W9-BEA-418
WESTERN'S CAMPUS
BOOKSTORE
USED BOOK
ON YOUR CAMPUS

WESTERN'S CAMPUS
28.80
BOOKSTORE
USED BOOK
ON YOUR CAMPUS

BASIC MICROBIOLOGY
WITH APPLICATIONS

THIRD EDITION

BASIC MICROBIOLOGY WITH APPLICATIONS

THOMAS D. BROCK
E. B. Fred Professor of Natural Sciences
University of Wisconsin, Madison

KATHERINE M. BROCK

DAVID M. WARD
Department of Microbiology
Montana State University, Bozeman

PRENTICE-HALL, ENGLEWOOD CLIFFS, NEW JERSEY 07632

Library of Congress Cataloging-in-Publication Data

BROCK, THOMAS D.
Basic microbiology with applications.

Bibliography: p.
Includes index.
1. Microbiology I. Brock, Katherine M.
II. Ward, David M. III. Title.
QR41.2.B76 1986 575 85-24333
ISBN 0-13-065244-X

FOR EMILY AND BRIAN
FOR NANCY, MOLLY, AND ABBY

Editorial/production supervision: Eleanor Henshaw Hiatt
Interior design: Photo Plus Art and Lorraine Mullaney
Cover design: Lorraine Mullaney
Manufacturing buyer: John Hall
Production coordination: Science Tech, Inc.
Art coordination: Science Tech, Inc.

Cover photograph: AIDS-infected small lymphocytes, magnification 33,000 x. Photograph by Centers for Disease Control, Atlanta, Georgia / Peter Arnold.

Printed in the United States of America

10 9 8 7 6 5 4 3 2 1

ISBN 0-13-065244-X 01

Prentice-Hall International (UK) Limited, *London*
Prentice-Hall of Australia Pty. Limited, *Sydney*
Prentice-Hall Canada Inc., *Toronto*
Prentice-Hall Hispanoamericana, S.A., *Mexico*
Prentice-Hall of India Private Limited, *New Delhi*
Prentice-Hall of Japan, Inc., *Tokyo*
Prentice-Hall of Southeast Asia Pte. Ltd., *Singapore*
Editora Prentice-Hall do Brazil, Ltda., *Rio de Janeiro*
Whitehall Books Limited, *Wellington, New Zealand*

Contents

PREFACE xv

1

INTRODUCTION 1

1.1 THE IMPACT OF MICROORGANISMS ON HUMAN AFFAIRS 2
1.2 THE DISCOVERY OF MICROORGANISMS 7
1.3 SPONTANEOUS GENERATION 10
1.4 THE GERM THEORY OF DISEASE 11
SUMMARY 15
KEY WORDS AND CONCEPTS 15
STUDY QUESTIONS 15
SUGGESTED READINGS 16

2

WHAT ARE MICROORGANISMS? 17

2.1 SIZES OF MICROORGANISMS 20
2.2 SEEING MICROORGANISMS 21
2.3 STRUCTURE OF CELLS 24

2.4 NAMES OF MICROORGANISMS 29
2.5 BACTERIA 29
2.6 ALGAE 34
2.7 FUNGI 35
BOX MUSHROOMS: DEADLY OR DELICIOUS? 38
2.8 LICHENS 38
2.9 PROTOZOA 40
2.10 VIRUSES 41
2.11 AN EVOLUTIONARY TREE OF THE LIVING WORLD 41
2.12 THE HISTORY OF MICROORGANISMS 43
SUMMARY 44
KEY WORDS AND CONCEPTS 46
STUDY QUESTIONS 46
SUGGESTED READINGS 47

3

CELL CHEMISTRY AND METABOLISM 48

3.1 A BIT OF CHEMISTRY 50
3.2 POLYMERS AND MONOMERS 52
3.3 CHEMICAL REACTIONS AND ENZYMES 58
3.4 NUTRITION AND BIOSYNTHESIS 62
3.5 OBTAINING ENERGY: CATABOLISM 64
3.6 BIOCHEMISTRY OF ENERGY GENERATION 66
3.7 METABOLIC DIVERSITY AMONG THE MICROORGANISMS 73
BOX AUTOTROPHS AT THE BOTTOM OF THE SEA 76
SUMMARY 79
KEY WORDS AND CONCEPTS 80
STUDY QUESTIONS 81
SUGGESTED READINGS 81

4

CARE AND FEEDING OF MICROBES 82

4.1 CULTURE MEDIA 83
BOX TRACKING THE LEGIONNAIRE'S KILLER 84
4.2 ENVIRONMENTAL REQUIREMENTS 86
4.3 STERILIZATION AND ASEPTIC TECHNIQUE 90
4.4 OBTAINING MICROORGANISMS IN CULTURE 94
4.5 SPECIAL PRECAUTIONS FOR WORKING WITH HARMFUL MICROORGANISMS 96

SUMMARY 98
KEY WORDS AND CONCEPTS 99
STUDY QUESTIONS 99
SUGGESTED READINGS 100

MICROBIAL GROWTH AND ITS CONTROL 101

5.1 GROWTH 101
5.2 MEASUREMENT OF GROWTH 104
5.3 CONTROL OF MICROBIAL GROWTH 109
BOX *SALMONELLA* SABOTAGE 110
5.4 CHEMICAL CONTROL OF MICROBIAL GROWTH 114
5.5 ANTIBIOTICS AND OTHER CHEMOTHERAPEUTIC AGENTS 115
5.6 KINDS OF CHEMOTHERAPEUTIC AGENTS AND THEIR ACTIONS 118
5.7 MEASURING ANTIMICROBIAL ACTIVITY 122
5.8 ANTIBIOTIC RESISTANCE 124
5.9 THE SEARCH FOR NEW ANTIBIOTICS 127
SUMMARY 128
KEY WORDS AND CONCEPTS 129
STUDY QUESTIONS 129
SUGGESTED READINGS 130

6

MOLECULAR BIOLOGY OF CELLS AND VIRUSES 131

6.1 THE MAJOR MACROMOLECULES: DNA, RNA, AND PROTEIN 132
6.2 PROTEIN SYNTHESIS 136
BOX IS THE GENETIC CODE UNIVERSAL? 140
6.3 REGULATION OF METABOLISM 141
6.4 ORGANIZATION OF DNA IN EUCARYOTIC CELLS 144
6.5 VIRUS REPRODUCTION 145
BOX VIROIDS AND PRIONS: THE SIMPLEST KILLERS 147
SUMMARY 148
KEY WORDS AND CONCEPTS 149
STUDY QUESTIONS 149
SUGGESTED READINGS 150

7

GENETICS AND GENETIC ENGINEERING 151

7.1 MUTATION 152
7.2 GENETIC EXCHANGE AND GENETIC RECOMBINATION 155
7.3 BACTERIAL TRANSFORMATION 157
7.4 TRANSDUCTION 160
7.5 PLASMIDS AND CONJUGATION 160
7.6 TRANSPOSONS: GENES THAT MOVE 166
7.7 GENETIC ENGINEERING 167
BOX HAZARDS OF GENETIC ENGINEERING 168
7.8 USES OF GENETIC ENGINEERING 174
SUMMARY 175
KEY WORDS AND CONCEPTS 176
STUDY QUESTIONS 176
SUGGESTED READINGS 177

8

INFECTIOUS DISEASE 179

8.1 HUMAN TISSUES AND THEIR NORMAL MICROBIAL INHABITANTS 182
BOX INTESTINAL GAS 187
8.2 GERM-FREE ANIMALS 188
8.3 INFECTIOUS DISEASES 190
8.4 HOW DO MICROORGANISMS CAUSE DISEASE? 192
8.5 HOST RESPONSES DURING INFECTION 197
8.6 DYNAMICS OF THE HOST-PARASITE INTERACTION 204
8.7 CLINICAL MICROBIOLOGY 208
8.8 ANTIMICROBIAL THERAPY 211
BOX PENICILLIN AT THE CROSSROADS: "IT LOOKS LIKE A MIRACLE." 212
SUMMARY 214
KEY WORDS AND CONCEPTS 214
STUDY QUESTIONS 215
SUGGESTED READINGS 216

9

IMMUNOLOGY 217

9.1 THE IMMUNOLOGICAL RESPONSE 218
9.2 ANTIBODY-MEDIATED IMMUNITY 223

9.3 THE COMPLEMENT SYSTEM AND THE IMMUNE SYSTEM 227
9.4 MECHANISM OF ANTIBODY FORMATION 229
9.5 CELL-MEDIATED IMMUNITY 232
9.6 IMMUNODEFICIENCY DISEASES 236
BOX NUDE MICE 237
9.7 ALLERGY AND HYPERSENSITIVITY 237
9.8 AUTOIMMUNE DISEASE 240
9.9 NONMICROBIAL IMMUNOLOGICAL PHENOMENA 241
9.10 IMMUNIZATION AGAINST DISEASE 244
9.11 MONOCLONAL ANTIBODIES 248
9.12 DIAGNOSTIC TESTS USING ANTIBODIES 250
SUMMARY 252
KEY WORDS AND CONCEPTS 252
STUDY QUESTIONS 253
SUGGESTED READINGS 255

10
PUBLIC HEALTH MICROBIOLOGY 256
10.1 SOURCES OF INFECTIOUS MICROORGANISMS 257
BOX TYPHOID MARY 259
10.2 HOW DISEASE SPREADS 259
10.3 HEALTH OF THE POPULATION 264
10.4 THE STATUS OF WORLD HEALTH 268
10.5 HOW DISEASES AFFECT HUMAN POPULATIONS 271
10.6 PRINCIPLES OF INFECTIOUS DISEASE CONTROL 275
10.7 PROCEDURES FOR PREVENTION OF INFECTIOUS DISEASE 278
BOX THE FALL OF SMALLPOX 280
10.8 HOSPITAL INFECTIONS 282
10.9 EMBALMING AND DISPOSAL OF THE DEAD 287
SUMMARY 287
KEY WORDS AND CONCEPTS 288
STUDY QUESTIONS 288
SUGGESTED READINGS 289

11
BACTERIAL DISEASES 291
11.1 STREPTOCOCCUS 291
11.2 STAPHYLOCOCCUS 294
11.3 NEISSERIA 296
11.4 HAEMOPHILUS 299

11.5 BORDETELLA 299
BOX RISK VERSUS BENEFIT IN VACCINATION 302
11.6 CORYNEBACTERIUM 300
11.7 CLOSTRIDIUM 303
11.8 BACILLUS 305
11.9 MYCOBACTERIUM 306
11.10 ENTERIC BACTERIA 310
11.11 YERSINIA 313
11.12 FRANCISELLA 314
11.13 BRUCELLA 314
11.14 PSEUDOMONAS 316
11.15 VIBRIO 316
11.16 CAMPYLOBACTER 317
11.17 SPIROCHETES 318
11.18 MYCOPLASMA 320
11.19 RICKETTSIA 320
11.20 CHLAMYDIA 322
11.21 LEGIONELLA 324
SUMMARY 324
KEY WORDS AND CONCEPTS 325
STUDY QUESTIONS 326
SUGGESTED READINGS 327

12

FUNGAL AND PROTOZOAL DISEASES 328

12.1 FUNGAL DISEASES 329
12.2 SUPERFICIAL MYCOSES 329
12.3 SUBCUTANEOUS AND SYSTEMIC MYCOSES 332
12.4 FUNGAL ALLERGIES 336
12.5 FUNGAL TOXINS 336
12.6 PROTOZOAL DISEASES 338
BOX WITCHCRAFT AND ST. ANTHONY'S FIRE: THE FUNGAL CONNECTION 339
12.7 DISEASES CAUSED BY FLAGELLATED PROTOZOA 339
12.8 DISEASES CAUSED BY AMOEBOID PROTOZOA 342
12.9 SPOROZOA AND MALARIA 343
SUMMARY 347
KEY WORDS AND CONCEPTS 348
STUDY QUESTIONS 348
SUGGESTED READINGS 349

13

VIRUSES AND VIRAL DISEASES 350

13.1 TECHNIQUES FOR VIRUS CULTIVATION 352
13.2 VIRUS REPRODUCTION 353
13.3 CONTROL OF VIRUS GROWTH 357
13.4 SMALLPOX 358
13.5 INFLUENZA 359
13.6 THE COMMON COLD 363
13.7 CHILDHOOD VIRUS DISEASES 364
BOX MEASLES AND RUBELLA EPIDEMICS ON COLLEGE CAMPUSES 367
13.8 HERPES INFECTIONS 368
13.9 ACQUIRED IMMUNODEFICIENCY SYNDROME (AIDS) 370
13.10 VIRAL DISEASES TRANSMITTED BY THE FECAL-ORAL ROUTE 372
13.11 RABIES 374
13.12 VIRUSES AND CANCER 376
SUMMARY 379
KEY WORDS AND CONCEPTS 380
STUDY QUESTIONS 381
SUGGESTED READINGS 382

14

ENVIRONMENTAL AND GLOBAL MICROBIOLOGY 383

14.1 MICROORGANISMS AND ECOSYSTEMS 385
14.2 MICROORGANISMS IN NATURE 388
BOX THE LIMITS OF LIFE 389
14.3 THE CARBON CYCLE 390
BOX MICROORGANISMS AND THE GREENHOUSE EFFECT 392
14.4 PETROLEUM MICROBIOLOGY AND THE CARBON CYCLE 392
14.5 THE NITROGEN CYCLE 395
14.6 THE SULFUR CYCLE 398
14.7 MINING MICROBIOLOGY 401
SUMMARY 405
KEY WORDS AND CONCEPTS 405
STUDY QUESTIONS 406
SUGGESTED READINGS 406

15

MICROBIOLOGY OF WATER AND WASTEWATER 407

15.1 WATER SUPPLY 408
15.2 PATHOGENIC ORGANISMS TRANSMITTED BY WATER 409
15.3 MICROBIOLOGICAL ASSAY OF WATER 411
15.4 WATER PURIFICATION 413
15.5 DRINKING WATER STANDARDS 416
BOX WATER PURIFICATION: ROBERT KOCH SHOWS THE WAY 417
15.6 MICROBIOLOGY OF WATER PIPELINES 418
15.7 WATER POLLUTION 420
15.8 SEWAGE-TREATMENT SYSTEMS 422
15.9 ALGAE AND WATER POLLUTION 431
SUMMARY 432
KEY WORDS AND CONCEPTS 433
STUDY QUESTIONS 433
SUGGESTED READINGS 435

16

FOOD MICROBIOLOGY 436

16.1 FOOD SPOILAGE 436
16.2 FOOD-BORNE DISEASES 441
16.3 ASSESSING MICROBIAL CONTENT OF FOODS 447
16.4 FOOD PRESERVATION 448
16.5 FOOD SANITATION 456
16.6 DAIRY MICROBIOLOGY 457
16.7 MEAT MICROBIOLOGY 468
BOX ANTIBIOTIC-RESISTANT *SALMONELLA*: FROM CATTLE TO CONSUMER 471
SUMMARY 473
KEY WORDS AND CONCEPTS 474
STUDY QUESTIONS 474
SUGGESTED READINGS 475

17

AGRICULTURAL MICROBIOLOGY 476

17.1 THE SOIL 476
17.2 PESTICIDES 479
17.3 NITROGEN IN AGRICULTURE 482

17.4 PLANT DISEASES 486
17.5 MICROBIAL INSECTICIDES 488
17.6 MYCOTOXINS AND GRAIN STORAGE 490
17.7 ANIMAL DISEASES 491
BOX THE POULTRY SLAUGHTER: "IT COULD HAVE BEEN WORSE." 494
17.8 RUMINANTS AND MICROORGANISMS 497
17.9 ANTIMICROBIAL AGENTS IN ANIMAL FEEDS 499
SUMMARY 501
KEY WORDS AND CONCEPTS 502
STUDY QUESTIONS 502
SUGGESTED READINGS 503

18
INDUSTRIAL MICROBIOLOGY 504

18.1 INDUSTRIAL FERMENTATION 506
18.2 YEASTS IN INDUSTRY 511
18.3 ALCOHOL AND ALCOHOLIC BEVERAGES 513
18.4 ANTIBIOTIC FERMENTATION 519
18.5 VITAMINS AND AMINO ACIDS 521
18.6 MICROBIAL BIOCONVERSION 524
18.7 ENZYME PRODUCTION BY MICROORGANISMS 524
BOX SWEETENING A SOFT DRINK: BIOTECHNOLOGY SHOWS THE WAY 526
18.8 VINEGAR 526
18.9 CITRIC ACID AND OTHER ORGANIC COMPOUNDS 528
18.10 FOOD FROM MICROORGANISMS 529
SUMMARY 531
KEY WORDS AND CONCEPTS 531
STUDY QUESTIONS 531
SUGGESTED READINGS 532

GLOSSARY 533

CREDITS AND ACKNOWLEDGEMENTS 545

INDEX 547

LIST OF BOXES

MUSHROOMS: DEADLY OR DELICIOUS? 38
AUTOTROPHS AT THE BOTTOM OF THE SEA 76

TRACKING THE LEGIONNAIRE'S KILLER 84
SALMONELLA SABOTAGE 110
IS THE GENETIC CODE UNIVERSAL? 140
VIROIDS AND PRIONS: THE SIMPLEST KILLERS 147
HAZARDS OF GENETIC ENGINEERING 168
INTESTINAL GAS 187
PENICILLIN AT THE CROSSROADS: "IT LOOKS LIKE A MIRACLE." 212
NUDE MICE 237
TYPHOID MARY 259
THE FALL OF SMALLPOX 280
RISK VERSUS BENEFIT IN VACCINATION 302
WITCHCRAFT AND ST. ANTHONY'S FIRE: THE FUNGAL CONNECTION 339
MEASLES AND RUBELLA EPIDEMICS ON COLLEGE CAMPUSES 367
THE LIMITS OF LIFE 389
MICROORGANISMS AND THE GREENHOUSE EFFECT 392
WATER PURIFICATION: ROBERT KOCH SHOWS THE WAY 417
ANTIBIOTIC-RESISTANT *SALMONELLA*: FROM CATTLE TO CONSUMER 471
THE POULTRY SLAUGHTER: "IT COULD HAVE BEEN WORSE." 494
SWEETENING A SOFT DRINK: BIOTECHNOLOGY SHOWS THE WAY 526

Preface

The new edition of this popular textbook reflects the remarkable changes the field of microbiology has undergone in recent years. There has been an explosion of activity, brought about in part by the rise of biotechnology and genetic engineering. In the decade which has passed since the discovery of recombinant DNA techniques, we have witnessed the development of this field and seen the first new microbiological products, such as insulin and new vaccines. Dramatic change in the field of microbiology is also due to the recognition of new and important human problems which have microbial roots. On the cover of this book, for example, is an electron micrograph of cells infected with the AIDS virus; the existence of AIDS was not even suspected when the previous edition of this book was published a few years ago. Several other diseases have also been newly described, such as toxic shock syndrome and infant botulism. Diseases such as legionellosis, chlamydial nongonococcal urethritis, and genital herpes have become much better understood during this period. And smallpox, once one of the most devastating diseases of mankind, has been eradicated. In few fields of science have such revolutionary developments occurred as rapidly as in the field of microbiology.

So many areas of microbiology have changed so greatly since the previous edition that large amounts of this text have been completely rewritten. Although we have been careful to keep the features of this book which have proved so useful to instructors and students, we have pruned heavily to make room for new material, taking care not to lengthen the book unduly.

A new coauthor has been heavily involved in the preparation of this edition. His experience in teaching introductory microbiology to large classes of students who are mainly nonscience majors, and his experience in teaching using the previous edition, have provided an opportunity to analyze in detail the goals of this text. Special attention has been given to the backgrounds of the students and the possible uses of the book. The book assumes that students have little chemistry and biology background. In order to make the text most accessible, many of the chapters, particularly those with technically difficult material (such as metabolism in Chapter 3 and molecular biology in Chapter 6), begin relatively simply, providing entry into the material at a fairly elementary level. Sophistication increases as each chapter progresses, so that the book is up-to-date and complete for students who have a stronger background (for example, life science majors who may take only one microbiology course). It is assumed that the more difficult material in some of the chapters will not be used in courses taught to nonscience majors, where the background of the student is a problem. The organization between, as well as within, chapters has also been designed so that major concepts are well illustrated without the need to use all chapters of the book. For example, in a one quarter course there may be insufficient time to cover the individual chapters on microbial diseases (Chapters 11-13), but there are numerous examples of the causes of diseases, clinical aspects of diseases, and treatment and prevention of diseases within the preceeding chapters, which deal with the disease process in individuals and populations (Chapters 8-10).

The focus of the book is on basic microbiological principles, but always in the context of applications to areas of human concern. It is divided into three sections of approximately equal length which provide a *balanced* coverage of the entire field: **general microbiology** (kinds of microorganisms, physiology, genetics, growth and control), **medical microbiology** (host/parasite relations, immunology, epidemiology, infectious diseases), and **environmental microbiology** (ecology, water and food microbiology, agricultural and industrial applications). Our emphasis has been on the humanistic aspect of microbiology. The coverage of medical microbiology, for instance, has been directed toward a consideration of the natural history of infectious disease rather than toward the merely clinical side. We have made a special effort to show how microbiology relates to human experience in the broadest possible context, and in doing so, hope that the book will encourage the student to think across disciplines, a theme for courses which are a part of university core curricula. For example, we have raised such issues as the effect of social behavior on the incidence of sexually transmitted diseases, the contrasting status of health in developing versus developed regions of the world, and the risks involved in using new technologies (for example, genetic engineering, vaccine use, and antibiotic overuse).

There are so many new features in this edition that it can almost be considered a new book. We list some of the major new features here:

1. *Boxes*. Special-interest boxes have been written for each chapter. The purpose of these boxes is to give broad insight into the *human* side of

microbiology through the issues that impact society, current events, historical perspective, and practical human experience. Some boxes also introduce students to the dynamic aspect of science and to the scientific method through examples of recent discoveries. Each box is written so that it can, to a great extent, stand alone.

2. *Biotechnology focus*. The book has a strong biotechnological focus. This is exhibited in the introductory material in Chapter 1 and in the expanded materials on genetic engineering in Chapter 7, immunology in Chapter 9, and industrial microbiology in Chapter 18.
3. *New chapter organization*. Some significant reorganization of the chapters has been done, with new chapters added on molecular biology (Chapter 6) and immunology (Chapter 9) and deletion of the chapter on elementary chemistry (whose key points have been incorporated into Chapters 3 and 6).
4. *New figures*. A very large number of new figures have been drawn especially for this edition, and many of the figures retained from the earlier editions have been modified.
5. *Key words and concepts*. Two levels of key word emphasis have been used, *italics* and **bold face,** the latter being reserved for particularly central terms and themes. Major key words and concepts have been assembled at the end of each chapter to assist students in understanding which points are most relevant.
6. *Two-color format*. In order to provide better comprehension of figures and diagrams, a second color has been used.
7. *Larger page size*. This has enabled us to provide clearer figures and tables.

Finally, we note that this book was produced using modern computerized equipment that permitted rapid progression from manuscript to bound book. Because of this, we have been able to include in up-to-date fashion fast-breaking subject matter, such as information about AIDS. The production of the final copy for the printer was done completely in Madison, Wisconsin, where the senior authors are located. In this way, almost up to the moment that printing of the book began, modifications and improvements could be made. Since most books of this type take over a year to produce, the clear advantages of the use of this new technology can be appreciated. The production of this book was competently handled by the staff of Science Tech, Inc.

Several people deserve special thanks for their help in bringing this book to completion. First and foremost are the reviewers who read and commented critically on the manuscript, Jane Ann Phillips and Dr. David W. Smith. A number of other reviewers read individual chapters and made useful comments. Dr. Ruth Siegel of the Science Tech staff also read the book in its entirety and made numerous useful comments. The whole manuscript was typed on an IMB-PC by Wayne Schlegelmilch, whose expertise in converting handwritten text into typesetting code was amazing. Irene Slater typed the numerous tables in coded form suitable for the typesetter. The typesetting itself was competently handled by Tom Linley of Impressions, Inc. The new art was prepared primarily by Dunn's Studio, with additional art prepared by Ed Phillips and Maurice Newport. Art direction was by Katherine Brock.

Thomas D. Brock
Katherine M. Brock
David M. Ward

1

Introduction

This textbook deals with **microorganisms**, a large and diverse group of free-living organisms which exist as single cells or cell clusters. Microbial cells differ from the cells of animals and plants, which are unable to live alone in nature and exist only as multicellular organisms. In contrast to the cell of an animal or plant, a single microbial cell is generally able to carry out its life processes of growth, energy generation, and reproduction independently of other cells. The science that deals with the study of microorganisms is called **microbiology**. It is a branch of biology parallel with *botany*, the study of plants, and *zoology*, the study of animals. However, the procedures and practices by which microorganisms are studied are quite different from those used to study plants and animals, and microbiology has therefore developed as a science independent of botany and zoology.

Even though microorganisms are small and appear to be simple, they are of major concern to humans. The diseases that microbes cause have dramatically affected human existence. For example, "Black Death" (bubonic plague) killed one-quarter of the whole population of Europe in the Middle Ages! During the Irish potato famine of the 1840s, more than a million people starved because the Irish potato was attacked by a fungus. The Irish immigration to America began as a result of this famine, greatly enriching the cultural diversity of the United States. It is widely accepted that in World

War I more soldiers died from infectious diseases than from guns. In 1918 a major influenza epidemic spread through the world, and over 20 million people died. In recent times, there has been a dramatic rise in the incidence of sexually transmitted diseases, culminating in the emergence around 1980 of AIDS (acquired immunodeficiency syndrome). Such events are vivid reminders of the harmful effects of microorganisms. Indeed, through the diseases they cause, microbes have frequently altered the course of human history.

On the other hand, many microorganisms are beneficial. Without microbes, we would have no bread, no beer, no wine, no cheese; we would have to fertilize our crops more extensively; and we would be without the antibiotic drugs which have erased many important diseases.

Because there are so many ways in which microorganisms impact on human experience, microbiology is probably the most *applied* of the biological sciences. At the same time, microbiology is one of the most *basic* parts of biology, because microorganisms have provided the most suitable experimental materials for studies on the nature of life itself, studies now classified under the heading *molecular biology*. More recently, microbiologists and molecular biologists have joined forces to develop the new technology called *genetic engineering*, which has resulted from a merging of the basic and applied aspects of microbiology. These developments have led to the creation of a new field called *biotechnology*.

1.1 THE IMPACT OF MICROORGANISMS ON HUMAN AFFAIRS

The goal of the microbiologist is to understand how microorganisms work, and through this understanding to devise ways that benefits may be increased and damages curtailed. Microbiologists have been eminently successful in achieving these goals, and microbiology has played a major role in the advancement of human health and welfare.

Microorganisms and health care One measure of the microbiologist's success is shown by the statistics in Figure 1.1, which compares the present causes of death in the United States to those at the beginning of the century. At the beginning of the century, the major causes of death were infectious diseases; currently, such diseases are of only minor importance. Control of infectious disease has come as a result of our comprehensive understanding of disease processes. As we will see later in this chapter, microbiology had its beginnings in these studies of disease.

However, in order to fully appreciate the development of the field of microbiology, one must imagine a time when antibiotics and vaccines were not available, and when many of the diseases currently under control were running rampant. Physicians were unaware that the disinfection of hands after examination of a patient was necessary to prevent the spread of the disease-causing agent to the next patient examined. Modern means of food preservation did not exist and foods often deteriorated drastically because of

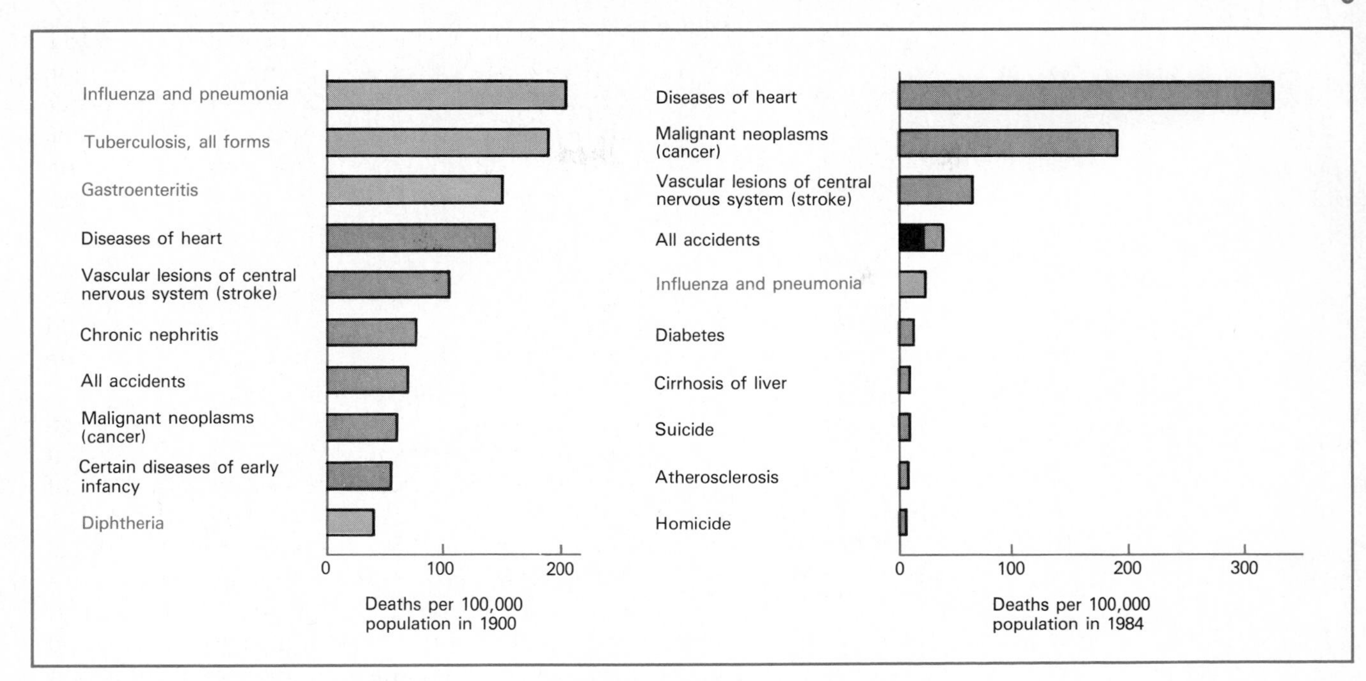

Figure 1.1 Death rates for the 10 leading causes of death: 1900 and 1984. Infectious diseases were the leading causes of death in 1900, whereas today they are much less important. Microbial diseases are shown in color. Automobile accident fatalities are shown in black.

the growth of microorganisms. The importance of water treatment had not been discovered, and supplies of drinking water were often mixed with water containing harmful human sewage. Although our current control of infectious microbes is excellent, this control has not come inexpensively. Many billions of dollars are spent yearly to maintain the human population relatively free of infectious disease. Table 1.1 summarizes the major health-related expenditures in the United States and the number of individuals involved in the effort to insure a good standard of public health.

A disquieting note when we consider the successes illustrated by Figure 1.1: although we now live in a world where microbes are mostly under control, for the individual dying slowly of AIDS, or the cancer patient whose immune system has been devastated as a result of treatment with an anti-cancer drug, microbes are still the major threat to survival. Although such tragic situations barely appear in our health statistics, they are of no less concern. Further, microbial diseases still constitute the major causes of death in many of the less developed countries of the world. Although eradication of smallpox from the world has been a stunning triumph for medical science, millions still die yearly in developing countries from such pervasive illnesses as malaria, cholera, African sleeping sickness, and severe diarrheal diseases.

The point has now been clearly made that microorganisms are serious threats to human existence. But we must emphasize that most microbes are *not* harmful to humans. In fact, most microorganisms cause no harm at all. More interestingly, a large number of microorganisms display *beneficial* ac-

TABLE 1.1 Health-related activities involving microbiology

Item	Annual total
Total U.S. expenditures for health	$355.4 billion
Hospitals	$35.6 billion
Drug expenditures	$23.7 billion
Biologicals (such as vaccines)	$2.3 billion
Pharmaceuticals (such as antibiotics)	$16.7 billion
Expenditures for public health	
Federal government (epidemic control)	$1.2 billion
State/local governments (reportable diseases)	$9.9 billion
Pollution abatement (surface/ground waters)	$22.2 billion
Sewage treatment	$14.9 billion
Drinking water treatment systems	$11.5 billion
Medical research expenditures	$6.2 billion
Labor force in health-related occupations	
Health-diagnosing occupations (e.g., physicians)	735,000 people
Health assessment and treatment (e.g., nurses)	1,900,000 people
Health technologists and technicians	1,111,000 people
Health service occupations (e.g., aides)	1,739,000 people

Data for 1982 or 1983 from the Statistical Abstract for the United States for 1985.

tivities, carrying out processes that are of immense value to human society. Even in the health-care industry, microorganisms play beneficial roles. For instance, the pharmaceutical industry is a multi-billion dollar industry built in part on the large-scale production of antibiotics by microorganisms. We will have much to say about antibiotics in this book, but note here that worldwide antibiotic production is over *100,000 tons* per year! A number of other major pharmaceutical product lines are also derived, at least in part, from the activities of microbes (Table 1.1).

Microorganisms and agriculture Our whole system of agriculture depends in many important ways on microbial activities, as summarized in Table 1.2. A number of major crops are members of a plant group called the *legumes*, which live in close association with special bacteria which form tiny nodules on their roots. In these root nodules, atmospheric nitrogen (N_2) is converted into fixed nitrogen compounds that the plants can use for growth. In this way, the activities of the root nodule bacteria reduce the need for costly fertilizer. As seen in Table 1.2, billions of acres and billions of dollars are annually committed to the cultivation of these leguminous crop plants.

Also of major agricultural importance are the microorganisms that are essential for the digestive process in cattle and sheep. These important farm animals have a special digestive organ called the *rumen* in which microorganisms carry out the digestive process. As shown in Table 1.2, many billions of dollars of meat and milk production are linked to these rumen microbes. Without these microbes, cattle and sheep production would be virtually impossible.

In addition to benefits to agriculture, microorganisms also cause harmful effects. Animal and plant diseases due to microorganisms have major eco-

TABLE 1.2 Agricultural-related activities involving microbiology

Crop	Value	Acres	Number of farms
Plant crops (root-nodule bacteria)			
Soybeans	$10.4 billion	69 million	508,000
Hay, silage	$2.3 billion	60 million	300,000
Snap beans	$107 million	204,000	—
Peas	$113 million	303,800	—
Animal products (rumen bacteria)			
Dairy	$16.3 billion	—	200,000
Beef cattle	$31.6 billion	—	1,279,000
Sheep	$608 million	—	99,000
Plant diseases (sprays and dusts for crops)	$180 million	11.7 million	—
Total agricultural (farming, fish, forestry)	$60.9 billion	(110 billion pounds)	—

Data for 1982 or 1983 from the Statistical Abstract of the United States for 1985.

nomic impact. An example of this is the large amount of money spent annually for sprays and dusts used to control diseases of crop plants (Table 1.2).

Microorganisms and the food industry Once food crops are produced, they must be delivered in wholesome form to consumers. Some of the roles of microbes in the food industry are outlined in Table 1.3. We note first the food spoilage that wastes vast amounts of money every year. The canning, frozen-food, and dried-food industries exist to prepare foods in such ways that they will not spoil. As shown in Table 1.3, food preservation is an almost $30 billion industry.

However, not all microbes have harmful effects on foods. Dairy products manufactured, at least in part, via microbial activity include cheese, yogurt, and buttermilk, all products of major economic value (Table 1.3). Sauerkraut,

TABLE 1.3 Food-related activities involving microbiology

Category	Annual value	Annual production
Foods treated to prevent microbial growth	$29.8 billion	
Canned foods		17.8 billion pounds
Frozen foods		6.1 billion pounds
Dried foods		2.2 billion pounds
Foods made with the aid of microorganisms		
Dairy (total dairy income)	$18.8 billion	
Cheese		6.2 billion pounds
Corn sweeteners		16.3 billion pounds
Meat (from ruminant animals)	$1.2 billion	41.4 billion pounds
Baked goods (produced with yeast)	$17.7 billion	
Alcoholic beverages	$60 billion	
Beer		196 million barrels
Wine		849.5 million gallons
Distilled spirits		181 million gallons

Data for 1982 and 1983 from Statistical Abstract for the United States, 1985.

TABLE 1.4 Fuels and minerals whose production is affected by microorganisms

Product	Annual value	Amount produced
Fuels		
Petroleum (crude)	$90 billion	133 billion gallons
Natural gas (methane)	$45.6 billion	18.5 billion cubic feet
Minerals		
Copper (bacterial leaching)	$280 million	380 million pounds
Sulfur (sulfur bacteria)	$434.7 million	7.9 billion pounds
Diatomite (algae)	$107.6 million	1.22 billion pounds

Data for 1981 and 1982 from the Statistical Abstract of the United States for 1985.

pickles, and some sausages also owe their existence to microbial activity. Baked goods are made using yeast. Even more pervasive in our society are the alcoholic beverages, also based on the activities of yeast, which build a $60 billion industry.

All of these applications of microbes are of ancient origin, but microbiology has not rested on the past. Consider, for instance, the microbe's contribution to a carbonated soft drink. The major sugar in many soft drinks is *fructose*, produced from corn starch via microbial activity. As seen in Table 1.3, over 16 billion pounds of corn sweeteners are produced each year! In diet soft drinks, the artificial sweetener *aspartame* is a combination of two amino acids, both produced microbiologically. Finally, the *citric acid* added to many soft drinks to give them tang and bite is produced in a large-scale industrial process using a fungus.

Mineral- and energy-related activities involving microorganisms Our complex industrial society is energy-driven, and here also microorganisms play major roles. Natural gas, methane, is a product of bacterial action. It is harvested in vast amounts, as listed in Table 1.4. A few other mineral and energy products are also the result of microbial activity, but of even greater interest is the relationship of microorganisms to the petroleum industry. Crude oil is subject to vigorous microbial attack, and drilling, recovery, and storage of crude oil all have to be done under conditions that minimize microbial damage. As seen in Table 1.4, the petroleum industry is almost a $100 billion industry!

Further, human activity will result in the complete consumption of available fossil fuels during the next few decades, so that we must seek new ways to supply the energy needs of society. In the future, microorganisms may provide alternative energy sources. Photosynthetic microorganisms can harvest light energy for the production of *biomass*, energy stored in living organisms. These novel forms of biomass and existing waste materials such as domestic refuse, surplus grain, and animal wastes, can be converted to "biofuels," such as methane and ethanol, by other microorganisms. Other microbial products may be used as "chemical feedstocks," the chemicals from which synthetic materials are manufactured, and which are now also commonly

TABLE 1.5 Major United States corporations with microbiological and biotechnological interests

Company	Sales, billions $$	Areas of major interest	Activities in biotechnology
Exxon	88	Oil, chemicals	+
Mobil	55	Oil	
duPont	35	Chemicals	+
Std. Oil Indiana	28	Oil, chemicals	+
Shell	19.7	Oil, chemicals	+
Phillips Petroleum	15.2	Oil, chemicals	+
Proctor and Gamble	12.4	Consumer products, food	
Dow Chemical	11	Chemicals, agriculture	+
Allied Corp.	10.3	Conglomerate	+
Beatrice Foods	9.0	Foods, consumer products	
General Foods	8.0	Foods	
PepsiCo	7.8	Beverages	
3-M	7.0	Chemicals, minerals	+
Coca-Cola	6.9	Beverages	
Consolidated Foods	6.5	Foods	
Monsanto	6.3	Chemicals, agriculture	+
W.R. Grace	6.2	Fertilizers, chemicals, agriculture	+
Anheuser-Busch	6.0	Brewing	+
Nabisco Brands	5.9	Foods	+
Johnson and Johnson	5.9	Health care	+
General Mills	5.5	Foods	
Ralston Purina	4.9	Foods	
Colgate-Palmolive	4.9	Consumer products	
Archer-Daniels-Midland	4.3	Chemicals, high-fructose syrup	+
Borden	4.3	Foods	
CPC International	4.0	Chemicals, food	+
Bristol-Myers	3.9	Pharmaceuticals	+
Pfizer	3.7	Pharmaceuticals, chemicals	+
H.J. Heinz	3.7	Foods	
Pillsbury	3.7	Foods	
American Cyanamid	3.5	Pharmaceuticals, chemicals	+
United Brands	3.5	Foods	
Owens-Illinois	3.4	Chemicals	+
Carnation	3.3	Foods	
Campbell Soup	3.3	Foods	
Merck	3.2	Pharmaceuticals	+
SmithKline Beckman	3.1	Pharmaceuticals	+
Warner-Lambert	3.1	Pharmaceuticals	+
Eli Lilly	3.0	Pharmaceuticals	+
Abbott Laboratories	2.9	Pharmaceuticals	+
National Distillers	2.3	Alcoholic beverages	
Upjohn	1.9	Pharmaceuticals	+
Rohm and Haas	1.9	Chemicals, agriculture	+
Baxter Travenol	1.8	Pharmaceuticals	+
Schering-Pough	1.8	Pharmaceuticals	+
Squibb	1.8	Pharmaceuticals	+

Sales data are from Rand McNally Commercial Atlas. Biotechnology interests are from Genetic Engineering and Biotechnology Firms Worldwide Directory, 1985.

TABLE 1.6 Biotechnology in the U.S.

State	Number of biotechnology companies
California	185
New York	120
New Jersey	115
Massachusetts	84
Illinois	50
Connecticut	35
Maryland	35
Ohio	35
Pennsylvania	34
Wisconsin	27
Texas	25
Florida	23
Colorado	21
Michigan	17
Missouri	16
Washington	16
Minnesota	15
Indiana	14
Virginia	14
Iowa	12
North Carolina	12
Rest of country	94

Data compiled from *Genetic Engineering and Biotechnology Firms, Worldwide Directory*, 4th edition, 1985.
The numbers reflect all categories of companies with an interest in biotechnology, primarily research companies, but also including publishers of biotechnology-related materials, consultants, and investment firms.

derived from petroleum. Human activity is also decreasing the supplies of other substances, such as metals, and microorganisms are being used increasingly in metal recovery from low grade ores (see Table 1.4).

Microorganisms and biotechnology But the above recital of the benefits of microbiology is only the beginning. One of the most exciting new areas of microbiology at present is that called **biotechnology**. In the broad sense, biotechnology entails the use of living microorganisms in large-scale industrial processes, but by "biotechnology" today we usually mean the application of novel genetic procedures, generally involving microorganisms. Biotechnology is frequently equated with **genetic engineering**, the discipline that concerns the artificial manipulation of genes and their products.

Genes from human sources, for instance, can be broken into pieces, modified, and added to or subtracted from, using microbes and their enzymes as precise and sophisticated tools. It is even possible to make completely artificial genes using genetic engineering techniques. And once the desired gene has been selected or created, it can be inserted into a microbe where it can be made to reproduce and make the desired gene product. For instance, human insulin, a hormone found in abnormally low amounts in people with diabetes, has now been produced microbiologically with the human insulin gene engineered into a microbe.

The enormous impact which microbiology and biotechnology have on the United States economy is shown by the information presented in Table 1.5 on the previous page. There we see that many of the largest United States corporations have established biotechnology programs. Further, the excitement of biotechnology has brought about the establishment of an amazing number of new companies. These companies, whose broad distribution is illustrated in Table 1.6, are impelled by the potential of biotechnology to invest large amounts of money in research on new applications of microorganisms. We can anticipate over the next generation truly startling discoveries from these energetic and innovative enterprises.

The overwhelming influence of microorganisms in human society is clear. We have many reasons to be aware of microorganisms and their activities. As the eminent French scientist Louis Pasteur, one of the founders of microbiology, expressed it: "The role of the infinitely small is infinitely large."

We have now briefly delineated the practical significance of microbiology. But this book begins with a careful discussion of the basic properties and activities of microorganisms. Only after a solid understanding of basic microbiology has been established, is it possible to discuss the applications. Thus, in this book we will be moving from the basic to the applied, and from the simple to the complex. But before we begin, let us consider briefly some of the events which led up to our current understanding of microorganisms and their activities.

1.2 THE DISCOVERY OF MICROORGANISMS

One of the most significant advances, the discovery of microorganisms themselves, occurred as a result of basic research done without any preconceived practical goal, but merely because of an interest in using microscopes to see

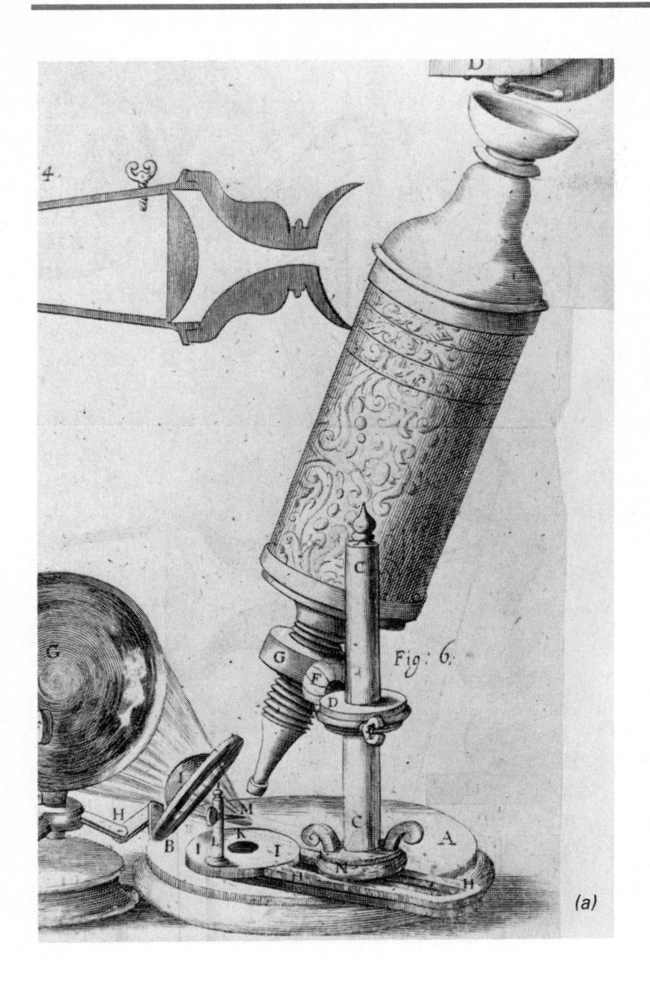

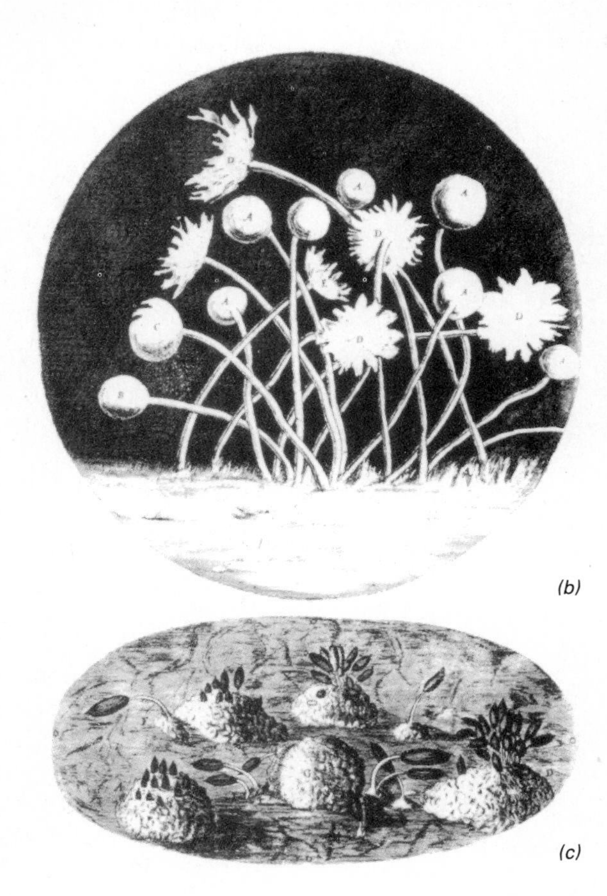

Figure 1.2 The early microscopic observations of Robert Hooke. (a) Robert Hooke's microscope, as illustrated in his great book *Micrographia*, published in 1665. This is a compound microscope having two lenses, one near the eye and the other near the object. (b) Hooke's drawing of a blue mold growing on the surface of leather; the round structures contain spores of the mold. (c) A mold growing on the surface of an aging and deteriorating rose leaf.

the very small. Although the existence of creatures too small to be seen with the eye had long been suspected, their discovery was linked to the invention of the **microscope**. Robert Hooke, using elegantly ornate microscopes (Figure 1.2*a*), described the fruiting structures of molds in 1664 (Figure 1.2*b*). However, the first person to see microorganisms in any detail was the Dutch amateur microscope builder Antoni van Leeuwenhoek, who used simple microscopes of his own construction (Figure 1.3). Leeuwenhoek's microscopes were extremely crude by today's standards, but by careful manipulation and focussing he was able to see organisms as small as bacteria. Leeuwenhoek's reaction to viewing the microbial world for the first time is reflected in a letter he wrote to the Royal Society of London in 1674: ". . . Tho my teeth are kept usually very clean, nevertheless when I view them in a Magnifying Glass, I find growing between them a little white matter as thick as wetted flower . . . and then to my great surprise perceived that the aforesaid matter contained very many small Animals, which moved themselves very extravagantly. The biggest sort . . . darted themselves thro the water or spittle, as a Jack or Pike goes thro the water . . . I took in my mouth some very strong wine-Vinegar, and closing my Teeth, I gargled and rinsed them very well with fair water,

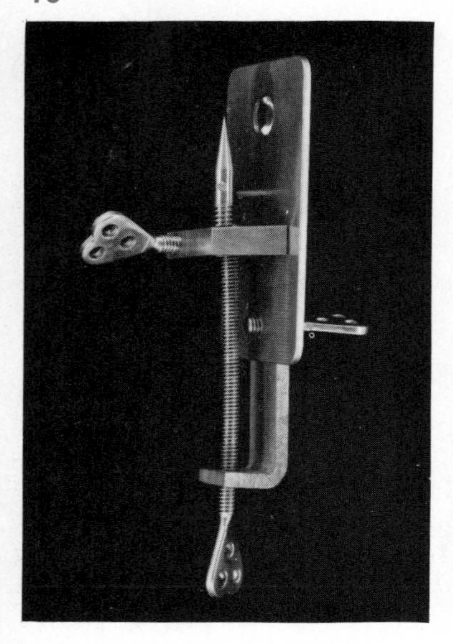

Figure 1.3 A replica of Leeuwenhoek's microscope. The object to be viewed was placed on the pointed tip at the end of the screw and was moved back and forth by turning the screw.

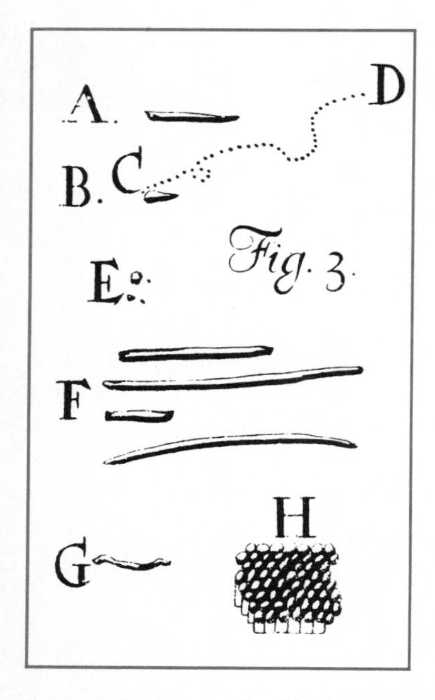

Figure 1.4 Leeuwenhoek's drawings of bacteria, published in 1684. Even from these crude drawings we can recognize several kinds of common bacteria.

but there were an innumerable quantity of Animals yet remaining in the scurf upon the Teeth . . . The number of these Animals in the scurf of a man's Teeth, are so many that I believe they exceed the number of Men in a kingdom."

Drawings of some of Leeuwenhoek's "wee animalcules" are shown in Figure 1.4. His observations were confirmed by other workers, but progress in understanding the nature of these tiny organisms came slowly until the nineteenth century. Only then did improved microscopes become available, enabling scientists to penetrate deeper into the mysteries of the cell. In fact, microbiology did not develop as a separate discipline until almost the end of the nineteenth century. This long delay occurred because, in addition to microscopy, certain other basic techniques for the study of microorganisms needed to be devised. Investigation of two perplexing questions led to the development of these techniques and laid the foundation of microbiological science: (1) Does spontaneous generation occur? (2) What is the nature of contagious disease? By the end of the century, both questions were answered, and microbiology was firmly established as a distinct and growing field of science.

1.3 SPONTANEOUS GENERATION

The basic idea of spontaneous generation can easily be understood. If food is allowed to stand for some time, it putrefies, and when the putrefied material is examined microscopically, it is found to be teeming with bacteria. Where do these bacteria come from, since they are not seen in fresh food? Some people said they developed from seeds or germs that had contaminated the food from the air, whereas others said that they arose spontaneously.

Spontaneous generation would mean that life could arise from something nonliving, and many people could not imagine something so complex as a living cell arising spontaneously from dead materials. The most powerful opponent of spontaneous generation was the French chemist Louis Pasteur (Figure 1.5), whose work on this problem was the most exacting and convincing. Pasteur first showed that there were structures present in air that closely resembled the microorganisms seen in putrefying materials. He did this by passing air through guncotton filters, the fibers of which stop solid particles. After the guncotton was dissolved in a mixture of alcohol and ether, the particles that it trapped fell to the bottom of the liquid and were examined on a microscope slide. Pasteur found that in ordinary air there existed constantly a variety of solid structures ranging in size from 0.01 millimeter (mm) to more than 1.0 mm. Many of these structures resembled the spores of common molds, the cysts of protozoa, and various other microbial cells. As many as 20 to 30 of them were found in 15 liters of ordinary air, and they could not be distinguished from the organisms found in much larger numbers in putrefying materials. Pasteur concluded, therefore, that the organisms found in putrefying materials originated from the organized bodies present in the air, which are constantly being deposited on all objects. If this conclusion was correct, it meant that food treated to destroy all the living organisms contaminating it would not putrefy. In fact, another Frenchman, Nicholas Appert, had already devised a method for food preservation based on heat treatment, but did not understand the principle upon which his method worked.

Pasteur used heat to eliminate contaminants, since many workers had shown that if a nutrient infusion was sealed in a glass flask and heated to boiling, it never putrefied. The proponents of spontaneous generation had criticized such experiments by declaring that fresh air was necessary for spontaneous generation and that the air inside the sealed flask was affected in some way by heating so that it would no longer support spontaneous generation. Pasteur skirted this objection simply and brilliantly by constructing a swan-necked flask, open to the air, now called the *Pasteur flask* (Figure 1.6). In such a flask, putrefying materials could be heated to boiling; after the flask was cooled, air could reenter but the bends in the neck prevented particulate matter, bacteria, or other microorganisms from getting in. Material sterilized in such a flask did not putrefy, and no microorganisms ever appeared so long as the neck of the flask remained intact. If the neck was broken, however, putrefaction occurred, and the liquid soon teemed with living organisms. This simple experiment effectively settled the controversy of spontaneous generation.

Figure 1.5 Louis Pasteur.

Killing all bacteria or germs is a process we now call **sterilization** and the procedures that Pasteur and others used were eventually carried over into microbiological research. Disproving the theory of spontaneous generation thus led to the development of effective sterilization procedures, without which microbiology as a science could not have developed.

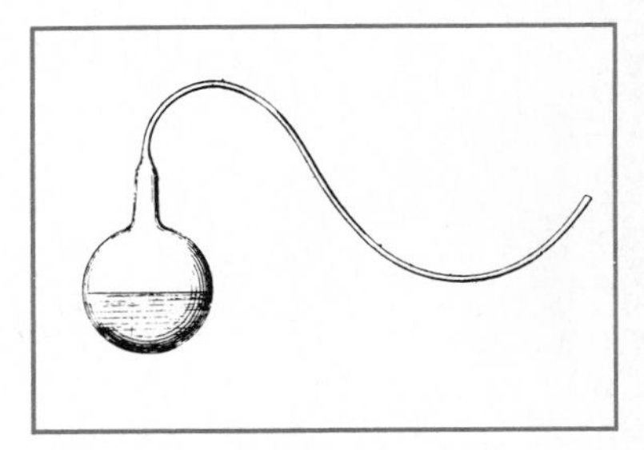

Figure 1.6 One of Pasteur's swan-necked flasks.

It was later shown that flasks and other vessels could be protected from contamination by cotton stoppers, which still permitted the exchange of air. The principles of sterile technique, developed so effectively by Pasteur, are the first procedures learned by the novice microbiologist. Food science also owes a debt to Pasteur since his principles are applied in the canning and preservation of many foods (described in Chapter 16).

1.4 THE GERM THEORY OF DISEASE

Proof that microorganisms could cause disease provided a great stimulus for the development of the science of microbiology. As early as the sixteenth century, it was thought that something that induced disease could be transmitted from a diseased person to a healthy person. Many diseases spread through populations and were called *contagious;* the unknown thing that did the spreading was called the *contagion.* After Leeuwenhoek's discoveries, it was more or less widely held that microorganisms might be responsible for contagious diseases, but proof was lacking. In 1846, Miles Joseph Berkeley provided the first clear demonstration that a microorganism caused a disease by showing that a mold (fungus) was responsible for the Irish potato blight. Ignaz Semmelweis and Joseph Lister showed that use of simple rules of hygiene could limit the spread of disease, thus providing some evidence for the importance of microorganisms in causing human diseases. But it was the work of Robert Koch, a German physician (Figure 1.7), that placed the germ theory of disease on a firm footing.

Figure 1.7 Robert Koch.

Koch's early work In his early work, published in 1876, Koch studied anthrax, a disease of cattle which sometimes also occurs in humans. Anthrax is caused by a bacterium now called *Bacillus anthracis,* and the blood of an infected animal teems with cells of this large bacterium. Koch established by careful microscopy that the bacteria were always present in the blood of an animal that had the disease. However, the mere association of a bacterium with a disease does not prove that it causes the disease; it might instead be a result of the disease. Therefore, Koch demonstrated that it was possible to take a small amount of blood from a diseased animal and inject it into another animal, which in turn became diseased and died. He could then take blood from this second animal, inject it into another, and again obtain the characteristic disease symptoms. By repeating this process up to 20 times, successively transferring small amounts of blood containing bacteria from one animal to another, he proved that some living agent did indeed cause anthrax. The twentieth animal died just as rapidly as the first, and in each case Koch could demonstrate by microscopy that the blood of the dying animal contained large numbers of the bacterium.

Koch carried this experiment further. He found that the bacteria could also be cultivated in nutrient fluids outside an animal's body, and even after many transfers in culture, the bacteria could still cause the disease when reinoculated into an animal. Bacteria from a diseased animal and bacteria in culture induced the same disease symptoms upon injection. The experiment thus proved that the bacterium *Bacillus anthracis* was the cause of the disease anthrax. This was the first proof that an infectious disease could be caused by a specific bacterium. Koch's work was important in elucidating the nature of anthrax, and even more important, it provided the experimental basis for the isolation and culture of a number of other infectious agents.

Koch's pure culture methods In order to prove that a microorganism is the cause of a disease, one must be sure that this organism alone is present in culture; that is, the culture must be *pure*. A **pure culture** is defined as a group of cells, all of which have arisen via successive cell divisions from a *single* parent cell. With materials as small as microorganisms, ascertaining purity is not easy, because even a very tiny sample of blood or other body fluid may contain several kinds of organisms that may all grow together in culture. Koch realized the importance of pure cultures and developed several ingenious methods of obtaining them, of which the most useful was that involving the isolation of single colonies. Koch observed that when a solid nutrient surface, such as a potato slice, was exposed to air and then incubated, bacterial colonies developed, each having a characteristic shape and color. He inferred that each colony had arisen from a single bacterial cell that had fallen on the surface, found suitable nutrients, and began to multiply. Because the solid surface prevented the bacteria from moving around, all the offspring of the initial cell remained together, and when a large enough number were present, the mass of cells became visible to the naked eye. Koch assumed colonies with different shapes and colors were derived from different kinds of organisms. When the cells of a single colony were spread out on a fresh surface, many colonies developed, each with the same shape and color as the original.

Koch realized that this discovery provided a simple way of obtaining pure cultures, since streaking mixed cultures on solid nutrient surfaces spread the various organisms so far apart that the colonies they produced did not mingle. Because many organisms could not grow on potato slices, Koch devised a *semisolid medium,* in which gelatin was added to a nutrient fluid, such as blood serum, in order to solidify it. When the gelatin-containing fluid was warmed, it liquefied and could be poured onto glass plates; after cooling, the solidified medium could be inoculated. Later, *agar* (a material derived from seaweed) was found to be a better solidifying agent than gelatin, and this substance is widely used today to obtain colonies.

Koch and tuberculosis Probably the most important single discovery of a bacterial disease agent was Koch's discovery of the causal agent of tuberculosis, *Mycobacterium tuberculosis.* Tuberculosis has been one of the great scourges of mankind. It was still the second greatest cause of death at the beginning of the twentieth century (see Figure 1.1), and remains to this day a leading cause of death in developing countries. Also called "consumption" or "phthisis," the disease has been known for centuries. Tuberculosis can involve many parts of the body, including the bones and the skin, but the organ most commonly affected is the lung. In advanced cases there is considerable degeneration of the lungs, difficulty in breathing, spitting of blood, and emaciation; death is often cruelly slow in coming as the patient gradually wastes away. Two of the most famous heroines of grand opera, Violetta in *La Traviata* and Mimi in *La Bohème,* die of tuberculosis, and their lingering deaths provide the essential pathos for the last acts of these operas.

In contrast to some other infectious diseases, in the nineteenth century the communicability of tuberculosis was by no means clear. Although the bacterium is a necessary condition for the development of the disease, it alone is not sufficient, since the general health of the person, heredity, and social factors strongly influence whether an infection leads to disease. Many people infected with the bacterium never develop the illness. But once it had been clearly established that *some* infectious diseases could be caused by bacteria, it was natural to turn to tuberculosis and look for its causal agent. Before Koch, the most important work on tuberculosis had been done by a French surgeon, Jean-Antoine Villemin, who showed that material from tubercular lung tissue could be used to initiate an infection in rabbits or guinea pigs and could be transmitted from one infected animal to another without any decrease in ability to cause disease. His work, published in 1868 before Koch began his studies on anthrax, showed clearly that tuberculosis was infectious but did not show what the infectious agent was. In a brilliant series of studies conducted between 1880 and 1882, Koch succeeded in obtaining cultures of the organism on an artificial medium and these cultures retained their infectivity for experimental animals. Koch also developed a specific staining procedure that permitted microscopic examination of tissues for the presence of the bacteria; such examination is still an important diagnostic tool.

The magnitude of Koch's achievement is hard to appreciate without knowing some of the difficulties inherent in working with *M. tuberculosis.* First, it is a slow-growing bacterium. Whereas other disease-causing bacteria usually produce visible colonies on culture media after a day or two of incubation, *M. tuberculosis* requires 10 days to 2 weeks, and even then the colonies are small and hard to see. Other workers were probably unsuccessful in culturing the organism because they were impatient, discarding their cultures too early. The second peculiarity of the organism is the difficulty with which it can be stained by dyes for microscopy. When examining infected tissues, some staining procedure is essential because the bacteria are virtually impossible to see among the large amount of deteriorating tissue cells and other debris. Most bacteria can be stained easily with dyes, but *M. tuberculosis* is refractory to staining by normal methods. Koch discovered through a lucky accident (as Pasteur said in other circumstances, "Chance favors the prepared mind") that the bacteria could be stained if the dye solution was made alkaline with potassium hydroxide; staining was also faster if the material was heated during the staining process. After this treatment, both the bacteria and the tissues were heavily stained. The rest of the procedure required a decolorization treatment that removed the stain from the tissues under conditions that did not decolorize the bacteria.

Koch's postulates Although the discovery of the tubercle bacillus was of major significance, even more important was that the experimental procedures that Koch developed provided a framework for the study of *any* infectious disease. We now call these experimental requirements *Koch's postulates*:

1. The organism must always be found in animals suffering from the disease and must not be present in healthy individuals.
2. The organism must be cultivated in pure culture away from the animal body.
3. Such a culture, when inoculated into susceptible animals, must initiate the characteristic disease symptoms.
4. The organism must be reisolated from these experimental animals and cultured again in the laboratory, after which it must still be the same as the original organism.

Koch's postulates not only supplied a means of demonstrating that specific organisms cause specific diseases, but also provided a tremendous impetus for the development of the science of microbiology by stressing laboratory culture.

The development by Koch of the proper procedures and criteria for the isolation and study of pathogens opened the way for subsequent discoveries in many other laboratories around the world. In the 20 years following the formulation of Koch's postulates, the causal agents of a wide variety of contagious diseases were isolated. These discoveries led to the development of successful procedures for the prevention and cure of many contagious diseases and contributed to the development of modern medical practice. The impact of Koch's work has been felt worldwide.

SUMMARY

The purpose of this chapter has been to present a brief overview of the significance of microbiology and its role in human affairs, and to provide some brief insights into the historical background of the field. Although we have concentrated in our historical treatment on Louis Pasteur and Robert Koch, it should be emphasized that these two eminent scientists were primarily the leaders of rather large research groups. The activities of the members of these research groups were no less important than those of the leaders. By the end of the nineteenth century, the field of microbiology had become a well-established discipline, with many students and researchers. Many of the concepts which are still in use had their origins in the research carried out a century ago.

We have seen in this chapter the enormous economic impact which microorganisms and their activities are having in human affairs. Whether we are concerned with harmful or beneficial activities of microorganisms, large amounts of money are involved, and large numbers of people are employed. With the anticipated rise in biotechnology, we can predict that even greater impacts of microbiology will be felt in the future. There are, indeed, strong reasons for studying microbiology!

KEY WORDS AND CONCEPTS

Microorganism
Microbiology
Biotechnology
Genetic engineering
Microscope
Spontaneous generation
Sterilization
Germ theory
Pure culture
Koch's postulates

STUDY QUESTIONS

1. How do microorganisms differ from plants and animals?
2. In what way are microorganisms similar to plants and animals?
3. List and discuss several health-related problems which involve microorganisms.
4. Describe one way in which microorganisms are beneficial in a health-related activity.
5. What is biotechnology and what are the potential economic benefits of this new development?
6. Describe the set of experiments that Louis Pasteur performed which showed that *spontaneous generation* does not exist.
7. How did Pasteur's work on spontaneous generation relate to the development of *sterilization* methods for microbiology?

8. Outline the experiment Robert Koch performed that showed that the bacterium *Bacillus anthracis* is the causal organism of the disease *anthrax*.
9. Why was Koch's development of *pure culture* methods so important for the development of microbiology?
10. How are pure culture methods used in the demonstration by *Koch's postulates* that a specific microorganism is the cause of a specific infectious disease?

SUGGESTED READINGS

BROCK, T.D. 1975. *Milestones in microbiology.* American Society for Microbiology, Washington, D.C.

BROCK, T.D., D.W. SMITH, and M.T. MADIGAN. 1984. *Biology of microorganisms, 4th edition.* Prentice-Hall, Englewood Cliffs, N.J.

BULLOCH, W. 1938. *The history of bacteriology.* Oxford University Press, London.

DOBELL, C. 1932. *Antony van Leeuwenhoek and his "little animals".* Reprint, 1960, Dover Publications, New York.

HOOKE, R. 1665. *Micrographia.* Royal Society of London. Reprint, 1961, Dover Publications, New York.

2

What Are Microorganisms?

Microorganisms are so small that at first glance one might think that they were unimportant. That is not the case, however; we may say that they are small but powerful. Just how small are they? How can we see them? What are they composed of? What do they do? To answer some of these questions, we will look in this and the next chapter at the structure and function of microorganisms.

All living organisms are composed of **cells**, which are the basic units of living things. Complex organisms such as plants and animals are composed of very large numbers of cells. For example, the human body contains about 10 trillion cells. We call such organisms multicellular, and in such organisms most cells are found in *tissues*, each of which carries out certain specific functions. All of the different cells and tissues then collectively make up the living plant or animal. However, in multicellular organisms any single cell is unable to live by itself and depends on the others for survival.

In contrast to multicellular organisms, most microorganisms are single-celled organisms and can independently carry out life processes resulting in growth and reproduction. Each cell is thus an independent organism, able to grow without the support of other cells. Some representative microbial cells are shown in Figure 2.1.

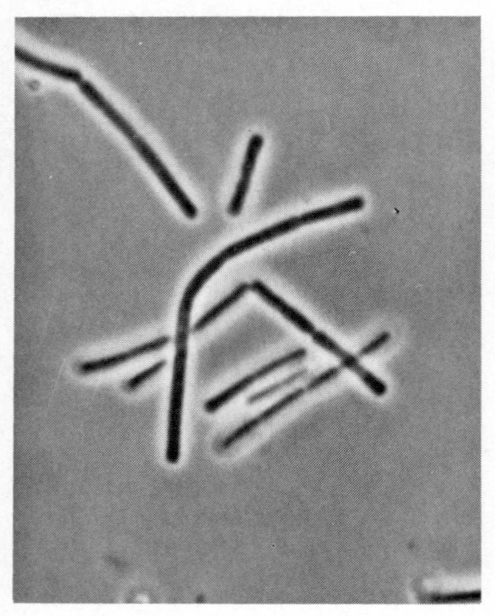
(a)

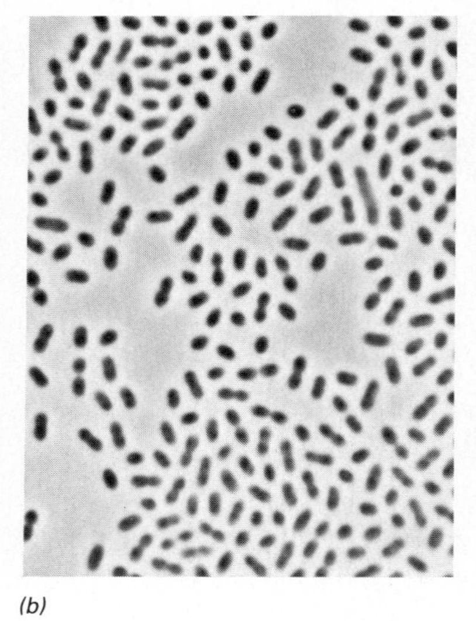
(b)

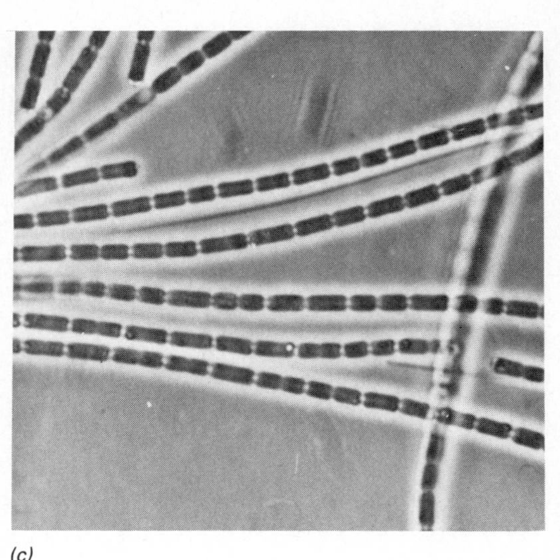
(c)

Figure 2.1 Typical representatives of the microbial world. All photomicrographs were taken by a phase-contrast microscope (see Section 2.2). (a) Large rod-shaped bacterium, *Bacillus cereus*. Magnification, 1600 ×. (b) Short rod-shaped bacterium, *Acetobacter aceti*. Magnification, 1600 ×. (c) Filamentous cyanobacterium, *Anabaena*. Magnification, 1600 ×. (d) Yeast cells dividing by budding. Magnification, 1900 ×. (e) Fungus filaments. Magnification, 500 ×. (f) A freshwater alga. Magnification, 300 ×. (g) A protozoan, *Paramecium*. Magnification, 125 ×.

What is the structure of a cell? All cells have a barrier separating inside from outside which is called the **cell membrane**. It is through the cell membrane that all food materials and other substances of vital importance to the cell pass in, and it is through this membrane that waste materials and other cell products pass out. If the membrane is damaged, the interior contents of the cell leak out, and the cell usually dies. As we shall see, some drugs and other chemical agents damage the cell membrane and, in this way, bring about the destruction of cells.

The cell membrane is a very thin, highly flexible layer and is structurally weak. By itself, it usually cannot hold the cell together, and an additional stronger layer, called the **cell wall**, is usually necessary. The wall is a relatively rigid layer which is present outside the membrane and protects the membrane and strengthens the cell. Plant cells and most microorganisms have such rigid cell walls. Animal cells, however, do not have walls; these cells have developed other means of support and protection.

Inside a cell is a complicated assembly of substances and structures which is called the **protoplasm**. These materials and structures, bathed in water, carry out the functions of the cell. The word *metabolism* is often used to refer to the collection of chemical processes that cells carry out. Microbial metabolism will be discussed in Chapter 3. In the present chapter, we will discuss the structures of cells and will present a survey of the diversity of types of microorganisms.

Upon careful study of the structures of cells, two basic types have been recognized which differ greatly in structure. These are called **procaryote** and **eucaryote** and these two types of cells are so different that the separation of procaryotic from eucaryotic cell lines is thought to have been a major event in biological evolution. **Bacteria** are the only procaryotes. There are several groups of eucaryotic microorganisms, including **algae**, **fungi**, and **protozoa**. In addition, all higher life forms (plants and animals) are constructed of eucaryotic cells. Figure 2.2 summarizes the major types of cellular life forms.

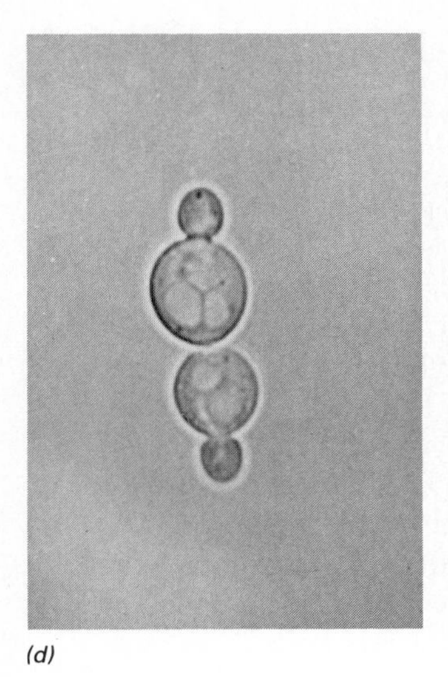
(d)

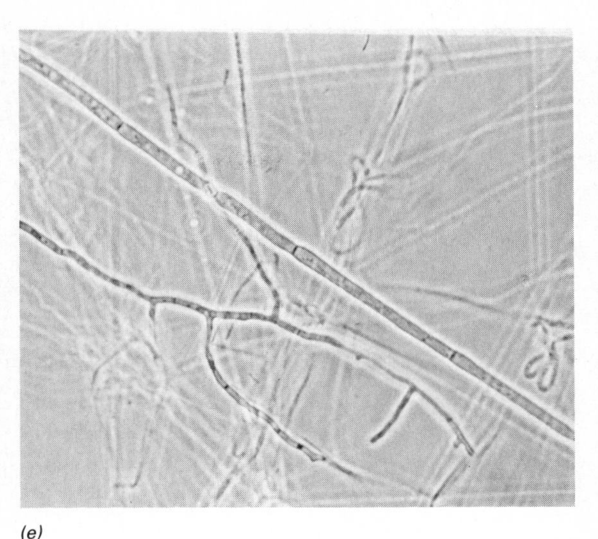
(e)

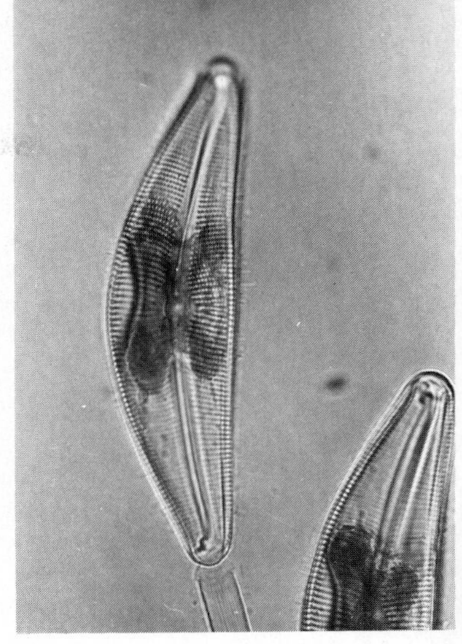
(f)

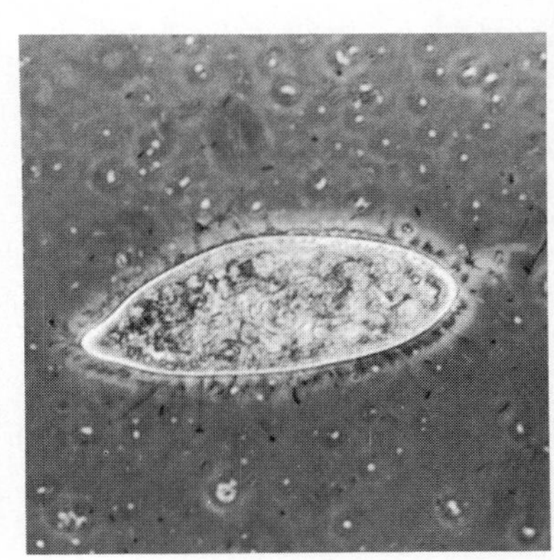
(g)

Viruses are not cells. They have very simple structures, and unlike cells, they have no metabolism of their own. Viruses cannot reproduce independently of cells, but must infect cells which then reproduce the virus. Are viruses living? The answer to this question depends on how we define life and will be discussed later in this book. Because they cause important diseases, we will consider virus structure and diversity briefly in this chapter and the ability of viruses to infect cells and cause disease in Chapters 6 and 13. An example of one kind of virus is shown in Figure 2.3. In this figure, a large number of virus particles are seen, all arranged in a uniform pattern. Viruses vary considerably in size and shape. Some virus particles are much smaller than those shown in Figure 2.3, whereas others are larger. The shapes of virus particles vary also, with some virus particles resembling tiny crystals and others being fairly irregular in shape. Viruses are known which infect all kinds of living organisms: animals, plants, algae, fungi, protozoa, and bacteria. Although many viruses cause disease in the organisms they infect, virus infection does not always lead to disease; many viruses cause inapparent in-

	Procaryotic	Eucaryotic
Differentiated higher forms		Animals Plants
Microorganisms	Archaebacteria Eubacteria	Algae Fungi Protozoa

Figure 2.2 The major types of cellular life forms. Note the distinction between the procaryotic and eucaryotic cell types. All multicellular organisms are eucaryotic.

fections. Note that the virus particles shown in Figure 2.3 look very different from cells.

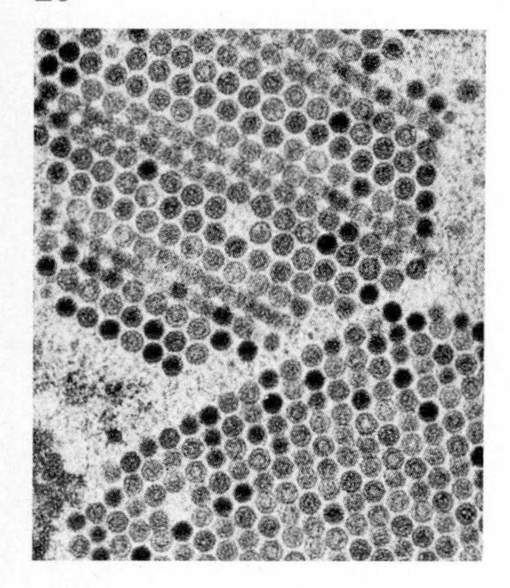

Figure 2.3 Particles of a virus within an infected cell. This is *adenovirus,* which causes various respiratory infections. Each individual virus particle has the shape of a tiny hexagon. Each virus particle is about 0.07 μm (70 nm) in diameter. Hundreds of individual virus particles are aggregated into large clusters.

2.1 SIZES OF MICROORGANISMS

Microorganisms are very small. Before discussing microbial size, we must consider a new unit of measurement, the *micrometer,* abbreviated μm. Consider first a dimension of one millimeter (abbreviated mm), which is about the size of the period at the end of this sentence. A μm is one thousand times smaller than a millimeter. We cannot see something which is only one μm in size. In fact, the human eye cannot perceive objects smaller than about 200 μm. Thus, bacteria of typical size, about 1 to 2 μm long, are completely invisible to the naked eye. To illustrate how minute a bacterium is, consider that 1,000 bacteria could be placed end to end on top of the period at the end of this sentence.

Viruses are even smaller than bacteria, ranging in size from 0.02 μm to 0.3 μm. A common unit of measure for viruses is the *nanometer* (abbreviated

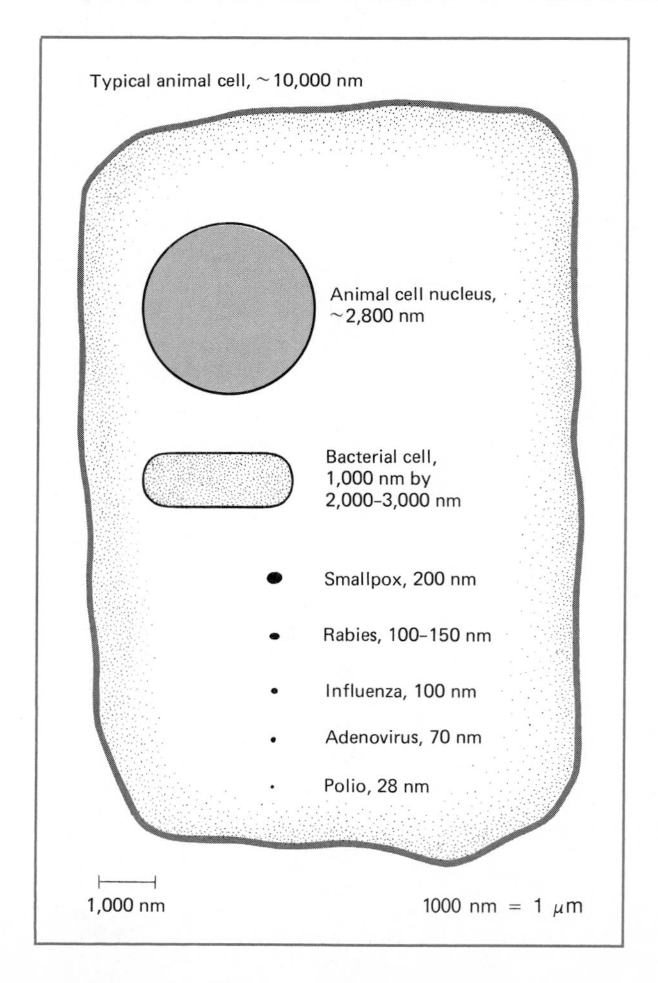

Figure 2.4 Relative sizes of cells and viruses.

nm), which is 1,000 times smaller than a μm (and one million times smaller than a millimeter).

The relative sizes of different cells and viruses are shown in Figure 2.4. As can be seen, viruses are smaller than procaryotic cells, which in turn are usually smaller than eucaryotic cells.

2.2 SEEING MICROORGANISMS

It is impossible to see a single microbial cell without a special device called a **microscope**. The microscope is the most important tool of the microbiologist. The standard type of microscope used in microbiological study is called the **light microscope**, because illumination is brought about by the use of visible light. Since the nineteenth century, microscopes have become increasingly sophisticated, and the image seen has improved accordingly. This improvement in microscopes is illustrated by the various views of yeast cells shown in Figure 2.5.

The key attributes of a microscope are magnification and resolution. *Magnification* refers to the ability of the microscope to enlarge objects and *resolution* refers to the ability of the microscope to separate into distinct images the various parts of the object being viewed. In crude or poorly designed microscopes, magnification may occur without resolution, resulting in rather fuzzy images. There is a practical limit to resolution in the light microscope, determined by the wavelength of the illuminating light. To resolve objects smaller than 0.2 μm it is necessary to use something other than visible light to view the specimen. A beam of electrons has a smaller wavelength than visible light and thus the **electron microscope** can be used to achieve resolution at higher magnification than can a light microscope (see Figure 2.5*d*).

The type of microscope most commonly used today is called a *compound light microscope* because it has two lenses. The *objective* lens is placed close to the object to be viewed and the *ocular* lens, or *eyepiece*, is placed close to the eye (Figure 2.6). The total magnification of a compound microscope is calculated by multiplying the magnification of the objective by the magnification of the eyepiece. Thus, with an objective lens magnifying 40 fold (40×) and an eyepiece magnifying 10 fold (10×), the total magnification is 400 fold (400×). The highest magnification usually used in light microscopy is about 1000 fold, achieved by using a 100× objective and a 10× ocular. The 100× objective is a type called an *oil-immersion* objective, because a drop of oil is placed between the lens and the object. The oil reduces light scattering and increases resolution. The specimen to be viewed is placed on the microscope stage and the light is directed through the specimen by the *condenser*, whose function is to focus the light precisely on the specimen.

The simplest procedure for observing a microorganism is to make a *wet mount*. This involves placing a drop of water containing the microbial cells

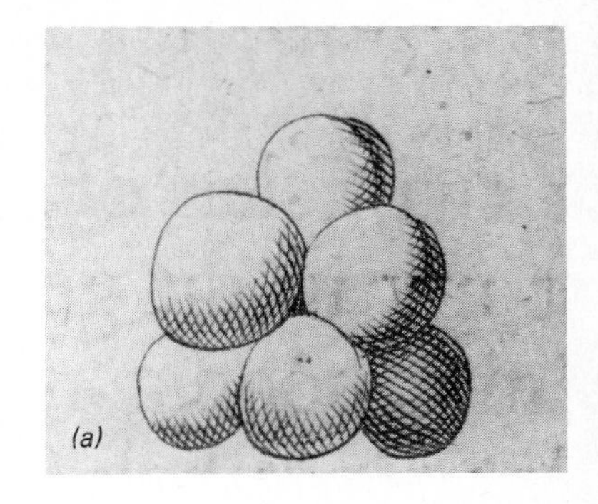

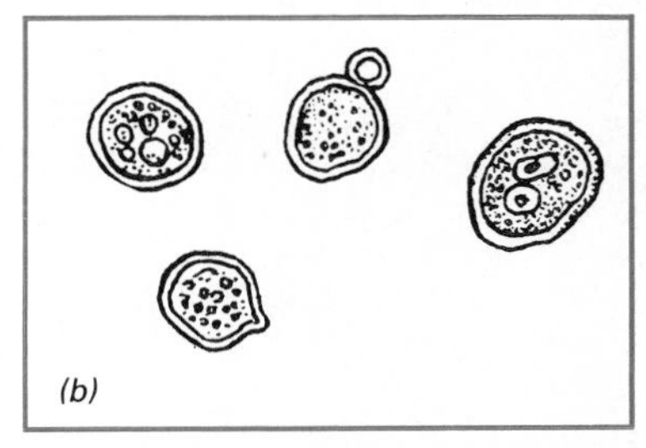

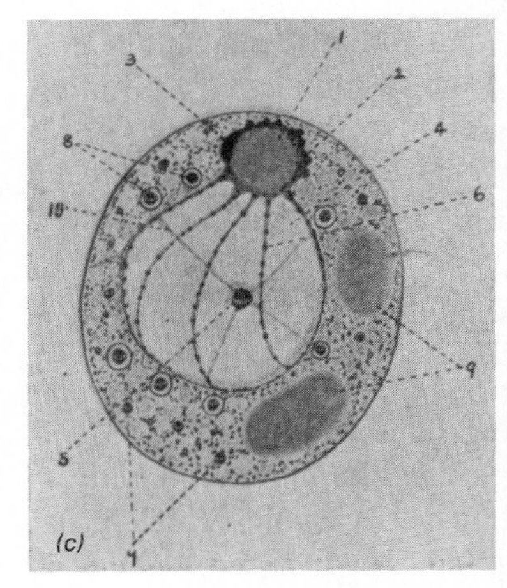

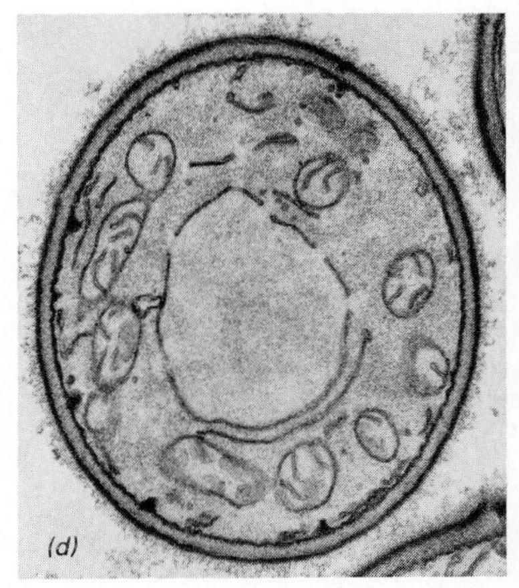

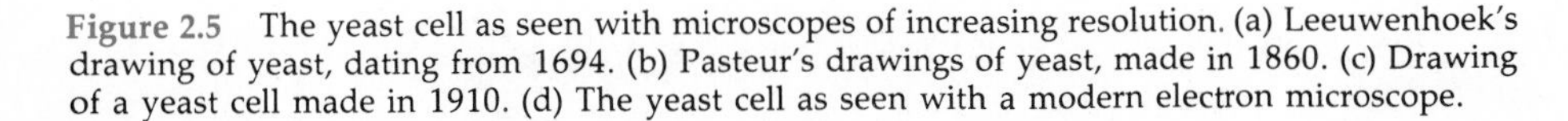

Figure 2.5 The yeast cell as seen with microscopes of increasing resolution. (a) Leeuwenhoek's drawing of yeast, dating from 1694. (b) Pasteur's drawings of yeast, made in 1860. (c) Drawing of a yeast cell made in 1910. (d) The yeast cell as seen with a modern electron microscope.

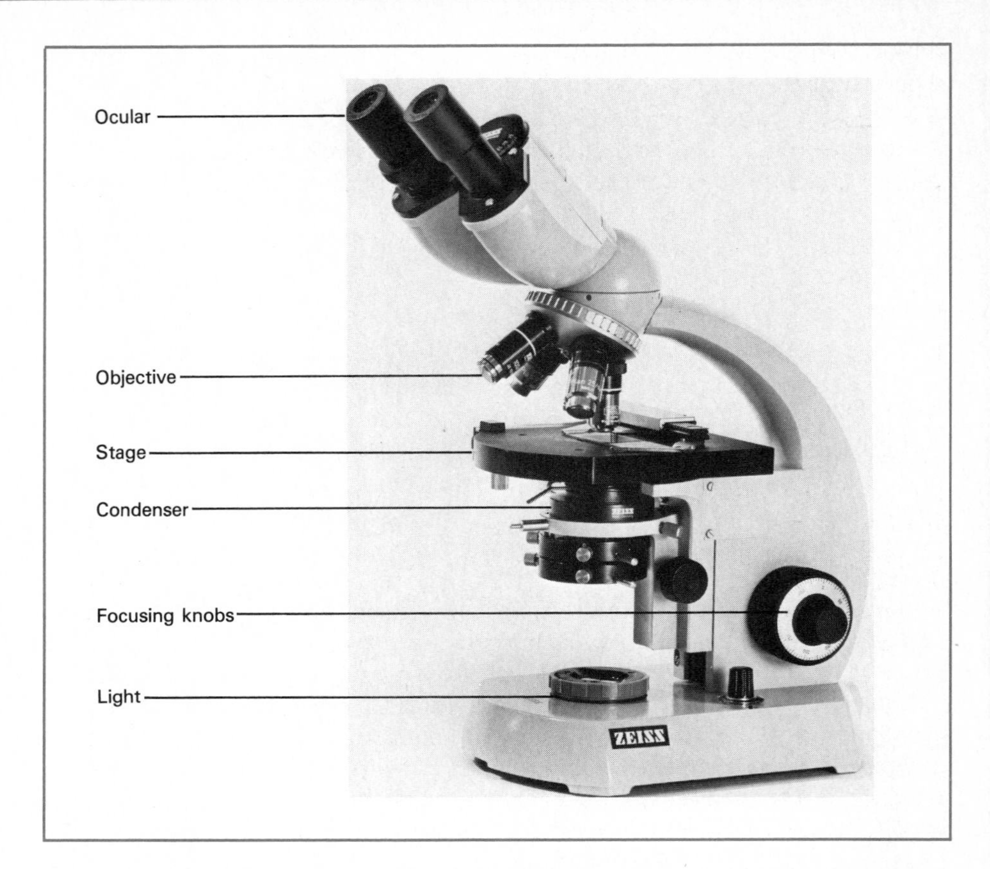
Ocular
Objective
Stage
Condenser
Focusing knobs
Light
ZEISS

Figure 2.6 A light microscope, with the various parts labeled.

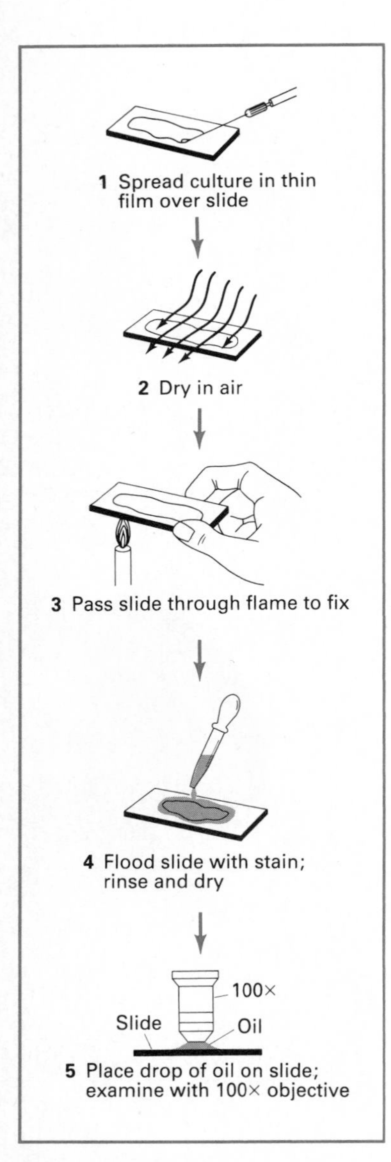
1 Spread culture in thin film over slide
2 Dry in air
3 Pass slide through flame to fix
4 Flood slide with stain; rinse and dry
100×
Slide
Oil
5 Place drop of oil on slide; examine with 100× objective

Figure 2.7 Staining cells for microscopic observation.

on a glass slide and covering the sample with a thin square of glass called a *cover slip*. Bacterial specimens are often dried before examination, although this usually leads to some distortion. The drop of water containing the specimen is spread thinly over the slide, and the water is allowed to evaporate. The dried slide is passed several times quickly through a flame to fix the cells to the surface, and the slide is then rinsed in water to remove salts. Usually, dried material is stained before examination, as described below.

Staining A major difficulty in viewing microorganisms is the lack of contrast between the cell and the surrounding medium. Unless a cell is pigmented (algae, for instance, are often green because of the presence of chlorophyll) they are virtually invisible. The simplest way of increasing the contrast of unpigmented cells is through the use of *dyes,* which color the cells and hence make them darker than the surrounding medium. A variety of dyes are used in microbiology; the most common are *methylene blue, crystal violet,* and *safranin red.*

In the simplest staining procedure, a drop of the dye is added to cells in a wet mount. However, for most studies, staining is done with dried preparations (Figure 2.7). The slide containing dried and washed organisms is flooded for a minute or two with a dilute solution of a dye, then rinsed several times in water, and blotted dry. It is usual to observe dried stained preparations of bacteria with an oil-immersion lens. To do this, a drop of oil is added directly to the surface of the dry slide and the oil-immersion objective is carefully lowered into position.

One of the most widespread and useful staining procedures is the **Gram stain**, named for the Danish bacteriologist Hans Christian Gram, who developed it in 1884. The Gram stain is called a *differential* staining procedure because bacteria of different types are stained different colors. This has permitted a division of bacteria into two groups, called **Gram positive** and **Gram negative**. After Gram staining (Figure 2.8), the Gram-positive bacteria appear purple, and the Gram-negative bacteria appear red. This difference in reaction to the Gram stain arises because of differences in the cell wall structure of Gram-positive and Gram-negative cells. The Gram-positive cell has a single thick wall layer, through which the decolorizing solvent (alcohol) does not readily penetrate, whereas the Gram-negative cell has a wall with several thinner layers through which the decolorizer readily penetrates. The Gram stain is one of the most useful staining procedures in the bacteriological laboratory; it is almost essential in identifying an unknown bacterium to know first whether it is Gram-positive or Gram-negative.

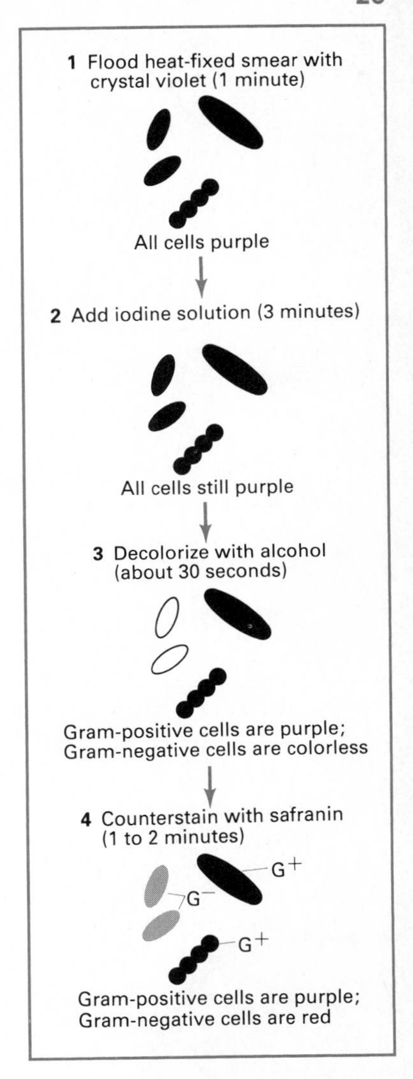

Figure 2.8 Steps in the Gram stain procedure.

Special light microscopes In addition to the conventional light microscope, several other light microscopes have been developed for research and specialized applications. The *phase-contrast microscope* uses a special optical procedure so that cells appear with greater contrast than when observed with a normal light microscope. Because the cells have excellent contrast without staining, they can be viewed in the living state. Examples of phase-contrast images were shown in Figure 2.1.

A *dark-field microscope* is an ordinary light microscope with the condenser system modified so that light is directed at the specimen from the sides. Direct light does not reach the objective, and the only light seen is that scattered from the cells. The cells thus look light on a dark background. The principle is the same as that involved when dust particles are seen in a shaft of sunlight. Dark-field microscopy makes possible the observation in the living state of particles and cells so tiny that they are invisible in a conventional light microscope. In medical bacteriology, dark-field microscopy has been used most often to study the very thin cells of the causal agent of syphilis, *Treponema pallidum*. These tiny spiral-shaped organisms move rapidly and can readily be seen in the dark field microscope (see Figure 11.16).

The *fluorescence microscope* is used to see specimens that fluoresce. Fluorescence, a property of certain chemicals and dyes, results when a chemical subjected to light of one color gives off light of another color. Some microorganisms contain naturally fluorescing substances (for example, the *green* pigment, chlorophyll, fluoresces *red* when illuminated with blue light). Cells can also be treated with certain dyes to make them fluoresce. The fluorescence microscope is widely used in medical bacteriology, where specific disease-causing microbes can be visualized by use of fluorescent dyes (see Figure 9.21).

The electron microscope Light microscopy is unsuitable for seeing viruses or for studying the detailed structures of cells, and the *electron microscope* must be used. An electron beam is used instead of light rays, and magnets are used as lenses. The specimen is mounted on a thin film of plastic and placed in a vacuum. The image is seen on a small screen resembling a television screen, and for a permanent record photographs are taken. Electron microscopes are not only much more expensive and complicated to use than light microscopes, but also must be carefully aligned and maintained if they are to yield good results.

Only very thin objects can be examined with an electron microscope, and if whole cells are observed only the outline of a specimen can be seen. Electron microscopy can provide useful information about cell shape and may often reveal structural features on the outsides of microorganisms. However, to really use the electron microscope to study interior structures, very *thin sections* must be cut through cells. Such sections are prepared by embedding cells in plastic and making very thin slices (0.1 μm or less) with a glass or diamond knife. In such sections, cellular structures can be seen with greatly increased resolution.

Another type of electron microscope, designed especially for looking at surfaces of cells, is the **scanning electron microscope**. In this device, the electron beam is directed down on the surface of the specimen (coated with a thin layer of metal), and the electrons scattered from the cells are caught and made to form an image.

2.3 STRUCTURE OF CELLS

We have stated briefly that two types of cells are recognized, called *procaryotic* and *eucaryotic*. Procaryotic microorganisms are all classified in the broad group called the *bacteria*. Eucaryotic microorganisms include the *algae*, *fungi*, and *protozoa*. These two types of cells are distinguished in a number of ways, of which the most important is the structure of the **nucleus**. Eucaryotes have a **true nucleus**, whereas procaryotes lack a true nucleus but have a **nuclear region**. A term occasionally used when discussing structures of organisms is **morphology**, which means, literally, the study of form. In the present section, we present a short discussion of the morphology of the procaryotic and eucaryotic cell, and show how these two types of cells differ.

The procaryotic cell Careful study with light and electron microscopes has revealed the detailed structure of procaryotic cells. Starting at the outside of a typical procaryotic cell and proceeding inward, the following structures are seen (Figure 2.9*a* and *b*): (1) cell wall; (2) cell membrane; (3) internal membranes; (4) ribosomes; (5) nuclear region. What are these structures and what are their functions?

The **cell wall** confers strength on the cell and determines the cell shape. The cell wall also acts like armor, protecting the cell from many external influences. If the cell wall is removed, the membrane usually breaks and the cell bursts, a process called **lysis**.

The **cell membrane** is a very thin structure through which all of the food materials pass in and through which waste products and other products of

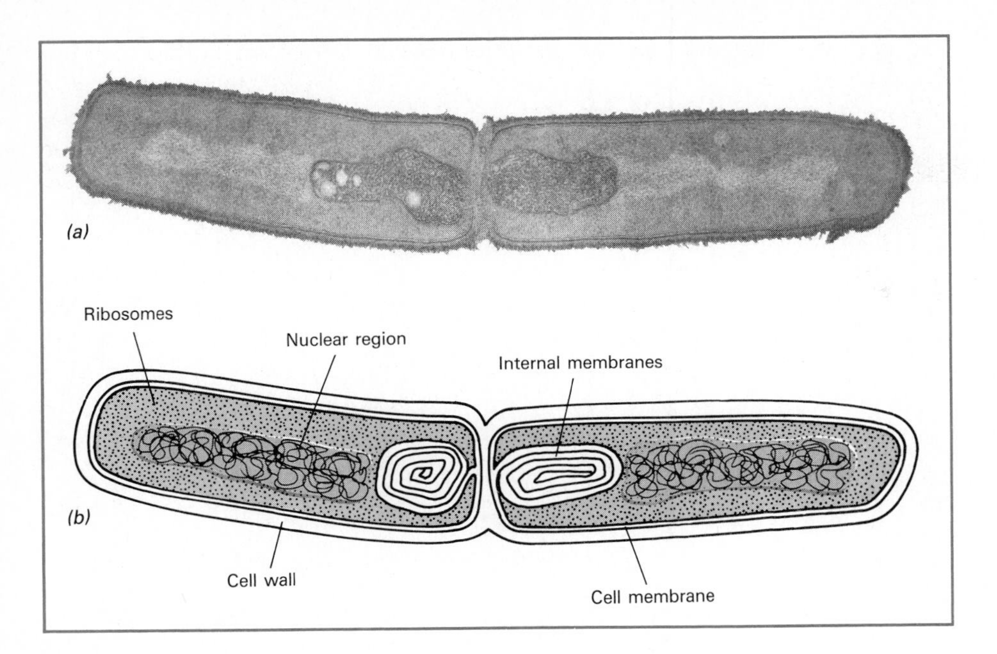

Figure 2.9 The interior structures of a procaryotic cell. (a) An electron micrograph of a thin section through a dividing cell of *Bacillus subtilis.* (b) Diagram showing the major components.

metabolism pass out. Connected to the cell membrane and extending into the interior of the cell are usually many other thin membranes, called *intracellular membranes,* which serve as surfaces onto which other substances of the cell attach and upon which many important cell functions take place. These intracellular membranes serve as organizers for many cell functions, which otherwise would take place randomly within the cell. For example, procaryotes which are able to use light energy and carry out the process of photosynthesis contain a type of internal membrane called the *photosynthetic membrane.*

The cell also contains **ribosomes**, small particles composed of *protein* and *ribonucleic acid* (RNA), which are just barely visible in the electron microscope. A single cell may have as many as 10,000 ribosomes. Synthesis of cell proteins takes place upon these structures. We discuss ribosomes, proteins, and RNA in Chapter 6.

As we have noted, the **nuclear region** of the procaryotic cell is primitive, in contrast to that of the eucaryotic cell to be discussed in the next section. Procaryotic cells do not possess a true nucleus, the functions of the nucleus being carried out by a single long strand of **deoxyribonucleic acid (DNA)**. DNA, the genetic material of the cell, is the key substance bearing information that determines production of proteins and other cellular substances and structures. This nucleic acid is present in a more or less free state within the cell. The procaryote is thus said to have a nuclear region (the place where the DNA is present) rather than a nucleus. During cell division, the DNA within a cell is copied exactly. The cell then splits in two by a process called *binary fission,* each resultant daughter cell receiving an identical DNA molecule. (The cells in Figure 2.9*a* are undergoing fission.)

Many, but not all, procaryotic microorganisms are able to move. Movement of a procaryotic cell is usually by means of a structure called a **flagellum**

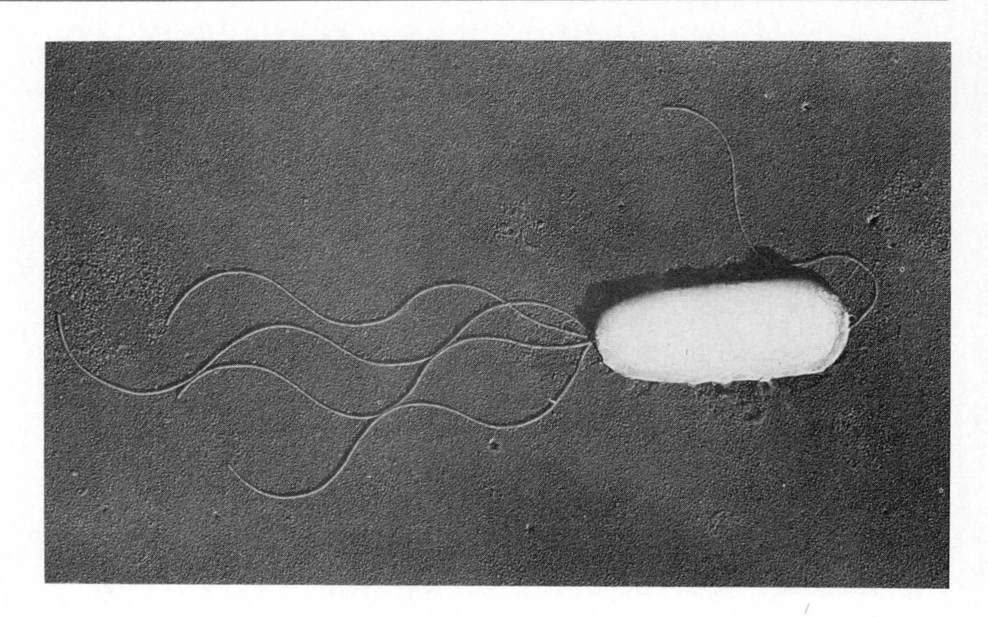

Figure 2.10 Electron micrograph of a bacterial cell, showing flagella. Magnification, 17,000 ×.

(plural, *flagella*). Each flagellum (Figure 2.10) consists of a single rigid coiled tube of protein. The rotation of flagella propels the cell through the water. Flagella are too small to be seen in the light microscope without special staining techniques, but are readily visible in the electron microscope.

Various other structures are found in some, but not all, procaryotes, including *spores* and *capsules*. These structures are described later in this chapter (see Section 2.5).

The eucaryotic cell Eucaryotic cells are larger and more complex in structure than procaryotic cells. The key difference is, of course, that eucaryotes contain true nuclei. Eucaryotic cells also contain within them distinct structures called *organelles* within which important cellular functions occur (Figure 2.11). These structures are lacking in procaryotes, in which similar functions, if they occur, are not restricted to special organelles.

The most obvious eucaryotic organelle is the **nucleus**, a special membrane-surrounded structure within which the genetic material of the cell, the DNA, is located. The DNA in the nucleus is organized into **chromosomes**, which are invisible except at the time of cell division. Before cell division occurs, the chromosomes are duplicated and then condense, become thicker and then undergo division as the nucleus divides. The process of nuclear division in eucaryotes is called *mitosis* and is a complex but highly organized process. Two identical daughter cells result from the division of one parent cell. Each daughter cell receives a nucleus with an identical set of chromosomes.

Both binary fission in procaryotes and mitosis in eucaryotes are examples of *asexual reproduction*. Many eucaryotic microorganisms also carry out the process of *sexual reproduction*. During sexual reproduction, specialized reproductive cells from two different parents (usually called male and female in analogy with higher organisms) fuse, so that the genetic information within

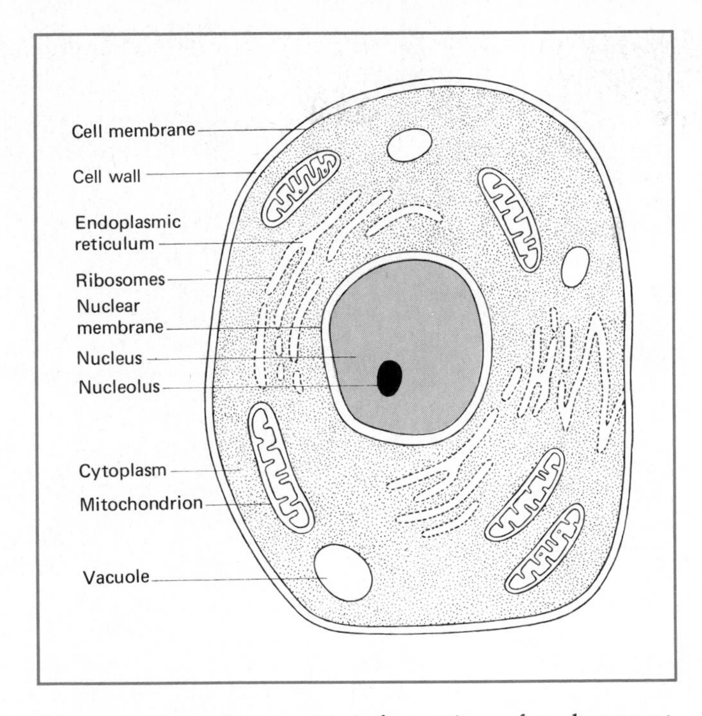

Figure 2.11 The structure of a eucaryotic cell.

the offspring is derived in part from the genetic information of each parent. Thus, each offspring is not exactly like either parent.

As we have noted, some of the functions in eucaryotes are carried out in special structures called **organelles**. One organelle found in most eucaryotes is the **mitochondrion** (plural, *mitochondria*). Mitochondria are the organelles within which the energy-generating functions of the cell occur. The energy produced in the mitochondria is then used throughout the cell.

The algae are eucaryotic microorganisms which carry out the process of *photosynthesis*. In these organisms, an additional organelle is found: the **chloroplast**. The chloroplast, the site where chlorophyll is localized, is green, and it is within this organelle that the light-gathering functions involved in photosynthesis occur.

The *cell membrane* of the eucaryotic cell differs chemically but not functionally from that of the procaryotic cell. The *cell walls* of eucaryotes are usually much thicker than the cell walls of procaryotes and also have different chemical structures. Among the eucaryotes, algae and fungi have cell walls, but protozoa do not.

There are several types of internal membrane systems within a eucaryotic cell. One system is called the *endoplasmic reticulum*, and it is on this system that ribosomes are attached and protein synthesis occurs. Membranes are also found surrounding special regions within the cell called *vacuoles*.

Many eucaryotic cells are motile, and two types of organelles of motility are recognized: flagella and cilia (Figure 2.12). **Flagella** are long filamentous structures, each of which is attached to one end of the cell; they move in a whiplike manner. It should be emphasized that the flagella of eucaryotes are quite different in structure from those of procaryotes, even though the same

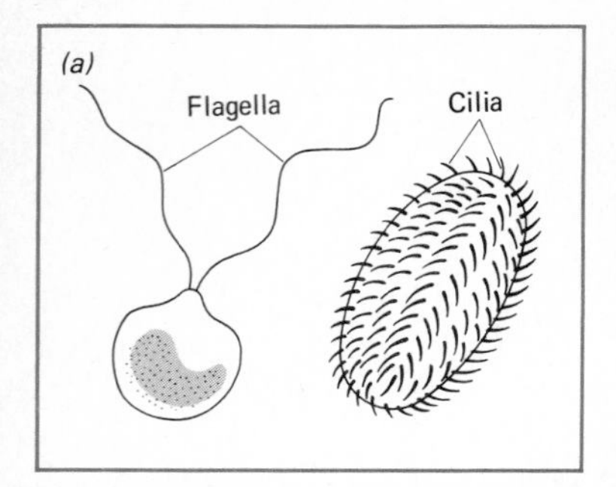

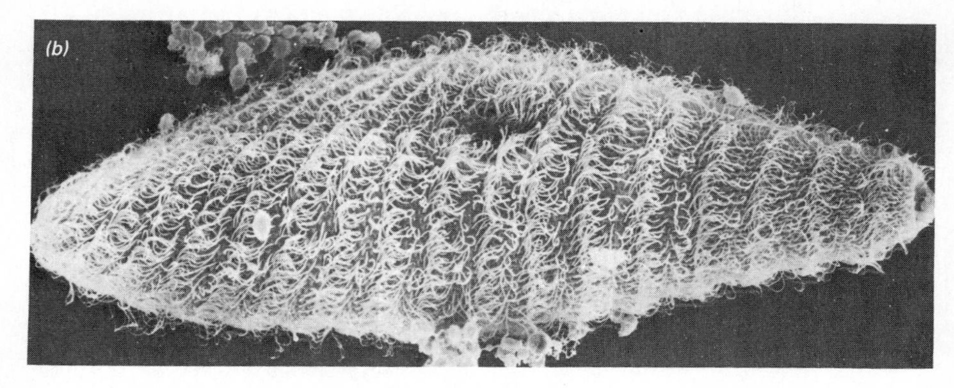

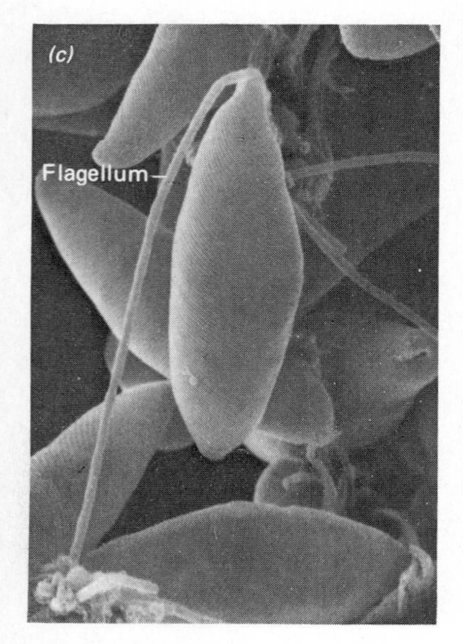

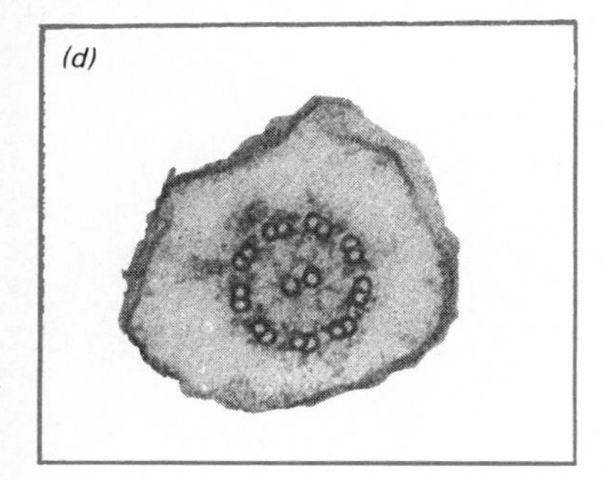

Figure 2.12 Cilia and flagella, organs of motility in eucaryotic cells. (a) Diagram comparing structure of cilia and flagella. (b) *Paramecium* cell, showing large number of cilia. This is a scanning electron micrograph. Magnification, 1800 ×. (c) *Euglena*, a flagellate. Magnification, 1800 ×. (d) An electron micrograph of a cross section of a eucaryotic flagellum, showing the complex internal structure.

name is used for both. The flagellum of a eucaryote is composed of two central protein fibers surrounded by nine outer fibers, with each of the latter composed of two subfibrils (Figure 2.12*d*). Eucaryotic flagella are large enough to be seen with the light microscope, and their movements can be easily followed.

Cilia (singular, *cilium*) are similar to eucaryotic flagella in structure but are shorter and more numerous. A single cell may have over 10,000 cilia. These organelles operate like oars and propel the cell through the water. Protozoa are the only microorganisms which possess cilia, but ciliated cells are also common in higher organisms.

To further emphasize the vast structural differences between procaryotes and eucaryotes, a comparison between the cells of these two kinds of organisms is given in Table 2.1. Although procaryotes and eucaryotes differ greatly in structure, it should be emphasized that they are quite similar in chemical composition and metabolism. Evidence is strong that procaryotes and eucaryotes have had a common origin.

TABLE 2.1 Characteristics of procaryotic and eucaryotic cells

Characteristic	Procaryote	Eucaryote
Size of cell	Small, 0.5–2.0 μm	Larger, 2–200 μm
Nuclear body	No nuclear membrane; no mitosis	True nucleus; nuclear membrane; mitosis
DNA	Single molecule; not in chromosomes	Several or many chromosomes
Organelles	None	Mitochondria, chloroplasts, vacuoles, others
Cell wall	Relatively thin; usually peptidoglycan	Thick or absent; chemically different
Manner of movement	Flagella of submicroscopic size; single protein fiber	Flagella or cilia of microscopic size; complex pattern of fibers

Viruses Viruses are simpler in structure than even procaryotic cells. The simplest viruses contain nothing more than genetic material (DNA or RNA) surrounded by a **protein coat**. An example of the assembly of these parts and the formation of a complete virus particle is shown in Figure 2.13. The more complex viruses also have a primitive membrane surrounding the protein coat and the genetic material. Viruses cannot carry out metabolism or reproduction by themselves. When a virus particle infects a cell, the control of the cell's metabolism can be drastically altered, causing the cell to make more virus particles. The host cell is usually damaged and often destroyed as a result of virus infection. We discuss viruses and virus diseases in detail in Chapters 6 and 13.

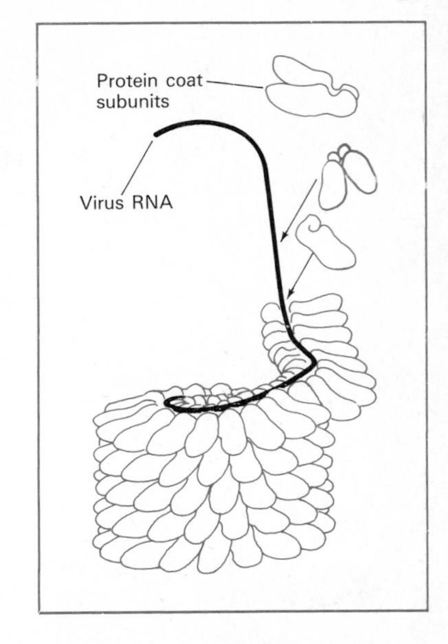

Figure 2.13 Structure of a simple virus particle, tobacco mosaic virus.

2.4 NAMES OF MICROORGANISMS

Because of the diversity of microorganisms, it is useful to give names to different organisms. Microbiologists use the *binomial system* of nomenclature, first developed for plants and animals by the Swedish botanist Carl von Linné (Linnaeus) in the eighteenth century. In the binomial system, each organism has two names. The first name is the **genus** name, and the second is the **species** name. Thus, the name of a common yeast is *Saccharomyces cerevisiae, Saccharomyces* being the genus and *cerevisiae* the species. There may be several species with the same genus name, for example, *Saccharomyces cerevisiae, S. pastorianus,* and *S. carlsbergensis*. We can refer to the genus *Saccharomyces* without designating a specific organism, and when doing so, we are referring to a whole group of related organisms.

The names of microorganisms are usually derived from Latin or Greek and hence may be unfamiliar. This is unfortunate, but there is no way to avoid learning at least some names of microorganisms. After all, we cannot discuss the cast of characters without calling them by name. Sometimes, the meaning of a name can be deduced from its Latin or Greek roots. Thus, *Saccharomyces cerevisiae,* the beer yeast, got its name in the following way: yeasts convert sugar to alcohol, and *saccharo* means "sugar"; a yeast is a fungus, and *-myces* derives from the Greek meaning "fungus"; *cerevisiae* derives from the Latin word meaning "beer."

We have now explored how differences in the structures of the major types of microorganisms help us to distinguish and define procaryotic and eucaryotic cells and viruses. In the remainder of this chapter we will highlight some of the types of microbes found within each group.

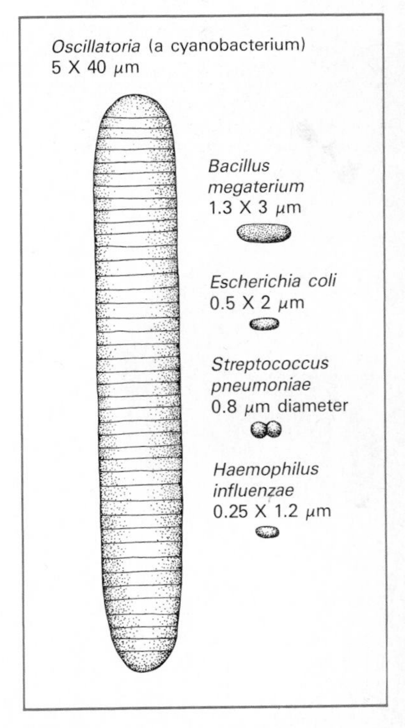

Figure 2.14 Comparison of sizes of a variety of procaryotic microorganisms.

2.5 BACTERIA

Bacteria vary in size from cells as small as 0.1 μm in width to those more than 5 μm in diameter (Figure 2.14). Among the bacteria, differences occur not only in form and structure, but also in function. Since function (metabolism) is the subject of the next chapter, we will consider here bacterial diversity primarily as reflected in structure.

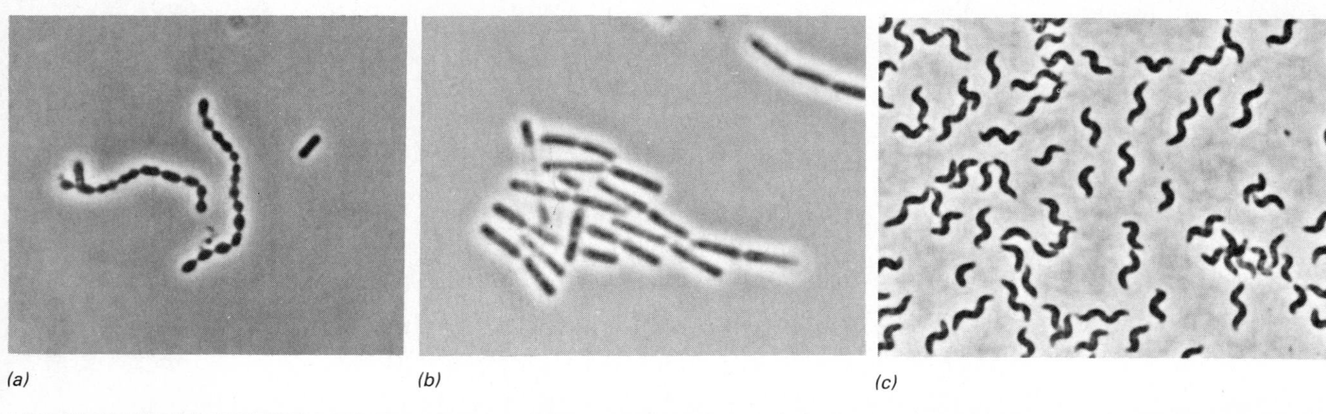
(a) (b) (c)

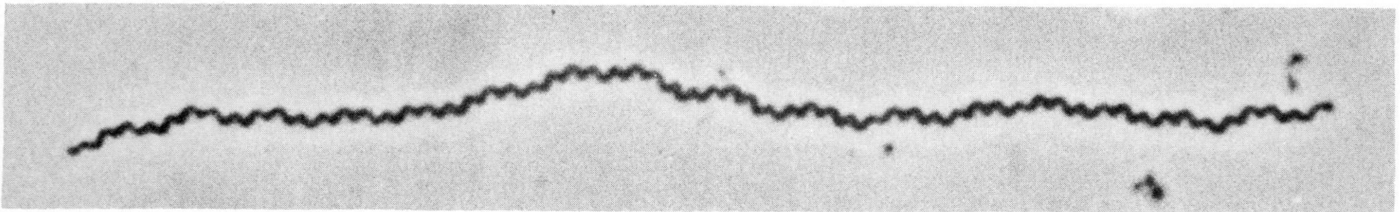
(d)

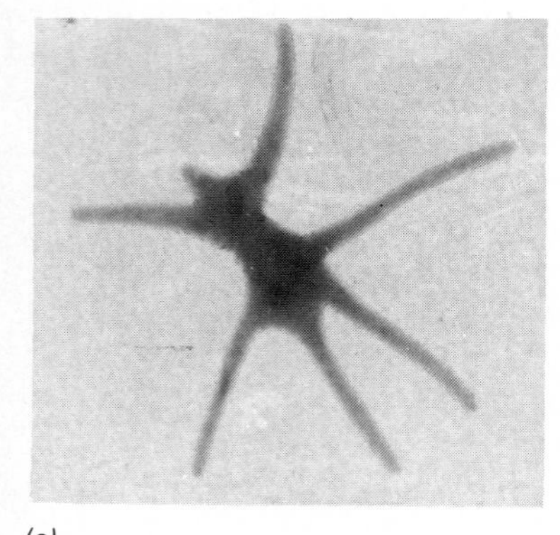
(e)

Figure 2.15 Shapes of bacteria. (a) A coccus-shaped organism, *Streptococcus.* Magnification, 1500 ×. (b) A rod-shaped organism, *Bacillus.* Magnification, 1500 ×. (c) A spiral-shaped bacterium, *Rhodospirillum.* Magnification, 1500 ×. (d) A spirochete, *Spirochaeta plicatilis.* Magnification, 1500 ×. (e) A bacterium with a number of appendages, *Ancalomicrobium adetum.* Magnification, 6300 ×.

Several distinct shapes of bacteria can be recognized, which have been given names. Examples of some of these various bacterial shapes are given in Figure 2.15. A bacterium which is spherical or egg-shaped is called a *coccus* (plural, *cocci*) (Figure 2.15*a*). A bacterium with a cylindrical shape is called, simply, a *rod* (Figure 2.15*b*). Some rods are curved, frequently forming spiral-shaped patterns, and are called *spirilla* (Figure 2.15*c*).

In many microorganisms, the cells remain together in groups or clusters, and the arrangements in these groups are often characteristic of different organisms. For instance, cocci or rods may occur in long chains (as shown for cocci in Figure 2.15*a*). Some cocci form thin sheets of cells, while others occur in three-dimensional cubes or irregular grapelike clusters. Several groups of bacteria are recognized by their unusual shape. Examples include coiled bacteria, the *spirochetes* (Figure 2.15*d*), and bacteria which possess extensions of their cells as long tubes or stalks, the *appendaged bacteria* (Figure 2.15*e*).

Bacteria of another group, the *spore-forming bacteria,* produce structures called **endospores** within their cells (Figure 2.16). Endospores are very resistant to heat and cannot be killed easily even by boiling. When spore-forming bacteria are present, heat sterilization of foods and other perishable products is difficult. A knowledge of the nature and properties of spores is of considerable importance in applied microbiology. Endospores can be seen easily with a phase-contrast microscope; they appear as bright shining objects in contrast to the dark appearance of the bacterial cell. Spore-forming bacteria are found most commonly in the soil, and virtually any sample of soil will have some bacterial spores present. Since soil particles contaminate nearly

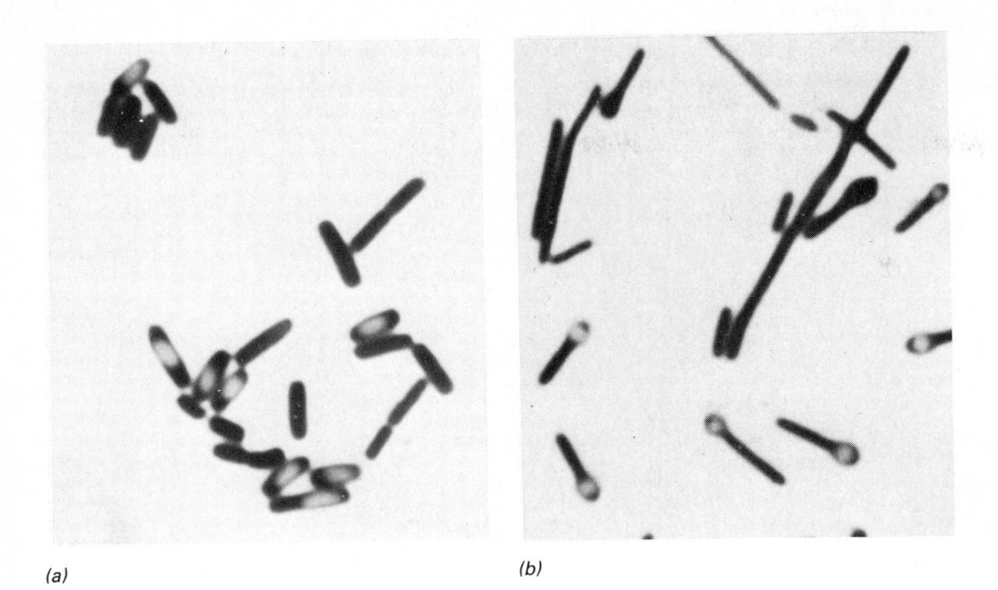

(a) (b)

Figure 2.16 Bacteria containing endospores; the spores are the light structures. (a) *Clostridium bifermentans*, subterminal. (b) *Clostridium tetani*, terminal.

all materials, it must be assumed that spores are present in any item to be sterilized. Bacteria that are actively growing do not form spores, but when growth ceases due to starvation or some other cause, spore formation may be initiated. Spores are more resistant than normal cells to drying, radiation, and drugs, as well as to heat. Spores are able to remain alive but dormant and inactive for many years; however, they can convert back to normal cells in a matter of minutes, given proper conditions. This process, called *spore germination*, involves outgrowth of a new cell from within the spore (Figure 2.17). As soon as germination begins, resistance to heat and to other harmful agents is lost.

Cell shape and arrangement, in combination with Gram-stain reaction, presence or absence of endospores, and whether metabolism is aerobic or anaerobic, are features used to separate many of the major groups of rod- and coccus-shaped bacteria (Table 2.2). We will consider some of these bacterial groups further at various points in the book.

Although size, shape, and cell arrangement are relatively constant for a single organism, variations do occur, often influenced by environment. Some

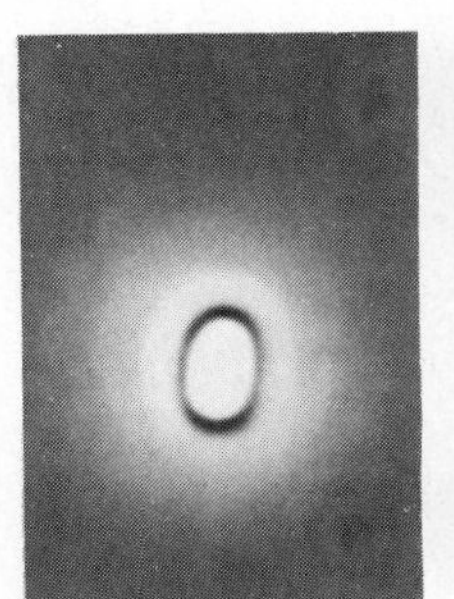
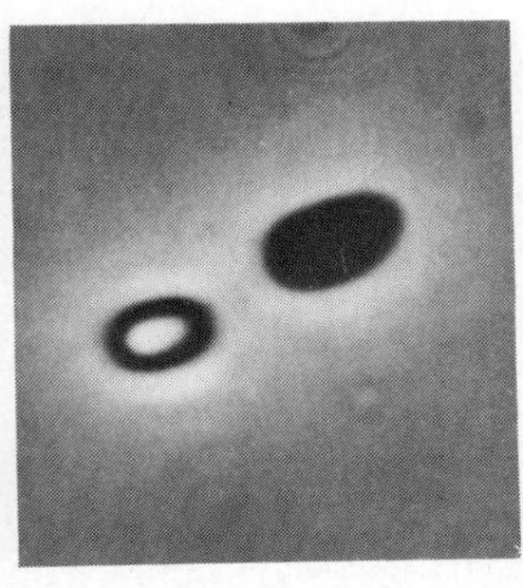
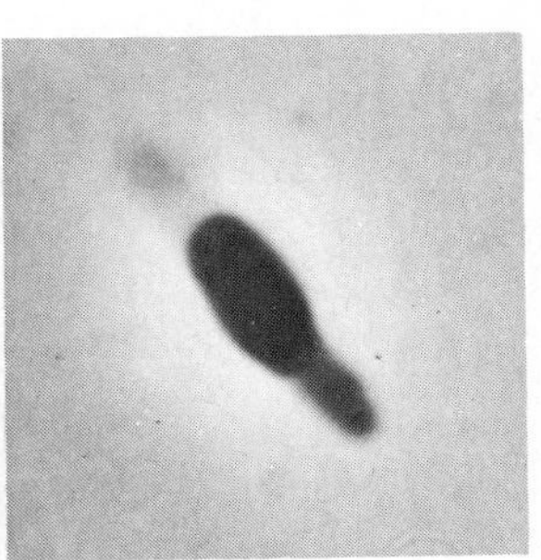
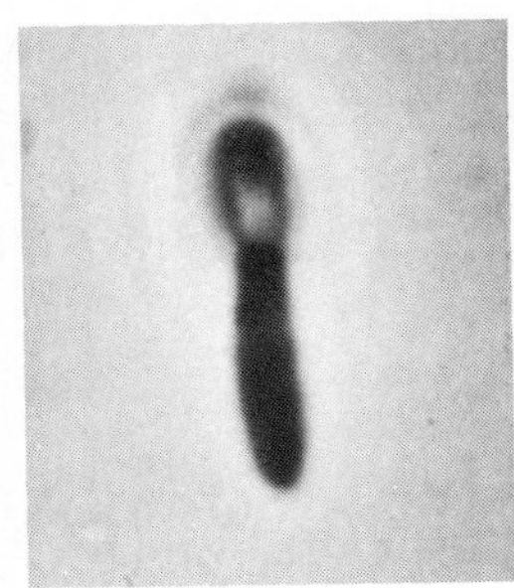
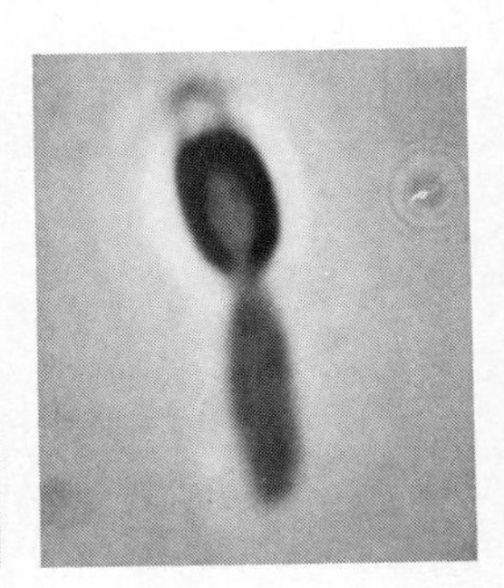

Figure 2.17 Stages in endospore germination.

TABLE 2.2 Some major groups of rod- and coccus-shaped bacteria

Shape	Gram reaction	Oxygen requirement	Endospore formation	Representative genus	Important example
Rod	+	+	+	*Bacillus*	Anthrax
	+	−	+	*Clostridium*	Gas gangrene
	+	+/−	−	*Lactobacillus*	Yogurt
	−	+	−	*Pseudomonas*	Hospital infections
	−	+/−	−	*Escherichia*	Normal in intestine
			Cell arrangement		
Coccus	+	+/−	Packets or clusters	*Staphylococcus*	Boils
	+	+	Packets or clusters	*Micrococcus*	Normal in soil
	+	+/−	Chains	*Streptococcus*	Sore throat
	−	+	Chains or packets	*Neisseria*	Gonorrhea

+, must have oxygen.
+/−, can live with or without oxygen.
−, are harmed by oxygen.

organisms are more variable than others. For instance, in the genus *Arthrobacter*, cells of a given species are sometimes rods and at other times cocci. Cells with a variable morphology are called *pleomorphic,* meaning "many-shaped." A great many organisms show some pleomorphism with age, so that accurate description of cell morphology must be done using young cells.

Bacteria called *actinomycetes* form long filaments which are often extensively branched. Among these are *Nocardia* (Figure 2.18*a*) and *Streptomyces* (Figure 2.18*b*). These bacteria form irregular masses of filaments, collectively referred to as a *mycelium* (analogous to that of fungi, discussed below). Streptomycetes also have the ability to form spores (sometimes called *conidia*),

Figure 2.18 Filamentous bacteria, forming mycelium. (a) *Nocardia corallina.* Magnification, 1000 ×. (b) *Streptomyces.* Magnification, 525 ×.

(a)

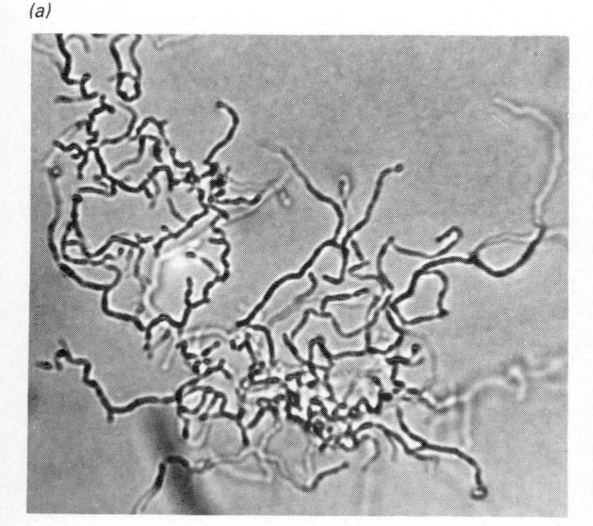

(b)

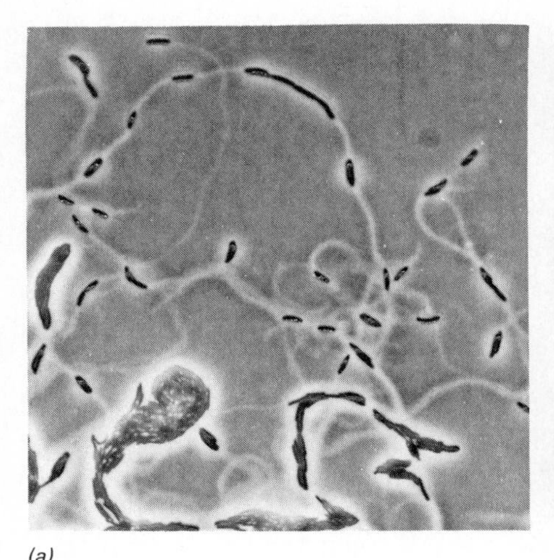

(a)

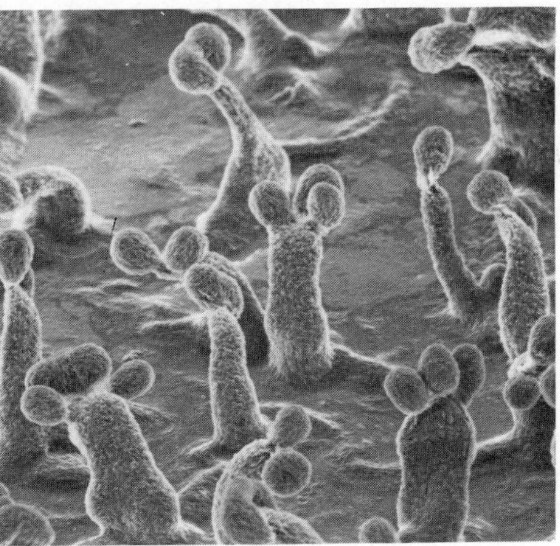

(b)

Figure 2.19 Gliding bacteria which form fruiting bodies. (a) *Myxococcus fulvus*, single cells swarming on agar, showing the characteristic slime tracks. Magnification, 370 ×. (b) *Stigmatella aurantiaca*, scanning electron micrograph of fruiting bodies. Magnification 200 ×.

which are resistant to adverse environmental conditions such as dryness (but usually not as resistant to heat as endospores). As the streptomycete mycelium develops, some filaments grow into the air, and spores are formed on these aerial filaments. Actinomycetes are common soil microorganisms and they are especially important as sources of antibiotics (see Chapter 5).

Perhaps the most complex bacteria are found among the *gliding bacteria*. These organisms move without flagella, gliding over surfaces and leaving tracks of slime (Figure 2.19*a*). Some gliding bacteria, the *myxobacteria*, have complex life cycles. Individual cells of myxobacteria form slimy aggregates which then produce raised structures called *fruiting bodies* (Figure 2.19*b*). Within the fruiting body, cells turn into myxospores, although these myxospores are not heat resistant like endospores. Myxospores germinate and form gliding rods, thus completing the life cycle.

Some bacteria possess thick coatings outside their cell walls which are called **capsules** (Figure 2.20). This is a gummy material secreted onto the surface of the cell, forming a compact layer. Capsules can be visualized in the light microscope by use of a special negative staining procedure (illustrated in Figure 2.20). Many disease-causing bacteria have capsules that help them adhere to surfaces or resist host defenses (see Chapter 8).

Some bacteria move by means of flagella. Two types of flagellar arrangement exist: polar and peritrichous (Figure 2.21). Polarly flagellated organisms have a single flagellum or a group of flagella at one or both poles of the cell, whereas peritrichously flagellated organisms have flagella attached at many points on the surface of the cell. Although special staining procedures are available to demonstrate the presence of flagella during light microscopy, these structures can also be readily revealed with the electron microscope (see Figure 2.10).

The diversity among bacteria is well demonstrated by differences in shape and structure. Still more variety will be found among the bacteria when we consider their metabolism in the next chapter.

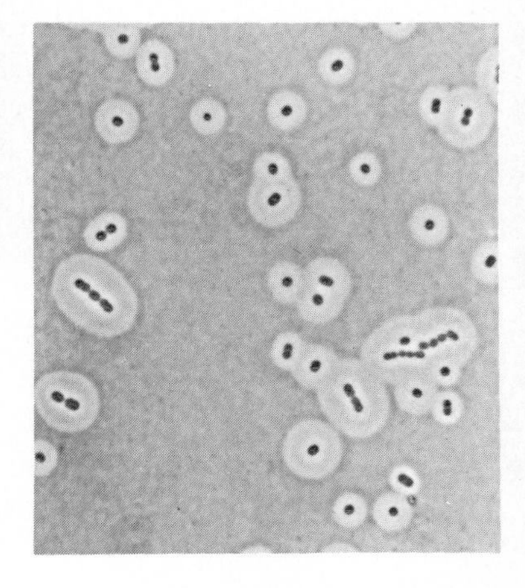

Figure 2.20 Capsulated bacteria of the genus *Acinetobacter*. The specimen was treated with india ink (a procedure called *negative staining*) to reveal the capsule.

(a)

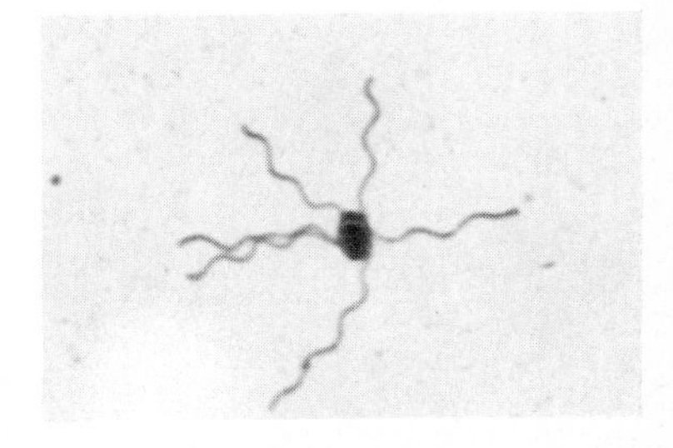

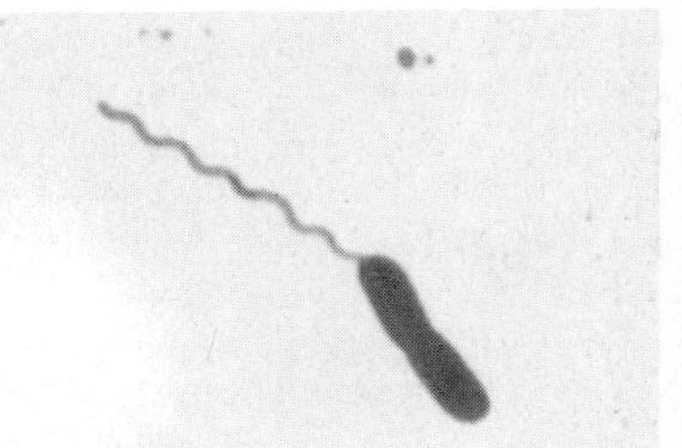

(b)

Figure 2.21 Bacterial flagella, as revealed by a special staining procedure. (a) Peritrichous flagellation. (b) Polar flagellation.

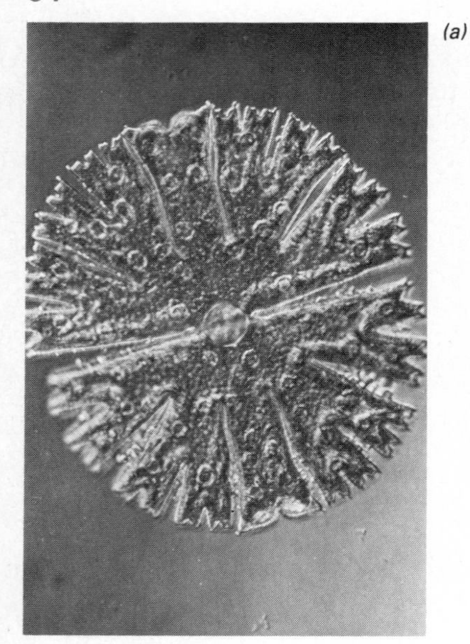

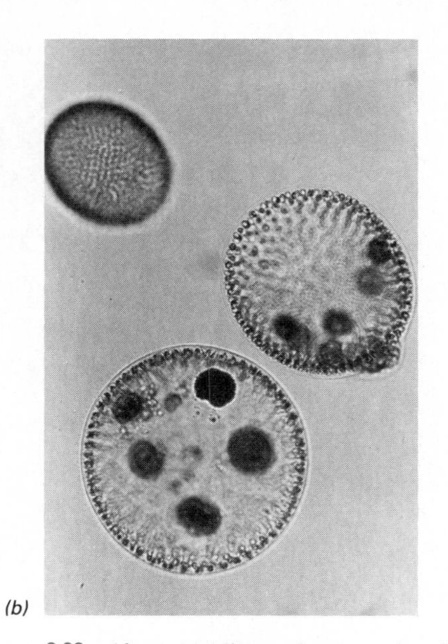

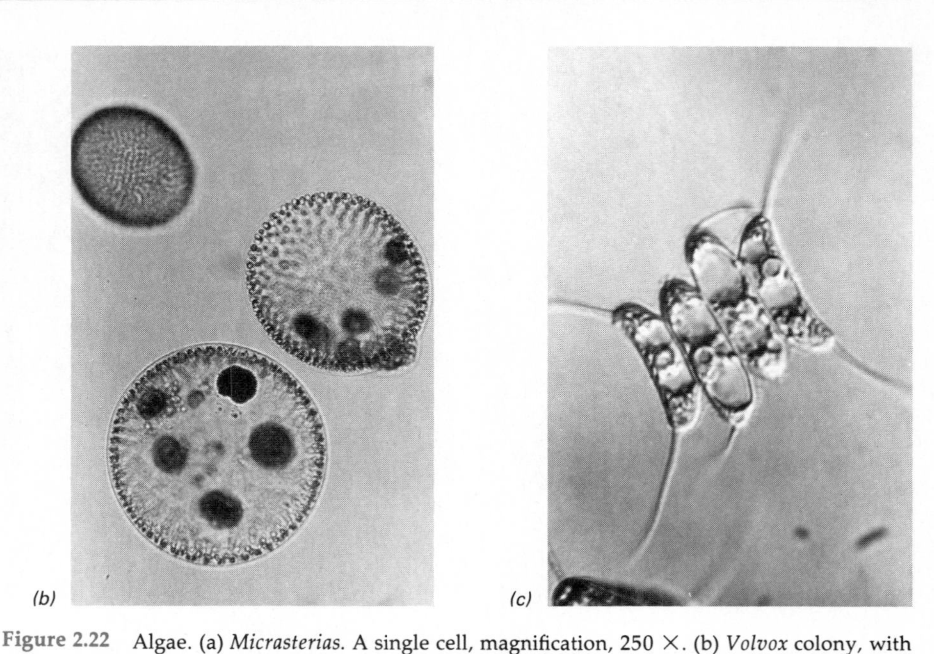

Figure 2.22 Algae. (a) *Micrasterias*. A single cell, magnification, 250 ×. (b) *Volvox* colony, with a large number of cells. Magnification, 1000 ×. (c) *Scenedesmus*, a packet of four cells. Magnification, 100 ×. (d) *Ulothrix*, a filamentous alga. Note the chloroplasts. Magnification, 350 ×.

2.6 ALGAE

Algae contain **chlorophyll**, a green pigment that serves as a light-gathering substance and makes it possible for them to use light as a source of energy. Because of chlorophyll, most algae are green, and they can be recognized first by their green color. However, a few kinds of common algae are not green but appear brown or red because in addition to chlorophyll other pigments are present that mask the green color.

A characteristic structure of algae is the **chloroplast**, one of the organelles so common in eucaryotes. The chloroplast is the site of the chlorophyll pigments, and can often be recognized microscopically within an algal cell by its green color. As we have noted, the chloroplast is the site of photosynthesis in algae. The chloroplasts of eucaryotic algae are especially easy to see in Figure 2.22*d*.

Algae can be either *unicellular* (Figure 2.22*a*) or *colonial*, occurring as aggregates of cells (Figure 2.22*b-d*). When the cells are arranged end-to-end, the alga is said to be *filamentous* (Figure 2.22*d*). Among the filamentous forms, both unbranched filaments and more complicated branched filaments occur. In many cases, the filamentous algae occur in such large masses that they can be seen easily without a microscope. Small ponds or streams or even fish tanks frequently show such massive accumulations of algae.

The identification of algae makes use primarily of the microscope. The main characteristics used in identification are (1) color and kind of pigments; (2) cellular organization, whether unicellular, colonial, filamentous, or a more complex form; (3) motility, and if motile, the shape, number, and position of the flagella, or whether motile by gliding motility; (4) structure and appearance

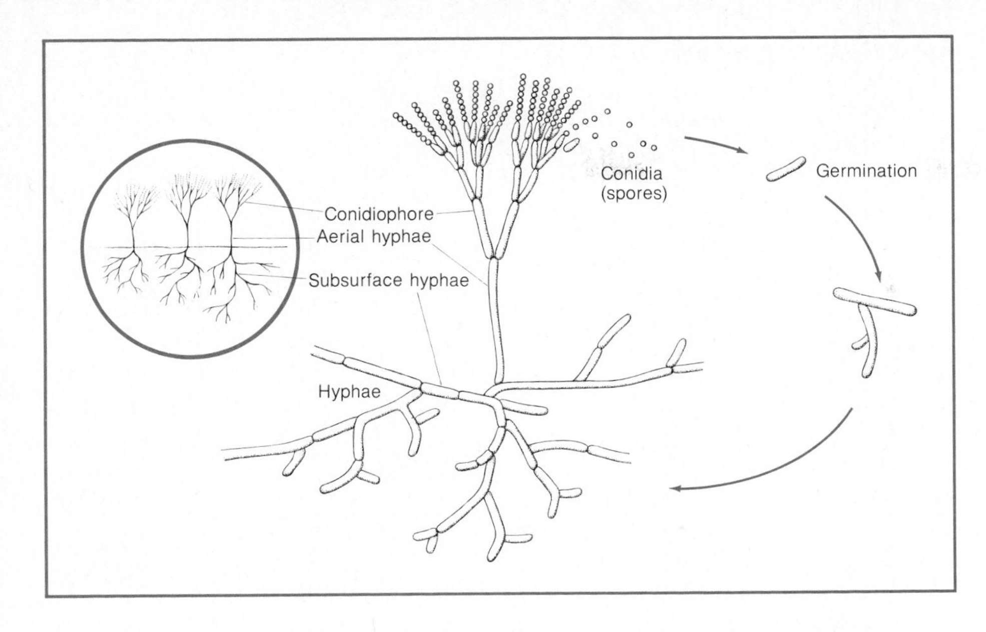

Figure 2.23 Mold structure and growth.

of the cell wall (some groups have complex walls composed of silica or calcium carbonate, and the markings on the walls are distinctive); (5) presence and type of sexual reproduction, and the kinds of structures involved in or formed during sexual reproduction; and (6) habitat, especially whether fresh-water or marine, since some kinds of algae are found only in one or the other of these habitats.

2.7 FUNGI

Fungi can be distinguished from algae because the fungi do not have chlorophyll and thus are not green. Fungi can be differentiated from bacteria by the fact that fungal cells are much larger, and vacuoles, nuclei, and other intracellular organelles typical of eucaryotic cells can usually be seen inside, even with the ordinary light microscope. Although the fungi are a large and rather diverse group of eucaryotic microorganisms, three groups of fungi are of major practical importance: the molds, the yeasts, and the mushrooms.

Molds The molds are filamentous fungi (Figure 2.1*e* and 2.23). They are widespread in nature and are commonly seen on stale bread, cheese, or fruit. Each filament grows mainly at the tip, by extension of the terminal cell. A single filament is called a **hypha** (plural, *hyphae*). Hyphae usually grow together across a surface and form rather compact tufts, collectively called a **mycelium**, which can be easily seen without a microscope. The mycelium arises because the individual hyphae form branches as they grow, and these branches intertwine, resulting in a compact mat, or felt. From the mycelium, other branches may reach up into the air above the surface, and on these aerial branches spores (also called **conidia**, singular, *conidium*) are formed. Conidia are round structures, often highly pigmented, that are resistant to

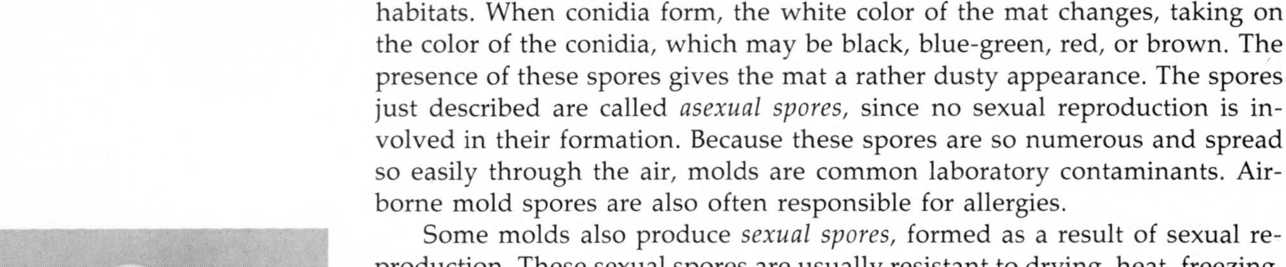

drying, are very lightweight, and permit the fungus to be dispersed to new habitats. When conidia form, the white color of the mat changes, taking on the color of the conidia, which may be black, blue-green, red, or brown. The presence of these spores gives the mat a rather dusty appearance. The spores just described are called *asexual spores,* since no sexual reproduction is involved in their formation. Because these spores are so numerous and spread so easily through the air, molds are common laboratory contaminants. Airborne mold spores are also often responsible for allergies.

Some molds also produce *sexual spores,* formed as a result of sexual reproduction. These sexual spores are usually resistant to drying, heat, freezing, and some chemical agents. However, these sexual spores are not as resistant to heat as bacterial endospores. Either an asexual or sexual spore of a fungus can germinate and develop into a new hypha and mycelium.

The classification of molds is based on hypha and spore characteristics. If the fungus has hyphae with cross walls it is classified in a different group than if it has hyphae without cross walls. The morphology of the spore-bearing structures and the arrangement of the spores, both asexual and sexual, also provide a major basis of classification. Some of the most important microbial industrial products are the result of mold activity. For instance, *penicillin,* the most important antibiotic in medicine, is produced by a mold (see Chapters 5 and 18). A number of filamentous fungi are also important disease-causing agents of animals and plants (see Chapters 12 and 17).

Yeasts The yeasts are unicellular fungi. Yeast cells are usually spherical, oval, or cylindrical, and cell division takes place by **budding** (Figure 2.24). In the budding process, a new cell forms as a small outgrowth of the old cell; the bud gradually enlarges, and then separates. Yeasts do not form filaments or a mycelium, and the population of yeast cells remains a collection of single cells. Yeast cells are considerably larger than bacterial cells and can be told from bacteria by their size and by the obvious presence of internal cell structures. Yeasts also exhibit sexual reproduction by a process called *mating,* in which two yeast cells fuse. Within the fused cell, called a *zygote,* spores form. For the most part, yeasts spread from place to place as ordinary vegetative cells rather than as spores.

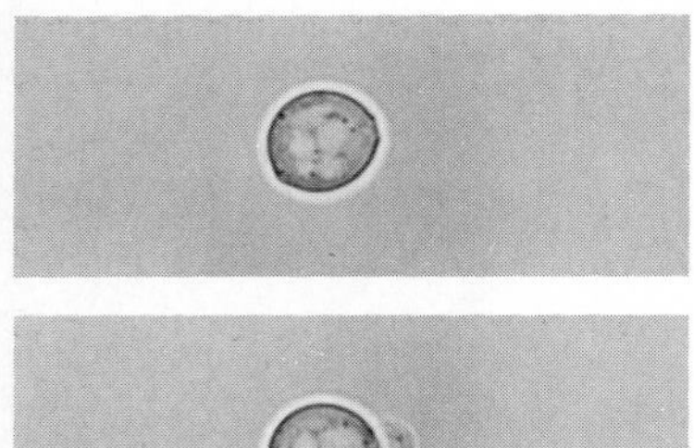
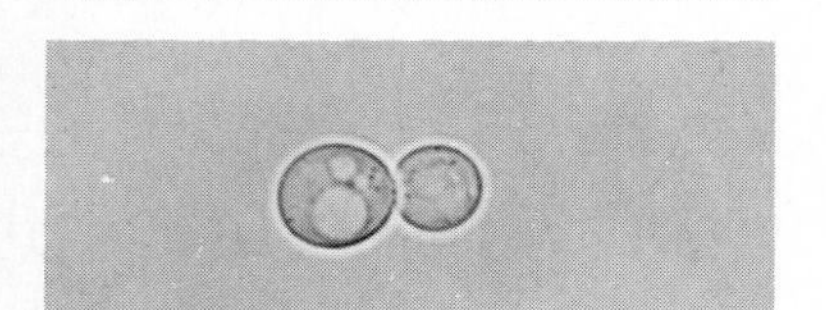
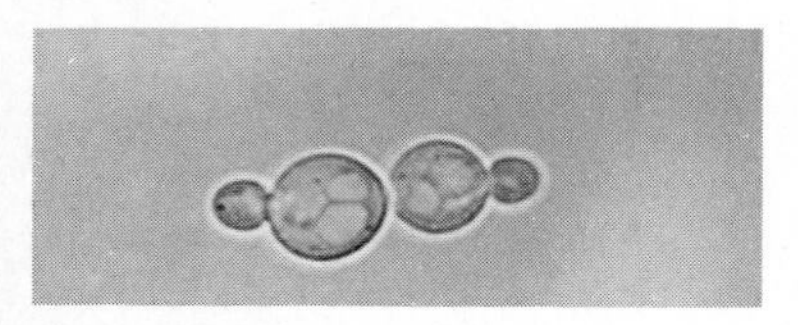
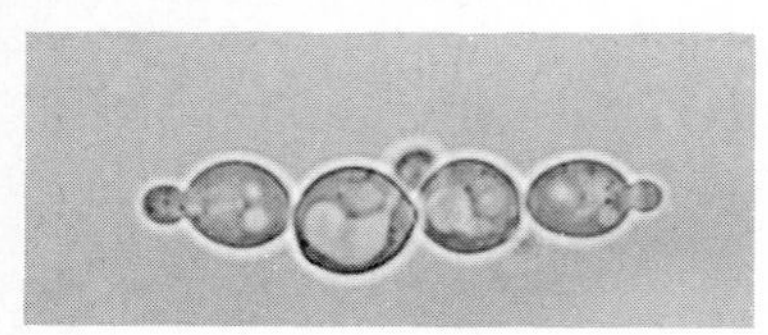
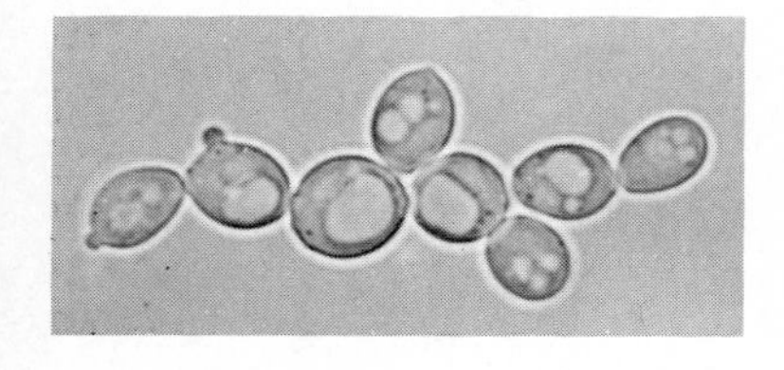

Figure 2.24 Yeast: development of buds from a single cell.

Classification of yeasts is based partly on the kinds of sexual spores formed and partly on the basis of nutrition and biochemistry. The common yeast, *Saccharomyces cerevisiae,* used in making bread, beer, and whiskey and other spirits, is one of the most important organisms affecting human society; we shall discuss its industrial uses extensively in Chapter 18. Most other yeasts are of little human importance, although a few cause human diseases (see Chapter 12).

Mushrooms The mushrooms are a group of filamentous fungi that form large complicated structures called **fruiting bodies** (the fruiting body is commonly called the "mushroom"). The fruiting body is formed through the growing-together of a large number of individual hyphae (Figure 2.25). If the mushroom fruiting body is cut and examined under the microscope, the individual hyphae can be seen pressed tightly together.

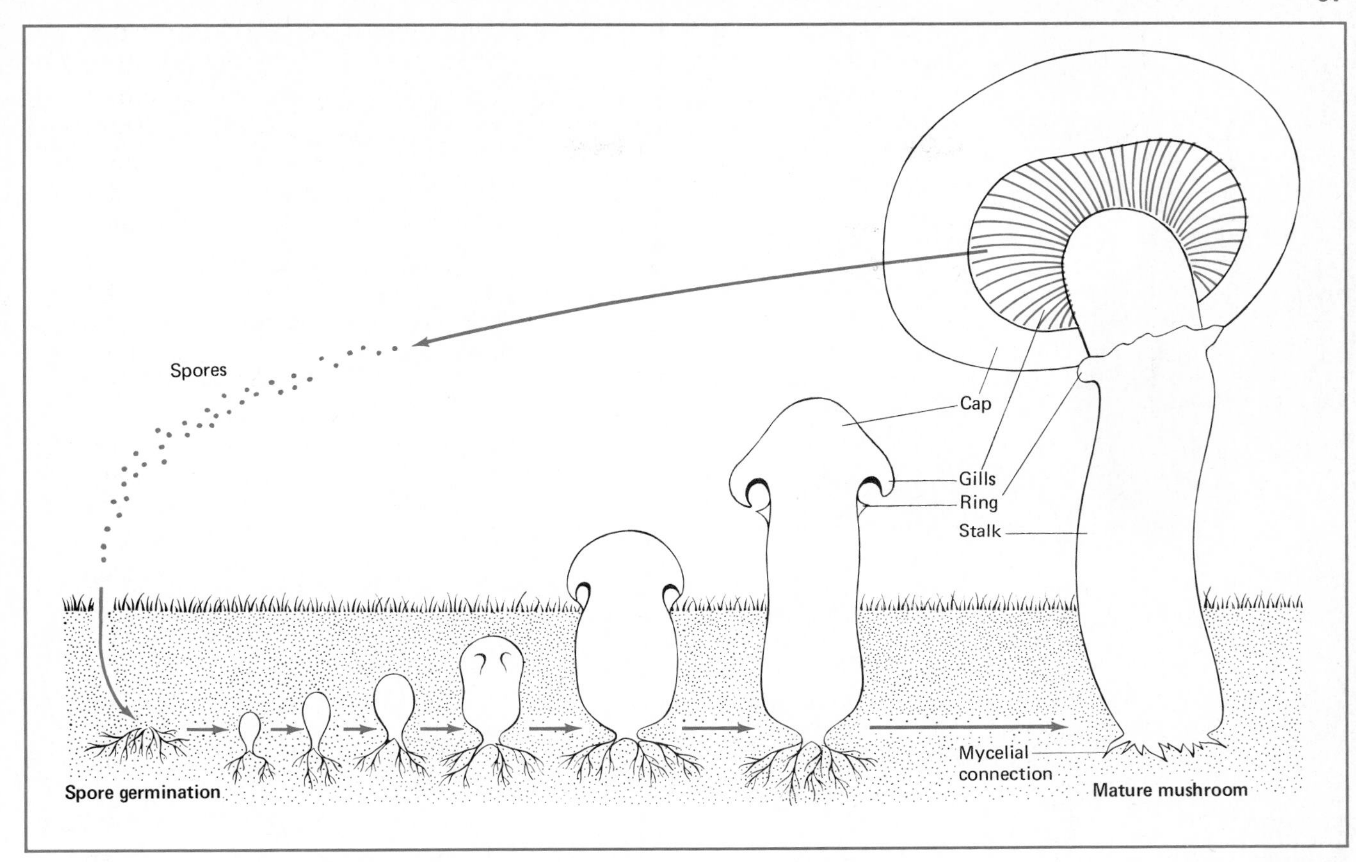

Figure 2.25 Mushroom life cycle, showing how the fruiting body develops from underground hyphae.

During most of its existence, the mushroom fungus lives as a simple mycelium, growing within soil, leaf litter, or decaying logs. When environmental conditions are favorable, the fruiting body develops, beginning first as a small button-shaped structure underground and then expanding into the full-grown fruiting body that we see above ground. The nutrients for growth come from organic matter in the soil and are taken up by the hyphal filaments which, like the roots of a plant, feed the growing fruiting body. Sexual spores called *basidiospores*, are formed, borne on the underside of the fruiting body, either on flat plates called *gills* or within deep *pores*. The spores are often colored and impart a color to the underside of the fruiting body cap. Some mushrooms, called *puff balls*, produce their spores within spherical fruiting bodies instead of on gills or pores, the spores puffing out through cracks or holes that develop in the fruiting body as it dries. The spore is the agent of dispersal of mushrooms and is carried away by the wind. If it alights in a favorable place, the spore will germinate and initiate the growth of new hyphae, mycelium, and fruiting body.

Classification of mushrooms is based on the size, shape, and color of the fruiting body, the arrangement of the gills or pores, and the color of the

THE MUSHROOM: DEADLY OR DELICIOUS?

Mushrooms are prized delicacies at the dinner table. Although their food value is low, they add a gourmet touch to meats and vegetables, enhancing taste through subtle nuances of flavor and texture. A skilled mushroom hunter, visiting a forested area favorable for mushroom growth, can often return after a few hours with a basket of fresh, delicious mushrooms. Wild mushrooms are highly prized in Europe, the Orient, and parts of North America. Most highly valued are the morels, the chantrelles, and the truffles, but many other species also are good to eat.

Unfortunately, only a few mushrooms possess these highly prized qualities. Most mushrooms are bland or bitter. A few are *deadly* poisonous. How to tell the deadly from the delicious? The collector must understand mushroom structure intimately, and be able to identify both genus and species.

The mushroom hunter's task can be most dramatically illustrated by examples from the genus *Amanita*. The genus contains the delicious edible species *Amanita caesarea* (said to be a favorite of Julius Caesar), but also contains one of the most poisonous of all species, *Amanita verna,* known as the *destroying angel*. Fatal in over 50 percent of cases, the destroying angel causes severe abdominal pains, vomiting, and diarrhea, followed by total physical collapse. A related species, the fly amanita (*A. muscaria*) is also fatal in large doses, and in smaller doses causes psychological effects, including general excitation and hallucinations.

Although there are many more edible than poisonous mushrooms, the mushroom hunter is strongly advised to learn to recognize on sight a few indisputable edible forms and to eat *only* these. Despite the delicious flavor of *Amanita caesarea* (Julius Caesar had tasters try his food first), most mushroom hunters shun the genus *Amanita* completely, to reduce the chances of getting a poisonous specimen. In no other area of microbiology is a mistake in classification so likely to have serious consequences!

spores. There is a wide variety of mushroom species, of which some are edible, while a few are poisonous (see the Box above and also Section 12.5). Commercial mushrooms are cultivated in vast beds containing a mixture of sterilized soil and horse manure. The mushrooms are harvested just as the mushroom buttons are expanding, since full-grown fruiting bodies are tougher and less tasty than immature fruiting bodies. Commercial production of mushrooms is discussed in Chapter 18.

2.8 LICHENS

Lichens are leafy or encrusting growths that are widespread in nature and are often found growing on bare rocks, tree trunks, house roofs, and on the surfaces of bare soils (Figure 2.26). The lichen is one of the best examples of **mutualism**, a trait commonly found in the biological world. The lichen plant, usually called a *thallus,* consists of two organisms, a fungus and an alga, living together to their mutual benefit. In some mutualisms, the two organisms can also live separately, but in the case of the lichen, the mutualistic relationship is generally necessary for existence in nature. Lichens are extremely interesting associations because they demonstrate so clearly the value and importance of cooperation in the biological world. Lichens are usually found in environ-

Figure 2.26 A lichen growing on a branch of a dead tree.

ments where other organisms do not grow, and it is almost certain that their success in colonizing such extreme environments is due to the mutual interrelationships between the alga and fungus partners.

The lichen thallus usually consists of a tight association of many fungus hyphae, within which the algal cells are embedded (Figure 2.27). The shape of the lichen thallus is determined primarily by the fungal partner, and a wide variety of fungi are able to form lichen associations. The diversity of algal types is much smaller, and many different kinds of lichens may have the same algal component. The algae are usually present in defined layers or clumps within the lichen thallus.

Both the fungus and the alga are always found living together in nature. The alga is photosynthetic and is able to produce organic matter from carbon dioxide of the air. Some of the organic matter produced by the alga is then used as nutrient by the fungus. Since the fungus is unable to photosynthesize, its ability to live in nature is dependent on the activity of its algal symbiont.

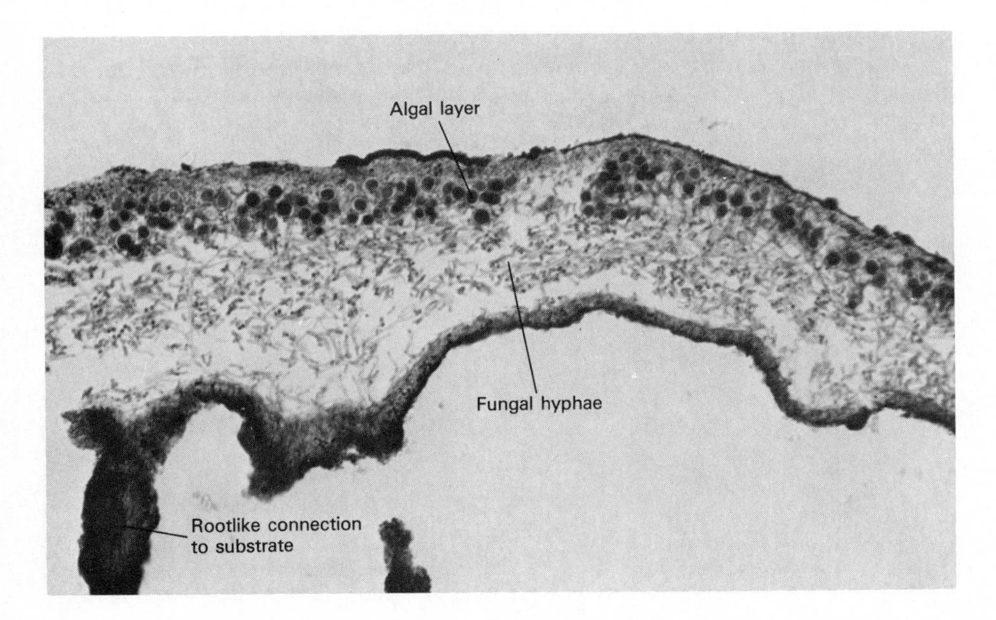

Figure 2.27 Photomicrograph of a cross section through a lichen.

The fungus clearly benefits from associating with the alga, but how does the alga benefit? The fungus provides a firm anchor within which the alga can grow protected from erosion by rain or wind. In addition, the fungus absorbs from the rock or other substrate upon which the lichen is living the inorganic nutrients essential for the growth of the alga. Another role of the fungus is to protect the alga from drying; most of the habitats in which lichens live are dry (rock, bare soil, roof tops), and fungi are in general much better able to tolerate dry conditions than are algae.

Although lichens live in nature under rather harsh conditions, they are extremely sensitive to air pollution and quickly disappear from large cities when heavy air pollution occurs. One reason for the great sensitivity of lichens to air pollution is that they absorb and concentrate materials from rainwater and air and have no means for excreting them, so that lethal concentrations are reached. Lichens have been extensively used as indicators of air pollution and may provide some early warnings of impending chronic air pollution problems. Studies made on the richness of the lichen flora in cities and their surroundings show that the number of species of lichens found on various habitats decreases as one moves from the countryside to the center of the city.

2.9 PROTOZOA

Protozoa are unicellular, colorless, generally motile, eucaryotic microorganisms that lack cell walls (Figure 2.28). They are distinguished from bacteria by their greater size and eucaryotic nature, from algae by their lack of chlorophyll, and from yeasts and other fungi by their motility and absence of the cell wall. Protozoa usually obtain food by ingesting other organisms or organic particles. They do this by surrounding the food particle with a portion of their flexible membrane and engulfing the particle (Figure 2.29), or by swallowing the particle through a special structure called the gullet.

Depending on the group, the type of protozoal movement varies. The **amoebae** move by what is called **amoeboid motion**; the cytoplasm of the

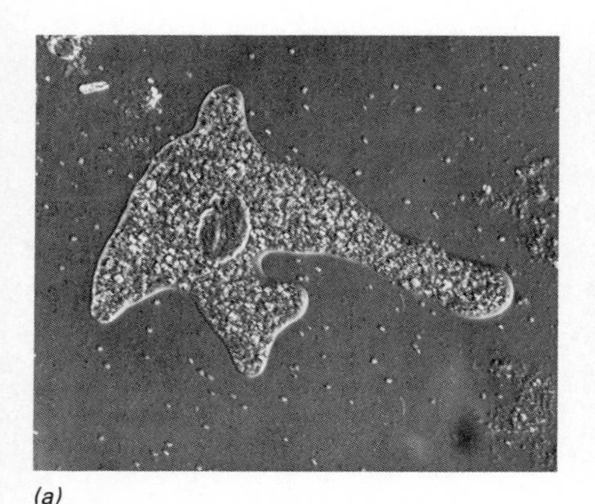

(a)

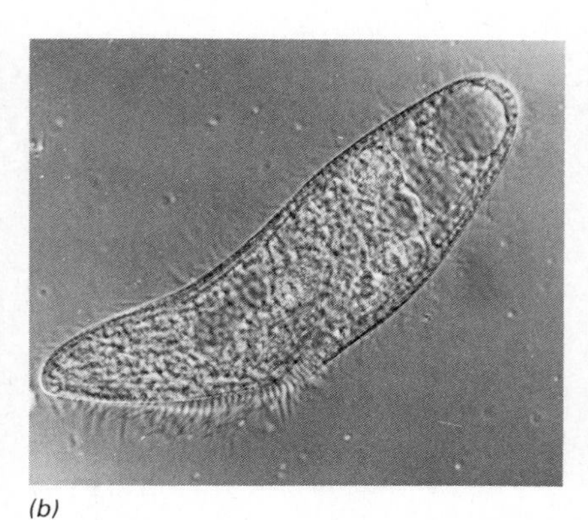

(b)

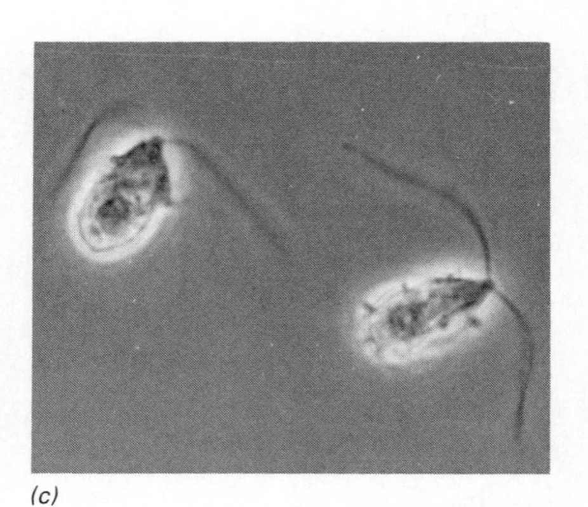

(c)

Figure 2.28 Protozoa. (a) An amoeba, *Amoeba proteus.* Magnification, 125 ×. (b) A ciliate, *Blepharisma.* Magnification, 120 ×. (c) A flagellate, *Dunaliella.* Magnification, 1900 ×.

cell flows forward in a lobe of the cell, called a *pseudopodium* (false foot), and the rest of the cell flows toward this lobe. Amoeboid motion requires a solid substrate and is rather slow. A few amoebas are disease-causing agents, but most are harmless and live in soil or water.

The **flagellates** move by use of flagella, usually by a single long flagellum attached at one end of the cell. The *trypanosomes*, a subgroup of flagellates, cause African sleeping sickness and a number of other diseases of humans and animals.

The **ciliates**, a third group of protozoa, move by the action of a large number of smaller appendages called *cilia*, which are attached all over the surface of the cell (see Figure 2.12). The ciliates are the most complicated protozoa structurally, and in addition they have complicated sexual reproduction mechanisms as well as the regular cell division process.

A fourth group of protozoa, the **sporozoans**, contains parasites of man or animals; the agent that causes malaria, *Plasmodium vivax*, is a member of this group. These nonmotile organisms do not engulf food particles but absorb dissolved food materials directly through their membranes. Despite their name, sporozoans do not form true spores as do bacteria, algae, and fungi.

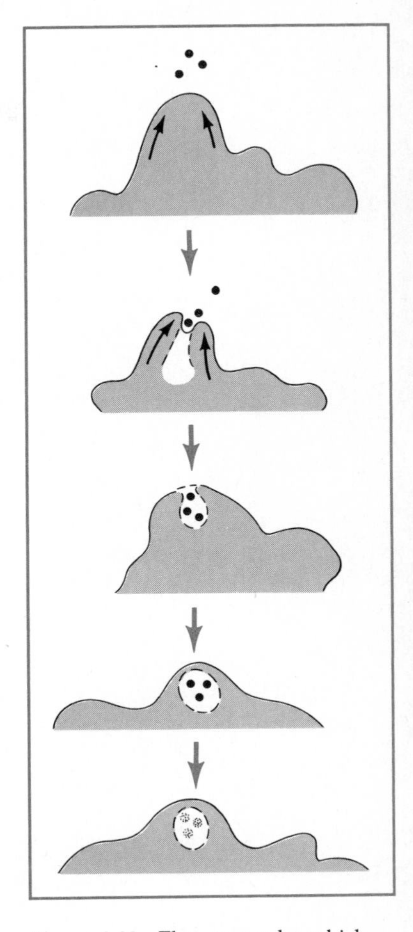

Figure 2.29 The process by which an amoeba engulfs small particles.

2.10 VIRUSES

Viruses are not cellular, have no metabolism of their own, and are generally considered not to be living microorganisms. Viruses depend upon growth in other cells for their existence. The cell which supports the growth of a virus is called the **host**. We will discuss the reproductive cycle of viruses in Chapters 6 and 13. In this section, we consider briefly the various types and structures of viruses.

Viruses exhibit various shapes, sizes and structures (Figure 2.30). One of the more important differences is the presence of either DNA or RNA as the genetic material. Thus, we can refer to viruses as either **DNA viruses** or **RNA viruses**. Another major structural feature found in some viruses is the membrane-like *envelope* which surrounds the viral protein coat. A virus without an envelope is said to be a *naked virus*. Because of their very small size, identification of viruses depends on the use of electron microscopy and studies of molecular structure. The manner of assembly of a virus from nucleic acid and protein was illustrated in Figure 2.13.

Another way in which the diversity of viruses is exhibited is in the type of host infected. Viruses infect various species of plants, animals, fungi, and bacteria. The names of some viruses reflect the type of host infected. For example, the tobacco mosaic virus infects tobacco plants, among others. Other viruses are named after the diseases they cause. For example, the smallpox virus causes smallpox and poliovirus causes polio (infantile paralysis). A number of diseases caused by viruses will be considered in Chapter 13.

2.11 AN EVOLUTIONARY TREE OF THE LIVING WORLD

We have discussed the structures of procaryotic and eucaryotic microorganisms and have given a brief glimpse of the diversity of the microbial world. However, underlying the diversity in cell structure that we have presented

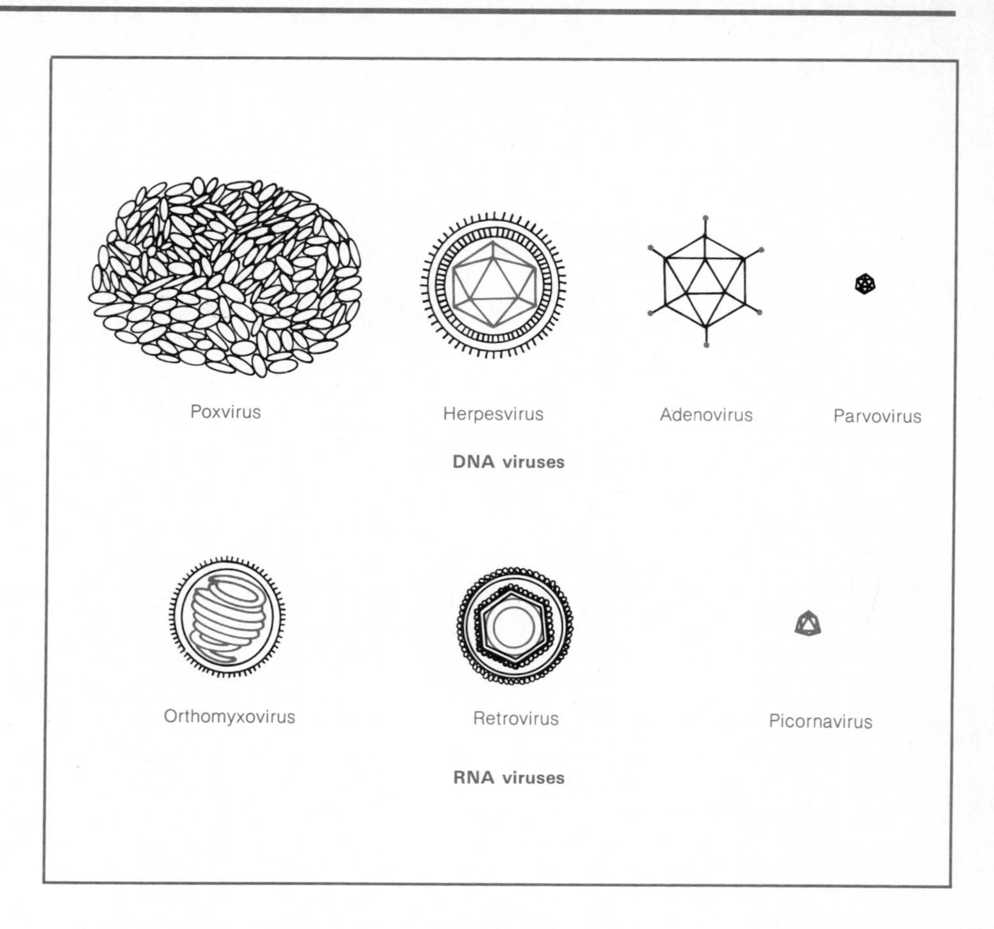

Figure 2.30 Animal viruses of various sizes, shapes, and types of nucleic acid.

is a fundamental chemical diversity based on the genetic characteristics of living organisms; this chemical basis of genetics will be discussed in Chapters 6 and 7. By careful study of the underlying chemical structure of the genetic material, it is possible to develop an overall scheme for how life evolved. One important cellular structure whose genetic analysis is especially favorable for evolutionary analysis is the *ribosome,* a structure basic to cell metabolism. The chemical analysis of the RNA of the ribosome has been used to discern evolutionary differences between organisms. After surveying several hundred representative bacteria, algae, fungi, protozoa, plants, and animals, the results displayed in Figure 2.31 were obtained. It can be seen that the bacteria can be divided into two major groups. One group, called the **eubacteria**, contains most of the types of bacteria which we have described in this chapter. Another group, called the **archaebacteria**, contains some unusual bacteria which either make methane gas (the *methanogens*), or live in extremely salty environments (the *halophiles*), or live in hot sulfurous environments (the *sulfur-dependent archaebacteria*). From a chemical-genetic point of view, the archaebacteria appear to be as different from the eubacteria as either bacterial group is from eucaryotic life forms. Viewed in this way, the bacteria exhibit far greater diversity than that which is found among all eucaryotic life forms, including algae, fungi, protozoa, plants, and animals.

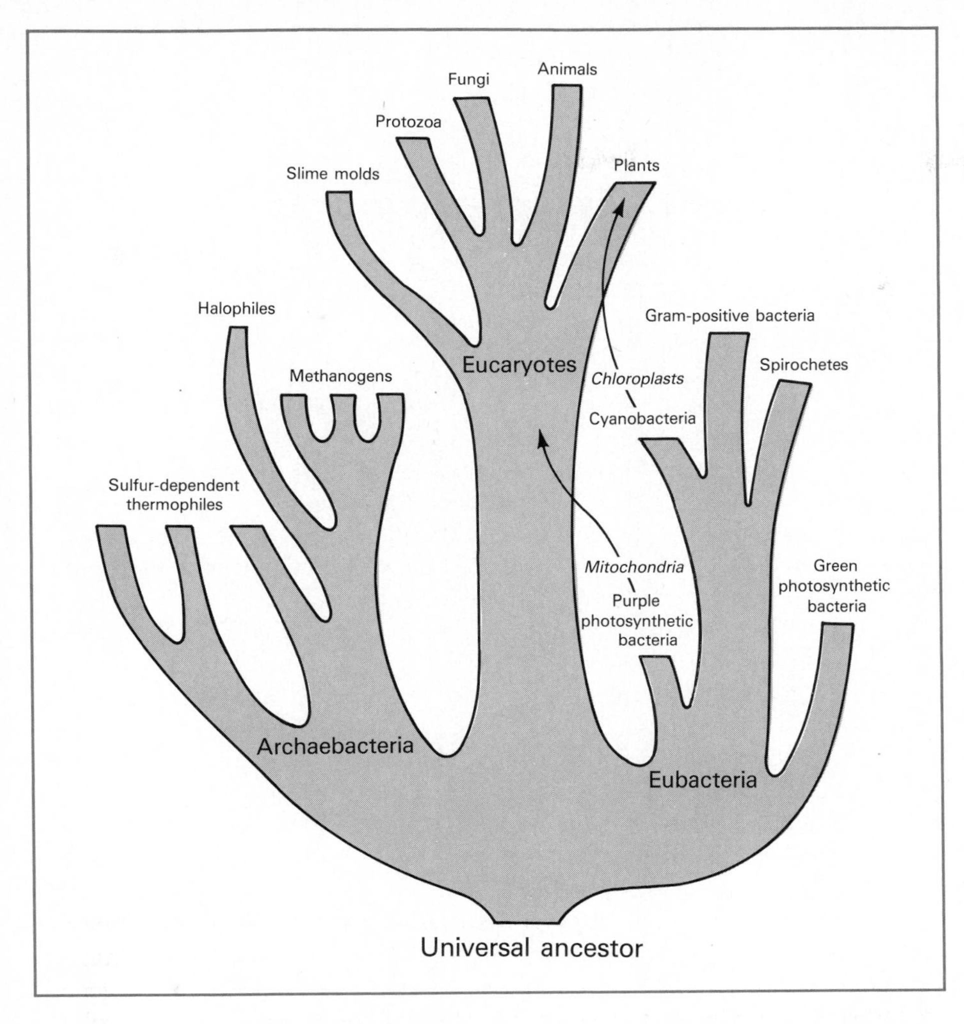

Figure 2.31 Evolutionary tree of the living world, based on the structure of RNA of the ribosomes.

Another interesting result illustrated in the figure is that eucaryotes do not appear to derive directly from procaryotes, but both derive from some common *universal ancestor*. But even more, the eucaryotic cell appears to have originated by the *merging* of two separate entities, each with its own origin. Mitochondria and chloroplasts, the two major organelles of eucaryotes, have apparently arisen from a *procaryotic* lineage, whereas the main eucaryotic cell component, the *nuclear component*, appears to have arisen further back in time from the universal ancestor. Thus, the eucaryotic cell is a *chimera*, with two origins, one for the nuclear component, the other for the cytoplasmic (organellar) component.

2.12 THE HISTORY OF MICROORGANISMS

To understand the history of microorganisms one must consider a time scale calibrated in billions of years (Figure 2.32). The earth itself is about 4.5 billion years old. We obtain evidence about the history of life by looking for the

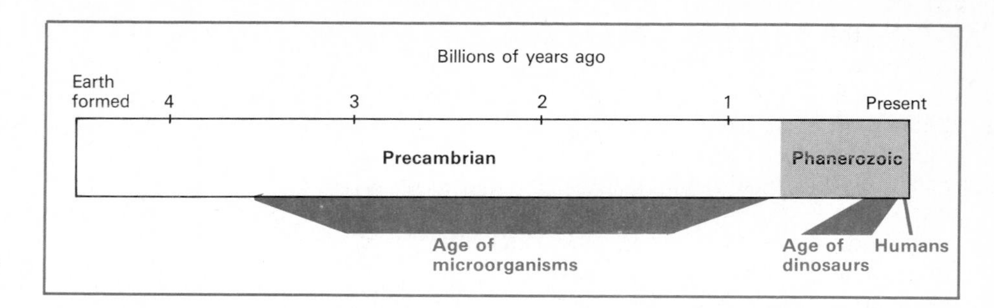

Figure 2.32 A time line for earth history, showing the time of appearance of several groups of living organisms.

presence of life forms (fossils) in ancient rocks. Fossil evidence of humans does not date back more than a few million years, a recent moment in the history of the earth. By looking in older rocks, we can find fossil evidence that the earth has not always been as we know it today. The *age of dinosaurs*, for example, was a time 65–200 million years ago, when reptiles dominated the earth. Simpler animal life forms, such as trilobites, are found in even older rocks, but rocks formed earlier than 600 million years ago do not appear to contain visible fossils. Geologists call the period of time during which animals have been visible in rocks the *Phanerozoic period.* The period before 600 million years ago, when no visible fossils are seen, is called the *Precambrian period.*

Although Precambrian rocks appear to be devoid of fossils, careful examination of thin sections of such rocks under the light microscope occasionally reveals a surprising diversity of *microfossils* (Figure 2.33). The best interpretation of these microfossils is that bacteria or algae flourished in the Precambrian period and became fossilized. Examination of a variety of rocks of Precambrian age has revealed well-preserved microfossils in rocks formed as long ago as 3.5 billion years. Microfossil-bearing rocks became increasingly abundant throughout the Precambrian period, and during the latter half of this period increasingly complicated organisms appeared. Rocks formed more recently than one billion years ago show evidence of primitive plant-like organisms and it is presumed that it is from these simple forms that the complex plant and animal life of the Phanerozoic period evolved. However, the establishment of higher plants and animals did not lead to the disappearance of microorganisms, which have continued to diversify throughout the Phanerozoic period.

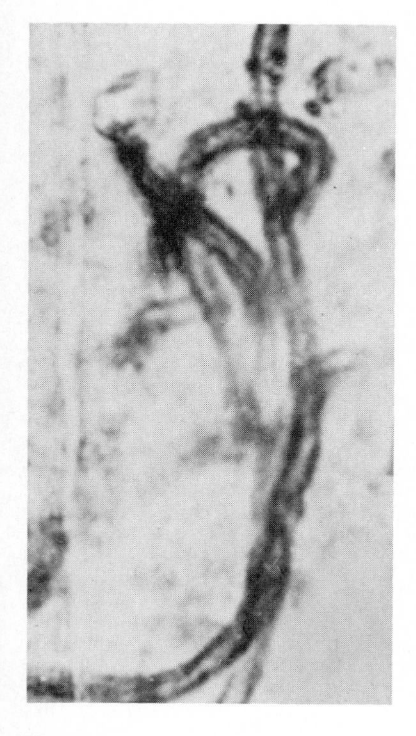

Figure 2.33 Microfossils found in Precambrian rocks dating back 900 million years. Even older rocks show evidence of microfossils. The filament is about 3 μm in diameter.

We thus see that the history of life on earth is a lengthy one, extending back nearly as long as the history of the earth itself. Considering the whole of earth history, we find that microbes, rather than higher life forms, have dominated the earth the longest. In the same sense that we refer to the age of the dinosaurs, we can also refer to the *age of microorganisms* (Figure 2.32).

SUMMARY

In this chapter we have discussed the structures of microorganisms. We showed that microorganisms can be separated into two major groups, called **procaryotes** and **eucaryotes**. The procaryotes are the bacteria, and the eucaryotes include the algae, fungi, and protozoa. The cells of plants and animals are also eucaryotic.

All cells have a barrier separating inside from outside which is called the **cell membrane**. Most bacteria, and all algae and fungi, have a **cell wall**, a rigid layer outside the membrane which protects the membrane and provides strength to the cell. Inside the cell is a complicated assembly of substances and structures which is called **cytoplasm.**

We discussed the sizes of microorganisms and showed how their structures are studied with the **microscope**. For detailed study of internal structure, an **electron microscope** must be used. The study of bacteria with the light microscope often involves the use of special staining procedures which increase contrast of cells and make them more readily visible. One special staining procedure, the **Gram stain**, also is a *differential stain*, since it permits the division of bacteria into two groups: Gram positive and Gram negative.

Procaryotic and eucaryotic cells have a number of structural differences. The most important is that eucaryotes have true **nuclei**, whereas procaryotes lack true nuclei but contain instead a region of the cell where the genetic material (DNA) is present, called the **nuclear region**. Eucaryotes also contain distinct structures called **organelles** within which certain cellular functions occur. Two important organelles in eucaryotes are **mitochondria**, involved in energy generation, and **chloroplasts**, which are found in eucaryotes capable of carrying out photosynthesis. An important function in many microorganisms is **motility**, the ability to move. In most cases, special organs of motility are present, termed **flagella** and **cilia**.

Viruses are simpler in structure than cells. The simplest viruses contain nothing more than genetic material (DNA or RNA) surrounded by a protein coat. The more complex viruses also contain a primitive membrane, but no virus is able to reproduce itself outside of a living cell.

A wide variety of **bacteria** have been described. A major criterion for the classification of bacteria is the shape and structure of the cell. Three main bacterial shapes are the **coccus, rod,** and **spirillum**. An important group of bacteria produce **endospores**, which are heat-resistant structures that are difficult to destroy in the sterilization process.

Among the eucaryotic microorganisms, those containing chlorophyll and chloroplasts are classified as the **algae**. The **fungi** consist of eucaryotic microorganisms that do not have chlorophyll but do have cell walls. Three important groups of fungi are the **molds, yeasts,** and **mushrooms.** The yeasts are unicellular organisms, but the other fungi form long, filamentous structures called **hyphae**. Hyphae usually grow together across a surface and form compact mats, collectively called a **mycelium.** Many fungi form spores, sometimes called **conidia,** which permit dispersal through the air. Mushrooms are a group of filamentous fungi that form structures called **fruiting bodies.** A **lichen** is a biological unit containing both a fungus and an alga living together for their mutual benefit. The lichen is one of the best examples in the biological world of cooperation between organisms. **Protozoa**, the microorganisms most closely resembling animals, are unicellular, colorless eucaryotes which lack cell walls and which obtain their food by engulfing other organisms.

An overall classification of living organisms based on chemical analysis of genetic characteristics has permitted the construction of an evolutionary tree upon which all living organisms can be placed. The analysis shows that the procaryotic world can be divided into two major divisions, called **eubacteria** and **archaebacteria**. These two groups differ as much between themselves as do all of the eucaryotic organisms. Both eucaryotes and procaryotes appear to be derived at a very early time from a common organism called the **universal ancestor.** Also, the organelles of eucaryotes appear to have arisen from a procaryotic lineage, whereas the nuclear portion of the eucaryote has had a separate origin.

The knowledge of cell structure learned in this chapter will be of value when discussing the functional characteristics of microorganisms in the next chapter, as well as in the discussion of genetic characteristics in subsequent chapters. We will also find a knowledge of cell structure useful for discussing the infectious diseases caused by microorganisms (see Chapters 8 through 13).

KEY WORDS AND CONCEPTS

Cell
Protoplasm
Procaryote
Eucaryote
Gram stain
Morphology
Cell wall
Lysis
Cell membrane
Ribosome
Nuclear region
Flagellum (procaryote)
Nucleus
Chromosome
Organelle
Mitochondrion
Flagellum (eucaryote)
Cilium
Genus
Species
Bacteria
Coccus
Rod
Spirillum
Endospore
Capsule
Algae
Chlorophyll
Chloroplast
Fungi
Hyphae
Mycelium
Lichen
Mutualism
Protozoa
Amoeba
Flagellate
Ciliate
Sporozoan
Virus
Protein coat
Host
DNA virus
RNA virus
Archaebacteria
Eubacteria

STUDY QUESTIONS

1. Trace the pathway of light through the light microscope. What is the function of the *ocular*? of the *objective*? of the *condenser*?
2. Why are staining methods useful in the study of bacteria?
3. What is the *Gram stain*? Why is it of greater value than a simple stain?
4. What is the color of a *Gram-positive* organism after the Gram stain is completed? of a *Gram-negative* organism?

5. Draw a diagram of a "typical" bacterial cell. Indicate the following procaryotic structures and describe the function of each briefly: *cell wall, cell membrane, nuclear region, ribosome, flagellum.*
6. Draw a diagram of a "typical" eucaryotic cell. Indicate the following eucaryotic structures and describe the function of each briefly: *cell wall, cell membrane, nucleus, mitochondrion, chloroplast, flagellum, cilium.*
7. Compare and contrast the procaryotic and eucaryotic cell. How are they similar? different?
8. Cells range in size from 0.1 μm to greater than 10 μm. Why do we give the same name *cell* to things with such different sizes?
9. How would you differentiate the following major groups: *bacteria, algae, fungi, protozoa, animal cell, plant cell*?
10. How would you most simply distinguish between a *fungus* and an *alga*?
11. Three major groups of fungi are *molds, yeast,* and *mushrooms*. How would you most simply tell whether a fungus you were examining belonged to one of these three groups?
12. Why are *viruses* not considered cells?

SUGGESTED READINGS

BOLD, H.C. and M.J. WYNNE. 1978. *Introduction to the algae: structure and reproduction.* Prentice-Hall, Inc. Englewood Cliffs, NJ.

BROCK, T.D., D.W. SMITH, and M.T. MADIGAN. 1984. *Biology of microorganisms, 4th edition.* Prentice-Hall, Inc., Englewood Cliffs, NJ.

BUCHANAN, R.E. and N.E. GIBBONS, Eds. 1974. *Bergey's manual of determinative bacteriology, 8th edition.* Williams and Wilkins, Baltimore.

FAWCETT, D.W. 1981. *The cell.* W.B. Saunders, Philadelphia.

STARR, M.P., H. STOLP, H.G. TRÜPER, A. BALOWS, and H.G. SCHLEGEL. 1981. *The prokaryotes—a handbook on habitats, isolation, and identification of bacteria. 2 volumes.* Springer-Verlag, New York.

3

Cell Chemistry and Metabolism

Microbial cells are built of chemical substances of a wide variety of types, and when a cell grows, all of these chemical constituents increase in amount. The basic chemical elements of a cell come from outside the cell, from the cell's **environment**, but these chemical elements are transformed by the cell into the characteristic constituents of which the cell is composed. Life forms always have chemical constitutions that are different from those of the environment.

The chemicals from the environment of which a cell is built are called **nutrients**. Nutrients are taken up into the cell and changed into cell constituents. This process by which a cell is built up from the simple nutrients obtained from its environment is called **anabolism**, which means, literally, *building up*. Because anabolism results in the chemical synthesis of new materials, it is often called **biosynthesis**.

Biosynthesis is an energy-requiring process, and each cell must thus have means of generating energy. Some organisms, the *phototrophic organisms,* use light as a source of energy, but most organisms obtain their energy from chemical substances. Chemicals used as energy sources are broken down into simpler constituents, and as this breakdown occurs, energy is released. The process by which chemicals are broken down and energy released is called

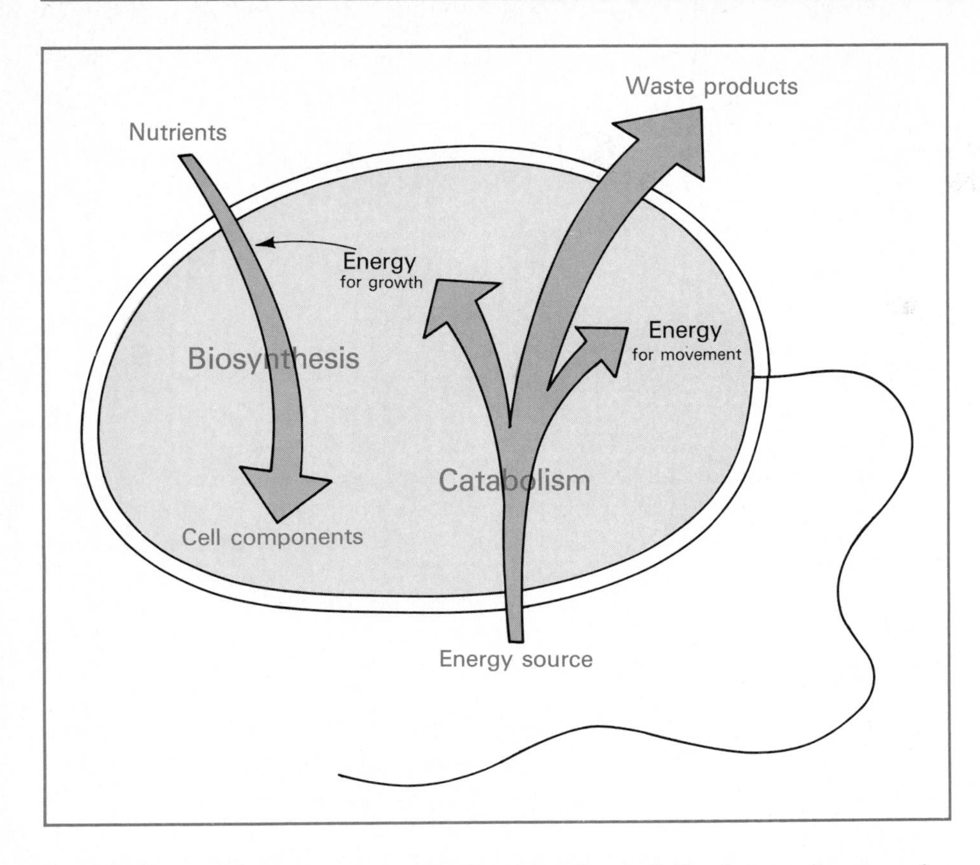

Figure 3.1 A simplified view of the major features of cell metabolism.

catabolism, which means, literally, *breaking down*. Cells also need energy for other functions, such as cell movement (motility).

We thus see that there are two basic kinds of chemical transformation processes in cells, the building-up processes called *anabolism* and the breaking-down processes called *catabolism*. Both anabolism and catabolism occur at the same time in cells. Together, anabolism and catabolism are collectively referred to as **metabolism**, derived from the Greek word *metabole* which means *change*. Although when we look at it under the microscope a cell appears to be a fixed and stable structure, it is actually a dynamic entity, continually undergoing change, as a result of all the chemical reactions which are constantly taking place.

A simplified overview of cell metabolism is shown in Figure 3.1, which depicts how catabolic reactions supply the energy needed for cell functions. In this chapter we will consider some of the biosynthetic and catabolic processes used by microorganisms.

A knowledge of cell metabolism helps in understanding the biochemistry of microbial growth, and this makes it possible to control the growth of unwanted microorganisms by the use of antibiotics or other chemicals. Also, since different microorganisms have different biosynthetic and catabolic abilities, biochemistry helps in understanding microbial disease processes. Note in Figure 3.1 that a cell usually converts the energy source partially to waste products (sometimes called *metabolic products*). Another important reason to study cell metabolism is that some of these products, while waste products as far as the cell is concerned, have useful human applications. A good example is ethanol (commonly called alcohol), a product of the metabolic reactions of yeast, which is used in beverages and as an automotive fuel. Another use for study of biochemical reactions is that they aid in the identification of microorganisms.

3.1 A BIT OF CHEMISTRY

All chemicals are comprised of orderly assemblages of the fundamental units of matter—atoms. Each atom consists of a nucleus containing protons and neutrons, surrounded by orbiting electrons (Figure 3.2). Most atoms do not exist independently in nature but in combination with other atoms. A *molecule* or *compound* is a specific association of atoms held together by strong forces called *chemical bonds*. When adjacent atoms in a chemical compound share electrons the chemical bond is particularly strong and is said to be a *covalent bond*. An example of a compound is water, H_2O, which consists of one oxygen atom that shares electrons with each of two hydrogen atoms (Figure 3.3). Other examples of compounds are: hydrogen sulfide, H_2S; ammonia, NH_3; and carbon dioxide, CO_2.

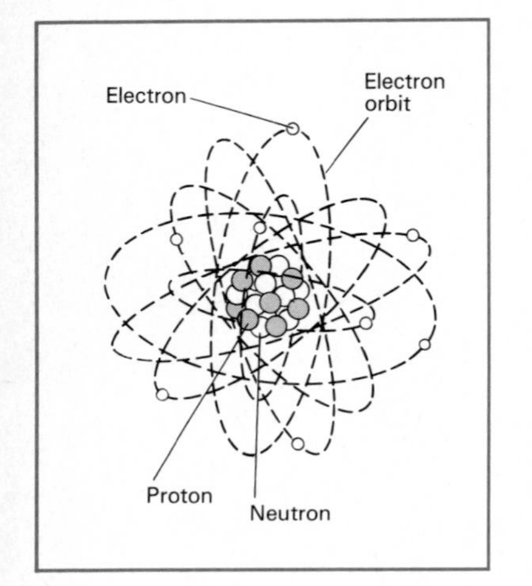

Figure 3.2 The structure of the atom.

Although covalently linked compounds are relatively stable, they still are able to interact with other compounds by way of weaker types of interactions. Weak bonds between molecules can form by several means and play important roles in life processes. Many of the key types of reactions occurring in living organisms are only possible through the making and breaking of weak bonds between organic molecules of cells. An extremely important type of weak bond in living organisms is the **hydrogen bond**.

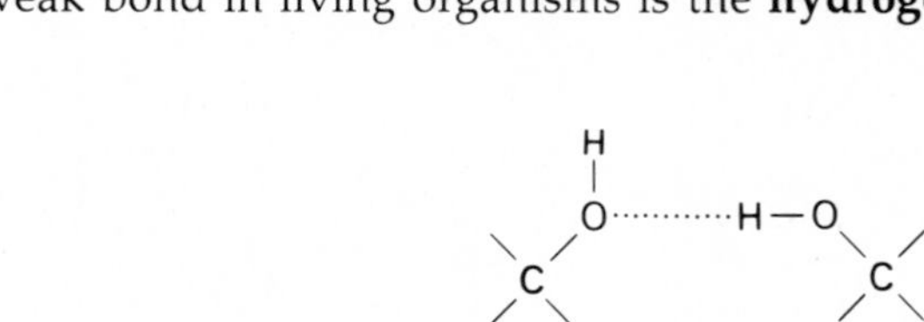

Hydrogen bond between two oxygen atoms
Hydrogen bond between oxygen and nitrogen atoms

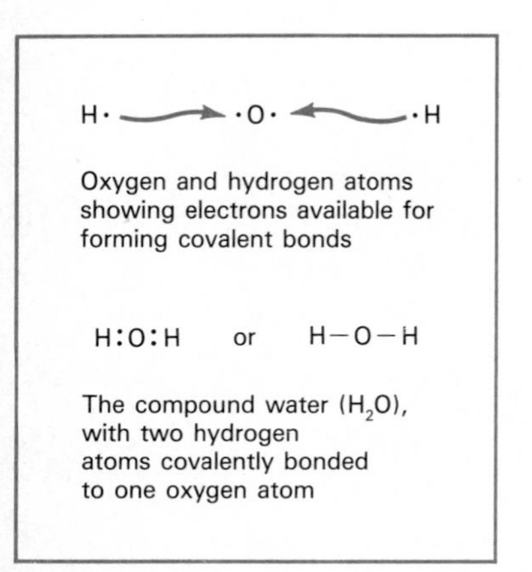

Figure 3.3 The formation of a compound, water (H_2O), by covalent bonding of atoms of the different elements hydrogen (H) and oxygen (O).

A hydrogen bond arises when a hydrogen atom covalently linked to one atom interacts with an acceptor atom on another molecule. In biochemistry, the most important hydrogen bonds involve hydrogen atoms attached to oxygen (O—H) or nitrogen (N—H). The acceptor atom also is usually a nitrogen or oxygen atom. Some examples of hydrogen bonds formed between organic molecules are shown. Hydrogen-bond formation is very important in the association of large organic molecules in living cells.

Molecules that will form hydrogen bonds with other molecules will also form them with water molecules. When a hydrogen-bond forming molecule is placed in water, some of the hydrogen bonds between water molecules are broken, and hydrogen bonds between water molecules and the added compounds are formed. Molecules that associate in this manner with water molecules are called **hydrophilic** (*hydro*, "water"; and *philic*, "loving"), and the atoms or groups of atoms in these molecules that undergo hydrogen-bond formation are often called *hydrophilic groups*. Substances such as sugars, which have many —OH groups, are hydrophilic and dissolve readily in water. On the other hand, many substances, such as cooking oils, gasoline, and kerosene, lack hydrophilic groups and are water insoluble. Atoms or groups of atoms in molecules that do not associate with water are called **hydrophobic** (literally, "water-fearing"). When a hydrophobic substance is placed in water, there is no tendency for hydrogen-bond formation between the substance and water. The water molecules remain firmly bonded together, and the hydrophobic groups associate with each other. This association between hydrophobic groups is in effect a type of bonding and has been called *hydrophobic bonding*. Hydrophobic bonding is very important in cells, because structures such as the cell membrane, high in fatty substances that are hydrophobic, are held together by hydrophobic bonds.

Organic compounds Most of the compounds of living systems contain carbon atoms. Carbon-containing compounds are called *organic compounds*, since the first carbon-based compounds were found in living organisms (*organic* refers to life or life processes). A single carbon atom has the ability to form *four* covalent bonds, and such bonds can form with a wide variety of other elements, such as hydrogen (H), oxygen (O), nitrogen (N), sulfur (S), and phosphorus (P). Further, carbon is one of a few elements that has the ability to form bonds not only with other elements but with itself (Figure 3.4). This capacity greatly increases the variety of compounds that can be formed. A vast number of carbon-based compounds are known, and more are being discovered or synthesized each year. It has been said that life is possible only because carbon can form such a tremendous variety of compounds.

Because of the large number of possible compounds, and the large number of atoms that can be present in a single compound, writing formulas for organic compounds presents some difficulties. It is not sufficient merely to give the number of atoms of each kind present in an organic compound, as we would for other elements, because the same number of atoms may combine in a variety of ways to make a number of different compounds. For instance, the sugars glucose and fructose both have the formula $C_6H_{12}O_6$. The most accurate way of writing organic structures is to indicate all bonds between atoms, but this is cumbersome and space-consuming. In order to be able to write structures conveniently, certain simplifications are used. In practice,

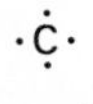

·Ċ·

The carbon atom with its four electrons in the outer shell

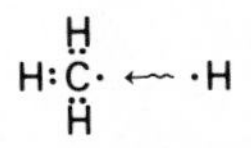

H:C· ← ·H (with H above and below C)

Covalent bonding to four hydrogen atoms to make a compound, methane, CH_4

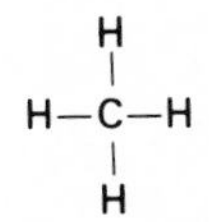

H—C—H (with H above and below C)

Conventional structure of methane

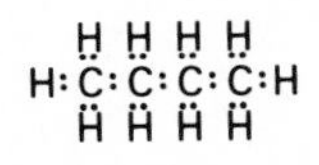

H:C:C:C:C:H (with H above and below each C)

Bonding of carbon to itself and to hydrogen to make a chain, in this case butane, C_4H_{10}.

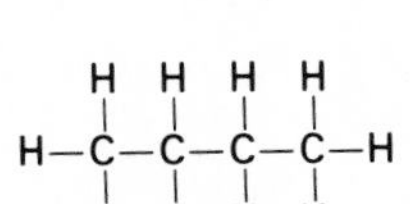

H—C—C—C—C—H (with H above and below each C)

Conventional structure of butane

Figure 3.4 Covalent bonding of carbon to other atoms and to itself, leading to the formation of organic compounds.

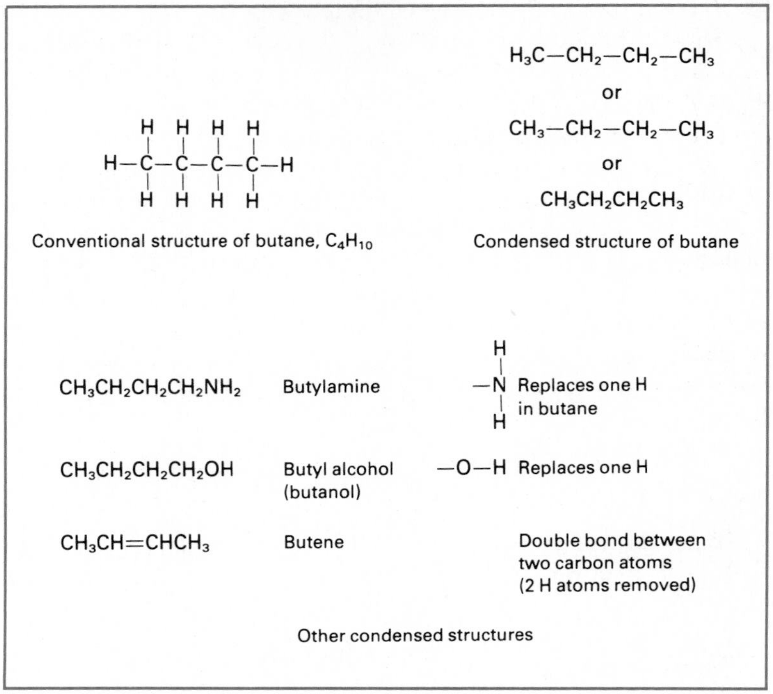

```
  H   H   H   H
  |   |   |   |
H—C———C———C———C—H
  |   |   |   |
  H   H   H   H
```

Conventional structure of butane, C_4H_{10}

$H_3C—CH_2—CH_2—CH_3$

or

$CH_3—CH_2—CH_2—CH_3$

or

$CH_3CH_2CH_2CH_3$

Condensed structure of butane

$CH_3CH_2CH_2CH_2NH_2$	Butylamine	$—N(H)_2$	Replaces one H in butane
$CH_3CH_2CH_2CH_2OH$	Butyl alcohol (butanol)	—O—H	Replaces one H
$CH_3CH{=}CHCH_3$	Butene		Double bond between two carbon atoms (2 H atoms removed)

Other condensed structures

Figure 3.5 Writing the structures of organic compounds.

what is done is to write condensed structures, in which key or important bonds are shown but other bonds are not. Figure 3.5 illustrates one way in which organic structures are often written.

We will now consider those classes of organic compounds which are particularly relevant to the biochemistry of a cell.

3.2 POLYMERS AND MONOMERS

In Chapter 2 we discussed the basic parts of a cell, such as the cell wall, membrane, and nuclear region. What is the chemical composition of these cell structures? Although there are differences in chemical composition of different cells, all cells use similar chemicals to make up the various parts. Table 3.1 lists the main chemical components of the major parts of a typical cell. Each structure is composed of very large molecules, called *macromolecules* (*macro-* means *large*). Macromolecules are built up of individual "building blocks," which are connected in specific ways. A single building block is called a *monomer* (*mono-* means *one*), and the macromolecule is called a *polymer* (*poly-* means *many*). There are only a few types of polymers important in cell biochemistry, and each is made of a characteristic set of monomers. In the rest of this section, we discuss the major macromolecules of cells, and the monomers of which they are built.

Sugars, polysaccharides, and cell walls Cell walls are usually composed of a type of polymer called a **polysaccharide**. In a polysaccharide,

TABLE 3.1 Chemical composition of major cell components

Component	Type of macromolecule	Type of monomer
Cell wall	Polysaccharide	Sugars
Enzyme	Protein	Amino acids
Nuclear region	Deoxyribonucleic acid (DNA)	Deoxyribose sugar, phosphate, nucleic acid bases (adenine, thymine, guanine, cytosine)
Ribosome	Ribonucleic acid (RNA)	Ribose sugar, phosphate, nucleic acid bases (adenine, uracil, guanine, cytosine)
	Protein	Amino acids
Membrane	Lipid-protein	Amino acids, fatty acids
Flagellum	Protein	Amino acids

sugar monomers are linked together. For instance, *cellulose*, a polymer found in the cell wall of plants and some algae, is a simple polysaccharide composed of a series of identical *glucose* molecules (Figure 3.6*a*). Some polysaccharides are chemically more complex than cellulose. A good example of a complex polysaccharide is the component of the bacterial cell wall which consists of alternating units of two sugars, as illustrated in Figure 3.6*b*. The sugars that are the building blocks of polysaccharides are often called *carbohydrates*, because they are composed of carbon, hydrogen, and oxygen atoms. The hydrogen and oxygen in carbohydrates are in the same proportion as in water, two hydrogens for each oxygen (Figure 3.7). Many sugars found in living organisms contain six carbon atoms and are called *hexoses*. An important pair of sugars are the two found in nucleic acids, called *ribose* and *deoxyribose* (Figure 3.7). These two sugars each have five carbon atoms and are called *pentoses*. Two other important polysaccharides in the biological world are *starch*, an energy-rich food storage product of plants, and *glycogen*, a food storage product in the muscles of animals. (The weariness of the long-distance runner is due at least in part to a deficiency in muscle glycogen.) It is interesting that all three of the major polysaccharides of the biological world, cellulose, starch, and glycogen, are composed solely of glucose monomers. It is the manner in which the glucose monomer units are connected in the polymer

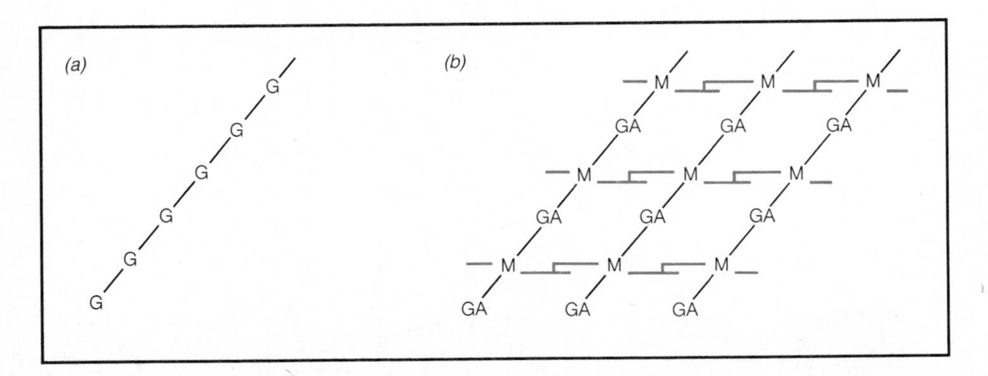

Figure 3.6 Cell wall polysaccharides. (a) Cellulose, the major cell wall constituent of plants, is composed of monomers of the sugar, glucose (G). (b) The bacterial cell wall is composed of two alternating sugars, N-acetyl glucosamine (GA) and muramic acid (M). The chains of sugars are cross-linked by short peptide (protein) segments (shown in color), leading to the formation of a complex polymer called *peptidoglycan*.

which accounts for the marked differences in the properties of these three polymers.

Proteins, peptides, and amino acids One of the most important kinds of biological polymers is **protein**. Proteins are found in flagella, ribosomes, and enzymes, among other components (Table 3.1). We noted in Chapter 2 the role of flagella in cell movement. Ribosomes play a central role in protein synthesis, and will be discussed in detail in Chapter 6. Enzymes are central participants in cell metabolism and will be discussed in the next section.

The building blocks of proteins are **amino acids**. There are about 20 different amino acids found in protein, and 100 or more amino acid building blocks in a single protein macromolecule. Amino acids contain carbon, hydrogen, oxygen, nitrogen, and sometimes sulfur. The structures of some amino acids are given in Figure 3.8. It should be noted that one end of all amino acids has the same structure, whereas the other end differs from one amino acid to another. Table 3.2 lists the 20 amino acids commonly found in proteins, and their abbreviated chemical symbols.

The covalent bonds that link amino acids together in a protein are called **peptide bonds**, and involve the constant end of the molecule. In the long

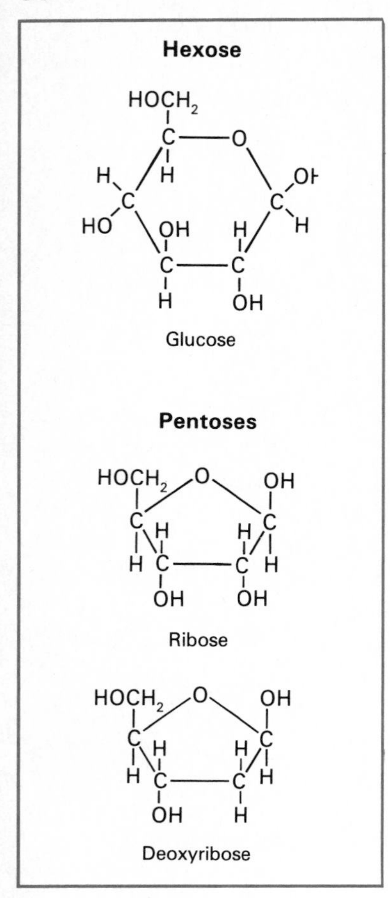

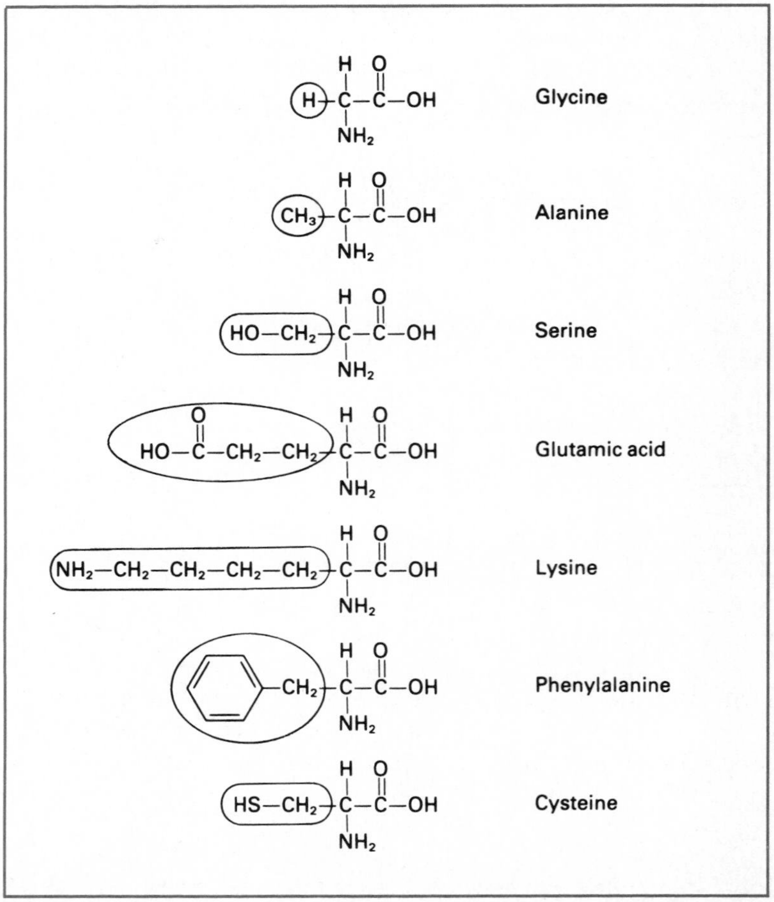

Figure 3.7 Structure of sugar monomers. Hexoses contain six carbon atoms, whereas pentoses contain five carbons. The pentoses shown are important components of nucleic acids.

Glycine: (H)–CH(NH_2)–COOH

Alanine: (CH_3)–CH(NH_2)–COOH

Serine: (HO–CH_2)–CH(NH_2)–COOH

Glutamic acid: (HO–C(=O)–CH_2–CH_2)–CH(NH_2)–COOH

Lysine: (NH_2–CH_2–CH_2–CH_2–CH_2)–CH(NH_2)–COOH

Phenylalanine: (C_6H_5–CH_2)–CH(NH_2)–COOH

Cysteine: (HS–CH_2)–CH(NH_2)–COOH

Figure 3.8 Structures of some amino acids. The unique end is circled.

TABLE 3.2 The 20 amino acids grouped by chemical characteristics*

Ionizable side chains	Nonionizable side chains	Hydrophobic side chains
Acid side chains	Glycine (gly)	Valine (val)
Aspartic acid (asp)	Alanine (ala)	Leucine (leu)
Glutamic acid (glu)	Serine (ser)	Isoleucine (ile)
Basic side chains	Cysteine (cys)	Phenylalanine (phe)
Lysine (lys)	Threonine (thr)	Tryptophan (trp)
Arginine (arg)	Tyrosine (tyr)	Methionine (met)
Histidine (his)	Asparagine (asn)	
	Proline (pro)	
	Glutamine (gln)	

*The abbreviations commonly used in writing protein structures are shown in parentheses.

polypeptide chains that are formed, the side chains of the amino acids project out from the polypeptide backbone. Many of the unique properties of specific proteins are due to the arrangements of side chains on the outside of the polypeptide chains. The structure of a relatively simple protein, the hormone insulin, is shown in Figure 3.9. Note that this protein consists of *two* polypeptide chains, cross-linked by means of sulfur atoms. One sulfur atom from the amino acid *cysteine* on each polypeptide chain is involved in this cross-linking, which occurs through a type of bond called a *disulfide bond*. Although the structure in Figure 3.9 is accurate, it does not show the actual orientation of the protein macromolecule. This is because polypeptides undergo a complex folding process in which various parts of the molecule interact. A diagram of a folded enzyme molecule, showing how various distant parts of the polypeptide chains are brought together, is illustrated in Figure 3.10. Figure 3.10*b* also shows the *active site* of the molecule, the site where the function of the enzyme occurs.

Nucleic acids and their building blocks Two of the most important polymers in living organisms are the nucleic acids, **ribonucleic acid (RNA)** and **deoxyribonucleic acid (DNA)** (Figure 3.11). RNA and DNA are similar in structure, but they have important differences. They are composed of long chains of sugar molecules which are connected by phosphate groups, and attached to each sugar molecule is a nitrogen-containing organic molecule called a *nucleic acid base*. The sugar in RNA is *ribose* and the sugar in DNA is *deoxyribose* (see Figure 3.7). The bases in RNA are **adenine**, **cytosine**, **guanine**, and **uracil** (Figure 3.12). The first three bases listed are also found in DNA, but instead of uracil, DNA has **thymine** (Figure 3.12).

We will have more to say about nucleic acids in Chapters 6 and 7. We note here that the ribosome, a structure which plays a central role in protein synthesis, is composed of protein and RNA. There are also several other kinds of RNA molecules in the cell. As has been noted earlier, DNA is the hereditary material of the cell.

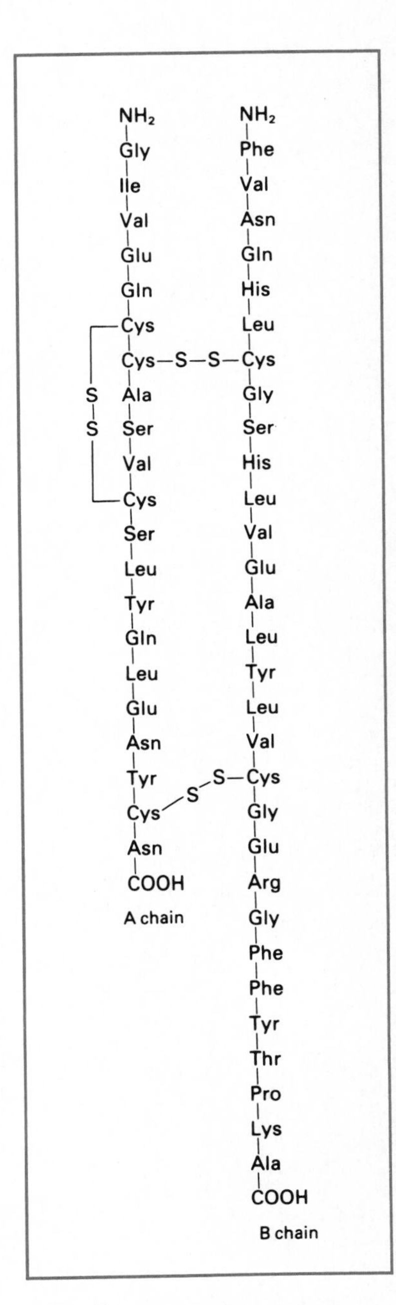

Figure 3.9 The amino acid sequence of a protein, insulin. Each amino acid is abbreviated as described in Table 3.2.

Lipids and the cell membrane The cell membrane is a complex structure composed of proteins embedded in a **lipid** envelope. The lipids in the

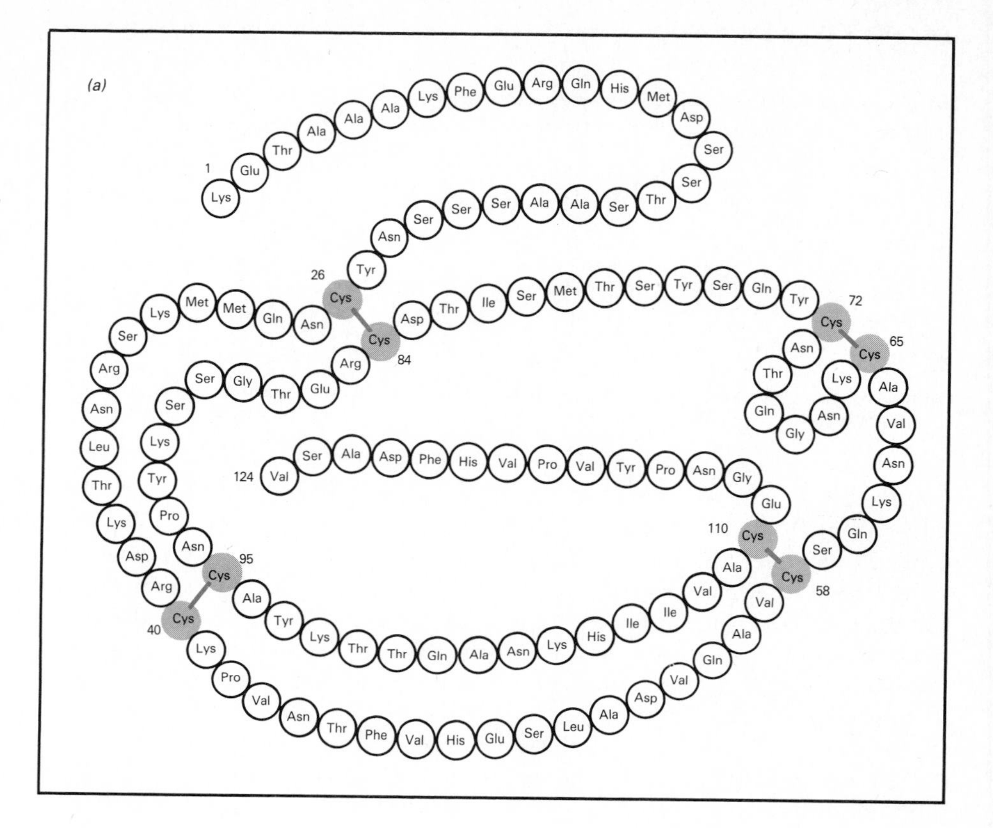

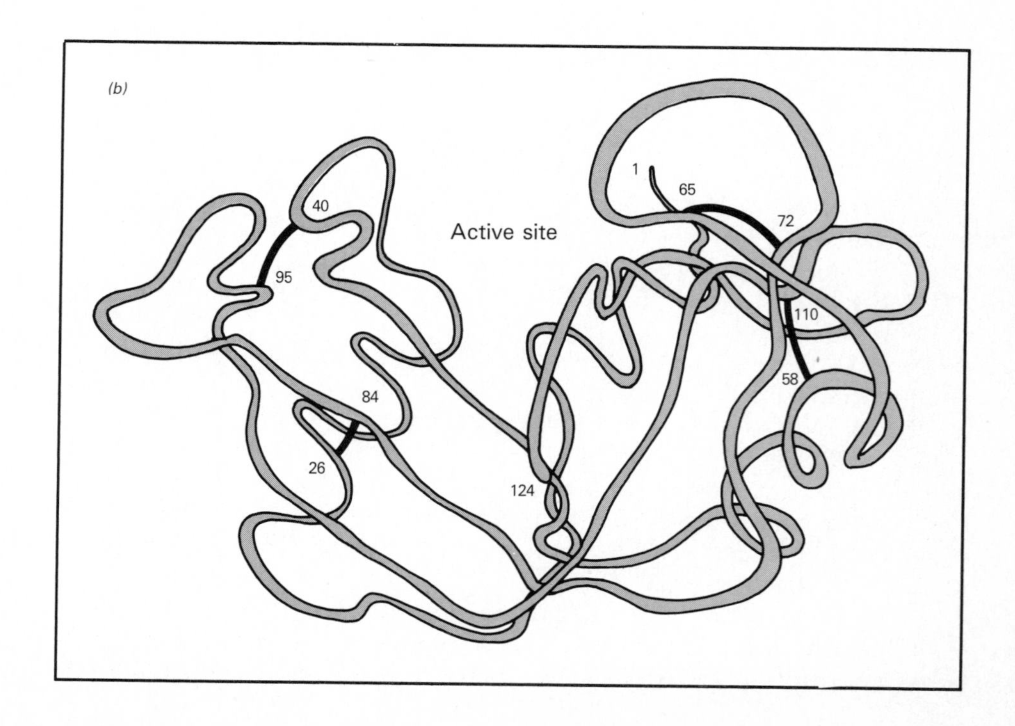

Figure 3.10 Structure of the enzyme ribonuclease. (a) Primary amino acid sequence, showing how sulfur-sulfur bonds between cysteine residues cause folding of the chain. (b) The three-dimensional structure of ribonuclease, showing how the macromolecule folds so that a site of enzymatic activity is formed, the *active site.*

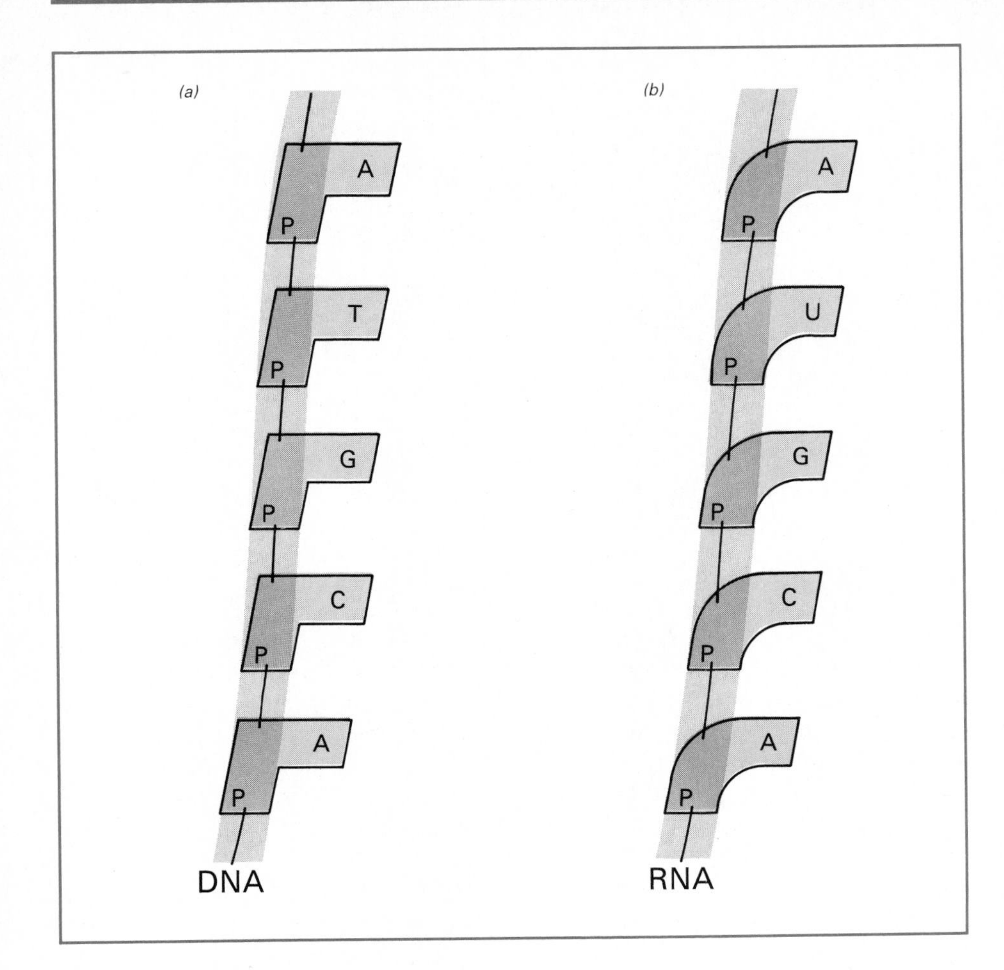

Figure 3.11 Simplified structures of the nucleic acids. (a) Deoxyribonucleic acid (*DNA*). (b) Ribonucleic acid (*RNA*). The nucleotide monomers are linked by bonds involving phosphate atoms (*P*), forming the "ribbon-like" backbone of the polymer. The nucleic acid bases protrude from the backbone. Abbreviations: *A*, adenine; *G*, guanine; *C*, cytosine; *T*, thymine; *U*, uracil.

cell membrane are not really polymers, but an array of molecules held together by weak bonds. Most membrane lipids, called *phospholipids*, consist of two chemically different parts. One end has an ionic charge, while the other end consists of two long uncharged carbon chains, called fatty acids. The charged end, being hydrophilic, tends to associate with water whereas the uncharged end, being hydrophobic, tends to associate with other uncharged ends rather than with water. The result is that in water, lipids tend to form a double layer (or lipid bilayer) in which hydrophilic ends face the outside and the hydrophobic ends face toward the inside, as is shown in Figure 3.13. The membrane is a lipid bilayer which extends completely around the cell, providing a barrier between the environment outside and the protoplasm inside of the cell. It is interesting that such an important structure of the cell is only held together by relatively weak bonds. There are numerous other examples in which weak bonds, rather than stronger covalent bonds, are important in the biochemistry of a cell. We will discuss some of these in Chapter 6.

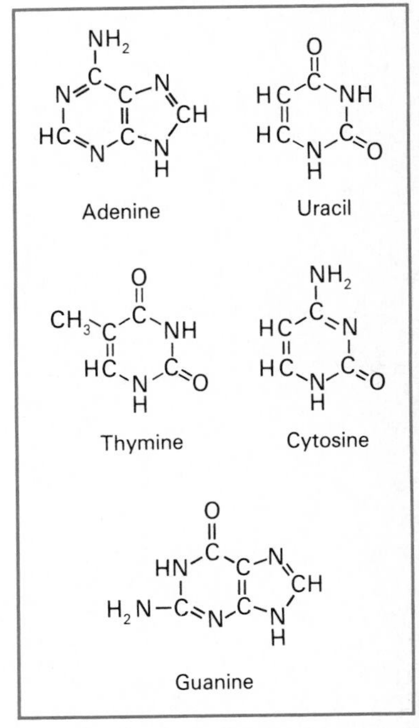

Figure 3.12 Structures of the nucleic acid bases.

Overall composition of a cell Now that we have discussed the various chemical constituents of a cell, we can return to a discussion of the cell itself.

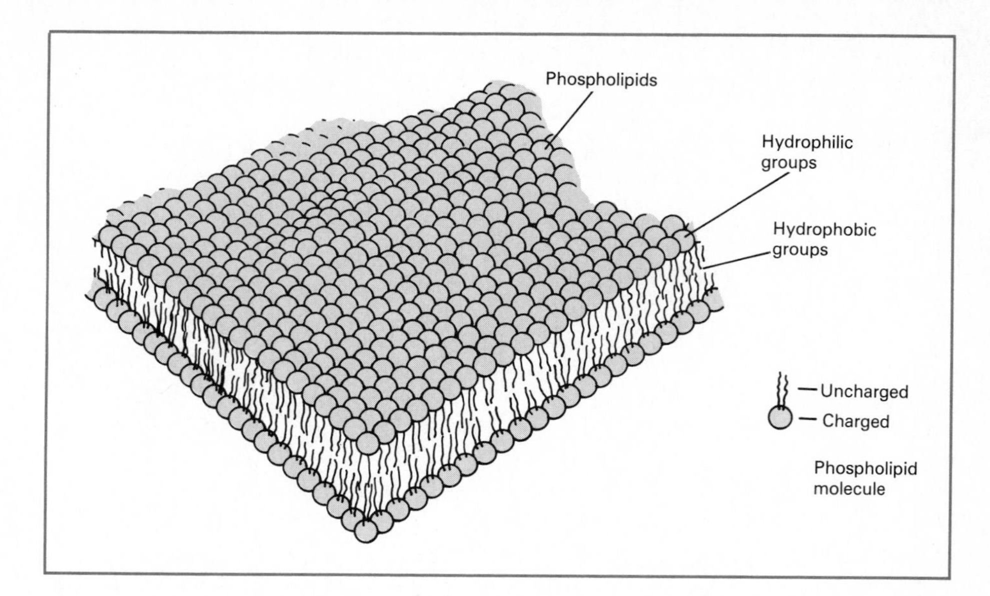

Figure 3.13 The assembly of a cell membrane from phospholipid molecules. In the cell, protein molecules (not shown) are embedded in the lipid layer.

Table 3.3 shows where and in what amounts the major polymers are found within a typical bacterial cell. Note that the dry weight of a cell is about half protein. There is also considerable RNA within the cell (about 20 percent of the total cell dry weight), most of it being associated with ribosomes. Lipids are the next most abundant cell component, making up about 10 percent of the weight of a cell. With the exception of DNA, there are a large number of copies of each type of macromolecule. Note that there are a thousand or more different types of protein in the cell. Some insight into why there is such variety among the proteins of the cell will be obtained in the rest of this chapter.

3.3 CHEMICAL REACTIONS AND ENZYMES

The metabolism of a cell is really the result of a complex series of chemical reactions which involve either the synthesis of monomeric and polymeric cell constituents from nutrients (biosynthesis or anabolism), or the breakdown of

TABLE 3.3 Approximate amounts of the different types of macromolecules in a typical bacterial cell

Macromolecule	Approximate percentage of total cell dry weight	Approximate number of molecules per cell	Different kinds of molecules per cell
Protein	55	2,000,000	Over 1000
RNA	22	300,000	Around 500
DNA	3	1 or few	1
Lipid	10	22,000,000	4
Carbohydrate	10	1,000,000	3

polymers and monomers as energy sources resulting in the release of energy (catabolism). The energy released with catabolism is then used to drive biosynthesis.

In a chemical reaction, chemicals interact to form new products. An example of a chemical reaction is the reaction in which hydrogen (H_2) and oxygen (O_2) combine with the formation of water:

$$2H_2 + O_2 \rightarrow 2H_2O + \text{(energy released)}$$

This reaction, which is equivalent to burning hydrogen gas in air, results in the liberation of a large amount of energy as heat. Reactions that give off energy, such as the example just given, are called *exothermic* reactions. Most catabolic reactions are exothermic and part of the energy given off is conserved by the cell and used in energy-requiring biosynthetic reactions. Energy-requiring reactions are called *endothermic* reactions.*

Activation energy and catalysts Although exothermic reactions may occur spontaneously, the reactants often remain together indefinitely without reacting. This is because chemical reactions require that the bonds between atoms in molecules be broken or partially loosened if they are to react with each other. Covalent bonds, the type formed in most molecules, are strong bonds and usually do not break spontaneously. Therefore, for most reactions to be initiated, energy must be applied. The energy necessary to break covalent bonds and get a reaction started is called **activation energy**. Once the reaction is initiated, if it is exothermic it can continue spontaneously, since the heat given off provides further activation energy. The concept of activation, which is extremely important in biochemistry, is illustrated in Figure 3.14.

A good example of the need for activation energy is shown by the reaction in which methane (natural gas) is burned:

$$CH_4 + 2O_2 \rightarrow CO_2 + 2H_2O + \text{(energy released)}$$

Methane can be present in air for long periods of time, provided no source of activation energy is present. If a spark or other small source of heat energy is added, some of the methane molecules are activated and react with oxygen in the air. The energy released from this initial reaction activates more methane molecules, leading to more reaction. A chain reaction is initiated and the reaction builds up to a very fast rate. In fact, an explosion may occur, caused by the sudden release of a large amount of energy.

A **catalyst** is an agent which reduces the activation energy and thus make a reaction occur more readily. A catalyst is defined as a substance that speeds up a chemical reaction but is not changed during the reaction. Good examples of catalysts used in chemistry are finely divided metals, such as nickel, palladium, and platinum. One of the most important chemical catalytic processes is the "catalytic cracking" of petroleum to produce gasoline from crude oil, in which a silica-alumina catalyst is used.

*Exothermic reactions are sometimes called *exergonic* and endothermic reactions are sometimes called *endergonic* (*ergo*- is a Greek combining form meaning *work*).

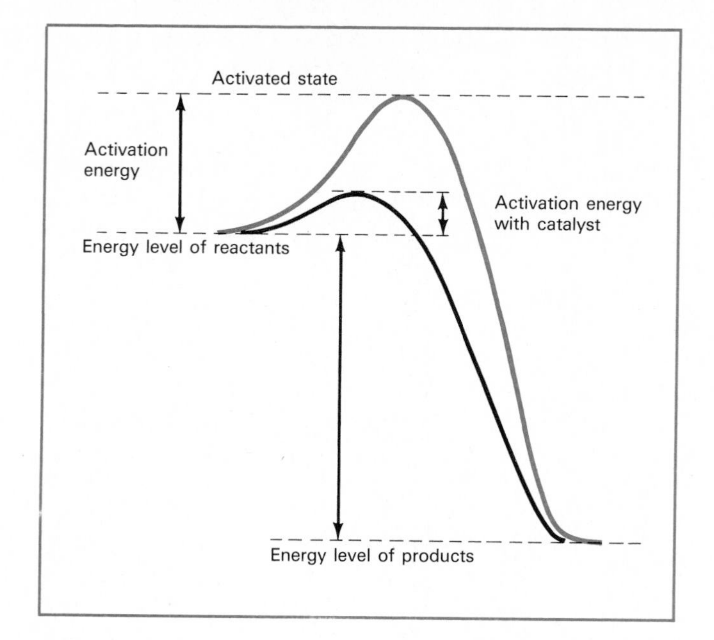

Figure 3.14 The concept of activation energy. Chemical reactions may not proceed spontaneously even though energy would be released, because the reactants must first be *activated.* Once activation has occurred, the reaction then proceeds spontaneously. Catalysts such as enzymes lower the required activation energy.

Enzymes Enzymes are the catalysts in living organisms. Without enzymes, life would be impossible, because most biochemical reactions have considerable activation energies and will not occur spontaneously. Enzymes are proteins (see Section 3.2). Because of weak bonding and disulfide bonding between the amino acids of the polypeptide chains, each enzyme protein folds into a large, generally spherical, macromolecule (see Figure 3.10) that has a number of reactive groups on the outside. These reactive groups interact with molecules present in cells and bring about catalysis. The molecule with which an enzyme reacts is called its **substrate** and the site on the enzyme where the reaction takes place is called the *active site* (see Figure 3.10*b*). Enzymes are very specific; a single enzyme will combine with only one or a very few kinds of substrates. Binding of substrate to enzyme results in the formation of an *enzyme-substrate complex.* As a result of the formation of this substrate complex, the substrate is activated, and chemical reaction can occur.

As an example of how an enzyme catalyzes a chemical reaction, let us consider the reaction in which sucrose is converted to glucose and fructose by addition of water (Figure 3.15). Even though this reaction is moderately exothermic and hence favorable, the rate of the reaction is extremely slow. This is shown by the fact that sucrose solutions in water (syrups) are stable for very long periods of time. However, an enzyme called *invertase* (or *sucrase*) which is very common in yeast and other organisms is able to catalyze this reaction and greatly speed it up. Because the reaction in which sucrose is converted to glucose and fructose involves the addition of water, it is called a *hydrolysis* reaction (*hydro-* refers to water and *lysis* means *breaking down*).

Several key aspects of enzymes deserve emphasis:

1. Enzymes are agents that very efficiently catalyze chemical reactions. They are active even when present in very low concentrations. It is

usually difficult to detect the enzyme molecule directly in a cell, but it is readily detected through the chemical reaction that it catalyzes.

2. Enzymes are very specific. An enzyme catalyzes only a single or a very few chemical reactions. Thus, for the large number of chemical reactions occurring in living organisms, many enzymes are necessary. It has been estimated that a single bacterial cell has over 1000 kinds of enzymes. This explains the large variety of proteins found in a cell, as noted in Table 3.3.
3. Enzymes, as other catalysts, are not changed as a result of the chemical reactions they catalyze. At the end of a reaction, the enzyme is still able to catalyze the same reaction again.

The catalytic function of an enzyme is determined by how the polypeptide chain is folded, and this folding depends almost solely on the amino acid sequence. Thus, the specificity of an enzyme molecule resides in its amino acid sequence. As we will learn in Chapter 6, the information for each amino acid sequence resides in the DNA of the cell. The DNA, the genetic material of the cell, specifies the amino acid sequence. It is a fascinating thought that the control of life processes is based on enzymes, agents whose only roles are to catalyze chemical reactions. The differences between one organism and another, between plant and animal, bacterium and protozoan, human and ape, are mediated by the different enzymes that each possess.

Enzymes are generally named for the substrate they react with or the reaction that they carry out. The suffix *-ase* is used to indicate that an enzyme is being named. Thus, an enzyme which hydrolyzes cellulose is called *cellulase,* an enzyme which hydrolyzes malt sugar, maltose, is called *maltase,* and an enzyme which hydrolyzes milk sugar, lactose, is called *lactase.*

Some enzymes are named for the kinds of functions they carry out. The general class of enzymes that remove hydrogen atoms from molecules are called *dehydrogenases* and the specific enzyme that removes hydrogens from glutamic acid is called *glutamic dehydrogenase*. Enzymes that introduce oxygen

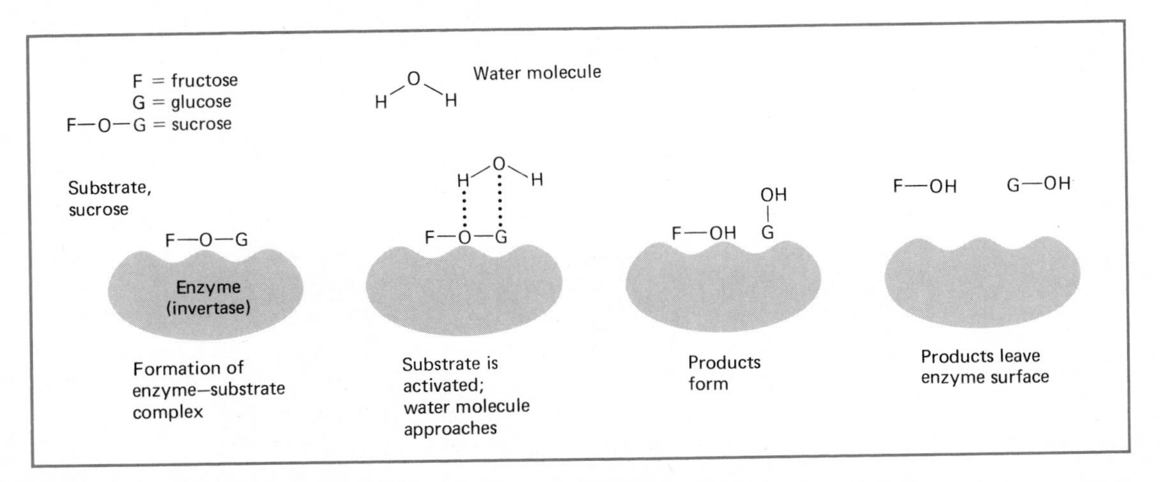

Figure 3.15 Mechanism of enzyme catalysis. The enzyme *invertase* catalyzes the hydrolysis (addition of water) of sucrose into its component sugars, fructose and glucose.

into molecules are called *oxidases;* the enzyme glucose oxidase introduces oxygen into glucose. Enzymes that catalyze the reduction of substances are called *reductases;* nitrate reductase causes the reduction of nitrate. The enzymes that break down nucleic acids are called *nucleases.* If the nucleic acid is ribonucleic acid, the enzyme is a *ribonuclease,* sometimes abbreviated RNase; with deoxyribonucleic acid, the enzyme is a *deoxyribonuclease,* abbreviated DNase. Enzymes that break down proteins are called *proteases.* Because thousands of enzymes are known, thousands of enzyme names exist. Fortunately, we shall mention only a few enzymes in this book.

3.4 NUTRITION AND BIOSYNTHESIS

Chemical elements are required by microorganisms as the building blocks of the biochemical constituents of cells. These elements must be present in the environment in order for the cell to be able to grow. However, the forms in which the elements are available to cells can vary considerably. Each cell contains enzymes that allow it to use nutrients only if they are provided in a specific form. An element may be present within an environment in a form which a cell cannot use, and thus be unavailable to the organism. Common sources of the major elements required by microorganisms are shown in Table 3.4.

Carbon is available either in organic form or as carbon dioxide. **Heterotrophic** organisms are defined as those which use organic compounds as carbon sources, and also often use the same organic compounds as energy sources. Many organisms can be grown in media which contain a single organic compound (for example the sugar glucose). Such organisms must be able to rearrange this organic compound into the vast array of chemical structures present in the cell. We will return to this point later in this chapter after considering the enzymatic pathways used by heterotrophic organisms in en-

TABLE 3.4 Common forms of the major elements needed for biosynthesis of cell components

Element	Usual form in the environment
C	Carbon dioxide (CO_2) Organic compounds
H	Water (H_2O) Organic compounds
O	Water (H_2O) Oxygen gas (O_2)
N	Ammonia (NH_3) Nitrate (NO_3^-) Nitrogen gas (N_2) Organic compounds (e.g., amino acids)
P	Phosphate (PO_4^{3-})
S	Hydrogen sulfide (H_2S) Sulfate (SO_4^{2-}) Organic compounds (e.g., cysteine)

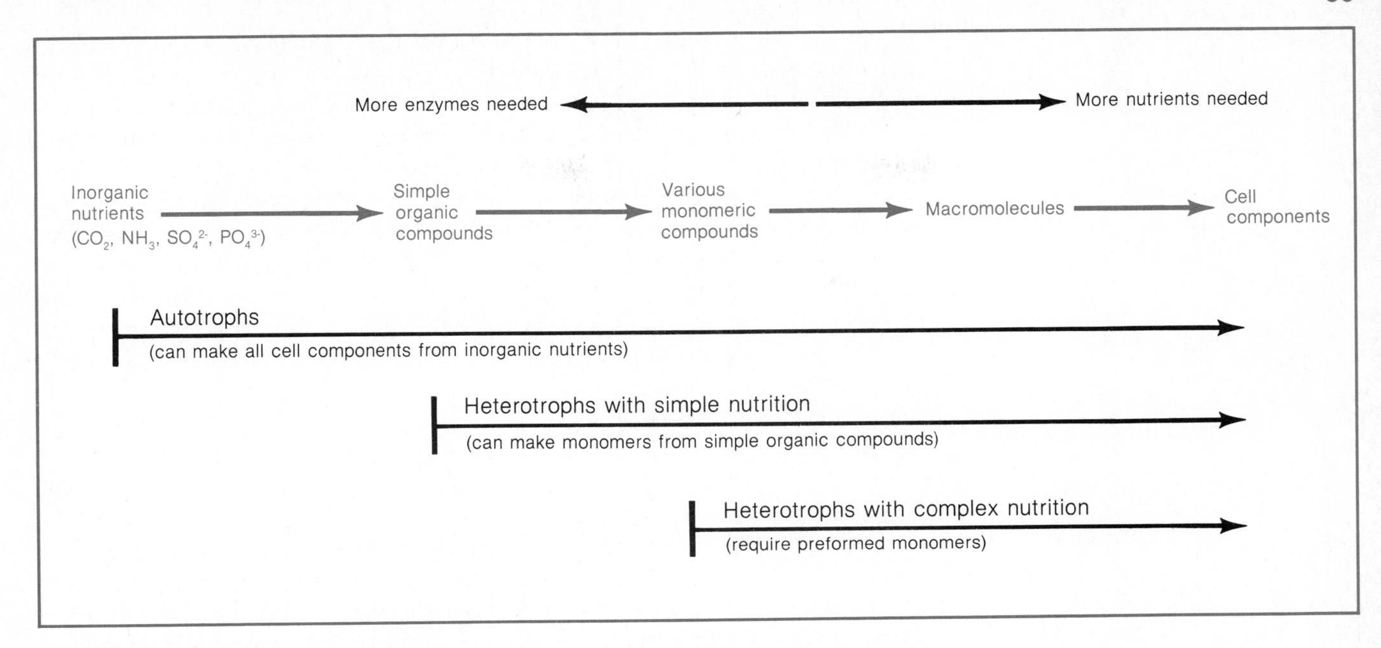

Figure 3.16 Differences in biosynthetic abilities of organisms based on the types of nutrients required.

ergy generation. One major group of organisms is able to obtain *all* of the carbon needed from carbon dioxide. Organisms of this group are called **autotrophs** (literally, *self-feeding*). We will discuss the unique characteristics of autotrophs later in this chapter.

Nitrogen is required in the cell in organic compounds such as amino acids and nucleic acid bases. It may be provided as organic nitrogen (for example, amino acids) in the environment, but is often available in inorganic form as nitrate (NO_3^-), ammonia (NH_3) or nitrogen gas (N_2). If nitrate or nitrogen gas is used as a source of nitrogen, either of these compounds is first converted to ammonia. Although many organisms are able to convert nitrate to ammonia, the conversion of nitrogen gas to ammonia, a process called **nitrogen fixation**, is a property of only a few bacteria. This process will be discussed in greater detail in Chapter 17, as it results in the enrichment of soils with nitrogen which would otherwise have to be supplied as fertilizer. Many heterotrophic bacteria and fungi are able to use ammonia or nitrate as a source of nitrogen. On the other hand, some microorganisms, such as protozoa, almost always require organic nitrogen compounds, being unable to use nitrate or ammonia. Humans, too, require organic nitrogen compounds and use proteins as the primary nitrogen source. Other elements required by organisms include sulfur (S), phosphorus (P), sodium (Na), potassium (K), calcium (Ca), magnesium (Mg), and iron (Fe).

There are great differences among microorganisms in their biosynthetic capabilities as determined by the collection of enzymes they contain. There is a corresponding variation among microorganisms in nutrient requirements (depicted in Figure 3.16). The simplest type of nutrition is that of the autotroph, which can be grown on completely inorganic compounds. However, an autotrophic organism would need to have a most complex array of biosynthetic enzymes to convert these simple nutrients into the complex biochemicals of the cell. Even among heterotrophic microorganisms there is a wide range of biosynthetic capabilities. Many heterotrophic microorganisms can be grown using a single organic compound for both carbon and energy source. Other heterotrophic microorganisms may require a complex collection of organic nutrients, such as amino acids and nucleic acid bases, because they lack the enzymes needed to make them from simpler compounds.

Once an organism has synthesized monomers from the nutrients in its environment, it must of course assemble them into the various polymers, or macromolecules of the cell components. The synthesis of macromolecules will be the subject of Chapter 6.

3.5 OBTAINING ENERGY: CATABOLISM

Energy is needed to drive the biosynthetic reactions which must occur as cells grow. In biological systems, this energy is provided in chemical form as *high-energy phosphate bonds*, shown as ~ *P* (sometimes called "squiggle P").* A typical compound that provides such ~P bonds is **adenosine triphosphate, ATP** (Figure 3.17). There are two high-energy phosphate bonds in ATP, and when they are broken, the energy liberated can be used to drive energy-requiring reactions. In most reactions, only the outer high-energy phosphate bond is used, ATP then being converted into *adenosine diphosphate*, ADP, plus *phosphate*, P:

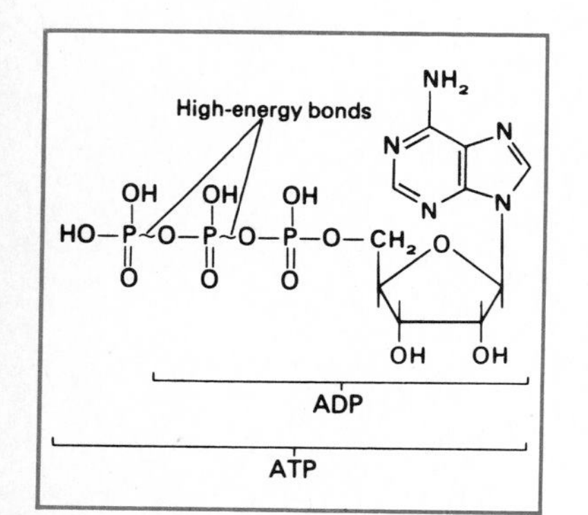

Figure 3.17 Structure of adenosine triphosphate, ATP, showing the location of the two high-energy phosphate bonds.

$$\text{ATP} \longrightarrow \text{ADP} + \text{P} + (\text{energy released})$$

By adding energy back to the ADP and phosphate, ATP can be regenerated:

$$\text{ADP} + \text{P} + (\text{added energy}) \longrightarrow \text{ATP}$$

The principal role of catabolic reactions is the release of energy from chemical compounds and its use in the synthesis of ATP from ADP + P. The ATP is then available for energy-requiring biosynthetic reactions.

Oxidation and reduction reactions In order to understand the reactions of catabolism, one must first have some understanding of two important types of reactions which occur in energy metabolism: oxidation and reduction. **Oxidation** is the removal of electrons from a compound. Because electrons are never free in a cell, oxidation is always coupled with **reduction**. An example of a coupled oxidation/reduction reaction is the oxidation of hydrogen with oxygen described earlier. We now rewrite this reaction to include the electrons that are being transferred:

*From a chemical viewpoint, it is inaccurate to use the term "high-energy phosphate bond," since the high energy does not exactly reside in the specific bond shown. However, the concept of the high-energy phosphate bond is a useful formalism in biochemistry.

$H_2 \rightarrow 2H^+ + 2e^-$ (*electron donating*)
$O + 2e^- \rightarrow O^{2-}$ (*electron accepting*)
$2H^+ + O^{2-} \rightarrow H_2O$ (*water formation*)

During this series of coupled reactions, hydrogen is oxidized and oxygen is reduced.

An example of a more complex oxidation/reduction reaction is the oxidation of glucose coupled to the reduction of oxygen:

$$\text{Glucose } (C_6H_{12}O_6) + 6O_2 \rightarrow 6CO_2 + 6H_2O$$

Here the oxidation of glucose yields electrons and the reduction of oxygen consumes electrons:

Oxidation: Glucose $\rightarrow 6CO_2$ + 12 electrons + $12H^+$
Reduction: $6O_2$ + 12 electrons + $12H^+ \rightarrow 6H_2O$

During oxidation, glucose is oxidized to carbon dioxide. Note that there is more oxygen and less hydrogen per carbon atom in the product *carbon dioxide* than in the substrate *glucose*. During the reduction, oxygen becomes reduced to water, a product containing more hydrogen than the substrate oxygen.

The simplest way to think about these oxidation and reduction reactions is in terms of the donation and acceptance of electrons. The energy source donates electrons and becomes oxidized; it can be called the **electron donor**. In the same reaction, the **electron acceptor** accepts the electrons and becomes reduced. In the last example shown above, glucose is the electron donor and oxygen is the electron acceptor.

A whole series of oxidation-reduction reactions may be coupled, with the oxidized product of one reaction serving as the electron donor to the next electron acceptor. The last electron acceptor in such a coupled system is called the *terminal electron acceptor*. A variety of compounds may serve as electron donor or electron acceptor. The combination of electron donor (energy source) and terminal electron acceptor determines the amount of energy available to the organism. We will consider the variety of types of catabolism in Section 3.7.

Although we have presented the oxidation of glucose to carbon dioxide as a single reaction, in reality, there are numerous chemical reactions involved in this process, each step being catalyzed by a separate enzyme. This is true for all biochemical oxidations. As a result, when oxidation occurs, the terminal electron acceptor seldom immediately receives the electrons directly from the electron donor. Instead the electrons are temporarily carried by a specific **electron carrier**. The most common electron carrier is **nicotinamide adenine dinucleotide** (abbreviated **NAD**) which is capable of accepting two electrons (Figure 3.18). When NAD accepts two electrons it becomes reduced (written as NADH), and in the oxidation of NADH two electrons are given up and NAD is formed again. A related electron carrier is *nicotinamide adenine dinucleotide phosphate* (*NADP*). In general NAD plays a role in catabolic oxidation/

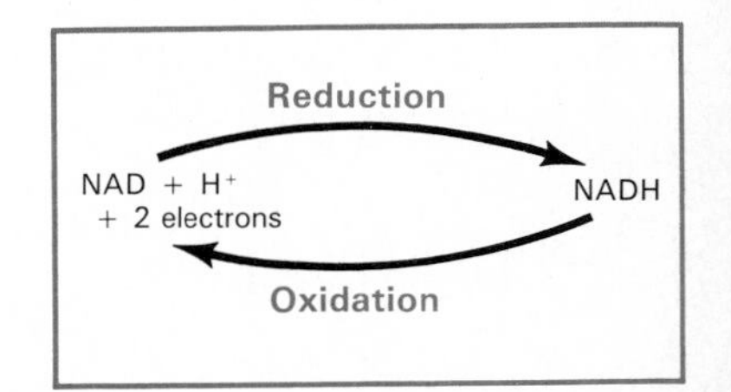

Figure 3.18 The oxidation-reduction reaction of the electron carrier nicotinamide adenine dinucleotide, NAD.

reduction reactions, and NADP acts in biosynthetic reactions. Other electron carriers important in the formation of ATP will be considered below.

3.6 BIOCHEMISTRY OF ENERGY GENERATION

Now we shall look more carefully at some of the main reactions of catabolism. Just how is the substrate (energy source) degraded in such a way that ATP is produced? The key to how a cell is able to trap the energy released during catabolic oxidation/reduction reactions is that the overall reaction is carried out in a large number of discrete steps. At each step, a small amount of energy is released, some of which can be trapped. (A violent explosion such as occurs when hydrogen gas is burned would be wasteful!)

There are two major mechanisms for energy generation from the oxidation of chemical compounds: **fermentation** and **respiration**. Respiration requires an external electron acceptor such as O_2 and a means of transporting electrons to the O_2. Fermentation, on the other hand, can occur in the absence of O_2. By means of the metabolic reactions of fermentation and respiration, energy can be liberated from chemical compounds and be coupled to the production of high-energy phosphate bonds.

Fermentation Many organisms oxidize certain organic compounds in the absence of oxygen or any other external electron acceptor. This is possible because the needed electron acceptor is produced as an intermediate during the breakdown of the energy source. The process by which an organic compound is broken down in the absence of an external electron acceptor is called **fermentation**. Only part of the organic energy source is oxidized to carbon dioxide; other products of a fermentation reaction include reduced organic compounds in which much of the chemical energy initially present in the energy source still remains. Numerous end products, called *fermentation products*, are possible, including alcohols (for example, ethanol), and acids (for example, lactic and acetic acids). Because fermentation products such as ethanol, lactic acid, and acetic acid are important in human affairs (one thinks, for instance, of beer, yogurt, and vinegar), a knowledge of how fermentation processes are carried out is of considerable practical importance.

Glycolysis The most common way in which a fermentable substrate such as glucose is broken down is by a process called **glycolysis**. Figure 3.19 shows the pathway of glycolysis used by yeast in the production of ethanol and carbon dioxide. Breakdown of glucose occurs by a series of reactions, each catalyzed by a specific enzyme. In the first few steps, phosphate from ATP is added, resulting in the activation of glucose. The six-carbon phosphorylated sugar formed, fructose diphosphate, is then split into two three-carbon phosphorylated molecules, glyceraldehyde phosphate. All subsequent reactions occur twice, once for each molecule of glyceraldehyde phosphate. The glyceraldehyde phosphate is oxidized, the electrons being transferred temporarily to NAD. During the next several enzymatic reactions, the resulting three-carbon molecule is rearranged and high-energy phosphate in-

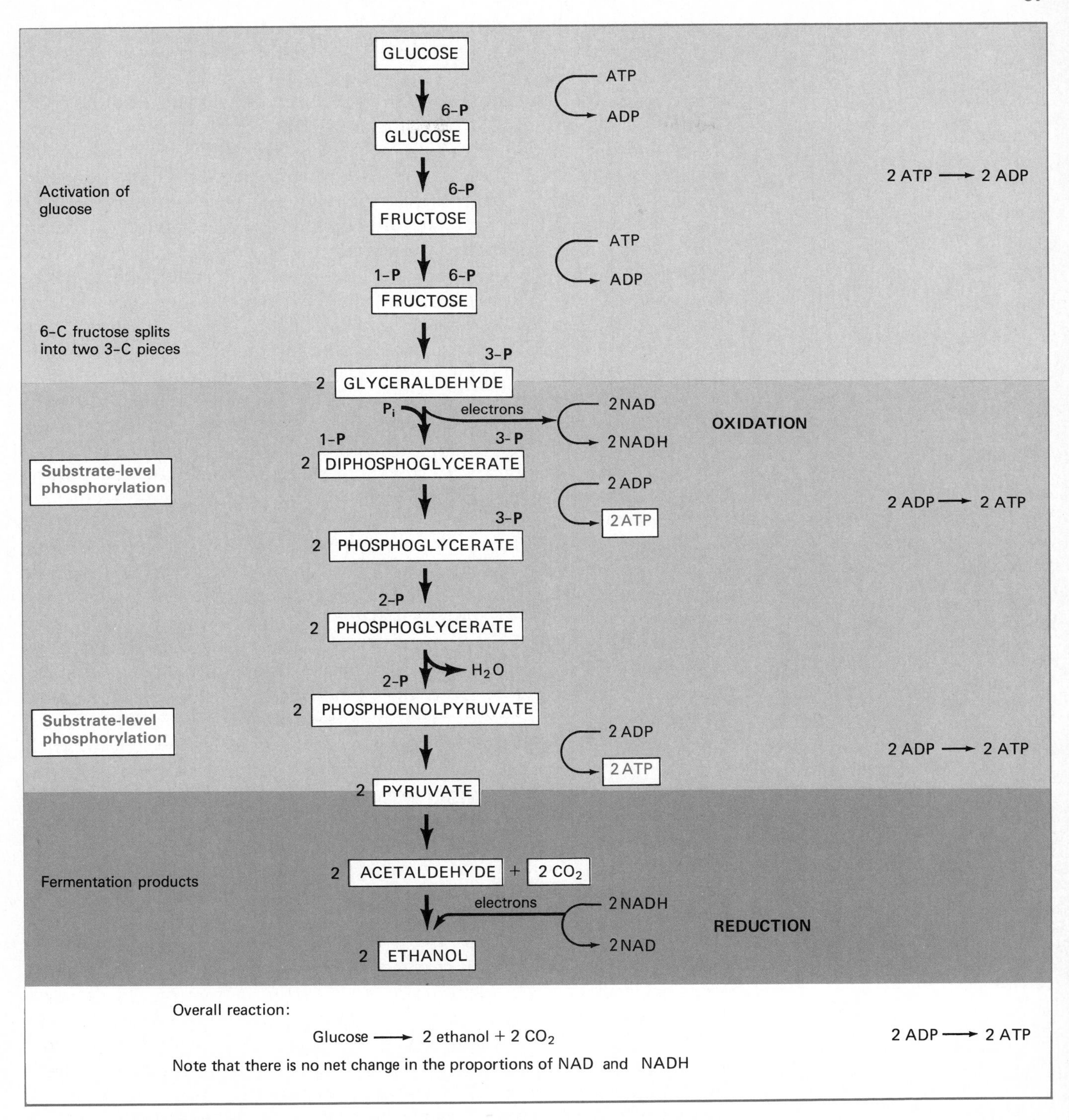

Figure 3.19 Glycolysis, the series of biochemical steps by which glucose is converted into ethanol and CO_2 by yeast under anaerobic conditions. Each arrow indicates a step catalyzed by a different enzyme.

termediates are formed which transfer their high-energy phosphate bonds to ADP, resulting in the formation of ATP. This process, in which a high-energy phosphate bond in ATP is made as a result of substrate oxidation, is called **substrate-level phosphorylation**. A key compound formed as a result of these reactions is *pyruvate*, which is both an important intermediate in the energy-generation process and an important starting material for biosynthesis. In energy generation, the pyruvate is reduced, accepting electrons from the NADH that had formed earlier at the glyceraldehyde-phosphate oxidation step.

How does substrate-level phosphorylation work? During the catabolism of glucose, phosphorylated organic components are produced, which, like ATP, have high energy ~P bonds. Such components are intermediates in the breakdown of the energy source to its product, and they are thus degraded further. In this process, ATP synthesis occurs. In Figure 3.19, the compounds *diphosphoglycerate* and *phosphoenol pyruvate* are high-energy phosphate compounds, and when they break down, ATP formation occurs.

The final steps in glycolysis, after substrate-level phosphorylation, involve the formation of fermentation products from pyruvate. In yeast (Figure 3.19). the fermentation products are ethanol (more reduced than glucose) and carbon dioxide (more oxidized than glucose). The NADH formed at an early stage of glycolysis is converted back to NAD in the final stages. Thus, the oxidation/reduction balance is preserved.

The net energy yield from alcohol fermentation by yeast is rather small, as only two ATP molecules are made per glucose molecule. Some of the energy from glucose remains in the ethanol formed, but this energy is not available to the yeast cell in the absence of O_2.

Since the energy yield is small, an organism which obtains energy by fermentation has to convert a large amount of its energy source to products to obtain sufficient ATP to drive biosynthetic reactions. Although the energy yield is small for the microbe, the yield of product in the fermentation is large, which is, of course, of major commercial importance. In the beverage industry, it is the alcohol which is the desired product, but in the baking industry, it is the CO_2, since this is the agent responsible for the rising of bread. The production of yogurt and cheese depends on the formation of the fermentation product, lactic acid. Further discussion of food fermentations is given in Chapter 16, and alcoholic fermentations are discussed in detail in Chapter 18. One final point about fermentation processes: They do not involve oxygen and hence occur under completely *anaerobic* conditions.

Respiration Organisms that respire are usually aerobic, using O_2 as a terminal electron acceptor. **Respiration** can be defined as an oxidation process in which O_2 is the terminal electron acceptor. A key difference from fermentation is that during respiration the energy source is oxidized completely to CO_2. The many electrons derived from this total oxidation are transferred to O_2 via a special system, the electron transport system, which we discuss below. As a result of the activity of the electron transport system, much more ATP is generated than could be formed during fermentation. During the flow of electrons through the electron carriers of the electron transport system, chemical bond energy is released that can be used to form ATP. This is accomplished by a process called **electron-transport phosphorylation**, in which ADP and phosphate are converted into ATP. In the transfer of one pair of electrons from the electron carrier NADH via the electron transport system to oxygen, three ATPs can potentially be made.

Let us now consider the details of the respiration process. Two distinct systems are involved. One accomplishes the complete oxidation of the energy source to CO_2, whereas the other brings about the synthesis of ATP during the electron flow to the electron acceptor, O_2. Let us consider the substrate oxidation process first.

An organism which respires an energy source such as glucose first converts it to pyruvate via the same enzymes used in glycolysis (Figure 3.19). The subsequent reactions involve the *tricarboxylic acid cycle* (*TCA cycle*) (Figure 3.20). The TCA cycle gets its name from the first few intermediates which

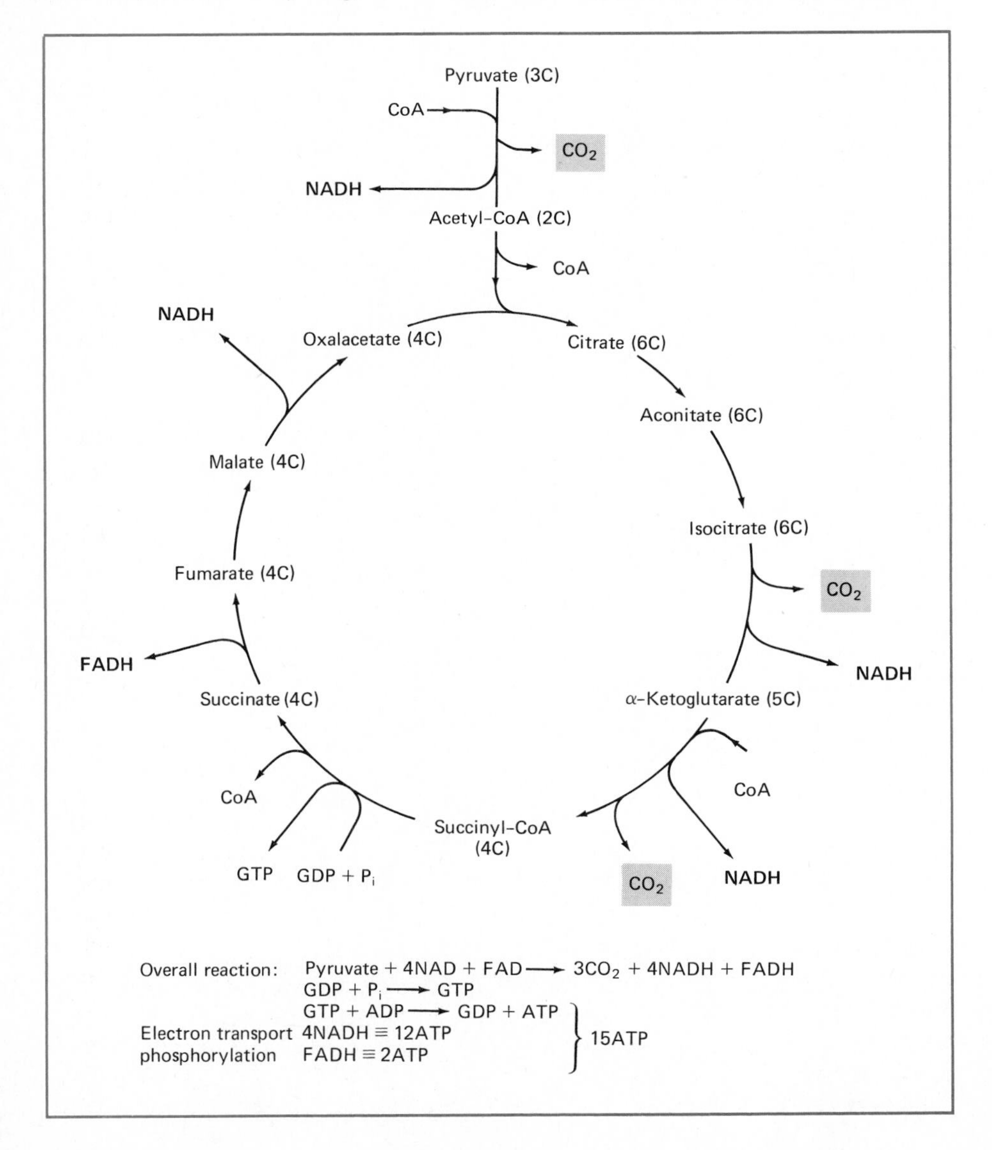

Figure 3.20 The tricarboxylic acid (*TCA*) cycle. The three-carbon compound pyruvate is oxidized completely to CO_2, and the energy released is conserved in NADH and FADH.

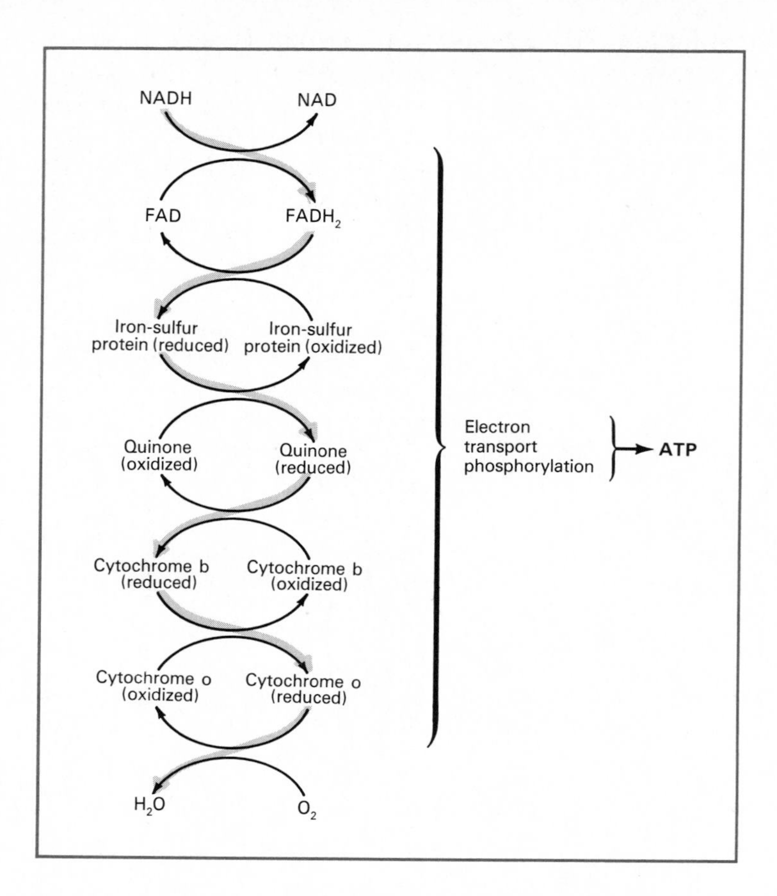

Figure 3.21 The electron-transport system showing the flow of electrons between NADH and the ultimate electron acceptor oxygen (O_2).

are organic compounds with three carboxyl (—COOH) groups. It is also sometimes referred to as the *citric acid cycle* after this first intermediate.* Pyruvate is first oxidized to CO_2 and acetyl-CoA, which is a high-energy form of acetic acid. Acetyl-CoA then enters the TCA cycle. Note that the formation of acetyl-CoA and several steps in the TCA cycle are coupled to the reduction of the electron carrier NAD to NADH. During the several enzymatic reactions of the TCA cycle, two molecules of CO_2 are produced. Note that with each oxidation step, NAD (or another electron carrier, FAD) is reduced to form NADH (or FADH).

Although a small amount of ATP is synthesized by substrate-level phosphorylation in glycolysis and the TCA cycle, by far the most ATP synthesis occurs as a result of the flow of electrons through the electron-transport system from NADH or FADH to oxygen. The several different electron carriers involved are shown in Figure 3.21. As NADH is oxidized, electrons flow to FAD, thus reducing it to FADH. The electrons next cascade in a series of oxidation/reduction reactions from FADH through several electron carriers including, in order, iron-sulfur proteins, quinone, cytochromes b and o, and ultimately the final electron acceptor, oxygen. As noted in Figure 3.21, ATP is formed as a consequence of this electron flow.

*The TCA cycle is sometimes called the *Krebs cycle*, after its discoverer, Sir Hans Krebs.

Since ATP synthesis occurs as a result of electron flow, this process is called **electron-transport phosphorylation**. This process is complex and before explaining it we must introduce several new concepts involving the role of the cell membrane.

We first note that the electron-transport system is located within a membrane. In bacteria the electron-transport system is located in the cell membrane, whereas in eucaryotic microorganisms it is located in membranes of the mitochondrion. The electron flow from one electron carrier to another within the membrane causes a transfer of protons (H^+) from the inside to the outside of the membrane (Figure 3.22). This occurs because some electron carriers carry protons as well as electrons. For example, FAD accepts two electrons and two protons and becomes reduced to FADH. When FADH is oxidized by another electron carrier, both the electrons and protons are released. In this process, protons are accepted from the inside of the membrane, but liberated on the outside of the membrane, so that a transfer of protons from inside to outside of the membrane occurs. Thus, as a result of electron flow through the electron transport system, a *proton gradient* is set up with OH^- (hydroxyl ion) on the inside and H^+ on the outside (Figure 3.22).

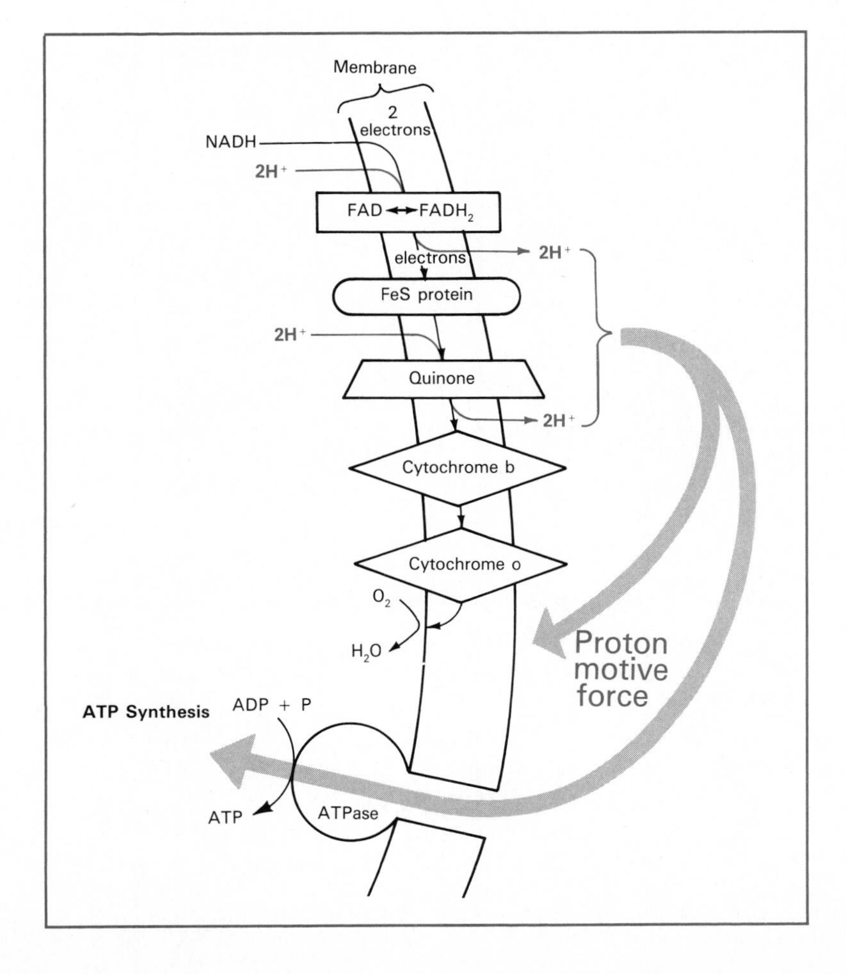

Figure 3.22 Electron transport and electron-transport phosphorylation. Top: Oxidation of NADH and the flow of electrons through the electron transport system leading to the transfer of protons (H^+) from the inside to the outside of the membrane. The tendency of protons to return to the inside is called *proton motive force.* Bottom: ATP synthesis occurs as protons reenter the cell. An ATPase enzyme uses the proton motive force in the synthesis of ATP.

How can the transfer of protons across a membrane result in ATP synthesis? Protons are positively charged ions which cannot by themselves move back across the membrane. We note that the separation of electric charge on two sides of an impermeable barrier in a battery results in an *electromotive force* which permits energy to be stored. In an analogous way, the separation of the protons from hydroxyl ions across a membrane results in a **proton motive force**. As we shall see, this proton motive force can be used to drive ATP synthesis.

The last step in electron-transport phosphorylation is the conversion of the energy of proton motive force into the chemical energy of ATP. This occurs through the action of a special enzyme also located in the membrane called *ATPase* (see lower part of Figure 3.22). This enzyme allows the controlled flow of protons back to the inside of the membrane, which releases the energy that had been stored as the proton motive force, driving the synthesis of ATP.

During the flow of electrons from each NADH to oxygen, a sufficiently large proton motive force is generated to form up to three ATP molecules. Considering the large number of NADH and FADH molecules oxidized during the complete oxidation of the energy source (see Figure 3.20), considerably more ATP is formed during respiration of glucose than during fermentation of glucose. Including the substrate-level phosphorylations which occur in glycolysis and the TCA cycle, 38 ATPs can be formed from a single glucose molecule during respiration. Respiration and fermentation are compared in Table 3.5.

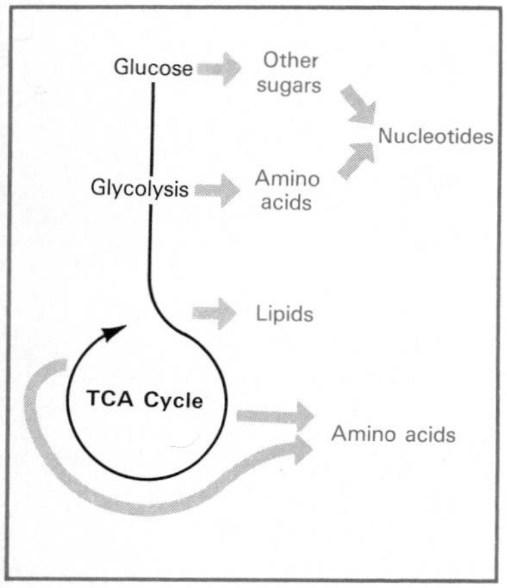

Figure 3.23 The use of catabolic pathways for biosynthesis of monomers.

Integration of metabolic pathways Many of the enzymatic pathways of catabolism are also used in the biosynthesis of the carbon structures or "backbones" of the various monomeric organic compounds found within the cell. As shown in Figure 3.23, sugars, nucleotides, lipids, and amino acids are largely derived from the intermediates of glycolysis and the TCA cycle. Although we will not discuss the enzymatic pathways by which the synthesis of specific monomers from specific catabolic intermediates occurs, it is important to note that biosynthesis and catabolism are linked through a number of common biochemical intermediates.

TABLE 3.5 A comparison of fermentation and respiration

Characteristic	Fermentation	Respiration
Substrates	Sugars; some amino acids	Many organic compounds
Products	Partially oxidized products	Completely oxidized product: CO_2
Oxygen requirements	None, anaerobic	O_2 required, aerobic
Pathway	Glycolysis	Glycolysis and tricarboxylic acid cycle
Amount of ATP	Low, 2 per glucose	High, 38 per glucose

3.7 METABOLIC DIVERSITY AMONG THE MICROORGANISMS

In Chapter 2 we discussed the variety among microorganisms based on differences in the structure of their cells. There is an equally impressive diversity among microorganisms based on types of energy metabolism. These differences, summarized in Table 3.6, are largely a reflection of the ways in which different organisms obtain energy. Up to now, we have discussed organisms which obtain energy by the aerobic oxidation or fermentation of sugars. Other types of respiration processes are known. Also, an organism may obtain energy from light or from the oxidation of inorganic chemicals.

Energy from organic chemicals Most bacteria, all protozoa, and all fungi are heterotrophic, obtaining their energy from organic compounds. There is an impressive variety in the types of organic compounds which may serve as electron donors (energy sources) for microorganisms in respiratory reactions. Certain bacteria are able to metabolize hundreds of different organic compounds. Members of the bacterial genus *Pseudomonas* are commonly recognized for this ability. Such microorganisms may be able to carry out the metabolism of unusual organic compounds not commonly found in living matter, such as pesticides, which are synthetic molecules potentially harmful to humans (see Chapter 17). This is, of course, a function of the types of enzymes a given organism possesses. The diversity among heterotrophic microorganisms is, in part, reflected by the abilities of different organisms to use different energy sources.

TABLE 3.6 Variety among microorganisms in energy metabolism

Type	Energy source	Electron donor	Electron acceptor	Types of microorganisms
Phototrophy	Light			
Oxygenic		H_2O	CO_2	Algae; cyanobacteria
Anoxygenic		H_2S	CO_2	Purple and green bacteria
Heterotrophy	Organic			
Aerobic respiration		Many organic compounds	O_2	Many bacteria; most fungi; most protozoa
Fermentation		Some organic compounds	—	Many bacteria; yeast (fungi); some protozoa
Anaerobic respiration	Organic			
Denitrification		Many organic compounds	NO_3^-	Denitrifying bacteria
Sulfate reduction		Few organic compounds	SO_4^{2-}	Sulfate-reducing bacteria
Methanogenesis*	Inorganic	H_2	CO_2	Methanogenic bacteria
Lithotrophy	Inorganic			
Nitrification		NH_3 or NO_2^-	O_2	Nitrifying bacteria
Sulfur oxidation		H_2S, S^0	O_2	Sulfur-oxidizing bacteria

*Also lithotrophy.

We have already considered that some microorganisms are able to carry out heterotrophic reactions under anaerobic conditions by fermentation. There is considerable variety in the types of compounds which different microorganisms are able to ferment. Although sugars are the most commonly studied fermentation substrates, other organic components such as amino acids, nucleic acid bases, and organic acids are also fermentable by some organisms. In addition, some fermenting microorganisms can use different enzymatic pathways to convert the same energy source to different products. For example, yeasts convert glucose to ethanol and CO_2 as shown in Figure 3.19, whereas lactic acid bacteria convert glucose to lactic acid, because they use a different enzyme to metabolize the glycolysis intermediate, pyruvate. Fermentation can be carried out by yeasts, some anaerobic protozoa, and many bacteria.

Whereas some bacteria grow exclusively by aerobic respiration, other bacteria are poisoned by oxygen and can only obtain energy by fermentation. Still other bacteria, called *facultative anaerobes,* can grow either by aerobic respiration or by fermentation. As mentioned in Chapter 2, many bacteria are classified according to these metabolic differences (see Table 2.2).

Anaerobic respiration We have defined respiration as the oxidation of a substrate using oxygen as an electron acceptor. We have also noted that respiration involves a set of electron carriers combined in an electron-transport system. Some organisms use inorganic electron acceptors other than oxygen in electron-transport processes. Because these processes are analogous to respiration but occur in the absence of oxygen, they are called **anaerobic respiration**. A number of bacteria carry out the anaerobic respiration process. Examples of alternate electron acceptors used by such bacteria are nitrate (NO_3^-), sulfate (SO_4^{2-}), and carbon dioxide (CO_2).

Bacteria which use nitrate as electron acceptor, called *denitrifying bacteria,* oxidize organic compounds and reduce the nitrate to nitrogen gas (N_2). The process is similar to that diagrammed in Figure 3.21, except nitrate is used instead of oxygen. As in aerobic respiration, energy of the organic molecule is conserved via electron-transport phosphorylation. When the process of *denitrification* occurs in the soil, it can result in the loss to the atmosphere of soil nitrate. Since nitrate is one of the common nitrogen sources for plants in the soil, nitrogen losses due to denitrification must be made up by addition of fertilizer (see Chapter 17). One important characteristic of soil denitrification is that it only occurs under anaerobic conditions, so that it will be found in soils where oxygen penetration does not occur, such as water-logged soils. Even well-tilled soils may undergo denitrification if they become flooded so that oxygen penetration is prevented.

The members of another group of bacteria, the *sulfate-reducing bacteria,* use sulfate instead of oxygen as an electron acceptor. *Sulfate reduction* results in the formation of hydrogen sulfide, a potentially toxic product. Sulfate-reducing bacteria are strictly anaerobic and are usually found in muds and marshes. They are responsible for the "rotten egg" odor which can sometimes be detected around lakes and coastal areas.

Another important group of bacteria carrying out an anaerobic respiration process are the *methane-producing* or *methanogenic* bacteria. These organisms reduce CO_2 to methane gas, CH_4. The methanogenic bacteria are particularly active in habitats where organic matter is being decomposed anaerobically, such as sewage-treatment plants (see Chapter 15), the bottoms of lakes, and wetlands. Another interesting habitat for methanogenic bacteria is the digestive organ in cattle and sheep called the *rumen*. The methane produced by bacteria in the rumen is belched by the animal and can represent a significant loss of energy in the conversion of feed to milk or meat. This is an important concern in beef and dairy cattle husbandry (see Chapter 17). Another interesting habitat for methanogenic bacteria is the large intestine of humans (see the Box in Chapter 8, page 187).

Energy from inorganic chemicals A number of bacteria are able to obtain energy by oxidizing inorganic compounds. Such organisms are called *lithotrophs* (*litho-* means *rock* and *trophic* refers to feeding, so that these organisms are, literally, the *rock-eating bacteria*). Examples of inorganic energy sources used by lithotrophic bacteria are ammonia (NH_3), hydrogen sulfide (H_2S), and iron (Fe^{2+}).

The bacteria which use NH_3 as energy source convert it to nitrate (NO_3^-), a process called *nitrification*. The nitrification process is aerobic, and so can be considered to be a type of respiration. However, the complete oxidation of ammonia to nitrate requires the cooperative action of *two* groups of bacteria. One group of bacteria oxidizes ammonia to nitrite (NO_2^-). The nitrite formed by the first group is oxidized by the second group to nitrate. Nitrification is an important process in the soil because it allows the conversion of ammonia-based fertilizers to a form, nitrate, which is easily utilized by plants. The nitrification process will be discussed in Chapter 17, when we consider agricultural microbiology.

Another group of lithotrophic bacteria, the *sulfur-oxidizing bacteria,* oxidize hydrogen sulfide, elemental sulfur (S^0), or metal sulfide ores such as iron sulfides (FeS_2, called pyrite) to sulfuric acid (H_2SO_4). These oxidations are also aerobic respiration processes. The oxidized product, sulfuric acid, contributes to the acidity of the environment. How this process can lead to environmental pollution will be considered in Chapter 14.

There are other inorganic compounds which can be metabolized by different bacterial groups. For example, some bacteria are able to oxidize hydrogen gas (H_2), while still others can obtain energy from the oxidation of ferrous iron (Fe^{2+}).

As we noted earlier, some organisms, called *autotrophs,* are able to obtain all of their carbon for cell synthesis from carbon dioxide (CO_2). Many of the lithotrophic bacteria are autotrophic. Such lithotrophic autotrophs are able to grow and make all their cell material in completely mineral environments, containing simply an inorganic energy source, carbon dioxide as carbon source, and the required mineral elements.

Perhaps the most interesting place where lithotrophic autotrophs live is in the deep-sea hydrothermal vents at the bottom of the sea floor, discovered in 1977 near centers of oceanic volcanic activity. Here, hydrogen sulfide flow-

AUTOTROPHS AT THE BOTTOM OF THE SEA

In 1977 scientists in the submarine Alvin exploring the ocean floor of the East Pacific Rise (between Baja California and the Galapagos) discovered an amazing new biological world. Around hot springs, or vents, on the ocean floor were dense communities of large tube worms, giant clams, crabs, fish, and some unusual animals which scientists had never seen before. What were they doing there? We know that photosynthesis, the source of food for animal life, occurs only near the ocean surface where light is available. Overall, the deep ocean floor is a biologically barren place, since light cannot penetrate that far and very little food ever sinks to such great depths. What were the food sources for these deep-sea vent communities? The water which flows out of the deep-sea springs contains reduced *inorganic* compounds, in particular hydrogen sulfide, H_2S. Therefore the logical source of food for these newly discovered communities could be organic matter produced by lithotrophic rather than by photosynthetic processes. Sulfur-oxidizing lithotrophic bacteria use H_2S as an energy source and reduce carbon dioxide, thus being able to grow within or in the vicinity of deep sea vents. Like algae at the ocean surface, these bacteria could provide food for the higher organisms found around the vents. Since 1977, deep-sea vent communities have been discovered in a number of other locations on the ocean floor. The existence of these communities shows that life on earth need not be driven solely by sunlight. Here is a good example of how an unusual type of bacterial metabolism, lithotrophy, can be important to higher life forms.

A scanning electron photomicrograph of microorganisms covering surfaces such as rocks which are exposed to hydrothermal fluids in the immediate vicinity of a vent.

ing into the ocean is oxidized by autotrophic sulfur bacteria. These bacteria, using hydrogen sulfide as energy source and carbon dioxide as carbon source, produce organic matter which is used by an entire community of novel deep-sea organisms (see the Box above).

Energy from light Up to now we have been discussing energy release as a result of the oxidation of chemicals, either organic or inorganic compounds. Green plants, and a number of microorganisms, are able to obtain usable energy from sunlight. These **phototrophic** organisms contain special pigments, the **chlorophylls**, which capture light energy. Chlorophyll resides in a special type of structure called the *photosynthetic apparatus*. The light energy trapped by chlorophyll within the photosynthetic apparatus is then used to generate a proton motive force, from which ATP can be synthesized. Most phototrophs are also autotrophs, able to obtain all of their carbon for cell synthesis from CO_2.

We should note here that when an organism uses CO_2 as carbon source, it must *reduce* the carbon to the level needed for biosynthesis of cell material.

Such a reduction process is energy-requiring, and the energy needed comes from ATP, which, we have just noted, can be made using light energy. The CO_2 reduction process also requires *reducing power*, which the organism must produce. The reducing agent which the phototrophic organisms use to reduce CO_2 is NADPH, which is also made within the photosynthetic apparatus. If ATP and NADPH are available, then phototrophic organisms are able to grow autotrophically using CO_2 as carbon source. The role of light is thus to provide the energy needed to make ATP and NADPH, the actual process of CO_2 reduction (sometimes called *CO_2 fixation*) occurring completely in the dark.

We have already noted that ATP can be formed in the photosynthetic apparatus as a result of the generation of a proton-motive force. Where does the reducing power come from? In green plants, algae, and cyanobacteria, reducing power is obtained from water, which is *split* into hydrogen (reducing power) and oxygen (O_2), using light energy. When water is split, the electrons are transferred through the electron transport chain to NADP, as illustrated in Figure 3.24*a*. As seen, the oxygen from water is released as molecular oxygen, O_2. The production of oxygen, so characteristic of green plant photosynthesis, is thus due to the formation of reducing power during the splitting of water. Green plant photosynthesis is thus commonly called **oxygenic photosynthesis**. We note that the oxygen in the atmosphere, so vital for the survival of animal (including human) life, is dependent upon the activity of green plant photosynthesis. On land, most of the photosynthesis occurs in higher plants, but in lakes and oceans, microorganisms are primarily responsible. Probably half of all the oxygen in the atmosphere is derived from microbial photosynthesis.

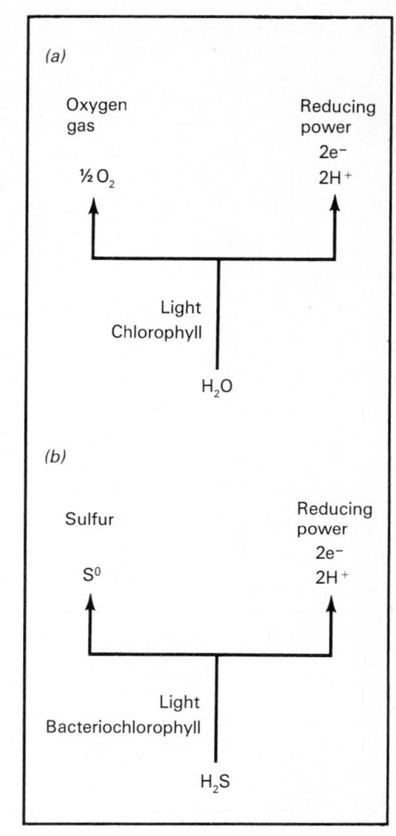

Figure 3.24 The formation of reducing power in photosynthesis. (a) Green plants, algae, and cyanobacteria produce reducing power by splitting water. Molecular oxygen, O_2, is produced. (b) The purple and green bacteria use hydrogen sulfide, H_2S, and form elemental sulfur, S^0.

Another important attribute of photosynthesis is that the carbon of CO_2 is converted into organic carbon. The overall reaction of photosynthesis can be formulated:

$$6CO_2 + 6H_2O \rightarrow C_6H_{12}O_6 + 6O_2$$

where organic carbon shown in the equation is in the form of a sugar. Animals are incapable of growing on carbon dioxide as sole carbon source and are dependent, ultimately, on green plant photosynthesis for survival. Thus, green plant photosynthesis plays a prime role in the development of the biological communities of the earth, as we discuss in Chapter 14.

Although most organisms carrying out oxygenic photosynthesis are eucaryotes (either green plants or algae), one procaryotic group, the *cyanobacteria* (sometimes called *blue-green algae*) also carry out oxygenic photosynthesis. The cyanobacteria are widespread in aquatic environments, and are frequently responsible for massive unsightly growths in lakes and streams as a result of nutrient enrichment of these waters by human activities. We discuss this problem, called *eutrophication*, in Chapter 14. The cyanobacteria are an interesting group of organisms, and frequently have complex morphologies. Some examples of cyanobacteria are given in the photomicrographs in Figure 3.25.

In addition to the organisms carrying out oxygenic photosynthesis, there are other procaryotes, the *purple* and the *green bacteria*, which carry out a

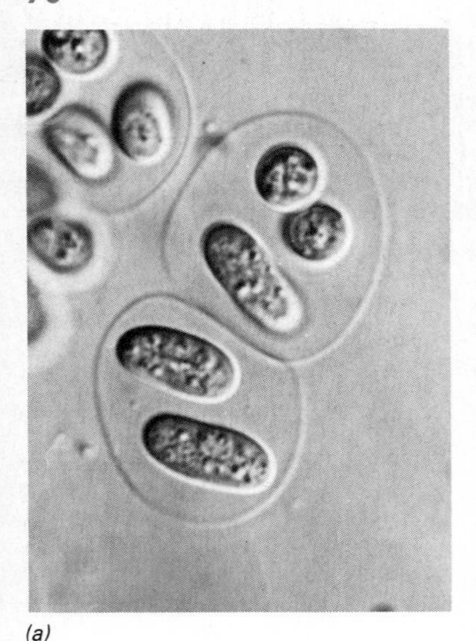

(a)

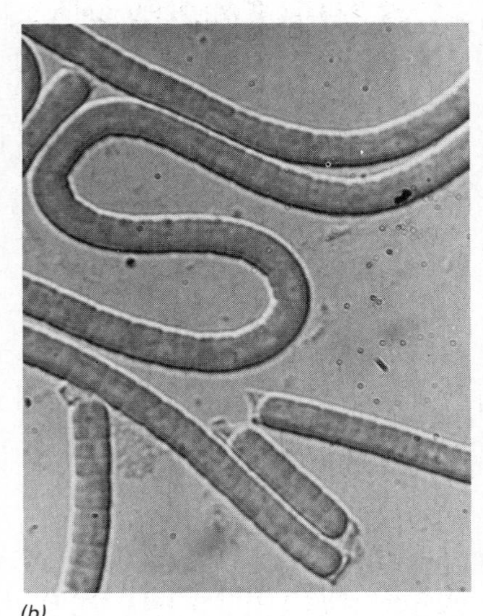

(b)

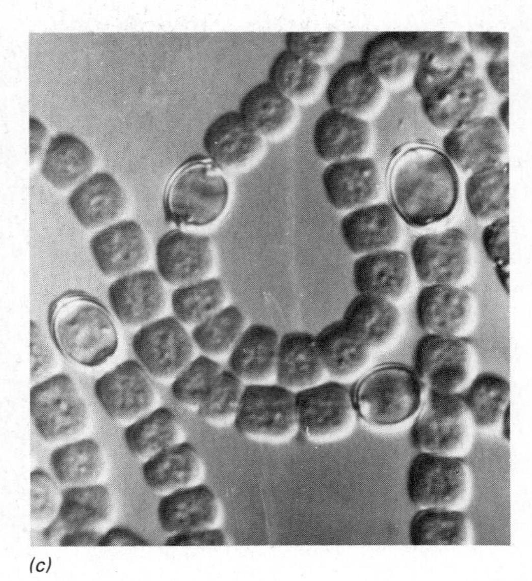

(c)

Figure 3.25 Some cyanobacteria. Magnification 1200 ×. (a) *Gloeocapsa,* a unicellular organism. The cells are encased in a sheath. (b) *Oscillatoria,* a filamentous organism. (c) *Anabaena,* a filamentous organism that forms chains of spherical cells.

photosynthetic process of somewhat different character. The purple and green bacteria contain pigments of a type called *bacteriochlorophyll,* which are similar to but chemically different from green plant chlorophyll. These bacteria do not obtain reducing power from the splitting of water, and do not form oxygen. The electron donor for the purple and green bacteria may be either a reduced sulfur compound such as hydrogen sulfide (H_2S), or an organic compound. If hydrogen sulfide is used, elemental sulfur may be formed, and the purple and green bacteria form characteristic sulfur granules, as illustrated in Figure 3.26. The overall process by which hydrogen sulfide is used as an electron donor in photosynthesis in purple and green bacteria is illustrated in Figure 3.24*b*. Since oxygen is not formed in bacterial photosynthesis, the process is often called *anoxygenic photosynthesis*. Purple and green bacteria growing phototrophically using hydrogen sulfide as electron donor are able to obtain all of their carbon for cell synthesis from CO_2, and are thus autotrophs. However,

(a)

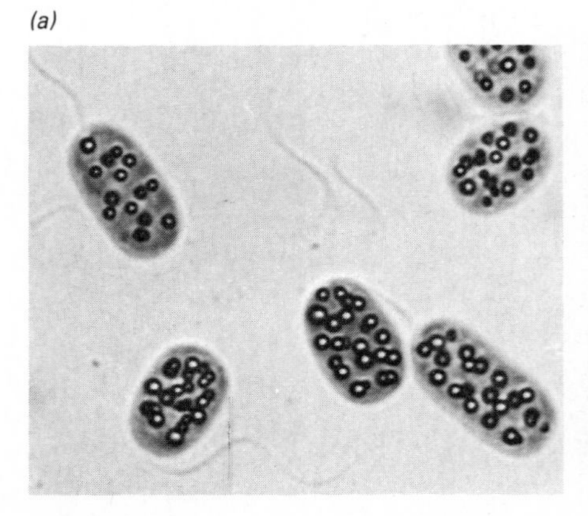

(b)

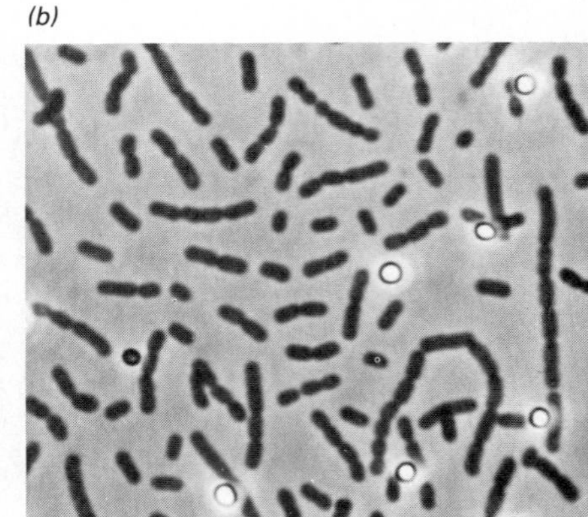

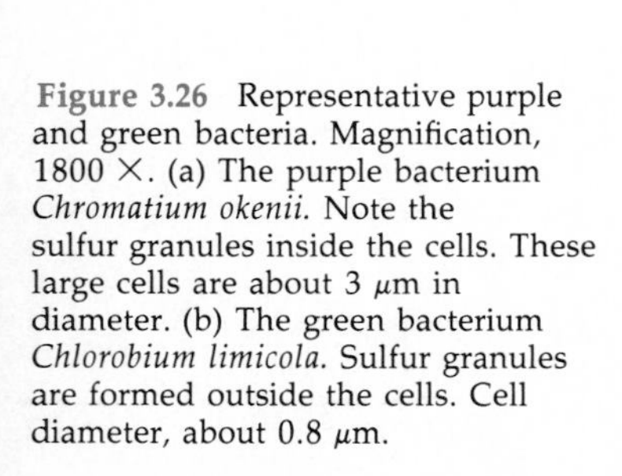

Figure 3.26 Representative purple and green bacteria. Magnification, 1800 ×. (a) The purple bacterium *Chromatium okenii.* Note the sulfur granules inside the cells. These large cells are about 3 μm in diameter. (b) The green bacterium *Chlorobium limicola.* Sulfur granules are formed outside the cells. Cell diameter, about 0.8 μm.

some photosynthetic bacteria use organic compounds as either electron donors or as carbon sources, while obtaining energy from light. Such organisms are not autotrophs, although they are phototrophic; they are sometimes called *photoheterotrophic* organisms.

SUMMARY

In this chapter, we have begun with a simple discussion of cell chemistry, followed by a discussion of cell metabolism. Cell metabolism includes two distinct types of processes, called **catabolism** and **anabolism**. Catabolic processes are involved in the release of energy from the oxidation of chemicals. In this energy release, some of the energy is conserved in the form of high-energy phosphate bonds in ATP. The catabolic processes are catalyzed by enzymes, and a number of discrete steps are involved in the breaking down of an organic energy source such as glucose.

We have made a distinction between two kinds of *catabolic processes* involved in energy generation: fermentation and respiration. In **respiration** processes, electrons from the substrate are transferred through an electron transport system to molecular oxygen, O_2. High-energy phosphate bonds in ATP are formed by a process of **electron-transport phosphorylation** which involves the participation of the electron-transport chain. During the transfer of electrons to oxygen, a **proton motive force** is generated, and this proton motive force can be used to bring about the synthesis of ATP.

In **fermentation**, an external electron acceptor such as oxygen is not involved. Rather, the organism uses some of the substrate molecules to synthesize an internal electron acceptor. In this way, some of the carbon of the substrate is reduced whereas the rest of the carbon is oxidized to CO_2. Organisms carrying out fermentation processes produce characteristic end products, such as ethanol, lactic acid, or acetic acid.

Energy generated as a result of catabolism is used in the biosynthetic reactions of anabolism. **Biosynthesis** results in the building up of the vast array of organic constituents of which the cell is constructed. We have noted that the major cell constituents are **polymers** or **macromolecules**, formed in specific ways from small-molecular weight building blocks called **monomers**.

In addition to the simple respiration and fermentation processes, some organisms have different energy-generation processes. In **anaerobic respiration**, a terminal electron acceptor other than oxygen is used, of which the most common are nitrate, sulfate, and carbon dioxide. Some organisms, called **lithotrophs**, are able to obtain energy from the oxidation of inorganic chemicals such as hydrogen sulfide, hydrogen (H_2), and ferrous iron (Fe^{2+}). These organisms are able to grow and synthesize organic matter in completely mineral environments.

Other important organisms are able to obtain energy from sunlight, using chlorophyll or bacteriochlorophyll. This process, called **photosynthesis**, is the basis of the food chain of the earth. Two types of photosynthesis are recognized: **oxygenic**, carried out by green plants, algae, and cyanobacteria, and **anoxygenic**, carried out by a few specialized groups of purple and green sulfur bacteria.

KEY WORDS AND CONCEPTS

Environment
Nutrient
Anabolism
Catabolism
Biosynthesis
Monomer
Polymer
Macromolecule
Polysaccharide
Amino acid
Protein
Ribonucleic acid (RNA)
Deoxyribonucleic acid (DNA)
Nucleic acid base
Lipid
Activation energy
Enzyme
Heterotroph
Autotroph
Oxidation
Reduction
Electron acceptor
Electron donor
Electron carrier
Fermentation
Respiration
Glycolysis
Tricarboxylic acid cycle
Electron transport phosphorylation
Substrate-level phosphorylation
Proton motive force
Anaerobic respiration
Denitrification
Sulfate reduction
Methanogenesis
Lithotrophic
Nitrification
Phototrophic
Photosynthesis

STUDY QUESTIONS

1. How does an *organic compound* differ from an *inorganic compound*? Why is it correct to state that not all organic compounds are associated with life?
2. For each of the following *polymers*, list at least one corresponding *monomer*: starch, insulin, cellulose, DNA, RNA.
3. List at least three ways in which RNA and DNA are similar. List two ways in which they are different.
4. Describe the manner by which the *hydrophilic* and *hydrophobic* portions of *phospholipid* molecules interact in the formation of the cell membrane.
5. Explain how *methane* and *oxygen* can remain together indefinitely without reacting, only to react violently in the presence of a spark.
6. What is the function of an *enzyme* in a chemical reaction? What are the functions of enzymes in the cell?
7. Give the appropriate name for each of the following enzymes: lactose-splitting enzyme, ribonucleic acid-decomposing enzyme, sucrose-splitting enzyme, enzyme which removes hydrogen atoms, enzyme which adds oxygen atoms.
8. How do *heterotrophs* differ from *autotrophs*?
9. Define *metabolism, catabolism, anabolism.* Which process, catabolism or anabolism, results in the release of usable energy?

10. What roles does ATP play in cell metabolism? NAD?
11. Define *oxidation* and *reduction.*
12. List three *electron carriers* that participate in cell respiration.
13. Show by diagram how electrons are passed through the *electron transport system* and finally to oxygen.
14. Compare and contrast *fermentation* and *respiration* with respect to pathways, oxygen requirements, energy yield, and end products.
15. Why is the cell yield higher with *aerobic* than with *anaerobic* growth?
16. What is *anaerobic respiration*? Show how nitrate can be used in this process.
17. What is a *lithotroph?* List three energy sources used by lithotrophs.
18. How do *phototrophic* organisms obtain reducing power for biosynthesis? How do they obtain ATP?
19. Compare and contrast *oxygenic* and *anoxygenic* photosynthesis.
20. Prepare an outline of the various types of energy metabolism in microorganisms, indicating both the energy sources and electron donors and acceptors.

SUGGESTED READINGS

BROCK, T.D., D.W. SMITH, and M.T. MADIGAN. 1984. *Biology of microorganisms, 4th edition.* Prentice-Hall, Englewood Cliffs, NJ.

GOTTSCHALK, G. 1985. *Bacterial metabolism, 2nd edition.* Springer-Verlag, New York.

HINKLE, P.C., and R.E. MCCARTY. 1978. *How cells make ATP.* Scientific American 238: 104–123. (An excellent overview of proton motive force as a means of energy storage in cells.)

STRYER, L. 1981. *Biochemistry, 2nd edition.* W.H. Freeman, New York.

4

Care and Feeding of Microbes

We discussed in Chapter 1 the work of Robert Koch on the role of microorganisms as causal agents of infectious disease. We showed that proof that a specific microorganism caused a specific disease depended on obtaining a pure culture of the organism. Pure cultures are required not only in disease investigation, but in any study where a microbe is suspected of playing a role. Thus, our understanding of what things microorganisms do and how they do them depends upon our ability to study microorganisms in the laboratory. Laboratory study requires that we be able to culture microorganisms, providing them with the proper foods and conditions so that they can grow. Once we have isolated a pure culture, we can then proceed to a study of the characteristics of the organism and a determination of its capabilities. As we have noted in Chapter 1, a pure culture is often obtained by single-colony isolation on an agar culture medium. It is assumed that the cells of the colony have all been derived from the initial growth of the single cell which had been placed on the agar. The word *clone* is often used to describe a population of cells all derived from a single cell. Because a pure culture is usually obtained by colony isolation, a pure culture and a clone are synonymous.

The laboratory study of microorganisms is not unusually difficult to understand. In practice, however, great care must be taken to ensure that pure cultures remain uncontaminated and that appropriate environmental and nutrient conditions are provided. When disease-causing (*pathogenic*) microorganisms are to be studied, special precautions must be observed to prevent unintended infection. We will present here the basic principles and procedures involved in laboratory study of microorganisms.

4.1 CULTURE MEDIA

We discussed in Chapter 3 the chemical constitution of cells and noted that a number of chemical elements must be obtained from the environment. The chemical compounds which supply these elements are called **nutrients**. The aqueous solution containing such necessary nutrients is called a **culture medium** (plural, **media**). The food materials present in the culture medium provide a microbial cell with those ingredients required to produce more cells like itself. We divide the nutrients of a culture medium into four major kinds: energy sources, cell structural components, minerals, and growth factors. There is a wide variety of food ingredients that provide these basic materials, but not all organisms require the same ones, nor can any one organism use all kinds.

Nutritional components As was pointed out in Chapter 3, there are many types of *energy sources* which microorganisms may use. It is, of course, necessary to have the correct energy source in order to obtain growth of any specific organism. For example, a heterotroph must be supplied with the correct organic compound, whereas a lithotroph must be supplied with the correct inorganic compound, and a phototroph must be supplied with light.

In Chapter 3 we also discussed the variation in the nutrients needed by different microorganisms for biosynthesis. Organisms with simple nutrition need only a few simple compounds as sources of the elements they need for biosynthetic reactions. For example, many microorganisms can be grown on a single energy and carbon source plus a few inorganic salts to provide nitrogen, phosphorus, sulfur, and other needed major elements. Other organisms with complex nutrition may require many of the important monomeric compounds necessary for biosynthesis of polymers. Nutrients of this sort, required in small amounts are called **growth factors**. It is not uncommon, for instance, for a microorganism to require amino acids or nucleic acid bases. If the organism one intends to grow has these requirements, it is, of course, necessary to provide these compounds as nutrients in the culture medium. It is usually the case that if a substance is required as a growth factor, the organism lacks the enzymatic machinery for the synthesis of that substance from other nutrients.

The best-known growth factors are **vitamins**. It should be emphasized that many microorganisms do not require growth factors, but in the ones that do, failure to provide them will result in a complete lack of growth. The most commonly required vitamins in microorganisms are thiamine (vitamin B_1), biotin, pyridoxine (vitamin B_6), and cobalamin (vitamin B_{12}). Other vitamins sometimes required are folic acid, nicotinic acid (niacin), lipoic acid, pantothenic acid, riboflavin, and vitamin K. Certain disease-causing bacteria can

TRACKING THE LEGIONNAIRES' KILLER

When 221 people at the 1976 American Legion convention in Philadelphia developed a severe lung disease of an unknown nature, microbiologists at the Centers for Disease Control in Atlanta, Georgia, were faced with one of their biggest challenges of modern times: obtain a pure culture of the culprit and fulfill Koch's postulates. But the bacterium, if indeed there was a bacterium, did not grow on any available culture medium. So the microbiologists did what Robert Koch had done before them: they inoculated guinea pigs with diseased lung tissue and used this animal host as a *living* culture medium. Luck was with them: some of the guinea pigs got sick and their tissues contained tiny bacteria of a new type. With the living guinea pig culture as a start, many new culture-medium formulas were tried, and many petri plates were streaked with infected tissue. Finally, persistence paid off and tiny colonies appeared on *one* of the culture plates. The trick, as it turned out, was simple, albeit unsuspected. The Legionnaires' killer required an uncommon growth factor, the amino acid *cysteine*, and also a larger than normal amount of iron. Otherwise, the culture medium was fairly standard. Once colonies were obtained, a pure culture was easily isolated, and Koch's postulates were fulfilled by inoculation of more guinea pigs. The culture was then used in antibiotic testing, and erythromycin was found to be effective in treatment. Using the new culture medium, nationwide studies were carried out, and the disease, now called *legionellosis*, was found to be widespread, especially in the elderly and the sick. Between 25,000 and 50,000 legionellosis cases occur in the United States yearly. The organism grows readily in air conditioning units and hot water heaters, and is transmitted to people in mists and small droplets. The organism, named *Legionella pneumophila*, has joined company with all the other bacterial pathogens which have been recognized since the days of Koch. Without the laboratory skill of the microbiologist, legionellosis would still be an unsolved mystery. The dry recital of bacteriological technique often hides a high drama from the world of medicine.

grow only when blood or *heme*, the red pigment of blood, is present in the culture medium.

A knowledge of growth factor requirements is of great importance for successful culture of microorganisms in the laboratory. There are some organisms, however, which have not been cultured successfully. It is likely that these organisms fail to grow because the culture media used lack one or more required growth factors. The isolation of the bacterium which causes legionellosis (Legionnaire's disease) provides a good example of this (see the Box).

Synthetic and complex media It is common to distinguish between two types of culture media, synthetic and complex. In a **synthetic medium**, (sometimes called a *defined medium*) every essential nutrient is provided by a pure chemical of known composition (Table 4.1). In actual practice, many of the medium ingredients serve double or even triple duty. Potassium phosphate, for instance, serves as a source of both potassium and phosphorus. Ammonium sulfate may serve as a source of both nitrogen and sulfur. Glucose is both the carbon and energy source for many organisms.

In a **complex medium** (Table 4.2), certain of the ingredients used are plant, animal, or microbial extracts which supply all the essential nutrients.

TABLE 4.1 Components of a typical synthetic culture medium.

Component	Amount	Function in Medium
K_2HPO_4	7.0 g	pH buffer; K, P source
KH_2PO_4	2.0 g	pH buffer; K, P source
$(NH_4)_2SO_4$	1.0 g	N, S source
$MgSO_4$	0.1 g	Mg, S source
Glucose	5.0 g	C and energy source
Agar	15.0 g	Solidifying agent
Distilled H_2O	1000 ml	
Adjust to pH 7.0		

Impurities in the ingredients supply other needed nutrients.

Examples of such materials are malt extract (from barley malt), meat extract (from beef muscle), and yeast extract (from baker's yeast). Other commonly used ingredients are tryptone and casitone, which are digests of the milk protein casein; peptone, a digest of beef muscle; and beef heart infusion, an extract of beef heart. These materials usually contain all essential organic and inorganic nutrients for many heterotrophic microorganisms and are often used in preparing versatile culture media capable of supporting the growth of a variety of microorganisms.

Liquid and solid media Culture media can be prepared for use either in a liquid state, or in a gel (solid) state. Liquid culture media are dispensed in tubes, flasks, or bottles, which can then be inoculated and incubated. Microbial growth in liquid cultures is often evenly dispersed throughout the liquid.

For many studies it is desirable to obtain microbial growth on or in a solid substrate. For this purpose, a liquid culture medium can be converted into a solid form by adding a gelling agent. Gelling agents frequently used are agar, gelatin, and silica gel.

Agar, the most commonly used gelling agent, is manufactured from certain seaweeds. Agar is not a nutrient for most microorganisms and, hence, can be added to a culture medium without significantly modifying the quality of the medium.

TABLE 4.2 Components of a typical complex culture medium

Component	Amount	Function in Medium
Tryptone	10.0 g	Source of amino acids (N, S) and P
Yeast extract	3.0 g	Source of vitamins and other growth factors
Glucose	10.0 g	Carbon and energy source
Distilled H_2O	1000 ml	
Adjust to pH 7.4		

Impurities in the ingredients supply other needed nutrients including minerals.

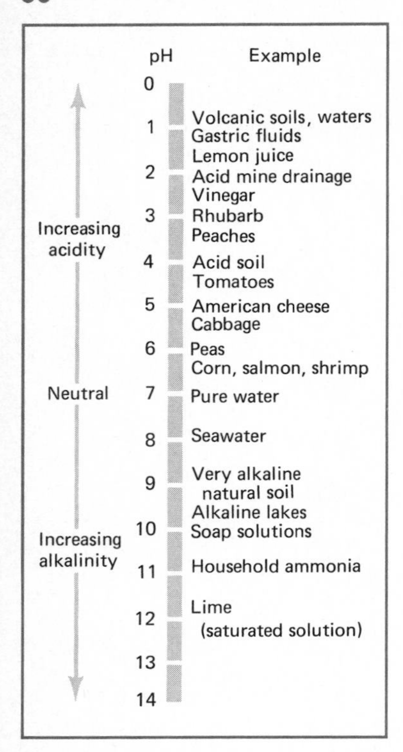

Figure 4.1 The pH scale.

4.2 ENVIRONMENTAL REQUIREMENTS

In addition to satisfying the nutritional needs of a microorganisms, environmental conditions, such as acidity (pH), oxygen tension, water activity, and temperature must be controlled.

Acidity and alkalinity (pH) The proper acidity or alkalinity must be provided for each organism. Acidity or alkalinity of a solution is expressed by its **pH value** on a scale in which neutrality is pH 7 (Figure 4.1). Those pH values that are less than 7 are *acidic*, and those greater than 7 are *alkaline* (or *basic*). Each pH unit stands for a tenfold change in acidity or alkalinity, so that vinegar (pH near 2) and household ammonia (pH near 11) differ in acidity by one billion times. Indicator dyes are often added directly to culture media and can then indicate not only the initial pH of the medium but also any change in pH that results from growth or activity of the microorganisms.

Each organism has a pH range within which growth is possible and usually also has a well-defined pH optimum at which the cells can best grow. Organisms that have an acid optimum pH are called **acidophiles** (acid-loving), and those with an alkaline optimum pH are called **alkalophiles**. It is an interesting point that a pH that may be highly detrimental to one organism may be the pH at which another organism thrives.

The pH of a culture medium is adjusted by adding an alkaline compound, such as sodium hydroxide, if the medium is too acid or an acidic compound, such as hydrochloric acid, if the medium is too alkaline. Because microorganisms usually induce changes in pH as they grow, it is often desirable to add to the culture medium a pH *buffer*, which acts to keep the pH relatively constant. Such pH buffers work only over a narrow pH range; hence, different buffers must be selected for different pH regions. For near neutral pH ranges (pH 6 to 7.5), sodium phosphate and potassium phosphate make excellent buffers (see Table 4.1).

Oxygen As discussed in Chapter 3, microorganisms vary in their need for, or tolerance of, oxygen. Microorganisms can be divided into several groups depending on the effect of oxygen, as outlined in Table 4.3. Organisms which lack a respiratory system will not be able to use oxygen as terminal electron

TABLE 4.3 Terms used to describe O_2 relations of microorganisms

Group	O_2 Effect
Aerobes	
Obligate	Required
Facultative	Not required, but growth better with O_2
Microaerophilic	Required, but at levels lower than atmospheric
Anaerobes	
Aerotolerant	Not required, and growth no better when O_2 present
Aerophobic (obligate anaerobes)	Harmful

acceptor. Such organisms are called **anaerobes**, but there are two kinds of anaerobes, those which can tolerate oxygen even though they cannot use it, and those which are killed by oxygen. Organisms killed by oxygen are called *obligate anaerobes*. Although the reason why obligate anaerobes are killed by oxygen is not clear, one idea is that obligate anaerobes are unable to detoxify some of the products of oxygen metabolism. When oxygen is reduced several toxic products, hydrogen peroxide (H_2O_2), superoxide (O_2^-), and hydroxyl radical ($OH\cdot$) are formed. Aerobes have enzymes which decompose these products, whereas anaerobes seem to lack these enzymes.

For the growth of many aerobes, it is necessary to provide extensive aeration. This is because O_2 is only poorly soluble in water, and the O_2 used up by the organisms during growth is not replaced fast enough by diffusion from the air. Forced aeration of cultures is therefore frequently desirable and can be achieved either by vigorously shaking the flask or tube on a shaker or by bubbling sterilized air into the medium through a fine glass tube or porous glass disc. Usually aerobes grow much better with forced aeration than when O_2 is provided by simple diffusion.

For anaerobic culture, the problem is to exclude oxygen. One of the more difficult techniques in microbiology is the maintenance of anaerobic conditions: oxygen is ubiquitous in the air. Obligate anaerobes vary in their sensitivity to oxygen, and a number of procedures are available for reducing the O_2 content of cultures—some simple and suitable mainly for less sensitive organisms, others more complex but necessary for the most fastidious obligate anaerobes.

Bottles or tubes filled completely to the top with culture medium and provided with tightly fitting stoppers will provide anaerobic conditions for organisms not too sensitive to small amounts of oxygen. It is also possible to add a chemical which reacts with oxygen and excludes it from the culture medium. Such a substance is called a *reducing agent* because it reduces oxygen to water. A good example is *thioglycollate,* which is added to a medium commonly used to test whether an organism is aerobic, facultative, or anaerobic (Figure 4.2). After thioglycollate reacts with oxygen throughout the tube, oxygen can only penetrate near the top of the tube where the medium contacts air. Aerobes grow only at the top of the tube. Facultative organisms grow throughout the tube, but best near the top. Anaerobes grow only near the bottom of the tube, where oxygen cannot penetrate. An indicator dye, such as *resazurin*, is usually added to the medium because the dye will change color in the presence of oxygen and thereby indicate the degree of penetration of oxygen into the medium.

To remove all traces of O_2 for the culture of very fastidious anaerobes, it is possible to place an O_2-consuming gas in a jar holding the tubes or plates. One of the most useful devices for this is the *anaerobic jar*, a heavy-walled jar with a gas-tight seal, within which tubes, plates, or other containers to be incubated are placed (Figure 4.3). The air in the jar is replaced with hydrogen gas (H_2), and in the presence of a chemical catalyst the traces of O_2 left in the vessel or culture medium are consumed, thus leading to anaerobic conditions.

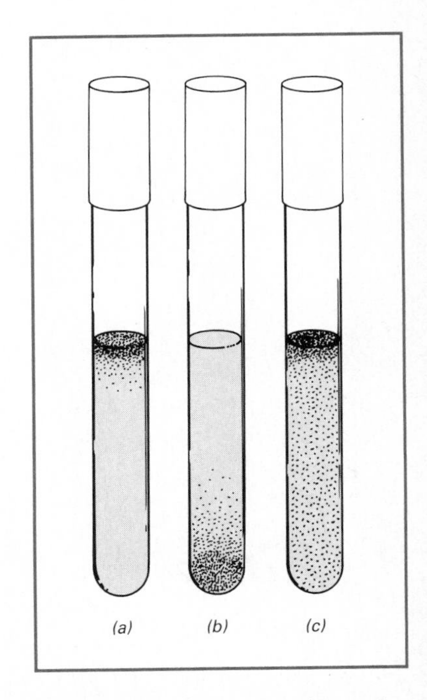

Figure 4.2 Aerobic, anaerobic, and facultative growth, as revealed by the position of microbial colonies within tubes of a culture medium. A small amount of agar has been added to keep the liquid from becoming disturbed. (a) Oxygen only penetrates a short distance into the tube, so that aerobes grow only at the surface. (b) Anaerobes, being sensitive to oxygen, only grow away from the surface. (c) Facultative anaerobes are able to grow either in the presence or absence of oxygen and grow throughout the tube.

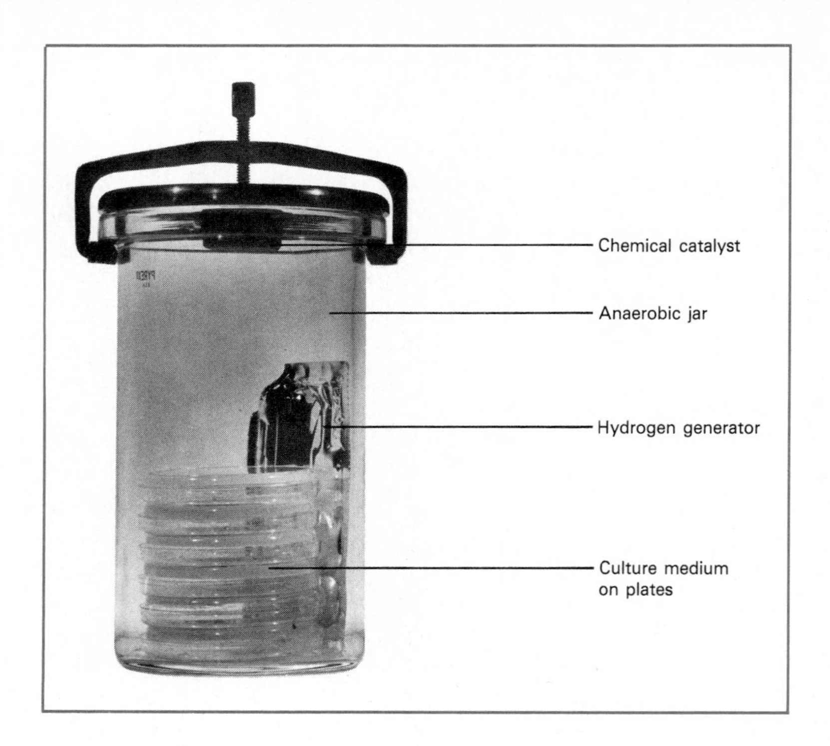

Figure 4.3 Special jar for incubating cultures under anaerobic conditions.

For the most fastidious anaerobes, such as the methanogens, it is necessary not only to carefully remove all traces of O_2 but also to carry out all manipulations of cultures in an anaerobic atmosphere, as these organisms are frequently killed by even brief exposure to O_2. A procedure pioneered by the microbiologist R.E. Hungate, called the *Hungate technique* (Figure 4.4), is sometimes used. In this technique, a tiny jet of O_2-free hydrogen or nitrogen gas is directed into the culture vessel during manipulations, thus driving out any O_2 that might enter. By the use of the Hungate technique, a variety of anaerobic bacteria that had not been cultured before have been isolated from nature, including some human pathogens. For extensive research on anaerobes, special boxes fitted with gloves, called *anaerobic glove boxes,* permit work with open cultures in completely anaerobic atmospheres.

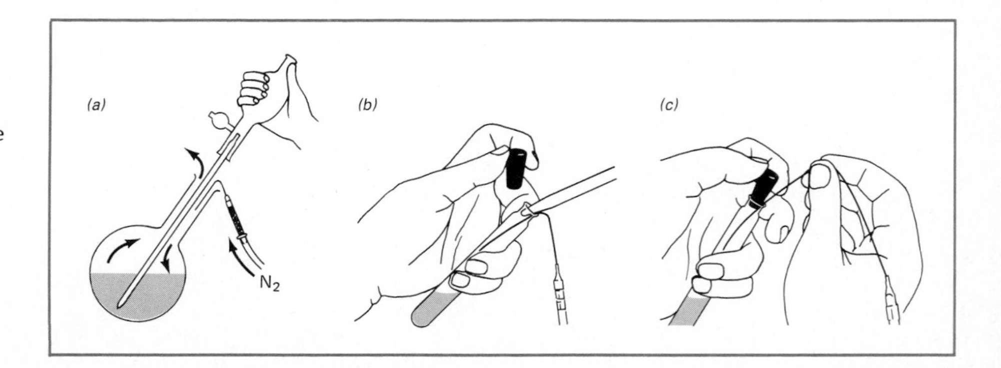

Figure 4.4 The Hungate technique for culturing fastidious anaerobes. (a) Medium in flask is gassed with nitrogen (N_2) while a portion is removed. Gassing prevents the entrance of oxygen. (b) Medium is transferred to tube to be inoculated. Inoculum is added. (c) The gassing needle is gradually removed from the tube as the stopper is inserted, preventing entrance of oxygen. The rubber stopper is impermeable to oxygen, so that anaerobic conditions remain inside the tube as long as the stopper is in place. To remove a sample at a later time, the gassing needle is again reinserted as the stopper is opened.

Temperature Each organism has a restricted range of temperatures within which it will grow. Three cardinal temperatures can be recognized, the *minimum*, the *maximum*, and the *optimum* temperature (Figure 4.5). The optimum temperature is usually nearer the maximum than the minimum. Because of this, it is better to grow an organism at a temperature somewhat below optimum, to ensure that any accidental increase in temperature does not lead to death.

Some microorganisms have an optimum temperature as low as 5°C, and others are known with optima as high as 105°C. No one organism spans more than a small part of this temperature range, however (Figure 4.6). Organisms that cause disease in man have their optima near body temperature, 37°C. Organisms with low temperature optima are called **psychrophiles** (which means cold-loving), and organisms with high temperature optima are called **thermophiles** (heat-loving). Those with optima between room temperature (18 to 25°C) and body temperature (37°C) are called **mesophiles** (*meso* means middle).

To obtain the appropriate temperature for microbial growth, *incubators* or *water baths* are necessary. It is important that the incubator be equipped with a *thermostat* so that the temperature remains relatively constant. Incubators and water baths suitable for most routine microbiological work may be purchased from various scientific and hospital supply companies. A simple temperature-controlled water bath can be constructed by placing a thermostatically controlled immersion heater in a glass aquarium (such as the type of heater used in a tropical fish tank).

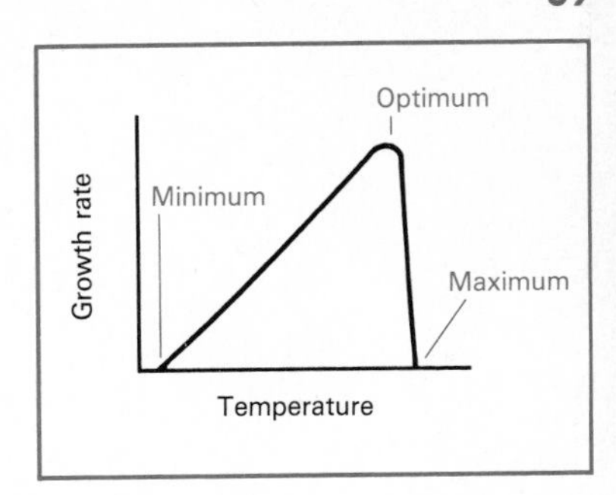

Figure 4.5 The typical growth response of a microorganism to temperature, showing the minimum, maximum, and optimum growth temperatures.

Growth on membrane filters Another way to provide a nutrient surface for microbial growth is to use a **membrane filter**. These filters are tough cellulose acetate or cellulose nitrate discs, so manufactured that large numbers of tiny holes are present. The filter itself does not act as a nutrient but is laid on agar medium or on a cellulose pad saturated with liquid culture medium

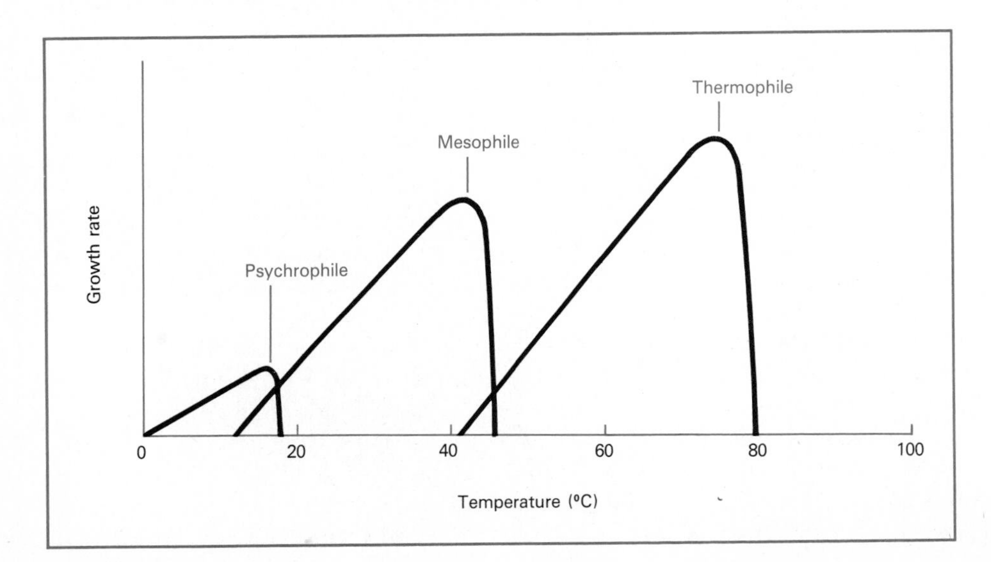

Figure 4.6 Temperature responses of different kinds of microorganisms.

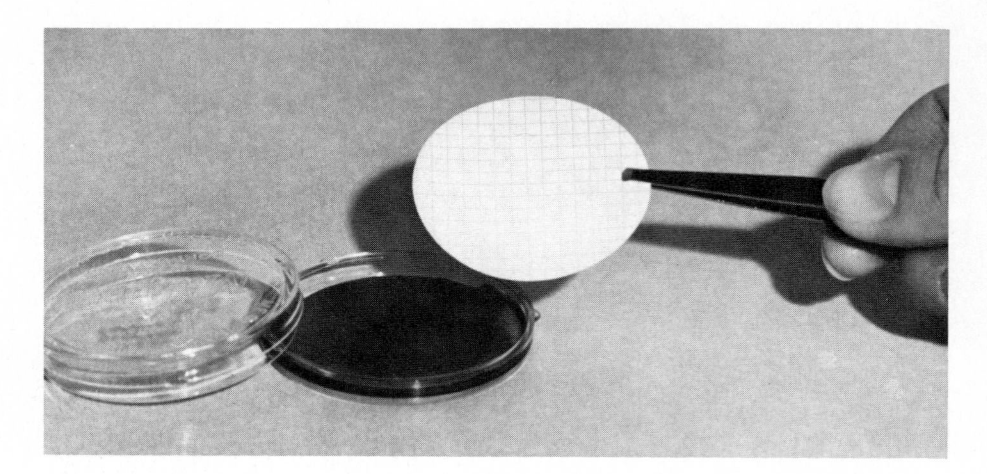

Figure 4.7 Use of a membrane filter as a surface for microbial growth. The filter with microorganisms is placed on the surface of a plate containing a culture medium.

(Figure 4.7). The medium seeps through the holes, forming a thin layer on the surface of the filter, and the microorganisms then grow in this thin film. Membrane filters remain solid at high temperatures (even up to the boiling point of water) and do not contribute any nutrients to the medium. Such filters are widely used in water-pollution studies, as will be described in Sections 5.2 and 15.3. One reason for using membrane filters is to concentrate microorganisms for culture from dilute suspensions. If only a low concentration of cells is present in a sample, a large amount of liquid can be filtered in order to obtain enough cells on the filter to obtain colonies.

4.3 STERILIZATION AND ASEPTIC TECHNIQUE

We have considered what must be put into a medium and how to prepare it. Before proceeding to the use of media to obtain cultures of microorganisms, we must first consider how to exclude unwanted or contaminant organisms. Microbes can be considered to be everywhere. Because of their small sizes, they are easily dispersed in the air and on surfaces (such as the human skin). Therefore, we must *sterilize* the culture medium soon after its preparation to eliminate microorganisms already contaminating it. It is equally important to take precautions during the subsequent handling of a sterile culture medium, to exclude from the medium all but the organisms desired. Thus, other materials which come into contact with culture media must also be sterilized. It is also necessary to sterilize contaminated materials, such as samples collected in hospitals, in order to control the spread of harmful bacteria.

Sterilization of media The most common way of sterilizing culture media is by heat. However, since heat may also cause harmful changes in many of the ingredients of culture media, it is desirable to keep the heating time as short as possible. Although most microorganisms are quickly killed by temperatures near boiling, bacterial endospores are very resistant to such heat and may survive hours of boiling. Since endospores are ubiquitous, being common in soil, dust, and air, all heat sterilization procedures are so designed

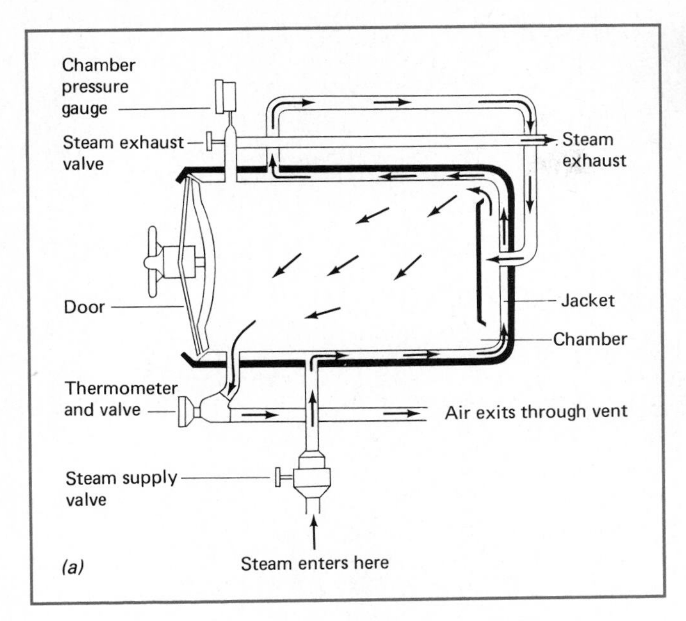

Figure 4.8 The autoclave, one of the most important tools of the microbiologist. (a) The flow of steam through an autoclave. (b) Typical autoclaves.

that the destruction of any contaminating endospores is ensured. The best procedure is to heat under pressure, since at pressures above atmospheric the temperature can be increased above 100°C, thus decreasing the time necessary to make the medium sterile. Devices for heating under pressure are called **autoclaves** (Figure 4.8). Like pressure cookers, autoclaves allow the buildup of pressure which raises the boiling point of water. It is then possible to heat above 100°C without the medium boiling over. The usual procedure is to heat at a pressure of 15 lb/sq in. (1.1 kg/sq cm), which yields a temperature of 121°C, for 15 to 20 minutes. Since it is not pressure but heat that sterilizes, it is important to be sure that the temperature within the medium itself has reached 121°C, before beginning to time the sterilization. In small laboratories, or if only a limited budget is available, a large pressure cooker, such as is used for home canning of foods, provides a very inexpensive and perfectly satisfactory autoclave as it works on exactly the same principle as described above.

Filter sterilization of solutions An especially valuable technique for sterilizing aqueous solutions containing heat-sensitive materials is filtration (Figure 4.9). The filter used must have holes too small for the passage of contaminating microorganisms while still allowing the passage of the liquid (Figure 4.10). The most common filter material is the *membrane filter*, discussed above. These were mentioned above for growing cells; here, it is the cell-free filtrate that is collected.

It should be noted that filtration only removes all organisms larger than the pore size of the filter. With most filters, viruses and very tiny bacteria easily pass through the filter. Thus sterilization by filtration is not as certain as sterilization by heat.

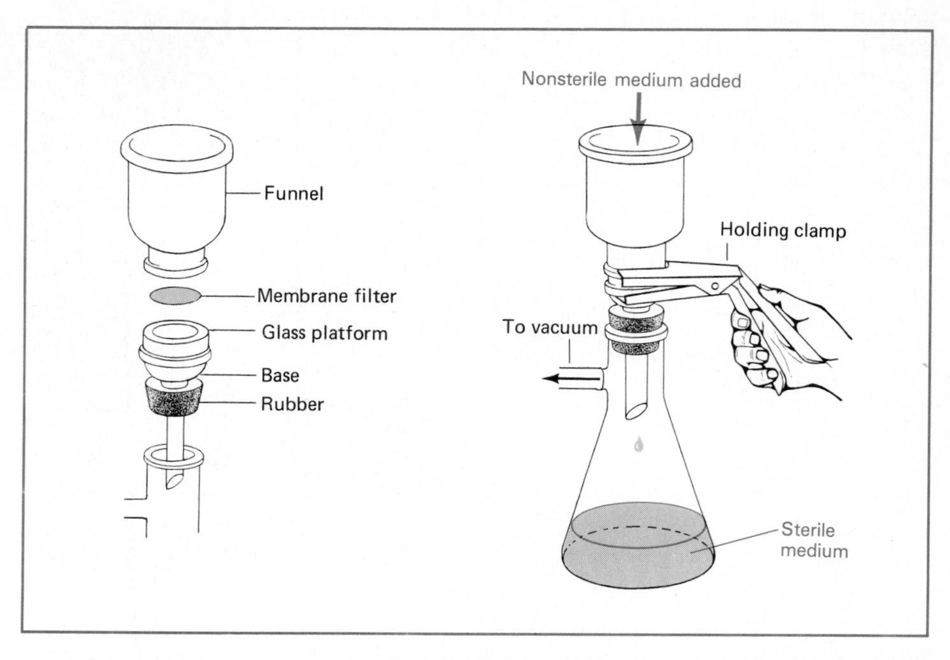

Figure 4.9 Use of a membrane filter for sterilization of a liquid.

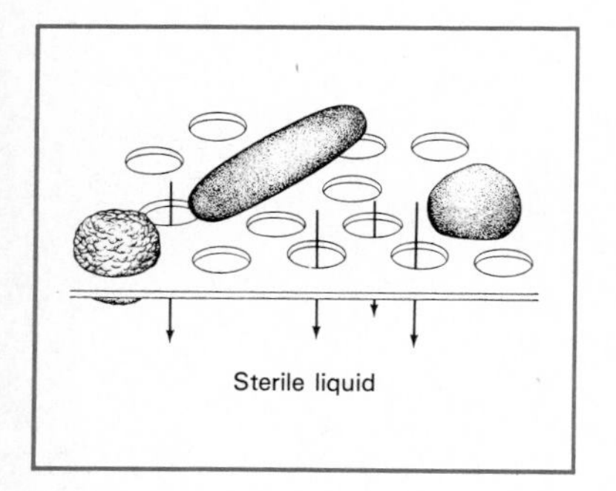

Figure 4.10 The membrane filter removes cells and other particles that are larger than the pores of the filter.

Closures Any container in which the contents are to be kept sterile must have an effective closure. Not only must the closure be effective in maintaining sterility, it must also be easy to handle so that it can be removed and put on again quickly during manipulations. In culture vessels where aeration is needed, the closure must also permit passage of air.

The traditional closure for a culture vessel is the *cotton plug*, made by twisting a piece of cotton and forcing it into the neck of the tube or flask. Air can pass through the plug, but organisms cannot. The cotton plug adapts to the shape of the vessel, permits reasonable passage of air, can be autoclaved, can be fairly easily manipulated, and is very inexpensive. However, it has the following disadvantages: it does not always prevent passage of contaminants down the wall of the vessel; it frays, and fibers of cotton may fall into the medium; vitamins and other organic materials present even in purified cotton may leach into the culture medium and affect results of nutritional experiments; it becomes wet if used in water baths; it can easily become dislodged from the vessel, especially if vigorous aeration on a shaker is carried out; and plugs are tedious and time consuming to prepare. A variety of alternatives to the cotton plug exist (Figure 4.11).

Sterilization of materials other than liquids The autoclave or pressure cooker can also be used for sterilization of glassware, pipettes, surgical instruments, and other items. A drying cycle after autoclaving helps to eliminate moisture which condenses on glassware from steam during autoclaving.

The most convenient way of sterilizing objects such as empty test tubes, petri dishes, pipettes, hypodermic needles, or instruments is by use of dry heat. At least 90 minutes heating at 160 to 170°C is necessary for sterilization, although when bulky items or large loads are being sterilized, more time may

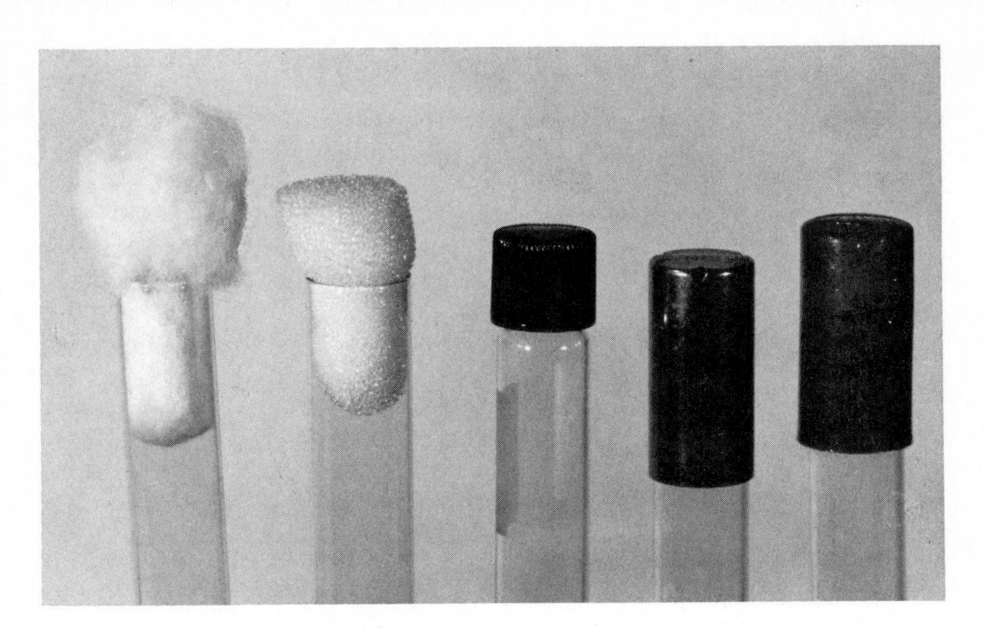

Figure 4.11 Several kinds of closures used to keep tubes sterile. Left to right: cotton, plastic foam, screw cap, metal cap, plastic cap.

be necessary. A higher temperature is necessary with dry heat than with steam because endospores are more resistant to heat when dry.

Objects that are sensitive to heat or that should not become damp, as happens in the autoclave, can often be sterilized by use of a gas. The most commonly used gas is ethylene oxide, although other gases such as methyl bromide, propylene oxide, or ozone may also be used. The items to be sterilized are placed in a gas-tight chamber, and the gas is admitted for the prescribed time. The chamber is then exhausted and the items aired for 8 to 10 days before use, since the gases are toxic to humans as well as microbes. This method is finding increasing use in hospitals for the sterilization of plastics and other heat-sensitive materials.

Microorganisms are easily destroyed by burning, a process called incineration. Incineration is used frequently in the laboratory to sterilize the loops and needles used in transferring cultures. The contaminated instrument is heated in a flame until it is red-hot. Incineration is used in hospitals and laboratories for the disposal of items that may have become contaminated with pathogenic organisms such as sputum cups, disposable paper sheets, surgical dressings, and infected laboratory animals.

Aseptic technique The prevention of contamination during manipulations of cultures and sterile culture media is called **aseptic technique**. Its mastery is required for success in the microbiology laboratory. Air-borne contaminants are the most common problem, since the air always has a population of microorganisms. When containers are opened, they must be handled in such a way that contaminant-laden air does not enter. This is best done by keeping the containers at an angle so that most of the opening is not exposed to the air. Operations should preferably be carried out in a dust-free room in which air currents are absent.

Figure 4.12 Aseptic transfer. (a) Loop is heated until red-hot. (b) Tube is uncapped and loop is cooled in air briefly. (c) Sample is removed and tube is recapped. Sample is transferred to a sterile tube. Loop is reheated before being taken out of service.

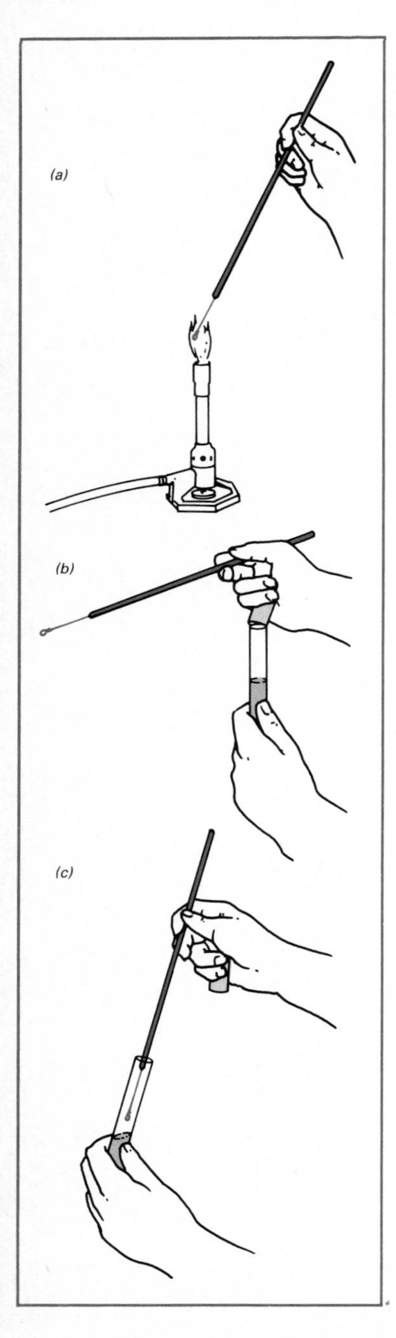

Aseptic transfer of a culture from one tube of medium to another is usually accomplished with an inoculating loop or needle which has been sterilized by incineration in a flame (Figure 4.12).

Transfers of liquid cultures are often made with sterile *pipettes*. Mouth pipetting should not be used for microbial cultures because of possible danger of infection; a special remote pipetting bulb should be used instead.

After the culture is transferred and the tubes closed, the pipette must be discarded directly into a jar or tray of antiseptic solution, such as phenol or mercuric dichloride. Once the pipette has been used with a culture, care must be taken that the tip does not touch the laboratory bench or anything else, as this will result in transferring organisms to the object touched.

Working with agar in petri plates requires special precautions because when the lid of the plate is removed, the surface of the agar is exposed to the air and contamination may easily occur. When inoculating a plate, the lid is raised on one side only and just high enough to allow the loop to enter (Figure 4.13). After incubation, well-isolated colonies should be obtained on one part of the plate (Figure 4.13*c*).

During manipulation of cultures, care must be taken to avoid creating any mist or droplets in the air, since droplets of culture medium laden with microbes can be a serious source of contamination. This is especially important if pathogenic or potentially pathogenic organisms are being studied.

4.4 OBTAINING MICROORGANISMS IN CULTURE

In nature, microorganisms almost always live in mixtures. However, before most characteristics of a particular microorganism can be determined, the organism must first be isolated in pure culture. How does one go about isolating a pure culture of an organism? The general procedure is to use culture medium and incubation conditions which encourage the desired organism and discourage other organisms. Once the organism has been increased in amount relative to the others, it is purified so that the culture contains only the single type of microorganism desired. The isolation and maintenance of pure cultures is one of the most important procedures in microbiology.

Enrichment culture Organisms vary widely in their nutritional and environmental requirements. It is possible, by choosing the appropriate growth conditions, to select from a mixture an organism of interest. The procedure of adjusting culture conditions to select a particular organism is called **enrichment culture** and is a procedure of considerable importance in microbiology.

The usual practice in enrichment culture procedures is to prepare a culture medium that will favor the growth of the organism of interest. The medium is then inoculated with a sample of material thought to contain the organism, such as soil, water, blood, or tissue. The appropriate conditions of aeration, temperature, and pH are provided and the inoculated medium is then incubated. After an appropriate period of time, usually several days, the en-

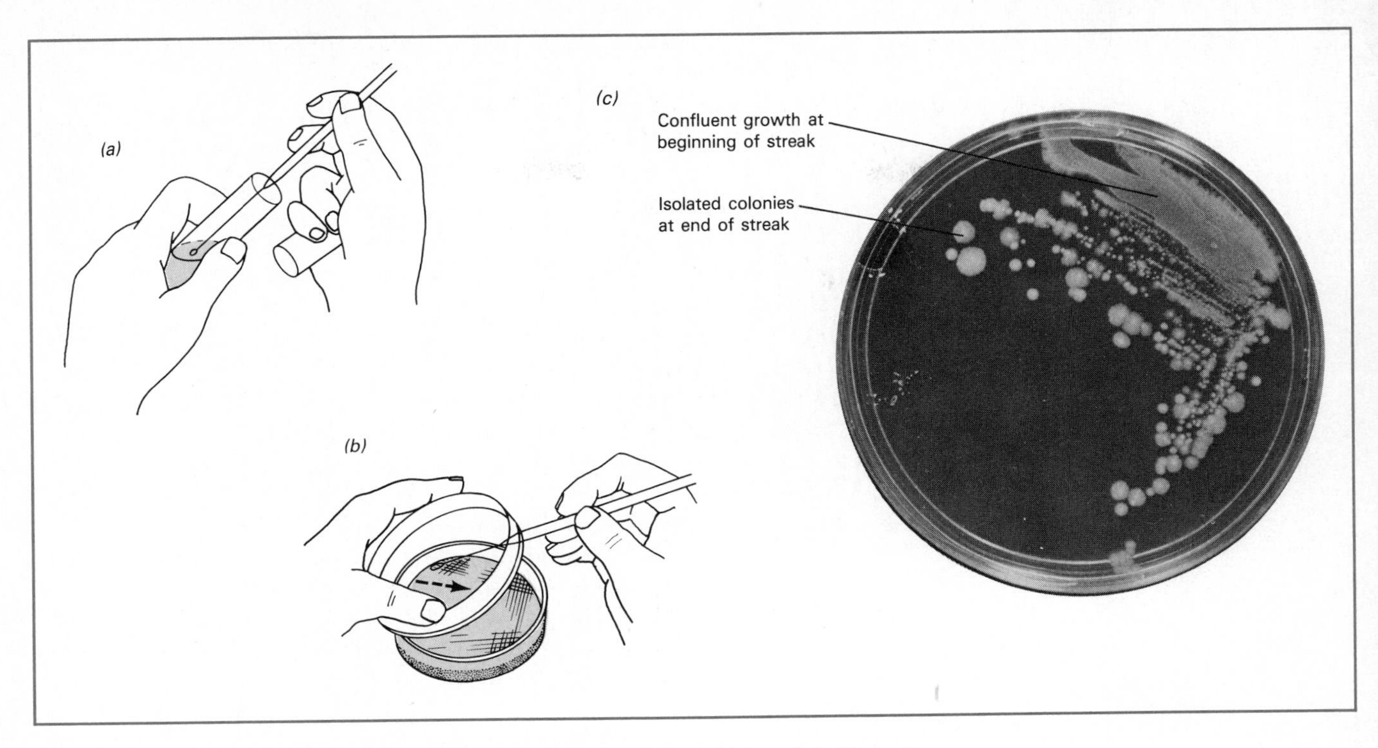

Figure 4.13 Method of making a streak plate to obtain pure cultures. (a) Loopful of inoculum is removed from tube. (b) Streak is made over a sterile agar plate, spreading out the organisms. (c) Appearance of the streaked plate after incubation. Note the presence of isolated colonies. It is from such well-isolated colonies that pure cultures usually can be obtained.

richment culture is examined visually and microscopically for evidence of microbial growth. Cultures in which growth has taken place are then used further for isolation of pure cultures by streaking on medium solidified with agar (see Figure 4.13).

Some examples of enrichment cultures follow:

1. Photosynthetic organisms can be enriched by using nutrient salts media devoid of any organic or inorganic energy sources (but with CO_2 as a source of carbon) and incubating the cultures in the light. Inocula from rivers, lakes, oceans, or the soil may be used.
2. Nitrifying bacteria can be enriched by preparing medium with ammonium as sole energy source and incubating in the dark. Inocula can be from soil, sewage, or mud.
3. Organisms using specific organic compounds as energy sources can be enriched by preparing culture media in which the organic compound in question is the sole energy source. As an example, if we were interested in an organism that can utilize the herbicide 2,4-D (2,4-dichlorophenoxyacetic acid), we could prepare a culture medium in which 2,4-D is the sole carbon and energy source.
4. If thermophilic organisms are desired, enrichment cultures can be prepared using high temperatures for incubation.

5. If anaerobic organisms are desired, enrichment cultures devoid of O_2 can be prepared.

These examples are just a few of many that might have been given. Using appropriate enrichment culture procedures, thousands of cultures have been obtained and most of our knowledge of microbial diversity has come from a study of pure cultures derived by enrichment procedures.

Selective and differential media *Selective culture media* are designed for the isolation of specific organisms. It has been found that some culture media can be prepared that allow the growth of single organisms or a group of closely related organisms and prevent the growth of virtually all others. Such selective media are quite useful in isolating an organism of interest. For instance, in water-pollution studies, the presence of either of the intestinal pathogenic bacteria *Salmonella* or *Shigella* is of great interest. Several selective media have been developed that, when inoculated with a water sample and incubated at the proper temperature, permit growth of only these two organisms. Selective media usually contain one or more substances inhibitory to the growth of organisms other than those to be selected.

A **differential medium** is one that allows several or many organisms to grow but contains dyes or other ingredients to produce variations in colony color in different organisms. Although more than one type of organism can grow on the differential medium, only a single type would be able to carry out the specific reaction which causes the color change. Differential media are widely used in the identification of unknown microorganisms in clinical, food, dairy, and other microbiological laboratories.

Maintaining pure cultures Once a pure culture is obtained, it must be kept pure. One of the most frequent ways in which erroneous results and conclusions are obtained in microbiology is by use of contaminated cultures. Cultures of organisms of interest that are maintained in the laboratory for study and reference are called *stock cultures*. The stock culture must be maintained so that it is free from contamination, retains viability, and remains true to type. Cultures which are infrequently used can be purchased from a commercial culture collection such as the American Type Culture Collection or the German Collection of Microorganisms. For long-term storage of cultures, they may be frozen in *liquid nitrogen*, at which low temperature viability is effectively maintained. Many cultures can also be dried by a *freeze-drying* process (*lyophilization*) and preserved almost indefinitely in the dried state.

4.5 SPECIAL PRECAUTIONS FOR WORKING WITH HARMFUL MICROORGANISMS

When working with disease-causing (pathogenic) microorganisms, it is not sufficient merely to keep cultures uncontaminated. One must be constantly on the watch to avoid infecting laboratory personnel or others. For work with the most dangerous pathogens, special bacteriological transfer hoods are available, inside of which all operations are carried out, thus confining the pathogen to a single location (Figure 4.14).

All utensils, pipettes, glassware, and other items that have come in contact with the pathogen must be sterilized before being discarded. Special disinfecting jars for pipettes and small instruments are desirable. All cultures containing the pathogen must be clearly marked, and separate incubators should be used rather than general-use incubators.

When a large amount of work is to be done with pathogens, special laboratories are often used in which the air flow can be controlled. The inside of such a laboratory should be at a lower air pressure than the outside so that air flows into the laboratory, not out. In this way, air-borne pathogens will remain within the laboratory. The air-exit system needed to maintain the lowered internal pressure should contain an incinerating device so that any pathogens leaving the laboratory in the exhaust air are killed.

Most important of all, work with pathogens should be done only by someone with proper training who is already skilled in their study. Fortunately, most microorganisms are nonpathogenic; so the student can learn the proper procedures of aseptic technique in complete safety.

It is important that the shipping and transfer of pathogenic organisms be carried out properly. The Centers for Disease Control, Atlanta, Georgia, has designed a special type of container for shipping dangerous materials (Figure 4.15). Pathogenic cultures or other harmful agents are to be placed inside a primary container that is wrapped in absorbent material. If breakage occurs, any liquid will be absorbed and will not leak out. The primary container and absorbent packing are then placed inside a secondary container that is in turn put inside an outer shipping container. Appropriate warning labels must be

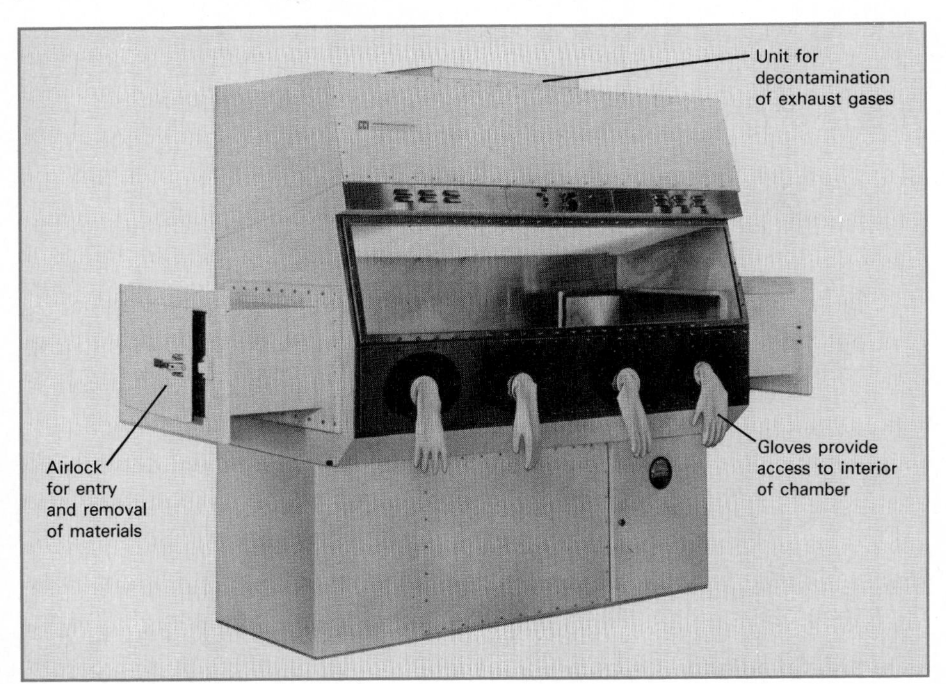

Figure 4.14 Special biohazard transfer hood used for working with especially dangerous pathogens.

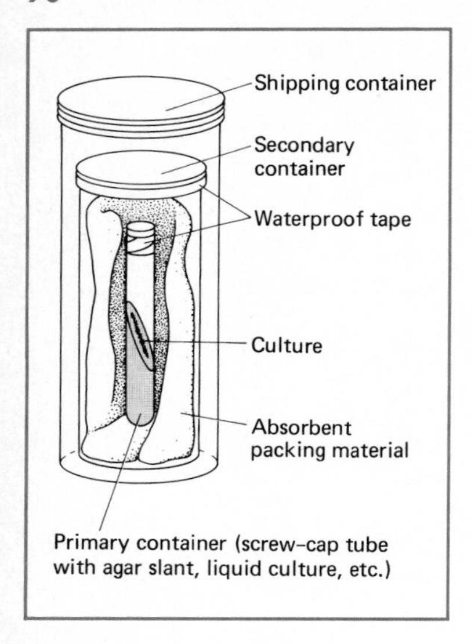

Figure 4.15 Special container to be used for shipping pathogenic cultures.

used, and receipt of delivery should be acknowledged. The people who handle pathogenic materials in the mail cannot be expected to know their danger and must be protected.

In the last decade techniques have been developed which allow the combining of genetic material (DNA) from two different organisms. The methods of *genetic engineering* will be discussed in Chapter 7. Although a genetically-engineered organism is not necessarily a harmful one, scientists became concerned about the possibility of creating a particularly harmful new microorganism in the laboratory, and worked to develop procedures to ensure the safety of laboratory personnel and the general public. As with pathogenic microorganisms, physical containment of the microorganism to the laboratory, or to a special part of a laboratory (such as a biohazard hood), is one way to control the spread of the potentially harmful agent. It is also possible to use special strains of microorganisms which have been "engineered" to be entirely dependent on laboratory culture and which cannot grow in the environment outside the laboratory (e.g., in humans, other animals, or plants), for high risk experiments. This is biological, rather than physical, containment, since the spread of the organism outside the laboratory is prevented by the inability of the organism to grow anywhere except on a specific culture medium in the laboratory. Usually a combination of physical and biological containment is recommended, but if the biological containment is adequate, the level of physical containment can be safely reduced.

SUMMARY

In this chapter, we have reviewed the procedures used to grow microorganisms in the laboratory. These are essential skills for a microbiologist since so much can be learned about a microorganism by obtaining it in purified form and studying it under defined laboratory conditions. A **culture medium** must first be carefully prepared based on the **nutritional** and **environmental** needs of the microorganism being studied. The energy source, organic nutrients, mineral and growth factor requirements, and the appropriate pH must be provided. Appropriate environmental conditions, such as temperature and oxygen concentration, must be maintained during the incubation of the inoculated medium. The medium, as well as all other materials used in manipulating microorganisms, must be **sterilized** to eliminate unwanted contaminating microorganisms. Proper manipulation of microbial cultures requires use of **aseptic technique**, which permits the microbiologist to grow only the microorganism of interest. Using aseptic technique, microorganisms can be obtained from their natural habitats, by encouraging their growth in **enrichment cultures** or on **selective** or **differential media**. The elimination of all but the single desired microorganism results in a **pure culture**, which then can be used for detailed study and identification. In order to ensure that the pure culture is always available for study, it is preserved as a **stock culture**, or sometimes in a **culture collection**, from which it may always be retrieved quickly without repeating the steps needed to isolate it from nature. When working with dangerous microorganisms, such as pathogenic ones, special precautions are taken to protect laboratory personnel, and to avoid spreading these organisms outside the laboratory.

KEY WORDS AND CONCEPTS

Clone
Culture medium
Nutritional requirement
Energy source
Growth factor
Vitamin
Environmental requirement
pH
Oxygen
Anaerobe
Temperature optimum
Psychrophile
Mesophile
Thermophile
Synthetic medium
Complex medium
Agar
Sterilization
Autoclave
Aseptic technique
Enrichment culture
Selective culture
Differential culture
Pure culture

STUDY QUESTIONS

1. What is meant by the word *clone*?
2. Name the major kinds of nutrients which a *culture medium* must provide.
3. What are *growth factors*? Give three examples.
4. What is the difference between a *simple* and a *complex culture medium*?
5. What do we call microorganisms which are adapted to low pH? Give an example of an acidic liquid. What do we call microorganisms which are adapted to high pH? Give an example of an alkaline solution.
6. What component of a culture medium keeps the pH stable?
7. Distinguish between *aerobic, anaerobic,* and *facultative* microorganisms. What methods can be used to culture extremely anaerobic microorganisms?
8. Where would you expect to find *psychrophilic* microorganisms? *mesophilic* microorganisms? *thermophilic* microorganisms?
9. Why is it necessary to *sterilize* a culture medium? Name a method suitable for sterilizing materials which are not sensitive to heat. Name a method used to sterilize materials which are sensitive to heat.
10. What is the objective of *aseptic technique*? Describe how you would use aseptic technique to transfer a culture from one tube or plate of culture medium to another.
11. What is an *enrichment culture*? How would you design an enrichment culture for oil-degrading microorganisms? for psychrophilic microorganisms?

12. What is a *pure culture*? Describe how you would obtain one from an enrichment culture.
13. Name two types of microorganisms whose manipulation in the laboratory requires special precautions. What steps can be taken to contain them?
14. What is the difference between *physical containment* and *biological containment*?

SUGGESTED READINGS

BROCK, T.D., D.W. SMITH, and M.T. MADIGAN. 1984. *Biology of microorganisms, 4th edition*. Prentice-Hall, Englewood Cliffs, N.J.

FRASER, D.W., and J.E. MCDADE. 1979. *Legionellosis*. Scientific American 241: 82–99. (A review of Legionnaire's disease.)

PHILLIPS, J.A. and T.D. BROCK. 1984. *General microbiology: a laboratory manual*. Prentice-Hall, Englewood Cliffs, N.J.

5

Microbial Growth and its Control

Microorganisms exert most of their effects through growth, and a knowledge of the processes involved in the growth of microorganisms is essential if we are to predict their activities and control them for our own benefit. A characteristic feature of microbial growth is the rapidity with which the population increases; very large numbers are reached quickly, often with harmful consequences. In this chapter, we discuss the measurement of microbial growth and examine some of the chemical and physical agents that are used to control it.

5.1 GROWTH

Growth is defined as an increase in the number of microbial cells or an increase in microbial mass. **Growth rate** is the change in cell number or mass per unit time. In unicellular microorganisms, growth usually involves an increase in cell number. A single cell continually increases in size until it is approximately double its original size; then cell division occurs, resulting in the formation of two cells the size of the original cell. During this cell-division cycle, all the structural components of the cell double. The interval for the formation of two cells from one is called a **generation**, and the time required

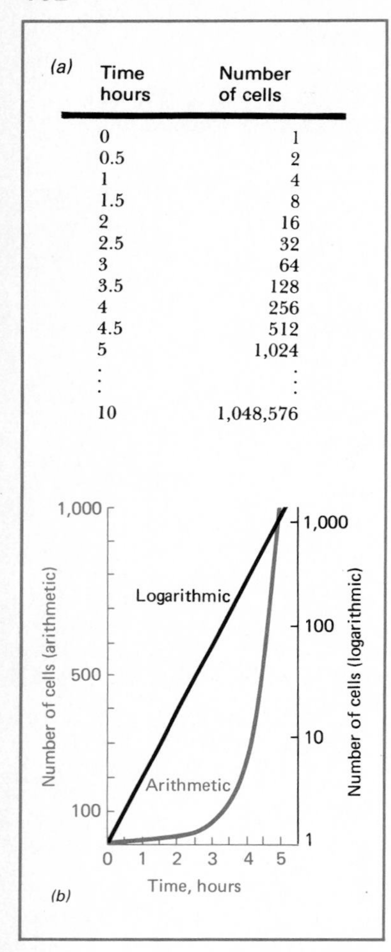
(a)

Time hours	Number of cells
0	1
0.5	2
1	4
1.5	8
2	16
2.5	32
3	64
3.5	128
4	256
4.5	512
5	1,024
⋮	⋮
10	1,048,576

Figure 5.1 The rate of growth of a microbial culture. (a) Data based on a doubling time of 30 minutes. (b) Above data plotted on an arithmetic and a logarithmic scale.

for this to occur is called the **generation time**. The generation time is thus the time required for cell number to double. Because of this, the generation time is also sometimes called the *doubling time*. Note that during a single generation, both the cell number and cell mass have doubled. Generation times vary widely among organisms. Most bacteria have generation times of 1 to 3 hours, but a few are known that divide in as little as 10 minutes. At the other extreme, some slow-growing protozoa and algae have generation times of 24 hours or more.

Exponential growth The progressive doubling of cell number results in a continually increasing rate of growth in the population, a process called **exponential growth**. Thus, one cell doubles to give rise to two cells, two cells give rise to four, four to eight, eight to sixteen, and so on. If this process continued unchecked, very quickly an enormous number of cells would be produced. For instance, beginning with a single cell with a doubling time of 30 minutes, at the end of 10 hours there would be over 1,000,000 cells (Figure 5.1*a*).

One of the characteristics of exponential growth is that the rate of increase in cell number is slow initially but increases at an ever faster rate. This results, in the later stages, in an explosive increase in cell numbers. A practical implication of exponential growth is that when a nonsterile product such as milk is allowed to stand under conditions such that microbial growth can occur, a few hours during the early stages of exponential growth are not detrimental, whereas standing for the same length of time during the later stages is disastrous.

Because of the rapid exponential growth of many microorganisms, large populations of cells develop quickly, and one is often forced to deal with very large numbers. For instance, cell numbers in the millions, hundred millions, and even billions occur quite often. Since it is difficult to handle such large numbers, the microbiologist makes use of scientific notation with employs exponents of 10. Thus we express 1,000,000 (one million) as 10^6; 10,000,000 as 10^7; 100,000,000 as 10^8; and 1,000,000,000 (one billion) as 10^9. To express a number such as 5,000,000, the figure is written as the unit integer multiplied by the proper power of 10: thus $5{,}000{,}000 = 5.0 \times 10^6$; $25{,}000{,}000 = 2.5 \times 10^7$; and $700{,}000{,}000 = 7.0 \times 10^8$. It is convenient to plot the increase in cell number as the logarithm of cell number because the resulting plot is a straight line (Figure 5.1*b*).

Growth cycle Populations of unicellular microorganisms exhibit characteristic growth cycles, which can be divided into several distinct phases: *lag phase, exponential phase, stationary phase,* and *death phase* (Figure 5.2).

When a microorganism is inoculated into a fresh medium, growth usually does not begin immediately but only after a period of time, called the **lag phase**, which may be brief or extended, depending on conditions. The lag phase usually occurs as the cells adjust to the new medium. This adjustment requires the production of enzymes needed for the utilization of the nutrients present in the environment and in the production of those monomers needed for biosynthesis which are not already present. If an exponentially growing

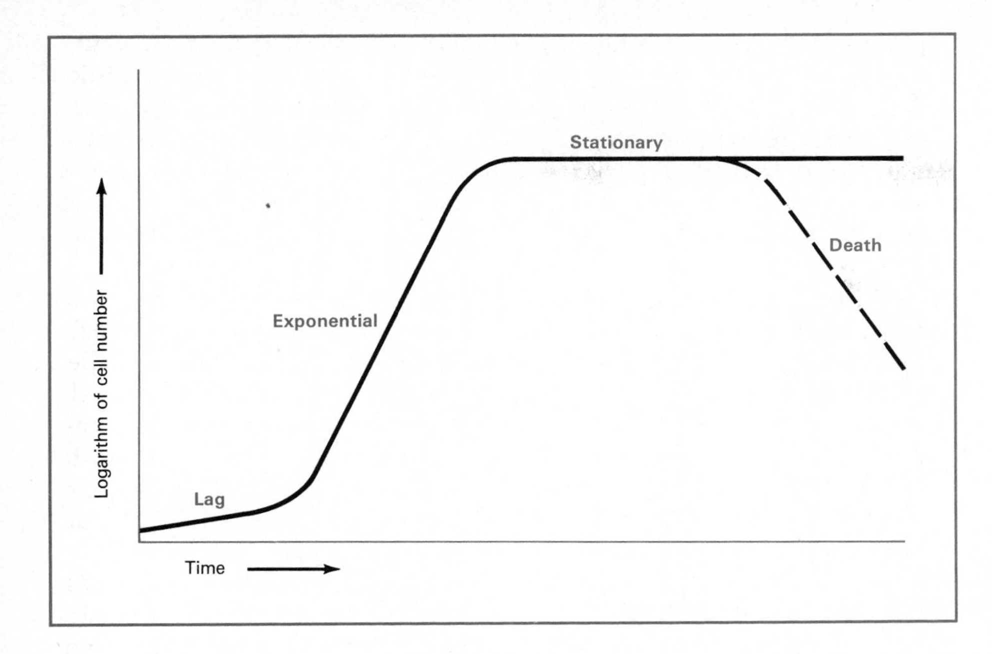

Figure 5.2 Typical growth curve of a microbial culture, showing the various phases.

culture is transferred to the exact same culture medium at the same temperature, a lag does not occur, since no adjustment by the cells is needed.

The *exponential phase* of growth has already been discussed. As noted, it is a consequence of the fact that each cell divides to form two cells, each of which also divides to form two more cells, and so on. Most unicellular microorganisms grow exponentially, but rates of exponential growth vary greatly. For instance, the organism causing typhoid fever, *Salmonella typhi*, grows very rapidly, with a generation time of 20 to 30 minutes, whereas the tubercle bacterium, *Mycobacterium tuberculosis*, grows slowly, with only one or two doublings per day. The rate of exponential growth is influenced by environmental conditions (temperature, composition of the culture medium) as well as by characteristics of the organism itself. In general, bacteria grow faster than eucaryotic microorganisms, and small eucaryotes grow faster than large ones.

Exponential growth cannot occur indefinitely. One can calculate that a single bacterium with a generation time of 20 minutes would, if it continued to grow exponentially for 48 hours, produce a population that weighed about 4,000 times the weight of the earth! This is particularly impressive since a single bacterial cell weighs only about one-trillionth of a gram. Obviously, something must happen to limit growth long before this time. What generally happens is that either an essential nutrient of the culture medium is used up or some waste product of the organism builds up in the medium to an inhibitory level and exponential growth ceases. The population has reached the **stationary phase**.

In the stationary phase there is no net increase or decrease in cell number. However, although no growth occurs in the stationary phase, many cell functions may continue, including energy metabolism and some biosynthetic processes. Certain cell metabolites, called *secondary metabolites*, are produced primarily in the stationary phase. Examples of such secondary metabolites include the antibiotics (for example, penicillin, streptomycin) and some enzymes. In some organisms, growth may even occur during the stationary phase; some cells in the population grow while others die, the two processes balancing out so that no net increase or decrease in cell number occurs. In endospore-forming bacteria, the endospore is produced after the culture has entered the stationary phase.

The stationary phase may extend indefinitely, or may be followed by the **death phase**, in which the organisms in the population die. Death occurs either because the organisms undergo starvation or because some toxic product accumulates and kills them. In some cases, not only do the organisms die, but the cells may also disintegrate, a process called *lysis*.

5.2 MEASUREMENT OF GROWTH

Growth is measured by following changes in number of cells or weight of cell mass. There are several methods for counting cell number or estimating cell mass, suited to different organisms or different problems.

Total cell count The number of cells in a population can be measured by counting under the microscope, a method called the **direct microscopic count**. Two kinds of direct microscopic counts are done, either on samples dried on slides or on samples in liquid. With liquid samples, special *counting chambers* must be used. In such a counting chamber, a special grid is marked on the surface of the glass slide, with squares of known small area (Figure 5.3). Over each square on the grid is a volume of known size, very small but precisely measured. The number of cells per unit area of grid can be counted

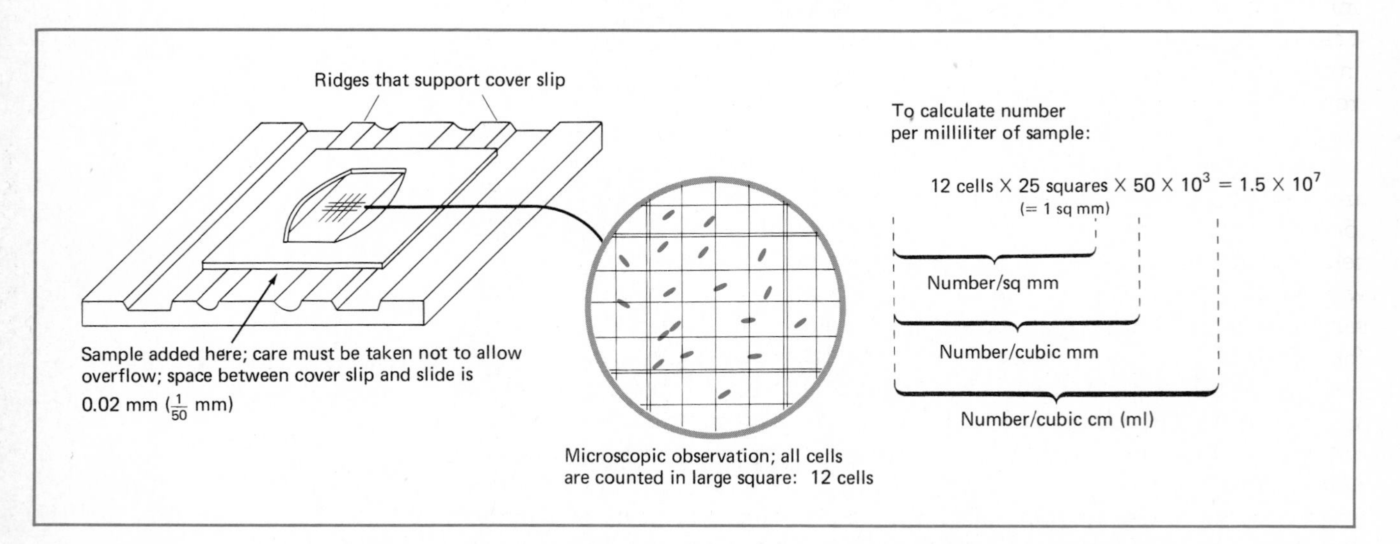

Figure 5.3 Direct microscopic count of cell number, using a counting chamber. The total ruled grid has 25 large squares and measures 1 square mm in area. When making microscopic counts, several large squares are counted and an average value obtained.

under the microscope, giving a measure of the number of cells per small chamber volume. Converting this value to the number of cells per milliliter of suspension is easily done by multiplying by a conversion factor based on the volume of the chamber sample.

The *dry-slide method* for direct microscopic counting (sometimes called the *Breed counting method*) is used primarily in dairy work. A known small volume of sample is spread uniformly on a 1-sq-cm area marked on a slide. After drying and staining, the cells in several microscopic fields are counted. The area of the microscopic field is determined; then, knowing this and the volume of sample spread in the centimeter-square area, the number of cells per unit volume is calculated. A description of the method of counting bacteria in milk is given in Chapter 16.

Direct microscopic counting is tedious but is a quick and easy way of estimating microbial cell number. However, it has certain limitations: (1) Dead cells are not distinguished from living cells. (2) Small cells are difficult to see under the microscope, and some cells are probably missed. (3) Precision is difficult to achieve. (4) A phase-contrast microscope is required when the sample is not stained. (5) The method is not suitable for cell suspensions of low density. With bacteria, if a cell suspension has less than 10^6 cells per milliliter, few if any bacteria will be seen.

To avoid the tedium of direct microscopic counting, *electronic cell counters* have been developed. These counters were first devised for counting of red blood cells in hospital laboratories but can also be used for counting microbial cells. In electronic counting, the electrical resistance of the fluid within a small hole is measured. As each cell passes through the hole, the resistance increases sharply, and the increase is recorded. A known small volume of sample is allowed to flow through the hole, and each cell is counted as it passes. An electronic cell counter is accurate and quick but is fairly expensive, so it is only valuable if a very large number of similar samples are being counted. It cannot distinguish between a live organism and a dead organism or between an organism and an inert particle; hence debris and precipitates must be absent from the sample being counted.

Viable count In the methods just described, both living and dead cells are counted. In many cases we are interested in counting only live cells, and for this purpose viable cell counting methods have been developed. A viable cell is defined as one that is able to divide and form offspring, and the usual way to perform a viable count is to determine the number of cells in the sample capable of forming *colonies* on a suitable agar medium. For this reason, the viable count is often called the **plate count**, or **colony count**.

There are two ways of performing a plate count: the spread plate method and the pour plate method (Figure 5.4). With the **spread plate method**, a volume of culture no larger than 0.1 ml is spread over the surface of an agar plate, using a sterile glass spreader. The plate is then incubated until the colonies appear, and the number of colonies is counted. It is important that the surface of the plate be dry so that the liquid that is spread soaks in. Volumes greater than 0.1 ml should never be used, since the excess liquid will not soak in and may cause the colonies to coalesce as they form, making

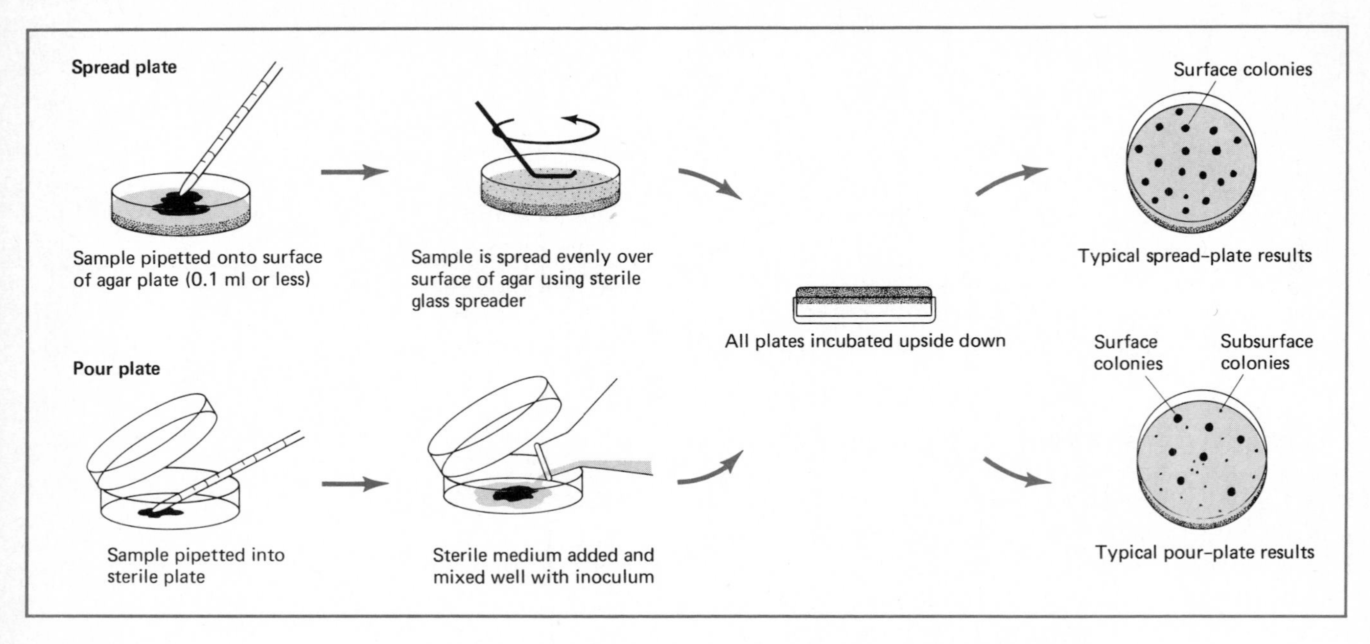

Figure 5.4 Two methods of performing a viable count (plate count).

them difficult to count. In the **pour plate method** (Figure 5.4), a known volume of 0.1 to 1.0 ml of culture is pipetted into a sterile petri plate; melted agar medium is then added and mixed well by gently swirling the plate on the table top. Because the sample is mixed with the molten agar medium, a larger volume can be used than with the spread plate; however, with the pour plate the organism to be counted must be able to briefly withstand the temperature of melted agar, 45°C.

With both the spread plate and pour plate methods, it is important that the number of colonies developing on the plates not be too large, since on crowded plates some cells may not form colonies and the count will be erroneous. It is also essential that the number of colonies not be too small, or the statistical significance of the calculated count will be low. The usual practice, which is most valid statistically, is to count only those plates that have between 30 and 300 colonies. To obtain the appropriate colony number, the sample to be counted must thus usually be diluted. Since one rarely knows the approximate viable count ahead of time, it is usually necessary to make more than one dilution. Several tenfold dilutions of the sample are commonly used (Figure 5.5). To make a tenfold dilution, one can mix 0.5 ml of sample with 4.5 ml of diluent, or 1.0 ml with 9.0 ml diluent. If a hundredfold dilution is needed, 0.05 ml can be mixed with 4.95 ml diluent, or 0.1 ml with 9.9 ml diluent, or of course two successive tenfold dilutions may be made. In most cases, such *serial dilutions* are needed to reach the final dilution desired. Thus, if a $1/10^6$ (1 to 1,000,000) dilution is needed, this can be achieved by making three successive $1/10^2$ (1 to 100) dilutions or six successive tenfold dilutions. The liquid used for making the dilutions is important. It is best if it is identical with the liquid used in making the solidified medium, although for economy,

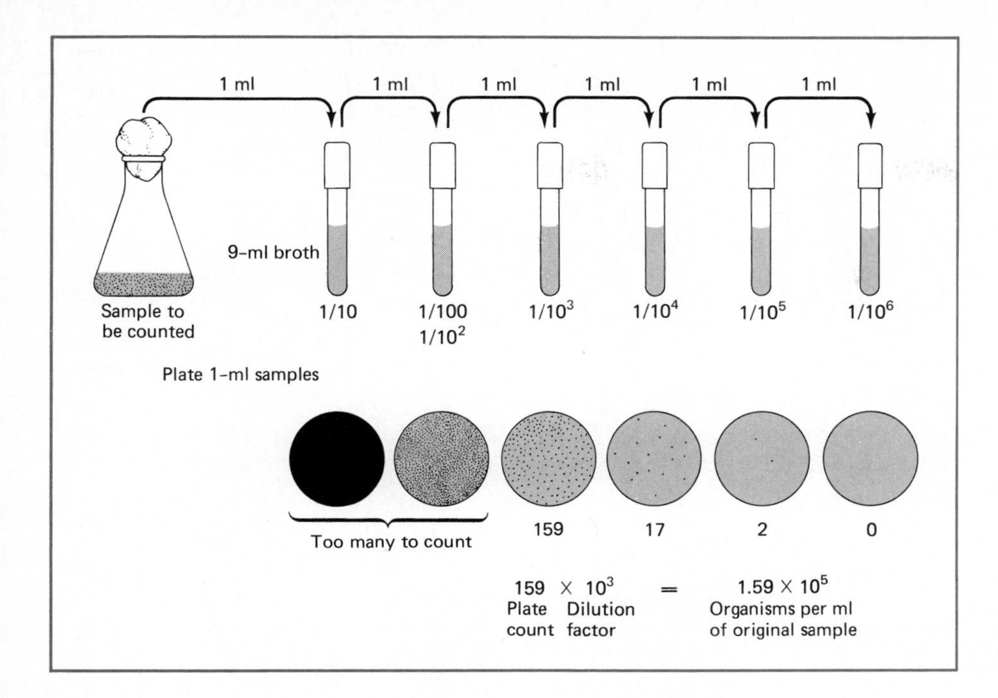

Figure 5.5 Procedure for viable count using serial dilutions of the sample.

it is often possible to use a sterile solution of inorganic salts or a phosphate buffer. It is important when making dilutions that a *separate* sterile pipette be used for each dilution, even of the same sample. This is because in the initial sample, which contains the largest number of organisms, not all the organisms will be washed out of the pipette when its contents are expelled. Organisms sticking to the pipette will be washed out in later dilutions and can cause serious error in the final count obtained.

The number of colonies obtained on a plate will depend not only on the inoculum size but also on the suitability of the culture medium and the incubation conditions used; also, it will depend on the length of incubation. The cells deposited on the plate will not all develop into colonies at the same rate, and if a short incubation time is used, less than the maximum number of colonies will be obtained. Furthermore, the size of colonies often varies. If some tiny colonies develop, they may be missed during the counting. It is usual to determine incubation conditions (medium, temperature, time) that will give the maximum number of colonies and then to use these conditions throughout. Viable counts are usually subject to large error, and if accurate counts are desired great care must be taken and many duplicate plates must be prepared. Note that two or more cells in a clump will form only a single colony, so that a viable count may be erroneously low. To more clearly state the result, viable counts are often expressed as the number of *colony-forming units* rather than as the number of *viable cells* (since a colony-forming unit may contain one or more cells).

Despite the difficulties involved with viable counting, the procedure gives the best information on the number of viable cells, so it is widely used. In food, dairy, medical, and aquatic microbiology, viable counts are used routinely (see especially Chapters 15 and 16). Standard methods have been developed that are suitable for viable counts of different materials, and these methods are clearly described in appropriate manuals. In these fields of applied microbiology, identical methods are used in all laboratories, so that the results in one laboratory are comparable to those in another.

Probably the most useful aspect of viable counting is the sensitivity of the method. Samples containing only a very few cells can be counted, thus permitting sensitive detection of contamination of products or materials.

For some work, *membrane filters* are used instead of agar plates, as mentioned in Chapter 4. The principle of counting is the same as in a plate count. An appropriately diluted sample is filtered, and the filter is placed on a suitable culture medium. After incubation, the colonies that develop on the filter are counted. Membrane filtration is most useful for counting when the sample contains only a very small number of organisms, since a relatively large volume of sample can be passed through the filter. The method is used in this way in aquatic microbiology, especially for obtaining bacterial counts of drinking water and other relatively unpolluted waters (see Chapter 15).

Some organisms do not readily form colonies on agar plates or membrane filters but will initiate growth in liquid medium. To count such organisms, a technique called the *most probable number* (or MPN) method has been developed, which permits an estimate of viable numbers after incubation in liquid medium. With this method, the sample is diluted to the point at which some portions of the sample, but not all, contain a cell. If a series of tubes is inoculated with identical portions taken at this dilution, some will contain a cell whereas others will not; after incubation, some will show growth whereas others will not. By counting the fraction of tubes showing growth, one can estimate the viable count, using statistical tables that have been developed. As will be described in detail later (in Chapter 15), MPN methods are used most commonly in aquatic microbiology.

Cell mass For many studies it is desirable to estimate the weight of cells rather than the number. Net weight can be measured by centrifuging the cells, and weighing the pellet of cells obtained. Dry weight is measured by drying the centrifuged cell mass before weighing, usually by placing it overnight in an oven at 100 to 105°C.

A simpler and very useful method for obtaining a relative estimate of cell mass is by use of *turbidity* measurements. A cell suspension looks turbid because each cell scatters light. The more cell material present, the more the suspension scatters light and the more turbid it will be. Turbidity can be measured with an electrically operated device called a *colorimeter*, or *spectrophotometer*. With such a device, the turbidity is expressed in units of *absorbance*. For unicellular organisms, absorbance is proportional to cell number as well as cell weight, and turbidity readings can thus be used as a substitute for counting. To perform cell counts in this way, a standard curve must be prepared for each organism studied, relating cell number to cell mass or absorbance. Turbidity is a much less sensitive way of measuring cell density than is viable counting but has the virtues that it is quick, easy, and does not destroy the sample. Such measurements are used widely to follow the rate of growth of cultures, since the same sample can be checked repeatedly.

5.3 CONTROL OF MICROBIAL GROWTH

The control of microbial growth is necessary in many practical situations. Control can be effected either by *killing* organisms or by *inhibiting* their growth. The complete killing of all organisms is generally called *sterilization* and is brought about by use of heat, radiation, or chemicals. Sterilization means the complete destruction of all microbial cells present; once a product is sterilized, it will remain sterile indefinitely if it is properly sealed.

Heat sterilization One of the most important and widely used agents for sterilization is heat. We have already discussed the use of heat to sterilize culture media in Chapter 4, and here we shall discuss some of the principles and practices of heat sterilization. The lethal temperature varies among microorganisms: some are very sensitive and are killed at temperatures as low as 30°C, whereas others are quite resistant and require temperatures of boiling or higher. When a population is heated to a lethal temperature, death of all cells in the population does not occur at the same time, but a certain percentage of organisms dies during each given time period. However, the higher the temperature, the faster the rate of killing (Figure 5.6). The **thermal death time** is defined as the time required at a given temperature to kill *all* organisms in the population.

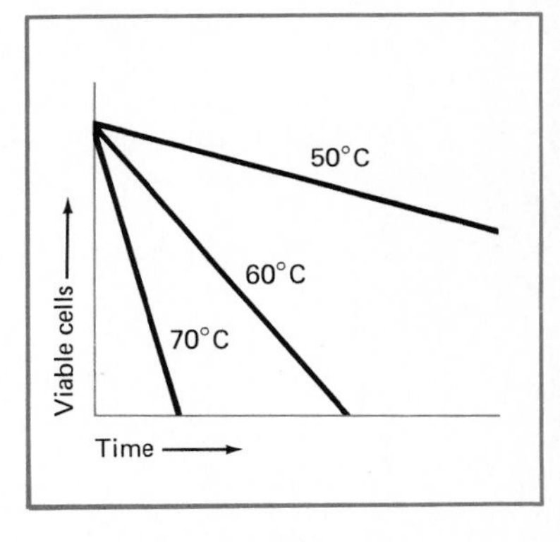

Figure 5.6 Effect of temperature on the viability of cells in a microbial population.

The thermal death time depends on the number of organisms present, since it obviously takes longer to kill a large number of organisms than a small number. Other factors to be considered are the species of microorganisms involved and the nature of the product being sterilized. In general, sterilization is more rapid at low pH than at neutral or higher pH, because most microorganisms are sensitive to low pH. For this reason, acidic foods such as tomatoes, fruits, and pickles are much easier to sterilize than are more neutral foods such as corn and beans. Moisture also affects the thermal death time. Dry cells are usually more resistant to heat than moist ones; for this reason, heat sterilization of dry objects always requires higher temperatures and longer times than does the sterilization of moist objects. Sugary and salty products also are usually quite resistant to heat sterilization, probably because the sugar or salt has a dehydrating effect on the microbial cells.

The single most important factor affecting the rate of heat sterilization is the presence or absence of bacterial endospores. Because of the high heat resistance of bacterial spores, very much longer heating times are necessary to sterilize a product containing spores than one containing only vegetative cells. If only vegetative cells are present, most products can be sterilized by heating at 60 to 70°C for a few minutes, whereas if spores are present, temperatures as high as 121°C are usually required. To achieve temperatures this high, since they are above boiling, an autoclave is required (see Figure 4.8). One of the most important practical processes involving heat sterilization is the *canning* of foods, as discussed in Chapter 16.

SALMONELLA SABOTAGE

Within a two-week period in the spring of 1985, about 18,000 persons in northern Illinois and surrounding states experienced severe intestinal disease due to infection with a bacterium called *Salmonella typhimurium*. The outbreak was traced to milk provided by a single dairy plant operated by a large grocery store chain. The bacterium was easily cultivated from patients and from the milk, using a culture medium selective for *Salmonella*. All of the cultures obtained were of a single strain which was highly resistant to the antibiotics ampicillin and tetracycline. The outbreak was the largest in the United States associated with a single *Salmonella* strain and illustrates how widely a consumed product, once contaminated, can spread its evil. *Salmonella* is found in dairy cattle and in raw milk, but is killed by the pasteurization process. Careful engineering study at the suspected dairy plant showed that a defective valve had allowed unpasteurized milk to contaminate pasteurized milk. The economic impact of this outbreak on the company was enormous: millions of dollars were forfeited in lost revenue, legal fees were huge, and the milk processing plant closed its doors permanently. This story illustrates two points: the pasteurization process in a large dairy is complex; and antibiotic-resistant bacteria (see Figure 5.16) are pervasive in modern society.

Pasteurization Pasteurization is a process using mild heat to reduce microbial populations in milk and other foods. The process is named for Louis Pasteur, who first used heat for controlling the spoilage of wine. Pasteurization is not synonymous with sterilization, since not all organisms are killed. Pasteurization is used to kill any harmful microorganisms that might be present, since they are usually fairly sensitive to heat and rarely form endospores. Because it kills spoilage organisms, pasteurization is also used to improve the keeping qualities of the product. Pasteurization is used commonly for milk, a rather heat-sensitive product, the flavor of which is altered greatly by heat sterilization but much less so by pasteurization. Beer, wine, cider, and other beverages are also pasteurized (see Chapter 16).

Pasteurization can be done by heating the substance at 63 to 66°C for 30 minutes and then quickly cooling. However, for pasteurizing large quantities of a liquid such as milk, a process called *flash pasteurization* is preferable. The product is passed through a heater where its temperature is raised quickly to 71°C, held there for 15 seconds, and then quickly cooled. Flash pasteurization is more satisfactory than bulk pasteurization, since the product is heated for a shorter period of time and its flavor is therefore less altered. Flash pasteurization has the additional advantage that since it is carried out on a continuous basis, it is more adaptable to large-scale operation (but see the Box above).

Incineration Incineration is a process of sterilization involving very high temperatures, the contaminating organisms actually being ignited and burned to death. Incineration is chiefly used for sterilizing heat-resistant items such as metal inoculating loops and needles, and for disposing of burnable contaminated materials from hospitals and laboratories. Incineration is also a suitable way of sterilizing air, which is passed over copper coils or pipe heated to a high temperature.

Use of low temperature to control growth Most organisms grow very little if at all in the cold; for this reason low temperatures are frequently used for the storage of perishable food products, thus slowing the rate of microbial growth. A few organisms are able to grow at refrigeration temperatures, however, and may cause spoilage.

Storage for long periods of time is possible at temperatures below freezing. Freezing greatly alters the physical structure of many food products and therefore cannot be used for everything, but it is widely and successfully used for the preservation of meats and many vegetables and fruits. Freezers providing temperatures of −20°C are most commonly used for storing frozen products. At such temperatures, storage for weeks or months is possible, but even at these temperatures some microbial growth may occur, usually in pockets of liquid water trapped within the frozen mass and nonmicrobial chemical changes in the food may still occur. For very long-term storage, lower temperatures are necessary, such as −70°C (dry ice temperature) or −195°C (liquid nitrogen temperature), but maintenance at such low temperatures is expensive, and consequently they are not used for routine food storage.

It should be emphasized that many microorganisms are not killed upon freezing, so that foods to be frozen should be of the best quality and free of spoilage. If a contaminated product is frozen, much of its microbial content will remain alive, and when it is thawed many of the organisms will still be present to cause problems.

Drying Since all microorganisms require water for growth, microbial growth often can be controlled by removing water from the product. However, drying often does not kill organisms already present in the product but only prevents them from growing, so that it is essential that the drying process be used on products or materials that are free of undesirable organisms.

Most microorganisms can grow only if water in the liquid state is present. Some fungi can grow under conditions of humid air even in the absence of liquid water. They apparently are able to extract water directly from the air. These fungi are common in the humid tropics and often cause serious damage (*mildew*) to cloth and leather goods, as well as to cameras and other optical instruments (Figure 5.7).

We stated in Chapter 2 that endospores, conidia, and other types of spores formed by microorganisms are not easily killed by drying. Vegetative cells of some microorganisms are also rather resistant; in general, small cells are more resistant to drying than large cells, and procaryotes are much more resistant than eucaryotes. Some disease-causing organisms resistant to drying are *Mycobacterium tuberculosis*, the causal agent of tuberculosis, and *Staphylococcus aureus*, a causal agent of pimples, boils, superficial wound infections, toxic shock syndrome, and pneumonia. On the other hand, the spirochete that causes syphilis, *Treponema pallidum*, is so sensitive to drying that it dies almost instantly when exposed to air. For this reason, the organism is transmitted only by intimate contact, such as sexual intercourse.

Figure 5.7 Development of fungal growth (*mildew*) on a pair of leather boots stored in damp warm conditions. The boot on the left was stored for 6 months under humid tropical conditions.

Growth of organisms in foods can be prevented by drying the food material. Although all foods can be dried, some foods retain their flavor and texture upon drying better than others. Milk, meats, fish, vegetables, fruits, and eggs are all preserved by drying. The practical aspects of preserving food in this way will be discussed in Chapter 16.

Radiation Some types of radiation cause death of living organisms and are of value in the sterilization and control of microorganisms in various materials. We distinguish between two kinds of radiations: ionizing and nonionizing (Figure 5.8). *Electromagnetic* radiations that have effects on living organisms include ultraviolet, visible, and infrared radiation. The latter two are usually beneficial and are used as sources of energy by photosynthetic organisms, whereas ultraviolet radiation is usually lethal. **Ionizing radiations** include X rays, cosmic rays, and emanations from radioactive materials, such as gamma rays.

Both intensity and wavelength can affect the killing power of **ultraviolet radiation**. Sources of short wavelength are more lethal than sources of long wavelength. Ultraviolet radiation from the sun that reaches the earth's surface is of long wavelength and is only weakly lethal for microorganisms. Artificial ultraviolet sources, such as germicidal lamps, are of shorter wavelength and are much more effective in killing microorganisms. The lethal effect of ultraviolet radiation is due to its effect on deoxyribonucleic acid (DNA). At low doses, ultraviolet radiation may only cause minor damage to DNA, usually resulting in only a few base changes in the DNA. Such a change in the DNA is called mutation, a topic which we will discuss in detail in Chapter 7.

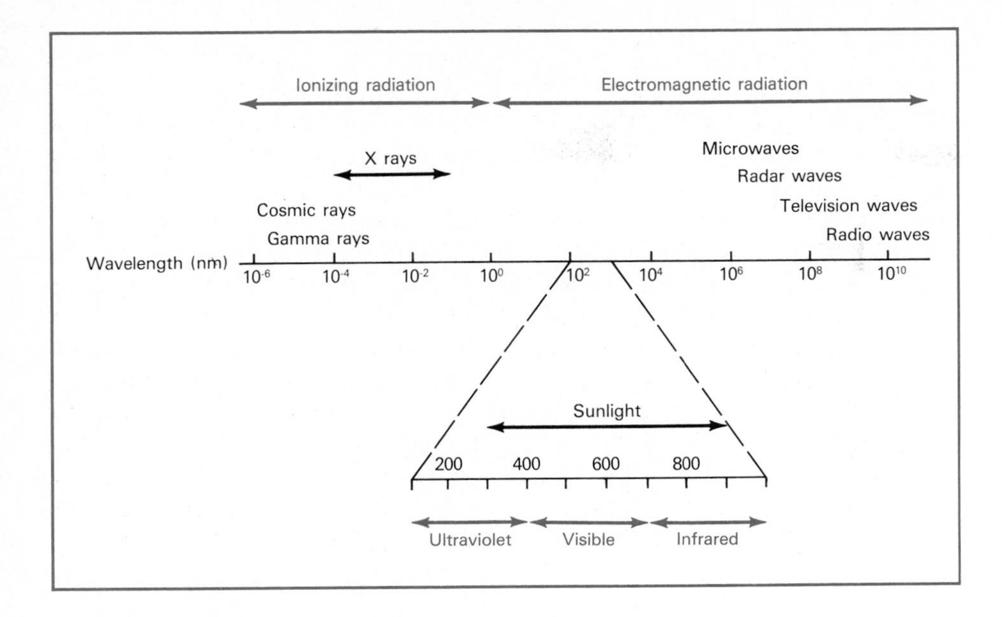

Figure 5.8 Wavelengths of radiation.

Germicidal ultraviolet radiation passes very poorly through many kinds of glass and not at all through opaque objects, so that its practical uses are limited. It can be used to sterilize air or surfaces, such as toilet seats or counter tops. It is often used in hospitals and in microbiology laboratories, since it can be applied for a few hours or overnight before the room is used. Ultraviolet radiation is also now used to reduce the microbial load of sewage. Because radiation from germicidal lamps is very damaging to the eyes, it should never be allowed to enter the eyes, either directly or after reflection from surfaces. Germicidal ultraviolet lamps are available commercially from various lamp manufacturers.

Ionizing radiation kills not only by directly affecting cell constituents but also by inducing in the water surrounding the cells the formation of highly toxic substances called *free radicals*; these react with and inactivate sensitive cell constituents. All kinds of cell constituents are affected, but death usually results from effects on DNA.

Ionizing radiation may be used to sterilize materials such as drugs, foods, and other items that are heat sensitive. In contrast to ultraviolet radiation, ionizing radiation is penetrating and hence can be used on products even after they have been packaged. Sources of ionizing radiation are X-ray machines, radioisotopes such as cobalt-60, and neutron piles in atomic energy installations. The most convenient commercial sources for radiation sterilization are cobalt-60 radioisotopes, which require no maintenance or input of energy once installed.

Strict precautions must be used in handling of ionizing radiation, since it is highly lethal for humans as well as microorganisms. Protective lead shielding is essential, and the doses received by operators must be carefully monitored. Since ionizing radiation has been shown to be effective in the sterilization of foods, its use may become wide-spread in the future.

5.4 CHEMICAL CONTROL OF MICROBIAL GROWTH

An **antimicrobial agent** is a chemical that kills or inhibits the growth of microorganisms. Such a substance may be either a synthetic chemical or a natural product. Agents that kill organisms are often called "cidal" agents, with a prefix indicating the kind of organism killed. Thus, we have **bactericidal**, **fungicidal**, and **algicidal** agents. A bactericidal agent, or bactericide, kills bacteria. It may or may not kill other kinds of microorganisms. Agents that do not kill but only inhibit growth are called "static" agents, and we can speak of **bacteriostatic**, **fungistatic**, and **algistatic** agents.

The distinction between a static and a cidal agent is often arbitrary, since an agent that is cidal at high concentrations may be static at lower concentrations. To be effective, a static agent must be continuously present, since if it is removed or its activity neutralized, the organisms present in the product may initiate growth if conditions are favorable.

Antimicrobial agents can vary in their **selective toxicity**. Some act in a rather nonselective manner and have similar effects on all types of cells. Others are far more selective and are more toxic to microorganisms than to animal tissues. Antimicrobial agents with selective toxicity are especially useful as *chemotherapeutic agents* in treating infectious diseases, as they can be used to kill disease-causing microbes without harming the host. They will be described in Section 5.5.

Disinfectants and antiseptics Cidal agents with a broad spectrum of target organisms are usually called *germicides*. Germicides are sometimes divided into two groups; **antiseptics** and **disinfectants**. An antiseptic is a substance that can kill most microorganisms (but not usually their spores) and is sufficiently harmless that it can be applied to skin or mucous membranes (Table 5.1). It is rarely safe enough to be taken internally. Antiseptics that are used on the skin may be swabbed on with cotton. Soaps containing antiseptics leave a residue of the compound on the skin after the soap is rinsed off, and this active residue may continue to affect microbial viability for some time, thus increasing the efficacy of the treatment. A disinfectant is an agent that kills microorganisms (but not necessarily their spores) and is distinguishable from an antiseptic by the fact that it is not safe for application to living tissue. Disinfectant use is restricted to inanimate objects, such as tables, floors, or dishes. Disinfectants used for surfaces are generally applied as solutions, by swabbing with sponges or mops. Dairy equipment and medical instruments may be dipped or rinsed in disinfectant solutions.

The most common effect of a germicidal agent is the inactivation of proteins of the microbial cell. Since human tissue also contains vital proteins, it is easy to see why so many compounds of both static and cidal type are not suitable for use on humans. Another common effect of disinfectants is an attack on the cell membrane so as to cause loss of cell constituents, or lysis. Again, there is no specificity in this attack, both microbial and human cell membranes being similarly affected.

TABLE 5.1 Antiseptics and disinfectants

Agent	Use
Antiseptics	
Organic mercurials	Skin
Silver nitrate	Eyes of newborn to prevent gonorrhea
Iodine solution	Skin
Alcohol (70% ethanol in water)	Skin
Bis-phenols (hexachlorophene)	Soaps, lotions, body deodorants
Cationic detergents (quaternary ammonium compounds)	Soaps, lotions
Hydrogen Peroxide (3% solution)	Skin
Disinfectants	
Mercuric dichloride	Tables, bench tops, floors
Copper sulfate	Algicide in swimming pools, water supplies
Iodine solution	Medical instruments
Chlorine gas	Purification of water supplies
Chlorine compounds	Dairy, food industry equipment
Phenolic compounds	Surfaces
Cationic detergents (quaternary ammonium compounds)	Medical instruments; food, dairy equipment
Ethylene oxide	Temperature-sensitive laboratory materials such as plastics

5.5 ANTIBIOTICS AND OTHER CHEMOTHERAPEUTIC AGENTS

An important goal in medicine is the control of the growth of microorganisms which cause disease within the human body. In order to prevent harm to the host tissue, an antimicrobial agent with *selective toxicity* is required. A **chemotherapeutic agent** is a substance which is able to kill or inhibit the growth of microorganisms selectively within the tissues of a host. **Antibiotics** are chemical substances produced by certain microorganisms which are active against other microorganisms. Antibiotics constitute a special class of chemotherapeutic agents, distinguished most importantly by the fact that they are natural products (products of microbial activity) rather than synthetic chemicals (products of human activity). We discuss later in this chapter some of the microorganisms which produce antibiotics.

A very large number of antibiotics have been discovered, but probably less than 1 percent of them have been of great practical value in medicine. Those few which have been useful have had a dramatic impact on the treatment of many infectious diseases. In addition to antibiotics, a few useful chemotherapeutic agents have also been synthesized chemically. Further, some antibiotics can be made more effective by chemical modification; these are said to be semisynthetic.

The production by microorganisms of antibiotics, a process called *microbial antagonism*, is illustrated in Figure 5.9. Antibiotic-producing microorganisms are especially common in the soil. The medically useful antibiotics are produced by two microbial groups: (1) fungi, especially those of the genus *Penicillium*, which produce antibiotics such as penicillins and griseofulvin,

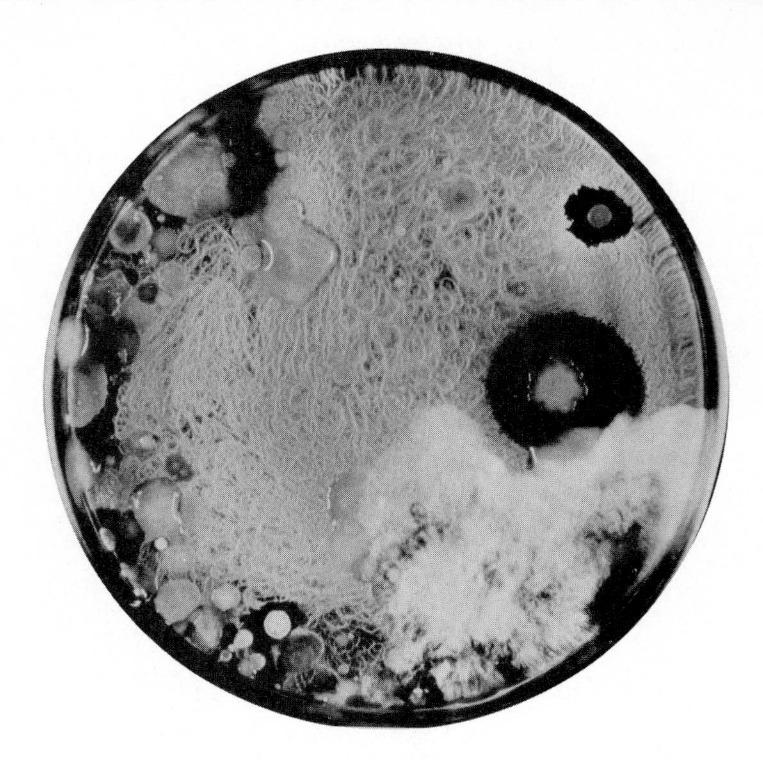

Figure 5.9 Crowded colonies of soil microorganisms, showing antagonism. The clear zones around several colonies are due to antibiotic action.

and the genus *Cephalosporium*, which produce cephalosporins; (2) bacteria, especially those of the actinomycete group, primarily of the genus *Streptomyces*, which produce antibiotics such as streptomycin, chloramphenicol, tetracycline, and erythromycin. It is among members of the genus *Streptomyces*, a genus of organisms widespread in soil, that most of the antibiotics have been discovered.

The sensitivity of microorganisms to antibiotics varies. Gram-positive bacteria are usually more sensitive to antibiotics than are Gram-negative bacteria, although conversely, some antibiotics act only on Gram-negative bacteria. An antibiotic that acts upon both Gram-positive and Gram-negative bacteria is called a **broad-spectrum antibiotic**. In general, a broad-spectrum antibiotic will find wider medical usage than a *narrow-spectrum antibiotic*, which acts on only a single group of organisms. A narrow-spectrum antibiotic may, however, be quite valuable for the control of certain kinds of diseases (Figure 5.10). Some antibiotics have an extremely limited spectrum of action, being toxic to only one or a few bacterial species. The physician needs and utilizes a wide variety of antibiotics and selects carefully the one needed for a particular patient with a particular kind of infection.

After an antibiotic is administered, it can interact in several different ways before it encounters and injures microorganisms. It is, of course, essential that the compound be able to move from the site of entry to the location of the microorganisms. The ability of the body to alter an antibiotic or prevent its distribution may influence how it should be administered. For example, some antibiotics are destroyed by stomach acid, and cannot be given orally unless protected. Other antibiotics cannot be absorbed through the intestinal wall

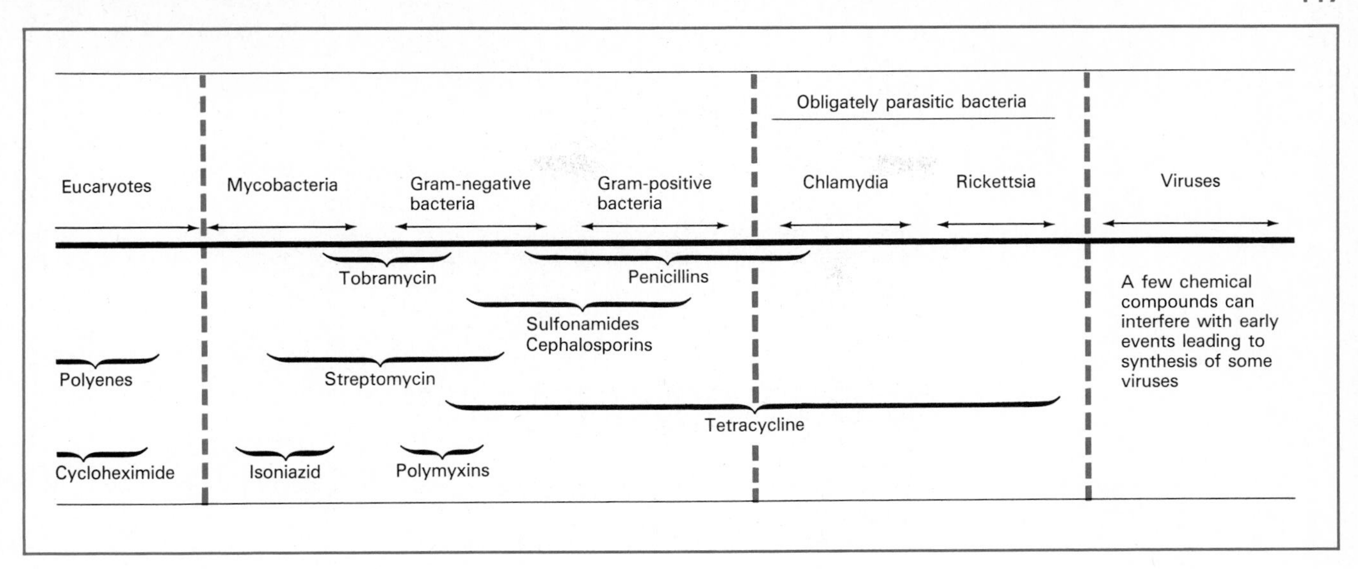

Figure 5.10 Ranges of action of antimicrobial agents.

into the bloodstream, so that they cannot be given orally at all. The exact amount of an antibiotic required to kill or inhibit a microorganism in laboratory culture can be determined readily, but the dose required to achieve the same effectiveness within the body may be different.

Although antibiotics are usually selective in their action, their toxicity for animals and humans varies, and a knowledge of antibiotic toxicity is essential for the wise use of antibiotics in medicine. Some antibiotics are so toxic that they can never be used in humans, others have a limited toxicity and may be used with care, while others are essentially nontoxic. This is in part related to the degree of difference between the cells of the invading microorganism and the cells of the host. Antibiotics which act against specific components of bacteria (for example the unique cell wall), are usually not very toxic for eucaryotic host cells. On the other hand, it is more difficult to selectively injure protozoan or fungal cells in the human body because both the infecting microbial cells and host cells are eucaryotic. A variety of toxic reactions can be observed within the host, including intestinal disturbances, kidney damage, and deafness.

However, even antibiotics which are nontoxic to humans may elicit allergic responses in some people. For example, 5 to 10 percent of the human population is allergic to penicillin and must not be treated with this antibiotic. Thus, a wide variety of factors determine the usefulness of an antibiotic. The ratio between the minimum toxic dose for the host and the minimum effective dose (to kill the microorganism) is called the **therapeutic index**. Antibiotics which are particularly useful generally have a high therapeutic index; that is, they are far more toxic to the microbe than to the host. A good example of an antibiotic with a high therapeutic index is penicillin. Other antibiotics are less useful because the doses required to harm the microbe also harm the host; they have a low therapeutic index. Such antibiotics may still have a useful role, however, if they are the only agents which act against a particular pathogen.

5.6 KINDS OF CHEMOTHERAPEUTIC AGENTS AND THEIR ACTIONS

Chemotherapeutic agents can be grouped based on chemical structure or on mode of action. In bacteria, the important targets of chemotherapeutic action are the cell wall, the cell membrane, and the biosynthetic processes of protein and nucleic acid synthesis. Some chemotherapeutic agents work because they mimic important growth factors needed in cell metabolism. Examples of each type are given in Table 5.2.

Cell-wall inhibitors Some of the most important chemotherapeutic agents affect cell wall synthesis in procaryotes. The structure of the peptidoglycan of the bacterial cell wall was shown in Figure 3.6. We note that the rigid layer of the bacterial cell wall is composed of polysaccharide chains that are cross-linked with peptides, in a configuration called *peptidoglycan*. The penicillin antibiotics inhibit the cross-linking reaction and hence prevent the formation of a complete cell wall. The wall thus becomes weak and the cell often bursts, a process called *lysis*. Human cells lack peptidoglycan and hence are unaffected by normal concentrations of these antibiotics, which are therefore quite nontoxic. Because they affect cell-wall synthesis, such antibiotics are bactericidal only to growing cells, since as bacteria grow they must continue to synthesize new cell wall in order to maintain cellular integrity. Nongrowing cells are not seriously affected, since new cell-wall material is not being synthesized. Because of this, nongrowing cells may remain alive for long periods in the presence of the antibiotic.

TABLE 5.2 Chemotherapeutic agents and their modes of action

Target in cell	Type of agent	Examples
Bacterial cell wall	Penicillins	Penicillin G, ampicillin, methicillin
	Cephalosporins	Cephalothin, cephamycin
Cell membrane		
Procaryotes	Polymyxins	Polymyxin B
Eucaryotes	Polyenes	Nystatin, amphotericin
Protein synthesis		
Procaryotes	Nitroaromatic	Chloramphenicol
	Aminoglycosides	Streptomycin, tobramycin
	Tetracyclines	Tetracycline, chlortetracycline
	Macrolides	Erythromycin
	Lincomycins	Lincomycin, clindamycin
Eucaryotes	Glutarimide	Cycloheximide
Nucleic acid synthesis	Rifamycins	Rifampin
Growth factor analog	Sulfonamides	Sulfanilamide
Unknown	Pyridine	Isoniazid
	Benzofuran	Griseofulvin

The two major types of cell-wall-inhibiting antibiotics are the **penicillins** and the **cephalosporins**. Penicillin was the first antibiotic to find widespread medical use and today is still one of the most useful antibiotics. Penicillin is not a single compound but a whole family of antibiotics, all related chemically but differing in detail and varying considerably in the range of organisms attacked. Most penicillins primarily affect Gram-positive organisms; others are active against both Gram-positive and Gram-negative bacteria, while a few are active only against Gram-negative bacteria. Some penicillins are destroyed by stomach acids and can only be given by injection, whereas others are acid stable and can be taken orally. The penicillin first discovered, now called penicillin G, is produced by the fungus *Penicillium chrysogenum*, and it was from the name of this genus that the antibiotic name was derived.

The core of the penicillin molecule is a substance called 6-*aminopenicillanic acid* (APA), and attached to this core are the side chains that confer specific activities. When APA is modified chemically to add different side chains, semisynthetic penicillins, such as methicillin and ampicillin, can be produced. Such new compounds may have different properties, such as altered antibacterial spectrum, resistance to acid, or longer shelf life. The most commonly used penicillins include penicillin G, ampicillin, and methicillin. Penicillins may cause allergic reactions in some individuals.

Members of the fungal genus *Cephalosporium* produce a family of antibiotics called *cephalosporins*, which act in a manner similar to that of the penicillins. Cephalosporins are particularly useful against bacteria which have become resistant to penicillin (see below), or in cases where a patient is allergic to penicillin. All cephalosporins are derivatives of a common chemical core which is similar, though not identical, to the penicillin core structure. Many cephalosporins are available, each with different attributes based on differences in the side chains attached to the core. Cephamycins are structurally similar to the cephalosporins but are produced by some members of the bacterial genus *Streptomyces*.

Cell-membrane inhibitors All cells are surrounded by a membrane. As discussed in Chapter 3, the membranes of all cells have a similar structure. It is thus very difficult to find antibiotics which destroy microbial membranes selectively. Two types of antibiotics which act against membranes are known, **polymyxins** and **polyenes**.

Polymyxins are produced by a species of the bacterium *Bacillus*. They act like detergents to destroy membranes rich in a type of lipid found commonly in Gram-negative bacteria, and are thus especially effective against this type of microorganism. They are unfortunately somewhat toxic to humans, however, and hence are only used medically after less toxic antibiotics have proved unsuccessful. The most common use for polymyxin is in the control of *Pseudomonas* infections, since this Gram-negative bacterium is often unaffected by other antibiotics.

Polyenes, a large group of antibiotics produced by *Streptomyces*, are active against fungi. *Nystatin* was one of the first polyenes to be discovered, and it has found some use in the control of fungal infections of the intestine. It is too toxic to be given by injection, but when given orally it remains in the intestinal tract and hence does not cause undue toxicity. Other polyene antibiotics include *filipin, candicidin,* and *amphotericin B.* These antibiotics act against a specific type of lipid found only in membranes of eucaryotes and have no effect on bacteria.

Protein synthesis inhibitors The process of protein synthesis will be discussed in detail in Chapter 6. As mentioned in Chapter 3, both procaryotic and eucaryotic cells contain ribosomes, where protein synthesis occurs within the cell. Although there are many similarities in the mechanism of protein synthesis in procaryotic and eucaryotic cells, the ribosomes of procaryotes and eucaryotes are different, and this provides a basis for selective toxicity.

Several classes of antibiotics which inhibit either procaryotic or eucaryotic protein synthesis are known. *Chloramphenicol,* a broad-spectrum antibiotic produced by a *Streptomyces* species, was the first antibiotic also to be produced by chemical synthesis. It inhibits protein synthesis in both Gram-positive and Gram-negative procaryotes. Chloramphenicol does have some toxicity in humans and has caused anemia and death when used for prolonged periods at high doses. Its most common medical use today is in situations where other antibiotics have proved ineffective.

Aminoglycoside antibiotics also act against bacterial ribosomes, bringing about inhibition of protein synthesis. *Streptomycin,* the first aminoglycoside, was discovered in the mid-1940s and is named for *Streptomyces,* the genus name of the bacterium which produces it. Since then several other aminoglycoside antibiotics have been discovered or have been produced by chemical modification, including *neomycin, gentamicin, kanamycin, amikacin,* and *tobramycin.* Streptomycin is now used mainly in the treatment of tuberculosis, and is usually given in combination with another antituberculosis agent. When given orally streptomycin does not pass through the intestinal wall into the bloodstream, so that for use in tuberculosis therapy, it must be administered by injection. When given for prolonged periods, streptomycin may cause deafness, a general problem associated with the use of aminoglycoside antibiotics. The newer aminoglycosides are more useful against Gram-negative bacteria. Tobramycin is especially useful against *Pseudomonas aeruginosa* which is resistant to the other aminoglycosides.

The **tetracyclines** are another family of antibiotics produced by *Streptomyces* which act by preventing bacterial protein synthesis. They include the parent antibiotic *tetracycline* and two derivatives, *chlortetracycline* and *oxytetracycline.* The tetracyclines are much less toxic to humans than are the aminoglycosides. In addition, they are absorbed into the bloodstream when given orally and hence are easy to administer. The tetracyclines are broad-spectrum antibiotics and are useful against a wide variety of Gram-positive and Gram-negative bacteria, although in contrast to streptomycin, they are not useful in the treatment of tuberculosis. Despite their lack of human toxicity, the tetracyclines must still be administered cautiously, since when given orally they alter the normal bacteria of the intestine and may cause intestinal disturbances.

Two other families of antibiotics produced by *Streptomyces* that act against the ribosomes of bacteria to inhibit protein synthesis are the **macrolides** and the **lincomycins**. The best-known and most widely used macrolide, *erythromycin*, is active against most Gram-positive organisms and is relatively nontoxic to humans. It is most often used as a substitute for penicillin in patients who are allergic to the latter drug and is also especially useful in treatment of whooping cough, legionellosis, and mycoplasmal pneumonia. *Lincomycin* and its semisynthetic chemical derivative *clindamycin* are also relatively nontoxic and are useful against some of the Gram-positive cocci as well as the Gram-negative anaerobe, *Bacteroides*.

Cycloheximide is also produced by a *Streptomyces* species. It inhibits protein synthesis in eucaryotes but has no effect in procaryotes; hence it is of no use in treating bacterial infections. Although effective against fungi and therefore of potential use for fungal infections, it is quite toxic to humans and has only been used medically in rare emergencies. It has found some use in agriculture in the control of fungus diseases and has been marketed especially for control of cherry tree diseases and turf diseases on golf courses.

Nucleic acid synthesis inhibitors A few antibiotics work by inhibiting the synthesis of nucleic acids. An example is *rifampin*, which blocks the production of RNA in bacteria. Since the enzyme that brings about RNA synthesis in eucaryotic cells is different from the parallel enzyme in bacteria, rifampin is a selective inhibitor and is relatively nontoxic. It is produced by chemical alteration of a natural antibiotic produced by a streptomycete bacterium, thus it is a semisynthetic antibiotic. Although it is a broad-range inhibitor of many bacteria, rifampin is rarely used because bacteria rapidly develop resistance to it.

Growth factor analogs There are several compounds which have antimicrobial properties because their structure resemble growth factors or metabolic intermediates. The organism is inhibited because the analog is used in place of the required growth factor, but is unable to replace the growth factor in function.

The **sulfa drugs**, or **sulfonamides**, a group of synthetic chemicals, are quite useful in chemotherapy and are often used instead of or in conjunction with antibiotics. The sulfa drugs, among the first chemotherapeutic agents discovered, contain a basic sulfonamide structure plus chemical side chains. The sulfonamides are analogs of the growth factor ***p*-aminobenzoic acid** (PABA), a necessary part of the vitamin **folic acid** (Figure 5.11). Many bacteria synthesize their folic acid from PABA and are inhibited by the sulfa drugs. However, if an organism does not synthesize folic acid from PABA but requires folic acid preformed in the environment, then its growth is not inhibited by sulfa drugs. Humans require preformed folic acid and are not affected by the sulfa drugs; hence these drugs can be used selectively in the control of human infections. Various sulfa drugs are used to control different Gram-positive or Gram-negative bacteria. They continue to find widespread medical use even since the discovery of antibiotics, and a wide variety of sulfa drugs is currently marketed.

Isoniazid (INH), is one of the most effective drugs for the treatment of tuberculosis. It is an excellent example of a limited-spectrum chemothera-

H_2N–C₆H₄–SO_2NH_2

(a) Sulfanilamide

H_2N–C₆H₄–COOH

(b) p-Aminobenzoic acid

(c) Folic acid

Figure 5.11 (a) The simplest sulfa drug, sulfanilamide. It is an analog of *p*-aminobenzoic acid (b), which itself is part of the growth factor folic acid (c).

peutic agent, as it is quite specific in its action against the causal agent of tuberculosis, *Mycobacterium tuberculosis*, and ineffective against other microorganisms. The drug is given orally, usually for extended periods of time. It is the usual practice to administer INH together with another antituberculosis drug, such as the antibiotics streptomycin or rifampin, or the synthetic chemical *p*-aminosalicylic acid.

Antibiotics of unknown mode of action Some antibiotics have been found to be effective in chemotherapy, but so far the exact way in which the antibiotic works is not known. *Griseofulvin* is one example of such an antibiotic produced by the fungus, *Penicillium griseofulvum*. It is active against other fungi and is used medically for the control of fungal infections of the skin. In contrast to most other antibiotics that inhibit fungi, it is not very toxic to humans. However, griseofulvin is not effective when applied directly to the skin but must be given orally. The drug then passes through the bloodstream and accumulates in the skin tissues. To cure a fungal infection of the skin, the drug must be taken over long periods of time at fairly high doses, a procedure possible only because of its low toxicity.

5.7 MEASURING ANTIMICROBIAL ACTIVITY

Antimicrobial activity is measured by determining the smallest amount of agent needed to inhibit the growth of a test organism, a value called the **minimum inhibitory concentration** (MIC). In one method of determining the MIC, a series of culture tubes is prepared, each tube containing medium with a different concentration of the agent, and all tubes of the series are inoculated. After incubation, the tubes in which growth does not occur (indicated by absence of visible turbidity) are noted, and the MIC is thus determined (Figure 5.12). This simple and effective procedure is often called the *tube dilution technique*. The MIC is not an absolute constant for a given agent, since it is affected by the kind of test organism used, the inoculum size, the composition of the culture medium, the incubation time, and the conditions of incubation, such as temperature, pH, and aeration. If all conditions are rigorously standardized, it is possible to compare different antibiotics and determine which is most effective against a given organism or to assess the activity of a single agent against a variety of organisms. Note that the tube

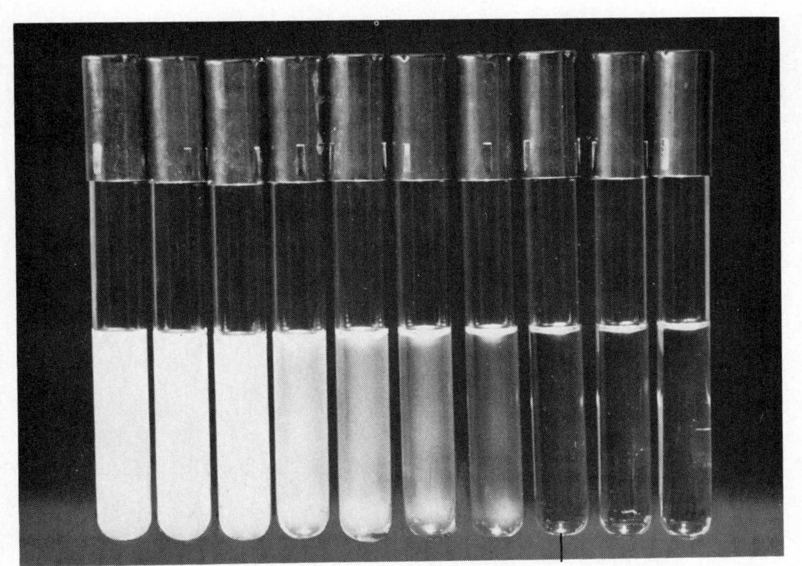
Minimum inhibitory concentration

Figure 5.12 Antibiotic assay by tube dilution, permitting detection of the *minimum inhibitory concentration, MIC.* A series of increasing concentrations of antibiotic is prepared in the culture medium. Each tube is inoculated and incubation allowed to proceed. Growth occurs in those tubes with antibiotic concentration below the MIC.

dilution method does not distinguish between a cidal and a static agent, since the agent is present in the culture medium throughout the entire incubation period.

Another commonly used procedure for studying antimicrobial action is the **agar diffusion method** (Figure 5.13). A petri plate containing an agar medium evenly inoculated with the test organism is prepared. The test plate may be inoculated by pouring an overlay of agar containing the test organism, or by swabbing the medium surface with a broth culture of the test organism. Known amounts of the antimicrobial agent are added to filter-paper discs which are then placed on the surface of the agar. During incubation, the agent diffuses from the filter paper into the agar; the further it gets from the filter paper, the smaller the concentration of the agent. At some distance from the disc the MIC is reached. Past this point growth occurs, but closer to the disc growth is absent. A **zone of inhibition** is thus created, and its size can be measured with a ruler; the diameter will be proportional to the amount of antimicrobial agent added to the disc.

The size of the zone is affected by the sensitivity of the test organism, the culture medium and incubation conditions, the rate of diffusion of the agent, and the concentration of agent in the filter-paper disc. The interpretation of the significance of wide and narrow inhibition zones is therefore not at all simple. An agent that produces a wide zone is not necessarily more active than an agent that produces a narrow zone, since the diffusion rate in agar of different agents varies widely. For comparing two agents, the tube dilution method is preferable. The agar diffusion method is simpler to set up, however, and is widely used for the assay of antibiotics and in clinical medicine.

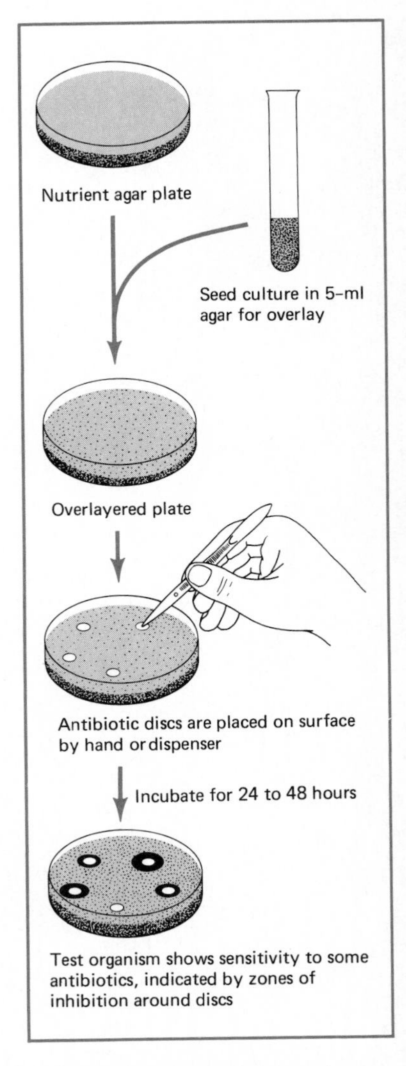
Nutrient agar plate
Seed culture in 5-ml agar for overlay
Overlayered plate
Antibiotic discs are placed on surface by hand or dispenser
Incubate for 24 to 48 hours
Test organism shows sensitivity to some antibiotics, indicated by zones of inhibition around discs

Figure 5.13 Agar diffusion method for assaying antibiotic activity.

Testing cultures for antibiotic sensitivity In medical practice, the sensitivity to antimicrobial agents of cultures isolated from patients is often needed

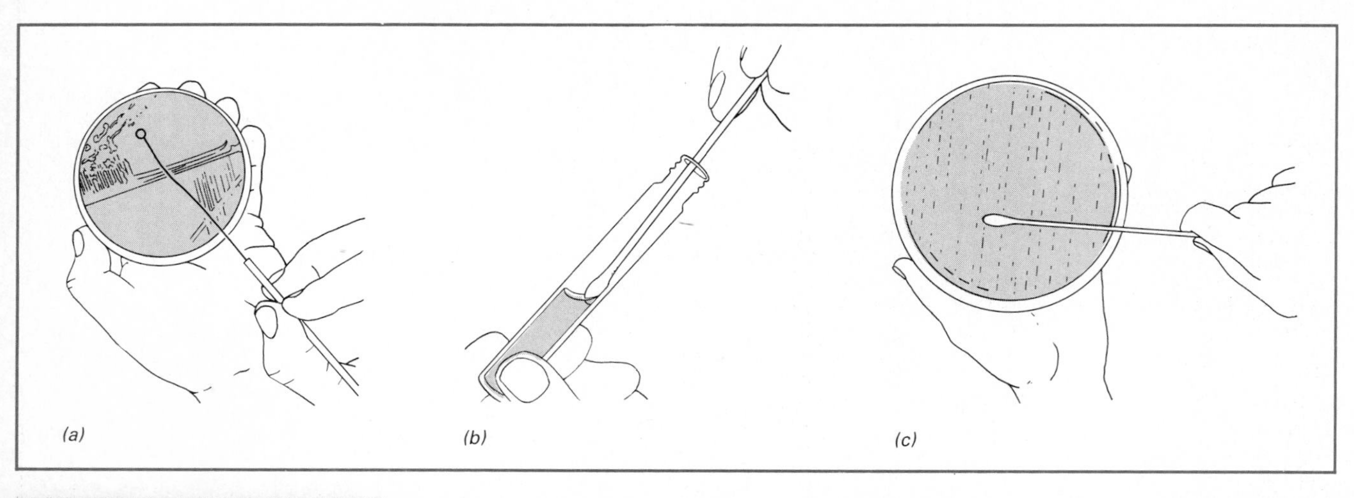

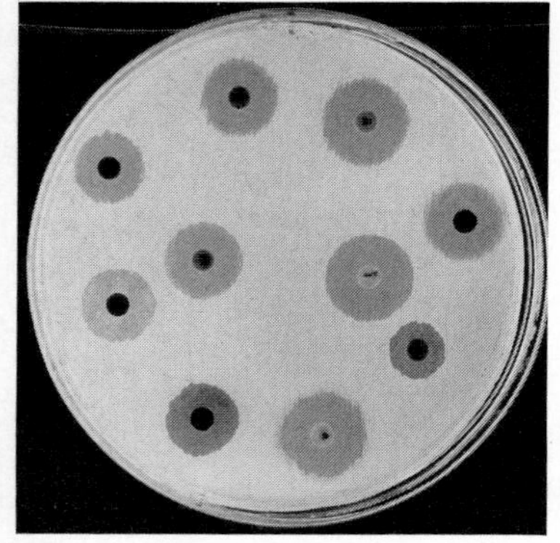

Figure 5.14 The Kirby-Bauer procedure for determining the sensitivity of an organism to antibiotics. (a) A colony is picked from an agar plate. It is inoculated into a tube of liquid culture medium and allowed to grow to a specified density. (b) A swab is dipped in the liquid culture. (c) The swab is streaked evenly over a plate of sterile agar medium. (d) Discs containing known amounts of different antibiotics are placed on the plate. After incubation, inhibition zones are observed. The susceptibility of the organism is determined by reference to a chart of zone-sizes (see Table 5.3).

to help determine the treatment of the patient. The sensitivity of a culture can be most easily determined by an agar diffusion method. Federal regulations of the Food and Drug Administration (FDA) now control the procedures used for sensitivity testing in the United States. The recommended procedure is called the *Kirby-Bauer method,* after the workers who developed it (Figure 5.14). A plate of suitable culture medium is inoculated by spreading an aliquot of culture evenly across the agar surface. Filter-paper discs containing known concentrations of different agents are then placed on the plate. The concentration of each agent on the disc is specified so that zone diameters of appropriate size will develop to indicate sensitivity or resistance. After incubation, the presence of inhibition zones around the discs of the different agents is noted. Table 5.3 presents typical zone sizes for several antibiotics. Zones observed on the plate are measured and compared to standard data, to determine if the isolate is truly sensitive to a given antibiotic.

5.8 ANTIBIOTIC RESISTANCE

It is obvious from the discussion above that not all antibiotics act against all microorganisms. Some microorganisms are resistant to some antibiotics. Antibiotic resistance can be an inherent property of a microorganism, or it can be acquired. There are several reasons why microorganisms may have inherent resistance to an antibiotic: 1) The organism may lack the structure which an antibiotic inhibits. For instance, some bacteria, such as mycoplasmas, lack a typical bacterial cell wall and are resistant to penicillins. 2) The organism may be impermeable to the antibiotic. 3) The organism may be able to change the

TABLE 5.3 Typical zone sizes for some common antibiotics

Antibiotic	Amount on disc	Inhibition zone diameter (mm)		
		Resistant	Intermediate	Sensitive
Ampicillin[a]	10 μg	11 or less	12–13	14 or more
Ampicillin[b]	10 μg	28 or less	—	29 or more
Cephoxitin	30 μg	14 or less	15–17	18 or more
Cephalothin	30 μg	14 or less	15–17	18 or more
Chloramphenicol	30 μg	12 or less	13–17	18 or more
Clindamycin	2 μg	14 or less	15–16	17 or more
Erythromycin	15 μg	13 or less	14–17	18 or more
Gentamicin	10 μg	12 or less	13–14	15 or more
Kanamycin	30 μg	13 or less	14–17	18 or more
Methicillin[c]	5 μg	9 or less	10–13	14 or more
Neomycin	30 μg	12 or less	13–16	17 or more
Nitrofurantoin	300 μg	14 or less	15–16	17 or more
Penicillin G[c]	10 Units	28 or less	—	29 or more
Penicillin G[d]	10 Units	11 or less	12–21	22 or more
Polymyxin B	300 Units	8 or less	9–11	12 or more
Streptomycin	10 μg	11 or less	12–14	15 or more
Tetracycline	30 μg	14 or less	15–18	19 or more
Trimethoprim-sulfamethoxazole	1.25/23.75 μg	10 or less	11–15	16 or more
Tobramycin	10 μg	12 or less	13–14	15 or more

[a]For Gram-negative organisms and enterococci.
[b]For staphylococci and highly penicillin-sensitive organisms.
[c]For staphylococci.
[d]For organisms other than staphylococci. Includes some organisms, such as enterococci and some Gram-negative bacilli, that may cause systemic infections treatable by high doses of Penicillin G.

antibiotic to an inactive form. This is very frequently the case in penicillin resistance. A group of enzymes called *penicillinases* or *β-lactamases* destroy the β-lactam ring of penicillin, thus rendering it inactive. Penicillinases are produced by a wide variety of bacteria but some penicillins are less susceptible to penicillinase action. A major objective in the creation of semisynthetic penicillins is to decrease the sensitivity to penicillinase action.

The ability of microorganisms to develop resistance to antibiotics has been a major problem in medicine, as well as a major impetus to continue the search for new and better antibiotics.

Acquired antibiotic resistance involves a change in the genetic information within a cell. Such genetic change can occur by at least two different mechanisms. One mechanism involves a small change, or *mutation*, of the DNA of the organism, causing in turn a change in the structure of the target molecule so that the antibiotic can no longer bind or carry out its function. Figure 5.15 shows mutants with increased resistance to an antibiotic growing within the zone of inhibition on a test plate. The mutants are able to grow at a higher concentration of the antibiotic so that a higher minimum inhibitory concentration would be required to control their growth. Development of resistance to one antibiotic in this way generally does not result in concomitant resistance to another antibiotic unless the two antibiotics are closely related chemically.

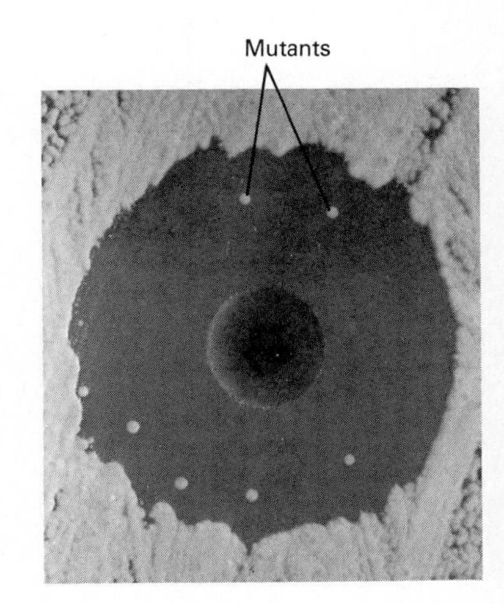

Figure 5.15 Development of antibiotic resistant mutants within the zone of inhibition. Close-up view of an inhibition zone.

The other major way in which genetic change can occur to confer acquired antibiotic resistance upon an organism is by the transfer of DNA from an antibiotic-resistant microbe to one which is sensitive. If, for instance, the DNA encoding the information needed to synthesize the enzyme penicillinase were acquired, the recipient microorganism could suddenly develop resistance to penicillin. Such has been the case for the bacterium which causes gonorrhea. In fact, it is relatively common to observe that a microorganism suddenly acquires resistance to several antibiotics at once by acquiring DNA from a resistant donor cell. How such large amounts of DNA can move between microorganisms will be discussed in Chapter 7. This mechanism of acquiring resistance to antibiotics is especially important since a microorganism can acquire resistance to many antibiotics at the same time. Careless and inappropriate use of antibiotics may result in rapid emergence of multiply resistant microorganisms.

Antibiotic resistance and antibiotic use There is a growing concern that the uncontrolled use of antibiotics is leading to the rapid development of antibiotic resistance in disease-causing microorganisms. Parallel to the history of discovery and clinical use of the many known antibiotics has been the emergence of bacteria which resist their action. This is, in fact, a major reason why we continually seek new antibiotics (see below) and attempt to modify existing ones through chemical alterations. There are numerous examples of the association between the use of antibiotics and the development of resistance. The example in Figure 5.16*a* shows a correlation between the number of tons of antibiotics used and the percentage of bacteria resistant to each antibiotic which were isolated from patients with diarrheal disease. In general, the more an antibiotic was used, the more bacteria were resistant to it.

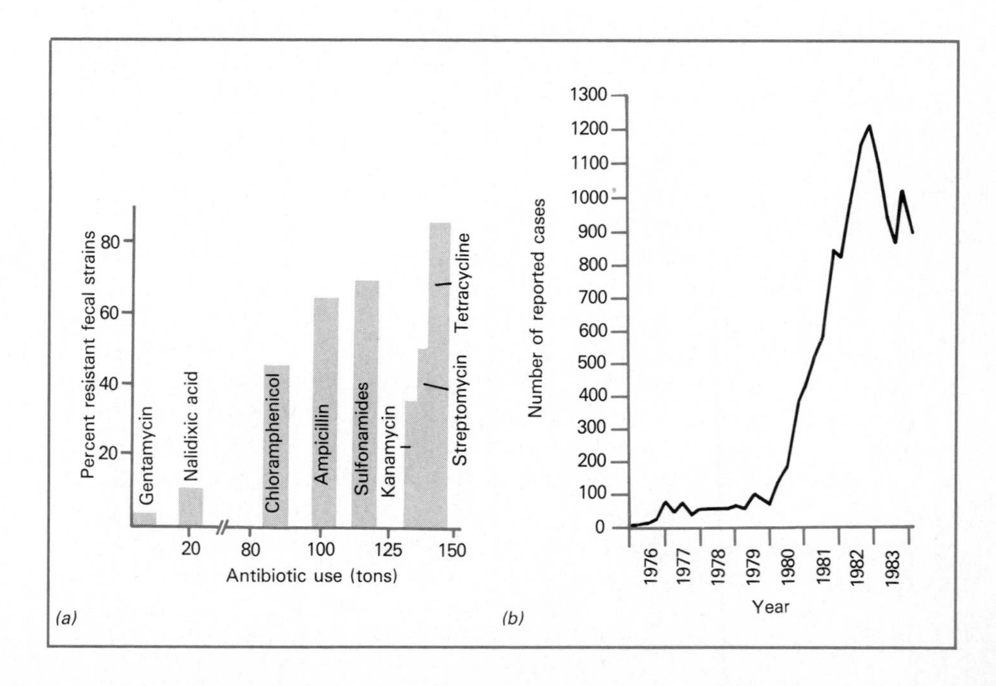

Figure 5.16 The emergence of antibiotic-resistant bacteria. (a) Relationship between antibiotic use and the percentage of bacteria isolated from diarrheal patients resistant to the antibiotic. Note that those antibiotics which have been used in the largest amounts, as indicated by the amount of antibiotic produced commercially, are those for which antibiotic-resistant strains are most frequent. (b) Increase in the incidence of bacteria causing gonorrhea which are resistant to penicillin.

There are many examples of diseases in which the drug recommended for treatment has changed due to the increased resistance of the microbe causing the disease. A classic example is the development of resistance to penicillin in *Neisseria gonorrhoeae*, the bacterium which causes gonorrhea (Figure 5.16*b*). The explanation for these observations is probably an ecological one. To grow in the presence of an antibiotic an organism must develop resistance to it. Thus, resistant microorganisms are selected for by the presence of the antibiotic in the environment. Surveys have shown that antibiotics are used far more often than is necessary. For instance, during the 1960s antibiotic production in the United States more than tripled although the population grew only 11 percent. The increased use of antibiotics is in part a result of poor prescription practices by physicians, who prescribe antibiotics when unnecessary. Antibiotics are also used extensively in agricultural practices both as growth-promoting substances in animal feeds and as prophylactics (to prevent the occurrence of disease rather than to treat an existing one). Several recent food poisoning outbreaks have been blamed on the use of antibiotics in animal feeds. By overloading the various environments with antibiotics, we create more stress for microorganisms in general. Rapid development of drug resistance is the result. The dairy plant disaster discussed earlier in this chapter (see the Box on page 110) was due to a bacterium which had multiple antibiotic resistance.

5.9 THE SEARCH FOR NEW ANTIBIOTICS

Antibiotics were originally discovered by chance through studies on antagonism between various microorganisms, but during the past 40 years, most new antibiotics have been discovered through intense and careful search. Seeking new antibiotics has occupied the time of a large number of microbiologists in pharmaceutical companies. Many pharmaceutical firms have invested millions of dollars in the search for new antibiotics and have developed a variety of techniques for such searches. New antibiotics are still needed, in part because resistance to existing antibiotics develops and in part because a number of infectious diseases are still not effectively controlled by those antibiotics currently available.

The most widely used procedure for searching for a new antibiotic is called the *screening approach*: A large number of possible antibiotic-producing organisms are isolated from natural environments, and each is tested against a variety of other organisms to see if antagonisms exist. The battery of test organisms is usually large and includes Gram-positive and Gram-negative bacteria, *Mycobacterium* (the causal agent of tuberculosis), yeasts and other fungi, and perhaps protozoa.

A convenient procedure for screening is to streak a possible antibiotic producer along one side of an agar plate and incubate the plate for a few days to allow the organism to grow and produce antibiotic. Then test bacteria are streaked at right angles to the first streak, and the plate is reincubated and examined for zones of inhibition (Figure 5.17). Since different antibiotics act upon different groups of organisms, it is desirable that any antibiotic

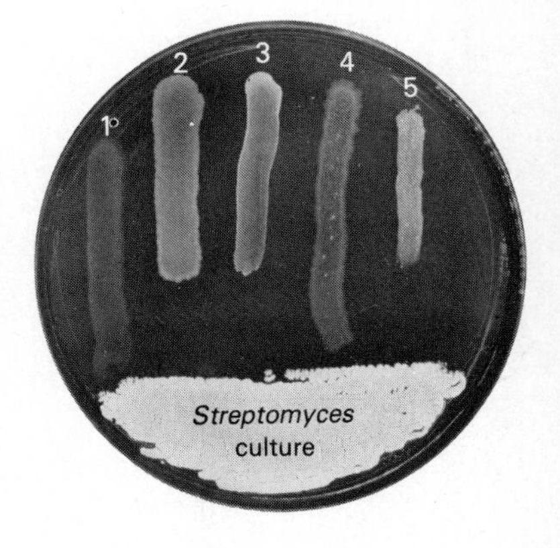

Figure 5.17 Method of testing an organism for the production of antibiotics. The producer (*Streptomyces*) was streaked across one-third of the plate and the plate was incubated. After good growth had been obtained, the test bacteria were streaked perpendicular to the *Streptomyces*, and the plate was further incubated. The failure of organisms 2, 3, and 5 to grow near the *Streptomyces* indicates that the *Streptomyces* produced an antibiotic active against these bacteria. Test organisms: 1) *Escherichia coli*; 2) *Bacillus subtilis*; 3) *Staphylococcus aureus*; 4) *Klebsiella pneumoniae*; 5) *Mycobacterium smegmatis*.

screening program use a wide range of test organisms to make certain that few antibiotic-producing organisms are missed. Since environment can greatly affect antibiotic production, obviously there must be appropriate medium and culture conditions permitting such production.

Once antibiotic activity is detected in a new isolate, the microbiologist will determine whether the agent is new or identical with an existing one. With so many antibiotics already known, the chances are good that the antibiotic will not be new. Often simple chemical methods will permit the characterization and identification of the antibiotic. If the agent appears to be new, larger amounts are prepared. For further testing, the antibiotic must be purified and preferably crystallized. Finally, it is tested for therapeutic activity—first in infected animals, later perhaps in humans.

Most screening programs concentrate on antibiotics produced by members of the genus *Streptomyces*, a group of organisms that has yielded most of the medically useful antibiotics now known, including chloramphenicol, streptomycin, neomycin, erythromycin, cycloheximide, cefamycin, nystatin, tobramycin, lincomycin, and tetracycline. *Streptomyces* occur predominantly in the soil, and species producing antibiotics are found in soils throughout the world.

Organisms often produce more than one antibiotic and sometimes as many as five or six. Multiple antibiotic production presents special problems, as activity against a test organism may be due to the concerted action of the several agents. The different antibiotics may all be closely related chemicals, or they may differ markedly. They may all be active against a single test organism, or they may show different spectra of activity. The presence of multiple antibiotics presents further problems and challenges for the microbiologist and chemist: Which antibiotic in the mixture is likely to be medically useful? How can its production be favored over the others? And how can it be purified? Antibiotics research presents endless puzzles and fascination for the microbiologist. When a new antibiotic is discovered, the interesting work has only just begun.

SUMMARY

In this chapter we have examined microbial growth and its control. Unicellular microorganisms grow **exponentially**, and often at **generation times** on the order of minutes. Because of this, very large populations of cells can develop quickly. However, growth does not continue indefinitely and cell populations eventually enter a stationary phase, sometimes followed by a death phase. The stationary phase ensues when essential nutrients are exhausted, or when toxic waste products accumulate. Growth may be measured by **direct microscopic counting** of cells, by **viable counting** procedures which require growing the cells to be counted, or by estimating cell mass.

A number of processes or agents are used to control microbial growth. **Sterilization** is the killing of all microorganisms and heating is one of the most effective sterilizing methods. Microbial growth can be prevented by use of low temperatures or drying, but these treatments do not result in sterilization. Sterilization can also be brought about by ultraviolet radiation or ionizing radiation.

A wide variety of chemicals, called **antimicrobial agents**, can be used to control microbial growth. **Cidal** agents kill microorganisms, whereas **static** agents inhibit growth without killing. Antimicrobial agents vary in their **selective toxicity**. **Disinfectants** are not selective and are used to sterilize inanimate surfaces, whereas **antiseptics** are less toxic to humans than to microorganisms and may be applied to the skin. **Chemotherapeutic agents** are usually much more toxic to microorganisms than to humans, and have been used extensively in treating human diseases. **Antibiotics** are antimicrobial agents produced by microorganisms which act against other microorganisms. A number of antibiotics have been used to control infectious diseases. However, use of antibiotics has caused the development of **antibiotic resistance** in microorganisms, which complicates the control of infections and creates a need to search for new antibiotics.

KEY WORDS AND CONCEPTS

Growth rate
Generation time
Exponential growth
Direct microscopic count
Viable count
Sterilization
Thermal death time
Pasteurization
Ionizing radiation
Ultraviolet radiation
Antimicrobial agent
Selective toxicity
Bactericidal
Bacteriostatic
Disinfectant
Antiseptic
Chemotherapeutic agent
Antibiotic
Minimum inhibitory concentration
Antibiotic resistance

STUDY QUESTIONS

1. The bacteria with the fastest *growth rates* double in about ten minutes. Calculate the time required for one such bacterial cell to produce a population of over one million cells. How many generations are required?
2. Draw a graph which shows the growth of a population of microbial cells after being inoculated into fresh medium. Label each phase of the *growth cycle.*
3. What is the practical significance of the fact that unicellular microorganisms grow *exponentially*?
4. Microbial growth may be measured by either *direct count* or *viable count*. Which method is simplest? Which method is more sensitive?
5. Express the following numbers as exponents of 10: 1,000; 100,000; 1,000,000,000; 5,000,000.

6. Name three different methods used to *sterilize* materials.
7. What is *thermal death time*? How would the presence of bacterial endospores affect the thermal death time?
8. Which preservation methods are used for the following products: canned garden vegetables, jams, milk, cereals?
9. What is the difference between a *bactericidal* and a *bacteriostatic* agent?
10. Which type of *antimicrobial agent* exhibits the most *selective toxicity*? Which exhibits the least *selective toxicity*?
11. *Antibiotics* vary in the way they affect cells. Name at least four different *modes of action* of antibiotics. For each, give an example of an antibiotic which works in this way to inhibit microorganisms.
12. Name an antibiotic which inhibits bacteria but not eucaryotes, and tell why it does not inhibit eucaryotic microorganisms.
13. Why is it more difficult to find antibiotics which are effective against protozoa and fungi, than to find antibacterial antibiotics?
14. What factors determine the *therapeutic index* of an antibiotic?
15. Briefly describe two different ways to determine the *minimum inhibitory concentration* of an antibiotic. Explain why this is not necessarily the same concentration needed to inhibit microorganisms within the body.
16. Why does *antibiotic resistance* develop? How does it develop (two ways)?
17. Describe how you would go about searching for a new antibiotic.
18. Name three types of microorganisms which are especially important antibiotic producers.

SUGGESTED READINGS

ABRAHAM, E.P. 1981. *The β-lactam antibiotics*. Scientific American 244: 76–86.

AMERICAN PUBLIC HEALTH ASSOCIATION. 1985. *Standard methods for the examination of water and wastewater, 16th edition*. American Public Health Association, Washington, D.C. (A useful reference providing methods commonly used in measuring the numbers of microorganisms present in water samples.)

BROCK, T.D., D.W. SMITH, and M.T. MADIGAN. 1984. *Biology of microorganisms, 4th edition*. Prentice-Hall, Inc., Englewood Cliffs, N.J.

GRAYSON, M. (EDITOR). 1982. *Antibiotics, chemotherapeutics and antibacterial agents for disease control*, John Wiley & Sons, Inc.

LENNETTE, E.H., A. BALOWS, W.J. HAUSLER, JR., and H.J. SHADOMY. 1985. *Manual of clinical microbiology, 4th edition*. American Society for Microbiology, Washington, D.C. (A good reference for information on measurement of antibiotic activity and sensitivity.)

PHILLIPS, J.A. and T.D. BROCK. 1984. *General microbiology: a laboratory manual*, Prentice-Hall, Inc., Englewood Cliffs, N.J.

6

Molecular Biology of Cells and Viruses

In Chapter 3 we discussed the structure of proteins and noted that the function of a protein is determined by its three-dimensional structure, which is itself determined by its **amino acid sequence**. The information that specifies the amino acid sequence of a protein resides in DNA, a polymer composed of four nucleic acid bases. The information in DNA is associated with the arrangement of these bases, the **nucleotide sequence**. In this chapter, we will examine the structure of DNA, to see how it is organized into **genes** in which the essential information for the synthesis of proteins is stored.

As mentioned in Chapter 3, it is the collection of proteins that determines the nature of an organism. Since DNA stores the information coding for a cell's complete complement of proteins, it is essential during growth and division, that a cell replicate its DNA, enabling its offspring to produce the same proteins. The process of **protein synthesis** is also of central importance to growth and reproduction. This process involves the interaction of the major cell **macromolecules**, **DNA**, **RNA**, and **protein**, as shown in Figure 6.1. The study of these molecules and processes is called **molecular biology**.

Although viruses are not cells, they are composed of similar macromolecules. Their reproduction within a host cell also involves replication and protein synthesis. After considering the molecular biology of cells, we will see how viruses use cellular processes to accomplish their own reproduction.

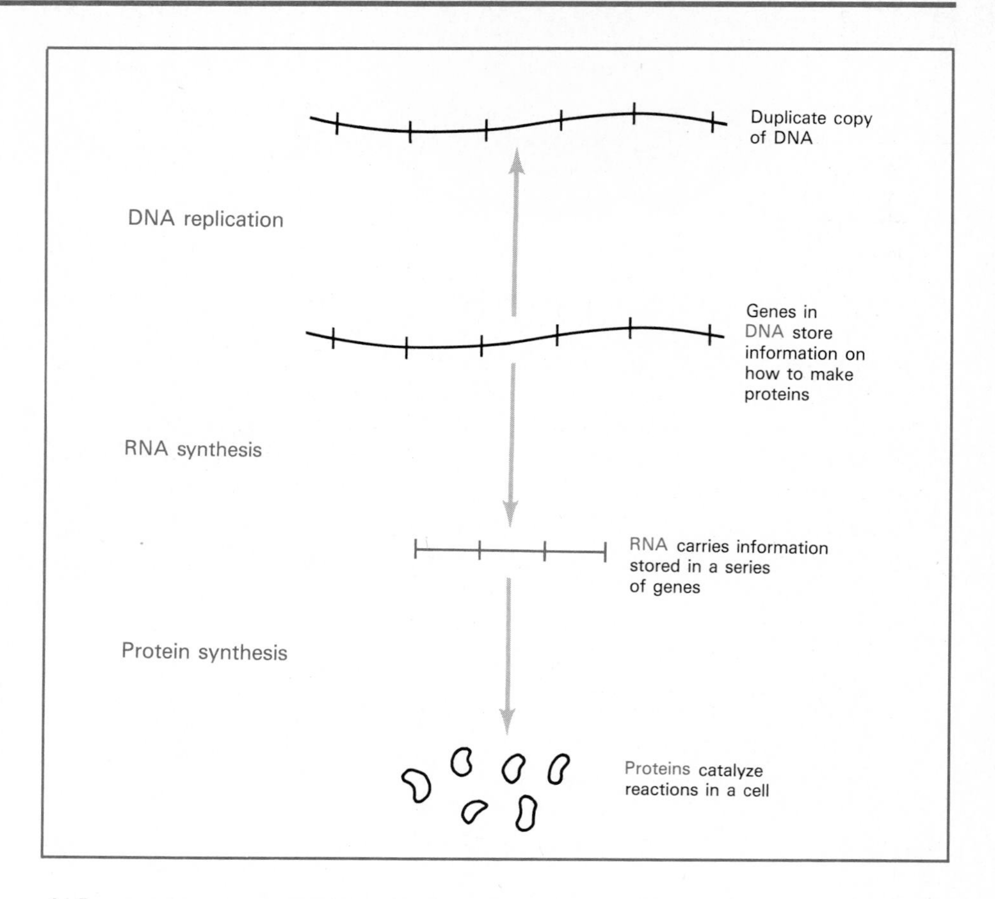

Figure 6.1 Central processes of information transfer.

6.1 THE MAJOR MACROMOLECULES: DNA, RNA, AND PROTEIN

In this section we discuss the chemistry of the major macromolecules of the cell. For each macromolecule, the way in which the monomer building blocks are connected in the polymer is presented.

Proteins As discussed in Chapter 3, proteins are polymers composed of a variety of amino acid monomers that are joined together by covalent bonds called **peptide bonds**. Since there are 20 different amino acids, a very large number of different proteins can be made by arranging the various amino acids in different sequences. The order of the amino acids which make up a specific protein is called the *primary sequence* (see Figure 3.9). The tendency of some amino acids to form cross-linking bonds with other amino acids within the polymer causes the chain of amino acids to fold. The protein chain folds into a three-dimensional structure which accounts for the specific function of the protein (see Figure 3.10). The twisted and folded three-dimensional protein structure is thus actually determined by the primary sequence of the amino acids in the protein.

DNA As we have noted, only four different nucleic acid bases are found in DNA: adenine (A), guanine (G), cytosine (C), and thymine (T). The

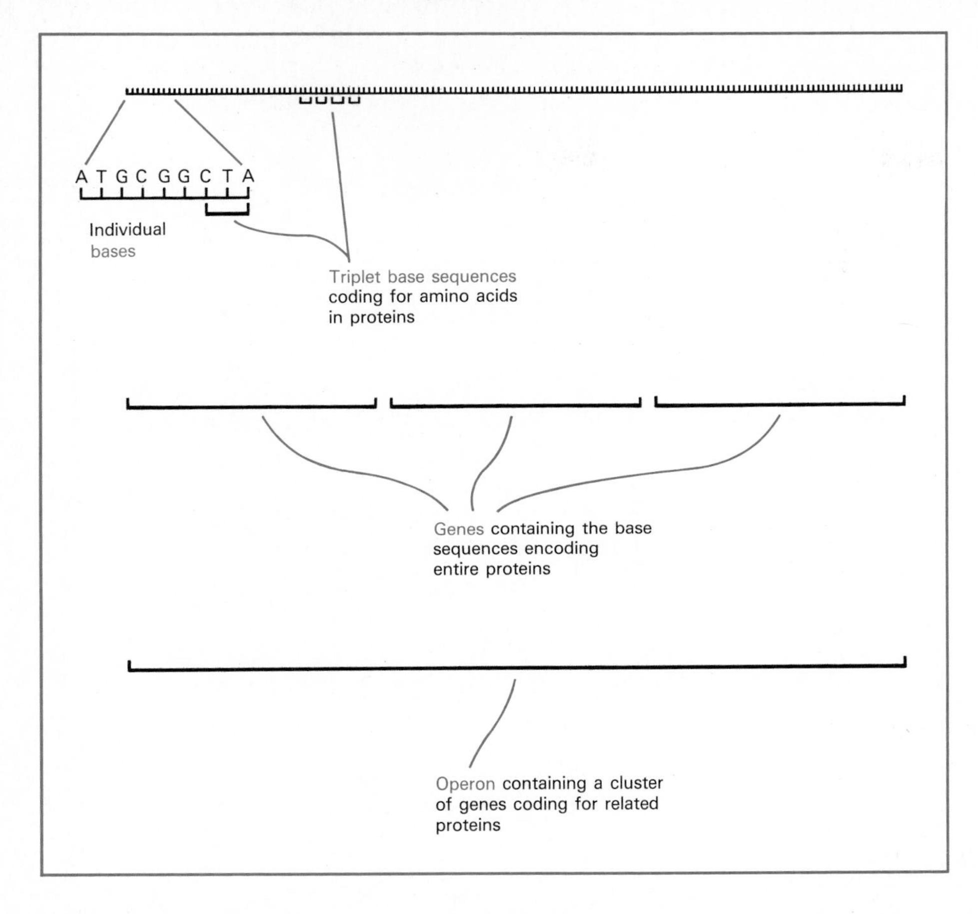

Figure 6.2 The organization of a functional DNA sequence.

information is stored in DNA in the sequence of bases along the polymer chain. A group of three bases in DNA codes for a specific amino acid and the code is thus called a **triplet code**. Since this code describes the way in which inherited traits are stored in the DNA of all cells, it is called the **genetic code**.

Information is arranged on the DNA molecule in such a way that a string of several hundred nucleotides contains the consecutive triplet codes corresponding to the individual amino acids making up the primary sequence of an entire protein as shown in Figure 6.2. A unit of DNA which is meaningful in the sense that an entire protein is encoded, is called a **gene**. In procaryotes, genes coding for related proteins are often organized into clusters, called **operons**. Proteins are not the only products encoded by genes, but they are the main products which carry out the many reactions occurring within a cell.

In reality, DNA does not exist as a single polymer strand, but as a *double-stranded* molecule. By understanding the structure of double-stranded DNA, we will be able to see how the information stored in DNA can be retrieved and used to synthesize the proteins a cell needs.

The chemistry of the four bases is such that weak bonds (called hydrogen bonds) can form between A and T and between C and G, as shown in Figure 6.3. These pairs are said to be *complementary* and to form **complementary**

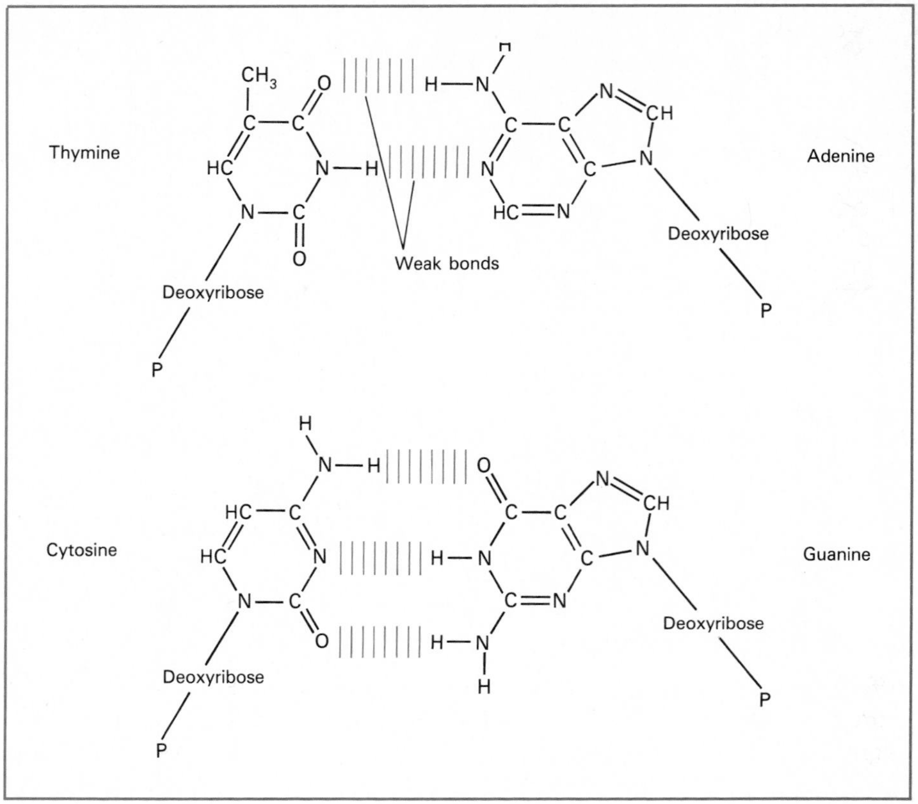

Figure 6.3 Specific pairing of the nucleic acid bases. Adenine (A) pairs with thymine (T) and guanine (G) pairs with cytosine (C).

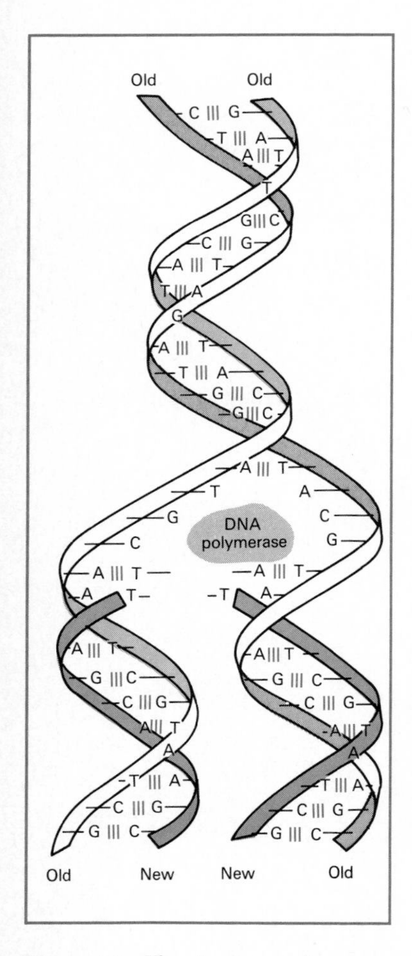

Figure 6.4 The structure and manner of replication of DNA. The enzyme DNA polymerase catalyzes the addition of new residues to the growing polynucleotide chain.

bonds. (No other pairs can form bonds.) Because of this complementary bonding, a single strand of DNA forms the pattern or **template**, against which a second strand of DNA can be produced. The entire second strand is thus complementary (a mirror image) to the first. The DNA double strand generally forms a coil, a configuration called a *double helix*. During cell division, the DNA double helix is duplicated and each daughter cell receives an exact copy of the complete double-stranded DNA molecule. The process by which this occurs is called **replication**. The production of an exact replica copy of the double-stranded DNA is accomplished by the splitting apart of the double strand, so that each strand can serve as a template against which the synthesis of new second strands can occur (Figure 6.4). Because of complementary bonding between the A and T and the C and G base pairs, the new strands are identical to the original opposing strands. The enzyme which catalyzes this process is called *DNA polymerase*.

It is possible to separate the two complementary strands of the double-stranded DNA by increasing the temperature to break down the weak bonds between complementary bases. If the mixture of single strands so formed is allowed to cool, complementary strands will reform, or **hybridize** (Figure 6.5), as their complementary base sequences match up and reform the weak hydrogen bonds. Single strands of DNA from different sources (e.g., two

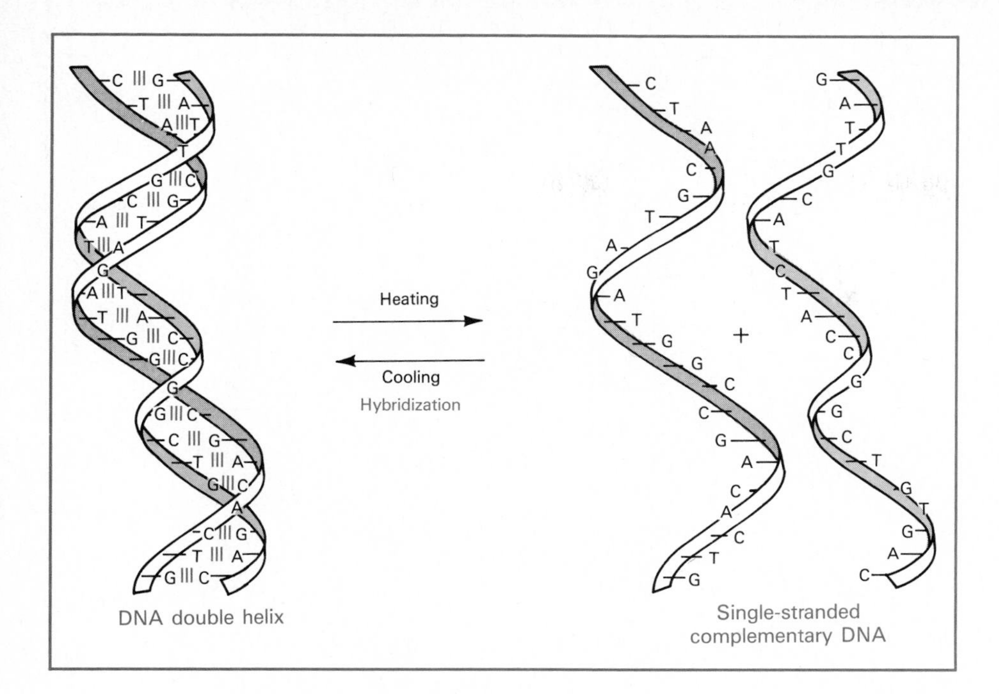

Figure 6.5 Separation of the two strands of the DNA double helix by heating and reformation upon cooling. The procedure shown can be used to create artificial hybrids and to search for complementary sequences in DNA strands.

different organisms) can be combined in such a hybridization experiment and if there is sufficient similarity in the base sequences, they will hybridize. The degree of hybridization is a measure of similarity of the two different DNAs.

RNA As illustrated in Figure 3.11, the chemistry of ribonucleic acid, RNA, is very similar to that of DNA. RNA differs from DNA in the type of sugar molecule and in the substitution in RNA of the base uracil (U) for the thymine (T) found in DNA. Like thymine, uracil forms weak bonds only with adenine (A).

RNA molecules are synthesized on a DNA template; their base sequences thus reflect the information encoded by the base sequences of the DNA strands from which they are made. The process by which the information from DNA is transferred into RNA is called **transcription**. Figure 6.6 shows an example of how transcription occurs. The two strands of a double-stranded DNA molecule separate so that one single strand can serve as the template upon which the RNA is synthesized. Transcription is catalyzed by the enzyme *RNA polymerase*. Each segment of DNA used as a template for transcription is preceded by a *promoter* region where the RNA polymerase binds. The transcription of the genes within the operon begins at this promoter site, and the transcription of all of the genes of the operon are thus controlled together. If the RNA molecule produced in this way carries the information for the protein coded for by a particular gene or operon, the RNA is called **messenger RNA** (or **mRNA**). mRNA molecules play an important role in carrying the information coded for by genes during the process of protein synthesis.

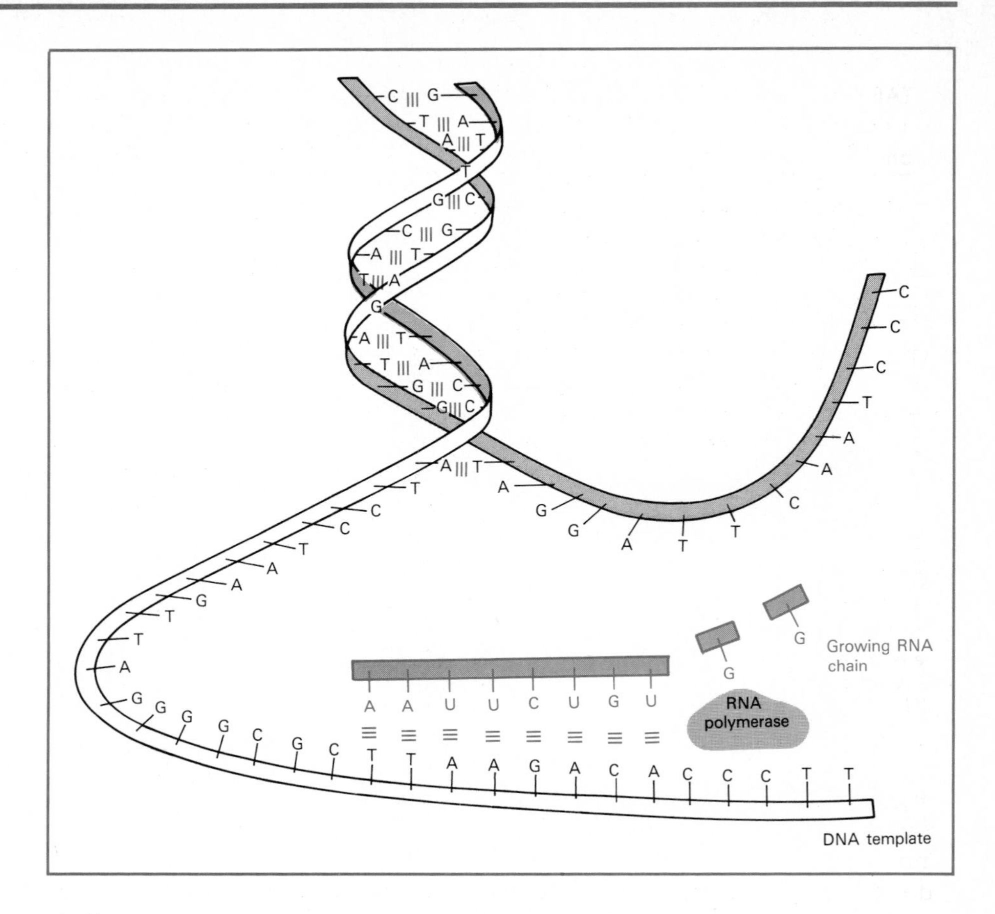

Figure 6.6 Transcription of messenger RNA by the enzyme RNA polymerase on the DNA double helix. Only one of the two strands, the coding strand, is transcribed.

6.2 PROTEIN SYNTHESIS

The process of protein synthesis is complex and involves many of the macromolecules and processes we have just discussed. In a sense, the process of converting the information stored in DNA into the sequence of amino acids in a protein can be thought of as a translation between two different languages. In one language, that of the base sequence of the DNA, the individual letters (the bases A, T, G, and C) are put together in groups of three to form meaningful words (triplet codes for amino acids). These words are arranged in sentences (genes) which code for the primary amino acid sequences of specific proteins. In the other language, amino acids are the words, which when linked in the appropriate order make up the sentences (proteins).

The first step in protein synthesis is the **transcription** of a messenger RNA (mRNA) from a gene or operon within the DNA of a cell (see Figure 6.6). The message may contain the encoded information for one or several related genes. Each triplet of bases in the DNA encodes a specific amino acid in the protein and is transcribed into the complementary triplet sequence in the mRNA. The mRNA triplet sequences are called **codons** as they actually code for the appropriate amino acid. The 64 possible codons of mRNA and the amino acids they represent are shown in Table 6.1. There are also special codons for the start (AUG) and stop (UAA, UAG, and UGA) of the message.

TABLE 6.1 The genetic code as expressed by triplet base sequences of mRNA*

Codon	Amino Acid	Codon	Amino Acid	Codon	Amino Acid	Codon	Amino Acid
UUU	Phenylalanine	CUU	Leucine	GUU	Valine	AUU	Isoleucine
UUC	Phenylalanine	CUC	Leucine	GUC	Valine	AUC	Isoleucine
UUG	Leucine	CUG	Leucine	GUG	Valine	AUG (start)†	Methionine
UUA	Leucine	CUA	Leucine	GUA	Valine	AUA	Isoleucine
UCU	Serine	CCU	Proline	GCU	Alanine	ACU	Threonine
UCC	Serine	CCC	Proline	GCC	Alanine	ACC	Threonine
UCG	Serine	CCG	Proline	GCG	Alanine	ACG	Threonine
UCA	Serine	CCA	Proline	GCA	Alanine	ACA	Threonine
UGU	Cysteine	CGU	Arginine	GGU	Glycine	AGU	Serine
UGC	Cysteine	CGC	Arginine	GGC	Glycine	AGC	Serine
UGG	Tryptophan	CGG	Arginine	GGG	Glycine	AGG	Arginine
UGA	None (stop signal)	CGA	Arginine	GGA	Glycine	AGA	Arginine
UAU	Tyrosine	CAU	Histidine	GAU	Aspartic	AAU	Asparagine
UAC	Tyrosine	CAC	Histidine	GAC	Aspartic	AAC	Asparagine
UAG	None (stop signal)	CAG	Glutamine	GAG	Glutamic	AAG	Lysine
UAA	None (stop signal)	CAA	Glutamine	GAA	Glutamic	AAA	Lysine

*The codons in DNA are complementary to those given here. Thus U here is complementary to the A in DNA, C is complementary to G, G to C, and A to T.
†AUG codes for N-formylmethionine at the beginning of mRNA.

The genetic code is a somewhat redundant code since most of the amino acids are encoded by more than one codon. It is remarkable that with only a few minor exceptions (see the Box, page 140), the same three bases in mRNA codons code for the same amino acids in all organisms.

The next and most crucial step in protein synthesis is called **translation**; this is the step during which the nucleotide sequence of the mRNA is translated into the corresponding amino acid sequence of the new protein. The whole translation process takes place on the ribosome. An important class of RNA molecules which plays a crucial role in the translation process is called **transfer RNA** (or **tRNA**). As shown in Figure 6.7, because of complementary base pairing within the molecule, the tRNA folds into a "cloverleaf" structure. Through the agency of transfer RNA, the genetic code is translated into the amino acid sequence of a protein. In a sense tRNA molecules are "bilingual" since they can recognize the triplet nucleotide sequence encoding a specific amino acid as well as the amino acid encoded. On one arm of the tRNA molecule is a region called the **anticodon**, which is complementary to the codon in the mRNA (Figure 6.7). The three bases of the codon thus pair with a tRNA molecule which contains the complementary anticodon. At one end of the tRNA molecule, the appropriate amino acid is attached. Thus, the appropriate tRNA molecule brings to the mRNA the amino acid encoded by the proper codon. One or more specific tRNA molecules are present for each amino acid (see Table 6.1).

Translation of the mRNA occurs on the ribosome surface (Figure 6.8). As the mRNA is translated, the tRNA molecules carrying the appropriate amino acids move to the complementary codons on the mRNA. After a peptide bond

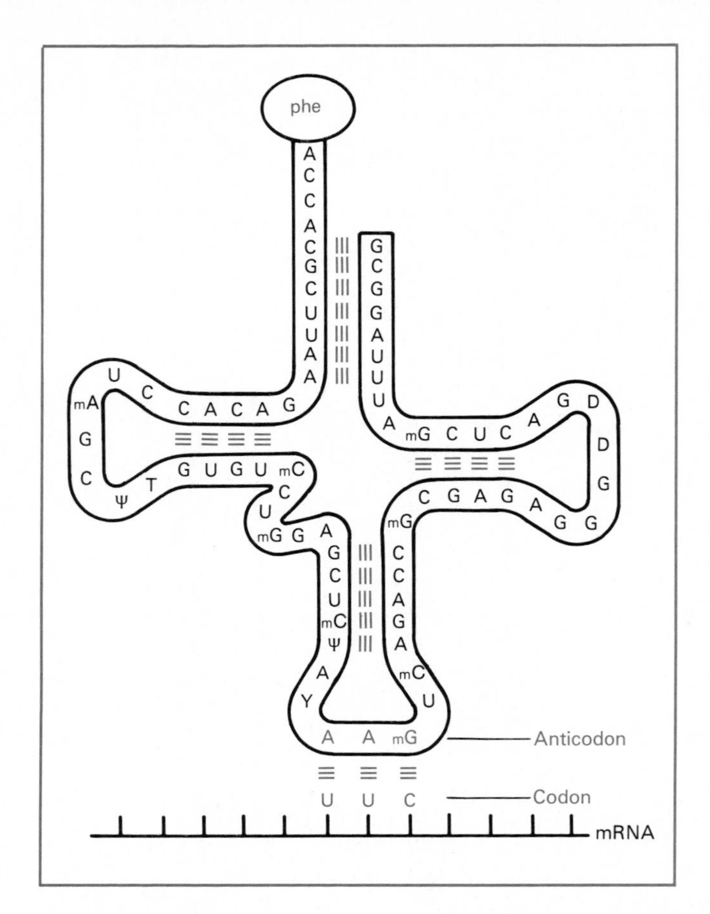

Figure 6.7 The structure of the transfer RNA (tRNA) molecule and the manner in which the anticodon of tRNA associates with the codon on mRNA by complementary base pairing. The amino acid corresponding to this codon (UUC) is phenylalanine which is bound to the opposite end of the tRNA molecule. Many tRNA molecules contain unusual bases such as methyl cytosine (mC) and pseudouridine (Ψ).

is formed between two amino acids, one tRNA molecule leaves the ribosome. The tRNA complementary to the next codon then comes to the site, carrying its amino acid. After the next peptide bond is formed, the preceding tRNA leaves the ribosome. The sequential association of tRNA molecules with codons on the mRNA and the formation of peptide bonds between adjacent amino acids gradually leads to the synthesis of the complete amino acid chain of the protein. When the chain is complete (as indicated by a *stop* codon), the ribosome, mRNA, and amino acid chain separate and the polypeptide chain folds into the three-dimensional structure of the protein.

In order for the series of codons on the mRNA to have meaning, it is essential that the translation be started at the correct base of the message. Each DNA sequence is said to have a proper **reading frame**. If translation were to begin one or two bases out of frame, the entire sequence would be translated incorrectly, changing the meaning of the encoded message. As illustrated in Figure 6.9, if translation were to begin at the second or third base in the sequence, the codons would all be read differently, resulting in a completely different amino acid sequence or none at all. How the reading frame may be altered by mutation will be described in Chapter 7.

Procaryotic and eucaryotic cells differ somewhat in the way in which they carry out the processes of transcription and translation. For instance, the

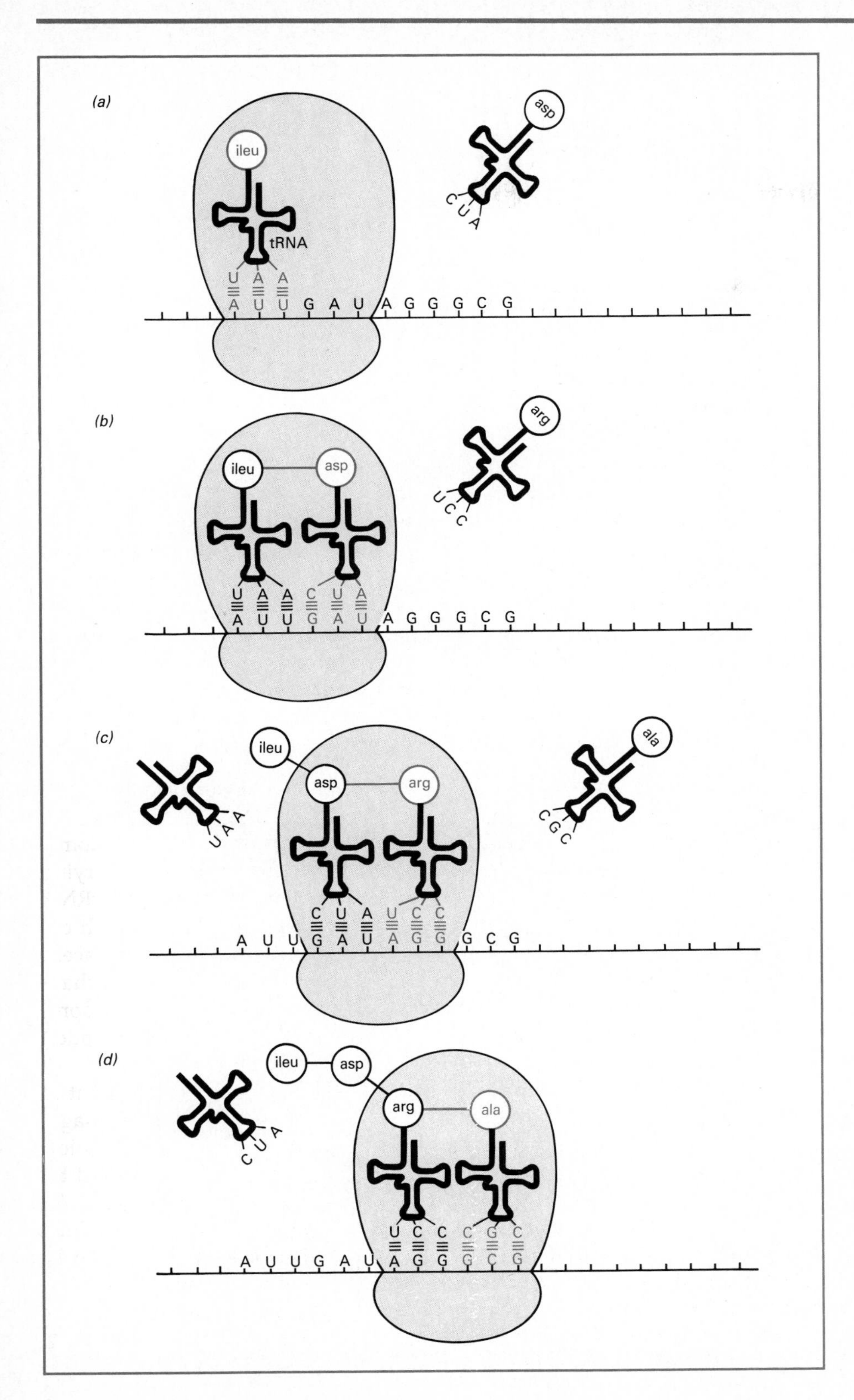

Figure 6.8 Translation of the information from messenger RNA (mRNA) into the amino acid sequence of protein. The whole process occurs on the surface of the ribosome. (a) Interaction between codon and anticodon brings into position the correct tRNA carrying the amino acid. (b) A peptide bond forms between amino acids on adjacent tRNA molecules. (c) and (d) As the ribosome moves to the next codon a new tRNA attaches and the old tRNA, now free of amino acid, leaves. The result is a growing chain of amino acids.

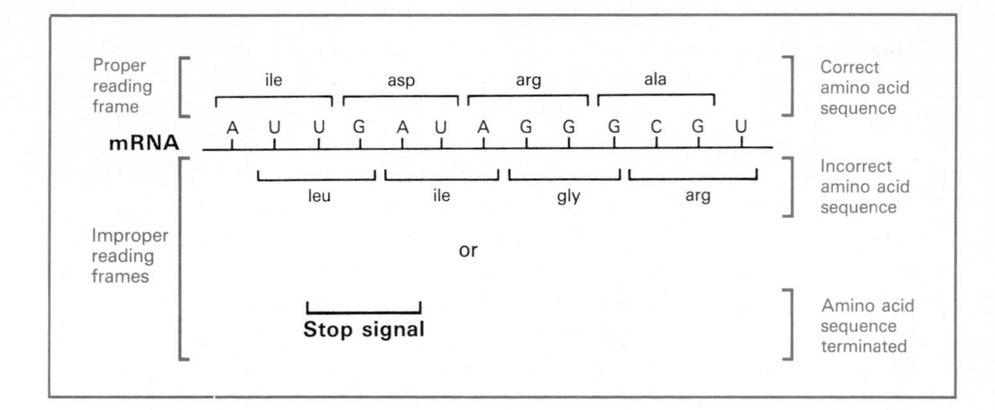

Figure 6.9 Translation of mRNA using an improper reading frame results in an incorrect amino acid sequence or an early termination of translation.

ribosomes of eucaryotic cells are larger and different enzymes are used to catalyze transcription. These differences are not significant enough to affect the way in which procaryotes and eucaryotes accomplish protein synthesis, but they do represent a basis for selective inhibition of one or the other type of cell. As discussed in Chapter 5, many antibiotics act by inhibiting transcription (RNA synthesis) or translation (protein synthesis) in procaryotic or eucaryotic cells. Some of these antibiotics and their modes of action were listed in Table 5.2. Rifampin, for example, inhibits transcription in bacteria, but not in eucaryotes, where different enzymes control the process. Many antibiotics, such as streptomycin, tobramycin, and the tetracyclines, act specifically against the smaller bacterial ribosome to inhibit translation but have no effect against the same process in eucaryotes. On the other hand, cyclo-

IS THE GENETIC CODE UNIVERSAL?

Since the genetic code was revealed in the 1960s, scientists have shown repeatedly that the same codons are used throughout the biological world. This fact suggests a common origin for all living things. However, despite the universality of the genetic code, some exceptions to it are known to exist.

Mitochondria, the organelles within eucaryotic cells which are responsible for energy production, contain some DNA that codes for certain proteins needed in energy generation. Mitochondria also contain components such as ribosomes and transfer RNA molecules which are needed for protein synthesis. Translation of genetic information inside mitochondria is not exactly like that which occurs in most procaryotic or eucaryotic cells. The triplet sequence UGA, which normally serves as a stop codon (see Table 6.1), codes in mitochondria for the amino acid tryptophan. Similarly, the sequence AUA, normally coding for isoleucine, codes for the amino acid methionine in mitochondria.

Other minor differences in the genetic code have recently been reported to occur in several ciliate protozoa. The codons UAA and UAG, which are also normally stop codons, appear to code for amino acids in *Paramecium* and perhaps in other ciliates. Since these differences are minor, it still seems that there is in principle a universal genetic code. Apparently, the code is not so rigid as to preclude minor changes which have occurred during the course of evolution.

heximide acts against only the larger eucaryotic ribosome to inhibit translation but does not affect procaryotes.

6.3 REGULATION OF METABOLISM

The genetic material determines the potential reactions a cell may perform. The ability to use certain nutrients, for example, may be a property of one organism but not of another, depending on the presence or absence of information in the DNA to encode the enzymes that participate in the utilization of the nutrient. For cells with the genetic potential, the ability to use a nutrient depends on whether the appropriate enzymes are produced and are active. Two types of metabolic regulation are recognized, one affecting the *synthesis* of enzymes, the other affecting the *activity* of enzymes. We discuss these two types of regulation here.

Regulation of protein synthesis Some enzymes, called **constitutive** enzymes, are made all of the time. Other enzymes are made only some of the time, as determined by environmental signals, usually nutrients. When a protein is being made as a result of transcription and translation of a gene, the gene is said to be **expressed**. The expression of a gene requires some sort of a "switch" which determines whether the genes are or are not expressed. Proteins are called *inducible*, if they are synthesized only when an environmental factor is present, or *repressible*, if their synthesis is inhibited when the factor is present. For example, a cell may be capable of producing an amino acid via a biosynthetic sequence brought about by several enzymes. The information for the synthesis of these enzymes is in genes within the DNA of the organism. If the amino acid is absent from the environment in which the cell is growing, the amino acid must be synthesized and the genes are expressed (Figure 6.10). If the amino acid is present in the environment, the synthesis of all of the enzymes needed for its production ceases.

The characteristics encoded within the DNA of a cell represent the cell's genetic potential and this is called its **genotype**. The **phenotype** of a cell is the collection of only those characteristics which are *expressed* under a given set of environmental conditions. Although the genotype is constant, the phenotype can vary depending upon how the conditions of the environment influence which genes are expressed. In the example shown in Figure 6.10, the cell always possesses the same genotypic potential of producing the enzymes for synthesis of the amino acid. However, two different phenotypes are possible depending upon whether or not gene expression occurs.

As noted, a chemical in the environment may **induce** the gene to be expressed. Alternatively, a chemical may turn off or **repress** gene expression. In many cases, a whole cluster of genes whose functions are related is regulated, rather than each gene being regulated independently. What is the nature of the switch which controls gene expression? We noted earlier that transcription begins at unique sites on the DNA called *promoters*. Regulation of gene expression in bacteria is brought about by selective control of transcription. A region of the DNA called the **operator**, which is adjacent to the

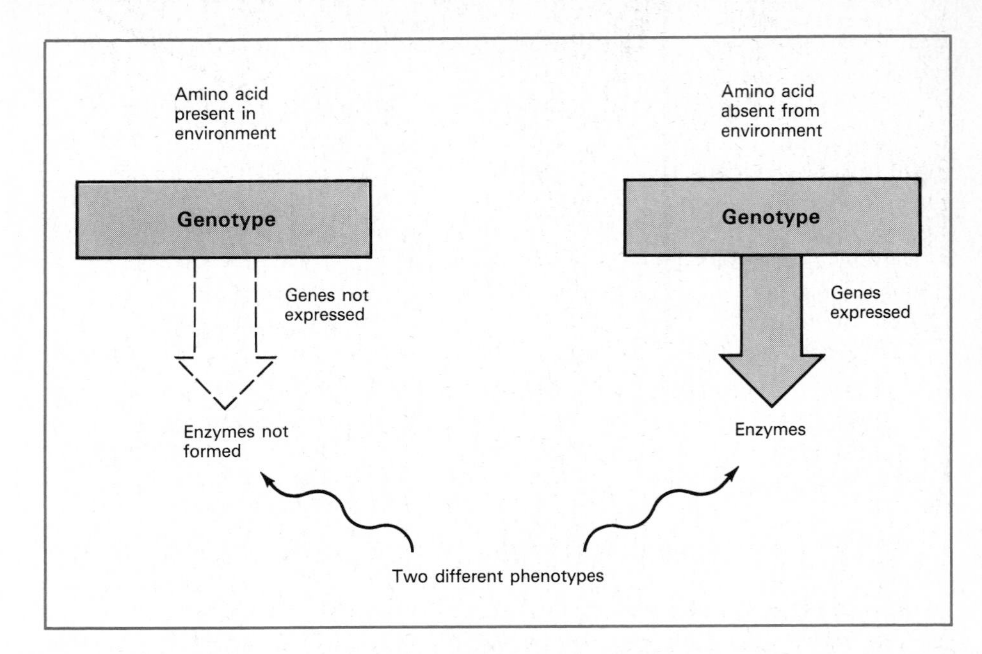

Figure 6.10 Expression of genes is influenced by the environment. The phenotype of the cell is thus determined by the interaction between genotype and environment.

promoter, interacts with a specific **repressor protein**. If the repressor protein is attached to the operator, transcription is blocked, whereas if the repressor is not attached, transcription can occur. Figure 6.11 shows the structure of an operon which is under **induction control**. The expression of the operon begins when a chemical called an *inducer* combines with the repressor protein, causing it to separate from the operator region. The RNA polymerase can then proceed to transcribe the gene, resulting in the production of mRNA which

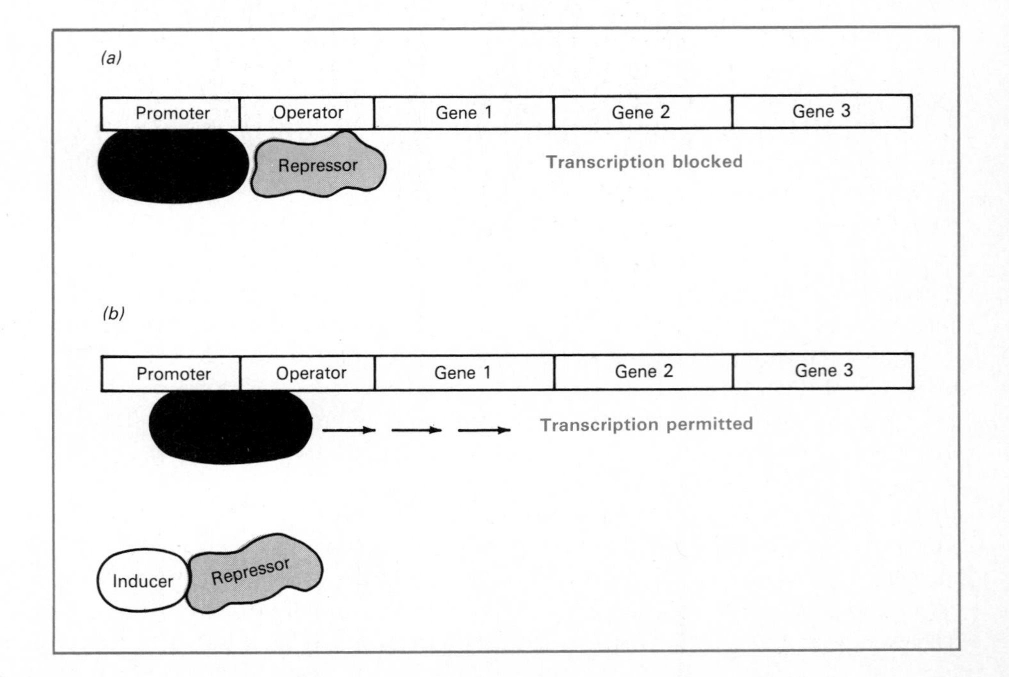

Figure 6.11 The process of enzyme induction. (a) A repressor protein binds to the operator region and blocks the action of RNA polymerase. (b) Inducer molecule binds to the repressor and inactivates it. Transcription by RNA polymerase occurs and a mRNA for that operon is formed.

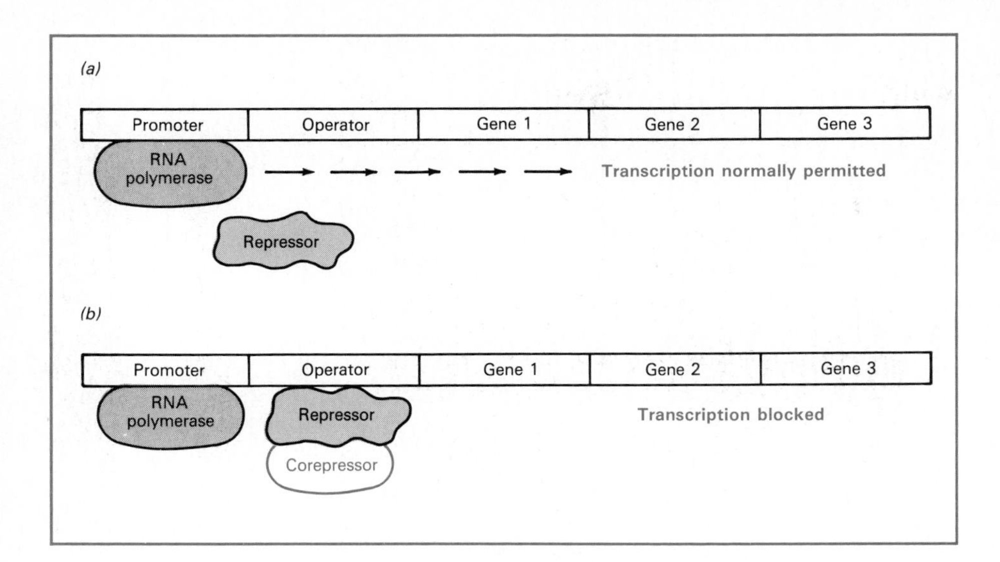

Figure 6.12 The process of enzyme repression. (a) Transcription of the operon occurs because the repressor is unable to bind to the operator. (b) After a corepressor (small molecule) binds to the repressor, the repressor now binds to the operator and blocks transcription. mRNA and the proteins it codes for are not made.

is in turn translated into protein. The binding between repressor protein and the operator and between the repressor protein and inducer is highly specific.

Genes which code for enzymes needed to metabolize energy sources are often under induction control. For example, *Escherichia coli* does not produce enzymes required for metabolism of lactose unless lactose is present in the environment. When lactose is present it binds to the repressor protein and expression of the genes encoding the enzymes for lactose decomposition begins.

Many operons are under **repression control** (Figure 6.12). Many of these operons code for enzymes involved in the production of monomers needed for biosynthesis. If, for example, a cell is growing in an environment which contains high levels of a certain amino acid, biosynthesis of this amino acid does not occur. How does the level of an amino acid control the expression of the operon encoding enzymes for its synthesis? In this case the specific repressor protein will bind to an operator region only if it *first* combines with a specific **corepressor**, usually the product which the encoded enzymes synthesize. When the repressor-corepressor complex binds to the operator, transcription of the genes does not occur. Thus, an excess of a compound shuts off the production of the enzymes which control its synthesis.

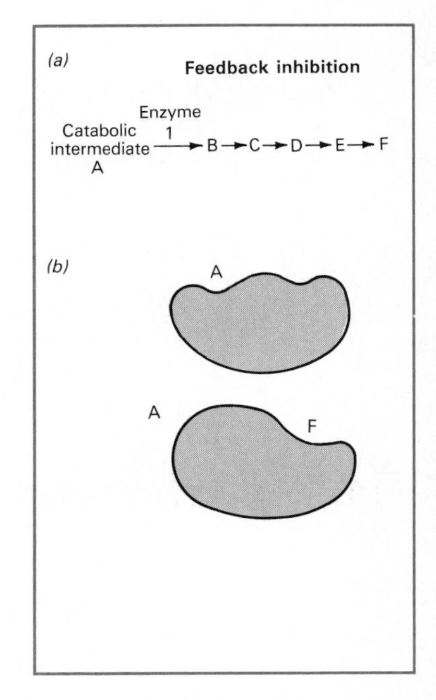

Figure 6.13 Feedback inhibition. (a) A biosynthetic pathway involving a series of enzymes. The final product of the pathway inhibits the first enzyme in the pathway, thus regulating its own production. (b) The manner by which the product of a pathway can bring about feedback inhibition. The product F combines with a site on enzyme 1 and causes a change in enzyme 1 so that it binds less effectively with its substrate.

Regulation of enzyme activity Another very specific control mechanism occurs in many cells, called *feedback inhibition*. This does not affect the synthesis of enzymes, but does affect their activity. In some metabolic reactions, the end product of a series of reactions can regulate the activity of the pathway by inhibiting one of the enzymes acting at an early step in the pathway. This can be shown by the reaction in Figure 6.13. Let *B*, *C*, *D*, and so on be a series of intermediates in a biosynthetic pathway leading from *A*, the substrate, to *F*, the final product (an amino acid, for instance). Each step from *A* to *F* is catalyzed by a specific enzyme. If there is a large amount of product, the early reactions of the pathway will be severely inhibited, and the amount of product will be correspondingly lessened so that the cell will not make excess product. When the product is once again in short supply, the inhibition will be less, and synthesis will proceed. This type of control seems to be an important mechanism to regulate the amount of product formed in a variety of biosynthetic pathways. If preformed product is added to a culture, the cells stop making it and use the added source instead. Note that feedback inhibition is an immediate type of control mechanism, acting on preexisting enzymes and leading to an immediate inhibition of the biosynthesis of the product. Induction and repression, on the other hand, do not act on preexisting enzymes but only on the synthesis of new enzymes; they are thus less immediate types of control mechanisms. Feedback inhibition can be viewed as a mechanism for finely regulating cell metabolism, whereas induction and repression are coarser mechanisms. Working together, these mechanisms result in an efficient regulation of cell metabolism, so that energy is not wasted by carrying out unnecessary reactions.

6.4 ORGANIZATION OF DNA IN EUCARYOTIC CELLS

One of the more recent and unexpected discoveries in molecular biology is that genes in eucaryotic organisms are not always organized in exactly the same way as the genes of procaryotic organisms. Often eucaryotic genes contain more information than that needed simply to encode a protein. **Intervening sequences**, or **introns**, interrupt the sequences which actually code for protein, which are called **exons** (Figure 6.14). Transcription of the gene

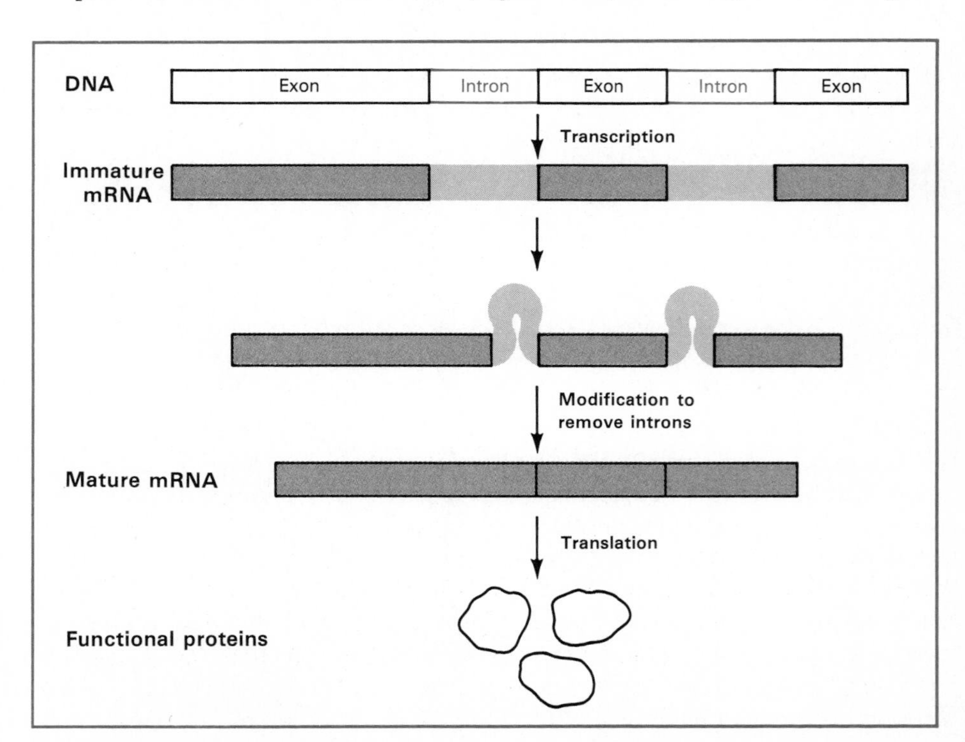

Figure 6.14 The structure of the eucaryotic gene. Noncoding segments called *introns* interrupt the coding sequence. The coding regions, called *exons*, are brought together during the maturation of the mRNA, a process called *RNA processing*.

plus intervening sequences leads to the production of an RNA molecule called a *primary transcript*. Before the message is translated it is modified into a "mature" mRNA by removal of the introns. Thus, in eucaryotes, another step is required during protein synthesis to splice together the exons and produce the essential parts of the message. The mature mRNA is then translated into the functional protein. These differences between procaryotic and eucaryotic gene organization and protein synthesis become very important when eucaryotic genes are transferred into bacteria by use of genetic engineering techniques, since bacteria are unable to carry out the splicing steps. This is one factor which complicates the expression of a eucaryotic gene in a procaryotic cell, as we will discuss in the next chapter.

6.5 VIRUS REPRODUCTION

Viruses are often considered to be nonliving particles. As mentioned in Chapter 2, viruses contain either DNA or RNA (but never both) and one or a few coat proteins. They are incapable of reproducing by themselves and depend on cells to carry out the synthesis of their essential macromolecules. How does a virus control a cell to accomplish its own replication?

A virus carries its own genetic information which encodes the proteins involved in its reproduction. One protein it needs is the one which makes up the coat of the intact virus particle. In order to reproduce, a virus must make coat protein and must also replicate its nucleic acid so that newly produced virus particles will have both genetic material and a protein coat. Thus, the processes which must occur in viral reproduction are the same as those which occur during growth and reproduction of cells (Figure 6.1). However, a virus cannot perform these reactions alone; it must depend on the machinery within the host cell to carry them out.

Lytic infection Let us consider first reproduction of DNA viruses (bacteriophages) of bacteria. As shown in Figure 6.15 a virus particle injects its DNA into the cell it infects. Before viral infection, the cell is busy replicating its own DNA, and transcribing and translating the information encoded in the DNA as it carries out biosynthesis, growth, and cell division. In effect, the viral DNA takes over the biosynthetic machinery of the host cell and uses it to produce the parts needed for production of new virus particles. Viral DNA replaces the host cell's DNA as a template both for replication (to produce more viral DNA) and transcription (to produce viral mRNA). Viral mRNAs are then translated using host cell ribosomes and tRNA-amino acid complexes, and viral proteins such as the coat protein are produced. In this way both the protein coat and the DNA of the virus are produced using the host cell's machinery. The parts of the virus are assembled into intact infectious virus particles within the host cell and are released upon cell lysis. Because the host cell is lysed in the process, this form of virus reproduction is called the **lytic cycle**.

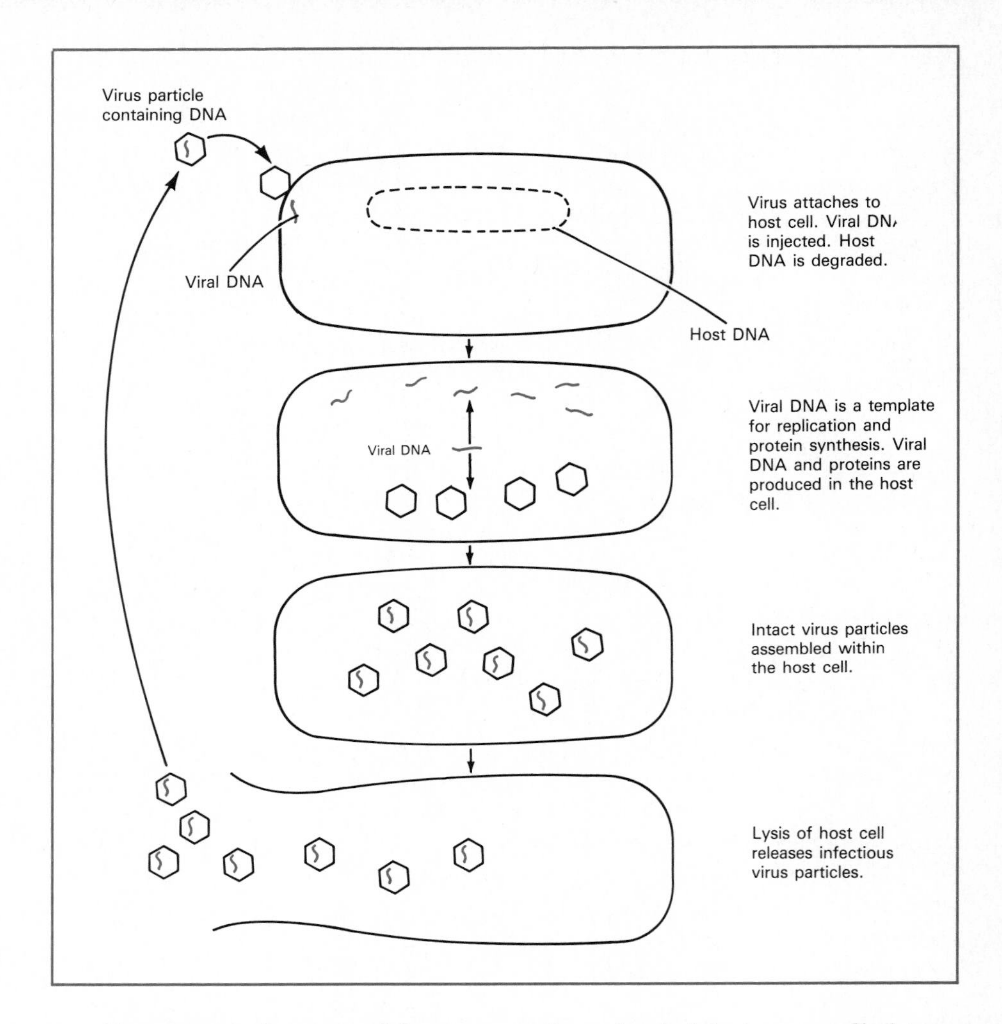
Virus particle
containing DNA
Viral DNA
Virus attaches to
host cell. Viral DN
is injected. Host
DNA is degraded.
Host DNA
Viral DNA
Viral DNA is a template
for replication and
protein synthesis. Viral
DNA and proteins are
produced in the host
cell.
Intact virus particles
assembled within
the host cell.
Lysis of host cell
releases infectious
virus particles.

Figure 6.15 Lytic infection of a host bacterial cell by a virus.

Temperate viruses and lysogeny Some bacterial viruses, called **temperate**, may be reproduced in another interesting way within the host cell. When the DNA of the virus enters the host cell, it may join with the host cell's DNA (Figure 6.16), rather than cause a lytic infection. Once the virus DNA is integrated within the host DNA, the viral DNA is referred to as a **provirus** or **prophage**. As growth and division of the host cell occurs, the replication of the host cell's DNA also leads to replication of the integrated viral DNA. Subsequent host cell divisions lead to a large population of bacteria, each of which carries the viral DNA integrated into the bacterial DNA. This type of viral infection is called a **lysogenic** or **latent infection**.

Environmental stimuli such as ultraviolet radiation, X rays, and chemicals which damage DNA can **induce** the provirus to change from the lysogenic to the lytic form of infection. The provirus then leaves the host DNA and, as described above, uses the host DNA replication and protein synthesis machinery to produce the parts of the virus. Assembly of intact virus particles precedes cell lysis, which in turn results in the release of infectious virus particles.

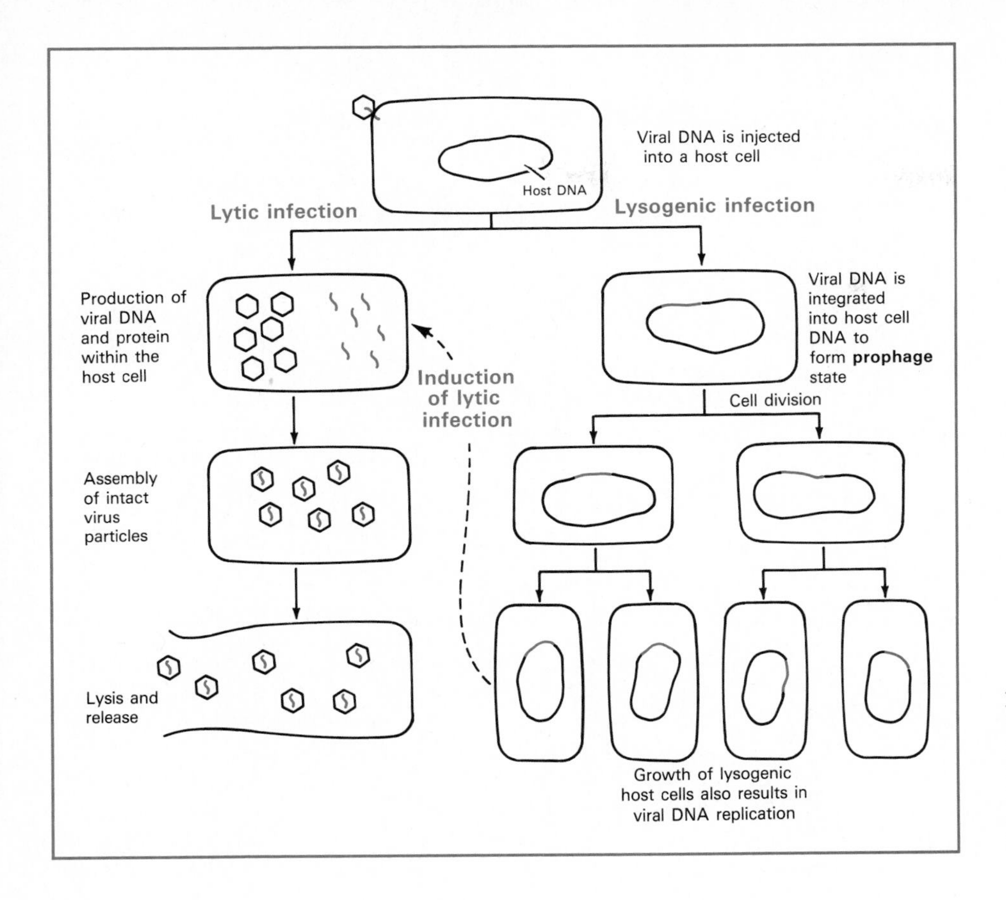

Figure 6.16 The lysogenic (latent) infection of a bacterial cell by a temperate virus. Upon infection, the alternatives are either a lytic or a lysogenic infection. The lytic cycle can also be initiated by induction of a lysogenized bacterial cell.

VIROIDS AND PRIONS: THE SIMPLEST KILLERS

Two different disease entities, viroids and prions, have recently been discovered to be simpler in structure than viruses, but like viruses each apparently depends on cells of a host for reproduction. *Viroids* are small pieces of RNA which are not complexed with any protein. The best studied viroid is the cause of a disease of potatoes. It is composed of 359 bases and thus has ten times less genetic material than the smallest known virus. In fact, there is only enough genetic information to code for one or two small proteins. Little is known about how viroids reproduce within the hosts they infect, but the fact that they contain RNA suggests a similarity to the RNA viruses.

Prions are small proteins which appear to be self-replicating yet devoid of any nucleic acid. It is estimated that a prion contains about 250 amino acids, and is possibly 100 times smaller than the smallest virus. Prions, like viroids, appear able to cause infectious diseases, such as scrapie in sheep and Creutzfeldt-Jakob disease in humans. Prions could represent the first exception to the rule that all self-replicating entities contain nucleic acid as the genetic material. One explanation for prion replication is that the host contains a latent gene which codes for the prion, and this gene is activated when the prion protein from the environment enters the cell.

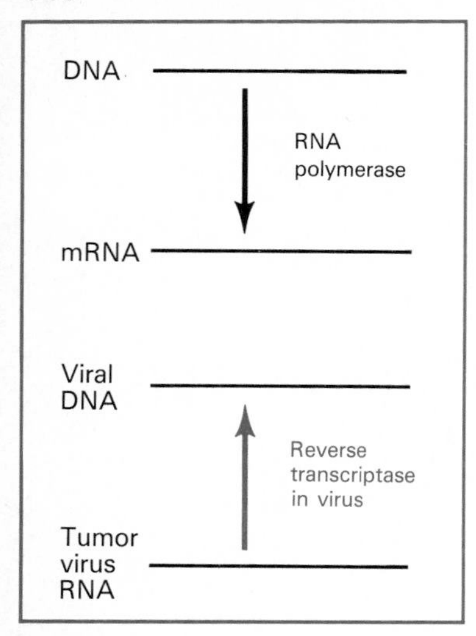

Figure 6.17 RNA tumor viruses have the enzyme *reverse transcriptase* which brings about the formation of DNA from RNA. The process of reverse transcription involves information transfer in the opposite direction to normal transcription.

RNA viruses There are several different types of RNA viruses. In some, such as poliovirus, the RNA of the virus serves directly as the mRNA for translation of viral proteins. In other RNA viruses, the details of how the information encoded in the viral RNA is used to produce viral protein are much more complex. The RNA tumor viruses are the most interesting because viral RNA serves as a template for DNA synthesis. This is the opposite of the usual process of transcription by which RNA is made from a DNA template (Figure 6.17). Another name for these RNA tumor viruses is, in fact, *retrovirus,* because the usual direction of nucleic acid synthesis is "reversed." The enzyme responsible for this reaction, called *reverse transcriptase,* is an important tool in genetic engineering (see Chapter 7). Retrovirus DNA formed from the infecting RNA can integrate with the host cell's DNA and form a provirus state similar to the prophage state of the lysogenic bacteriophages.

SUMMARY

In this chapter, we have considered **molecular biology**, the study of **macromolecules**, and how macromolecules are involved in fundamental processes that control biosynthesis, growth, and reproduction of cells and viruses. **DNA** is the macromolecule in which the information necessary to make the proteins is stored. The biochemical potential of a cell, its **genotype**, is a function of what proteins are specified by the cell's DNA. Each three bases in the DNA sequence codes for an individual amino acid in a protein. Different base triplets code for all of the essential amino acids, and collectively they make up the **genetic code**, which is nearly universal among all living things. Triplets are arranged into **genes**, which encode all of the amino acids of a protein. Genes are arranged into **operons** coding for related proteins, and the expression of each operon is controlled so that proteins are only made when required.

Growth and reproduction of the cell is dependent upon the important processes of **replication** of DNA, and **protein synthesis**. Replication is carried out by the enzyme DNA polymerase and accuracy is ensured by the **complementary bonding** between specific bases within the DNA. Protein synthesis is a more complex process which can be subdivided into **transcription**, the production of **messenger RNA** from DNA, and **translation** of the message into the protein. Translation itself is a complex event which occurs on ribosomes. During translation, **transfer RNA** molecules recognize the coding triplets on the mRNA and carry the correct amino acids to the ribosome. Once the primary amino acid sequence of the protein has been produced, interactions between the amino acids themselves cause the active three-dimensional protein structure to form, so that the protein can perform its function in the cell.

Viruses, like cells, must replicate their nucleic acid and produce viral proteins in order to reproduce. They are unable to carry out replication and protein synthesis by themselves, and infect cells in which these processes are carried out. In a **lytic infection**, a virus destroys the host cell, whereas in a **lysogenic infection**, the viral DNA integrates with the host cell DNA and the cell is not directly destroyed.

KEY WORDS AND CONCEPTS

Molecular biology
Macromolecule
Protein
Deoxyribonucleic acid (DNA)
Genetic code (triplet code)
Gene
Operon
Complementary bonding
Template
Replication
Hybridization
Ribonucleic acid (RNA)
Messenger RNA (mRNA)
Transfer RNA (tRNA)
Protein synthesis
Transcription
Translation
Reading frame
Regulation
Genotype
Phenotype
Induction
Repression
Feedback inhibition
Intron
Exon
Lytic infection
Lysogenic (latent) infection
Retrovirus
Reverse transcriptase

STUDY QUESTIONS

1. Name three different types of *macromolecules* found within cells. What type of monomer makes up each type of macromolecule?
2. Explain how the *primary amino acid sequence* of a protein can itself determine the active three-dimensional protein molecule.
3. How does a *triplet* of bases code for a particular amino acid?
4. Draw a sketch showing a sequence of twelve bases in single-stranded DNA. Using a second color, draw the strand which would be *complementary* to the first one.
5. What are the major differences in the chemistry of DNA and RNA?
6. Explain how single-stranded DNA can serve as a *template* for synthesis of either DNA or RNA.
7. *Transcribe* the DNA sequence you wrote in question 4 into the corresponding *messenger RNA* sequence.
8. What is the difference between a *codon* and an *anticodon*? On what types of RNA molecules are they found? How are they important in the process of *protein synthesis*?
9. Using the information found in Table 6.1, *translate* the mRNA sequence you wrote in question 7 into the corresponding amino acid sequence. Were there any "stop" messages in your RNA sequence?
10. If the *reading frame* were shifted one nucleotide to the right, what effect would this have on the amino acid sequence you determined in question 8?
11. Since the genetic code and protein synthesis are universal, why can some antibiotics be used to inhibit protein synthesis specifically in bacteria?

12. What is the difference between the *genotype* and *phenotype* of a cell? How does this relate to the ability of cells to *regulate* protein synthesis?

13. Explain the difference between *coarse control* and *fine control* in regulation of protein synthesis and activity.

14. Draw and label a diagram of an *operon* with its adjacent control regions. Show where RNA polymerase and the repressor or corepressor/repressor complex bind.

15. Explain how *intervening sequences* make protein synthesis more complicated in eucaryotic as compared with procaryotic cells.

16. Name two macromolecules or structures found within cells which viruses do not possess, and which make viruses dependent upon host cells for protein synthesis.

17. What are the steps involved in a *lytic infection* of a host cell by a virus?

18. What is the *provirus* state? Retroviruses are RNA viruses. What special enzyme allows them to form a provirus state?

SUGGESTED READINGS

BROCK, T.D., D.W. SMITH, and M.T. MADIGAN. 1984. *Biology of microorganisms, 4th edition*. Prentice-Hall, Inc., Englewood Cliffs, N.J.

CHAMBON, P. 1981. *Split genes*. Scientific American 244: 60–71. (A review of the organization of eucaryotic genetic information, focusing on introns and exons within genes.)

DIENER, T.O. 1981. *Viroids*. Scientific American 244: 66–73.

LEHNINGER, A.L. 1982. *Principles of biochemistry*, Worth Publ., Inc., New York, N.Y.

LEWIN, B.J. 1985. *Genes II*, John Wiley & Sons, Inc., New York, N.Y.

WATSON, J.D., J. TOOZE, and D.T. KURTZ. *Recombinant DNA, a short course*. Scientific American Books, New York. (A well-illustrated introduction to molecular biology and recombinant DNA technology.)

7

Genetics and Genetic Engineering

Genetics deals with the manner and mechanism by which characteristics are transmitted from an organism to its offspring. We have seen in Chapter 6 that genetic information is stored in DNA and that DNA replication ensures that during cell division each offspring receives a copy of the information contained in the DNA of the parent.

Changes can occur in the genetic material of the cell, and when they do, they may be observed as changes in the hereditary properties of the organism. A sudden change in a genetic character from one generation to the next is called a **mutation**. Mutations occur spontaneously, but environmental factors may cause relatively small, but sometimes quite significant change in DNA. Genetic change can also occur by means of **recombination**, which is the exchange of genetic information between organisms. These natural changes in genetic information are the driving forces behind the development of differences among living things which, in the long run, result in evolutionary change. As a result of **natural selection**, changes caused by mutation or recombination result in the survival or extinction of organisms.

The fields of molecular biology and genetics have advanced rapidly in the last few decades. We not only have a sophisticated understanding of natural genetic mechanisms, but it is now possible to bring about genetic change artificially by means of **recombinant DNA technology**. Since genes can be recombined in specific and predetermined ways, this new technology is also called **genetic engineering**. In this chapter, we will consider both natural and artificial genetic phenomena.

7.1 MUTATION

We discussed DNA replication in Chapter 6 and showed that when the DNA of the cell replicates during cell division the existing double-stranded molecule splits apart and each single strand acts as a template for the synthesis of a new complementary strand (see Figure 6.4). Recall that the complementarity between the pairs of bases adenine-thymine and guanine-cytosine permits the precise copying of one DNA strand. When this complementary strand is itself copied, its complement will be exactly the same as the original DNA strand. Mutations arise because of changes in the base sequence in the DNA that result in a DNA sequence different from that of the original. In most cases, mutations that occur in the base sequence of the DNA lead to genetic changes in the organisms; these changes are mostly harmful, although beneficial changes do occur occasionally.

Mutation can be either spontaneous or induced. **Spontaneous mutations** occur during replication, as a result of errors in the pairing of bases, leading to changes in the replicated DNA. In fact, spontaneous mutations in any specific gene occur about once in every 10^6 to 10^{10} replications. Thus in a normal, fully grown culture of organisms having approximately 10^8 cells/ml, there will probably be a number of different kinds of mutants in each milliliter of culture.

What is the result when a cell receives mutant DNA? The error in the DNA is transcribed into messenger RNA, and this erroneous mRNA in turn is used as a template and translated into protein. The triplet code that directs insertions of an amino acid via a transfer RNA will be thus be incorrect. Several possibilities may result. Since the genetic code is redundant, a single base change may have no effect. As seen in Figure 7.1, a change in the mRNA sequence from UAC to UAU would have no effect, since both sequences code for the same amino acid, tyrosine (see Table 6.1). Alternatively, a single base change may produce a codon that specifies a different amino acid. For example, a change from UAC to CAC would result in a change from tyrosine to histidine. If the change occurred at a critical point in the polypeptide chain, the protein could be inactive, or reduced in activity. Another possible result of a single base change would be the formation of a stop codon. For example, a change from UAC to the stop codon UAG would result in premature termination of translation of the message (Figure 7.1). The resultant protein would be incomplete and almost certainly not functional. Spontaneous mutation is by nature a sudden and unpredictable event and is relatively independent of the cell's environment. Mutational changes are permanent but are quite rare events.

Mutagens Some agents, such as X rays, ultraviolet irradiation, and certain chemicals, can cause mutations. These **mutagens** cause a change in the

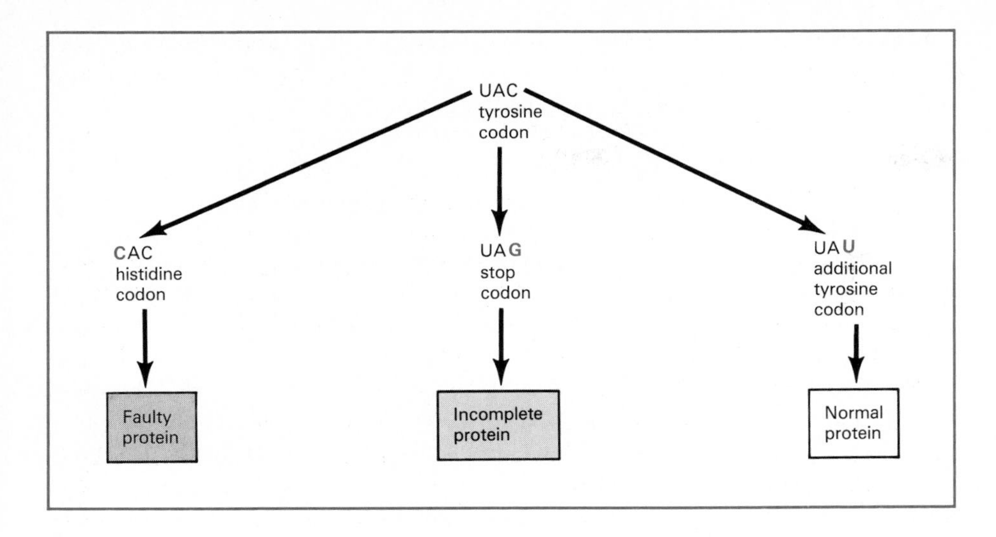

Figure 7.1 Possible effects of mutation: three different protein products from a single base change.

DNA that is similar to that which occurs in spontaneous mutation, but the rate at which these mutational events occur is much higher. Thus, if a large number of mutants are desired for certain studies, they can be obtained reasonably quickly by use of mutagens. The culture is treated with a mutagen and then allowed to grow in a medium that will select for the growth of the desired mutants.

Mutagens either act directly on bases in the DNA, changing them to other bases that are copied to produce altered DNA strands, or alter base-pairing properties so that the chance of an error during copying is increased. In either way, the result is the same: the production of an altered (frequently faulty) DNA.

Some types of mutagens alter the DNA during replication, so that addition or removal of nucleotides occur. Such changes are called *insertions* or *deletions*. As we illustrated in Figure 6.9, the DNA sequence must be read using the proper reading frame. Insertions and deletions alter the reading frame so that the translation of the mRNA results in a nonsensical protein. The example in Figure 7.2 shows how a change in the reading frame of a single base can change the sense of the message.

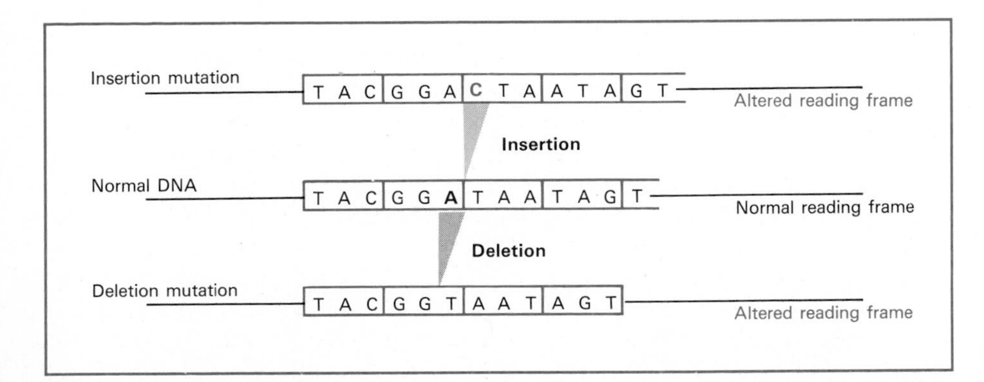

Figure 7.2 Shifts in reading frame caused by insertion or deletion mutations.

Since many of the chemical mutagenic agents used experimentally also occur naturally, it is possible that many spontaneous mutations are actually induced by natural mutagens. For instance, ultraviolet radiation from the sun can cause some mutations, and cosmic rays from outer space are similar to X rays in their mutagenic effects. Radioactivity is also mutagenic, and both natural and artificial radioactive materials can cause mutations under appropriate conditions.

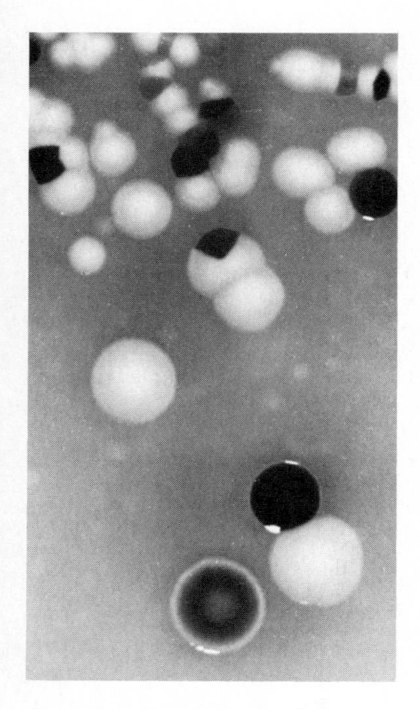

Figure 7.3 Development of pigmented mutants in a nonpigmented strain of bacteria, *Chromobacterium violaceum*. Mutations occurring when the colonies are still very small result in the formation of sectored colonies.

One of the most effective mutagens is nitrous acid, which is formed from nitrite (NO_2^-) in acid solutions. A number of bacteria can produce nitrite by the reduction of nitrate. Nitrite is also added to sausage and some other kinds of meat as a preservative. Nitrous acid reacts with adenine in DNA, converting it into the nucleic acid base hypoxanthine, which then pairs with cytosine during replication. Thus nitrous acid converts an adenine-thymine base pair into a guanine-cytosine pair. Other chemical mutagens include nitrosamines (formed in some foods by reaction of nitrite and amino acids), the nitrogen mustard gases (used in chemical warfare), the antibiotic mitomycin (another natural product), and the acridine dyes.

Kinds of mutants Typical mutations include loss of various synthetic capabilities (a requirement for the vitamin niacin, for example), loss of ability to utilize a certain sugar (unable to metabolize lactose, for example), and loss or gain of pigment (Figure 7.3). Mutations may affect an organism's ability to survive, which is of special importance in natural environments. For example, a mutation may affect the growth rate; faster growth will allow a mutant to become the dominant member of the population. As noted earlier, development of antibiotic-resistant mutants has had great practical consequence.

Selection procedures in genetics One of the noteworthy features of microbial genetics is that studies permit the use of very large population sizes, of the order of 10^8–10^9 cells/ml. Mutation frequencies in cultures for single genes are often 1 in 10^6 to 1 in 10^8 cells, so that any single milliliter of culture will always contain bacteria carrying mutations for almost every gene. Most of these mutants are unable to compete in conventional culture media with the wild type parent and are hence lost when the culture is transferred. However, if a sample of the culture is placed in a medium in which one of the mutants can grow and the parent cannot, then the mutant will be able to replace the parent. Culture conditions that permit growth of mutant but not of parent are said to be **selective**.

The extreme sensitivity of such selection procedures in microbial genetics should be emphasized. With the proper choice of mutant and culture conditions, it is possible to select a very small number of mutants from a very large population. As an example, mutants resistant to the antibiotic streptomycin are known (see Figure 5.15). Consider a culture containing a few streptomycin-resistant cells among a large population of streptomycin-sensitive cells. If a small sample of such a culture is plated on an agar medium containing streptomycin, none of the wild-type parent cells will grow, and only colonies of resistant mutants will appear. The streptomycin-containing medium is thus selective for the resistant mutants. As another example, consider a growth-factor-requiring strain capable of undergoing mutation to a type which does not require the growth factor. On a culture medium lacking the growth factor, cells of the parent strain (which require the factor) will not grow, whereas growth-factor-independent cells can form colonies.

The use of such selective conditions plays a major role in research in microbial genetics, because it makes possible the study of rare genetic events. This is of value not only in mutation studies but in studies on genetic recombination and genetic engineering, as discussed later in this chapter.

Mutagens and carcinogens Virtually every mutagen is also a **carcinogen**, capable of inducing cancer in animals or humans. Consequently, it is useful to carry out preliminary screening of compounds for carcinogenicity by assaying their mutagenicity in bacteria. Because of their ease of culture and study, and because the presence of mutations can often be easily detected, bacteria make ideal test organisms for determining the mutagenicity of chemicals.

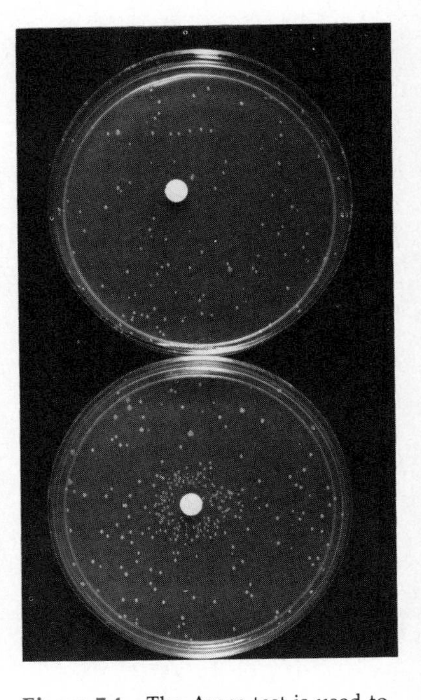

Figure 7.4 The Ames test is used to evaluate the mutagenicity of a chemical. Both plates were inoculated with a culture of a histidine-requiring mutant. The medium does not contain histidine, so that only revertant mutants can grow. Spontaneous revertants appear on both plates, but the chemical on the filter paper disc in the test plate has caused an increase in the mutation rate, as shown by the large number of colonies surrounding the disc. Revertants are not seen very close to the disk because the concentration of the mutagen is so high there that it is lethal.

Such a test is especially useful in determining whether new synthetic chemicals should be used in natural environments. The method used, called the **Ames Test**, is illustrated in Figure 7.4. The test bacterium used is one which requires a specific growth factor, such as the amino acid histidine. A sample of a culture is spread on the surface of a culture medium which does not provide the required amino acid. In such a medium, only mutants capable of producing the amino acid will grow and form colonies. These mutants are called *revertants* since they have reverted to the parent type. As shown on the control plate, some spontaneous revertants occur, resulting in a few colonies which grow dispersed over the entire plate. Since a huge number of bacteria were plated (around 10^9 cells per plate), it is not surprising to see some spontaneous mutants which have reverted to a type able to make the required amino acid and thus grow on the medium. A filter paper disc containing a mutagenic chemical has been placed on the test plate. The development of a large number of colonies around the disc indicates that the chemical has caused an increased rate of reversion. The Ames test thus permits a simple assessment of the mutagenicity of any chemical. A major virtue of the Ames test is that it is not only simple, but quick, sensitive, and inexpensive. If carcinogenicity is assessed in experimental animals, it requires months of time and thousands of dollars. Since mutagenicity shows a high degree of correlation with carcinogenicity, the Ames test provides a valuable preliminary assessment of one potential hazard of a chemical of interest.

7.2 GENETIC EXCHANGE AND GENETIC RECOMBINATION

Genetic recombination can be defined as a process in which genetic information from two parents is combined in one offspring. While mutation usually brings about only a very small amount of genetic change in a cell, genetic recombination usually involves much larger changes. Entire genes, or sets of

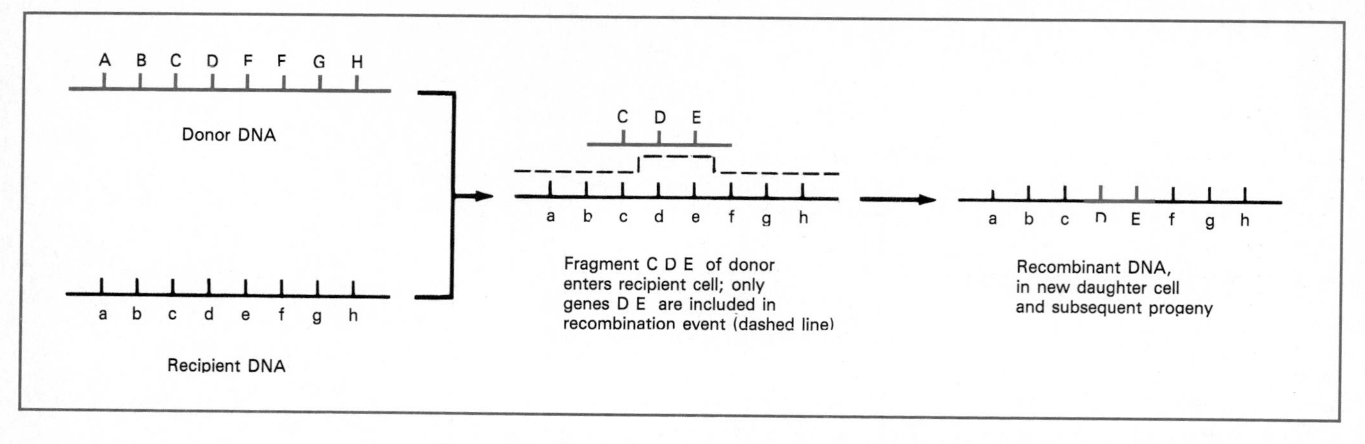

Figure 7.5 The basic process of bacterial recombination. Portions of the donor DNA are incorporated into the recipient DNA, resulting in a genetic recombinant. The letters denote specific genes. The two forms of a gene are noted by upper and lower-case letters. The new genes *replace* the corresponding genes of the recipient.

genes, are transferred between organisms. As a result, offspring are not exactly like either parent; they are **hybrids** and contain a combination of the traits exhibited by each parent. In eucaryotes, genetic recombination is a regular process which occurs as a result of sexual reproduction. Bacteria do not carry out an exactly analogous process, but they do have important ways of undergoing genetic exchange. There are distinct differences between genetic recombination in eucaryotes and procaryotes. In the present chapter, we discuss just the procaryotic processes, because these are of most microbiological interest from both a basic and applied point of view.

In bacteria, several processes have been recognized that bring about genetic recombination, but all involved the transfer of DNA from one cell, called the *donor,* to another cell, called the *recipient*. In the following discussion, the various mechanisms of DNA transfer are discussed and the events by which genetic recombination occurs are presented.

There are two ways in which a recipient cell can incorporate donor DNA: 1) by *replacement* of some its DNA with donor DNA; 2) by *insertion* of donor DNA without replacement. The first process is frequently called *homologous recombination* and the second process is called *insertion* or *integration*. We discuss homologous recombination first.

When a piece of donor DNA entering a cell undergoes homologous recombination, it *pairs* with the recipient DNA. For such pairing to occur, there must be some *homology* (similar or identical base sequences) between the two DNA molecules. As the result of this base pairing, the donor DNA replaces precisely the homologous DNA of the recipient. The process of genetic recombination is illustrated in Figure 7.5.

In the second mode of genetic exchange, the donor DNA becomes *inserted* into the recipient DNA, without replacement of recipient DNA. The donor DNA is said to become *integrated* into the recipient. We presented one type of DNA integration, that involving temperate viruses, in Chapter 6. Another example is presented in Figure 7.6. If the new piece of DNA is inserted *between*

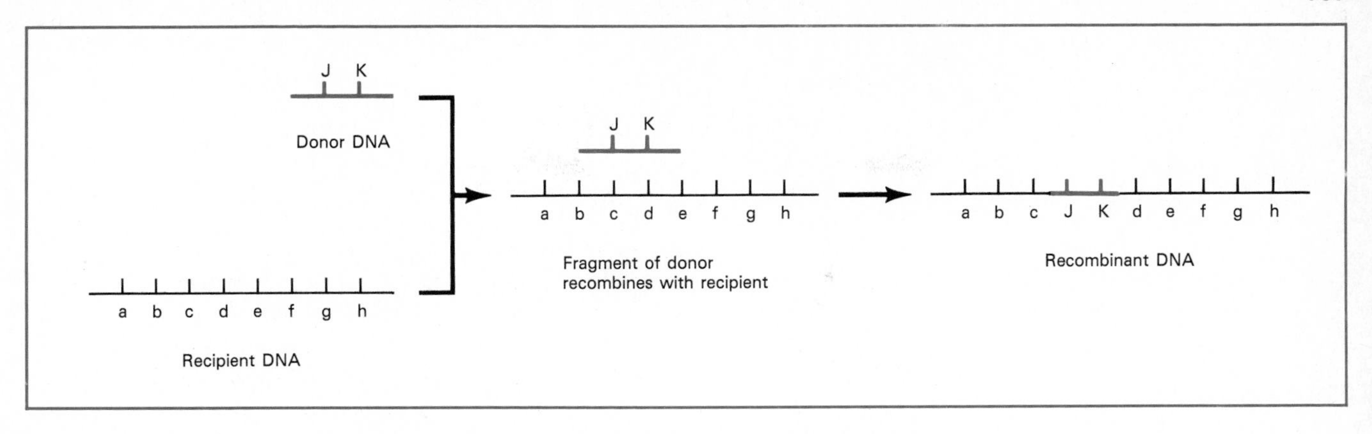

Figure 7.6 Insertion of a new piece of DNA into a recipient. The new genes are *added* to the existing genome.

recipient genes, then the recipient may acquire new functions controlled by the donor DNA without losing any preexisting functions. More often, however, the incoming DNA becomes inserted *in the middle* of a recipient gene, resulting in inactivation of the recipient gene.

Thus, in bacterial recombination, a part of the donor cell's DNA is transferred into the cell of the recipient. During subsequent cell replication, genetic elements from each DNA are incorporated into the DNA of the new progeny cell; this recombined DNA is then the genetic material for all subsequent progeny (Figure 7.5). The recipient is also called a **recombinant** since it is the cell in which recombination has occurred.

How does the DNA of the donor get transferred to the recipient? There are three distinct ways, which are illustrated in Figure 7.7: transformation, transduction, and conjugation. These processes are distinguished by the agency involved in the transfer of the DNA. In transformation, there is no special agent, but small fragments of free DNA participate. In transduction, also called virus-mediated genetic exchange, the virus particle is the agency of DNA transfer. The third mechanism, conjugation, involves cell-to-cell contact, and has some analogy to the sexual reproduction process in eucaryotes.

We discuss each of these methods further in the following pages.

7.3 BACTERIAL TRANSFORMATION

Transformation is a process by which free DNA is taken up directly by a recipient without the agency of any carrier. DNA fragments obtained from donor cells by lysis or extraction can be mixed with live recipient cells. With appropriate cells under proper conditions, the DNA fragments are taken up. The recipient cells grow, and if they express genes from the donor DNA, then transformation is known to have occurred. The process of transformation is diagramed in Figure 7.8. One prime requirement for transformation is of course that the donor DNA be able to enter the recipient. With the development of genetic engineering, scientists needed to find ways to increase the

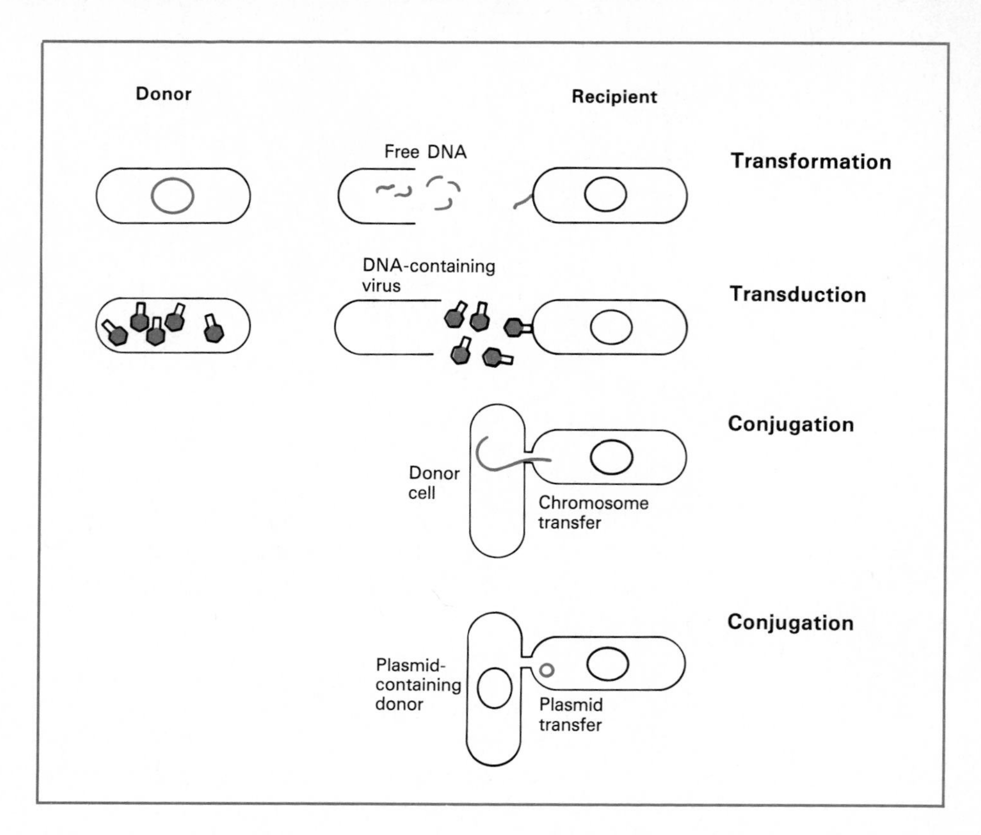

Figure 7.7 Processes by which DNA is transferred from donor to recipient bacterial cell. Just the initial steps in transfer are shown. For details of how the DNA is integrated into the recipient, see text.

efficiency of transforming cells. The application of these techniques to insert artificially recombined DNA into cells will be discussed later.

All items of genetic information may participate in transformation, although only a small amount of genetic information is involved at any one time. Examples of characters transformed include those specifying capsule formation, drug resistance, and nutritional requirements. The recombinant progeny must be detected in some way to know that recombination has indeed occurred. Detection of transformants is much more readily accomplished if

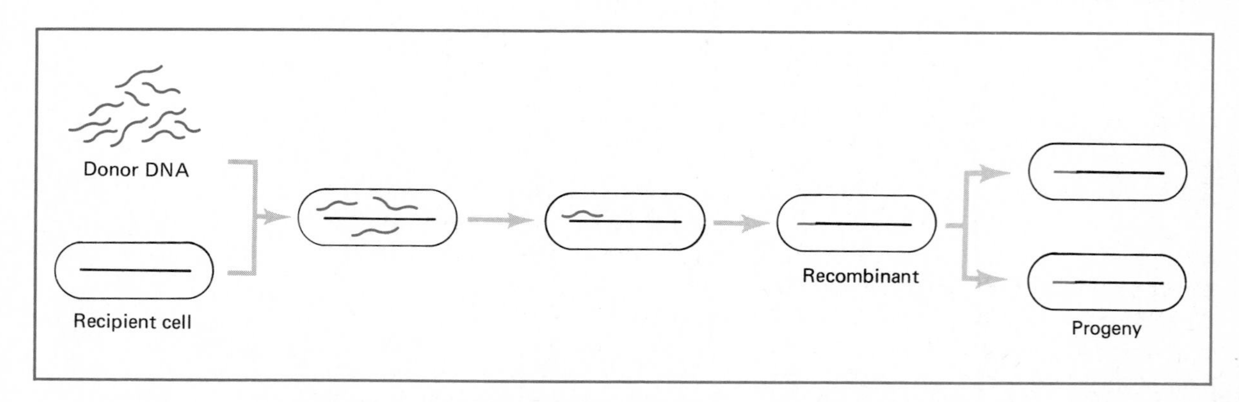

Figure 7.8 Bacterial transformation, the incorporation of free DNA into a cell and its integration into the cell genome.

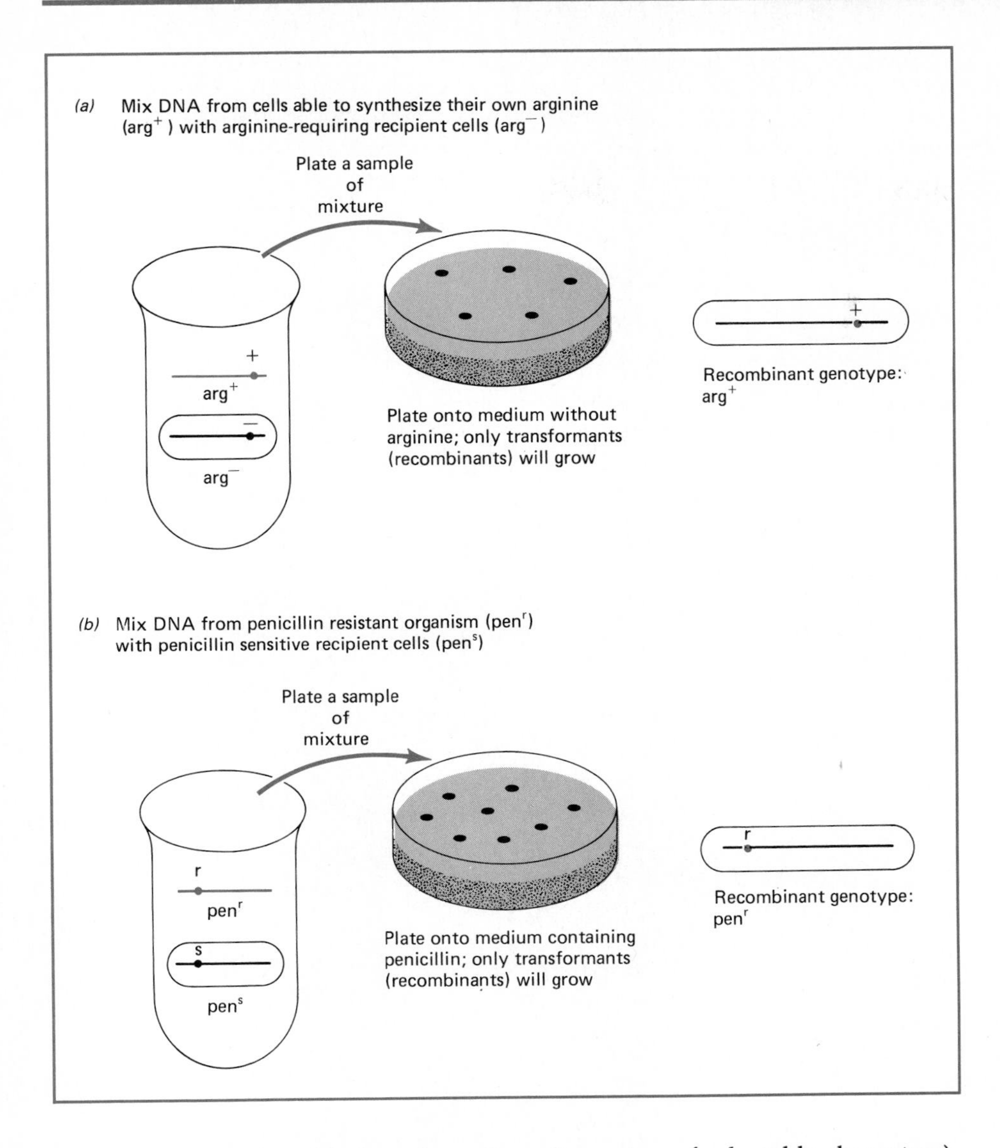
(a) Mix DNA from cells able to synthesize their own arginine (arg+) with arginine-requiring recipient cells (arg−)
Plate a sample of mixture
+
arg+
−
arg−
Plate onto medium without arginine; only transformants (recombinants) will grow
+
Recombinant genotype: arg+
(b) Mix DNA from penicillin resistant organism (penr) with penicillin sensitive recipient cells (pens)
Plate a sample of mixture
r
penr
s
pens
Plate onto medium containing penicillin; only transformants (recombinants) will grow
r
Recombinant genotype: penr

Figure 7.9 Laboratory procedure for the detection of genetic transformation.

the character is selectable (see above for a discussion of selectable characters). Two examples of the use of selectable characters to measure transformation are shown in Figure 7.9. In the first example, if recipient cells requiring arginine for growth (called arg^-) are mixed with donor DNA from cells that are capable of synthesizing their own arginine (arg^+), then the recombinant progeny become capable of arginine synthesis (arg^+). The way to find these arg^+ recombinants is to plate the transformed mixture onto a medium *without* arginine; only the recombinants (arg^+) are able to grow. This same method of observing recombinants works for a large number of nutritional characteristics. A similar method is useful in tracing the acquisition of resistance to a drug, such as penicillin. If the donor DNA comes from penicillin resistant (pen^R) cells, and the recipient is penicillin sensitive (pen^S), then after mixing and plating onto a medium containing penicillin, only the pen^R recombinants will be able to grow and form colonies. In the case of capsule or pigment formation, visual inspection of the recombinant colonies that arise after transformation may show which are obviously different. With such unselectable characteristics, rare transformation events cannot be detected. Since transformation is, in general, a rare event, it is difficult to use transformation to study the inheritance of unselectable characteristics.

7.4 TRANSDUCTION

The second way in which genetic recombination can occur in bacteria is by **transduction**, the virus-mediated transfer of genetic material from a donor cell to a recipient cell. Two ways in which transduction can occur during viral infections are shown in Figure 7.10. The reproduction of viruses was described in Chapter 6. Bacterial viruses usually cause a lytic infection resulting in the release of infectious virus particles (see Figure 6.15). During the viral infection the host cell DNA is fragmented into small pieces about the size of viral DNA. When the assembly of viral DNA and protein coats into infectious virus particles occurs, some of the host cell DNA fragments may become accidentally packaged into protein coats. After the lysis of the host cell, the virus particles which are released then infect other cells. However, those virus particles which carry host cell rather than viral DNA thus transfer genetic information from the first host cell (the donor) to the second (the recipient). If the transduced DNA recombines with the recipient cell's DNA, the recipient has acquired genetic information which it formerly did not have. This type of transduction can involve the transfer of any part of the host cell's DNA and is thus called **generalized transduction**.

As we discussed in Chapter 6, some viruses can produce a **latent**, or **lysogenic**, infection in which the viral DNA integrates with the host cell DNA (see Figure 6.16). When the latent virus is subsequently induced to enter a lytic cycle, the viral DNA leaves the host cell DNA. Occasionally the viral DNA takes with it a small amount of the DNA of the host cell. When the virus then infects another bacterial cell and its DNA integrates into the new host cell DNA, the DNA from the first bacterial host cell (the donor) is transferred to the second host cell (the recipient). This process is called **specialized transduction**, as it involves the transfer of a restricted group of host cell genes. In this case, the virus always integrates at a specific location on the host cell DNA, and the only genes transduced are those which lie near the integration site on the DNA.

7.5 PLASMIDS AND CONJUGATION

A third mechanism for genetic recombination in bacteria, called **conjugation**, involves direct contact between two bacterial cells of opposite mating types, as shown in Figure 7.11. Before we discuss conjugation in detail, we must discuss another type of genetic element that is involved in the conjugation process, the **plasmid**.

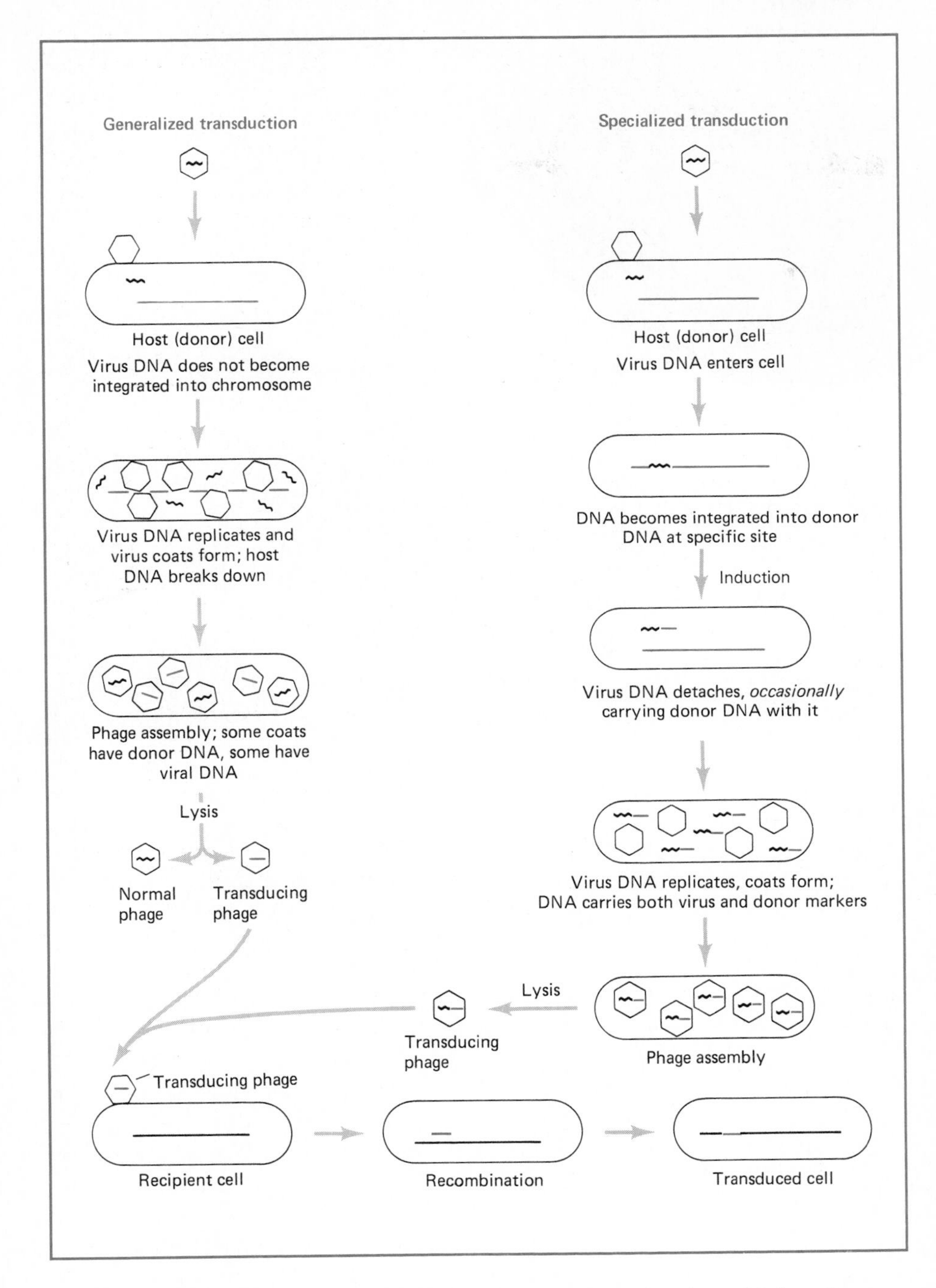
Generalized transduction
Host (donor) cell
Virus DNA does not become integrated into chromosome
Virus DNA replicates and virus coats form; host DNA breaks down
Phage assembly; some coats have donor DNA, some have viral DNA
Lysis
Normal phage
Transducing phage
Specialized transduction
Host (donor) cell
Virus DNA enters cell
DNA becomes integrated into donor DNA at specific site
Induction
Virus DNA detaches, occasionally carrying donor DNA with it
Virus DNA replicates, coats form; DNA carries both virus and donor markers
Phage assembly
Lysis
Transducing phage
Transducing phage
Recipient cell
Recombination
Transduced cell

Figure 7.10 Transduction, the transfer of genetic material from donor to recipient via virus particles.

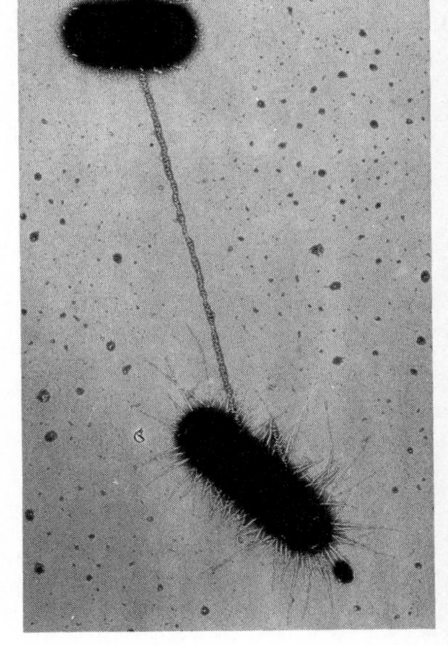

Figure 7.11 Direct contact between two conjugating bacteria is first made via a pilus. The cells are then drawn together for the actual transfer of DNA.

Plasmids Plasmids are small, circular DNA molecules which exist separately from the bacterial chromosome (Figure 7.12). Plasmids replicate independently from the rest of the DNA of a cell, as shown in Figure 7.13*a*. They contain genes that are not always essential for growth, so that under many conditions, they can be lost or gained without harm to the cell. Treatment of cells with certain drugs can result in loss of plasmids without any

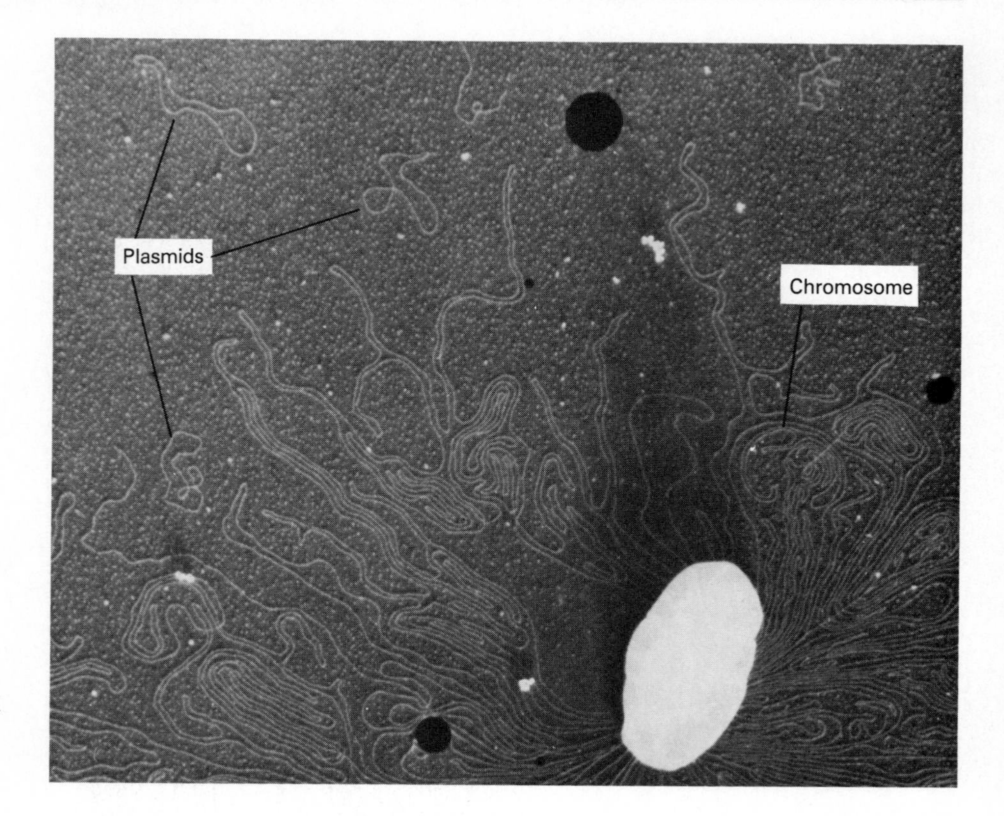

Figure 7.12 Bacterial plasmids, as shown in the electron microscope. The plasmids are the circular structures, much smaller than the main chromosomal DNA. The cell (large white structure) was broken gently so that the DNA would remain intact. (Photo courtesy of Drs. Huntington Potter and David Dressler, Harvard Medical School.)

other genetic change. This phenomenon is called **curing** (Figure 7.13*b*). Although they are nonessential genetic elements, plasmids can carry genes that confer selective advantage to cells under certain conditions. Examples of genes that can become incorporated into plasmids are the genes for resistance to heavy metals (mercury, copper, nickel, etc.) and genes for the production of toxins and other substances involved in the ability of a microorganism to cause disease.

The most dramatic examples of the importance of plasmids are those plasmids that contain genes for antibiotic resistance. Bacteria containing antibiotic-resistance plasmids present some of the most serious problems in the antibiotic therapy of infectious disease. Often, plasmids contain genes for resistance to several antibiotics; this is termed **multiple drug resistance**. Since these plasmids can be transferred at high frequency to other cells, antibiotic resistance can spread rapidly through the population. The emergence of bacteria containing plasmid-mediated multiple antibiotic resistance has been linked to the increasing use of antibiotics for the treatment of infectious diseases (see Section 5.8).

Conjugation How are plasmids transferred from a donor bacterial cell to a recipient cell? As shown in Figure 7.13*c*, plasmids may move from a donor cell to a recipient cell during the process called **conjugation**. The **F factor** (F stands for fertility) is one particular kind of plasmid which contains a gene coding for the production of a tube-like structure called a **pilus** (Figure

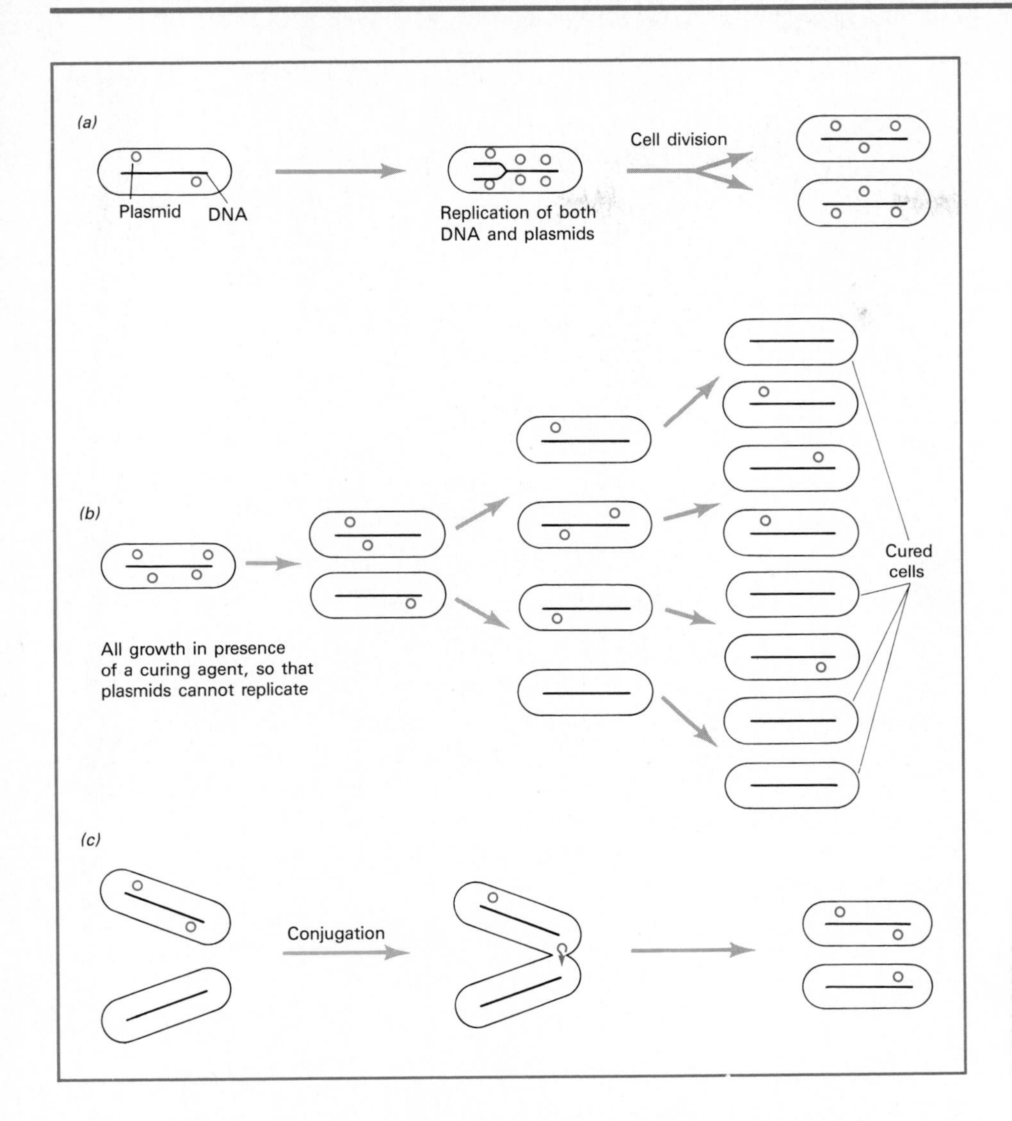

Figure 7.13 Replication and transfer of plasmids. (a) Replication of the plasmid independent of the chromosome. (b) Curing of the plasmid by drug treatment. (c) Transfer of a plasmid from cell to cell via conjugation.

7.11). The pilus of the donor cell is involved in attachment of donor to recipient cell, and once cell-to-cell contact has been made, the plasmid DNA can pass from donor to recipient.* The F factor can bring about not only its own transfer but also the transfer of other plasmids or part of the donor chromosomal DNA. As a result, the recipient cell acquires the genetic information encoded by the plasmid (antibiotic-resistance genes, for example). The recipient cell itself also becomes a donor cell since it receives the F factor gene which controls the ability to produce a sex pilus.

In a sense, plasmids by themselves might be considered living entities. They carry sufficient genetic information to control their own replication and movement between cells. They tend to accumulate genetic information which enhances the survival of the cells in which they are found (for instance, genes

*The exact mechanism by which the DNA passes from donor to recipient is not known but the DNA probably does *not* move through the pilus itself.

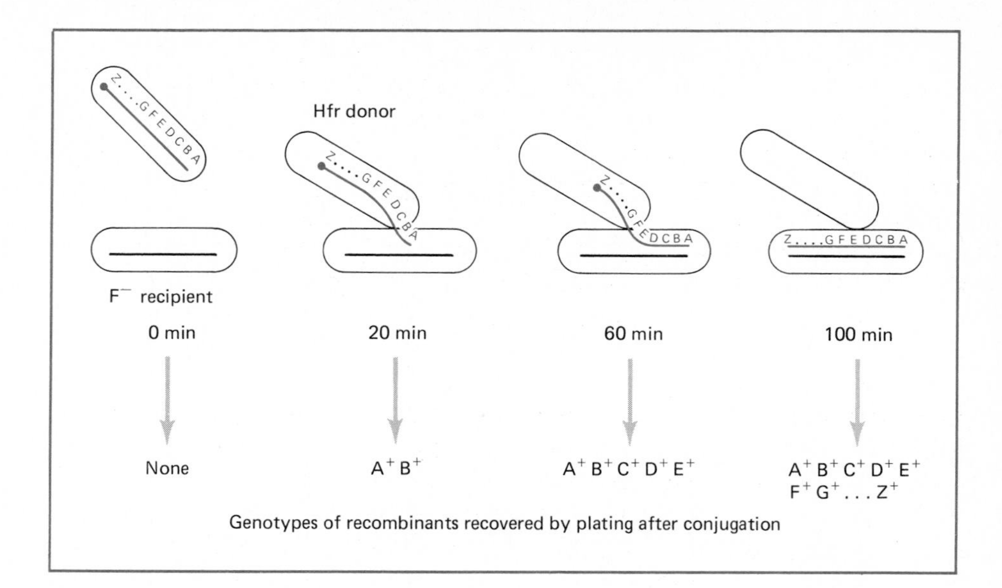

Figure 7.14 Bacterial conjugation, the transfer of genetic material via cell-to-cell contact. After the Hfr cell contacts the F^- cell, the DNA of the Hfr is transferred in sequential fashion into the F^- recipient.

that code for antibiotic and heavy metal resistance). However, like viruses, they are unable to reproduce without the aid of cells.

Some plasmids have the ability to integrate into the chromosomal DNA of the host cell. This is analogous to the integration of some viral DNAs into the DNA of the infected cell in lysogeny. When the F factor plasmid is integrated, conjugation results in transfer of part or all of the chromosome of the donor cell into the recipient cell, as shown in Figure 7.14. The recipient cell then contains DNA of its own plus DNA from the donor, and after recombination, the progeny will show traits from each parent. Because the integration of the F factor increases the rate of recombination of chromosomal genetic information between a donor and a recipient cell, donor strains in which the F factor is integrated are called **Hfr** strains (for *H*igh *f*requency of *r*ecombination).

Selection for recombinants formed by conjugation is accomplished by plating the mating mixture on culture media that allow growth of only the recombinant cells with the desired genotype. For instance, in the experiment shown in Figure 7.15, an Hfr donor which is sensitive to streptomycin (Str^S) and contains the genes coding for enzymes needed for synthesis of the amino acids threonine and leucine (T^+ and L^+) and for utilization of the energy source lactose (lac^+) is mated with an F^- recipient cell which lacks these genes but is resistant to streptomycin (Str^R). Selective media contain streptomycin so that only recipient cells can grow. The composition of each selective medium is varied depending on which genotypic characteristics are desired in the recombinant, as shown in the figure.

When conjugation occurs in Hfr cells, the circular chromosomal DNA is broken at a point specific for a given bacterial strain. Then, as shown in Figure 7.14, the donor DNA begins to enter the recipient cell. Since mating pairs do not remain attached indefinitely, it is more common to see recombinants with the genes that transfer early during conjugation than with genes which trans-

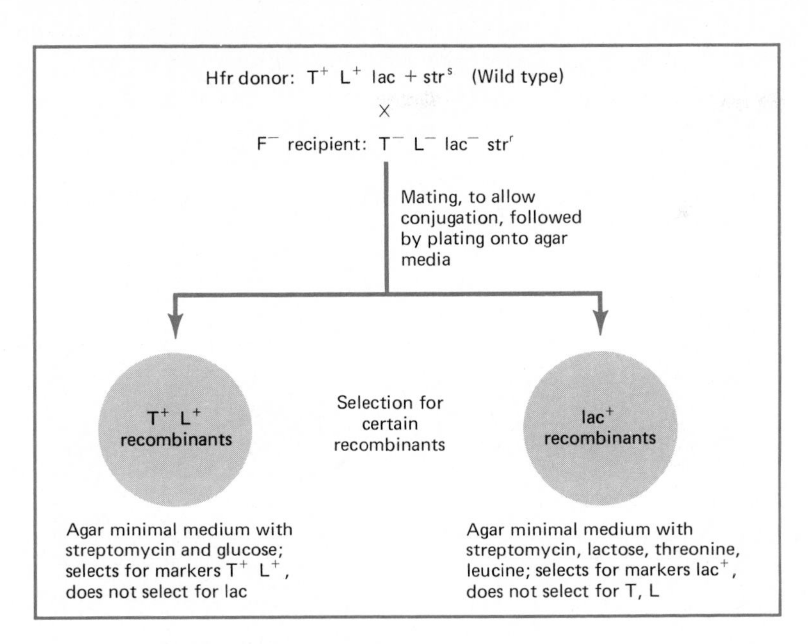

Figure 7.15 Laboratory procedure for the detection of genetic conjugation. Symbols for genes: T, requirement for threonine; L, requirement for leucine; lac, ability to utilize lactose; str^r, streptomycin resistant; str^s, streptomycin sensitive.

fer late. Knowing that the transfer of genetic material is always sequential, one can determine the order in which the genes on the chromosome are transferred. If a large number of matings are carried out using different genotypes and different Hfr strains, a **genetic map** for the organism can be formulated. A simplified version of the genetic map for *Escherichia coli* is shown in Figure 7.16. This is a greatly abbreviated map, since there are actually more than 650 known genes that could be shown in a complete map.

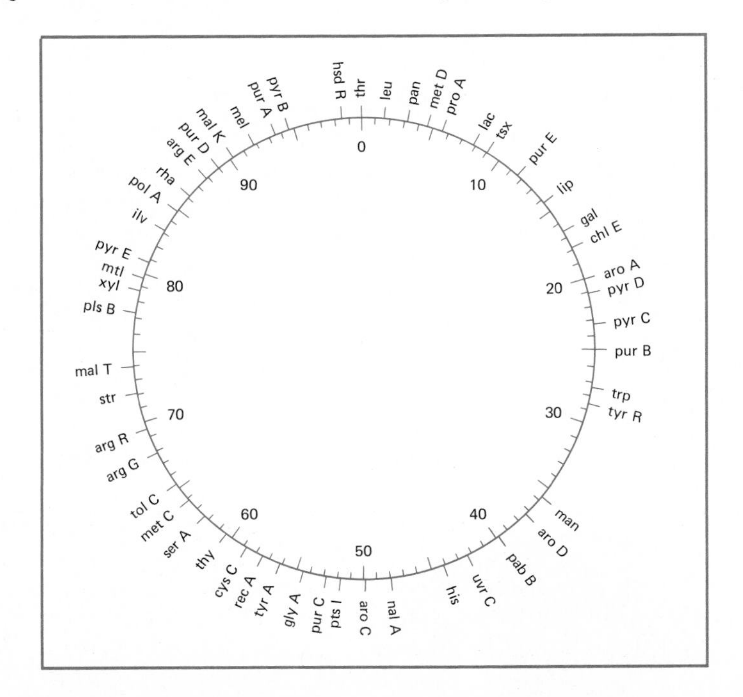

Figure 7.16 The genetic map of *Escherichia coli*. Only a small number of the mapped genes are shown.

7.6 TRANSPOSONS: GENES THAT MOVE

Although most genes are fixed in place on the DNA, some genes can move about on the chromosome or on plasmids. These movable genetic elements are called **transposons**. Their ability to move is due to the presence on the DNA of a specific base sequence called an **insertion sequence** (abbreviated *IS*), which is a short segment of DNA that contains genetic information for an enzyme system capable of cutting and reforming DNA pieces. Although an insertion sequence can move by itself, if *two* insertion sequences are flanking another piece of DNA containing genes, the whole structure:

insertion sequence—genes—insertion sequence

can move as a unit. The first such "moving genes" were discovered to occur in corn plants, but transposons are now known to occur in some yeast, protozoa, flies, and bacteria.

Examples of the types of gene rearrangement which may occur are shown in Figure 7.17. Gene movement may be another important way in which expression of genetic information is controlled. If a transposon moves to a

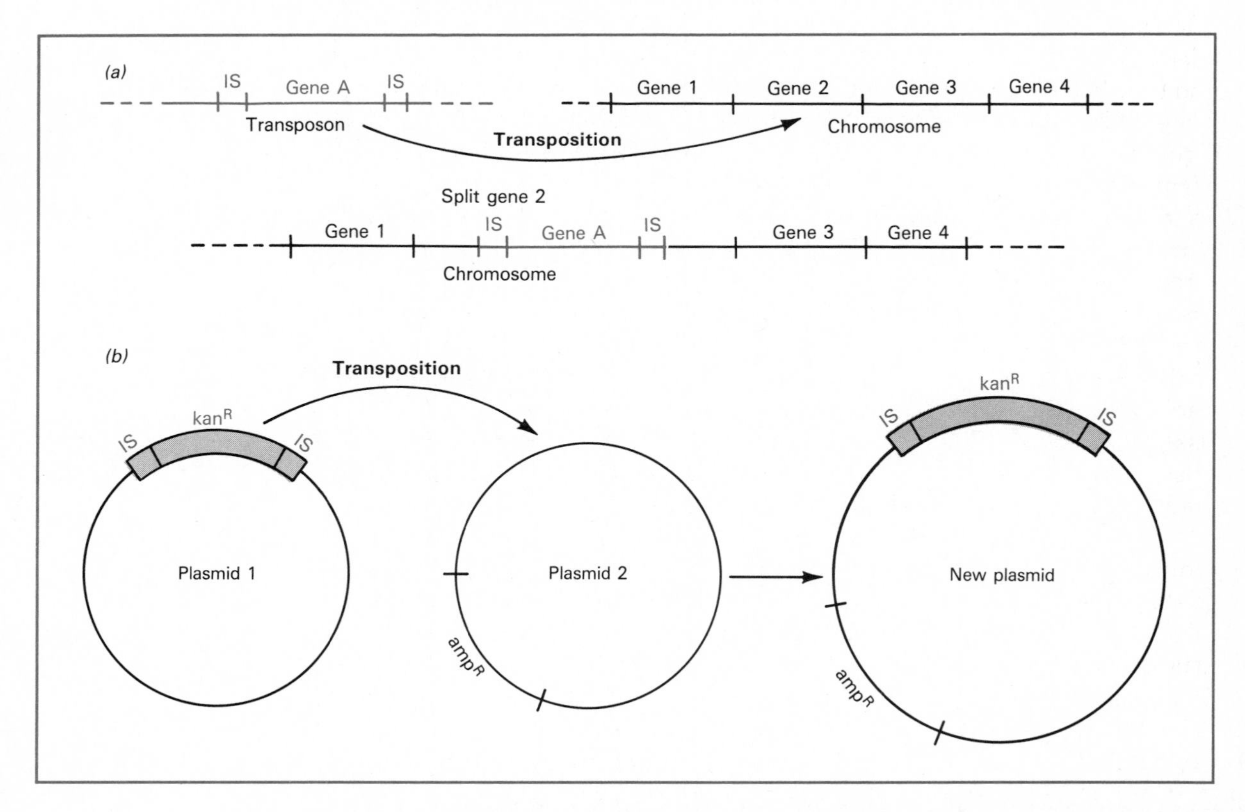

Figure 7.17 Transposons and their actions. (a) The transposon moves into the middle of gene 2. Gene 2 is now split by the transposon and is inactivated. Gene A of the transposon may be expressed. (b) Assembly of a plasmid containing multiple antibiotic resistance from two separate plasmids, one of which contains a transposon. amp^r, ampicillin resistance; kan^r, kanamycin resistance.

position within a gene, as shown in Figure 7.17*a*, the gene will be inactivated. The effect is similar to an insertion mutation (see Figure 7.2). Thus, movement of transposons into or out of genes is one way to inactivate or activate the gene.

Gene movement may also play a critical role in the assembly of plasmids. As shown in Figure 7.17*b*, two plasmids may recombine so that one acquires the genetic information of the other. This may explain how plasmids which code for multiple antibiotic resistance are produced. In the example shown in Figure 7.17*b*, plasmid 1 contains a gene which confers resistance to the antibiotic kanamycin; plasmid 2 contains a gene which confers resistance to the antibiotic ampicillin. Transposition from plasmid 1 to plasmid 2 results in the construction of a new plasmid which confers resistance to both antibiotics. A bacterium possessing this new plasmid will be resistant to both kanamycin and ampicillin. Thus, transposition can bring about significant change in the phenotype of an organism.

7.7 GENETIC ENGINEERING

The natural mechanisms for recombination provide a means of genetic exchange between closely related species. The universality of the genetic code and the commonality of the mechanisms for protein synthesis among all living things led scientists to predict that it should also be possible to artificially recombine DNAs from widely differing organisms. This would enable the use of microorganisms, which are easily grown both in laboratory culture and on an industrial scale, to produce products encoded by genes of higher life forms. It might also be possible to move genes from a microbe into a higher life form. These procedures, categorized as **recombinant DNA technology**, were first developed in the early 1970s. In the past decade, we have seen great increases in the sophistication by which we can artificially recombine genes, so that we are now seeing rapid development of this new **biotechnology** (see Table 1.5 and 1.6). As with any other emerging technology, there are both risks and benefits which arise with its development and application. How scientists perceived the potential problems of genetic engineering is considered in the Box on the next page.

The principles behind recombinant DNA technology are similar to those which occur in natural mechanisms of recombination when DNA from two related organisms is recombined. In normal genetic recombination, pieces of DNA from two separate sources are brought together and joined through action of normal cell enzymes. These same enzymes can be used to join pieces of DNA artificially. The DNA fragments are also formed in normal genetic recombination through action of enzymes, and some of these same enzymes can be used to create artificial DNA fragments. The lengths of the fragments that are used in genetic engineering are in general much shorter than those involved in normal genetic recombination, but the principles behind the recombination process are similar. The overall process can be broken down into several smaller essential steps, as described below and as shown in Figure 7.18.

HAZARDS OF GENETIC ENGINEERING?

The development of the technology for moving DNA from one organism into another organism, called *recombinant DNA technology,* has had significant impact on science, business, and the lives of humans in general. The ability to study specific fragments of foreign DNA in a well-understood microorganism could, for example, aid scientists in understanding the molecular basis of many of the diseases that afflict humans. Industry was interested in such technology, because of its tremendous potential for improving production. For the first time, a fast-growing microorganism could be used to produce the proteins encoded by genes of a higher organism. For example, insulin, normally made by humans and other animals, can be produced in large amounts by a bacterium into which the human insulin gene has been inserted. The biotechnology industry has attracted extensive capital investment because of its potential to create new products of predicted high demand. The same recombinant DNA technology, however, could also be used either accidentally or intentionally to create new organisms harmful to humans. For instance, it is now possible to create a potentially harmful organism by combining characteristics of a particularly harmful microbe (e.g., a pathogen) and a normal inhabitant of the human body.

However, the scientists who developed this new technology recognized the potential dangers and took the first steps toward the regulation of recombinant DNA technology. Perhaps they saw a parallel with the development of destructive nuclear weapons which followed the discovery of nuclear fission. Following the scientists' early lead, official guidelines for recombinant DNA research were adopted by an agency of the United States government, the National Institutes of Health. With these guidelines in place, research on recombinant DNA technology has developed at an impressive rate, and the tools of recombinant DNA research now find wide use in almost all phases of biomedical investigation (see Chapters 1 and 7).

1. *Isolating the desired gene.* First, the DNA of two different organisms must be obtained. It is impossible to move all of the DNA from one organism into another, and it is usually only of interest to move the DNA which encodes a particular product. This step involves breaking the donor organism's DNA into fragments small enough to be moved, but large enough to contain the desired gene or genes. Special enzymes, called **restriction enzymes**, are used to cut up the DNA into discrete fragments. Restriction enzymes have the unique ability to cut DNA at specific sequences without causing further degradation of the DNA. Each restriction enzyme recognizes a specific nucleotide sequence, so that different restriction enzymes cut DNA at different places. The fragmented DNA from a donor organism is usually moved into a recipient organism by attaching it to a small piece of carrier DNA, called a **vector**, which has also been cut by the same restriction enzyme. A special feature of *vector* DNA is that it can replicate by itself in the recipient, thus ensuring that the DNA of interest is not lost during cell division. An enzyme called *DNA ligase* is used to join the fragments into a new DNA molecule. The most widely used vectors are plasmids and temperate viruses.
2. *Moving recombinant DNA into host cells.* The vector bearing recombined foreign DNA is commonly moved into the recipient (host) cell using

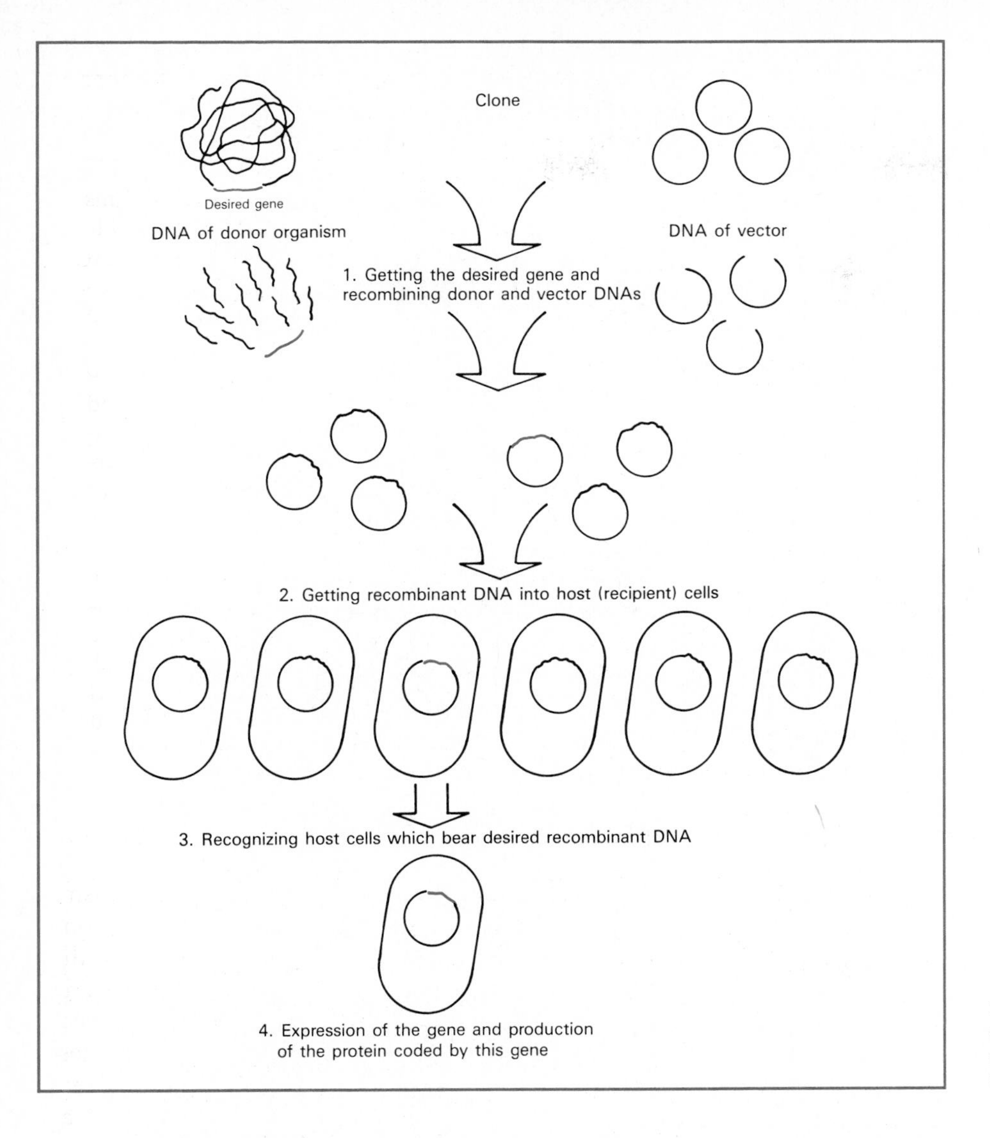
Clone
Desired gene
DNA of donor organism
DNA of vector
1. Getting the desired gene and recombining donor and vector DNAs
2. Getting recombinant DNA into host (recipient) cells
3. Recognizing host cells which bear desired recombinant DNA
4. Expression of the gene and production of the protein coded by this gene

Figure 7.18 Overview of the essential steps in genetic engineering: moving a gene from one organism to another.

transformation (discussed in Section 7.3). Since the vector can replicate in the host, the foreign DNA attached to the vector is also replicated.

3. *Recognizing host cells bearing recombinant DNA.* The recipient cell which bears the desired foreign gene must be selected. This is necessary since numerous other foreign genes might also recombine with vector DNA and be taken up by recipient cells. Selectable genetic characters (see Section 7.1) make it possible to readily recognize colonies of the desired host cell.
4. *Getting the recombined gene expressed.* Finally, expression of the foreign DNA must be accomplished so that the recipient cell produces large amounts of the desired product. This is one of the most crucial aspect of genetic engineering and is sometimes the most difficult problem.

Since a pure culture of cells all derived from a single cell is called a *clone* (see Chapter 4), if such a culture brings about the replication of a fragment of foreign DNA, the culture is said to contain a *DNA clone*. The whole technique for isolating and obtaining the replication of foreign DNA is thus sometimes called *DNA cloning*. Now that we have presented an overview of the process, we will consider the special tools of genetic engineering, and then proceed to a more detailed discussion of the procedures and the problems they involve.

Special enzymes for genetic engineering It is essential that the DNAs which are to be recombined be prepared in such a way that they can be joined together. Many cells contain enzymes which degrade DNA. A good reason to have such enzymes is probably that they confer protection from foreign DNA molecules (for example, viruses or transposons) which might injure the cell. An example of **restriction enzyme** action is shown in Figure 7.19*a*.

As we have noted, there are many different restriction enzymes, each of which recognizes a different nucleotide sequence. Those which recognize a sequence of four nucleotides are useful for cutting DNA into small pieces since there is a large number of such sequences in the DNA. Those which recognize sequences of more than four nucleotides are useful in cutting DNA into larger pieces since the occurrence of specific sequences of five or more nucleotides in DNA is more rare. A restriction enzyme does not destroy the DNA of the organism producing it, because the producing organism *modifies* its own DNA (usually by adding a $-CH_3$ group) at the site recognized by the restriction enzyme. Thus, a restriction enzyme selectively degrades foreign DNA. A restriction enzyme is designated by an abbreviation derived from the name of the source organism, followed by a number which indicates the specific bacterial strain. For example, EcoRI is a restriction enzyme from *Escherichia coli*, BamH1 is an enzyme from *Bacillus amylofaciens*.

The cutting action of many restriction enzymes results in the double-stranded DNA molecule being cut at the recognition site in such a way that each end contains a few nucleotides of single-stranded DNA. The single-stranded regions at each end are from opposite strands and are thus complementary. Because these complementary single-stranded ends will associate with each other, they are referred to as "sticky ends." If the DNA of both the donor and the vector are cut with the same restriction enzyme, the foreign DNA and the vector DNA will have the same sticky ends, as shown in Figure 7.19.

Although sticky ends tend to allow the association of foreign and vector DNA, they do not actually join the two pieces of DNA covalently. As shown in Figure 7.19*b*, this is accomplished by **DNA ligase**, which is normally involved in DNA replication and repair of damaged DNA within cells. Thus, by the combined use of a restriction enzyme and DNA ligase, DNA from two separate organisms can be joined together in a single molecule.

Vectors for genetic engineering The purpose of the **vector** is to bring about the entry of the recombined DNA into a recipient cell. The types of vectors commonly used are in fact the same entities which are responsible for natural movement of genes between cells—plasmids and viruses.

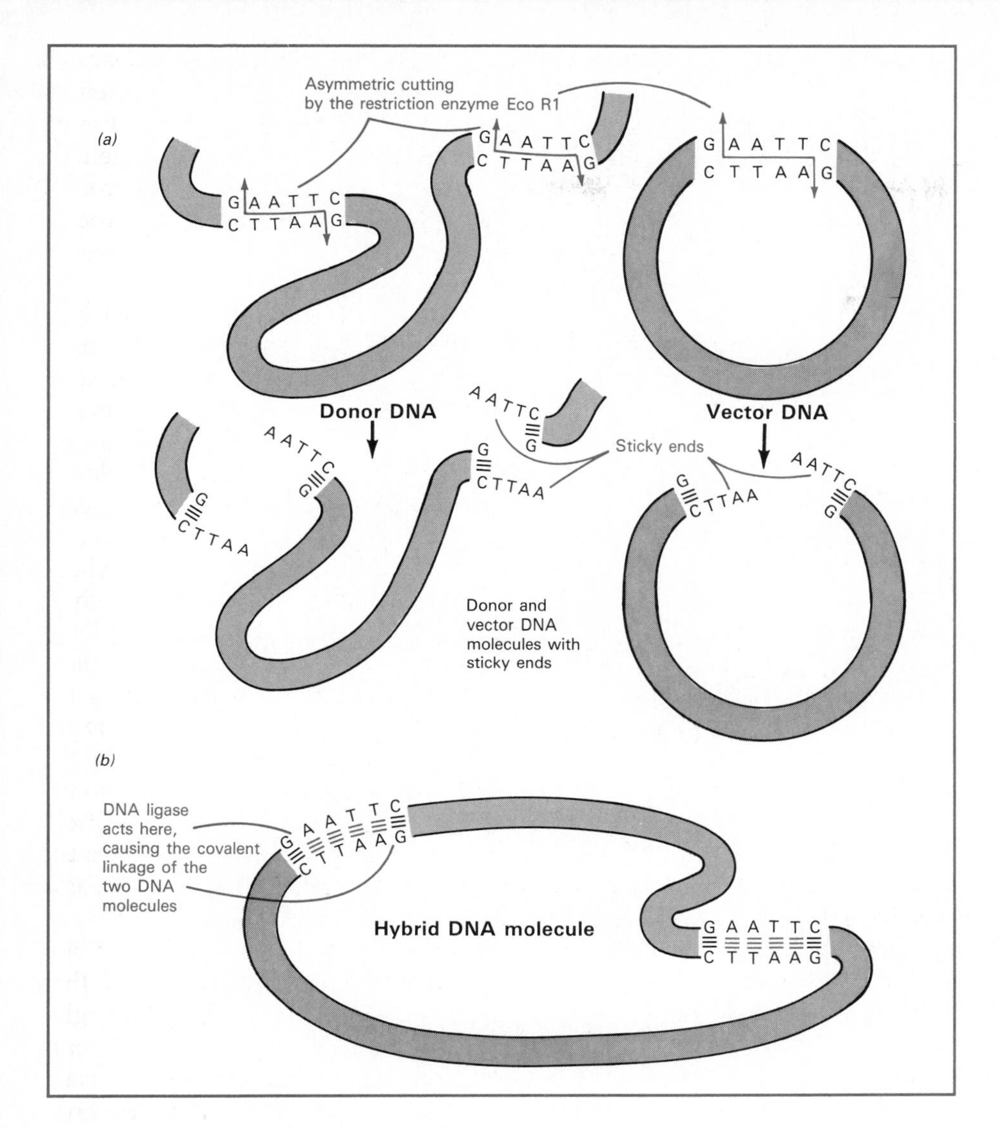

Figure 7.19 Use of special enzymes in genetic engineering. (a) The specific cutting of DNA by a restriction enzyme results in the formation of ends which contain small single-stranded complementary sequences ("sticky ends"). (b) The enzyme DNA ligase links pieces of DNA that have become associated by their sticky ends.

Vectors which are particularly useful in genetic engineering contain nucleotide sequences recognized by many different restriction enzymes, as this increases the number of different types of foreign DNA fragments with which they may recombine. Examples of a plasmid and a viral vector are shown in Figure 7.20. Plasmid vectors usually contain a single site for each restriction enzyme so that cutting simply opens the circular DNA molecule. Some virus vectors, such as bacteriophage λ (lambda), have two sites for each restriction enzyme. Cutting the vector thus yields three fragments, a left arm, a right arm, and a *stuffer*. When recombination occurs, the stuffer, which is not essential for viral reproduction, is replaced by the foreign DNA.

Another useful feature of a vector is the possession of restriction enzyme sequences within one of the important plasmid genes. As shown in Figure 7.20*a*, the commonly used plasmid vector called pBR322 contains tetracycline-

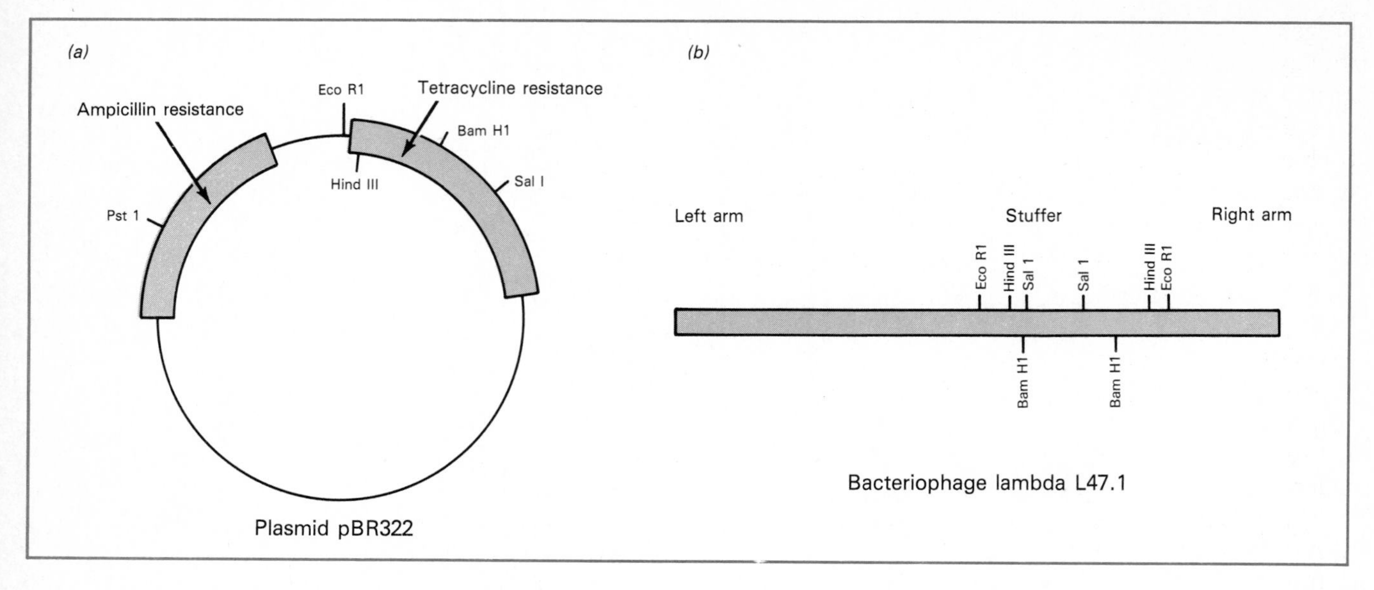

Figure 7.20 Vectors commonly used in genetic engineering. (a) The plasmid pBR322 contains antibiotic resistance genes and sites at which specific restriction enzymes act. (b) The bacterial virus λ is a temperate virus that contains a nonessential region (stuffer) in which foreign genes can be inserted.

resistance gene sequences which are recognized by three different restriction enzymes, BamH1, HindIII, and SalI. When pBR322 and foreign DNA that have been cut by one of these enzymes are recombined, the foreign DNA becomes inserted within the tetracycline-resistance gene, thus inactivating it. Acquisition of sensitivity to tetracycline then permits selection of recombinant plasmids.

Hosts for Vectors The nature of the host cell depends on the goal of the particular genetic engineering experiment being performed. Bacteria are used most commonly as host cells. They are easy to grow and can make products encoded by the DNA of other bacteria or of higher organisms. The specific bacterium used depends on the type of vector used, since the vector must be able to function within the host. Most of the bacterial plasmids and viruses used as vectors are able to infect *Escherichia coli*, and this is therefore the most common bacterial host. To avoid possible hazards, the *E. coli* strains used as hosts in genetic engineering have been "crippled" so that they are unable to grow in the human gut (their natural habitat) and are entirely dependent on laboratory culture. This makes it possible to avoid the escape of potentially hazardous genetically engineered microbes by biological means (called *biological containment*) in parallel with the physical containment discussed in Chapter 4.

It is also possible to transfer genes into eucaryotic microorganisms. Yeasts, such as *Saccharomyces cerevisiae*, contain plasmids which may be used as vectors for introducing foreign DNA. The cells of higher organisms grown in culture, or higher organisms themselves may also serve as hosts for foreign genes. Certain plant and animal viruses are known to integrate into the DNA of the cells they infect, and thus may be used as vectors to carry foreign genes.

Finding a desired gene by hybridization One major problem in genetic engineering is the determination of whether a clone of cells contains the gene of interest. The most common procedure is to detect the desired gene by **hybridization**. This requires a piece of DNA, called the *hybridization probe*, which is complementary to part of the gene of interest. As described in Figure 6.5, two identical, or *homologous*, regions of nucleotide sequence will bind to each other because of complementary base pairing to form double-stranded DNA. To search for a specific gene in a mixture of genes, one needs to separate the fragmented DNA into single strands and then allow the single-strand fragments to react with a single strand of a specific hybridization probe, as shown in Figure 7.21.

Other approaches to obtaining a desired gene A desired gene can be produced by *chemical synthesis* (Figure 7.21). This is only possible when the protein encoded by the gene is relatively small and its primary amino acid sequence is known. Using the genetic code it is possible to predict the nucleotide sequence of the gene from the amino acid sequence of the protein. It is then possible to chemically link nucleotides together to form a particular DNA sequence. An artificial gene of this type has been made which codes for the protein human insulin and this gene has been recombined with the plasmid pBR322 and placed into a strain of *E. coli*. A commercial process for making human insulin with such a strain has been perfected.

Another method for getting a desired gene is to make it from the mRNA molecule which carries the message encoded by the gene (Figure 7.21). The mRNA is then used as a template against which DNA is synthesized using the enzyme **reverse transcriptase**. This is the same enzyme used by some RNA viruses to produce DNA before insertion into the host cell's chromosomal DNA (as described in Figure 6.17). The DNA produced in this way, called **complementary DNA** (or **cDNA**) can be recombined with vector DNA and transferred into a host recipient cell. This approach has been used successfully to transfer into *E. coli* genes for the production of human interferon

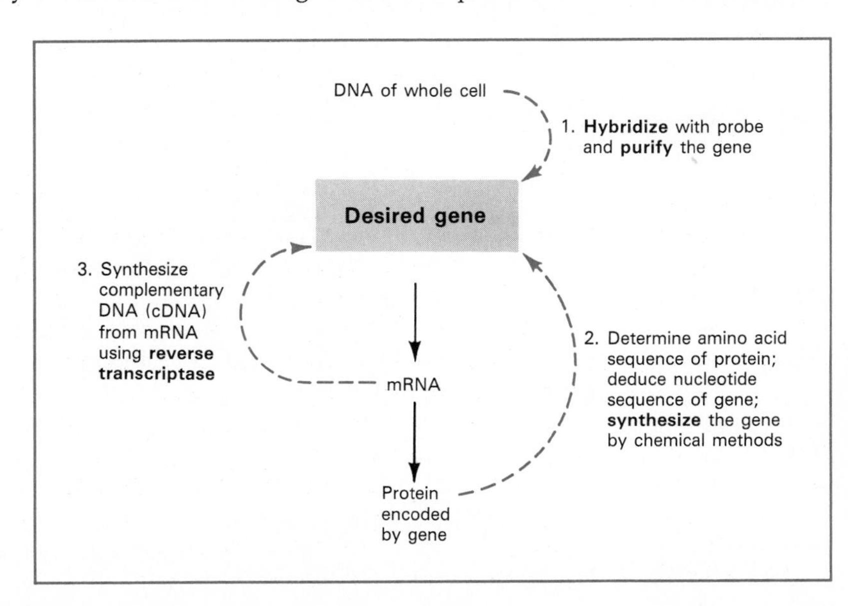

Figure 7.21 Alternate approaches to obtaining the desired gene for recombinant DNA technology.

(a protein involved in resistance to virus infection). It has also been used to prepare DNA from RNA viruses so that they can propagate in *E. coli*. The bacterium then expresses the viral genes, producing viral proteins which may be useful as vaccines. We will discuss vaccines and their production in Chapter 9.

A major difficulty arises in attempting to get bacteria to express mammalian genes. As illustrated in Figure 6.14, the genes of eucaryotic cells usually contain intervening sequences (**introns**) which do not code for the amino acid sequence of the gene product. Eucaryotic cells have a special mechanism for converting the primary transcript into the mature mRNA, but bacteria are unable to perform such modification. If the unmodified eucaryotic mRNA were translated directly, an incorrect (and inactive) protein would be made. To avoid this problem, instead of using a eucaryotic gene directly for cloning, the mRNA of the eucaryotic cell, from which the introns would have already been removed, is used. Reverse transcriptase is then used to copy information from the eucaryotic mRNA into DNA, and this DNA, now containing the gene as a single unit, is cloned in the bacterial host.

7.8 USES OF GENETIC ENGINEERING

A number of beneficial applications have come in the use of recombinant DNA technology to make products encoded by mammalian genes. Some examples are given in Table 7.1. Many of these products are produced in healthy humans, but may be required by people whose health is impaired. For example, diabetics are deficient in the protein insulin, which controls carbohydrate metabolism in the body. Insulin produced by pigs can be isolated and used to treat human diabetics, but its preparation is expensive, and it is less effective than human insulin in the body. By transferring the human insulin gene into *E. coli*, the bacterium is used as a "cellular factory" which can produce insulin. As a result, the production cost is lowered and a more effective product is obtained.

TABLE 7.1 Some proteins made by genetically engineered microorganisms

Protein product	Function
Mammalian source	
Interferon	Antiviral and anticancer agent
Insulin	Treatment of diabetes
Serum albumin	Transfusion applications
Growth hormone	Growth defects
Urokinase	Blood-clotting disorders
Parathyroid hormone	Calcium regulation
Viral source	
Human virus coat protein (Hepatitis B, cytomegalovirus, influenza)	Vaccines
Animal virus coat protein (foot-and-mouth disease)	Vaccines

People with growth deficiencies, such as dwarfs, have a genetic disorder and cannot produce human growth hormone. By using genetic engineering techniques, a new *E. coli* capable of producing human growth hormone has been developed. Interferon is a protein which is produced by human cells and which may have antiviral action. Because it is produced in the body in very small quantity, it has been difficult to obtain enough of it to evaluate its effectiveness in controlling viral diseases. By transferring the interferon gene into *E. coli*, it has now been possible to obtain much larger amounts of interferon, and at less cost, so that more complete studies of its activity can be carried out.

Genetic engineering has also been used in the production of vaccines. The immune system, and how it functions to control diseases, will be considered in detail in Chapter 9. One type of vaccine for viruses consists of the protein coat of the virus, which when injected into the human body induces immunity (as described in Chapter 9). The conventional way of preparing a virus vaccine is to produce large amounts of virus and then inactivate the virus particles by chemical treatment. There are potential hazards with this method, since complete inactivation of all virus particles may not occur. Such hazards can be avoided by cloning the virus gene for virus protein coat into a bacterial system. Production of virus protein can thus be obtained in the bacterium and the need to deal with live virus is eliminated. For example, the proteins of the virus which causes foot-and-mouth disease in cattle have been transferred into *E. coli*. These proteins produced by *E. coli* are now routinely used in many countries as a vaccine to immunize cattle against the disease.

SUMMARY

In this chapter we have discussed **genetics**, the study of the inheritable characteristics of microorganisms. Change in the genetic material of a cell involves alteration of the base sequence in the DNA, and may be either harmful or beneficial to the microorganism. **Natural selection** of beneficial changes is the driving force behind evolution. **Mutations** are small changes in the DNA sequence which occur as rare *spontaneous* events or more frequently when induced by **mutagenic** chemicals or radiation. Mutagens are often **carcinogenic**, and bacteria have become useful in testing for the potential of chemicals to cause cancer. **Genetic recombination** is a more dramatic type of genetic change than mutation and involves the transfer of genes or parts of genes between organisms, resulting in the formation of **hybrids** of the DNA deriving from each of two parent organisms. Recombination occurs in eucaryotic organisms during sexual reproduction. In procaryotes, three distinct mechanisms for the transfer of genes may occur. In each case, one cell acts as DNA **donor** and the other as DNA **recipient**. **Transformation** involves movement of naked DNA from one cell to another; **transduction** involves the transfer of DNA from donor to recipient via viruses; and in **conjugation** the movement of DNA between donor and recipient involves cell-to-cell contact. These exchanges of DNA between bacteria can be especially significant in the case of transfer of **plasmids**, small extrachromosomal pieces of DNA, which may code for **multiple antibiotic resistance**.

The understanding of how genetic information moves naturally between cells has led to the development of a new industry, called **recombinant DNA technology**. DNA from two different organisms can be recombined artificially so that new organisms with completely new characteristics can be formed. This is accomplished by cutting DNA from two sources with **restriction enzymes**, recombining them, and transferring the recombinant DNA into a **host** in which the genetic information is expressed. In this way, bacteria have been made to produce proteins encoded by human DNA, such as insulin, interferon, and human growth hormone. These new products of **biotechnology**, as well as such other new products as vaccines against viral diseases, demonstrate the practical aspects of genetics.

KEY WORDS AND CONCEPTS

Genetics
Mutation
Mutagen
Carcinogen
Recombination
Recombinant
Donor
Recipient
Transformation
Transduction
Conjugation
Plasmid
Multiple drug resistance
Transposon
Genetic engineering
Restriction enzyme
Vector
DNA ligase
Hybridization
Complementary DNA (cDNA)

STUDY QUESTIONS

1. Explain how *spontaneous mutation* and *natural selection* are involved in evolution.
2. Draw a diagram of a short DNA sequence and show how a mutation might occur. Use Table 6.1, to determine the effect of the mutation you diagramed on the amino acid encoded. Remember that the information in Table 6.1 gives base sequences of mRNA codons, rather than DNA base sequences.
3. Starting with the same DNA sequence you used in question 3, diagram an *insertion* or a *deletion mutation*. How has each of these changed the reading frame?
4. Explain how in the *Ames Test* bacteria are used to test for *mutagenicity* of a chemical. Why can it also be used to test for the *carcinogenicity* of a chemical?
5. What is the difference between *mutation* and *recombination* as ways in which genetic information can be altered?

6. What is meant by the term *hybrid*?
7. Discuss the differences between *transformation*, *transduction*, and *conjugation*. Which process involves viruses? Which process involves *plasmids*?
8. What is the difference between *generalized* and *specialized* transduction? Which mechanism involves lytic infection? Which involves lysogenic infection?
9. Though *plasmids* are important to bacteria, they are also important to humans. Explain.
10. Explain how *conjugation* can be used to construct a *genetic map*.
11. What is one unique property of *transposons*? Why might transposons be important in the assembly of *plasmids* encoding for *multiple drug resistance*?
12. Explain how *restriction enzymes* cut DNA to produce "sticky ends". How do "sticky ends" aid the *recombination* of foreign and *vector* DNA during *genetic engineering*?
13. What is a *vector*? What kinds of DNA can be used as a vector in *genetic engineering*?
14. In *genetic engineering* the goal is to place a gene of interest from one organism into another organism. However the "desired gene" is one among many other genes. Describe three different methods by which you might obtain the "desired gene".
15. Name three products normally made by humans which are now made by *recombinant* bacteria, as a result of *genetic engineering*.

SUGGESTED READINGS

BROCK, T.D., D.W. SMITH, and M.T. MADIGAN. 1984. *Biology of microorganisms, 4th edition*, Prentice-Hall, Inc., Englewood Cliffs, N.J.

CHILTON, M. 1983. *A vector for introducing new genes into plants*, Scientific American 248: 50–59. (Discussion of an approach using bacteria which cause plant disease to genetically engineer crop plants.)

FEDERAL REGISTER, volume 48, No. 106, June 1, 1983, Notices, Part III, Department of Health and Human Services, National Institutes of Health, *Guidelines for research involving recombinant DNA molecules: June 1983.* (The current guidelines governing the use of genetic engineering techniques, including physical and biological containment levels and methods.)

LEWIN, B.J. 1985. *Genes II*, John Wiley & Sons, Inc., New York, N.Y.

MANIATIS, T., E.F. FRITSCH, and J. SAMBROOK. 1982. *Molecular cloning, a laboratory manual*, Cold Spring Harbor Laboratory, Cold Spring Harbor, N.Y. (A complete manual for modern techniques used in recombinant DNA technology, including basic information on vectors, restriction enzymes, hybridization techniques, and many other modern methods of genetic engineering.)

PESTKA, S. 1983. *The purification and manufacture of human interferons,* Scientific American 249: 37–43. (Use of genetic engineering to develop bacteria producing human interferon.)

WATSON, J.D., J. TOOZE, and D.T. KURTZ. 1983. *Recombinant DNA, a short course.* Scientific American Books, New York.

8

Infectious Disease

Many microorganisms are capable of living and growing in or on higher organisms, including humans. The organism supporting the growth of a microorganism is called a **host**. The host provides environmental and nutritional factors necessary for the survival and growth of the microorganism. In some cases, the host also benefits from the association, generally because the microbe carries out some function that the host itself is unable to carry out. Those cases where *both* microbe and host benefit are called **mutualistic**. Several of these mutualistic relationships will be discussed in Chapter 17.

In many cases, only the microbe benefits from the association, and the host is either unaffected or is actually harmed. When only the microbe benefits, the relationship is termed *parasitic* and the microbe that benefits is called a **parasite**. In some cases, the parasite has little or no harmful effect on the host and its presence may be inapparent. Such situations are called *commensal*. In many cases the parasite brings about damage or harm to the host: such organisms are called **pathogens**. In most cases, the pathogen is also a parasite, and the harm is brought about as a result of growth of the parasite in the host.* A microorganism growing on dead materials outside the body is called

*Strictly speaking, a parasite need not harm its host, but the word *parasite* generally retains some of the concept of *host damage*. It is preferable to use the term *pathogen* when host damage is implied.

TABLE 8.1 Relationships between microorganisms and hosts

Term	Condition
Saprophytic	Microbe lives outside of host
Mutualistic	Both microbe and host benefit
Parasitic	Microbe alone benefits
Commensalistic	Microbe benefits and host not harmed
Pathogenic	Microbe benefits and host is harmed

a *saprophyte*. An overall summary of the terminology used in discussing host-parasite relationships is given in Table 8.1. **Infection** is the process in which a pathogen becomes established in a host and the condition brought about by a pathogen when growing in the host is thus called *infectious disease*.

It should be emphasized that infection is not the same as disease, since a potential pathogen may grow in a host without causing harm. Also, a microorganism can be pathogenic without growing within the host. This situation arises when an organism growing outside the body, such as in a food, produces a harmful product (a toxin) which the host ingests. The condition which results is an example of a *noninfectious disease* caused by a microorganism.

Further, not all diseases are caused by microorganisms. Examples of other noninfectious diseases include those caused by nutritional deficiencies, genetic defects, hormonal imbalances, problems of the immune system, higher organisms such as worms and insects, and other factors as shown in Table 8.2. Diseases caused by living parasitic organisms are often called **contagious** or **infectious diseases**, since the causal agent is transmissible from one individual to another.

During the course of infectious disease, the pathogen invades the host. Higher organisms are equipped with a variety of mechanisms of **host defense** which aid in protection against invasion by a pathogen. Pathogens vary in the degree to which they invade the body. The capacity of a pathogen to

TABLE 8.2 Examples of infectious and noninfectious diseases and their causes

Disease	Cause
Infectious diseases	
Influenza	Virus
Streptococcal sore throat	Bacterium
Athlete's foot	Fungus
Malaria	Protozoan
Noninfectious diseases	
Botulism	Toxic substance made by a bacterium
Rickets	Nutritional deficiency
Sickle cell anemia	Genetic defect
Rheumatic fever	Immune response against the body
Poison ivy	Allergic immune response

cause disease, termed **virulence**, is determined by specific mechanisms which aid in overcoming host defenses. The infectious disease process can thus be viewed as the conflict between pathogen virulence and host defenses (Figure 8.1). This dynamic process can be influenced by either the virulence of the pathogen or the strength of the host defense. Disease results whenever the pathogen virulence overcomes the host defense, a condition which may result either from weakened defense of the host or from a very high virulence of the pathogen (or, of course, from a combination of these factors).

In this chapter we will first consider the environments of the human body which are subject to microbial attack, and the microorganisms which normally inhabit these environments. We will then consider the means by which pathogens invade the human body, and the defensive responses which result during infection. Clinical aspects of infectious disease, including the role a microbiologist can play in diagnosis and treatment of disease, will also be considered. Despite the importance of microbes as disease agents, we should emphasize that most microorganisms do not cause disease at all. Most are

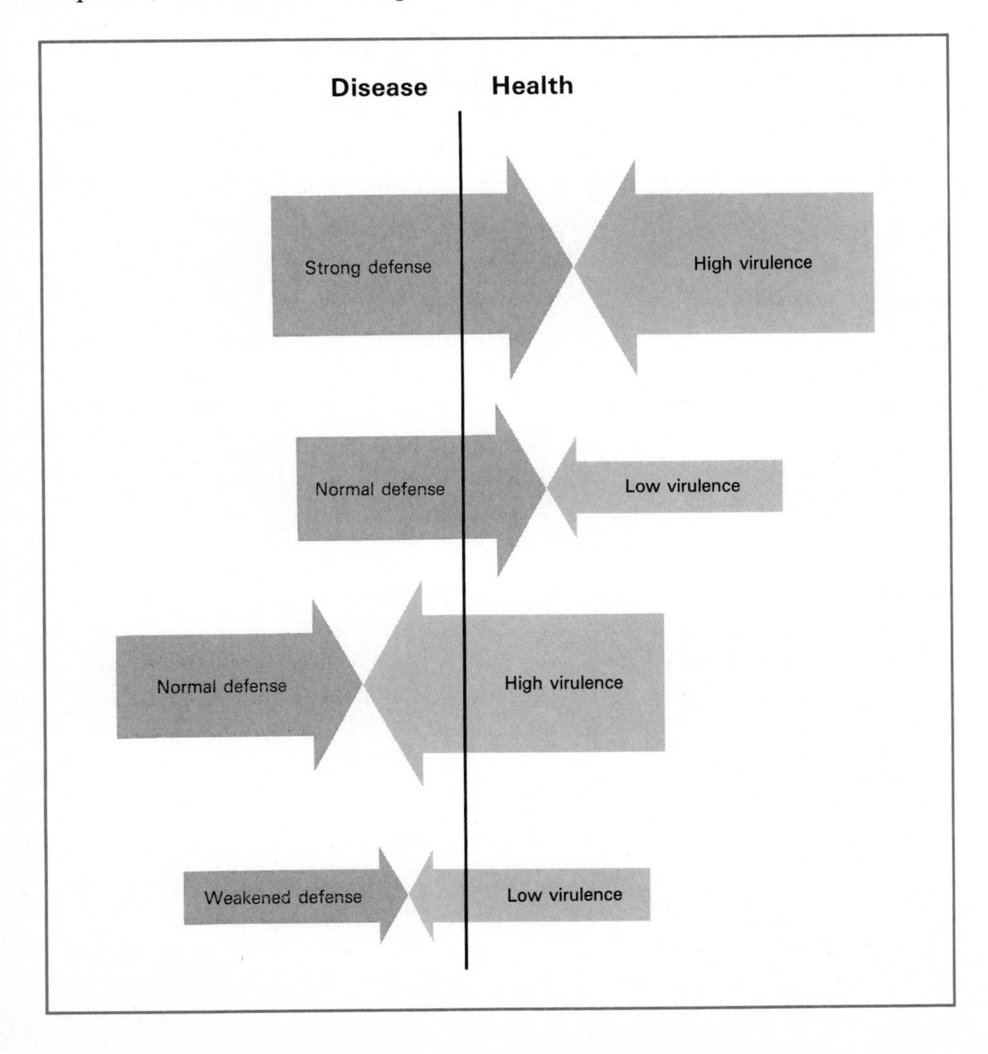

Figure 8.1 Outcome of a host-parasite interaction depending on the virulence of the pathogen and the defense of the host.

harmless and some are beneficial (or even essential) for the lives of their hosts.

8.1 HUMAN TISSUES AND THEIR NORMAL MICROBIAL INHABITANTS

The environments subject to invasion by pathogenic microorganisms include the exterior surfaces of the body and the soft tissues which line the various body cavities including the mouth, the gastrointestinal tract, the upper respiratory tract, and the linings of the genitourinary tract. Collectively, these tissues are referred to as **epithelial layers**. The cells of the epithelial layers are cemented together by extracellular polymeric substances in such a way that a physical barrier to invading microorganisms exists.

Epithelial surfaces are generally not sterile, but are colonized by microorganisms. These microorganisms represent the **normal flora** of the body. Microbes of the normal flora can be either beneficial or harmful. In some cases, the presence of a harmless microbe prevents infection by a pathogenic one. On the other hand, the normal flora may even play a role in disease if the host resistance is low. Members of the normal flora which can cause disease under the appropriate conditions of weakened host defense are called **opportunistic pathogens**.

Skin The skin is a protective barrier which plays a role in resistance of the host to infection. A cross-sectional view of skin (Figure 8.2) shows flat, scaly, dead cells in layers at the surface, and increasingly more rounded living cells below the surface. The platelike surface cells are filled with a protein

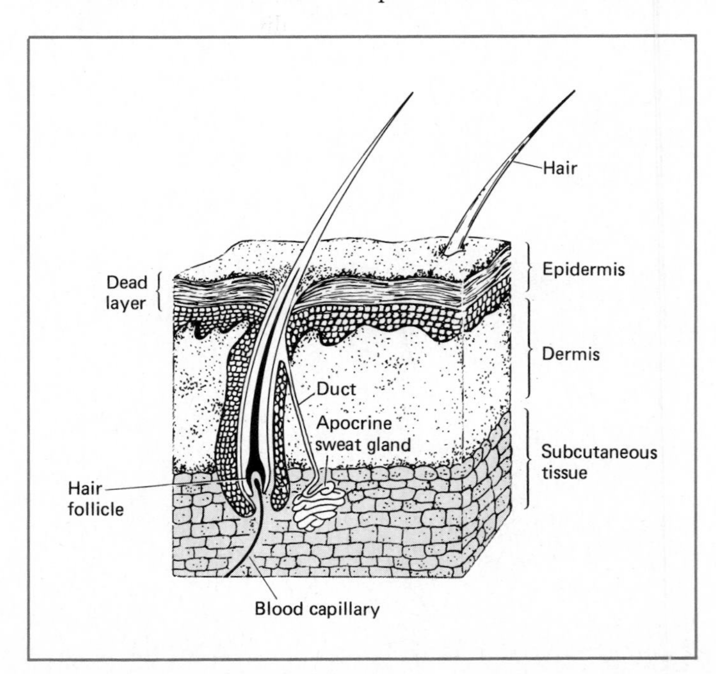

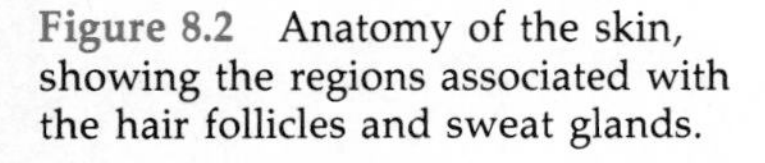
Figure 8.2 Anatomy of the skin, showing the regions associated with the hair follicles and sweat glands.

called keratin, which makes them more resistant and thus more protective to the deeper layers. Although the main function of the skin is probably protection of the body from fluid loss (drying), the hard horny surface of dead cells and keratin is also an effective barrier to microbial penetration. Much of the surface of the skin is not a favorable location for microbial growth as it is usually too dry. Most nonpathogenic microorganisms that live on the skin are restricted to the hair follicles and sweat glands, where moisture is present.

Underarm odor is caused by bacterial metabolism of sweat glands secretions. Underarm odor can thus be prevented either by destroying the bacteria present or by decreasing the flow of perspiration from the glands (for instance, by use of antiperspirants). Skin antisepsis, an important medical practice, is discussed in Chapter 10.

The pH of skin secretions in humans is usually acidic, with pH between 4 and 6, due to production of fatty acids. The microorganisms of the normal skin flora are hence usually mild acidophiles. The organisms are predominantly bacteria, including several species of *Staphylococcus* and *Corynebacterium*. The species *Corynebacterium acnes* is normally harmless but sometimes can cause *pimples* or *acne*. Most bacteria foreign to the skin die quickly when inoculated onto it. This is either because of their inability to tolerate the dryness and low pH of the skin, because their growth is inhibited by skin fatty acids, or because they cannot compete with the resident bacteria, which are better adapted to the skin environment.

The skin of the fetus is normally sterile but becomes infected at birth. The newborn baby is susceptible to infection by disease-causing organisms, such as the bacterium *Staphylococcus aureus* or the yeastlike fungus *Candida albicans*, at least in part because the baby lacks a normal flora that can compete with these pathogens. Within a week or two after birth, however, the infant has acquired its own normal skin flora, and its susceptibility to skin pathogens decreases.

Mouth The epithelial tissue within the mouth, like other moist cavity tissues (such as the esophagus and vaginal lining), is somewhat similar to skin tissue in structure, but lacks the dead keratin layer. The mouth is a favorable habitat for the growth of microorganisms, most of which attach firmly to epithelial cells. Though it might seem that the mouth is well aerated, the microenvironments within the mouth are often very anaerobic and thus a variety of anaerobic and facultatively anaerobic bacteria can be found as part of the normal flora.

Saliva contains many different microorganisms which slough off from the outermost epithelial layers, but it is not itself an especially good medium for the growth of microbes. Saliva contains antimicrobial agents which act as a chemical host defense and kill some invading microorganisms. The most important antimicrobial agent is **lysozyme**, an enzyme which digests the bacterial cell wall peptidoglycan causing lysis and death of the cell. Lysozyme is most effective against Gram-positive bacteria but also acts against Gram-negative bacteria under certain conditions. Lysozyme is also found in other secreted fluids, for example, in tears and mucus. Saliva contains an additional

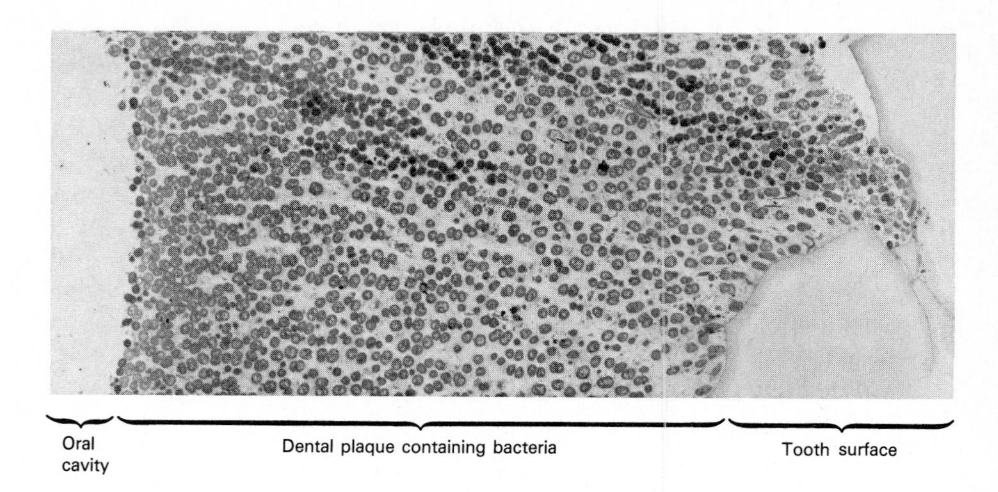

Figure 8.3 Electron microscope view of the bacteria of the dental plaque. This is a thin section made through the tooth surface. The thickness of the bacterial layer is about 40 μm.

antimicrobial defense system, called the **lactoperoxidase system**. In this system, the enzyme lactoperoxidase, together with two other substances found in saliva, thiocyanate (CNS^-) and hydrogen peroxide (H_2O_2), brings about the death of bacteria.

The surface of teeth provides a favorable environment for the growth of very large numbers of different microbes. On the surface of teeth there is a thin coating which consists of organic materials and densely packed bacterial cells (Figure 8.3). This coating, called **dental plaque**, adheres tenaciously to the surface of the teeth and if not removed by frequent brushing may develop into a thick covering. The bacteria in dental plaque are anaerobic, one of the most common ones being *Leptotrichia buccalis*. Associated with these anaerobic bacteria are facultatively anaerobic bacteria, including the important organisms *Streptococcus mutans* and *Streptococcus sanguis*, implicated as causal agents of tooth decay (*dental caries*). The crevices of the teeth, where food particles are retained, are the sites where tooth decay predominates. Diets high in sugars predispose individuals to tooth decay because the bacteria of the teeth produce lactic acid from the sugars; this acid attacks the hard surface of the tooth and causes removal of calcium from the tooth structure. Once the breakdown of the hard tissue has begun, microorganisms can penetrate more easily into the tooth and cause further breakdown. Susceptibility to tooth decay varies greatly among individuals and is also affected by diet, frequency of teeth brushing, and other factors. The incorporation of fluoride into the tooth structure makes it more resistant to attack by acid. Addition of fluorides to drinking water or dentifrices therefore aids in controlling tooth decay.

Gastrointestinal tract The anatomy of the gastrointestinal tract is shown in Figure 8.4. The gastric juice of the stomach is quite acid, with pH at 2 or below, and represents a chemical barrier to protect the intestinal tract, as many bacteria that enter the stomach in food are probably killed in this highly acidic environment. Despite the acidity of the stomach, the stomach wall of

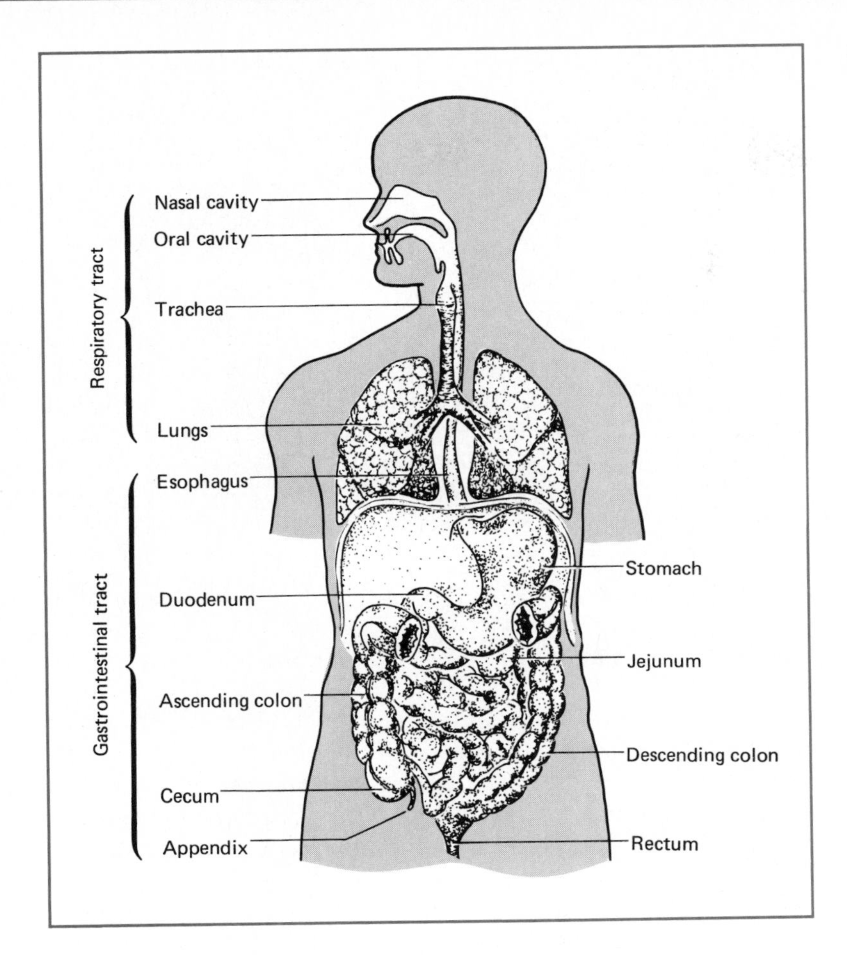

Figure 8.4 General anatomy of the gastrointestinal and respiratory tracts.

some animals is populated with nonacidophilic bacteria which live in acid-protected microenvironments, as shown in Figure 8.5.

As the food passes into the intestines, the pH gradually rises, and the lower small intestine and the large intestine are quite alkaline. The intestinal epithelium is specialized for absorption of nutrients released upon digestion of food in the stomach. Intestinal epithelial cells have numerous finger-like projections on the surface, called *microvilli*, which increase the surface area of the cell, enhancing absorption. Bacteria are present in large numbers on the walls of the intestines (Figure 8.6), and many are also found in the intestinal contents. Virtually all the bacteria of the intestines are either obligate or facultative anaerobes. The obligate anaerobic bacteria predominate and include many long, thin, Gram-negative rods which are as yet unnamed, as well as species of *Clostridium* and *Bacteroides*. Facultative anaerobes, including *Escherichia coli* and *Streptococcus faecalis*, are far less abundant.

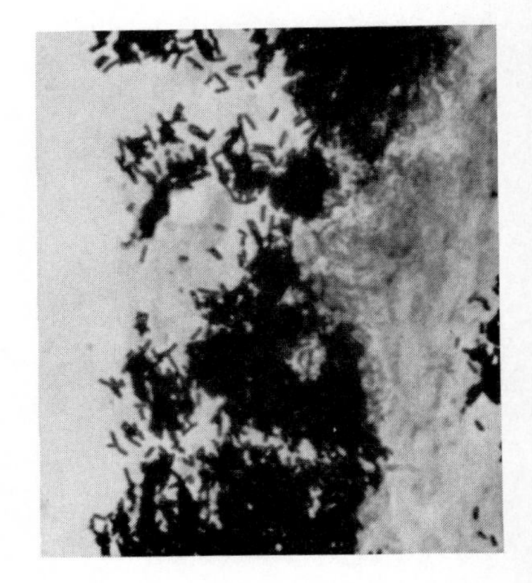

Figure 8.5 Dense layer of lactic acid bacteria attached to the wall of the stomach of a mouse. The stomach epithelium is at the right.

In the lower small intestine and colon, water is gradually removed from the digesting mass, and the contents are converted into *feces*, the excretory product. Bacteria make up much of the weight of feces. It is estimated that

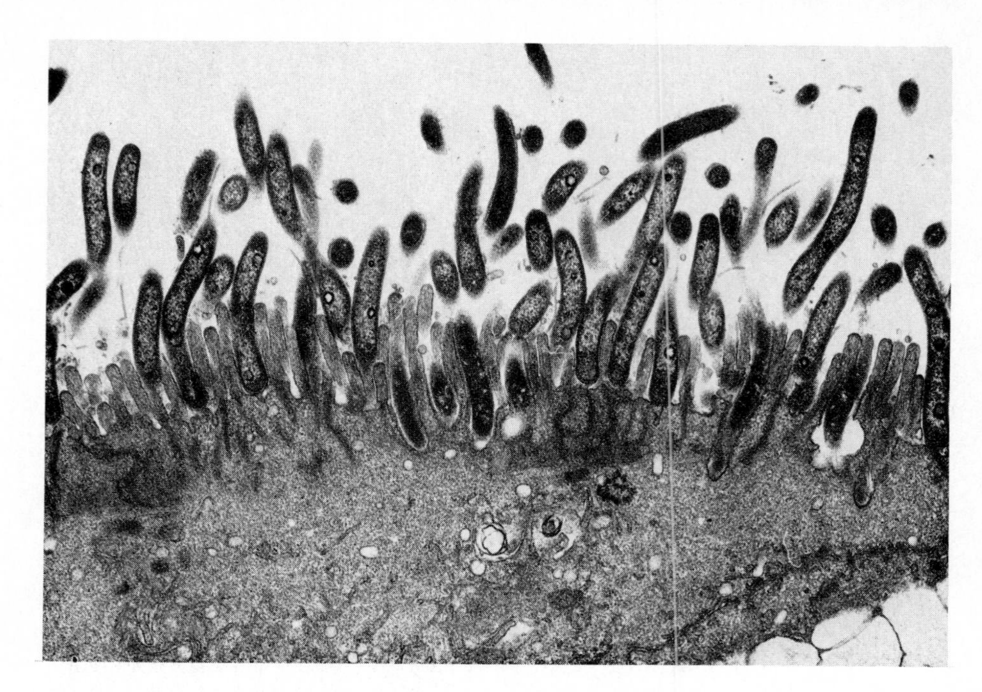

Figure 8.6 Electron microscope view of bacteria attached to epithelial cells of the large intestine of a monkey. In some cases the bacteria are embedded in pockets in the epithelial cells.

there are about 10^{11} bacterial cells per gram of feces, and perhaps as many as 400 different types.*

The bacteria that grow in the large intestine are responsible for the gases and odors produced in and expelled from the large intestine (see the Box, next page). The gases produced are primarily carbon dioxide, methane, and hydrogen, and the odors are primarily due to hydrogen sulfide, indole, and skatole.

The composition of the intestinal flora is influenced by diet. For example, breast-fed infants have a flora consisting largely of *Bifidobacterium*, a bacterium which requires a growth factor found in human but not in cow's milk. Infants fed cow's milk usually have a more complex bacterial flora. A major constituent of milk is the sugar lactose, and many of the bacteria in the large intestine can utilize lactose for growth. In adults, whose milk consumption is low, lactose-utilizing bacteria represent only a small part of the normal flora.

When certain antibiotics are given orally, they may enter the intestines and prevent the growth of intestinal bacteria. As the intestinal contents move downward, the inhibited intestinal bacteria are expelled, leading to the near-sterilization of the intestinal tract. Antibiotics given orally are thus often used to sterilize the intestinal tract before bowel or intestinal surgery. In the absence of this normal flora, however, exotic bacteria such as antibiotic-resistant *Staphylococcus aureus* or the yeast *Candida albicans* (naturally resistant to antibiotics affecting procaryotes) may become established. Normally, these organisms

*The significance of this large number of bacteria should be emphasized. If during use of toilet paper, 1 μg of feces should adhere to the fingers (a virtually invisible amount), about 10^5 bacteria would be present. Since a single cell of some of the most virulent pathogens may be capable of initiating infection, this invisible 1 μg of feces could easily be the source of a new infected host.

INTESTINAL GAS

The gas produced within the intestines, called *flatus*, is the result of the action of fermentative and methanogenic microorganisms. Some foods which are poorly absorbed in the stomach and intestines can be metabolized by fermentative bacteria resulting in the production of hydrogen (H_2) and carbon dioxide (CO_2). In many individuals, methanogenic bacteria then convert some of the H_2 and CO_2 to methane gas (CH_4). About one-third of the American population has an active methane-producing microbial flora in the intestines. Curiously, there is no pattern, such as inheritance, age, or diet, which can be associated with the presence in the intestinal tract of methanogenic bacteria.

Normal adult humans expel a few hundred milliliters of gas per day via the rectum. More than half of this gas is nitrogen (N_2) which enters the body in swallowed air and passes unchanged into the intestines, but the rest is microbially produced. Diet can have a dramatic effect on the amount and type of gas produced. If large quantities of beans are consumed, total gas production increases about tenfold. It is thought that a bean polysaccharide that is not digested by humans passes into the intestines, where fermentative bacteria convert it to H_2 and CO_2. When methane-producing bacteria are present, the H_2 and some of the CO_2 are converted to methane, so that CH_4 and CO_2 build up and are expelled. Although a high-bean diet can cause a gassy condition in most humans, some people inherently have a large amount of flatulence, even on a low-bean diet, probably due to their inability to digest or absorb certain fermentable sugars. A common reason for gassiness is poor absorption of the milk sugar lactose. If individuals with this abnormality do not eliminate lactose from their diets, their intestinal bacteria are stimulated to produce gas.

do not grow in the intestine because they cannot compete with the normal flora, but with the normal flora eliminated, they can take over. Occasionally, establishment of these exotic organisms can lead to a harmful alteration in digestive function. However, in most cases when antibiotic therapy is discontinued, the normal flora will eventually become reestablished.

Respiratory tract The gross anatomy of the respiratory tract was shown in Figure 8.4. Despite the constant inhalation of microorganisms present in the air, the lungs and lower trachea are usually sterile because of filtering and flushing actions within the upper respiratory tract. The epithelial layers of many of the tissues in the upper respiratory tract posses cilia and are covered with mucus. The cilia of upper respiratory epithelial cells propel mucus across the tissue, carrying foreign particles with it upward toward the mouth where the mucus is swallowed. This process, called **mucociliary flushing**, is thus an important physical feature of the host defense.

The tissues of the nose and throat are colonized by microorganisms. Resident normal flora include mainly aerobic and facultatively anaerobic bacteria such as *Staphylococcus*, *Streptococcus*, *Corynebacterium*, and *Moraxella*, but some anaerobic bacteria are also present.

Genitourinary tract The main anatomical features of the male and female genitourinary tracts are shown in Figure 8.7. In both male and female, the bladder itself is usually sterile, but the epithelium of the urethra is colonized by anaerobic Gram-negative rods and cocci.

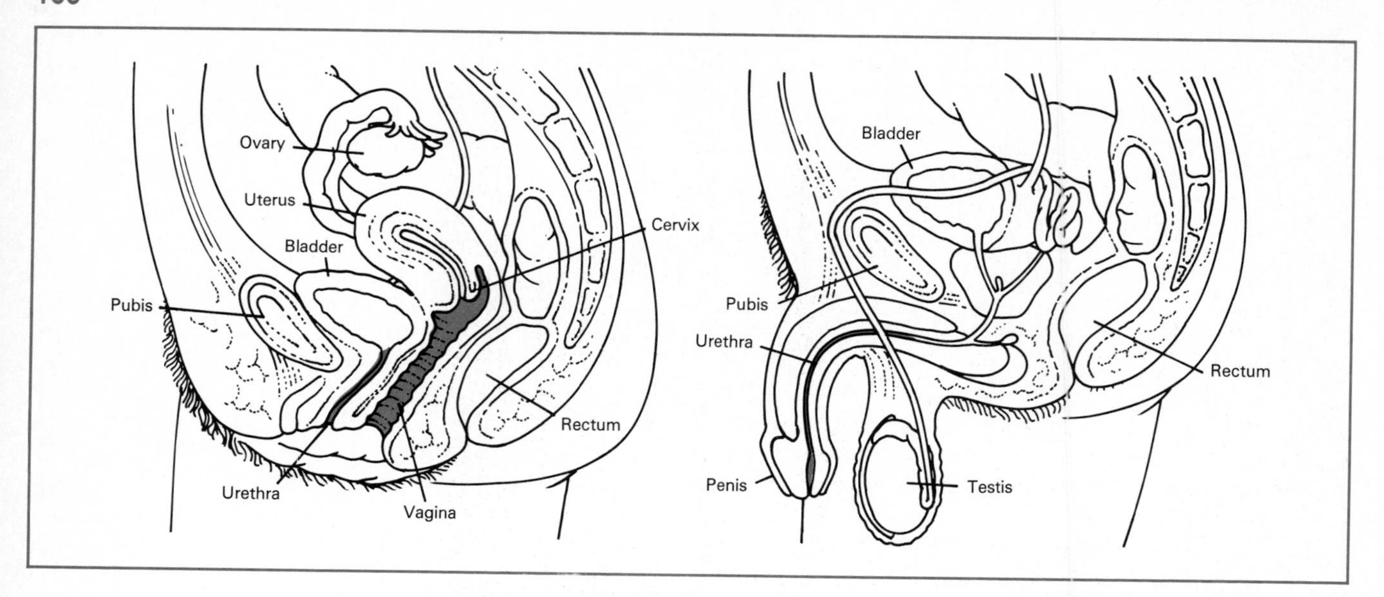

Figure 8.7 Anatomy of the genitourinary tracts of the human female and male, showing regions (color) where microorganisms often grow.

The vagina is another suitable microbial habitat and contains as normal flora a variety of anaerobic and facultatively anaerobic bacteria. Some yeasts, such as *Candida* and *Torulopsis,* are also usually present. The quantity and types of microorganisms in the vagina varies, however, due to menstrual cycle and to age-dependent hormonal changes. The vagina of the adult usually contains high numbers of a *Lactobacillus* species that produces lactic acid from a glycogen polysaccharide secreted by the vagina. Before puberty and after menopause, the vagina does not produce glycogen, and this *Lactobacillus* is absent.

8.2 GERM-FREE ANIMALS

As we have just learned, microorganisms are present in all normal animals. In order to evaluate the significance of the normal flora, it is necessary to develop special stocks of germ-free animals. A *germ-free animal* is one which has been bred and raised under totally aseptic conditions and completely lacks a microbial flora. Such germ-free animals can only be obtained in the laboratory. A fetus is usually germ-free as long as it remains inside the mother, and only at birth does it become infected. To obtain a germ-free animal, the fetus is surgically removed aseptically from the mother just before birth and placed in a sterile chamber called an *isolator*. All of the air, water, food, and other materials that go into the isolator must be sterile. The infant animal is fed sterile milk by hand from a bottle until it is old enough to feed itself. Careful attention must be given to be certain that the animals do not become contaminated (Figure 8.8). Although raising germ-free animals is a complicated business, it is now done routinely, and whole colonies of germ-free

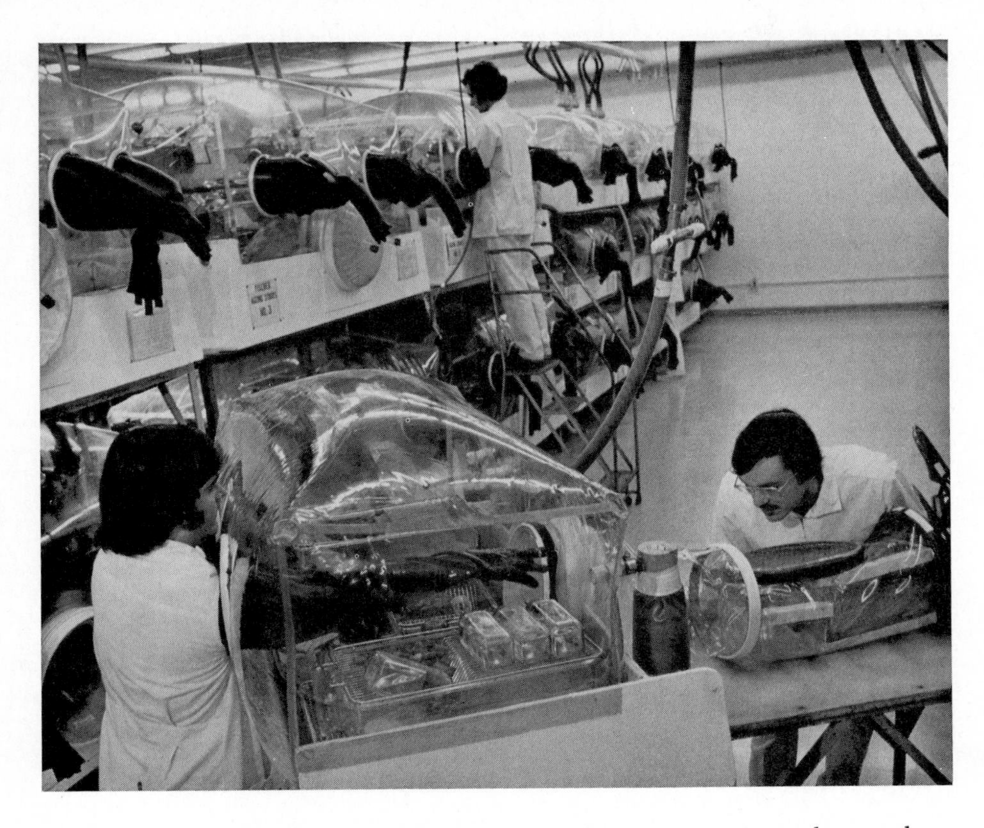

Figure 8.8 A chamber used to maintain and study germ-free animals.

animals have been established in which germ-free parents are used to produce germ-free offspring. Chickens, mice, rats, guinea pigs, and rabbits have all been raised in the germ-free condition and colonies established. Large animals such as sheep and cattle would of course be very difficult to raise germ-free, as very large isolators would be required. In some areas, germ-free mice can be purchased from laboratory animal supply companies.

To prove that an animal is germ-free, it must be shown that no microbial growth is obtained when culture media are inoculated with feces, urine, skin, hair, or other parts of the animal. Since no one culture medium will support the growth of all microorganisms, a wide variety of culture media must be used, and incubations must be carried out both aerobically and anaerobically.

Of what interest are germ-free animals? By studying the differences between germ-free and normal animals we can hope to find out whether the normal flora of the body is beneficial or harmful to the animal. The conclusion so far is that the normal flora can be both beneficial and harmful (Table 8.3). Among benefits, bacteria of the intestine produce vitamin K, an essential vitamin for mammals. Germ-free animals must be given vitamin K in their diet in order to survive, whereas conventional animals can live without such vitamin K supplements. Another benefit is that the normal flora helps the animal resist infectious disease. Conventional animals are much more resistant to infection than are germ-free animals: the bacteria of the normal flora probably keep pathogenic bacteria from becoming established by virtue of competition for nutrients and growth habitats and by producing inhibitory sub-

TABLE 8.3 Some beneficial and harmful aspects of the human normal flora

Beneficial	Harmful
Vitamin synthesis	Body odor
Synthesize products that inhibit pathogens	Intestinal gas
Outcompete pathogens	Tooth decay
Stimulate the immune system	Opportunistic infections
	Modify drugs

stances. As we have noted, when antibiotics are administered, a large proportion of the normal intestinal flora is often eliminated. In some cases, antibiotic-resistant pathogens presumably already present in the gut in very small numbers or entering as new colonizers are able to grow very well without the usual inhibition from the normal flora, and cause severe illness and even death. In such cases, antibiotic treatment to cure one disease has indirectly caused another disease. Also, the immune processes are much better developed in the normal than in the germ-free animal. Therefore, although animals can live without their normal flora, they seem to be better off when bacteria are present. On the other hand, the normal flora can at times be harmful. For example, the bacterium *Clostridium perfringens*, sometimes a member of the normal flora, can produce toxic substances that may retard the growth of the animal. The inhibition of such "normal" pathogens may explain stimulation of the growth of farm animals by feeding antibiotic-enriched feed (see Chapter 17). On balance, however, more benefit than harm is probably derived from the normal flora.

8.3 INFECTIOUS DISEASES

Although the vast majority of organisms associated with the body are harmless, the very restricted variety of harmful organisms which we have called **pathogens** often assume dominant importance in the study of microbiology. We discussed in Section 1.4 the history of the germ theory of disease and emphasized the importance of Koch's postulates for proving that a particular infectious disease is caused by a particular organism. Briefly, Koch's postulates require demonstration that an organism suspected of causing disease is constantly present when the disease exists, is absent from healthy individuals, can be isolated in pure culture, and can be reinoculated into healthy individuals and cause disease again. By the application of these postulates, many bacterial pathogens have been identified and studied, and great advances in medicine and public health have resulted.

In some cases, Koch's postulates cannot be applied because the organism cannot be cultured. The causal agent of leprosy, for example, has never been cultured away from living hosts, and of course, viruses can never be cultured in the absence of living cells. Despite inability to culture these agents, their causal role has been reasonably well-established by virtue of observation of their constant association with the disease state and because drug therapy or other treatments that cure the disease lead to a disappearance of the causal agent.

In humans, intentional inoculation is not acceptable and hence complete fulfillment of Koch's postulates is not possible. One way in which suspected causality has been proved is through laboratory accidents involving research workers studying the organism. Despite considerable precautions, accidental infections do occur in the microbiology laboratory, and if infection is followed by disease, this is strong evidence for causality.

Specificity of causal organisms In a number of cases, when a given microbial pathogen infects a host, a certain set of symptoms is elicited that can be distinguished from all other sets of symptoms and which can therefore be recognized as a specific disease. For example, virulent strains of *Corynebacterium diphtheriae,* when infecting susceptible humans, always give rise to the set of symptoms that we call *diphtheria.* Diphtheria is caused by no other organism, and *Corynebacterium diphtheriae* causes no other diseases. In this case, a *specific disease* is produced by a *specific pathogen.* Other examples of organisms causing specific diseases are *Clostridium tetani* (tetanus), *Brucella abortus* (brucellosis), and *Treponema pallidum* (syphilis). In Chapters 11, 12, and 13, a number of such specific diseases will be discussed.

For many infectious diseases, less specificity is involved. The common cold, for example, can be caused by any one of about one hundred different viruses. Pneumonia is the name of a disease involving infection of the lung and may be due to one of many microorganisms, including bacteria (such as *Streptococcus pneumoniae, Haemophilus influenzae, Staphylococcus aureus,* and *Mycoplasma pneumoniae*) and some viruses. Each of these microorganisms causes a lung infection which exhibits similar symptoms. The kidney may also be infected by a number of species of bacteria, including *Escherichia coli, Staphylococcus aureus, Streptococcus faecalis,* and *Pseudomonas aeruginosa,* among others. All these organisms cause a disease in the kidney that is called *nephritis* (from *nephro,* referring to "kidney," and *itis,* "inflammation"). In these cases, it is not possible from an analysis of the symptoms to make any good guess as to the causal organism of the infection in any particular host, so culture isolations must be performed. Other diseases of this sort are endocarditis (infection of the lining around the heart), meningitis (infection of the membranes called the *meninges* surrounding the spinal chord), and peritonitis (infection of the abdominal cavity).

Finally an individual organism may be able to cause more than one clinical disease. Thus, *Mycobacterium tuberculosis* causes pulmonary tuberculosis, but it also causes infections of the skin, bones, and internal organs, although such infections are usually sufficiently characteristic of *M. tuberculosis* that they can be recognized and distinguished clinically from infections of the same organs by other pathogens.

In addition to the wide range of infections discussed above, some diseases are caused by products which microorganisms produce rather than by an infection involving the microbe. An excellent example is the disease *botulism,* a type of food poisoning discussed in detail in Chapter 16. A chemical produced by the bacterium *Clostridium botulinum* is extremely toxic to the central nervous system. The disease botulism results when the toxic chemical is ingested. In fact, the pathogen *C. botulinum* rarely grows in the body, doing its damage as a result of growth outside the host. Thus, botulism is *noninfectious*, even though a microbe is needed to produce the toxic chemical.

We can approach infectious diseases either through a discussion of individual pathogens and the diseases they cause or through a discussion of the disease conditions to which various organs of a host may succumb. The latter approach is followed by pathologists, who are interested primarily in the diseased organ and only indirectly in the organism causing the disease. The microbiologist, on the other hand, emphasizes the infectious agent and approaches infectious diseases through a study of the microorganisms involved.

8.4 HOW DO MICROORGANISMS CAUSE DISEASE?

The host-pathogen relationship is a complex one that involves actions on the part of both the parasite and the host. One aspect of this host-pathogen relationship is the harm caused to the host. It is this harm which is recognized as disease. However, before the pathogen induces harm, it must initiate infection. What are the events in the infection process, and how does the host respond? In most cases, a definite sequence of events occurs during infection and disease:

1. Transfer of the pathogen to the host (*transmission*).
2. Entry of the pathogen into the host tissues and increase in numbers of the parasite (*invasion*).
3. Injury to the host, leading to *disease symptoms*.
4. Nonspecific response of the host to the presence of the pathogen (*innate resistance*).
5. Specific response of the host immune system against the pathogen (*acquired immunity*).

Transmission of the pathogen Transmission, the first step in infection, is a complex process that depends on the ability of the pathogen to spread from one host to another. Successful dispersal of the pathogen is a function of the location in the body where infection occurs, and the degree of incidence of infection in the host population. We shall reserve a detailed discussion of mechanisms of dispersal until Chapter 10.

Invasion and growth in the host If the pathogen is to cause disease, it must penetrate the host tissues, a process called **invasion**. During invasion there is a continuous interaction between pathogen and host involving features of the pathogen which allow it to counteract the specific physical and chemical host defenses. The skin and other types of epithelium provide the major physical barriers to microbial invasion. Very few, if any, organisms gain entry through the unbroken skin. However, in burns where much skin is lost, a major cause of death is infection. Wounds, especially large dirty ones, are common sources of infection. It is an interesting historical fact that in many wars there have been more deaths among soldiers as a result of wound infections than because of direct effects of weapons.

As discussed above, the mucus that covers the epithelium of the throat and respiratory tract is a barrier to the entrance of microbes. *Mucociliary flushing* in the ciliated epithelium layers of the upper respiratory tract tends to remove organisms which enter via the mouth or nose.

Many pathogens have the ability to attach specifically to certain host cells or cell types. Streptococci attach to the mucosal epithelium of the upper respiratory tract and thus are not eliminated by mucociliary flushing. Other bacteria selectively colonize the gastrointestinal epithelium. In some cases, the pathogen can remain localized at the epithelium and initiate damage by producing toxic substances, but in most cases, the pathogen penetrates the epithelium and either grows deeper in the tissue or spreads to other parts of the body where growth is initiated.

Certain pathogens produce specific enzymes that destroy the integrity of tissues. One example is the enzyme *hyaluronidase,* which acts on *hyaluronic acid,* a polysaccharide-like tissue cement holding cells together. Pathogens are unable to spread easily through normal tissues, but if they produce the enzyme hyaluronidase they can spread more readily. Another cementing substance in the body is the protein *collagen,* and some pathogens produce the enzyme *collagenase,* which also makes spreading easier.

After initial entry, the pathogen often becomes localized and multiplies forming a small **focus of infection** (Figure 8.9). In the skin, such foci are seen as boils, pimples, or carbuncles. These foci are usually filled with *pus,* a mixture of microorganisms, body fluids, and decomposing host cells. Infections that result in pus formation are called *pyogenic infections* (*pyo* meaning "pus"). Access to the interior of the body generally occurs in those areas where lymph glands are near the surface, such as the nose and throat, the tonsils, or the lymphoid follicles of the intestine. Inside the body, localization usually occurs in the lymph nodes, liver, spleen, or kidney. Selective colonization is the rule: only rarely is there extensive growth in a variety of organs or tissues.

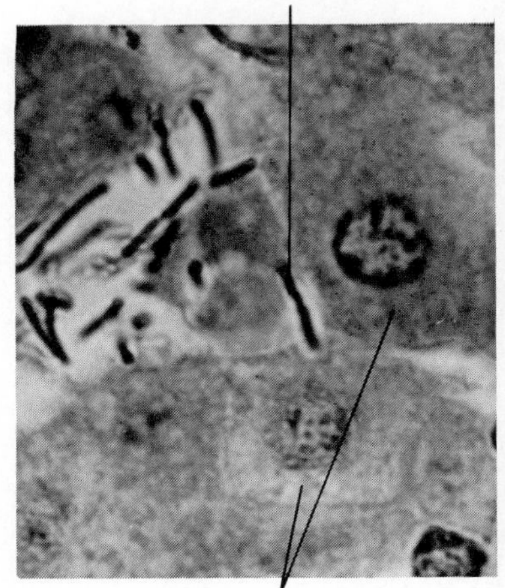

Figure 8.9 A pathogenic organism in infected tissue. *Bacillus anthracis,* the causal agent of anthrax.

A microbe occasionally passes from a focus of infection into the blood stream; when living bacteria are detected in the blood, the condition is usually called *bacteremia.* Other similar terms are used to describe the presence in the blood of viruses (*viremia*) or of toxins produced by microorganisms (*toxemia*). Blood itself is not an especially favorable medium for microbial growth, and when organisms are found there, they have usually come from some other place.

In almost all cases, damage to the host involves *growth* of the microbe in the body first. The organism must of course be able to grow under the environmental conditions (temperature, pH, etc.) found in the body. Nutrients are obtained from body fluids and tissues; the body is a rich source of microbial nutrients in the form of carbohydrates, amino acids, nucleic acids, and lipids.

Many pathogens produce enzymes which break down polymers such as polysaccharides or proteins; the monomers released are then used as nutrients.

However, not all microbial nutrients may be plentiful in the body at all times. Vitamins and growth factors are not necessarily in adequate supply in all tissues, and minerals may also be in short supply. Among the minerals, one frequently deficient is iron, a mineral required in fairly large amounts by most microorganisms. Ferric iron is highly insoluble at neutral pH values, and the host has a specific iron-binding protein called *transferrin*, which carries iron through the body. The affinity of this protein for iron is so high that there may be insufficient free iron for microbial growth. However, many bacteria produce specific iron-binding chemicals that enable them to grow even when concentrations of iron are low.

Some pathogens grow within cells of the host (intracellular) rather than between the cells comprising a tissue (intercellular). There are many chemical differences between the cytoplasm within host cells and the fluids surrounding such cells, and this may be the basis for the *intracellular infection* observed with some pathogens (Figure 8.10). Some bacteria, for example, members of the genera *Rickettsia* and *Chlamydia*, are actually obligate intracellular parasites and only grow within host cells. Because of their dependence on the intracellular environment, these bacteria generally cause diseases which are transmitted via insects or by direct contact between an infected and an uninfected host. Important diseases caused by rickettsia and chlamydia, such as typhus, Rocky Mountain spotted fever, and sexually transmitted chlamydial disease, will be discussed in Chapter 11.

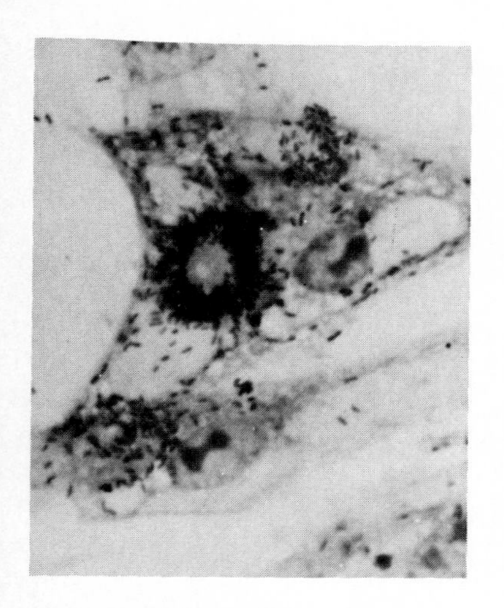

Figure 8.10 Intracellular growth of a pathogenic bacterium, *Rickettsia rickettsii*, the causal agent of Rocky Mountain spotted fever.

Many protozoa also cause intracellular infection. Although they may exist outside of the host, these protozoan pathogens spend a portion of their life cycle within cells of the host. A good example is the malaria pathogen *Plasmodium* infecting red blood cells, which is described in detail in Chapter 12. Viruses are by nature obligate intracellular parasites and their infection of host cells (described in Chapters 6 and 13) sometimes results in tissue injury in the host.

Harm to the host The mere presence of a microbe in the body rarely, if ever, causes damage to the host. The harm that is caused by pathogens is generally due to production of specific factors called **toxins**, which affect the host in some detrimental way. Two kinds of toxins are produced: (1) *exotoxins*, which are excreted or released from the pathogen; and (2) *endotoxins*, which generally remain bound to the microorganism and are released in significant amounts only when the pathogen dies and disintegrates (Figure 8.11).

Exotoxins are proteins. Since the exotoxin is released from the pathogen, action can take place at sites distant from the focus of infection. Table 8.4 lists some of the better known exotoxins produced by bacteria. The bacterial disease *diphtheria*, caused by *Corynebacterium diphtheriae*, can illustrate some of the properties of diseases involving exotoxins. The bacterium grows almost exclusively in the throat, never invading other parts of the body. The diphtheria toxin is released from this focus of infection and spreads throughout the body. Since all the symptoms of diphtheria are due to action of the toxin, elimination of the pathogen by treatment with antibiotics will not necessarily

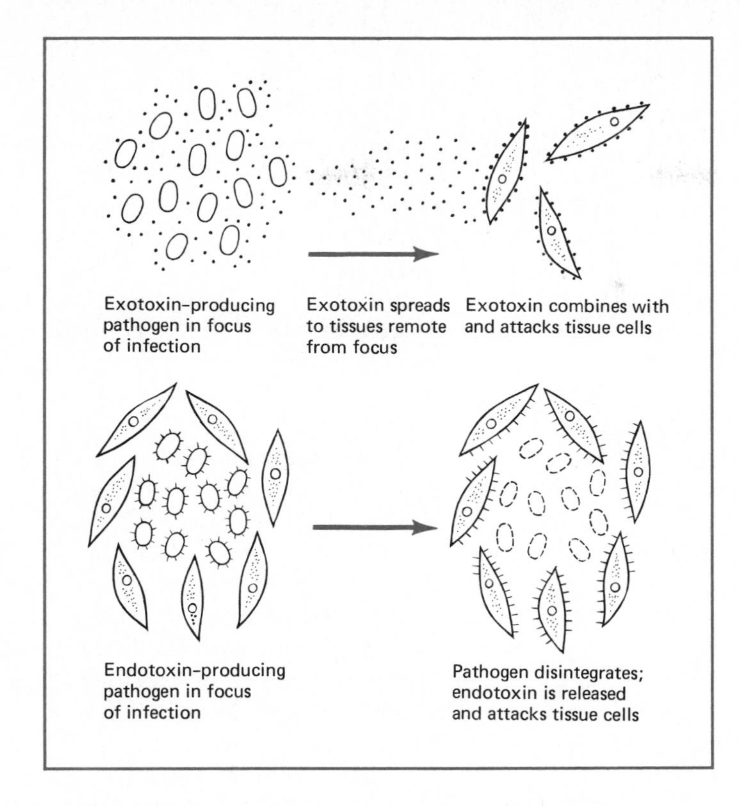
Exotoxin-producing pathogen in focus of infection
Exotoxin spreads to tissues remote from focus
Exotoxin combines with and attacks tissue cells
Endotoxin-producing pathogen in focus of infection
Pathogen disintegrates; endotoxin is released and attacks tissue cells

Figure 8.11 Differing actions of an exotoxin and an endotoxin.

eliminate disease symptoms, if the toxin has already spread throughout the body. Thus, effective treatment of diphtheria involves use of both an appropriate antibiotic to inhibit the pathogenic bacterium, and a treatment to neutralize the toxin.

TABLE 8.4 Some exotoxins produced by pathogenic bacteria

Bacterium	Disease	Toxin	Action
Clostridium botulinum	Botulism	Neurotoxin	Paralysis
C. tetani	Tetanus	Neurotoxin	Paralysis
C. perfringens	Gas gangrene	Alpha-toxin	Hemolysis
		Theta-toxin	Affects heart muscle
		Kappa-toxin	Digests collagen
		Lambda-toxin	Digests proteins
Corynebacterium diphtheriae	Diphtheria	Diphtheria toxin	Inhibits protein synthesis
Staphylococcus aureus	Pyogenic infections (boils, pimples)	Alpha-toxin	Hemolysis
		Leukocidin	Destroys leukocytes
	Food poisoning	Enterotoxin	Induces vomiting, diarrhea
Streptococcus pyogenes	Pyogenic infections, tonsillitis, and scarlet fever	Streptolysin O	Hemolysis
		Streptolysin S	Hemolysis
		Erythrogenic toxin	Causes scarlet-fever rash

The disease *tetanus* is caused by another exotoxin-producing bacterium, *Clostridium tetani*. This organism is an obligate anaerobe and will not grow in the living tissues of the body, which are aerobic. However, when oxygen supply to tissues is cut off as a result of deep puncture wounds, anaerobic conditions can develop, and *C. tetani* can grow. The toxin released by the bacterium causes paralysis at sites remote from the site of infection.

Another clostridium, *C. botulinum*, can cause death even without infecting the body. This organism, the cause of a very serious food poisoning (see Chapter 16), grows and produces its exotoxin in improperly preserved foods. When the food is eaten, the toxin spreads through the body from the intestine and causes paralytic death. The botulinum toxin is one of the most lethal substances known: one hundred nanograms of it is sufficient to kill an average human.

Endotoxins are cell-bound toxins, usually part of the outer cell layers of Gram-negative bacteria. They are complex chemically, containing both lipid and polysaccharide, with the lipid portion probably responsible for toxicity. In general, the toxicity of endotoxins is lower than that of exotoxins. However, endotoxins are of considerable medical significance and are responsible for many deaths. It is likely that small amounts of endotoxin are released even from living bacteria, but only when cells die and disintegrate are large amounts released. Symptoms induced by endotoxins include fever, diarrhea, hemorrhagic shock, and other tissue damage. Another important property of endotoxins is their ability to stimulate natural host defense mechanisms when present in the body in small amounts (as discussed in Chapter 9). Since certain bacteria of the normal flora, such as *Escherichia coli*, produce endotoxin in small amounts, natural stimulation of host defense mechanisms may occur continuously at a low level.

Virulence *Virulence* refers to the relative ability of a pathogen to cause disease. Virulence is determined primarily by two properties of an organism: its **invasiveness** and its **toxigenicity** (Figure 8.12). A highly invasive organism is able to grow well in the body and to set up a widespread generalized infection. Even if it is only weakly toxigenic, it may cause host damage because of the large numbers of cells present. On the other hand, a weakly invasive organism, even though it may grow only poorly in the body, may still be highly virulent if it produces an extremely potent toxin. An organism which is both highly invasive and highly toxigenic is usually a dangerous and highly virulent pathogen.

An example of a weakly invasive pathogen that is highly virulent is *Corynebacterium diphtheriae*. Even though this bacterium infects and grows only in the throat, it can cause death because the toxin it produces is extremely potent and can spread throughout the body. An example of a weakly toxigenic organism that is still highly virulent is *Streptococcus pneumoniae*, one of the most common agents causing bacterial pneumonia. This organism is not known to produce a toxin, yet it is able to damage the host and even cause death, because it is highly invasive and grows in the lungs in such large numbers that the functions of this organ are impaired. An organism which is both highly invasive and highly toxigenic is the bacterium *Yersinia pestis*, which

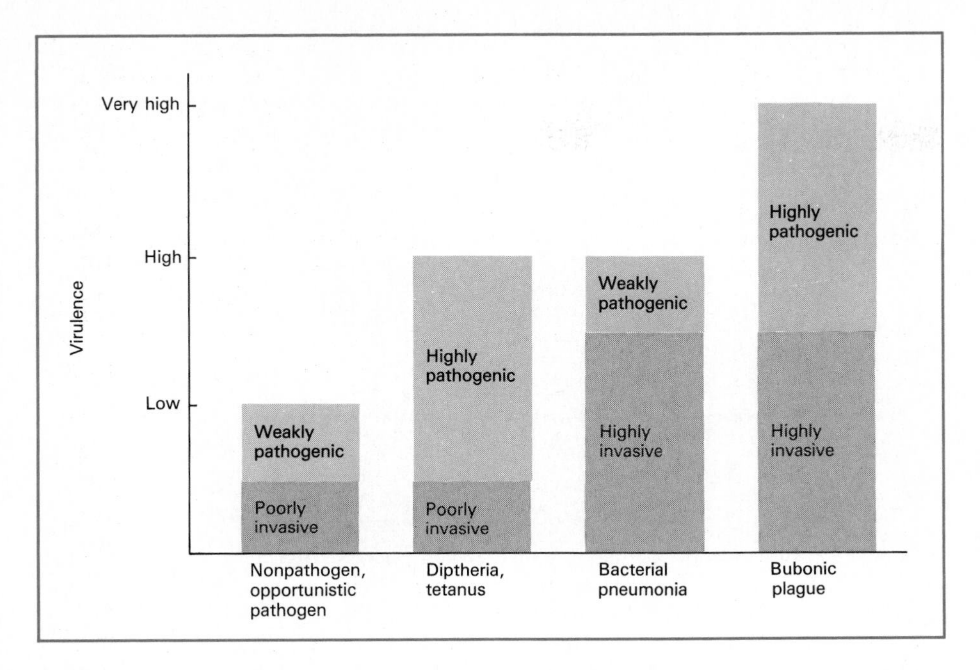

Figure 8.12 Virulence is determined by the additive effects of invasiveness and toxigenicity. Weakly invasive organisms, such as those causing diphtheria and tetanus, may still be virulent because they produce potent toxins. Weakly toxigenic organisms, such as *Streptococcus pneumoniae* causing bacterial pneumonia, may still be virulent because they are highly invasive.

causes bubonic plague. Major outbreaks of this disease during the Middle Ages killed one-fourth of the population of Europe, attesting to the high virulence of this pathogen.

There are many microorganisms which are so noninvasive and nontoxigenic that they do not ordinarily cause disease. The normal flora of the human body provide many examples of such organisms. It must be remembered, however, that disease may result from a decreased host defense as well as from a highly virulent pathogen. Thus, organisms that are usually harmless can cause disease if the host defense is impaired. Recall that these organisms are called opportunistic pathogens; they cause disease only in hosts with lowered resistance.

8.5 HOST RESPONSES DURING INFECTION

Mammals have a complex system that comes into play when foreign material enters the body. A microbial infection represents a major onslaught, and this system is of vital importance to host defense. Two types of responses may occur. One type, called **innate resistance**, involves defenses which are *nonspecific* and thus act against all invading pathogens. This type of resistance is already present in the body at the time infection occurs and its action is simply concentrated at the focus of infection in response to injury. The other type of host response is a highly *specific* one carried out by the host **immune system**. The protection provided by the immune response is acquired during the infection and thus is called **acquired immunity**. How the immune system can respond specifically to pathogens will be considered separately in Chapter 9.

TABLE 8.5 Components of whole blood

Component	Percent of total blood volume*	Number per ml (millions)	Percent of fluid volume
Cells	55		
Red cells (erythrocytes)		5000	
White cells (leukocytes)		5–10	
Platelets	<1	0.15–0.4	
Fluid	45		
Water			90
Proteins			9
Sugars and other organic compounds			<1

*A human has about 5 liters of whole blood.

We will now consider the types of innate resistance which limit the general growth of pathogens in the body, but before we do so we must first consider the blood and lymphatic circulatory systems of the body, since most of the innate resistance mechanisms originate in these systems.

Blood and lymph system Blood consists of cellular and noncellular components (Table 8.5). The most numerous cells in the blood are the *red blood cells,* which carry oxygen from the lungs to the tissues. The *white blood cells,* or *leukocytes,* although much less numerous, play important roles in nonspecific resistance, as will be discussed later. Small cell-like constituents which are numerous in blood, called *platelets,* are involved in the temporary control of leakage of blood from damaged vessels.

The fluid remaining after the cellular components are removed is called *plasma* (Table 8.6). It consists primarily of water in which a variety of salts and proteins are dissolved. Some of the plasma proteins play an important role in acquired immunity, whereas others are involved in nonspecific resistance. A plasma protein called *fibrinogen* is the clotting agent of the blood. When blood is removed from the body, the blood clotting system is activated and fibrinogen undergoes a complex set of reactions and becomes *fibrin,* of which the clot is composed. Clotting can be prevented by adding *anticoagulants* such as potassium oxalate, potassium citrate, or heparin. When plasma is allowed to clot, the fluid components left behind, called *serum,* consist of all the proteins and other dissolved materials of the plasma except fibrinogen.

TABLE 8.6 Blood fraction terminology

Term	Cells	Fibrinogen	Fluid
Whole blood	+	+	+
Plasma	−	+	+
Serum	−	−	+

Since serum contains many of the substances involved in specific immunity it is frequently prepared either for use in injections or for diagnostic studies.

Blood is pumped by the heart through a network of arteries and capillaries to various parts of the body and is returned through the veins (Figure 8.13). The circulatory system carries not only nutrients (including O_2) but also the components of the blood which are involved in host resistance to infection. At the same time, the circulatory system facilitates the spread of pathogens to various parts of the body.

Lymph is a fluid similar to blood but which lacks red cells. There is a separate circulatory system for lymph, called the **lymphatic system** within which lymph flows (Figure 8.13). Fluids in tissues drain into lymphatic capillaries, then into **lymph nodes**, found at various locations throughout the system, which filter out microorganisms and other particulate materials. Specialized white blood cells found in abundance in the lymphatic system, called **macrophages**, actually carry out the filtering action, as will be described below. In addition to filtering foreign particles, lymph nodes may be sites of infection, since organisms that are collected there by the filtering mechanisms may then proliferate. Lymphatic fluid eventually flows into the circulatory system via the thoracic duct.

Lymphocytes, another special type of cell found within the lymphatic system, are involved in the immune response. Their role in antibody formation will be discussed in Chapter 9.

Phagocytosis *Phagocytes* (literally, "cells that eat") are an important part of innate resistance against microbial invasion. Some of the leukocytes found in whole blood are phagocytes, and they are also found in various tissues and fluids of the body. Phagocytes are usually actively motile by amoeboid action. They are attracted to microbes, engulf, kill, and digest them. There are two major types of phagocytic cells (Figure 8.14). **Polymorphonuclear leukocytes** (abbreviated PMN) are actively motile cells containing a multilobed nucleus. They are short-lived cells found predominantly in the bloodstream and bone marrow. They appear in large numbers during the acute phase of an infection and therefore their presence in the blood can be used as an indicator of infection.

The other type of phagocyte is the **macrophage** (or **monocyte**) which appears quite different under the microscope from the PMN and which is less motile. There are two types of macrophages: **wandering cells**, which are found free in the blood and lymph; and fixed macrophages, or **histiocytes**, which are found embedded in various tissues and have only a limited mobility. Histiocytes constitute the active phagocytes of the lymphatic and reticuloendothelial systems, which we shall discuss later.

A phagocyte is attracted chemically to an invading microbe, which it rapidly engulfs and digests (Figure 8.15). Phagocytes work best when they can trap microbial cells against surfaces such as blood vessel walls, blood clots, or connective tissue fibers. As we noted earlier, the action of phagocytes is not restricted to microbes; any foreign particulate material can be ingested and may be digested and thus eliminated from the body.

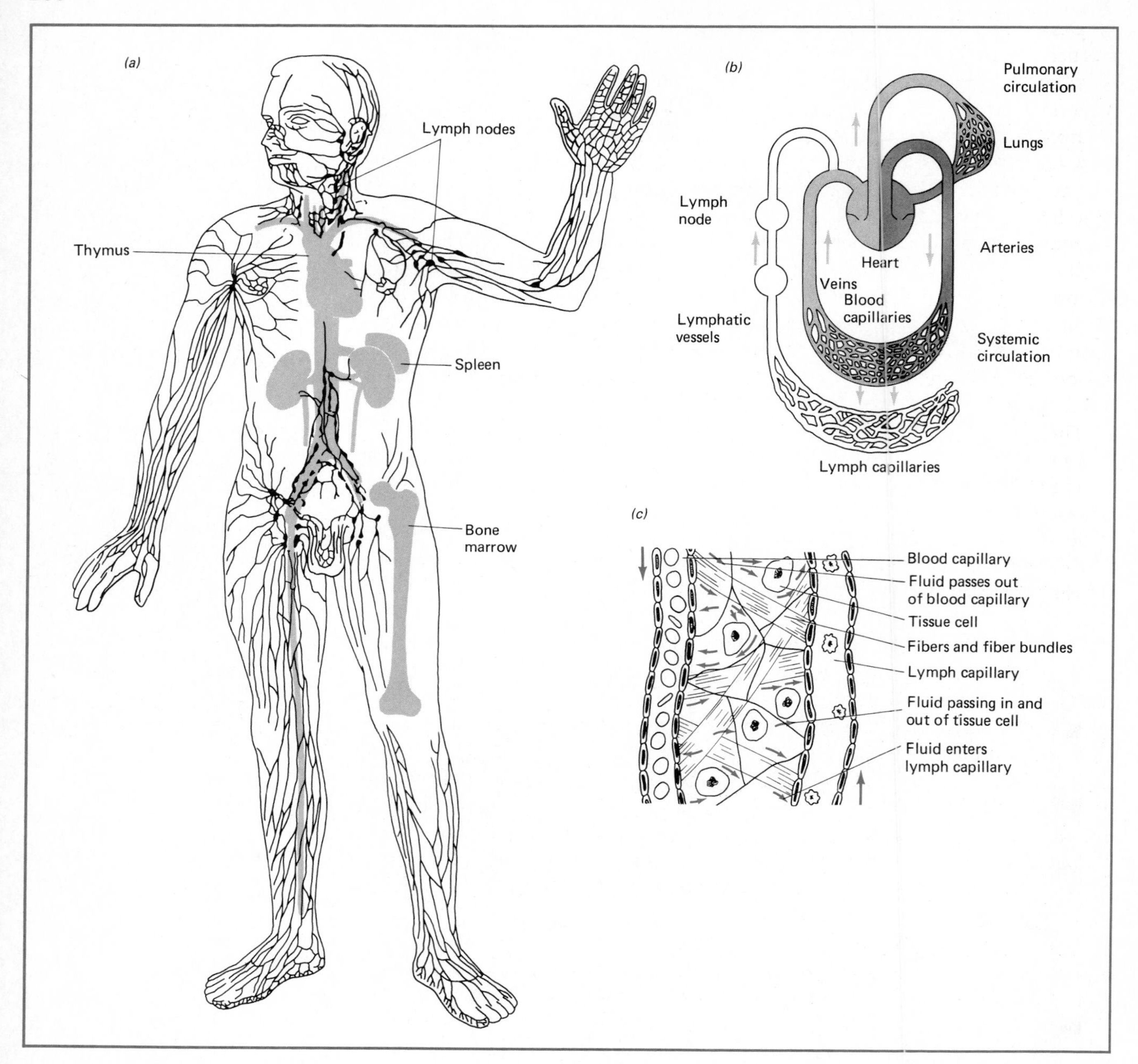

Figure 8.13 The blood and lymph systems. (a) Overall view of the major lymph systems, showing the locations of major organs. The lymph nodes are heavily concentrated at the joints and trunk, where they filter out foreign particles and bacteria. (b) Diagramatic relationship between the lymph and blood systems. Blood flows from the veins to the heart, then to the lungs where it becomes oxygenated, then through the arteries to the tissues. (c) Connection between the blood and lymph systems is shown microscopically. Both blood and lymph capillaries are closed vessels but are permeable to water and salts. Proteins of the blood also slowly move out of the capillaries. The tissue fluid is the means of transport between blood and lymph. Changes in capillary permeability, as a result of infection or other source of inflammation, can increase greatly the movement of fluid into the tissue spaces.

During many kinds of infection, an increased number of phagocytes are produced and disseminated through the bloodstream to the site of infection. Thus, infection may result in an increase in the number of phagocytic cells in the bloodstream, and a quantitative count of the number of such cells in a sample of blood can therefore be used as an indication of infection. Diseases often associated with an increased leukocyte count are appendicitis, meningitis, mononucleosis, pneumonia, gonorrhea, and pyogenic infections such as boils. In such infections the total white-cell count may be doubled, tripled, or even quadrupled.

During the course of infection, there is a shift in the type of white cell that predominates. During the active stage of infection, PMNs predominate. During later stages, monocytes, which play a major role in ingesting dead or dying bacteria, are more common. Thus, a *differential count* showing the proportions of various kinds of leukocytes aids in determining the progress of infection.

The liver, spleen, and lymph glands are unusually well supplied with fixed macrophages. As the blood and lymph pass through these organs, microorganisms are filtered out and ingested. The system of fixed macrophages in these organs is called the **reticuloendothelial system** This system is quite efficient in clearing foreign particles from the blood and lymph. Within hours after microbes enter, they are usually completely eliminated from the blood and can be found in the organs of the reticuloendothelial system, where they usually are quickly killed.

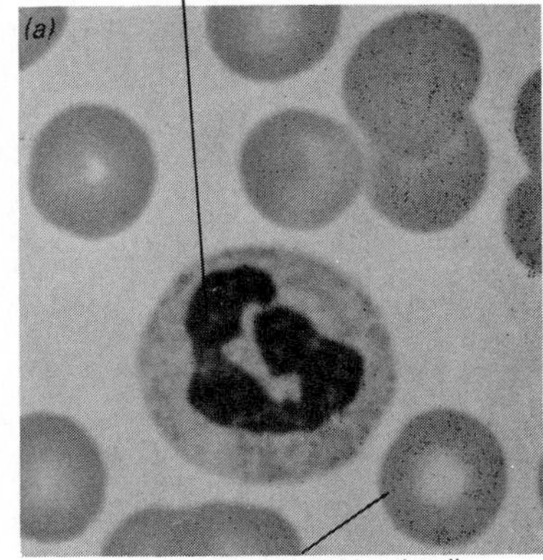

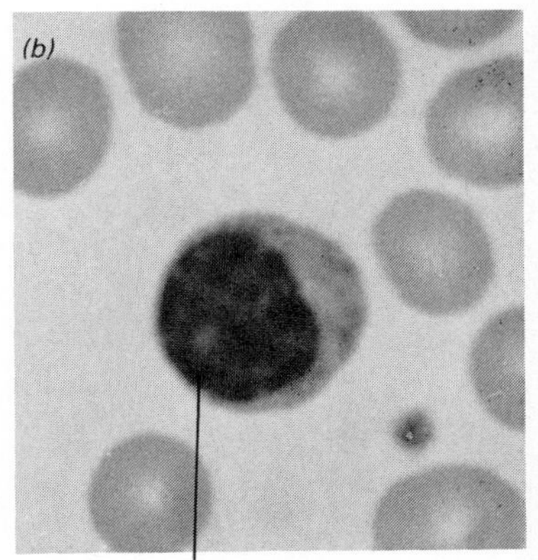

Figure 8.14 Two major types of phagocytic cells. (a) Polymorphonuclear leukocyte (PMN).
(b) Macrophage. The cells are about 10 μm in diameter.

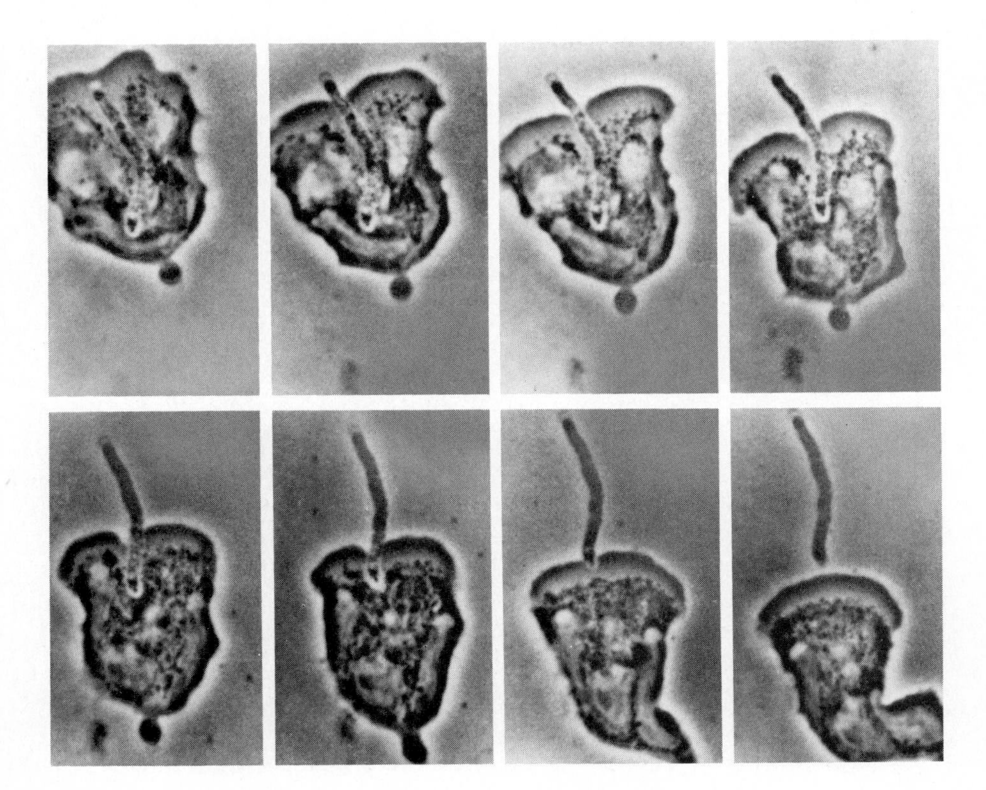

Figure 8.15 The process of phagocytosis, as viewed with a phase contrast microscope. The bacterial cell is *Bacillus megaterium*, one of the larger species of bacteria. The phagocyte is about 10 μm across.

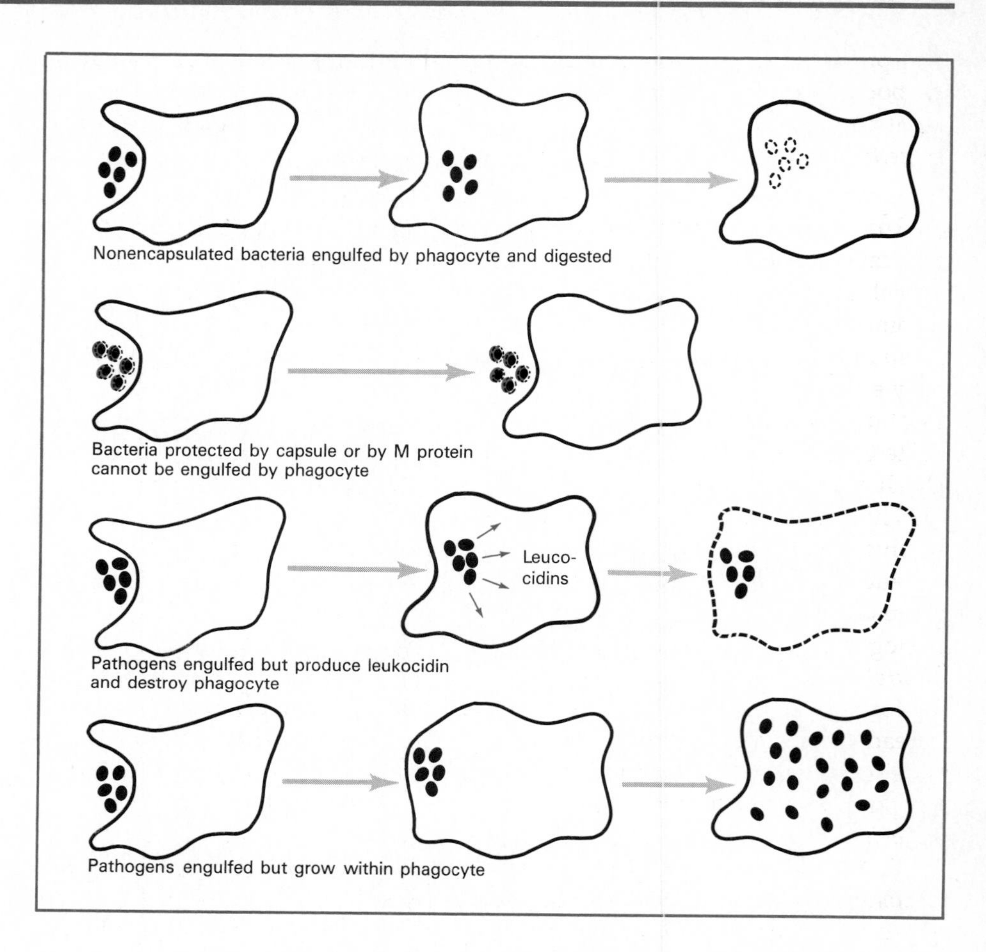
Nonencapsulated bacteria engulfed by phagocyte and digested
Bacteria protected by capsule or by M protein cannot be engulfed by phagocyte
Leuco-
cidins
Pathogens engulfed but produce leukocidin and destroy phagocyte
Pathogens engulfed but grow within phagocyte

Figure 8.16 Phagocytosis of bacteria and mechanisms by which evasion of phagocytosis can occur.

Pathogens frequently have systems that counteract phagocytic action (Figure 8.16). One way in which this is done is by the production of a *capsule* or *slime layer* that prevents the phagocyte from eating the cell. When a capsule surrounds a bacterial cell, the phagocyte may be unable to stick to and engulf it. The importance of the capsule as a virulence mechanism for some pathogens is well-documented. For example, encapsulated *Streptococcus pneumoniae* cells are usually highly virulent, but mutants which lack the capsule cannot cause the disease pneumonia. Pathogenic streptococci also possess proteins on their surfaces which have antiphagocytic action. The *M protein* of *Streptococcus pyogenes* is a major virulence factor in the disease condition *streptococcal sore throat*. Some microbes produce substances, called *leukocidins*, which destroy phagocytes. When a leukocidin-producing microbe is engulfed, its leukocidin kills the phagocyte while the microbe is unharmed. Pathogens such as *Staphylococcus* and *Streptococcus* that cause pyogenic infections (boils, pimples) often produce leukocidins; the pus itself contains large numbers of dead phagocytes as well as bacterial cells.

In some cases, pathogens are readily phagocytized but neither kill nor are killed by the phagocytes. These pathogens are able to remain alive within the phagocytes and even grow. Within the phagocytes, such pathogens are well-protected and are some of the most difficult organisms to eliminate from the body by drug treatment. Examples are *Mycobacterium tuberculosis* (causal agent of tuberculosis), *Salmonella typhi* (causal agent of typhoid fever), and *Brucella melitensis* (causal agent of undulant fever).

Inflammation The host responds to foreign matter or trauma by a process called *inflammation* (Figure 8.17). Causes of inflammation include microbial infections, cuts, abrasions, burns, and sharp blows. The inflamed site becomes swollen and red and is usually painful to the touch. The symptoms of inflammation develop because the blood vessels near the site of the foreign body enlarge and the blood capillaries increase in permeability, allowing cells and fluids to escape from the bloodstream and enter the tissues. Phagocytic white blood cells migrate from the capillaries and accumulate at the inflamed site. If infection exists, the microbial cells present may be ingested.

Fluid components of the blood also enter the tissues where they are involved in clot formation around the inflamed region; the invading microbe is hence localized. Although inflammation is often painful, it is beneficial because it leads to localization and destruction of the infecting organism. Some pathogenic bacteria, especially the streptococci, produce an enzyme called *streptokinase* that dissolves these clots, enabling the bacteria to spread out of the focus of infection. Other bacteria produce the enzyme *coagulase*, which, instead of dissolving clots, causes them to form. Pathogens producing coagulase (mainly staphylococci) thus often cause localized infections such as boils and pimples in the region of the clotted blood. These organisms also produce *leukocidin* and this toxin kills phagocytes that move into the clot area.

One of the factors that start the inflammatory response is the chemical *histamine*, which is released by cells in an injured region and causes an increase in capillary permeability. Certain drugs, called *antihistamines*, coun-

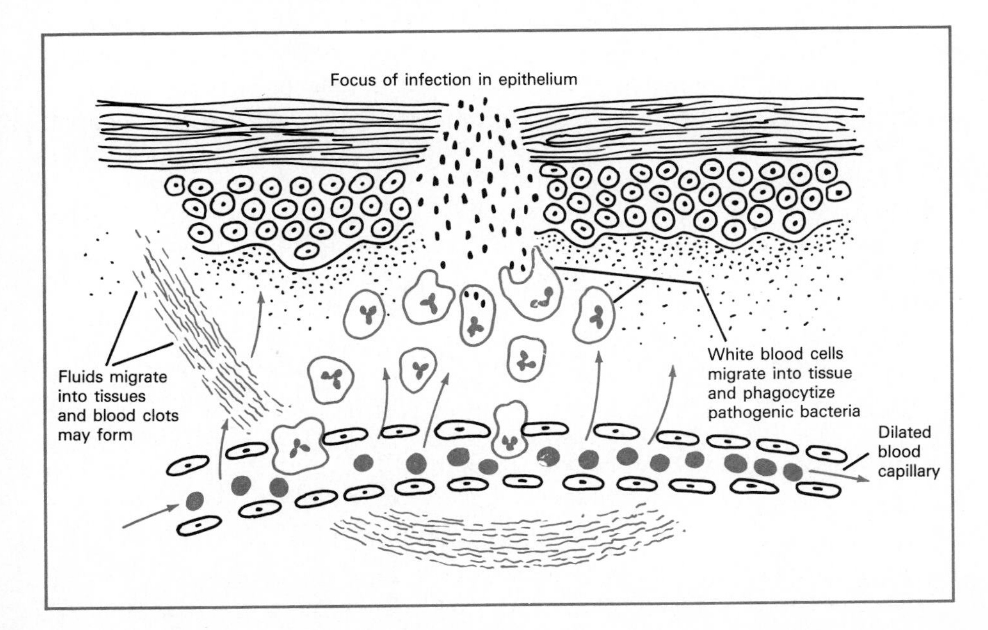
Focus of infection in epithelium
Fluids migrate
into tissues
and blood clots
may form
White blood cells
migrate into tissue
and phagocytize
pathogenic bacteria
Dilated
blood
capillary

Figure 8.17 Events in inflammation. Blood capillary dilates and leukocytes (PMNs) migrate into the tissue and engulf pathogens. Fluids also leak into the tissues with serum components such as fibrinogen participating in the formation of blood clots.

teract some of the effects of histamine and can be used to neutralize some of the pain associated with inflammation. However, antihistamines also prevent the benefits of inflammation, so that they must be used with caution.*

Pathogens vary in the degree to which they bring about inflammation. Some pathogens which are not especially invasive induce considerable inflammation and in this way cause more host damage than would be predicted from their invasive powers.

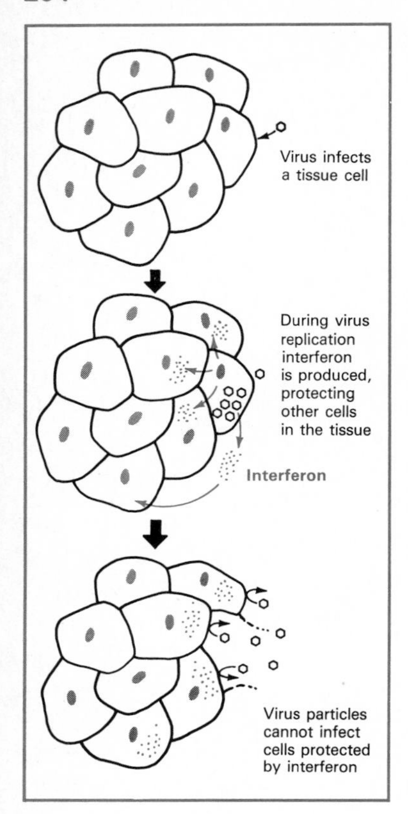
Virus infects a tissue cell
During virus replication interferon is produced, protecting other cells in the tissue
Interferon
Virus particles cannot infect cells protected by interferon

Figure 8.18 Interferon, produced after initial viral infection, protects uninfected cells from viral attack.

Interferon *Interferons* are antiviral proteins, produced by many animal cells in response to virus infection, that interfere with viral multiplication (Figure 8.18). Interferon has little or no harmful effect on uninfected cells because it specifically inhibits RNA synthesis directed by viruses, stopping production of viral proteins. Infected cells excrete interferon which increases the resistance of adjacent tissue cells to viral attack. Upon lysis of the infected cell, virus particles are unable to replicate successfully in the other (protected) cells in the surrounding area.

Because of the lack of toxicity and the potential protection which interferon could provide against viral infection, there has been great interest in obtaining it in large quantity. Unfortunately the amounts produced in the body are very small, and because interferon is species-specific, one must obtain human interferon to combat human viral infections. However, as described in Chapter 7, human interferon can now be produced by genetically engineered bacteria. Trials to test its effectiveness as an antiviral agent show great promise.

8.6 DYNAMICS OF THE HOST-PARASITE INTERACTION

The interaction between pathogen and host should be visualized as a continuous conflict. The pathogen first encounters the physical and chemical barriers of the host epithelial tissues. It must evade host defenses and penetrate the epithelium of the host. Once the pathogen has invaded the epithelium, it employs the mechanisms of pathogenicity, such as toxins, to elicit injury in the host.

Mechanisms of innate resistance are activated as a dynamic response to injury. Thus, as the pathogen invades and causes tissue destruction, the release of histamines causes blood capillaries to dilate, beginning the inflammatory response near the site of injury. Leukocytes migrate to the focus of infection to phagocytize pathogens, but in many cases the pathogen is resistant to phagocytosis (perhaps because of a capsule or antiphagocytic protein) or produces leukocidins to kill the leukocytes. Fluid components also leak into the tissues and participate in the formation of blood clots, but many pathogens produce substances to dissolve these clots. As dead tissue cells and leukocytes accumulate, pus may be generated. These materials, as well as some of the pathogenic microorganisms, drain into lymphatic capillaries where they may be phagocytized by macrophages in lymph nodes. Similarly, mi-

*Antihistamines are also used in treatment of allergies. This is a distinct, and perhaps more rational, use of antihistamine therapy.

TABLE 8.7 Examples of host defense and corresponding virulence mechanisms

Host defense mechanism	Pathogen virulence mechanism
Physical and chemical barriers	
Integrity of epithelial tissue	Toxins that lyse cells (hemolysins)
Tissue cements	Hyaluronidase, collagenase
Mucociliary flushing	Attachment mechanisms
Chemicals in secreted fluids	
Stomach acid	Acid tolerance
Lysozyme	Resistant cell wall structures
Lactoperoxidase system	
Innate resistance	
Phagocytic cells	Capsules, inhibitory proteins, leukocidins, intracellular growth
Blood clotting	Enzymes that dissolve clots (streptokinase)
Interferon	
Acquired immunity	
Antibodies	Immunoglobulin-degrading enzymes
T cells	

crobes which can disseminate through blood may be filtered by other organs of the reticuloendothelial system.

Ultimately, the lymphocytes within the lymphatic system encounter pathogenic microbes and an immune response is elicited. As will be described in Chapter 9, the lymphocytes produce both proteins called *antibodies* and cells called activated *T cells,* which are able to react with and inactivate specific chemicals on the surface of the pathogen, reducing virulence or killing the pathogen itself. Innate defense mechanisms can then participate in the clearing of the pathogen from the host and the disease ends.

The interactions between pathogen and host defenses are obviously complex. Strategies of host defense and virulence mechanisms used by pathogens to counter these defenses are summarized in Table 8.7.

Scenario of streptococcal sore throat It is instructive to consider as an example a specific disease to see how one particular pathogen is equipped to evade defenses and cause infection. The interaction between *Streptococcus pyogenes* and the human host causing streptococcal sore throat can be used as an example.

Streptococcal sore throat is an infectious disease of the upper respiratory tract caused by the bacterium *Streptococcus pyogenes.* The pathogen is transmitted to the uninfected host by direct mouth-to-mouth contact, by indirect contact via contaminated surfaces (such as a drinking glass), or through aerosol droplets which pass from an infected to an uninfected individual. The bacterium possesses a protein on its outer surface, called the *M protein,* which is involved in attachment to epithelial cells of the upper respiratory tract and excretes the enzyme hyaluronidase which breaks down the tissue cement so that it can invade the host tissue. It can also produce toxins which act to destroy cells. One toxin produced by the bacterium, called *hemolysin,* lyses

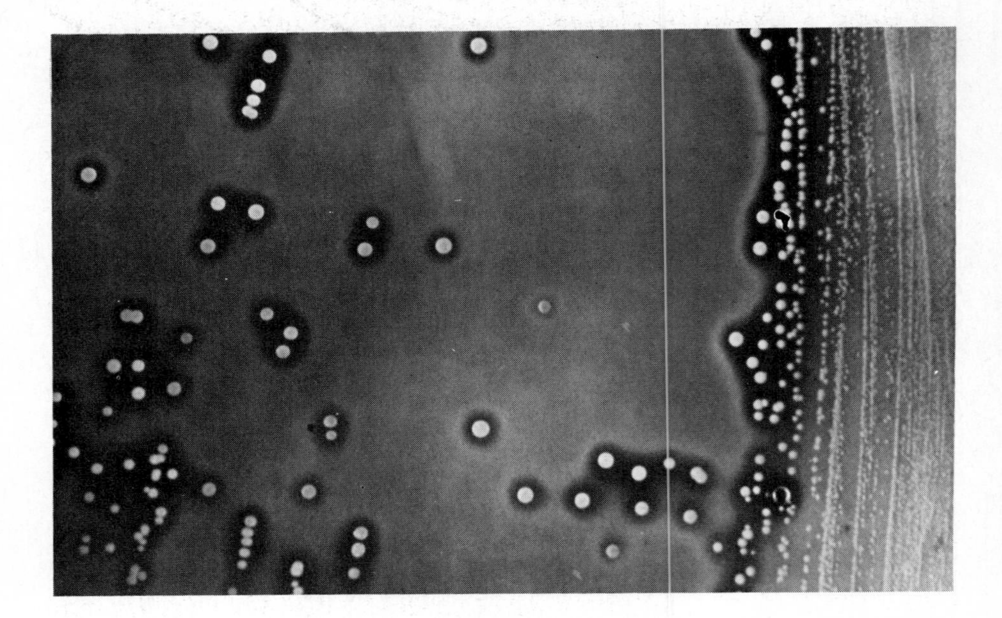

Figure 8.19 Demonstration of hemolysis on a blood agar plate. A small sample of blood was mixed with the melted agar before the plate was poured. The zone of clearing around each colony is due to lysis of red blood cells.

red blood cells. The production of this toxin can be used for the identification of the pathogen in throat cultures, benefiting diagnosis of streptococcal sore throat, as illustrated in Figure 8.19.

Once the invasion begins, *S. pyogenes* grows in the host tissue and injury results. The destruction of tissues stimulates the host to initiate a second line of defense involving the innate resistance mechanisms. Inflammation leads to the movement of phagocytic leukocytes into the infected area. *S. pyogenes* can evade phagocytes in several ways. It possesses a capsule, but apparently the M protein plays a more important role in countering the action of phagocytes. The protein itself is somehow able to prevent ingestion by leukocytes. *S. pyogenes* also produces leukocidins which are toxic to white blood cells. Inflammation also leads to the formation of blood clots, but *S. pyogenes* produces an enzyme, called streptokinase, which dissolves the blood clots. The result of this interaction between host defense and pathogen virulence is that dead cells and debris accumulate as pus at the focus of infection, a characteristic symptom of the disease. Other symptoms such as a sore throat result from damage to nerve endings and increased pressure due to inflammation.

The debris, containing some *S. pyogenes* cells, drains into the lymphatic capillaries and then into lymph nodes. Lymphocytes, representing a third line of defense (the immune system), can respond to the chemicals on the surface of *S. pyogenes* cells. The result is the production of an antibody which acts specifically against the M protein, inactivating it. The antibody is effective only at the focus of infection, and it is carried there through the circulatory system. Once the M protein on the cells at the focus of infection is inactivated, phagocytes are no longer inhibited and can destroy *S. pyogenes*. Thus, the disease comes to its natural end.

Clinical aspects of infectious disease The outward effect of the underlying conflict between invading pathogen and host defense is the expres-

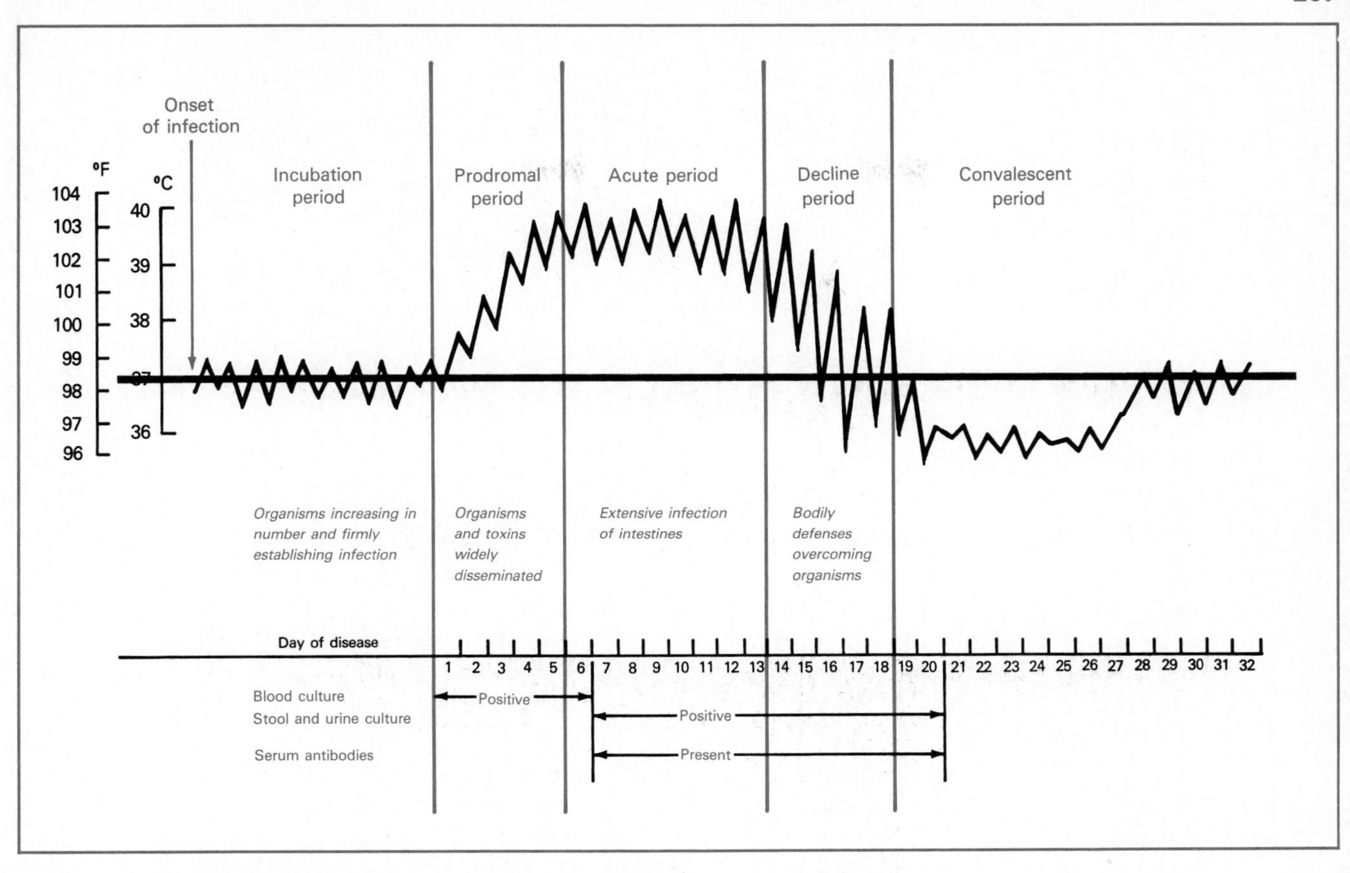

Figure 8.20 Stages of the disease typhoid fever, showing the relationship between features of the host/parasite interaction, symptoms (fever), and clinical observations.

sion of symptoms which are typical for a given disease. During the course of a disease, various stages can be identified relating to the appearance of different symptoms, as shown for the disease typhoid fever in Figure 8.20.

1. *Infection,* the time at which the organism invades the host.
2. *Incubation period,* the time between infection and the appearance of disease symptoms. Some diseases have incubation periods as short as a few hours, whereas others have incubation periods as long as several weeks (Figure 8.21). The incubation period for a given disease is determined by inoculum size, virulence of pathogen, resistance of host, and distance of site of entrance from focus of infection.
3. *Prodromal period,* a short period following incubation in which the first symptoms, such as headache and feeling of illness, appear.
4. *Acute period,* when the disease is at its height, with such overt symptoms as fever and chills.
5. *Decline period,* during which disease symptoms subside, and the host's temperature falls.

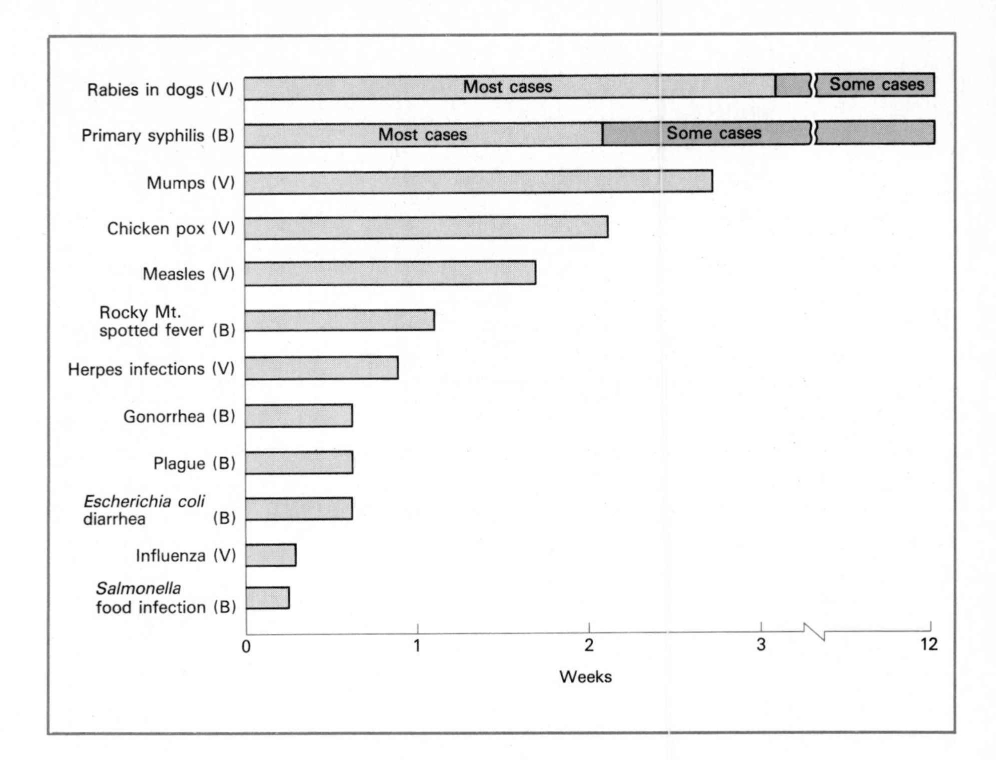

Figure 8.21 Average incubation period for some well-known diseases. The causative agent of each disease listed is either a bacterium (B) or a virus (V).

6. *Convalescent period,* during which the patient regains strength and returns to predisease status.

During the later stages of the infection cycle, the immune mechanisms of the host become increasingly important, and in most cases recovery from the disease requires the action of these immune mechanisms.

In cases of serious illness, it is usually not desirable for the disease to run its normal course. By identifying the pathogen and using drug therapy to eliminate it, the risk of serious complications or death can be reduced, and the host may return more rapidly to a condition of normal health.

8.7 CLINICAL MICROBIOLOGY

The most important role of the microbiologist in medicine is in the isolation and identification of the causal agents of infectious disease. This is a major area of microbiology called *clinical microbiology.*

If, on the basis of careful clinical examination, evidence suggests an infectious disease, samples of infected tissues or fluids are collected for microbiological analysis. Depending on the kind of infection, materials collected may include blood, urine, feces, sputum, cerebrospinal fluid, or pus. A sterile swab may be passed across a suspected infected area (Figure 8.22). Small bits of living tissue may be surgically removed (a procedure known as biopsy). In all cases, the sample must be carefully taken under aseptic conditions so that contamination is avoided. Considering the large numbers of normal flora present on many body surfaces, contamination can be a quite serious problem. Once taken, the sample is analyzed as soon as possible. If it cannot be analyzed immediately, it is usually refrigerated to slow down deterioration.

Collection of urine for microbiological analysis requires special precautions to ensure that the specimen is not contaminated with organisms from the urethral opening (see Figure 8.7). Urine samples can be collected by insertion of a sterile rubber tube up the urinary canal into the bladder (a procedure called *catheterization*), but this procedure in itself presents some dangers of introducing infection and is also often painful. An alternative is to collect an uncontaminated sample voided by the patient. In the male, the penis is cleansed by repeated washing with soap and water, followed by sponging with an antiseptic solution. The first portion of urine voided is discarded, and the next portion (so-called midstream urine) is collected directly into a sterile tube or bottle. In the female, the vulva is cleansed three or four times with soapy water, always wiping back toward the anus. In both cases, as soon as the sample is collected the container is covered with a sterile cap and is transported quickly to the laboratory. When the sample cannot be analyzed immediately, it must be refrigerated.

Blood for microbiological analysis is usually taken from a vein with a sterile hypodermic needle and syringe. The site of puncture is first disinfected so that bacteria of the skin will not be carried into the bloodstream as the puncture is made. The blood is transferred immediately from the syringe into an appropriate sterile medium, which must also contain a substance such as sodium polyanethol sulfonate (SPS) which functions as a stabilizing agent and anticoagulant. Blood is normally sterile; hence any indication of the presence of bacteria is presumptive evidence for infection. For this reason, samples must be taken with full precautions to avoid contamination, since contaminants would mislead the diagnostician.

Stomach fluid is obtained by inserting a rubber or plastic tube into the stomach via the mouth or nose and removing a sample by suction. *Feces* for microbiological analysis are usually not collected aseptically. The patient voids into a bed pan, and a sample of feces is transferred to a sterile container for transport to the laboratory.

The removal of *spinal fluid* for diagnostic studies requires strict aseptic conditions, not only to avoid contaminating the sample but, more importantly, to avoid introducing contaminating bacteria into the spinal column, since this region is very easily infected. In meningitis, which is an infection of the membranes covering the spinal cord and brain, bacteria are often in the spinal fluid. Since the spinal fluid is normally sterile, the presence of bacteria is almost always an indication of infection.

The causal agent of pulmonary tuberculosis (*Mycobacterium tuberculosis*) grows in the lungs. As a result of ciliary action of the mucous membranes, organisms are carried with sputum into the throat, and *sputum cultures* are often prepared for diagnosis of tuberculosis. The patient is requested to cough deeply, and the sputum is deposited in a sterile conical glass tube. The specimen should be examined visually to determine if the secretions are indeed sputum rather than saliva. Sputum, being heavier, will settle under saliva in the tube. Since some sputum is swallowed, the organism can also often be

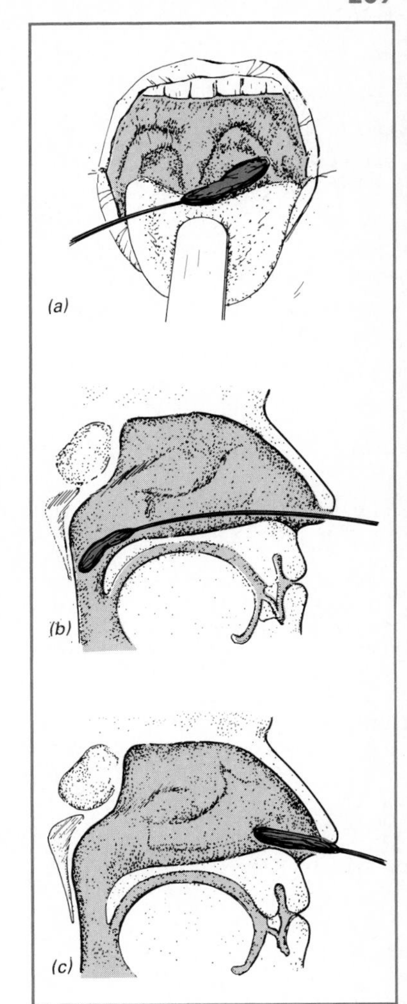

Figure 8.22 Methods for obtaining specimens from the upper respiratory tract. (a) Throat swab. (b) Nasopharyngeal swab passed through the nose. (c) Swabbing the inside of the nose.

found in stomach fluids, and for some diagnostic purposes these are preferable to sputum, since they are less likely to be contaminated with normal flora inhabiting the respiratory tract. The region surrounding the lungs is called the *pleura*, and *pleural fluid* may occasionally be removed for diagnosis of tuberculosis. This is done by entering the pleura with a long, specially designed needle through an incision in the patient's back. Although the sputum, stomach, and pleural fluids are most commonly examined for *M. tuberculosis*, diagnosis for other organisms infecting the lungs may also be done.

Many patients currently are given antibiotics as soon as infection is suspected. However, if microbiological analysis is to be carried out, samples should be taken before antibiotic therapy is begun so that growth of the pathogen is not prevented and the viability of the organism is not reduced by antibiotic present in the sample.

Microbiological analysis Once the sample is collected, it is then subjected to microbiological analysis as shown in Figure 8.23. Samples must be analyzed to determine the numbers and kinds of organisms present. For certain diseases, it is of value to make a **direct microscopic examination** of the sample. This procedure is useful: (1) when the causal agent has distinctive structural or staining properties that make it quite different microscopically

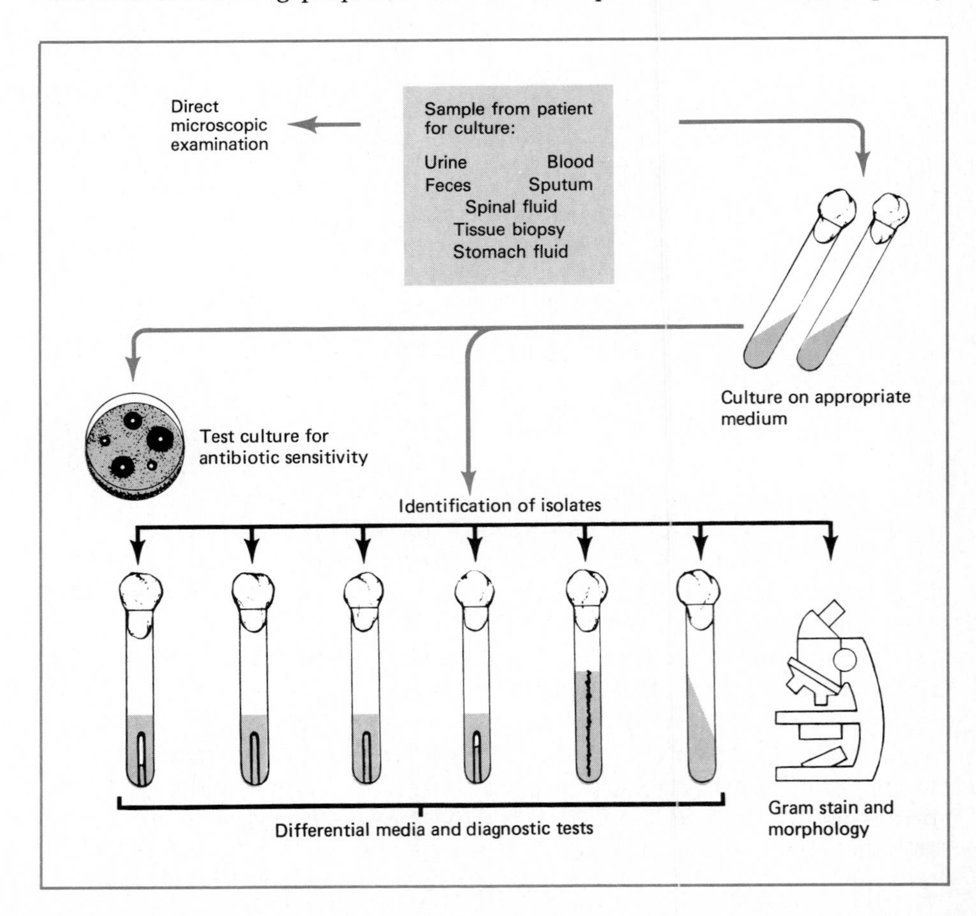

Figure 8.23 Flow sheet of a diagnostic procedure for isolation and identification of a human pathogen.

from the normal flora of the body; (2) when it is present in relatively large numbers; or (3) when it is present in a part of the body that is normally sterile.

Cultural studies are very important in diagnosis but must be performed and interpreted carefully. Successful culture requires selection of the appropriate culture medium and conditions of growth. Rich culture media containing blood are often used, as they provide relatively nonselective conditions in which a variety of pathogens will grow. If positive cultures are obtained from normally sterile fluids such as blood and spinal fluid, this usually indicates infectious disease; but cultures from areas of the body with an extensive normal flora, such as the skin, throat, or intestine, will always yield organisms even in a healthy person. The microbiologist must be able to distinguish members of the normal flora from potential disease agents. In many cases, *selective* and *differential* media, as described in Chapter 4, are used to culture specific pathogens suspected to be present. Once the organism is isolated, it may be obtained in pure culture and characterized taxonomically by its shape, staining properties, and biochemical properties as determined in differential media and other more specialized tests.

One of the most frequently performed procedures in clinical microbiology is determination of the **antibiotic sensitivity** of a culture (described in Chapter 5). Knowledge of the antibiotic sensitivity of a culture permits selection of the proper antibiotic for treatment of the infectious disease.

One note of caution is necessary regarding the collection and handling of samples, and of the cultures derived from them. Any sample or culture is potentially infectious, presenting a certain danger for the clinical microbiologist. The fact that a diagnostic procedure is being conducted implies some suspicion that a pathogen might be present. Consequently, great care must be taken in handling samples and cultures to ensure that laboratory and hospital personnel are not infected. Once samples have been processed, they should be sterilized before being discarded. Cultures present the greatest hazards, since the number of microbial cells has been built up to a high level, thus increasing the possibility of accidental infection. Continuous attention to all aspects of aseptic technique, as outlined in Chapter 4, is essential in all phases of clinical microbiology.

8.8 ANTIMICROBIAL THERAPY

The use of antimicrobial agents in the treatment of infectious diseases is called **chemotherapy**. The development, chemical nature, mode of action, and assay of antimicrobial agents was discussed in Chapter 5. Antibiotics have had a major impact on the management of patients with infectious disease. Currently, a wide variety of antibiotic agents are available, and effective agents are available for virtually all bacterial diseases. A few agents effective against fungi and protozoa are also known. So far, control of viral infections has not been possible with antibiotics, although several chemicals have been discovered that inhibit the action of certain viruses (see Chapter 13). We shall discuss in Chapter 11 the antibiotics that are used in the therapy of specific bacterial

PENICILLIN AT THE CROSSROADS: "IT LOOKS LIKE A MIRACLE"

Penicillin was discovered in London in 1929 by Alexander Fleming. It was the product of a mold that had contaminated an agar plate containing staphylococcal colonies, and caused inhibition of growth of the bacteria near the mold. Fleming's discovery languished until 1939 when Howard Florey (later Sir Howard Florey) began a study of penicillin and its action at Oxford University. It is hard to realize now how difficult work on penicillin was, and how unlikely Florey's success. The assay methods and testing procedures which today we take for granted did not exist then, and medical practice was oriented toward vaccines rather than chemotherapeutic agents. We owe an eternal debt of gratitude to Florey, and to his key collaborator Ernst Chain, for pushing the development of this antibiotic against great odds.

By the middle of March, 1940, after intense effort, Florey's group had managed to accumulate 100 mg of a brown powder consisting of rather impure penicillin. With this material the first animal tests were to be run. First, it was necessary to assess the toxicity of the antibiotic itself, and to determine the best method of administration. Fortunately, penicillin proved nontoxic to animals and tissues. The next step—an absolutely crucial one—was testing against a fatal infection in living animals. On Saturday the 25th of May, 1940, Florey performed the key experiment, surely a turning point in medical history. The day was one of Britain's lowest points in World War II—the Germans were across the English Channel and the British Army was trapped at the French city of Dunkirk. In Florey's experiment, eight white mice were infected with virulent streptococci and then four of the mice were injected with 5 mg of penicillin. The next day, Sunday, Florey rushed to the laboratory to examine the mice. The four control mice which had been infected but left untreated had all died, whereas three of the treated mice were perfectly well and the fourth, although sick, survived another two days. Jubilant, Florey called his collaborator and future wife Margaret Jennings on the telephone: "It looks like a miracle."

Then came the realization of the enormity of the next step: a human weighs 3000 times more than a mouse and the precious brown powder was in short supply. With the Battle of Britain raging in the air and the future of the whole British empire in doubt, Florey and his group pushed doggedly ahead with the purification of penicillin. (Fearing a German invasion, Florey destroyed all the research records to keep the secret of penicillin from the enemy's grasp, but he carefully preserved the penicillin-producing culture by smearing spores into the linings of some of his clothes, where the spores would remain dormant but alive for years.) Failing to interest a British drug company in the penicillin

diseases. Here we will discuss a few general principles.

No one agent is effective against all pathogens, so that care must be taken to select the appropriate drug. Extensive studies are made with each new antimicrobial agent before it is used in medicine, and these studies provide information needed to decide whether a particular antibiotic should be used for treating a particular disease. However, since pathogens may develop resistance to antimicrobial agents, it is essential to test the antibiotic sensitivity of the pathogen isolated from each patient. We have already described the methods by which antibiotic sensitivity is determined in Chapter 5.

In many cases, treatment of an infectious disease requires more than the use of drugs. A drug will not affect a toxin that has been produced by the organism, and since symptoms are frequently a result of the action of toxins, treatment with a drug alone will not be enough. Neutralization of toxin by use of passive immunization procedures is often required. This will be described in Chapter 9. It should definitely not be assumed that a drug must project, Florey determined to produce the penicillin needed for a clinical trial in his own laboratory at Oxford. Always short of funds, the Florey team established their own factory, improvising equipment to permit the growth of the mold in large containers. Finally, a gram or two of active material was available, enough for testing on only a single sick person.

The patient selected was a 43-year-old policeman, Albert Alexander, who had been in a hospital for two months fighting a losing battle against a spreading streptococcal infection. Policeman Alexander's infection, which had started as a small sore at the corner of his mouth, had invaded the subcutaneous tissues of his face, and had reached his eyes and scalp. Treatment with a sulfa drug had been unsuccessful and his left eye had to be removed. The infection continued to spread to the right shoulder, and then, to the lungs. On 12 February, the policeman was given 200 milligrams of penicillin by injection, followed at 3-hourly intervals by 100 mg doses. Within 24 hours, he was obviously better and by 17 February his remaining eye had become almost normal. Now, however, the supply of penicillin was exhausted; the Oxford group resorted to the clever but desperate trick of collecting the patient's urine, extracting the penicillin from it, and reinjecting this recovered antibiotic. For a week or so, it looked like the man might recover, but then, on 15 March, he died. The case, therefore, was not a real success, but it did show that penicillin could be given to a human repeatedly with no ill effects, and offered a glimmer of hope.

Now, the Oxford group rushed feverishly to produce more antibiotic. Their second patient, a boy of fifteen, was gravely ill from a streptococcal bacteremia, and in this case the boy responded favorably to penicillin treatment and recovered fully. Four more patients were treated over the next several months. Finally large-scale production of penicillin began, clinical trials continued, with increasing evidence that penicillin was effective. But now work in England became too difficult because of the war and Florey came to the United States to enlist further aid. The value of penicillin for the treatment of war wounds was evident and it was easy to convince the Americans to cooperate.

The rest of the story is medical history. The U.S. Department of Agriculture together with United States pharmaceutical firms put together an efficient research and development team that brought penicillin from the laboratory setting of Oxford University to commercial production on a world scale. By the end of World War II, penicillin was an item of commerce, available to all who needed it, and its success had pointed the way toward a major research effort in the search for other antibiotics. Today, there are over 50 antibiotics which have significant medical use, including many different forms of penicillin. Medical practice will never be the same. As the culmination of the efforts of the Oxford group, Florey and Chain shared with Fleming the Nobel Prize for Physiology and Medicine in 1945. Florey, the clear leader of the effort, has since been called "the most effective medical scientist since Lister."

always be used to cure an infection. Drugs themselves often have toxic side effects, and if a cure can be effected without drug treatment, that is preferable. In very few infections is the drug alone responsible for a cure. Most of the defense and immunity mechanisms described earlier in this chapter and in Chapter 9 are essential to bring about a cure, even when a highly effective drug is used. Many drugs do not kill pathogens but only prevent their growth; it is the host's defenses that eliminate the pathogens from the body.

In addition to use in curing diseases, antimicrobial agents are sometimes used to prevent future infections in people who may be unusually susceptible. Such use is called **chemoprophylaxis**. The best example of chemoprophylaxis is the use of penicillin to prevent streptococcal sore throats in rheumatic fever patients, since in these patients streptococcal infections often lead to a recurrence of rheumatic fever symptoms. Another example of chemoprophylaxis is the use of drugs in major surgery. Antibiotics are given to patients who have just undergone surgery, since the resistance of postoperative surgical patients to infectious disease is often low and the inoculum of foreign organisms during surgery may be high. However, such use of antibiotics should never supplant well-performed aseptic surgery.

Antimicrobial agents have played an enormous role in the elimination of infectious diseases as major causes of death. Before the availability of effective antimicrobial agents, diseases such as tuberculosis, pneumonia, typhoid fever, meningitis, and syphilis were leading causes of death in humans. Today these diseases are virtually absent from human populations where antimicrobial therapy is available. The practice of medicine has been completely changed since the widespread availability of antibiotics and other antimicrobial agents (see Figure 1.1).

SUMMARY

Microorganisms that inhabit the environments provided by animal bodies may enter **mutualistic**, **commensalistic**, or **parasitic** relationships with their **hosts**. **Pathogenic** microorganisms are those which can harm the host. Most pathogens live within the host and cause **infectious diseases**, but a few pathogens live outside the host and cause harm indirectly through the **toxins** they produce. The **normal flora** which extensively colonizes many of the surfaces of the body usually consists of commensalistic or mutualistic microorganisms, some of which may become **opportunistic pathogens** if the host is compromised.

Infectious disease is a dynamic process which results when the **virulence** of the pathogen overcomes the **host defense**. The body itself possesses physical and chemical barriers, such as **epithelial layers**, enzymes destructive to microorganisms, and **mucociliary flushing**, to resist invading pathogens, but pathogens counter these defenses in a variety of ways. The **invasiveness** and **toxigenicity** of pathogens combine to determine their virulence. When the pathogen injures the host during infection, the host responds in a nonspecific way through the processes of **innate resistance**. During **inflammation**, blood fluid components and cells leak from capillaries into the **focus of infection**. Clotting walls off the infected area, and **white blood cells** enter to **phagocytize** invading microorganisms. Again, pathogens possess many mechanisms which help them evade the defenses provided by the inflammatory response. The **lymphatic** and **blood systems** also help the body to filter out foreign microorganisms. The presence of the pathogen may eventually stimulate a specific response by the **immune system** which finally limits infection.

Clinical microbiology exists to support physicians in identifying pathogenic microorganisms through **direct microscopic examination** and **cultural studies** of samples collected from diseased patients. Another important responsibility of the clinical microbiologist is to determine the **antibiotic sensitivity** of the isolated pathogen, so that appropriate **antimicrobial therapy** can be recommended.

KEY WORDS AND CONCEPTS

Pathogen
Opportunistic pathogen
Infectious disease
Contagious disease

Epithelial layers
Mucociliary flushing
Virulence
Invasiveness
Toxigenicity
Exotoxin
Endotoxin
Focus of infection
Innate resistance
Inflammation
Phagocytosis
Lymphatic system
Interferon
Acquired resistance
Immune system
Clinical microbiology
Chemotherapy
Chemoprophylaxis

STUDY QUESTIONS

1. What are the differences between *mutualism, commensalism,* and *parasitism*?
2. How can a *pathogenic* microorganism cause disease without growing within the *host*? Give an example of such a disease.
3. Name three *infectious diseases* and the microorganisms which causes them. Name three *noninfectious diseases* and their causes.
4. Upon what surfaces in the body are the microorganisms belonging to the *normal flora* found? What areas of the body are normally considered sterile?
5. Although many members of the *normal flora* are commensals, some are *mutualistic*. Name three benefits the host may derive from its normal flora.
6. What causes tooth decay? How does fluoride help protect teeth?
7. What causes body odor? How do antiperspirants help prevent it?
8. Give an example of a disease caused by one specific microorganism. Give an example of a disease caused by more than one type of microorganism. Give an example of a microorganism which can cause more than one disease.
9. Imagine that a pathogenic microorganism is inhaled into the upper respiratory tract. Describe what physical and chemical barriers of the *host defense system* would protect the host from the pathogen. What mechanisms might the pathogen use to counter these defenses?
10. What two features of pathogenic microorganisms determine their *virulence*?
11. What is the difference between *endotoxin* and *exotoxin*? Give an example of each type of *toxin*.
12. What is the difference between *innate* and *acquired* immunity?
13. What components of blood are involved in *phagocytosis*? What mechanisms do pathogenic microorganisms use to avoid being killed by phagocytosis? Where else in the body does phagocytosis occur?
14. Imagine a cut made by a dirty knife. Describe how the body will respond to this injury. Will the response be a specific or a nonspecific type of defense mechanism?

15. Explain how the white blood cell count can be of value in diagnosing certain diseases.

16. Which system in the body is involved with filtering blood and the fluids which drain from tissues? What type of cell actually does the filtering? Name two specific organs which are a part of this system.

17. How does *interferon* protect cells from infections caused by viruses? How is interferon obtained for experimental purposes?

18. Describe the events which occur when you have streptococcal sore throat. Make a list of features of the pathogen which determine its virulence.

19. What are the clinical stages usually observed during disease? In which stage are the symptoms most pronounced?

20. Why must the *clinical microbiologist* be concerned with the *normal flora* of the body when collecting samples? Give two reasons.

21. Name two important contributions a *clinical microbiologist* can make which help in both the diagnosis and treatment of disease.

22. Why is it often insufficient to treat a patient only with antibiotics in order to cure a disease?

SUGGESTED READINGS

DAVIS, B.D., R. DULBECCO, H.N. EISEN, and H.S. GINSBERG. 1980. *Microbiology, 3rd edition*, Harper & Row, Publishers, Philadelphia, Pennsylvania. (An excellent detailed treatment of the principles of infection and immunity.)

JAWETZ, E., J.L. MELNIK, and E.A. ADELBERG. 1982. *Review of medical microbiology, 15th edition*, Lange Medical Publications, Los Altos, California. (A general textbook on the microorganisms that cause disease, including discussion of the disease process.)

JOKLIK, W.K., H.P. WILLETT, and D.B. AMOS. 1984. *Zinsser microbiology, 18th edition*, Appleton-Century-Crofts, East Norwalk, Connecticut. (A standard reference text on medical microbiology, including chapters on the host-parasite relationship, and the normal flora.)

LENNETTE, E.H., A. BALOWS, W.J. HAUSLER, JR., and H.J. SHADOMY. 1985. *Manual of clinical microbiology, 4th edition*, American Society for Microbiology. (Step-by-step descriptions of methods used for the culture and identification of pathogens from clinical material, and for antibiotic sensitivity testing.)

9

Immunology

In Chapter 8 we discussed the relationship between pathogenic microorganisms and their hosts. An essential part of host defense against disease-causing microorganisms in higher animals is the **immune response**, which leads to a type of resistance which develops during infection. The immune system responds specifically to chemicals foreign to the body, including the chemicals which coat the surfaces of pathogenic microorganisms, and the toxins produced by pathogens. These foreign substances which elicit the immune response are called **antigens**. As the result of antigen stimulation, the immune system produces specific proteins, called **antibodies**, or specific cells, called **activated T cells**, which react with the antigen. Together with other components of the host defense, the antigen-antibody or antigen-T cell reactions limit diseases by neutralizing the action of pathogenic microbes and their toxins. Immunity to infection which develops as a result of antigen stimulation should be contrasted with the immunity arising from the processes of phagocytosis and inflammation discussed in Chapter 8. Those immune processes which can occur in the absence of antigen stimulation are called *innate*, whereas those resulting from antigen stimulation are called *acquired*. As we shall see, acquired immunity is a major and extremely important property of the animal host, with significance for many processes other than microbial infection.

Three major features characterize the immune system: specificity, memory, and tolerance. The **specificity** of the antigen-antibody or antigen-T cell interaction is unlike any of the other host-resistance mechanisms described in Chapter 8. Once the immune system produces a specific type of antibody or activated T cell, it is capable of producing more of the same antibody or T cell, more rapidly and in larger amounts. This capacity for **memory** is of major importance in resistance of the host to subsequent reinfection or in the protection to the host provided by vaccination. The third property, **tolerance**, exists because chemicals on the surface of body cells are also potentially antigenic and would be materially altered if antibodies and activated T cells to these body cells were produced during an immune response. Since the immune system can develop tolerance, an immune response to body cells does not normally occur. The immune system is thus able to distinguish "self" from "nonself" (foreign) materials. Some problems related to the ability of the immune system to distinguish self from nonself include blood transfusion complications, tissue transplant rejection, allergy, and autoimmune disease.

In this chapter we will consider the basic biology of **antibody-mediated** and **cell-mediated immunity**. We generally make a distinction between the words "immune" and "immunological." **Immune** refers to the ability of the body to resist infectious disease, whereas **immunological** refers to the processes of antibody formation and activated T cell formation, whether or not these lead to immunity to a particular infectious disease. There are also many important immunological situations not involving infectious disease.

9.1 THE IMMUNOLOGICAL RESPONSE

Antigens As we have noted, the immunological response is elicited by specific chemicals called **antigens** which stimulate the production of antibodies or activated T cells. Compounds which are part of the surfaces of pathogenic microorganisms are often antigenic. Antigens are usually macromolecules such as proteins or polysaccharides, although some lipids and nucleic acids are also antigenic. The immune response is not actually against the macromolecule as a whole, but against certain parts of the antigen, called **antigenic determinants**. Chemically, antigenic determinants include sugars, amino acid side chains, organic acids, and other organic constituents. The specificity of antigenic stimulation is high, and two sugars as similar as glucose and fructose act as distinct antigenic determinants. Although the chemical substances such as sugars and amino acids which function as antigenic determinants do not themselves induce an immune response, they will combine with antibodies or activated T cells. Small molecules such as sugars and amino acids which behave in this way are called *haptens*. In addition to responding to natural antigenic determinants, the immune system can respond to artificial antigenic determinants. For instance, many drugs and other chemicals are able to combine with proteins, the combination then acting as an antigenic determinant. Neither the chemical nor the protein is by itself antigenic, but together they cause an immune response. Many allergies occur because of the complexing of chemicals with body proteins.

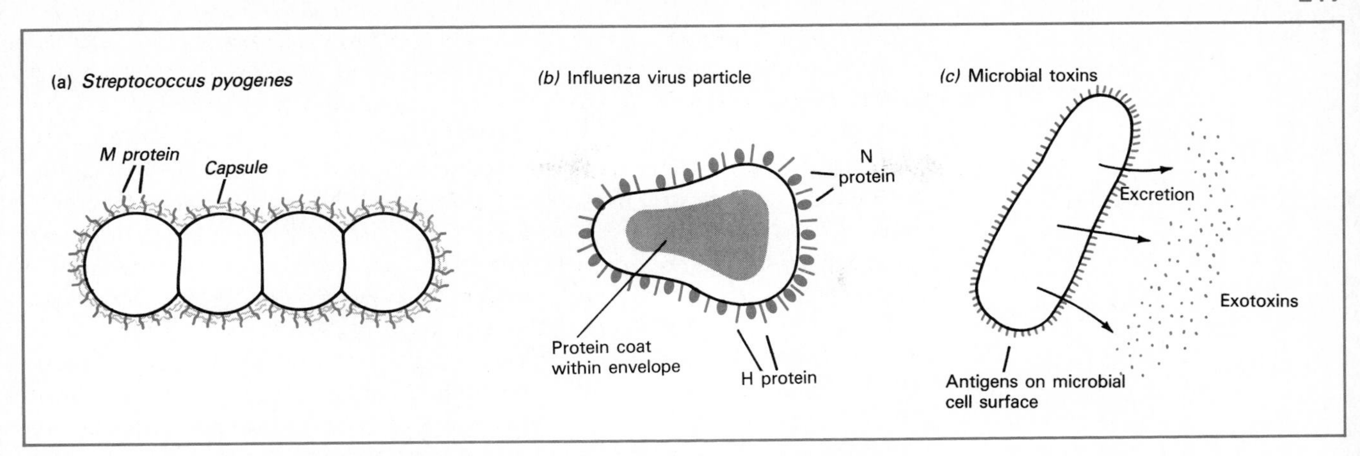

Figure 9.1 Examples of antigenic substances of pathogens. The antigens are shown in color. (a) Capsule and M protein on the surface of the bacterium *Streptococcus pyogenes*, a causal agent of sore throat. (b) H and N proteins on the envelope of influenza virus. (c) Toxins excreted by a pathogen may also be antigenic.

As we have noted, many microbial substances, being foreign, are antigenic. From the viewpoint of immunity, the antigens on the *outside* of pathogens are most significant, because these are the antigens of the living pathogen with which the immune system would interact. For example, the surface of the bacterium which causes streptococcal sore throat, *Streptococcus pyogenes*, contains a polysaccharide capsule and a fibrous protein called the M protein (Figure 9.1*a*). Both the capsule and the M protein are antigenic and can elicit antibody formation. From the point of view of the pathogen, these features are important virulence mechanisms which aid in countering the host defenses, or in injuring the host. As discussed in Chapter 8, the M protein of *S. pyogenes* is involved in the microbe's attachment to ciliated epithelial cells and its resistance to phagocytosis. Similarly, the influenza virus particle bears on its envelope two proteins, called the H and the N proteins (Figure 9.1*b*), which are antigenic. The H protein is involved in virus attachment whereas the N protein is involved in release of virus from the host cell. From the point of view of the host, however, these substances are recognized as being foreign to the body. As another example, toxins which pathogenic microorganisms produce may also be antigenic (Figure 9.1*c*). Since all of these substances are chemicals not normally found in the body, they have the potential to act as antigens.

It is often desirable to artificially induce an immunological response by injecting an antigen, a procedure called *vaccination*. **Vaccines** are preparations of antigenic materials that are able to bring about or enhance the immunological response. Vaccines generally consist of either microbial cells or their products, or virus particles or their components.

The chemicals of animal cells are also antigenic. An immune response is not mounted against the body's own cells, however. As we have noted, immunological tolerance is an essential feature of the immune system. It is what allows the system to respond selectively against foreign substances but not against body components. The cells of different individuals are antigenically different; one individual therefore exhibits an immune response against the cells of another individual. This is the basis for problems such as blood transfusion reactions and tissue transplant rejections.

Cell-mediated and antibody-mediated immunity The responses of the immune system to antigens are observed to be of two types: antibody-mediated and cell-mediated. Antibody-mediated immunity results from the formation of specific proteins called **antibodies**. Antibodies are specific proteins which react with the antigens which stimulate their formation. Thus, an antibody formed against the M protein of *S. pyogenes* will react specifically with the M protein and will usually not react with other antigens, even other antigens of *S. pyogenes* (Figure 9.2*a*). Similarly, antibody against the influenza virus N protein will not react with other influenza virus antigens (such as the H protein), or with the antigens of any other organism. Antibodies are found mainly in the blood serum and in secreted body fluids, such as saliva, mucus,

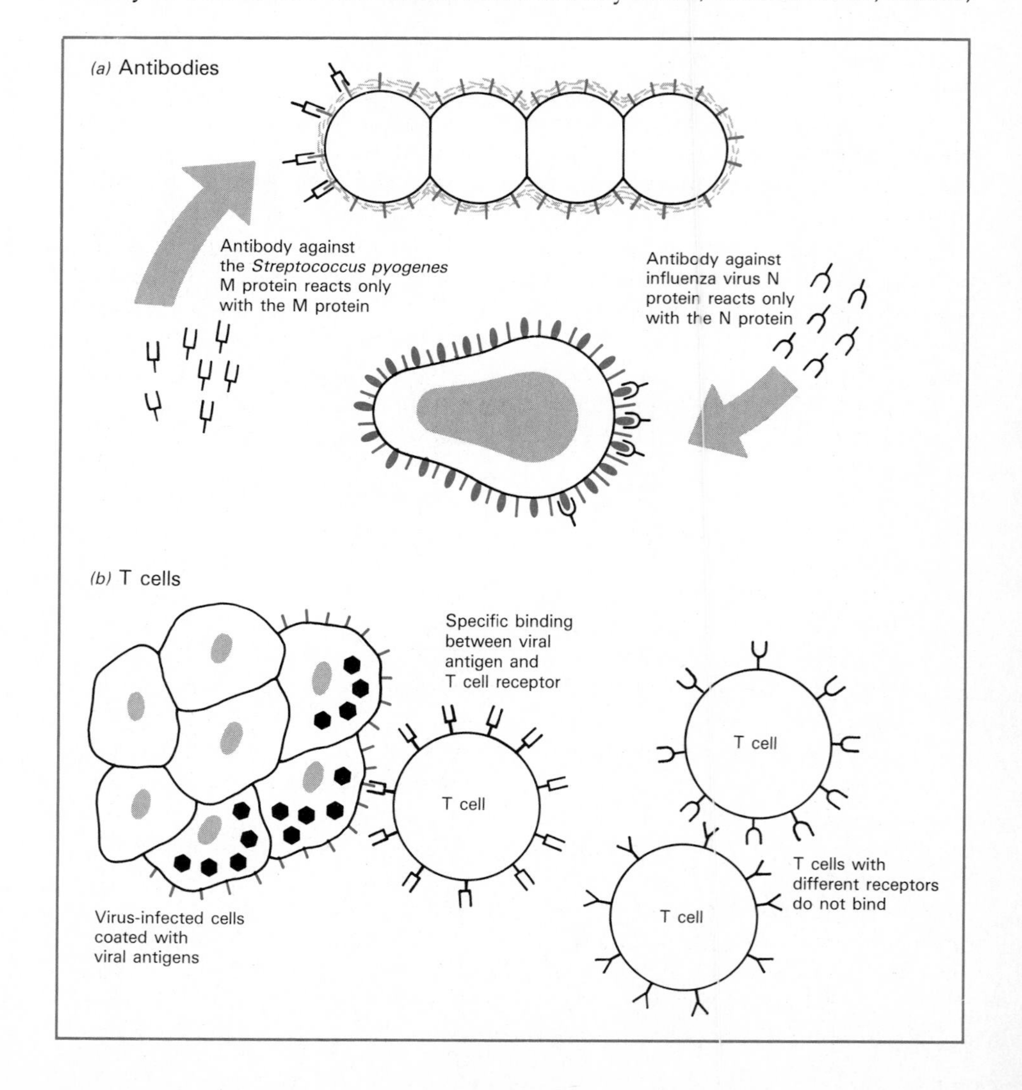
(a) Antibodies
Antibody against the Streptococcus pyogenes M protein reacts only with the M protein
Antibody against influenza virus N protein reacts only with the N protein
(b) T cells
Specific binding between viral antigen and T cell receptor
T cell
T cell
T cell
T cells with different receptors do not bind
Virus-infected cells coated with viral antigens

Figure 9.2 Two types of immune response. (a) Antibodies are small serum proteins which bind specifically to antigens. (b) Special cells of the immune system called T cells bear receptors which bind specifically to antigens.

and breast milk. A variety of antibody-antigen reactions may occur, as will be described below.

The other type of immune response is carried out by cells of a type called T cells. There are several different types of T cells, some which are helpers in the immunological response and some which respond specifically to antigenic stimulation. Those T cells which respond to antigenic stimulation are called **activated T cells**. They possess antibody-like proteins called *receptor proteins*, which coat their membranes. The reaction between receptor and antigen thus involves the binding of a T cell (Figure 9.2*b*). The degree of *specificity* between T cells and the antigens they recognize is similar to that between antibodies and antigens. A given type of T cell bears only one type of receptor, which can generally react with only a single type of antigen. Several subsequent reactions are possible after the initial antigen-T cell interaction, including direct killing of the microbial cell which bears the antigen. The reactions of T cells after binding to antigens will be described below.

A typical antibody-mediated immunological response is diagramed in Figure 9.3 as it occurs either during microbial infection or in response to the injection of an antigen (as in a vaccine). The amount of a specific antibody in the blood is called the **antibody titer**. When an antigen is first encountered, there is an initial delay of several days during which changes in the immune system occur, followed by a slow rise in the titer of the antibody against the antigen. This initial slow rise, called the **primary response**, is followed by a gradual decline in antibody titer. If subsequent exposure to the same antigen occurs, for example if another infection with the same pathogen occurs or if a second vaccine dose is given, there is a **secondary response**, in which antibody is produced more rapidly, to a higher level, and over a longer period of time. The secondary response illustrates the capacity for **memory** in the immune system. The boosting of the antibody level which occurs in the secondary response is taken advantage of during immunization procedures, which usually require more than one administration of vaccine. The second and subsequent injections of vaccine are called *"booster shots"* for this reason.

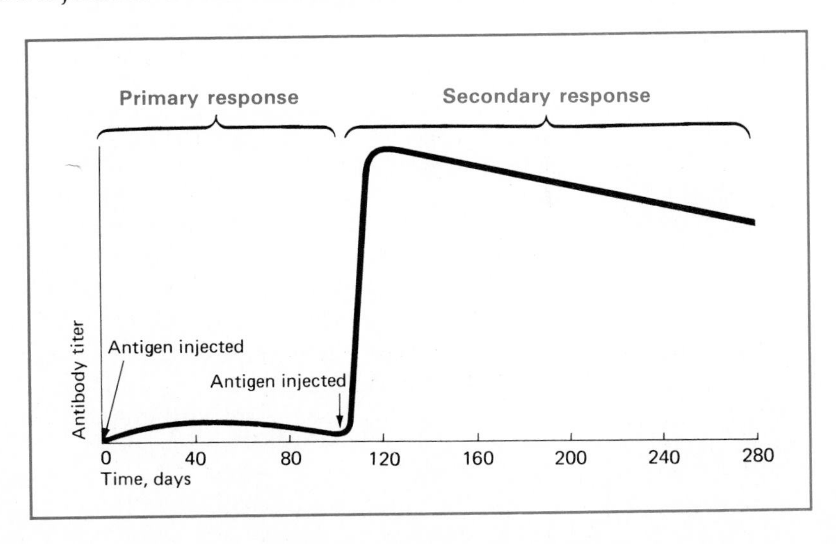

Figure 9.3 Typical antibody response after encounter with an antigen, showing the primary and secondary responses. A booster injection elicits a secondary response.

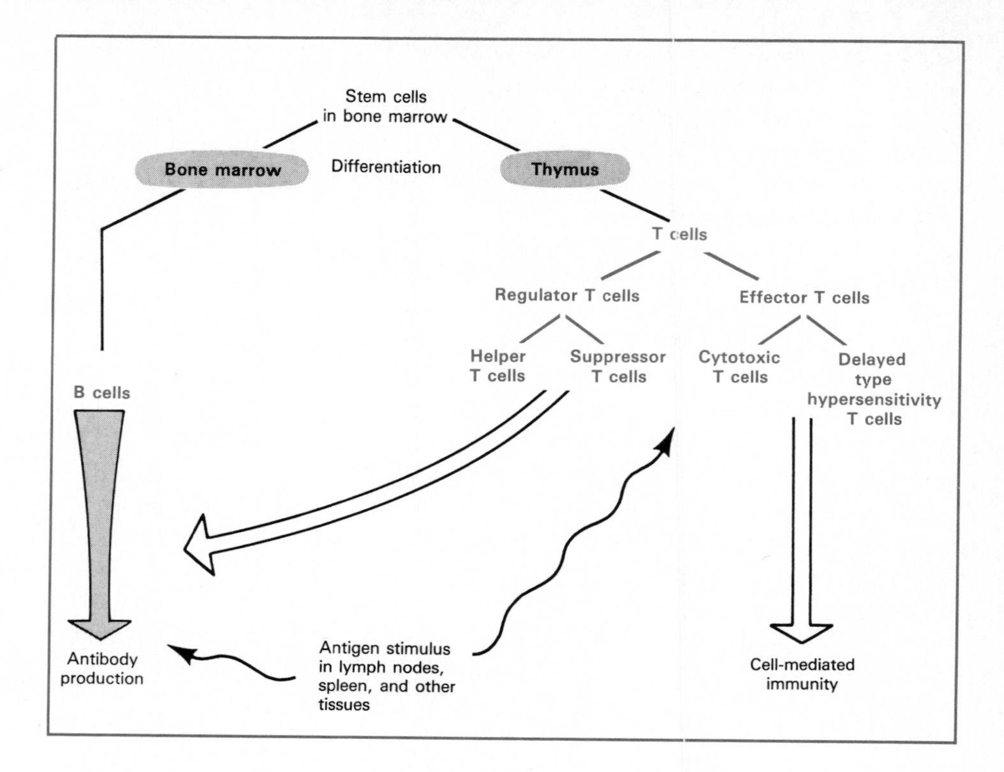

Figure 9.4 The major types of cells which function in the immune system. All the cell types shown originate from the bone marrow, but differentiate in different organs before moving to the various parts of the immune system.

A similar response occurs during cell-mediated immunity, with the primary and secondary responses involving an increase in the number of specific activated T cells, rather than an increase in antibody titer. As with the antibody response, the cell-mediated response may occur either during the course of infectious disease or after injection of vaccine.

The immune response of an individual is influenced by nutritional state, age, hormonal balance, and general well-being. Very young children and very old individuals usually respond poorly. Certain drugs, especially anti-inflammatory agents such as cortisone and other steroids, reduce the ability to respond. Radiation (such as X rays or atomic radiation) also severely reduces the immune response.

Cells of the immune system The immune system is controlled by cells of the lymphatic system (see Figure 8.13). Two major types of cells are involved, **B cells** which specialize in the production of antibody, and **T cells**, which carry out cell-mediated immunity. Both types of lymphatic cells originate in the bone marrow (Figure 9.4). During development of an individual, immature bone-marrow stem cells differentiate. Some differentiate within the bone marrow to become B cells. Other cells differentiate within the thymus gland and become T cells. (The "T" designation of these cells derives from their recognized origin within the thymus.) T cells differentiate further into the subclasses shown in Figure 9.4, each of which performs a different function to be described later. After this differentiation, both T and B lymphocytes are capable of recognizing and responding to antigens. They then migrate to lymph nodes, the spleen, and patches of subepithelial tissue in the tonsils, intestinal wall, and appendix. However, as described in Chapter 8, the lymphatic and blood circulations are interconnected (see Figure 8.13), so that some lymphocytes continually circulate via blood into the tissues and back into the lymphatic system. In the lymphatic and blood systems, and in the peripheral tissues of the body, T and B cells may encounter foreign antigens. Further differentiation of these cells occurs in response to stimulation by an antigen, as will be described below.

9.2 ANTIBODY-MEDIATED IMMUNITY

Antibody structure The proteins in blood serum which have antibody properties are called **immunoglobulins**. There are several different classes of immunoglobulins; the structure of the most common class, called *IgG* (or immunoglobulin G) is shown in Figure 9.5. An IgG antibody is a protein consisting of four separate amino acid chains. Two identical chains, called

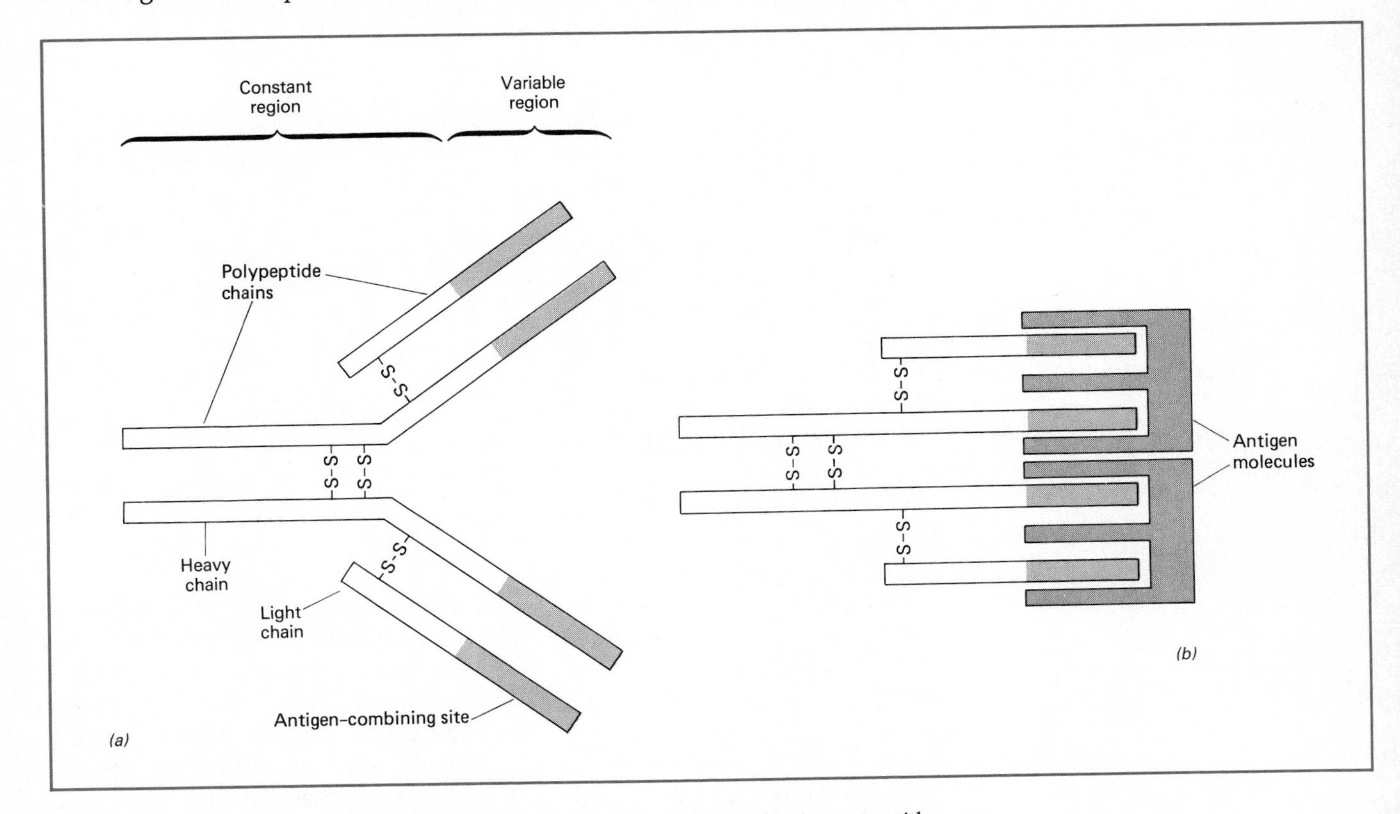

Figure 9.5 The structure of an antibody molecule and the manner in which it interacts with an antigen molecule. (a) Four separate proteins (2 light chains and 2 heavy chains) are held together by bonds between sulfur atoms (—S—S—). The variable regions of the antibody molecules contain the antigen-combining sites (in color). (b) Two antigens may interact with a single antibody molecule.

light chains, are shorter than two other identical chains, called *heavy chains*. The four chains are held together by covalent bonds between sulfur atoms, forming a structure which can be represented as "Y" shaped. One end of the antibody molecule contains the **antigen-combining sites**, where specific recognition of antigen occurs. Since different antibodies react with different antigens, this site on antibody molecules varies in structure, and is called the **variable region**. The other end of the antibody molecule does not interact with the antigen. For a given class of antibodies, this end of the molecule is constant in composition and is called the **constant region**.

How does an antibody recognize and react specifically with only a single antigen? As with other proteins, an antibody should be thought of as a macromolecule with a three-dimensional shape. The way in which an antibody recognizes and binds with an antigen is analogous to the way in which an enzyme reacts and binds with its substrate molecule. In Chapter 3 we discussed the specificity which enzymes have for their substrates (see Section 3.3); in Figure 3.10 we showed how the active site of an enzyme is a function of its three-dimensional configuration, which is determined primarily by its primary amino acid sequence. Different antibody molecules have variable regions with different primary amino acid sequences, and therefore with different three-dimensional structures. As a result, each different antibody is specific for and reacts with a different antigen.

Kinds of antibodies Studies of the chemistry of antibodies have shown that a variety of different antibody molecules are formed, each with its own function. The structures and functions of these various immunoglobulins are summarized in Table 9.1.

IgG is the major antibody found in the circulating body fluids, and plays a key role in antibody-antigen reactions in the blood, lymph, and tissue fluids. IgG molecules are *bivalent*, which means they have *two* antigen-combining sites, as illustrated in Figure 9.5. Immunoglobulin **IgM** has a much more complex structure than IgG and is much larger in size; it has ten antigen-combining sites. Perhaps because of its size, IgM does not enter the tissue fluids but remains confined to the blood and lymph systems. An important difference between IgM and IgG is that IgG antibody can pass through the placenta of the pregnant female (thus playing a role in immunity in the fetus), whereas IgM antibody cannot. IgM is the first antibody to appear after immunization; IgG appears later.

Immunoglobulin **IgA** is of interest because it is secreted into body fluids, such as the saliva, tears, breast milk and colostrum, gastrointestinal secretions, and mucous secretions of the respiratory and genitourinary tracts. IgA is also found in the blood, but the IgA secreted into body fluids has a modified structure, with a protein-carbohydrate complex attached to it that apparently serves as the "secretory piece," protecting the molecule from digestion by proteolytic enzymes which are also present in secreted fluids. The secretory piece can bind only to IgA because of the chemical nature of the constant region of this antibody class. The attachment of the secretory piece to the IgA molecule occurs in the mucosal cell through which the IgA passes on its way into the secretions. Because of its location in the secretions, IgA probably

TABLE 9.1 Kinds of antibodies and their properties

Designation	Properties	Structure*	Percent of total antibody	Distribution	Function of constant region
IgG	Major circulating antibody		80	Extracellular fluids and blood and lymph; crosses placenta	Binds complement
IgM	Complex structure; first antibody to appear after immunization		5–10	Blood and lymph only; does not cross placenta	Binds complement
IgA	Secretory antibody	Secretory piece	10	Secretions; extracellular and blood fluids	Binds secretory piece
IgD	Minor circulating antibody		1–3	Blood and lymph only	Binds to lymphocyte surfaces
IgE	Involved in allergic reactions		0.05	Blood and lymph; mast cells in tissues	Binds to mast cells

*See also Figure 9.5.

provides a first-line antibody attack against bacterial invaders, which as we have seen, become established initially on tissue surfaces (which are generally bathed in secretions). A breast-fed baby receives IgA antibodies from both colostrum and milk and usually has fewer problems with bacterial infections than a newborn fed cow's milk from a bottle.

The constant region of **IgE** enables this type of antibody to bind to cells called *mast* cells. This binding is important in allergic reactions, as will be discussed below.

The function of **IgD** type antibody is not certain. It appears to be bound to the surfaces of antibody-producing cells and may play a role in the production of antibodies of other classes.

Antigen-antibody reactions Many different kinds of antigen-antibody reactions occur. **Neutralization**, the simplest kind of reaction, occurs when an antibody combines with a soluble antigen, such as a toxin, thus blocking its toxicity (Figure 9.6). The antibody that combines with and neutralizes a toxin is sometimes called an **antitoxin**. Antibodies can play an important role in neutralizing specific features of pathogen invasiveness. For example, neutralization of the *Streptococcus pyogenes* M protein decreases its antiphagocytic

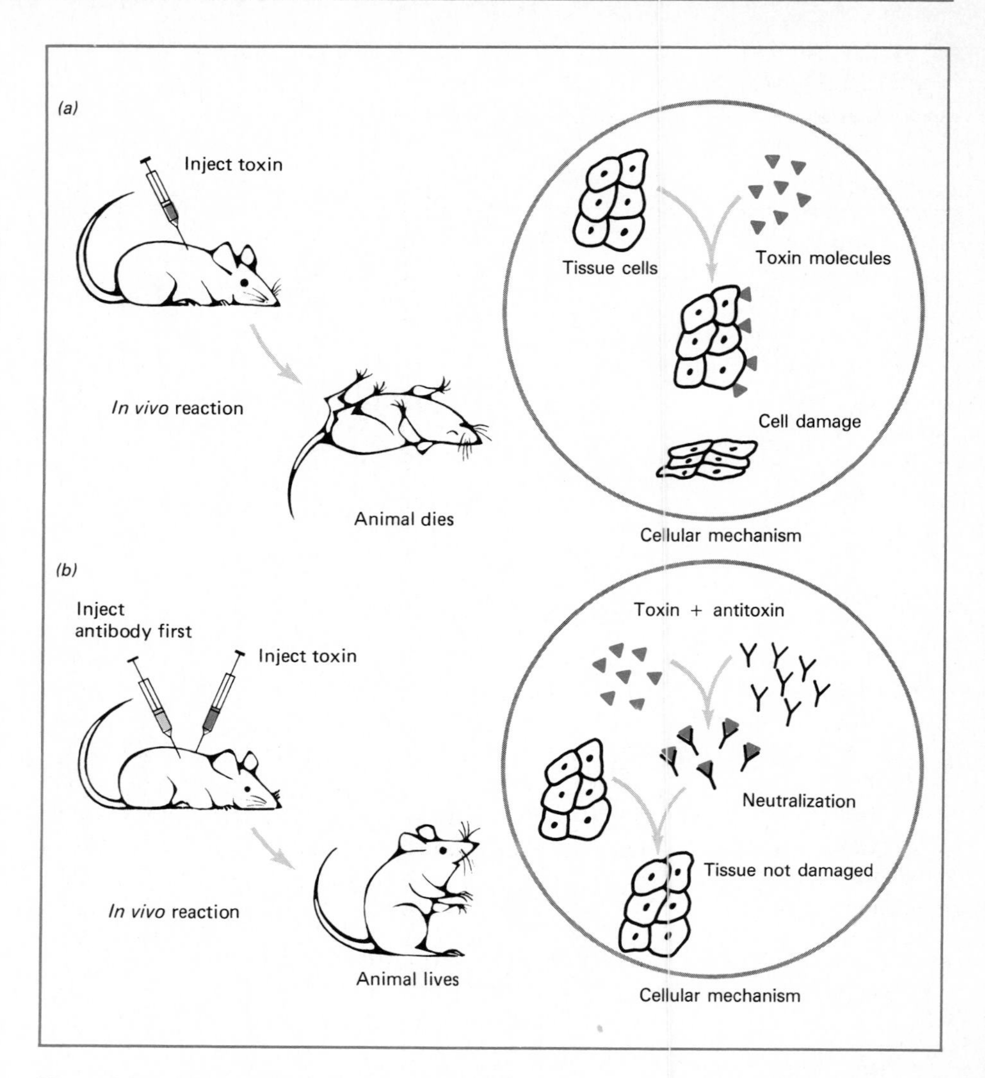

Figure 9.6 Neutralization of toxin by an antibody. (a) Action of an unneutralized toxin on an animal. (b) After neutralization of the toxin by specific antibody (antitoxin), the animal is protected.

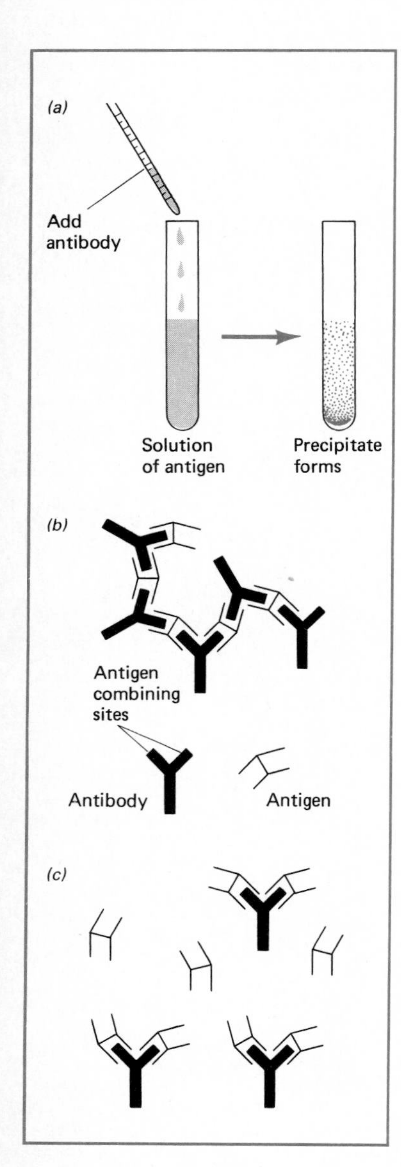

Figure 9.7 Antibody precipitation reaction. (a) Demonstration of precipitation in a test tube. (b) Mechanism of precipitation; precipitate forms when antibody and antigen are present at equivalent amounts. (c) No precipitation occurs when either antibody or antigen is present in excess.

action. Neutralization of viral surface proteins makes them unable to attach to host cells.

Precipitation occurs when a number of molecules of antibody and antigen react together, producing a clump, or aggregate (Figure 9.7*a*). The mechanism of precipitation for a bivalent antibody is illustrated in Figure 9.7*b*. Because there is frequently more than one antigen-combining site on each antibody molecule, antibodies can form a bridge between two or more antigens, so that a large aggregate containing alternating units of antibody and antigen develops. Precipitation requires that approximately equal numbers of antibody and antigen molecules react. If either antibody or antigen is in great excess, no precipitate will form, and the small antigen-antibody complexes will be soluble (Figure 9.7*c*). Precipitates are phagocytized more readily than are solutions, and precipitation thus leads to rapid clearing of antigens from the blood or other fluids. Addition of an excess of hapten can inhibit a precipitation reaction, and such inhibition is sometimes used in studies on the characterization of antigenic determinants.

Agglutination is like precipitation except that the antigen is part of a cell (such as an endotoxin) rather than a soluble component; the antibody molecules cause the cells to stick together and form clumps (Figure 9.8*a*), which are more readily phagocytized than are individual cells. Agglutination reactions can occur either as a result of combination of the antibody with cell-envelope antigens, in which case a compact agglutinate forms (Figure 9.8*b*), or the antibody can combine with flagellar antigens, in which case a much looser agglutinate forms (Figure 9.8*c*). Antibodies against flagella can also cause **immobilization** of motile cells, since motility is caused by the action of flagella. Immobilization is one of the few antigen-antibody reactions that can be detected with single cells.

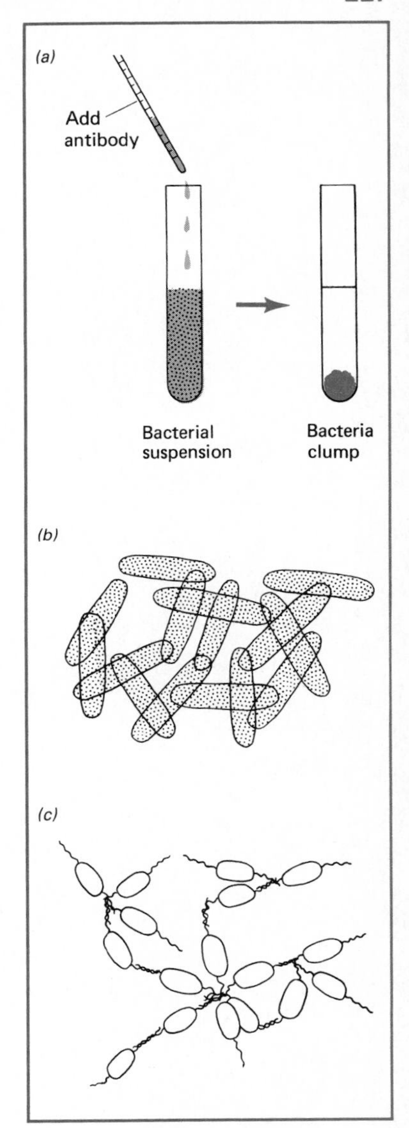

Figure 9.8 Antibody agglutination reaction. (a) Demonstration of agglutination in a test tube. (b) Appearance of agglutinated cells when antibody is directed against cell surface antigens. (c) Appearance of agglutinated cells when antibody is directed against flagellar antigens.

9.3 THE COMPLEMENT SYSTEM AND THE IMMUNE SYSTEM

Although antigen-antibody reactions may themselves help to counter the action of specific toxins or other pathogen virulence features, antibodies cannot by themselves destroy an invading pathogen. However, by reacting with a series of enzymes, called **complement**, the antigen-antibody complex *can* bring about the killing of a foreign cell. Enzymes of the complement system are continually present in the circulating fluids of the body and are not formed in response to infection. They are normally inactive, but are activated when antigen-antibody reactions occur. There is considerable economy in an arrangement such as this, since a wide variety of antibodies, each specific for a single antigen, can call into action the complement enzymatic machinery; thus the body does not need separate enzymes to counter the attack of each kind of invading agent.

Some reactions in which complement participates include: (1) bacterial lysis, especially in Gram-negative bacteria, when specific antibody combines with antigen on bacterial cells in the presence of complement; (2) microbial killing, even in the absence of lysis; and (3) phagocytosis, which may not occur during infection if the invading microorganism possesses a capsule or other surface structure that prevents the phagocyte from acting. However, when specific antibody combines with the cell in the presence of complement, the cell is changed in such a way that phagocytosis can occur. (This process in which antibody plus complement renders a cell susceptible to phagocytosis is sometimes called **opsonization**.)

Complement activation Complement is a system of 11 proteins, designated C1, C2, C3, and so forth. Activation of complement occurs only by antibodies of the IgG and IgM types (see Table 9.1). The antigen-antibody complex serves as a scaffold upon which the sequential assembly of the complement system can occur (Figure 9.9). When such antibodies combine with their respective antigens, they are altered in such a way that the first com-

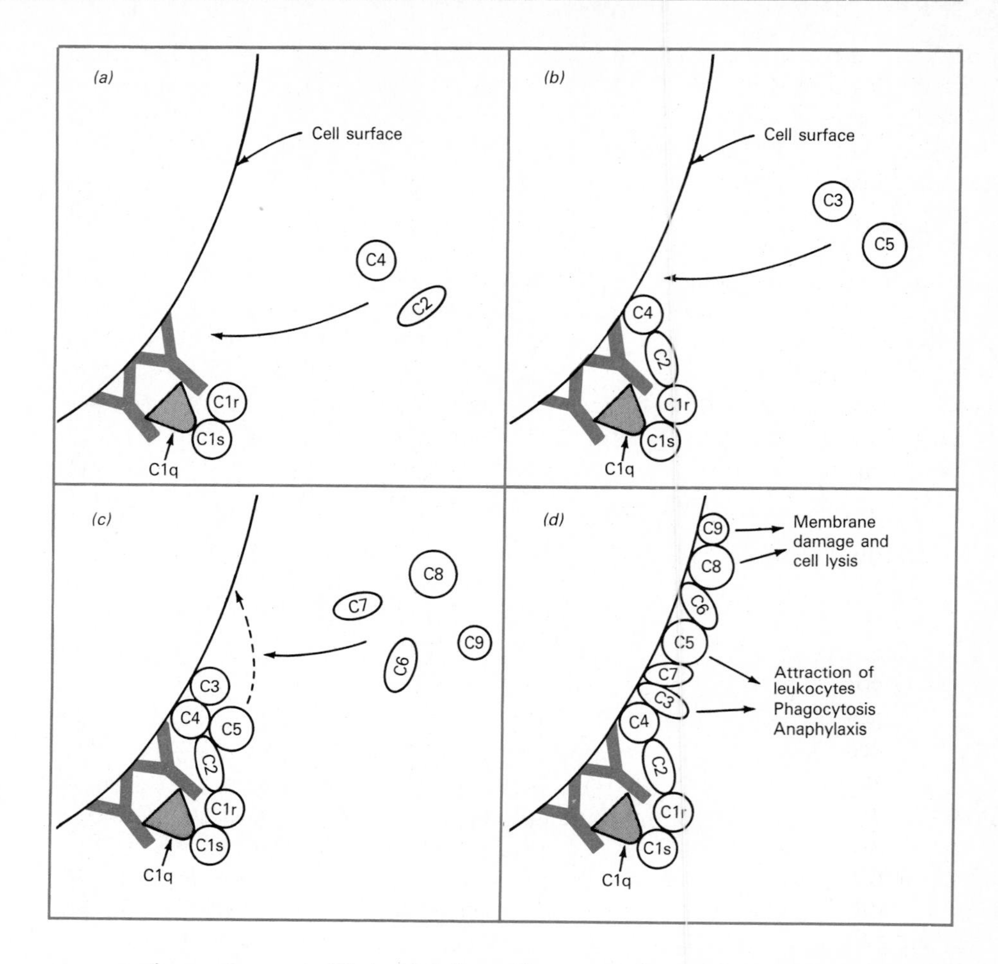

Figure 9.9 The major steps in the activation of complement. After antibody binds to cell surface antigen, sequential bonding of complement proteins occurs (parts a, b, and c). Some of the consequences of complement activation are shown in part (d).

ponent of complement, C1 (which is really a complex of three subunits called C1q, C1r, and C1s), combines with the antigen-antibody aggregate at the constant region of the antibody molecule. The C4 and C2 proteins then interact and form a complex, C4-C2, which links the growing complement sequence to the cell membrane. The complement protein C3 is the substrate for the C4-C2 complex, and when C3 becomes enzymatically activated, C5 can combine with it. Subsequent reactions with C5, C6, C7, C8, and C9 lead to the completion of the complement sequence of reactions.

A variety of biological effects result from the interaction between complement and antigen-antibody complexes. As outlined in Figure 9.9*d*, reactions at the C3 level result in attraction of phagocytes to invading agents, and in opsonization to promote phagocytosis. Reaction at C5 also leads to leukocyte attraction. The terminal series of reactions from C5 through C9 results in cell lysis and death. Lysis itself results from destruction of the integrity of the cell membrane, leading to the formation of holes through which cytoplasm can leak.

The complement system is involved in other immunological reactions besides those leading to the destruction of invading organisms. Allergic reactions and certain aspects of the inflammatory response also involve actions of the complement system. C3 and C5 fragments cause release of histamine from cells that store this substance. Histamine increases the permeability of capillaries, enabling leukocytes and fluid to escape into the tissue cells in an inflammatory reaction. One serious type of allergic reaction, **anaphylactic shock**, also sometimes involves components of the complement system. We shall discuss allergy later in this chapter.

9.4 MECHANISM OF ANTIBODY FORMATION

It has been estimated that an individual can form antibodies to over a million different antigens. How is it possible for higher animals to produce such a large variety of specific proteins? The problem of antibody formation can be divided into two separate questions:

1. How is antibody diversity generated?
2. How do the cells of the immune system interact to produce antibodies when stimulated by an antigen?

Genetic origin of antibody diversity To understand the **generation of antibody diversity**, we must consider the structure of an antibody again (see Figure 9.5). We consider here only the mechanism of formation of IgG antibodies. As noted, an IgG antibody is a protein composed of four amino acid chains. A protein is encoded by a specific gene in the DNA. The potential variety of antibody molecules is huge, but a separate gene does not preexist in the body for each antibody type. Rather, several genes, called C genes (for constant region) code for the constant end of an antibody molecule, one for each antibody class. The variable end of the molecule, where recognition and binding to different antigens takes place, contains different regions, each encoded by a separate gene. These regions are called the J (joining) and V (variable) regions; heavy chains also contain a third region called the D (diversity) region. The variation among antibodies occurs because there are many different J, D, and V genes which can be arranged in different combinations. These rearrangements occur via natural recombination events that take place in B and T cell lines that are maturing as part of the blood cell system in the bone marrow.

Before maturing in the bone marrow, immature B cells contain a complete set of C, J, D, and V genes for both heavy and light chains. There are about 150 V-region and 5 J-region genes for the light chain. Thus, about 750 combinations (150 $\times$ 5) between light chain V and J regions are possible. Similarly, there are about 80 V-region, 50 D-region, and 6 J-region genes for heavy chains; 24,000 combinations (80 $\times$ 50 $\times$ 6) are possible. Since different heavy and light chains may combine, over 18,000,000 (24,000 $\times$ 750) possible types of antibodies can be formed from only 200 or so genes. In this way, information storage is very efficient in genes coding for antibodies. As noted, antibody class is determined by different C-region genes corresponding to the different antibody classes. During the maturation of B cells, a large number of different types of B cells are derived, each containing the genetic infor-

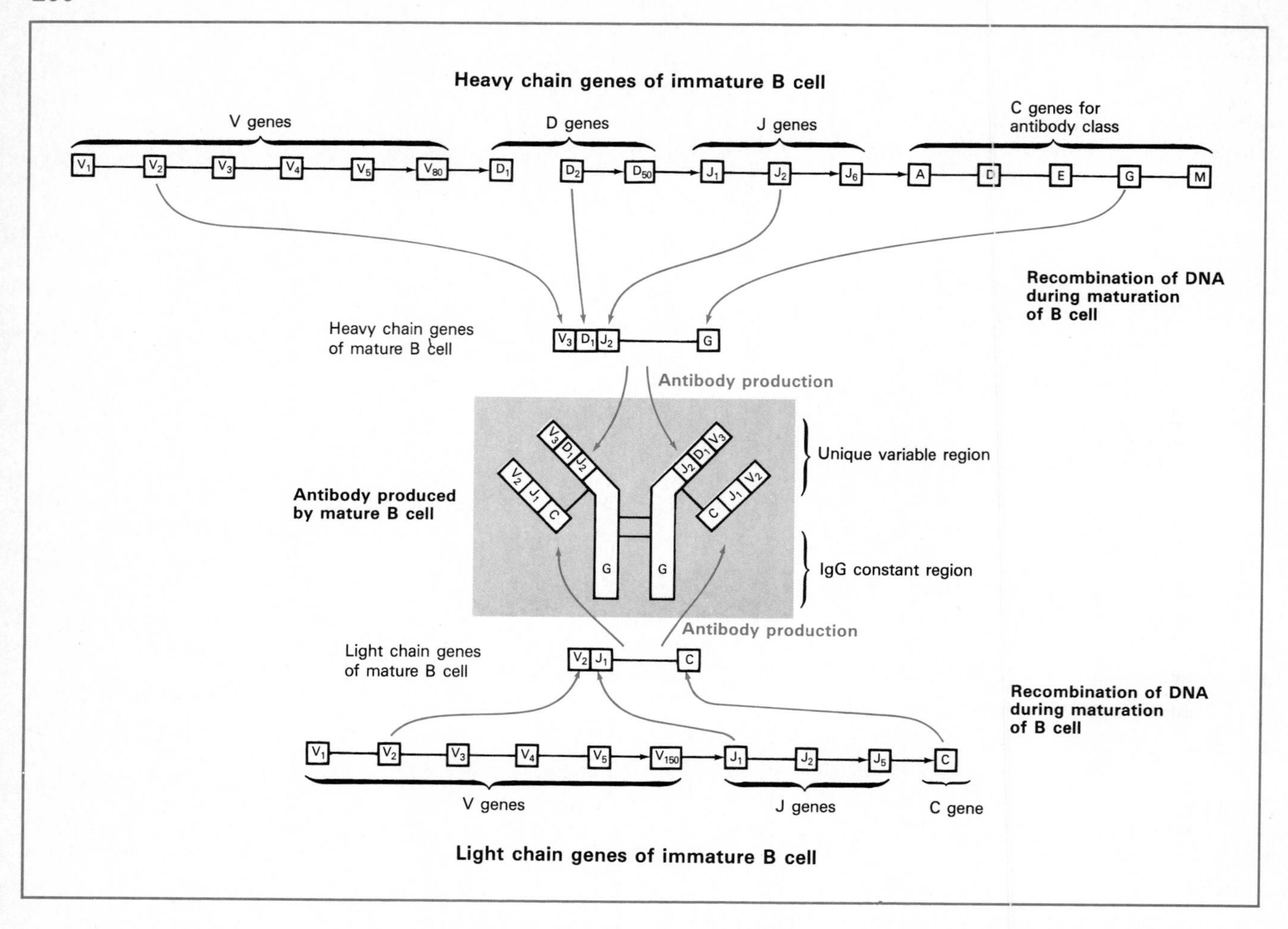

Figure 9.10 Steps in the formation of a specific antibody-forming cell type by gene rearrangement and recombination of different C, V, J, and D genes.

mation for the production of one unique antibody. This occurs by recombination of the different C, V, J, and D genes as shown in Figure 9.10. Maturation of antibody-forming cells by genetic recombination events results from transposition phenomena similar to those discussed in Chapter 7. Thus, a small number of different types of B cells is capable of maturing into millions of different antibody-forming cell lines. It should be emphasized that the origin of an antibody-forming cell line occurs in the complete absence of antigen stimulation. The role of the antigen in the maturation process is discussed below.

Maturation of antibody-forming clones The **production of antibody** in response to an antigen is a complex process involving B cells, certain types of T cells, and macrophages (phagocytic cells described in Section 8.5). The process is summarized in Figure 9.11.

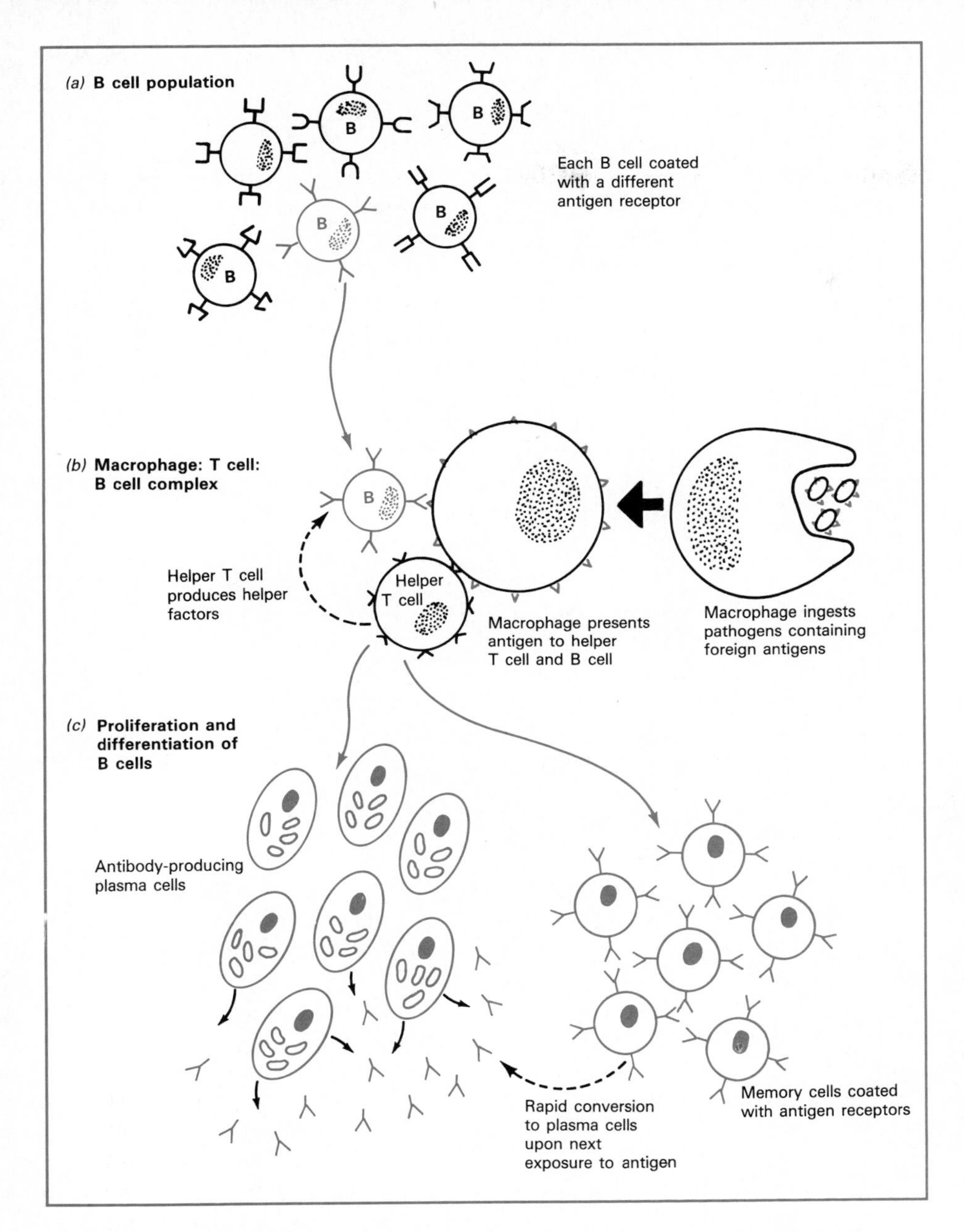
(a) B cell population
B
B
B
B
B
Each B cell coated with a different antigen receptor
(b) Macrophage: T cell: B cell complex
B
Helper T cell produces helper factors
Helper T cell
Macrophage presents antigen to helper T cell and B cell
Macrophage ingests pathogens containing foreign antigens
(c) Proliferation and differentiation of B cells
Antibody-producing plasma cells
Rapid conversion to plasma cells upon next exposure to antigen
Memory cells coated with antigen receptors

Figure 9.11 Events in the production of a clone of antibody-forming cells. (a) A large number of different B cells exist, each with the potential to respond to a different antigen. The origin of specific antibody-forming cell types was illustrated in Figure 9.10. (b) After a macrophage phagocytizes a foreign antigen, the macrophage interacts with a helper T cell and the B cell which responds to that specific antigen. (c) Helper factors produced by the helper T cell stimulate the B cell to divide and differentiate into either antibody-forming plasma cells, or memory cells. The concept of clonal selection is illustrated by the fact that only a single type of B cell (shown in color) proliferates and differentiates.

Before exposure to an antigen, B cells producing antibody against the antigen are present in only small numbers in the body (Figure 9.11*a*). The presence of the antigen causes the selective increase in this one type of B cell. The process by which this occurs is called **clonal selection**, since a single clone (or genetically unique type) of B cells is selected from the diverse pool of B cells. (We discussed clones in Chapter 7.)

Antigens usually become concentrated in the lymph nodes or in the liver and spleen where blood is filtered. *Macrophages* at these sites nonspecifically phagocytize foreign particles containing antigens and the macrophage surface becomes coated with the antigens of the particle which was phagocytized (Figure 9.11*b*). The role of the macrophage is to "present" the antigen to B and T cells. A special kind of T cell, called a **helper T cell**, binds to the antigen-coated macrophage and is stimulated to produce soluble substances which cause the appropriate B cell to divide and differentiate further into antibody-producing **plasma cells** (Figure 9.11*c*). The interaction of these three cell types: macrophage, B cell, and helper T cell, results in the *selective* proliferation of only those B cells which are capable of producing antibody against the antigen, and the modification of these cells into antibody-producing plasma cells. Plasma cells are short-lived cells which actively produce and excrete antibody for only about one week. Another type of differentiated B cell, called a **memory cell**, is also produced. Memory cells are much longer lived than plasma cells. They respond to subsequent exposure to the same antigen, differentiating rapidly into plasma cells which then produce antibody. Thus, it is the plasma cells which mediate the primary antibody response, and the memory cells which enable a more rapid secondary response.

The production of antibody is also regulated by a population of T cells called **suppressor T cells**. They bring about a decrease in antibody production once the antigen has been eliminated from the body.

In some cases IgM antibody can be produced by B cells without the involvement of T cells. T cell-independent antigens, to which B cells alone can respond, are usually large molecules often having repeating chemical units. Polysaccharide capsular antigens, which often possess repeating sugar units, are good examples of this type of antigen.

9.5 CELL-MEDIATED IMMUNITY

The cell-mediated immune response involves T cells of different types. Whereas antibodies are proteins which recognize and bind with specific antigens, T cells possess antibody-like molecules on their surfaces, and these react with antigens (see Figure 9.2*b*). Each individual T cell is capable of producing its own receptor molecule which reacts specifically with a single antigen. These **receptors** are not merely antibodies which have attached to T cells. Although the precise structure is not yet known, T cell receptors do appear to have regions analogous to the V, J, and C regions of antibodies.

Kinds of T cells and their reactions As shown in Figure 9.4, there are several classes of T cells, each of which functions differently in the immune system. These functions are outlined in Table 9.2. We have already considered the functions of helper T cells and suppressor T cells in regulating the production of antibodies by B cells. Two other types of T cells deserve further consideration, since these are the cells which bring about the elimination of foreign antigens from tissues during the cell-mediated immune response.

Cytotoxic T cells are also known as **killer T cells** because they not only recognize foreign antigens, but can also kill cells possessing such antigens. In addition to recognizing the antigen, killer T cells recognize the proteins of the major histocompatibility complex, the antigens which determine tissue

TABLE 9.2 Major types of T cells and their functions

Type	Function
Helper T cell	Stimulation of B cells during antibody production
Suppressor T cell	Act to decrease antibody production when antigen has been cleared from the body
Cytotoxic T cell (Killer T cell)	Recognize and destroy normal cells whose antigens are altered; attack virus-infected cells, cancer cells, foreign tissue
Delayed-type hypersensitivity T cell	Produce substances which attract and activate macrophages; important in delayed-type hypersensitivity (allergy)

type of animal cells. Thus, killer T cells are involved in recognition of normal cells of the body whose antigens have been altered. For example, when some viruses infect cells, the cell surface becomes coated with viral antigens. T cells may respond by recognizing both the viral antigen and the major histocompatibility complex proteins of the cell. Once the T cell recognizes and binds to the virus-infected cell, it is able to kill it (Figure 9.12). How a cytotoxic T cell kills the cells which possess foreign antigens is not known; however, the binding of such a T cell to the antigen on the surface of a target cell occurs first. Viral diseases in which cytotoxic T cells provide important protection include influenza, mumps, measles, rubella, rabies, and smallpox. Skin rashes characteristic of diseases like measles and smallpox probably result from destruction of virus-infected cells by cytotoxic T cells.

Cytotoxic T cells are also important in recognizing and destroying cancer cells. In **cancer**, normal cells become so altered that they divide much more rapidly than other cells (see Figure 13.16). Rapid division leads to tumor formation unless cancer cells are eliminated. Cancer cells also differ from normal cells in the surface antigens they possess. Killer T cells recognize both major histocompatibility antigens and specific cancer antigens and in some cases provide a mechanism of **immunosurveillance**, eliminating cancerous cells from the body.

An important process involving killer T cells is the rejection of cells which possess a foreign tissue type (different major histocompatibility antigens), a serious complication of **tissue transplantation**. The actions of T cell-mediated immune mechanisms are thus responsible for some of the difficulties involved in heart, kidney, and skin transplants, as will be discussed later in this chapter.

Another class of cells, called **natural killer cells**, also have the potential to kill normal cells with altered antigens, and may be quite important in immunosurveillance of some cancers. They are less specific than cytotoxic T cells in recognizing altered cells, so that their relationship to either T cells or to cells which kill nonspecifically, such as phagocytes, is uncertain.

T cells and delayed-type hypersensitivity A common response to antigen in cell-mediated immunity involves a class of T cells known as **delayed-type hypersensitivity T cells** (T_{DTH} cells). These T cells do not them-

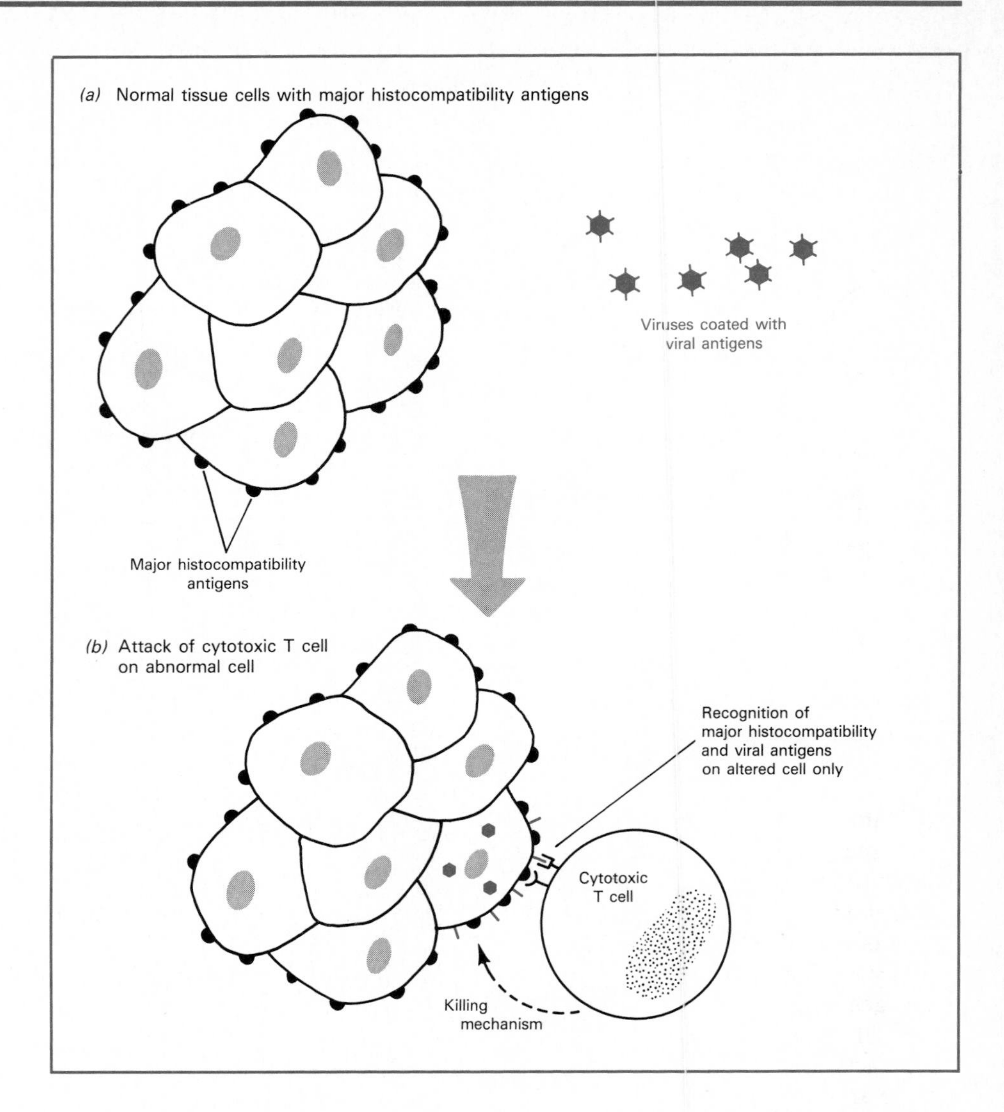

Figure 9.12 Cytotoxic T cell recognition and killing of altered cells. (a) A normal tissue cell is recognized as self by the presence on its surface of antigens called *histocompatibility antigens*. (b) After a cell becomes altered (in this case by infection with a virus) the cytotoxic T cell recognizes the viral-infected cell from the presence on its surface of the histocompatibility antigens *and* the viral antigens.

selves kill pathogens; they respond to antigen by excreting a variety of substances which collectively stimulate phagocytic macrophages (Figure 9.13). The activation of macrophages results in an increase in nonspecific killing of pathogens, thus benefitting the host defense.

The substances secreted by T_{DTH} cells are called **lymphokines**. They include *macrophage chemotactic factor*, which attracts macrophages toward the antigen-T cell complex; *migration inhibition factor*, which prevents the accumulating macrophages from leaving the area to which they are attracted; *macrophage activating factor*, which dramatically increases the ability of macrophages to kill the cells they phagocytize; and a type of *interferon*, which inhibits virus replication within nearby host cells (see Chapter 8). The specific reaction between T cells of this type and the antigens they recognize leads to a vigorous nonspecific attack on the infected tissue which is similar to an intense inflammatory response. Since the reaction involves the migration of

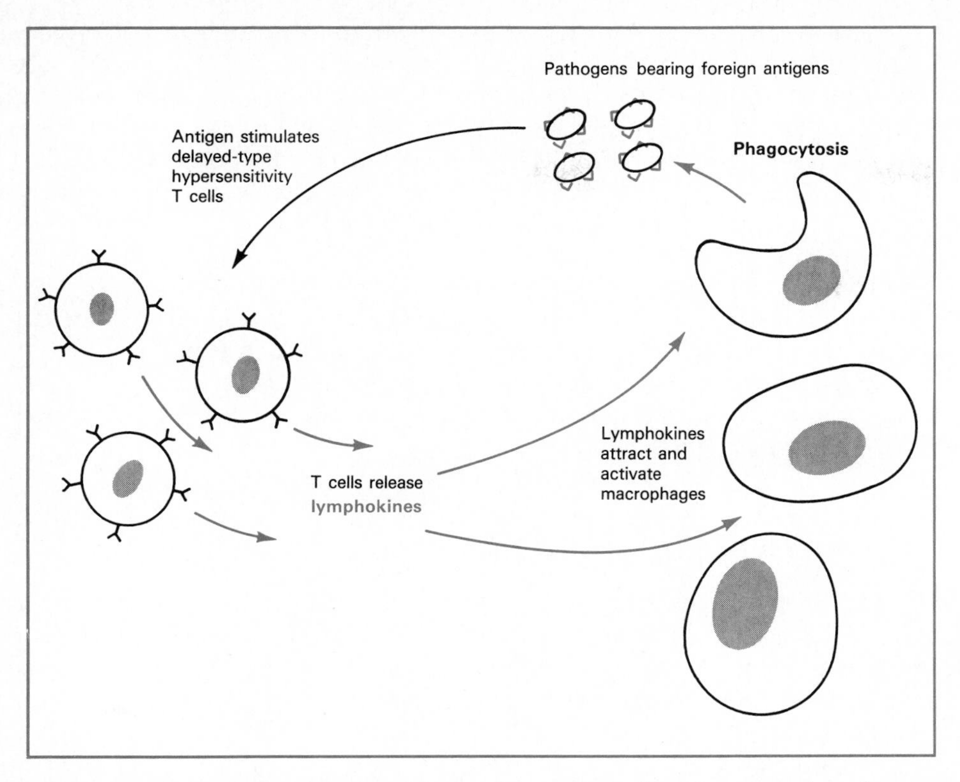

Figure 9.13 Action of delayed-type hypersensitivity T cells. These cells are stimulated by antigen to produce lymphokines which attract and activate macrophages. Macrophages then attack the cell bearing the antigen.

cells into the area of infection it is usually a somewhat slow reaction, hence the name "delayed." The word hypersensitivity, meaning "over-sensitive," relates to the intensity of the reaction which the antigen causes.

Delayed-type hypersensitivity is involved as an important component of host defense in many diseases, including diseases such as tuberculosis and leprosy in which the pathogen evades host defense by growing within macrophages. *Mycobacterium tuberculosis* can counter the normal killing mechanisms of macrophages. However, when T cells respond by secreting macrophage activating factor, the macrophages are stimulated to increase their killing power and are then able to kill the bacterium.

A classic demonstration of delayed-type hypersensitivity is the skin reaction called the *tuberculin* test, used for determining prior infection with *M. tuberculosis*. Antigens extracted from the pathogen, when injected subcutaneously into an animal which has had previous exposure to this bacterium, elicit a characteristic reaction of swelling and redness near the site of injection. Most other skin reactions, such as allergies, occur rapidly. The reaction to antigens of *M. tuberculosis*, a delayed-type reaction, develops only after 24 to 48 hours. During the development of the reaction, T cells in the region of the injection are stimulated by the antigen and release lymphokines that attract large numbers of macrophages. The characteristic skin reaction is an inflammatory reaction due to the destruction of the antigen by activated macrophages.

Mechanism of T cell proliferation As in the case of B cells, a large variety of different T cells, each with the potential to respond to different antigens, exists in the lymphatic system, blood stream, and tissues. The specificity of each T cell is determined by its surface-bound receptor which, as we have noted, bears some similarity to the structure of antibodies. As is the case with B cells, a specific T cell line arises from bone marrow stem cells by rearrangement of genes coding for different regions of the receptor molecule.

In response to an antigen, clonal selection of a specific T-cell type probably occurs in much the same way as with selection of a specific B cell (see Figure 9.11). Macrophages which have engulfed pathogens "present" the antigen to T cells. This interaction stimulates the proliferation of the T cell which bears the receptor recognizing the antigen. During the proliferation of either cytotoxic T cells or delayed-type hypersensitivity T cells, memory T cells are probably also produced. These act in the same way as memory B cells and cause a more rapid and dramatic secondary response upon subsequent exposure to the same antigen.

9.6 IMMUNODEFICIENCY DISEASES

The value of the immune system as a major defense mechanism against infectious disease is well illustrated by the consequences for individuals who lack components of the immune system because of genetic diseases. Such individuals are unusually susceptible to infectious disease and if not protected from infection die at an early age.

Agammaglobulinemia is an inherited disease involving inability to produce antibodies. Although persons with this condition possess normal cell-mediated immunity and resist most viral infections, they are extremely sensitive to diseases usually controlled by antibodies. Treatment of the condition involves passive injection of antibodies formed in another individual.

Persons with **Di George's syndrome** lack a thymus and because the thymus is essential for T cell formation they thus lack cell-mediated immunity (see the Box on the next page). Antibody-mediated immunity is also impaired because of the role played by T cells in antibody production. Although persons with Di George's syndrome are resistant to most bacterial infections because they produce antibody, they are very sensitive to viral infections, since these are usually controlled by the cell-mediated immune response. Treatment involves transplantation of the thymus gland.

Severe combined immunodeficiency is a serious disease in which bone marrow stem cells are abnormal; neither antibody- nor cell-mediated immunity is functional. Persons with this condition are susceptible to infection by essentially any microorganism, including microorganisms that are not normally pathogenic (opportunistic pathogens). Other than total isolation in a sterile environment, individuals with this immune-system disease survive only if transplantation of bone marrow to provide normal stem cells is successful.

One of the most dramatic and serious immune deficiency diseases is called **acquired immunodeficiency syndrome** (or **AIDS**). Since it was first recognized around 1980, AIDS has been a vivid reminder of the importance of the immune system, especially because of the high death rate among infected

NUDE MICE

A strange breed of mice has been discovered that lacks all body hair. Colonies of such *nude mice* are more difficult to maintain than normal mice because of their poor resistance to infectious agents. Whereas normal mice live for over 2 years, nude mice are difficult to keep alive for more than a few weeks unless they are specially protected from infection. It has been discovered that the immune system of nude mice is abnormal, due to the fact that they lack a thymus gland. Without a thymus gland, nude mice cannot make T cells and therefore lack cell-mediated immunity. Since the production of many antibodies requires helper T cells, nude mice also have an impaired antibody-mediated immune response. The lack of cell-mediated and antibody-mediated immunity explains the susceptibility of these mice to infectious agents; the condition of a nude mouse is like that of a human suffering from Di George's syndrome.

Nude mice have been used in the study of cancer because they do not reject transplanted human tumors. The growth of a human tumor can be studied as it develops in the mouse, and potential anticancer drugs can be tested for their ability to either prevent tumor growth or to reduce the size of tumors. The ability of tumors to grow unchecked in the nude mouse is also evidence that cell-mediated immunity is involved in immunosurveillance against cancer. In fact, most nude mice eventually die from their own spontaneous cancers, presumably because they lack the cells responsible for immunosurveillance.

individuals (see the discussion of AIDS virus in Section 13.9). This viral infection seems to affect mainly helper T cells so that both cell-mediated and antibody-mediated immunity are impaired. Those with AIDS usually succumb to opportunistic infections caused by organisms which are normally noninfectious. Some individuals with AIDS also develop relatively rare forms of cancer.

The beneficial aspects of the immune system are indisputable. However, there are many problems which occur as a consequence of having an immune system. The most important of these will be discussed below.

9.7 ALLERGY AND HYPERSENSITIVITY

The immune system responds to some antigens in a manner which is harmful to the body, a condition called *allergy* or *hypersensitivity*. In many cases, the antigens causing these effects are not even associated with pathogenic microorganisms. The antigens that cause allergies or hypersensitive reactions are called **allergens**. Two broad classes of hypersensitivity reactions have been identified:

1. **Immediate-type** hypersensitivity reactions, which are rapid and result in visible symptoms within a few minutes after the body is exposed to the allergen. The immediate-type hypersensitivity reactions are due to harmful consequences of antibody-antigen (allergen) reaction.
2. **Delayed-type** hypersensitivity reactions, which are slow and result in visible symptoms only after several hours or days. Delayed-type reactions are consequences of cell-mediated immune responses.

TABLE 9.3 Some common allergies

Immediate-type hypersensitivity	Delayed-type hypersensitivity
Hay fever	Poison ivy
Food allergies	Some insect bites
Penicillin and other drug allergies	Reaction to hair dyes
Hives	Reaction to cosmetics
Bee sting reactions	Reaction to jewelry metals
Serum sickness	

Some common allergies of each type of hypersensitive reaction are listed in Table 9.3.

Immediate-type hypersensitivity From 10 to 20 percent of the human population suffers from **immediate-type hypersensitivity** reactions. These reactions are caused by a variety of allergens such as pollens, animal dander, certain foods, and some drugs. These allergens stimulate the production of antibody of the IgE class, which binds to cells in the periphery of the body called *mast cells* (Figure 9.14). A first encounter with the allergen does not result in an allergic response, but an individual may become "sensitized." In subsequent encounters, allergen may react with the antibodies *already present* on mast cells, causing the cells to release chemicals which bring about the allergic reaction. The release of **histamines** and **serotonins** from mast cells then causes dilation of blood vessels and contraction of smooth muscle, initiating the symptoms of immediate-type hypersensitivity, including difficulty in breathing, flushed skin, copious mucus production, and itchy, watery eyes. The severity of the reaction depends on the site of the antibody-allergen reaction. In skin allergies, only local inflammatory reactions such as hives may appear. Allergies involving the upper respiratory tract are more generalized and severe breathing difficulty may result. Those involving the gastrointestinal tract result in symptoms such as diarrhea, whereas those involving the uterus can result in spontaneous abortion. **Antihistamines** are drugs that block the action of histamine and thus neutralize some of the effects of these types of allergies.

The most dramatic example of hypersensitivity is the phenomenon known as **anaphylactic shock**. As an example, if an animal that has previously been sensitized with small amounts of a foreign protein is injected a few weeks later with a large amount of the same protein, within a few minutes the animal may show signs of respiratory distress because of contraction of the bronchial muscles, which prevents the exhalation of air. The animal has an acute attack of asthma and usually dies within minutes. Anaphylaxis in humans, which can arise from a similar encounter with a foreign product, is signaled by the same acute asthma and is accompanied by flushing and itching, circulatory collapse, acute emphysema, and (if recovery occurs) a rash on the skin.

Anaphylactic shock is of most concern with drug treatments. Injection of drugs such as penicillin into unusually allergic individuals may cause death by anaphylaxis. Because of the seriousness of this drug reaction, it is extremely

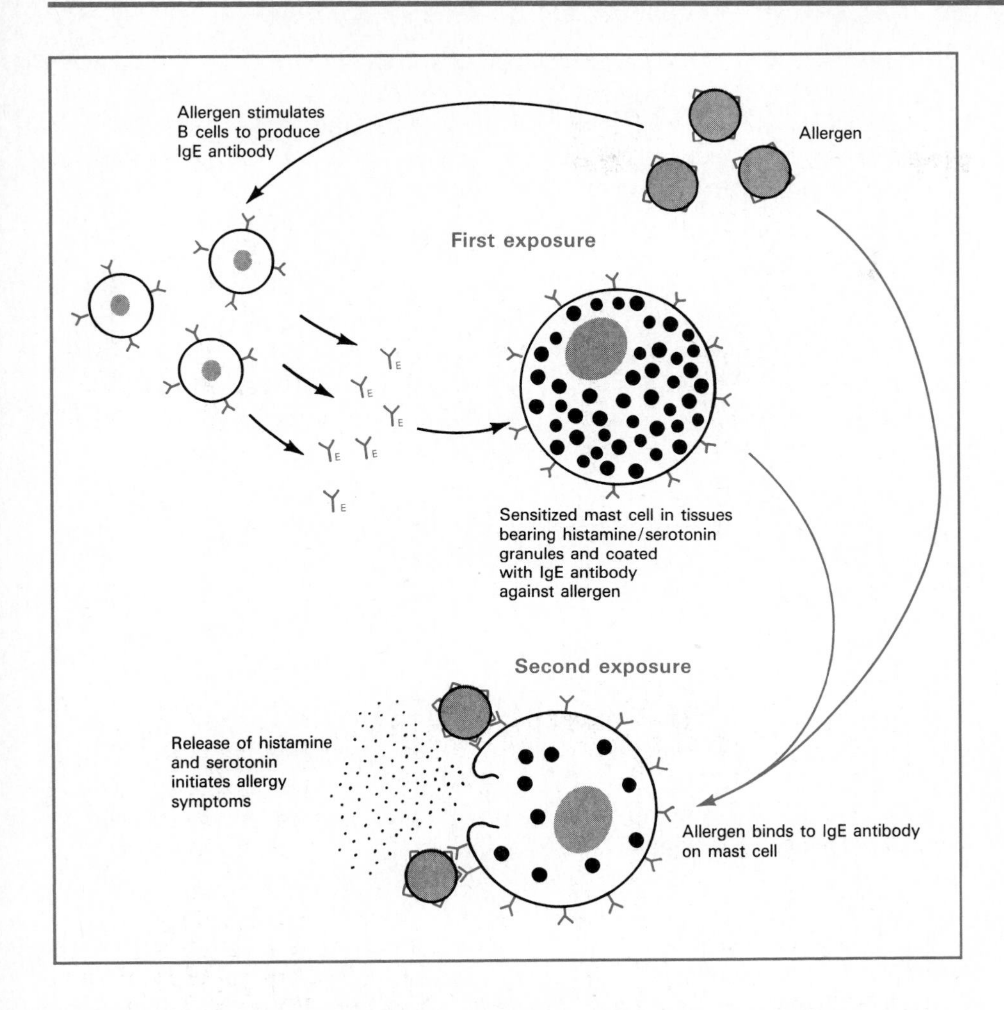
Allergen stimulates
B cells to produce
IgE antibody
Allergen
First exposure
Sensitized mast cell in tissues
bearing histamine/serotonin
granules and coated
with IgE antibody
against allergen
Second exposure
Release of histamine
and serotonin
initiates allergy
symptoms
Allergen binds to IgE antibody
on mast cell

Figure 9.14 Immediate-type hypersensitivity reactions. During the first exposure the antigen (allergen) causes formation of IgE antibody which binds to mast cells in the peripheral tissues. During subsequent encounters the allergen binds to sensitized mast cells, releasing histamine and serotonin, which are responsible for the symptoms.

important to know whether an individual is allergic to such drugs so that their use can be avoided. Venom from insects and snakes may also cause anaphylactic shock in some individuals. Fortunately, only a small percentage of people show these unusual reactions.

A number of other immediate-type hypersensitivity reactions involve a similar mechanism to that of anaphylaxis but are less severe and dramatic. For instance, **serum sickness** results when a relatively large amount of foreign serum (horse serum, used in certain immunological procedures, most commonly causes this problem) is injected and antibodies form against the foreign-serum proteins. The symptoms are usually a rash, enlargement of lymph glands, and swelling of body tissues, all resulting from increased permeability of the capillaries, another manifestation of histamine action.

Delayed-type hypersensitivity A number of allergies can be grouped within the **delayed-type hypersensitivity** reactions. These allergic reactions are elicited by delayed-type hypersensitivity T Cells (T_{DTH}) described in Section 9.5. The allergen combines with T_{DTH} cells, resulting in excretion of lymphokines, resulting in the attraction and activation of macrophages. The macrophages initiate tissue damage due to active phagocytosis.

Allergic reactions of this sort often involve contact between skin and the allergen. Thus **contact dermatitis** results when T_{DTH} cells respond to allergens which combine with proteins on the surfaces of normal body cells. Destruction of these altered cells by activated macrophages results in the symptoms. An excellent example is the hypersensitive reaction to poison ivy, in which contact with chemicals produced by the plant alter the normal antigens on certain skin cells. Many other chemicals also cause delayed-type hypersensitivity reactions, including ingredients of hair dyes, some industrial chemicals, some antibiotics which are applied to the skin, metals from jewelry, and some insect venoms.

Allergens in foods and **pollen** are usually proteins or polysaccharides. Allergies to pollen and spores are usually called *hay fever;* they tend to show pronounced seasonal incidence, developing when the particular pollen or spore is being released. In the case of *drug allergies,* the drug itself is not an antigen but becomes an antigenic determinant when it combines with serum protein. Both drug allergy and hay fever are types of hypersensitivity that seem to have hereditary factors.

The sensitivity of an individual to an allergen can often be determined by a skin test. A small amount of extract of the suspected allergen is injected under the skin; an allergic reaction will be seen as a reddening and swelling of the skin which may occur within a few minutes in the case of immediate-type hypersensitivity. Desensitization to an allergen can sometimes be achieved by carefully injecting gradually larger doses of allergen, resulting in the neutralization of existing antibody or its release from smooth muscle.

Allergic reactions are responsible for some of the symptoms of certain microbial infections. Upon initial infection, antigenic components of the microbes induce antibody or activated T cell formation. Subsequent infections then cause allergic-type reactions in the infected tissue or elsewhere in the body. Usually, the organisms that elicit allergic reactions cause chronic rather than acute infections. Examples of diseases with allergic manifestations are rheumatic fever and tuberculosis.

9.8 AUTOIMMUNE DISEASE

Self-recognition, exhibited as tolerance to the body's normal cell constituents, is one of the essential features of the immune system. Occasionally, problems can arise because of a breakdown in the self-recognition process. Autoimmune diseases are those conditions in which antibody- or cell-mediated immune responses are directed against normal cells within the body.

Some examples of autoimmune diseases are listed in Table 9.4. A variety of different mechanisms may be involved. The disease rheumatic fever only occurs as a complication following certain streptococcal infections, such as streptococcal sore throat. In this case, the body produces antibodies against

TABLE 9.4 Examples of autoimmune diseases

Disease	Cause
Rheumatic fever	Antibody against *Streptococcus pyogenes* also attacks heart tissue
Diabetes (insulin-dependent)	Antibody attacks pancreatic cells that produce insulin
Multiple sclerosis	Antibodies and T cells attack brain tissue
Rheumatoid arthritis	Antibodies attack cartilage
Male infertility (some cases)	Antibodies agglutinate sperm
Systemic lupus erythematosus	Massive antibody response to various cellular components

bacterial antigens during the normal host response against infection. These antibodies, however, may cross-react with antigens found on cells of the heart, leading to destruction of heart tissue. In other cases, autoimmunity results from some sort of breakdown in the self-recognition process itself. In insulin-dependent diabetes, for example, antibodies are produced which react with the pancreatic cells that produce insulin. In multiple sclerosis, both auto-antibodies and activated T cells are involved in destruction of brain tissue. Hopefully, a better understanding of the basic immune responses will lead to methods for controlling these important diseases.

9.9 NONMICROBIAL IMMUNOLOGICAL PHENOMENA

Several types of immunological phenomena have little connection with infectious disease, but are of interest because they can be understood in relation to the overall immune responses of the body. We discuss some of these in this section.

Blood groups Red blood cells contain antigens on their surfaces, and an understanding of blood cell antigens is important for blood transfusions and certain other medical concerns. Because of the self-recognition phenomenon discussed above, antibody formation against body substances usually does not occur. Thus, the antigenicity of red cells is only a problem during transfusions or other situations in which foreign red cells enter the body. There are at least 12 different red blood cell antigen systems. Two which are of particular importance deserve consideration in detail: ABO and Rh.

The best known blood group system is that called the **ABO system**. The ABO system is the one that is most frequently involved in transfusion problems, and is most often studied in blood matching of individuals. The key factors in the ABO system are the two blood group antigens designated as A and B. The antigenic determinants of these two antigens are polysaccharides. The ability of a person to produce these antigens is inherited, and a person can produce antigen A, antigen B, both, or neither. When both antigens are produced, the blood group is designated AB; when neither is produced, the blood group is designated O (Table 9.5).

TABLE 9.5 Antigens and antibodies of the ABO system

Blood group	Blood group antigen on red cells	Natural antibodies present in serum
A	A	Anti-B
B	B	Anti-A
AB	Both A and B	Neither
O	Neither	Anti-A, Anti-B

As noted, antibodies are not produced against the antigens of red blood cells of the body. However, antibodies against the A and B antigens are present normally in the blood of people who lack either or both antigens. Thus, type A individuals contain anti-B antibody; type B individuals contain anti-A antibody; type AB individuals contain neither antibody; and type O individuals contain both anti-A and anti-B antibodies. Antibodies against A and B antigens are called *natural antibodies*, since they appear to be produced even in the absence of antigenic stimulation. The explanation for this resides in the fact that the A and B polysaccharides, in addition to being present on red blood cells, are also found in certain common foods (the antigens in food and those on the red blood cell are said to *cross-react*). Since such foods are frequently eaten, antigenic stimulation against A and B substances occurs throughout life. Self-recognition insures that antibody is not formed against A or B if that antigen is also on the red blood cells, but antibody will be formed against either of these antigens if they are not on the red cells.

The reason for blood typing before blood transfusion is to prevent the blood-cell agglutination that would occur if blood containing a particular antigen were transfused into a person containing an antibody against this antigen. If agglutination occurred, the clumps of cells could lodge in blood vessels or arteries and block the flow of blood, causing serious illness or death.

The frequency of occurrence of the various blood groups varies in different populations. In the Caucasian population, which is of mixed ethnic background, blood groups A and O are the most common. Black African populations have a higher incidence of blood group B, whereas American Indians completely lack blood group B and are predominantly group O. Some South American Indians lack A and B antigens entirely. Frequencies of the ABO blood groups in various populations have been used in studying processes involved in human genetics, and in anthropological studies on racial origins.

Rh system One other blood group system, the **Rh system**, also has considerable medical importance. There are a number of Rh antigens, but for present purposes these can be simplified to Rh-positive and Rh-negative. An individual who possesses the Rh antigen is called Rh-positive; such a person does not produce antibodies against the Rh antigen, since one does not make antibodies against one's own antigens. An Rh-negative individual does not possess the Rh antigen and therefore will produce anti-Rh antibodies upon coming into contact with it. A serious blood disease **Rh disease of the newborn** (also called **erythroblastosis fetalis**) can occur if the mother is Rh-negative but carries Rh-positive children (Figure 9.15). Usually, the first pregnancy is normal, since the mother's immune system is not exposed to fetal red blood cells. During delivery, however, fetal red blood cells enter the mother's circulation, and the mother responds with the production of anti-Rh antibody. In subsequent pregnancies involving an Rh-positive fetus, the anti-Rh antibody (an IgG class antibody), can pass from the mother to the fetus through the placenta and destroy fetal red blood cells. Any further stimulation of the maternal immune system by exposure to Rh-positive fetal red blood cells will also lead to a secondary response in which a large amount of anti-Rh antibody is produced. Not all pregnancies of an Rh-negative woman

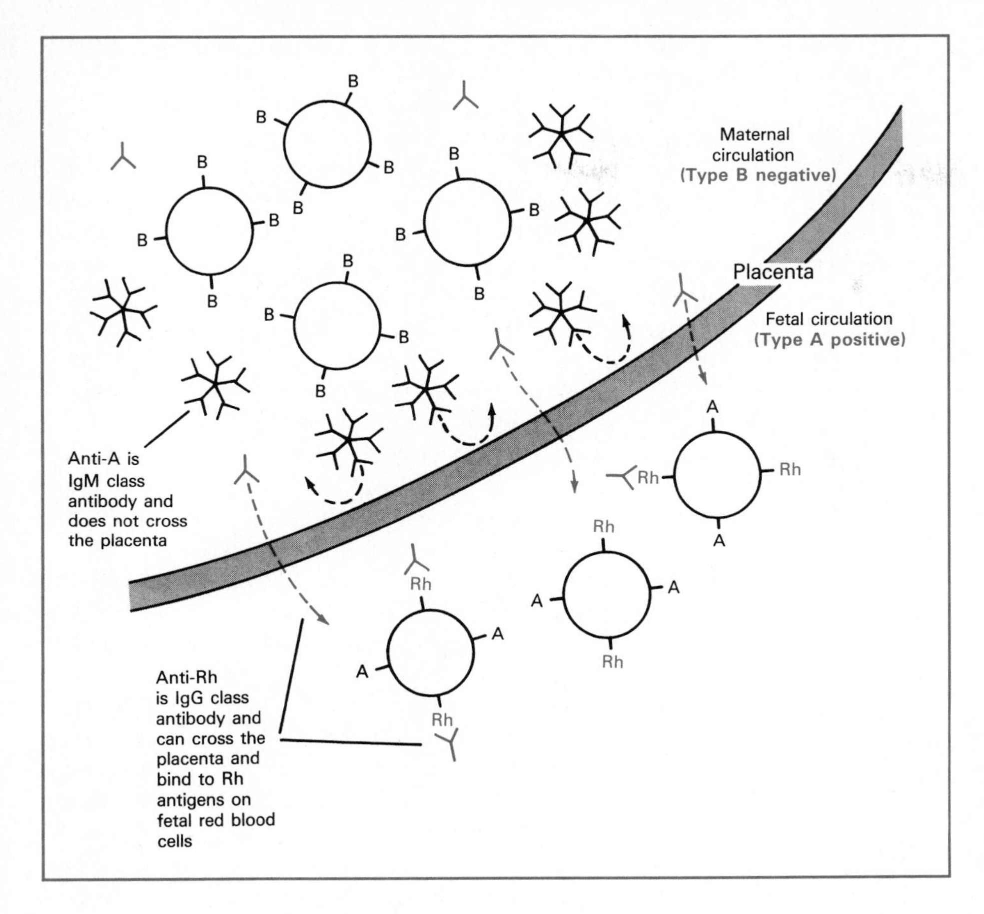
Maternal
circulation
(Type B negative)
Placenta
Fetal circulation
(Type A positive)
Anti-A is
IgM class
antibody and
does not cross
the placenta
Anti-Rh
is IgG class
antibody and
can cross the
placenta and
bind to Rh
antigens on
fetal red blood
cells

Figure 9.15 Rh disease of the newborn can result if an Rh-negative mother produces anti-Rh antibody. Unlike anti-A and anti-B antibody, which are type IgM, anti-Rh antibody is an IgG type and can cross the placenta, where it reacts with Rh-positive red blood cells of the fetus. The situation shown could only occur after the mother has been sensitized in an earlier pregnancy.

by an Rh-positive man lead to this complication, but it is wise to know that the possibility exists so that corrective measures can be taken.

Rh disease can be prevented by giving the mother serum containing anti-Rh antibody at delivery. The binding between Rh antibody and fetal red blood cells removes the Rh antigens obtained from her infant during delivery (parturition), so that the mother does not herself mount an immune response against the Rh antigen. Subsequent offspring will thus not be harmed, since no anti-Rh antibodies exist in the mother. However, this procedure must be repeated for every subsequent birth.

Blood typing In order to determine blood group, a typing procedure is performed in which serum containing known antibody is mixed with blood of the person to be typed, followed by observation of agglutination. ABO and Rh typing is usually done on microscope slides (Figure 9.16).

Before using blood, even of the same blood group, for transfusion, it is important to *cross-match* the blood with that of the recipient to ensure that there are no unforeseen incompatibilities. Cross-matching is done by mixing a drop of donor serum with a drop of fluid containing recipient cells and observing for agglutination.

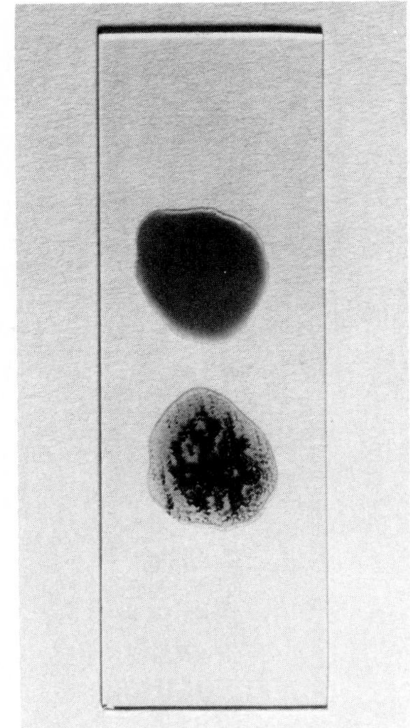

Figure 9.16 Procedure for blood typing. Blood of group B was mixed either with anti-A serum (top) or anti-B serum (bottom). Agglutination occurs only with anti-B serum.

Transplantation immunity The transplantation of tissues or organs from one individual to another is an increasingly important medical practice. However, in many cases, the transplanted tissue or organ is rejected, deteriorates, and dies. Rejection of transplanted tissues and organs is caused by a cell-mediated immune response. Cytotoxic T cells recognize the major histocompatibility complex of the cells of donor tissue as being foreign and attack the transplanted tissue. If transplanted tissues bear similar major histocompatibility antigens, there is a better chance that cytotoxic T cells will not reject the tissue. Identical twins, since they are of the same genotype, have identical histocompatibility antigens and hence can share organs or tissues without immunological rejection.

Homografts, such as skin grafts, involve transplantation of tissue from one part of the body to another. These can of course be performed without any immunologic complications, since the body does not react immunologically against itself. In order to transplant tissue from other donors, the tissue types of donor and recipient must be matched to ensure a reasonable probability of success. During such transplants, the immune system of the recipient is often temporarily suppressed by administration of drugs. **Immunosuppression**, however, makes the recipient very sensitive to infectious agents, so that antibiotics must be used as chemoprophylactics to combat possible microbial infections.

9.10 IMMUNIZATION AGAINST DISEASE

Knowledge of immune mechanisms has led to considerable advances in methods for the immunization against disease. Some of the major diseases that affect mankind have been controlled by appropriate immunization procedures. Immunization has had a long history, first being used to control smallpox in the early 19th century, and subsequently being greatly expanded through the work of Louis Pasteur. Before the days of antimicrobial therapy, immunization procedures were the only specific ways of controlling infectious diseases. Even today, immunization with vaccines provides the main approach to infectious disease prevention.

A person may be immunized in either of two ways (Figure 9.17). In **active immunization** an individual receives antigens and responds by producing antibody or activated T cells. Active immunization is a preventive procedure, carried out to protect an individual if the pathogen is ever encountered. In **passive immunization**, preformed antibodies are directly injected to control a disease which is in progress. Thus, active immunization is preventive whereas passive immunization is therapeutic.

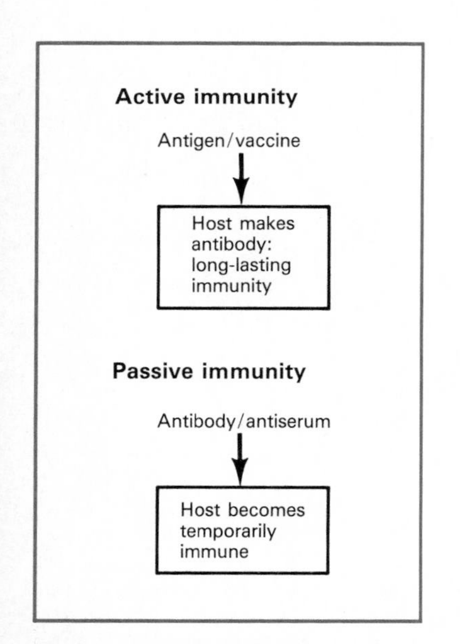

Figure 9.17 Contrast between active and passive immunity.

Passive immunization Passive immunity is conferred when a person is given an injection of serum containing antibodies taken from another person. The serum is usually taken from a *hyperimmune* person, which means that the serum has a very high antibody titer. Hyperimmune serums are also taken occasionally from animals and injected into humans. For instance, horses have been used for the production of serums containing antibodies against diphtheria, botulinum, and tetanus toxins. Antibodies can now also be produced in cell cultures as will be described in Section 9.11. A serum containing antibodies is called an **antiserum**, and an immune serum active against a toxin is called an **antitoxin**.

An immune serum is used to treat a person who is already suffering from a given disease and the serum is hence used as a therapeutic agent. For example, the botulinum toxin serves as an antigen, but it is so toxic that it causes death before the patient's immune system can respond. Because of this, passive administration of botulinum antitoxin is the only way to cure botulism once it begins.

This type of immunity is called **passive**, since the body does not produce the antibodies. Long-term immunity does not result, as the injected antibodies gradually disappear from the blood, but the antibodies present will probably be able to inactivate the disease agent and result in a cure. However, if a subsequent infection occurs, the person will generally be as sensitive to the disease as an unimmunized individual. Repeated use of immune serum must be avoided, as the individual will begin to develop antibodies against the proteins of the serum and exhibit a reaction against the serum. This type of reaction is sometimes called *serum sickness* (see Section 9.7).

Passive immunization is also used to prevent Rh disease of the newborn. In this case anti-Rh antibody is used to block the exposure of the antigen on fetal red blood cells to the maternal immune system (see Section 9.9).

Natural passive immunity develops in the fetus carried within its mother's womb. Antibodies pass across the placenta from mother to child and are present in the newborn infant. These antibodies confer immunity to many infections and help the infant avoid infection during the early months of life. However, these antibodies gradually disappear and are usually gone by the age of 6 months, at which time the susceptibility of the infant to infection increases. As we have noted, infants who are breast fed receive IgA antibodies contained in mother's milk, and are thus less prone to infections than infants who are fed cow's milk. Eventually, the immune system of the child develops and the child produces its own antibodies in response to antigens.

Active immunization and vaccines An immune response arises in an individual either as a result of natural exposure to antigens during the infectious disease process or as a result of artificial exposure to antigens as in immunization procedures. The antigen preparation used for immunization is called a **vaccine**, and the process is thus also known as **vaccination**. After vaccination, the individual develops either antibodies or activated T cells which provide protection should the antigen in its natural state (for instance, on the pathogen) be encountered subsequently. An important distinction between active and passive immunization is that in active immunization, continual formation of antibody or activated T cells occurs, and a secondary or booster response will occur if the person is later challenged with the same antigen. In passive immunization there is no production of antibody by the immunized individual. Vaccination is thus a way of preventing the spread of disease rather than a cure for a disease in progress.

Vaccination has been of major importance in controlling infectious diseases. Upon introduction of a specific immunization procedure, the incidence of the disease often drops markedly (Figure 9.18). For other examples of the success of vaccination in reducing disease incidence, see Figures 11.7 and 13.11, and the Box on page 280. A list of diseases for which vaccines are available is given in Table 9.6.

Vaccines consist of harmless products containing antigens from the pathogen or its products. Some vaccines are derived from toxins produced by the pathogen. Since the toxin itself cannot be injected, a chemically modified toxin called a **toxoid** is used. The toxoid still contains the appropriate antigenic determinants but is no longer toxic. One of the common ways of converting a toxin into a toxoid is by treatment with formaldehyde. If immunization requires the use of the pathogenic microbe itself, the organism must be inactivated in some way; usually, it is killed by treatment with heat, formaldehyde, or phenol. Vaccines against whooping cough and typhoid fever are routinely prepared from killed cultures of the causal agents.

Killed cells are not as effective as live cells as vaccines. Live cells not only contain the antigenic determinants in more immunogenic form, but they may also replicate to a small extent in the body, thus increasing the dosage. In some cases, it has been possible to develop an **attenuated** strain of the pathogen, still effective antigenically but no longer able to initiate infection. Attenuation may result from mutation of the pathogen, possibly by loss of ability to produce a capsule, toxin, or some other factor involved in virulence. When pathogens are kept in laboratory culture for long periods, their virulence often decreases or even disappears because faster-growing nonvirulent strains are selected.

There are many risks involved in the preparation and use of vaccines (see the Box on page 302). Some attenuated pathogens can recover their virulence by passage in an animal host or may revert to a virulent form by **spontaneous**

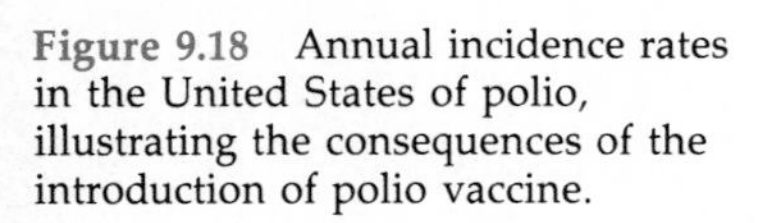

Figure 9.18 Annual incidence rates in the United States of polio, illustrating the consequences of the introduction of polio vaccine.

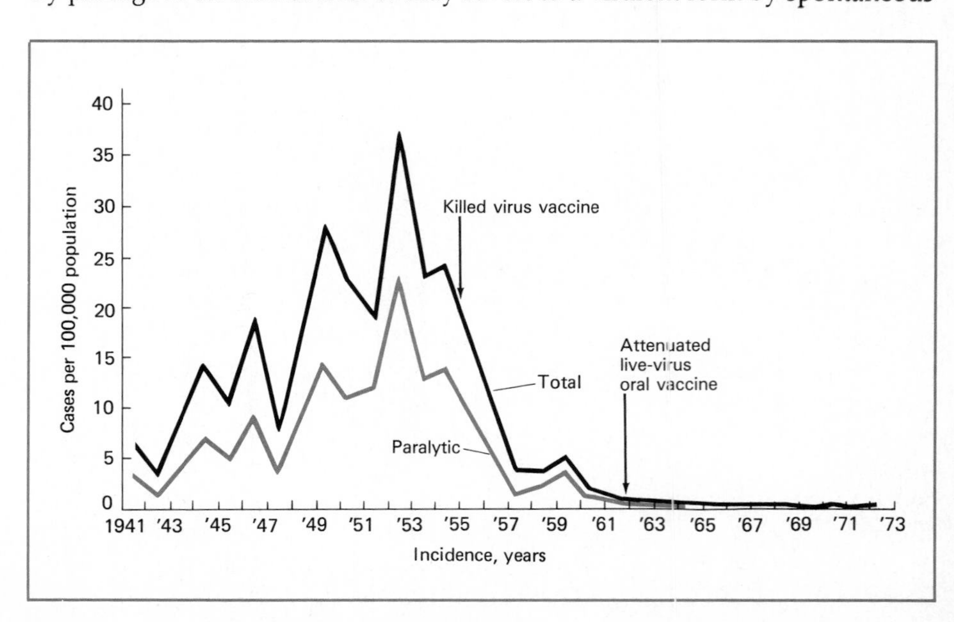

TABLE 9.6 Available vaccines for infectious diseases in humans

Disease	Type of vaccine used
Bacterial diseases	
Diphtheria	Toxoid
Tetanus	Toxoid
Pertussis	Killed bacteria
Typhoid fever	Killed bacteria
Paratyphoid fever	Killed bacteria
Cholera	Killed cells or cell extract
Plague	Killed cells or cell extract
Tuberculosis	Attenuated strain (BCG)
Meningitis	Purified bacterial polysaccharide
Bacterial pneumonia	Purified bacterial polysaccharide
Typhus fever	Killed bacteria
Viral diseases	
Smallpox	Attenuated virus
Yellow fever	Attenuated virus
Measles	Attenuated virus
Mumps	Attenuated virus
Rubella	Attenuated virus
Polio	Attenuated virus
Influenza	Inactivated virus
Rabies	Inactivated virus (human) or attenuated virus (dogs and other animals)

mutation during growth. Due to occasional errors in mass production, vaccines may become contaminated with pathogenic microorganisms. In rare cases, vaccines cause allergic reactions or other serious complications. It may be possible to produce safer vaccines by chemical synthesis of proteins corresponding to specific antigenic determinants. Since a protein rather than a whole attenuated pathogen is then injected, there is far less chance of accidental infection and adverse reaction. With the advent of recombinant DNA technology, it has become possible to use bacteria to mass-produce such specific proteins to serve as antigens. Thus the proteins rather than the pathogens can be used as vaccine. As mentioned in Chapter 7, the protein antigen (coat protein) of the virus causing foot-and-mouth disease has already been produced by recombinant bacteria and its efficacy as a vaccine is being tested in animals. There is great promise for future preparation of many more vaccines using recombinant DNA technology (see Chapter 13).

Immunization practices It is desirable to immunize against key infectious diseases at as early an age as possible, so that active immunity can replace the passive immunity received from the mother. However, very young infants have a rather poorly developed ability to form antibodies, so that immunization is not begun until a few months after birth. As noted in Figure 9.3, a single injection of antigen does not lead to a high antibody titer; it is therefore desirable to use a series of injections, so that a high titer of antibody is developed. The immunization schedule outlined in Table 9.7 provides for this spacing, as well as for periodic booster injections throughout life.

TABLE 9.7 Some recommended immunization procedures for infants and children*

Age	Immunization	Comments
During the first year of life	Diphtheria, pertussis, tetanus (DPT)	Given at 2, 4, and 6 months
	Oral polio (OP)	Given at 2, 4, and sometimes 6 months
During the second year	Mumps, measles, rubella (MMR)	Given at 15 months
	DPT booster	Given at 18 months
	OP booster	Given at 18 months
At 4–6 years	DPT booster, OP booster	May be given at 3 years on entering nursery school
At 14–16 years	Adult-type tetanus and diphtheria booster	Pertussis omitted

*These are recommended practices in some areas. Actual practice may vary in the details.

What happens inside the body during the time period that the injections are given? As noted in Section 9.4, antigen stimulation leads to the production of a line of activated B or T lymphocytes, and once this line of cells has been produced, continual cell division results in a small but significant population of these cells for considerable periods of time (usually several years). Periodic booster injections will stimulate further division of the activated lymphocytes, thus making it possible to maintain an active population of cells throughout adult life. A natural encounter with a pathogen will also result in a rapid booster response, leading to a further increase in production of antibodies or activated T cells that will attack the invading pathogen. If booster injections or natural exposures to the pathogen do not take place, the population of activated lymphocytes will eventually disappear, and the host will once again be susceptible to infection. It is not really known how long immunity lasts in the absence of antigenic stimulation; the length probably varies considerably from person to person.

Immunization procedures such as those listed in Table 9.7 are not only beneficial to the individual but are also effective public-health procedures, since disease spreads poorly through an immunized population. In fact, the disease smallpox has been eliminated worldwide as a result of an effective vaccination program (see the Box in Chapter 10, page 280).

9.11 MONOCLONAL ANTIBODIES

The production of antibodies either experimentally or clinically results in the formation of a series of closely related antibodies, each derived from a separate clone of B cells. For many purposes, it is desirable to obtain antibody produced by a single clone of B cells. Such antibodies, called *monoclonal antibodies*, have the advantage that they are monospecific, reacting to only a single antigenic determinant. For many research and diagnostic purposes, monoclonal anti-

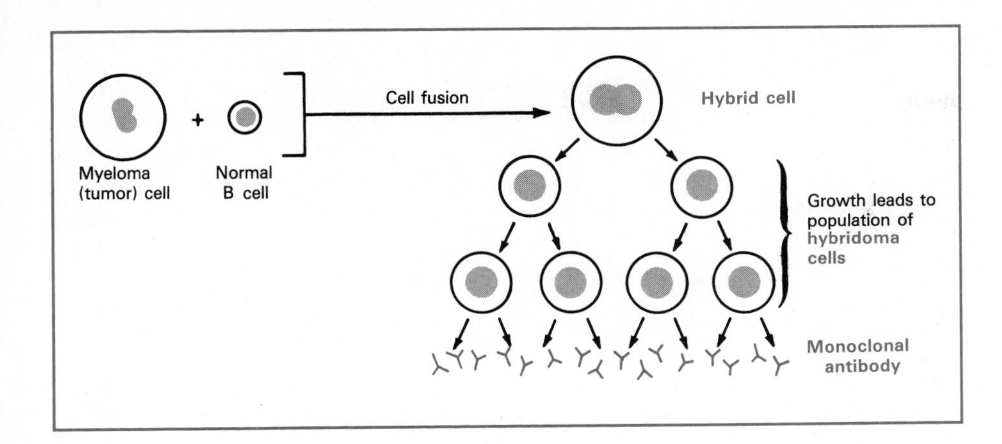

Figure 9.19 Monoclonal antibody production using the hybridoma technique.

bodies are a significant improvement over the mixed antibody preparations obtained from the serum of experimental animals. Monoclonal antibody production involves first the *selection* of a clone of antibody-producing B cells, and then the *culture* of these cells in such a way that large numbers of antibody-producing cells are obtained.

The technique used to culture B cells is called the **hybridoma technique**, because it involves the fusion of two different types of cells to form a hybrid (Figure 9.19). One type of cell is a B cell; the other cell is a type of cancer cell called a *myeloma* cell. Although it is ordinarily very difficult to grow human cells in laboratory culture, myeloma cells, like many other cancer cells, grow profusely in culture. By fusing a B cell with a myeloma cell, a long-lived (essentially immortal) hybrid cell can be obtained which grows rapidly (as do myeloma cells) and also produces antibody (as do B cells). The rapid growth of hybridoma cells leads to the development of a large population of cells, each of which is the progeny of a single type of B cell. All the antibody which is produced by the population of hybridoma cells is the same as that which the original B cell produced, and the antibody is therefore called **monoclonal antibody**. In a way, this procedure is analogous to the clonal selection which occurs when a B cell is stimulated by antigen in an animal, the important difference being that the "clone" is selected in the laboratory.

Because of its great specificity, many applications exist for monoclonal antibody in experimental and clinical medicine. Monoclonal antibodies are ideal for the selective detection and destruction of specific antigens. For example, it is now known that subtle antigenic differences exist between the surfaces of cancer cells and normal cells. Monoclonal antibodies which react against the antigenic determinants unique to cancer cells could be effective in treating cancer cells without causing harm to normal cells. This would be a vast improvement over current cancer therapy which involves the use of radiation or drugs which are not very selective killing agents and which thus cause significant side effects in the patient. Monoclonal antibodies may also be useful in determining tissue and blood types. They are also very useful in diagnostic reactions for the detection of specific chemical substances.

9.12 DIAGNOSTIC TESTS USING ANTIBODIES

The term *serology* refers to the use of antibodies (obtained from serum) for diagnostic procedures and laboratory research in either basic science or clinical medicine. There are many examples of the use of immune system components in diagnostic reactions. Since serological procedures may help determine the presence of pathogenic microorganisms which are difficult to culture or recognize by other means, serology plays an important role in a clinical microbiology laboratory. Other serological procedures are used as sensitive biochemical or pharmacological assays. A few of the more well known techniques are discussed below.

Radioimmunoassay (RIA) A sensitive assay for antigenic chemicals can be developed using radioactively labelled antigen. For example, insulin levels can be measured in the serum of diabetic patients by this method, as shown in Figure 9.20*a*. In a control reaction, radioactively labelled insulin is allowed to react with anti-insulin antibody. The formation of a precipitate results in the transfer of radioactivity from solution (unreacted insulin) to the precipitate. In the test reaction, a sample of serum to be assayed is also added. If insulin is present it will also react with anti-insulin antibody and there will be a decrease in the amount of radioactivity in the precipitate. The amount of the decrease is proportional to the amount of insulin.

ELISA technique The ELISA (or *e*nzyme-*l*inked *i*mmunoabsorbent *a*ssay) technique can also be used for the sensitive detection of specific antigenic chemicals (Figure 9.20*b*). In this test, antibody specific for a given antigen is coupled to an enzyme whose activity is easy to assay. In a control reaction, a known amount of antigen is added and a precipitate formed as a result of the antigen-antibody reaction. The precipitate also contains a defined amount of the enzyme conjugated to the antibody, and this enzyme activity can be measured. Antigen concentrations in test solutions can be determined by comparison of enzyme activity in the precipitate of the test system to that in the controls. The ELISA test is safer than the radioimmunoassay technique because no radioactive materials are required. It may also be used for the sensitive detection of antibody if the enzyme is linked to the antigen instead of the antibody.

Complement fixation technique An important property of the complement system (described in Section 9.3) is that the complement proteins are enzymatically altered and thus used up during the reaction with antigen-antibody complexes. The loss of complement as a result of an antibody-antigen reaction is called **complement fixation** and occurs whenever an IgG or IgM antibody reacts with antigen in the presence of complement. Complement fixation is thus a measure of the extent to which an antigen-antibody reaction has occurred. After an initial reaction to permit complement fixation to occur, an indicator system is added which consists of sheep red blood cells coated with antibody against them. The lysis of sheep red blood cells, caused by complement left over after the initial reaction, is easily observed without

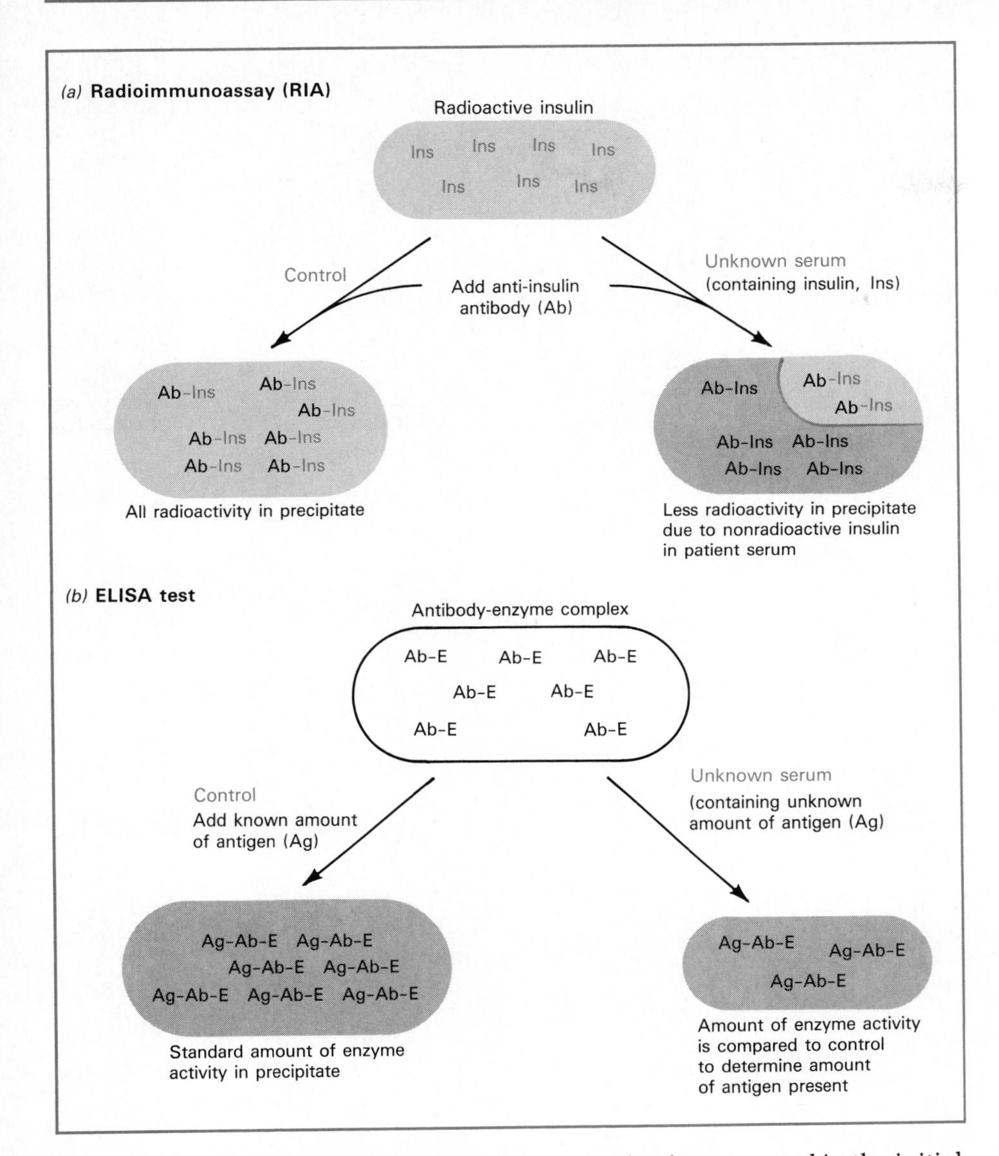

Figure 9.20 Two commonly used serologic diagnostic reactions. The control reactions are on the left, compared to reactions of a serum being tested on the right. (a) Radioimmunoassay (RIA). The color indicates radioactivity. (b) Enzyme-linked immunoabsorbent assay (ELISA).

the aid of a microscope. If complement was completely consumed in the initial reaction (indicating that a strong antigen-antibody reaction occurred) no lysis of sheep red blood cells will occur. If, on the other hand, complement was not completely consumed in the initial reaction (indicating a weak or absent antigen-antibody reaction), complement will be present and will react with the antibodies coating the sheep red blood cells to lyse them.

By measuring complement fixation, one has a means of determining the occurrence of an antigen-antibody reaction even if precipitation or some other visible reaction has not occurred.

Fluorescent antibody technique It is possible to conjugate a fluorescent dye to an antibody molecule; the result is a fluorescent antibody which can be used to detect the presence of specific pathogens. A fluorescent an-

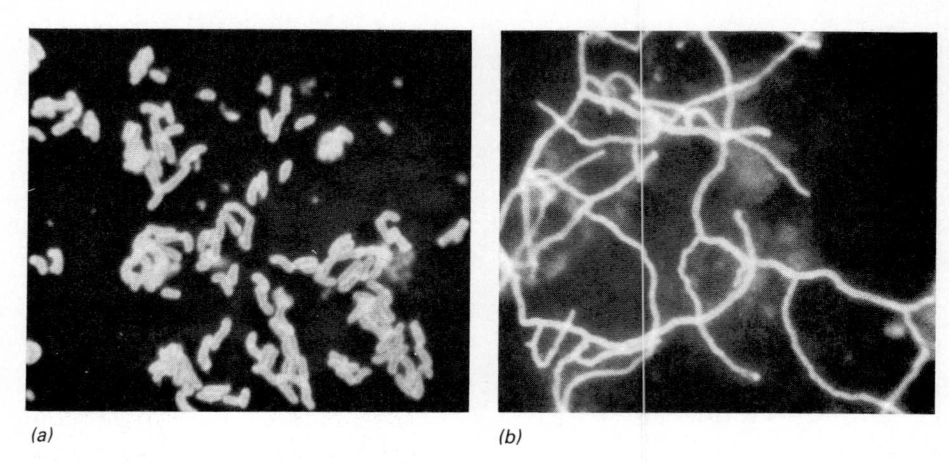

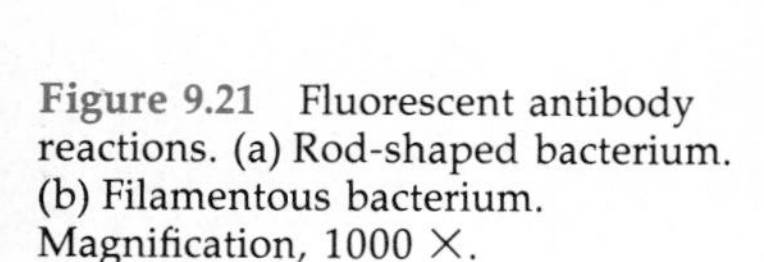

Figure 9.21 Fluorescent antibody reactions. (a) Rod-shaped bacterium. (b) Filamentous bacterium. Magnification, 1000 ×.

tibody specific for a microorganism is mixed with a sample suspected to contain that microorganism. The sample is then observed using a fluorescence microscope. If the organism is present, it will appear fluorescent due to the coating of antibody over its surface. Examples of such staining reactions are shown in Figure 9.21. A variation of the fluorescent antibody method is used in the diagnosis of syphilis.

SUMMARY

The **immune system** is a critical part of the host defense against infectious microorganisms, as evidenced by the difficulties suffered by those who have immune systems diseases, such as AIDS. Unlike innate resistance, the immune system provides a form of **acquired resistance**, which occurs in response to foreign chemicals in the body, including the macromolecules of pathogenic microorganisms. The hallmarks of the immune system are **specificity**, **memory**, and **tolerance of self**.

Antigens are chemicals foreign to the body which stimulate either an **antibody-mediated** or a **cell-mediated** immune response. An **antibody** is a type of protein, produced by **B cells** in response to the presence of foreign antigens, which reacts with the specific antigen which stimulated its production. Depending on the type, antibodies may **neutralize**, **precipitate**, or **agglutinate** the antigens with which they react. Antibody-antigen complexes may also react with **complement**, resulting in the death of foreign cells, or in an increase in the activity of phagocytic cells.

Activated T cells are also produced in response to specific antigens. Antibody-like **receptors** on T cell surfaces combine specifically with antigens. Several different types of T cells exist which participate in killing of foreign cells, in increasing the activity of phagocytic cells, and in formation of antibodies by B cells. Thus, both types of immune responses are involved in resistance to infectious disease.

Immunization is the artificial use of the immune response to control disease within individuals and populations. Preformed antibodies may be given to diseased people in **passive immunization**. Alternatively, an **atten-**

uated antigen preparation, called a **vaccine**, may be administered to **actively immunize** an individual before disease occurs. **Vaccination** programs have effectively controlled the spread of many important diseases.

There are several negative consequences of the immune system. **Allergies** may result from either antibody-mediated or cell-mediated immunologic reactions; blood transfusion and tissue transplant problems result from incompatibility in the antigens on normal human cells; breakdown in the immune system causes **autoimmune disease**.

Because of the specificity of the immune response, many **serologic** techniques have been developed to aid in diagnosis and treatment of disease. **Monoclonal antibodies** are becoming useful in testing of specific chemicals, and may be useful in anticancer therapy.

KEY WORDS AND CONCEPTS

Tolerance (self-recognition)
Antibody titer
Primary response
Secondary response
Antigen
Antigenic determinant
Antibody
Immunoglobulin
IgG, IgM, IgA, IgE, IgD
Neutralization
Precipitation
Agglutination
Complement activation
Opsonization
Clonal selection
Plasma cell
Memory cell
Helper T cell
Cell-mediated immunity
Activated T cell
Cytotoxic (killer) T cell
Immunosurveillance
Delayed-type hypersensitivity T cell
Lymphokines
Acquired immunodeficiency syndrome (AIDS)
Allergy
Immediate-type hypersensitivity
Anaphylactic shock
Serum sickness
Delayed-type hypersensitivity
Contact dermatitis
ABO system
Rh system
Immunosuppression
Autoimmune disease
Immunization
Passive immunity
Antiserum
Antitoxin
Active immunity
Vaccination
Attenuation
Toxoid
Monoclonal antibody
Serology

STUDY QUESTIONS

1. What is the difference between *innate resistance* and *acquired resistance*?
2. Give three examples of *antigens* on microorganisms. Give an example of an antigen produced by, but not actually *on* a microorganism.

3. What are the two major types of immune responses which may occur when foreign *antigens* enter the body? What cell types are responsible for each immunologic response?
4. Draw a diagram showing how the *antibody titer* changes during the first and second exposure to an *antigen*. Why is a *booster shot* given during *vaccination*?
5. Draw the general structure of *IgG antibody*, showing its *constant* and *variable regions*. Where is the *antigen combining site*? How is the structure of *IgG antibody* different from the structure of *IgM*?
6. What is unique about the structure of *IgA* antibody, and how is this related to its function? Where do you find this type of antibody?
7. Which type of antibody is important in *allergic* reactions? Why is it involved in these reactions?
8. In what different ways do antibodies and antigens react? How do such reactions aid in defense against pathogenic microorganisms? How are such reactions used in diagnostic tests?
9. What is *complement*? With what does complement react? What are some of the possible results of complement activation?
10. Explain how *opsonization* results in the elimination of pathogens.
11. Explain how a relatively small number of genes can code for a very large number of different antibodies.
12. Describe how *B cells, helper T cells,* and *macrophages* interact with the antigens of a pathogen to bring about *clonal selection* and the production of one type of antibody.
13. *Activated T cells* include *cytotoxic T cells* and *delayed-type hypersensitivity T cells*. What is the major difference in the way they help rid the body of pathogenic microorganisms?
14. Name three immune deficiency diseases. What is the problem in each of these diseases? Can each be treated?
15. Give an example of each of the major types of *allergy*: *immediate-*, and *delayed-type hypersensitivity*. Which type is antibody-mediated? Which type is cell-mediated?
16. Why are *antihistamines* used to combat some allergies?
17. What are *autoimmune diseases*? Give two examples.
18. What is your ABO blood group? Which antigens do your red blood cells possess? Which antibodies are in your blood serum? What ABO blood types can you receive during a transfusion?
19. *Rh disease of the newborn* involves reaction between maternal anti-Rh antibody and fetal Rh-positive red blood cells. Why doesn't a similar reaction occur between a mother and a fetus of different ABO blood groups?
20. What antigens control the tissue types of animal cells? Explain why tissues must be carefully typed before transplantation is attempted. What is *immunosuppression* and why is it used during organ transplants?
21. *Immunization* may be either *active* or *passive*. Explain the difference, and give an example of each type. Give an example of *natural passive immunization*.
22. What is *attenuation*? How is it accomplished?

23. Which diseases do we currently vaccinate against in the United States? When should children be vaccinated against these diseases?

24. What are *monoclonal antibodies*? Why are they more useful than antibodies produced by animals?

SUGGESTED READINGS

HOOD, L.E., I.L. WEISSMAN, W.B. WOOD, and J.H. WILSON. 1984. *Immunology, 2nd edition*. The Benjamin/Cummings Publishing Company, Menlo Park, California.

KIMBALL, J.W. 1983. *Introduction to immunology*. Macmillan Publishing Company, New York, New York.

LEDER, P. 1982. *The genetics of antibody diversity*. Scientific American 246: 102–115.

LENNETTE, E.H., A. BALOWS, W.J. HAUSLER, JR., and H.J. SHADOMY. 1985. *Manual of clinical microbiology, 4th edition*. American Society for Microbiology, Washington, D.C. (A good reference for information on serologic tests used in clinical microbiology.)

LERNER, R.A. 1983. *Synthetic vaccines*. Scientific American 248: 48–56. (A review describing the production of purified proteins useful as vaccines.)

MILSTEIN, C. 1980. *Monoclonal antibodies*. Scientific American 243: 66–74. (A review of the techniques used to produce antibodies in cell cultures.)

ROITT, I. 1984. *Essential immunology, 5th edition*. Blackwell Scientific Publications, Oxford, England.

ROITT, I., J. BROSTOFF, and D. MALE. 1985. *Immunology*. Gower Medical Publishing, Ltd., London, England.

ROSE, N.R. 1981. *Autoimmune diseases*. Scientific American 244: 80–103. (A review of diseases related to breakdown in the immune system.)

10

Public Health Microbiology

In the last two chapters we discussed the course of an infectious disease within an individual. In the present chapter, we discuss the spread and the control of diseases within populations. We have defined parasites as organisms capable of living at the expense of other living organisms, and have defined pathogens as parasites capable of causing harm. If a parasite grows in nature only in another host, then for this parasite to survive in nature, there must be some mechanism by which the parasite moves from host to host. Therefore, to understand infectious disease in populations, we must understand the **mode of transmission** of the parasite. A knowledge of how parasites are transmitted is very useful in developing effective control methods for infectious diseases. There are two major procedures for control of disease in populations: 1) prevent the transmission of the pathogen; 2) increase the resistance of the population through immunization.

When we consider infectious disease in populations, we frequently define diseases as *endemic* or *epidemic*. An **endemic** disease is one which is constantly present in a population, whereas a disease is said to be **epidemic** when it occurs in an unusually high number of individuals at the same time. A **pandemic** is a worldwide epidemic. The topic of the present chapter falls under the heading of "public health," because it deals with the health of the public at large. Public health is based on knowledge of the field of **epidemiology**, the study of the incidence and prevalence of disease in populations. Through the efforts of epidemiologists, we have learned how diseases affect populations, and have devised effective means of controlling disease. By organizing efforts to improve public health, through agencies of the federal and state governments, there has been vast improvement in the incidence of infectious disease in the United States (see Figure 1.1). Diseases such as tuberculosis and diphtheria, which were once among the leading causes of death, have now been brought under control. Diseases once *epidemic* in proportion have been reduced to insignificant problems. Smallpox, for example, has been completely eradicated. In this century, however, epidemic outbreaks of other diseases, such as influenza, and in the past decade, the emergence of new diseases, such as legionellosis and acquired immunodeficiency syndrome (AIDS), have been bitter reminders of our inability to control all diseases. In fact, infectious disease still remains a most significant problem in developing countries. How disease in the population is monitored and controlled will be considered in detail in this chapter.

10.1 SOURCES OF INFECTIOUS MICROORGANISMS

Obviously, an infected individual may serve as a source from which other individuals become infected. However, there are many other animate and inanimate sources from which an individual may contract infectious diseases (Figure 10.1). A site in which an infectious agent remains alive and from which it can be transmitted is called a **reservoir**.

For some diseases, the only reservoir is human beings, and close or intimate contact with a contagious individual is the only way infection occurs. Most viral and bacterial respiratory diseases, sexually transmitted diseases, staphylococcal and streptococcal diseases, typhoid fever, and diphtheria are in this category. It is also possible for some humans to harbor pathogenic microorganisms without exhibiting any disease symptoms. Such individuals, called **carriers**, are particularly dangerous reservoirs; the carrier state will be described further below.

Inanimate objects may also serve as reservoirs of pathogens. For example, bacteria of the genus *Clostridium* ordinarily live in the soil and only cause

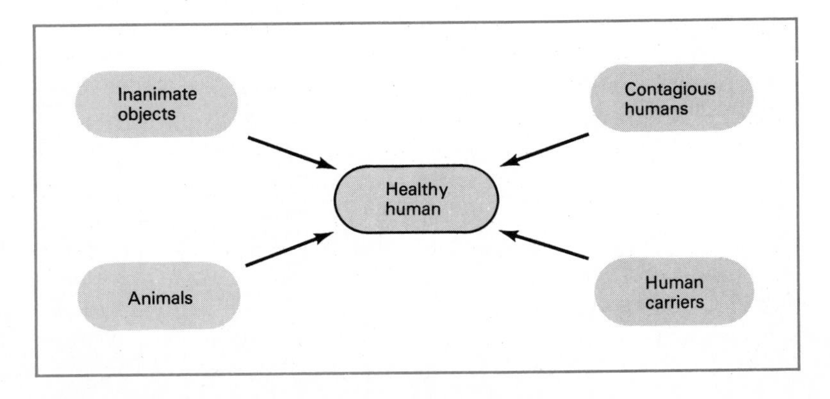

Figure 10.1 Reservoirs which may serve as sources of infectious agents.

disease when humans or animals come into contact with them. *Clostridium tetani* causes the disease tetanus when inoculated into an individual in a deep puncture wound. *C. botulinum*, the organism that causes the kind of food poisoning known as botulism, is even more restricted; it does not generally grow in the body, and hence in most cases is not an infectious agent. Rather, it spends its life in the soil and can become inoculated into foods where, under unusual circumstances, it can grow and produce a toxin that is harmful if ingested (see Chapter 16).

In some diseases, called **zoonoses**, animals serve as the reservoir, the disease being contracted by direct contact with the animal. Rabies, for example, is acquired when a rabid animal bites a human. In other cases, the infectious agent is carried from the infected animal to humans by arthropods, such as mosquitoes or ticks.

The type of reservoir is very important in determining how a disease is controlled, and in fact determines whether a disease can be controlled at all. One extreme is illustrated by the complete elimination of the disease *smallpox*. Humans are the only significant reservoir for the smallpox virus. Because there were no animal or inanimate reservoirs, and because there was an effective vaccine available, it was possible to eradicate the disease entirely (see Section 10.7). On the other hand, *Clostridium tetani* is primarily a soil inhabitant and only incidentally infects humans. Complete elimination of the bacterium from soil is, of course, impossible, so that it will never be possible to control tetanus through its reservoir.

Contagion during infection An individual is said to be **contagious** when he or she can transmit a pathogenic microorganism to another individual. Although this may occur at various stages during the disease process (see Chapter 8), the spread of disease between individuals is particularly important, if the diseased individual is contagious during the **incubation period** (the time between the initiation of the infection and the first appearance of disease symptoms). Some diseases have short incubation periods; others have long incubation periods (see Figure 8.21). From the point of view of spread through a population, the length of the incubation period is of considerable significance, because during this period the contagious individual is not aware of it and hence may infect numerous other people.

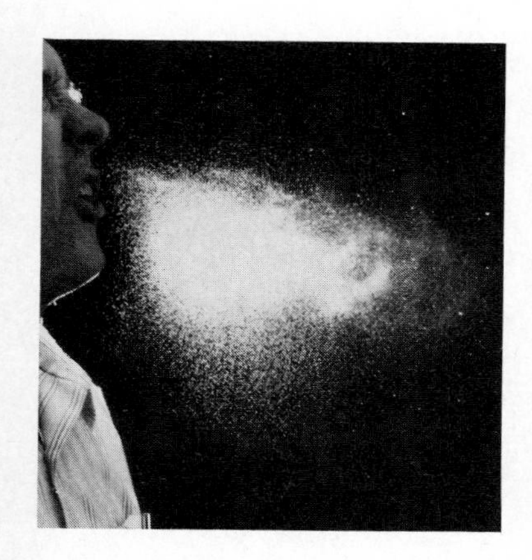

Figure 10.2 High-speed photograph of a sneeze.

For pathogens where humans are the only reservoir, the pathogen must leave the body and infect others, since only in this way is the pathogen able to maintain itself in the population. The most common routes of exit from the body are the *respiratory* and *gastrointestinal* tracts. Organisms living in the respiratory tract exit as a result of coughing or sneezing, which expels droplets containing mucus particles within which the pathogen is entrapped (Figure 10.2). Many pathogenic organisms are excreted with the feces; this includes both those that infect primarily the gastrointestinal tract and others that infect interior organs of the body and spread via the blood stream to the intestines. Another important route of exit, used by a few pathogens, is the *genital* tract, the organisms being emitted in secretions from the penis or vagina. Diseases in which the agents are transmitted in this manner are usually called **sexually transmitted** (or **venereal**) **diseases**.

TYPHOID MARY

The classic example of a chronic carrier was the woman known as "Typhoid Mary," a cook in the New York-Long Island area in the early part of this century. Typhoid Mary (her real name was Mary Mallon) was employed in a number of households and institutions, and as a cook she was in a central position to infect large numbers of people. Eventually, she was tracked down by the epidemiologist Dr. George Soper after an extensive investigation of a number of typhoid outbreaks. When Mary Mallon's feces were examined bacteriologically, they were found to contain large numbers of the typhoid bacterium *Salmonella typhi*. She remained a carrier for many years, probably because her gall bladder was infected, and organisms were continuously excreted from there into her intestine. Public-health authorities offered to remove her gall bladder, but she refused the operation. In order to prevent her from continuing to serve as a source of infection, she was imprisoned. After almost three years in prison, she won a court case and was released on the pledge that she would not cook or handle food for others and that she was to report to the local health department every three months. She soon disappeared, changed her name, and resumed cooking in hotels, restaurants, and sanitoria, leaving a wake of typhoid fever. Five years later, she was arrested as the result of the investigation of an epidemic at a New York maternity hospital. She was again imprisoned and she remained in prison until she died in 1938, 32 years after her first discovery as a typhoid carrier. This case shows the importance of carriers and that they can remain contagious for long periods of time. Fortunately, most disease-carrier states can now be eliminated with proper antimicrobial therapy, and current incidence is much lower.

Carriers Some individuals, called **carriers**, may harbor the infectious agent without exhibiting any disease symptoms. The basis for the carrier state can either be that an individual has a high level of resistance to a particular pathogen, or that the pathogen only causes a mild, subclinical form of the disease. The carrier condition may be either temporary or chronic. Carriers may be unaware of their condition and can therefore be extremely important reservoirs affecting the spread of disease. They can act as sources of infection for large numbers of people over a long period of time.

Carriers can be identified by routine surveys of populations, using cultural, radiological (for example, chest X ray), or immunological techniques. In general, carriers are only sought among groups of individuals who may be sources of infection for the public at large, such as food handlers in restaurants, groceries, or processing plants. Two diseases in which carriers have been of most significance are typhoid fever and tuberculosis, and routine surveys of food handlers for inapparent cases of these diseases are sometimes made. The classic example of a chronic carrier was the woman known as "Typhoid Mary," whose case is described in the Box above.

10.2 HOW DISEASE SPREADS

As we have noted, transmission of an infectious agent depends on the ability of the pathogen to leave an infected host and enter a new one; the pathogen must also survive *en route*. Various **modes of transmission** are summarized

TABLE 10.1 Modes of transmission of pathogens from one infected individual to another

Mode of transmission	Usual site of exit	Usual site of entry	Examples
Airborne	Upper respiratory tract	Upper respiratory tract	Influenza, common cold
Water-borne	Feces	Oral cavity	Cholera, polio
Food-borne	Feces	Oral cavity	Salmonellosis, typhoid fever
Direct human contact	Genitals	Genitals	Sexually transmitted diseases
	Mouth	Mouth	Streptococcal sore throat
Contact with inanimate objects		Skin, mouth, open wounds	Wound infections
Animal to human (zoonoses)	Mouth of animal	Broken skin	Rabies
via products	Beef, milk	Oral cavity	Brucellosis
via insects	Animal blood-meal	Human blood-meal	Bubonic plague, Rocky Mountain spotted fever
Human to human via insects	Human blood-meal	Human blood-meal	Malaria

in Table 10.1, together with the common sites of exit from and entry into the body.

It is essential that the pathogen be able to maintain viability during the time that it is outside the body. The exterior environment is much more rigorous than the host in many ways, and pathogens can usually survive outside the host only for short periods of time. Air especially is a harsh environment, and pathogens transmitted through this medium must be able to resist drying and the bactericidal action of bright sunlight if they are to be viable when they reach susceptible hosts. Water is somewhat more favorable as a dispersal agent, but the pathogen must contend with the lack of suitable nutrients for growth, and harmful features of the aquatic environment such as lower temperature, competition and predation from natural aquatic microorganisms, and toxic chemicals. In general, transmittal can be over greater distances via water than by air, although only a few pathogens are transmitted strictly by the water route. Some pathogens are so sensitive to the environment outside the host that they cannot survive for more than a few minutes. Those pathogens transmitted by the genital tract are usually in this category, and it is for this reason that they are transmitted only by direct intimate contact between the two hosts so that exposure to the outside world does not occur.

Note that a successful parasite is not necessarily a highly virulent one. From the point of view of the parasite, success means ability to reproduce, grow, and be transmitted from individual to individual, thus permitting long-term survival of the parasite. A highly virulent parasite, which may kill its host quickly, may not be transmitted, and hence may die with its host.

Airborne dispersal Many parasites spread through the air, usually attached to some sort of particle. A wide variety of microorganisms are found in the air but usually only in low numbers. The organisms found in the air outdoors are primarily harmless soil organisms, but indoors the organisms may be those commonly found in the human respiratory tract. During coughing and sneezing, or even talking, numerous droplets of moisture are expelled from the mouth. During a sneeze (see Figure 10.2), droplets laden with bacteria leave the mouth at a speed of over 200 miles per hour; the number of bacteria in a single sneeze may be as many as 10,000 to 100,000. In the air, the water in these droplets evaporates, leaving behind small particles of mucus to which the bacterial cells are attached. The bacteria in these lightweight particles come from the respiratory tract. Such bacteria are relatively resistant to drying and can remain alive for long periods of time. Thus, infection can arise from inhalation of contaminated dust; this has often been a serious source of infection in hospitals. However, infection by direct transmission of airborne droplets from one person to another is more common. This sort of direct transmission is more frequent when people are crowded together, and this is one explanation for increased respiratory infections in winter, when people remain indoors and have more direct contact with each other (see Section 10.5).

The average human breathes several million cubic feet of air in a lifetime, much of it containing microbe-laden dust. The speed at which air moves through the respiratory tract varies; it is fast in the upper respiratory tract and quite slow in the lower respiratory tract (Figure 10.3). As the air slows down, particles in it stop moving and settle, the larger particles first and the smaller ones later. In the tiny bronchioles, only particles below 3 μm can penetrate.

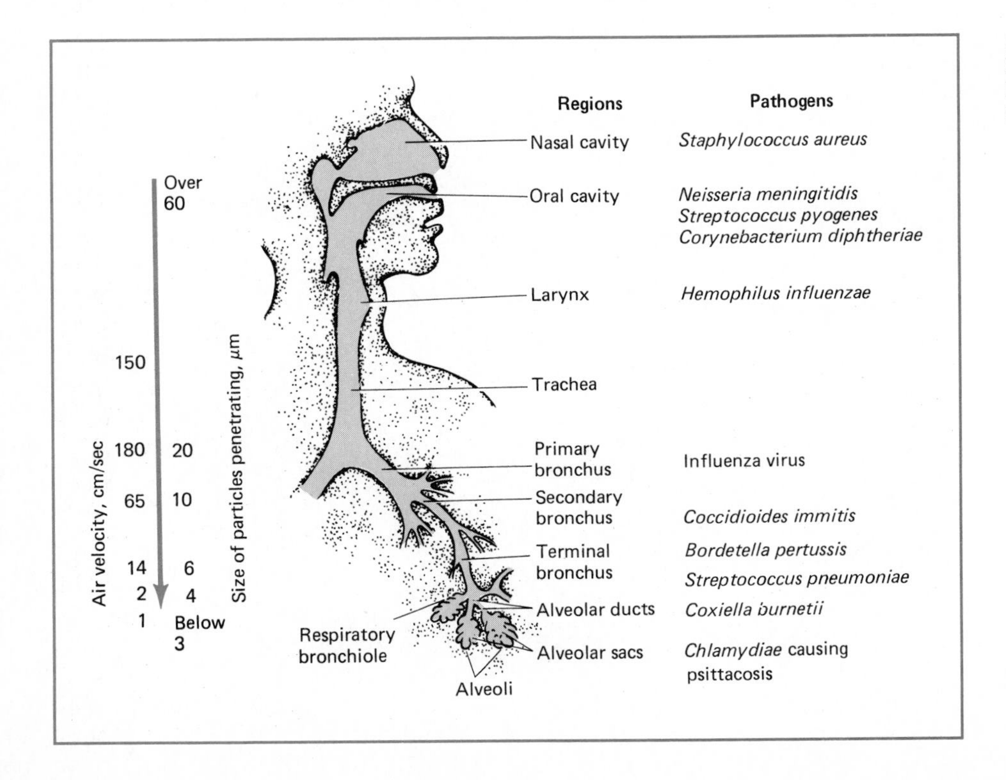

Figure 10.3 Characteristics of the human respiratory system and locations at which various organisms initiate infections.

Different pathogenic organisms reach different levels in the tract, thus accounting in part for the differences in the kinds of infections that occur in the upper and lower respiratory tract.

The kinds of diseases in which the causal organism is transmitted through the air include diphtheria, pneumonia, meningitis, psittacosis, the common cold, influenza, streptococcal sore throat, histoplasmosis, and tuberculosis. In all these diseases, the pathogen lives in the respiratory tract and is expelled into the air by the host during the period of infection. Some of these diseases will be discussed in Chapters 11, 12, and 13.

Water- and food-borne dispersal Many harmless microorganisms are found in lakes, rivers, and oceans. In water polluted by sewage, however, pathogens are frequently also present. Pathogens found in water usually come from feces of individuals with intestinal infections. These pathogens live and grow in the intestinal tract and are often passed out of the body in large numbers. Methods of water testing and of water treatment and purification designed to prevent the spread of these organisms are described in Chapter 15. Pathogens transmitted by polluted water include among others the causal agents of typhoid fever (*Salmonella typhi*), and cholera (*Vibrio cholerae*).

Intestinal pathogens are also transmitted in foods that have become contaminated by infected food handlers (discussed in Chapter 16). The major route of entry of water- and food-borne pathogens is ingestion, so that the mode of transmission of most of these diseases is said to be by the *fecal/oral route*.

Direct human contact As mentioned in Section 8.1, the human body is heavily colonized by microorganisms. Because of poor hygiene, it is also likely that the skin is also often contaminated with organisms which may be pathogenic. The contamination and decontamination of skin will be considered later in this chapter. *Staphylococcus aureus,* the bacterium which causes pimples and boils, is probably transmitted primarily by direct contact.

Sexually transmitted diseases Some pathogens die quickly outside the body because they are very sensitive to drying, light, or other environmental factors. These pathogens are usually transmitted only by more intimate contact between people during sexual intercourse or kissing. When the causal agent is transmitted via the genital organs, the disease is called a **sexually transmitted disease** (also called **venereal disease**). Classic examples of sexually transmitted diseases include syphilis and gonorrhea, but other diseases, such as genital herpes, chlamydial nongonococcal urethritis (NGU), and AIDS are of equal or greater importance. These will be described in Chapters 11 and 13.

Contact with inanimate objects Such things as pencils, toys, clothing, dishes, tables, or bedding can become contaminated by an infected individual. Pathogens that resist drying, such as *Staphylococcus aureus* and *Mycobacterium* *tuberculosis*, will remain viable on such objects and are the ones most frequently transmitted in this way.

Wound infections Most organisms cannot penetrate the unbroken skin but can infect a wound. The organism may come from another person or from the soil. Examples of wound infections include pus-forming infections due to staphylococci and streptococci, gas gangrene, caused by *Clostridium perfringens*, and tetanus, caused by *Clostridium tetani*. The most dangerous wounds are those in which contaminated material is driven deep into tissue that has been extensively destroyed. In such dead tissue, oxygen supply is cut off, anaerobic conditions develop, and obligate anaerobes such as *C. tetani* can grow.

Transmission involving animals Some pathogens are primarily animal parasites and only occasionally are transmitted to man. Diseases of this type, called **zoonoses**, may be passed to humans by animal bite or through the intermediary of an insect or other arthropod (Figure 10.4). Zoonoses which involve direct contact with animals or animal carcasses include rabies, caused by the rabies virus, and rabbit fever (tularemia), caused by the bacterium *Francisella tularensis*. Some zoonoses involve contact with animal products. Examples include anthrax, involving the transmission of the bacterium *Bacillus anthracis* through beef; and brucellosis, involving the transfer of bacteria of the genus *Brucella* via beef or milk. The control of a zoonosis is difficult because it requires the discovery and elimination of all infected animals. This can lead to significant economic loss if livestock are infected. Such control is nearly impossible in the wild. For instance, rabies is widespread in foxes, skunks, squirrels, bats, and other animals, and the virus is transmitted from them to pets (dogs and cats) and from pets to humans. One can reduce the incidence of rabies in humans by immunizing pets, but this does not eliminate the virus from wild animal populations.

Diseases involving transmission from animals to man via insects include plague, caused when fleas transmit the bacterium *Yersinia pestis* from rodents to man; and Rocky Mountain spotted fever, caused when ticks transmit the bacterium *Rickettsia rickettsii* from wild animals to man. Insects also play an essential role in transmission of pathogens from human to human in many important diseases. Mosquitoes, for example, carry the protozoa causing malaria (*Plasmodium*) and African sleeping sickness (*Trypanasoma*), and the virus causing yellow fever. When insects transmit an infectious agent from one animal or human to another, they are referred to as **vectors**. Control of diseases involving vectors may be better brought about by eradication of the vector rather than the pathogen itself. One of the most famous cases of control of an infectious disease via vector control was the elimination of the mosquito that carried the yellow fever virus during the building of the Panama Canal early in the twentieth century. This control measure permitted the successful building of the Canal.

Insects may also carry microorganisms from place to place. Flies, for example, are attracted to feces and from there can transfer contaminating microorganisms to other places.

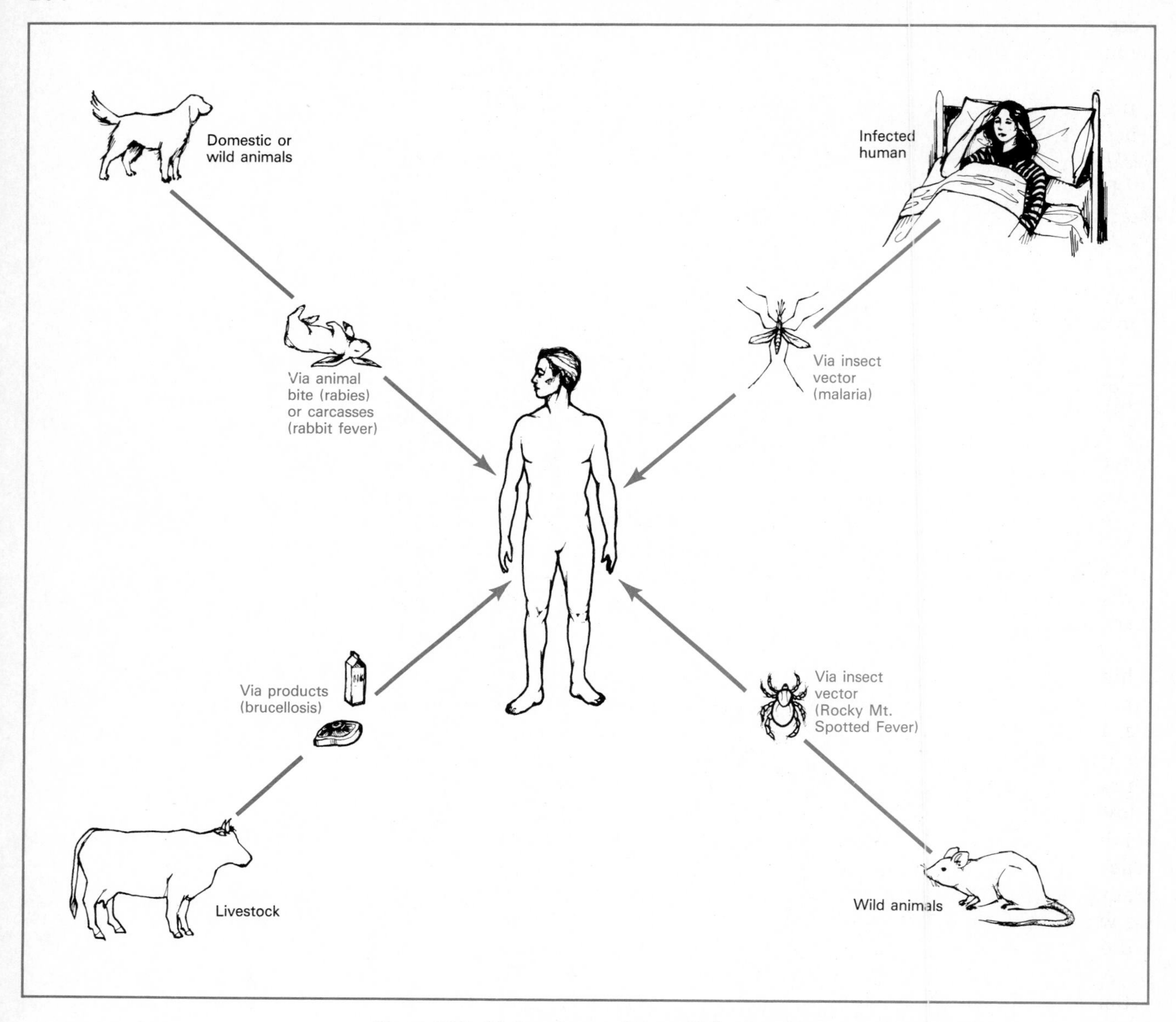

Figure 10.4 Modes of transmission of infectious disease involving animals.

10.3 HEALTH OF THE POPULATION

If we are to understand and control disease, we must have some idea of its prevalence in the population. The prevalence of disease is determined by obtaining statistics of illness and death, and from these data, a picture of the public health in that population can be obtained. Public health varies from region to region and has varied with time, so that a picture of the public health at a given moment provides only an instantaneous picture of the situation. By continuing to examine health statistics for many years, it is possible to assess the value of various public-health protective measures in influencing the incidence of disease.

Statistics of disease incidence in the United States are compiled by the Centers for Disease Control, an agency of the federal government. This agency issues a weekly publication on health statistics, called *Morbidity and Mortality Weekly Report,* and an annual summary at the end of each year.

There will always be a margin of error in health statistics because of inaccurate reporting. Statistics of long-term changes in incidence of disease are especially unreliable, since the ability of physicians to diagnose specific diseases depends on their education and awareness of these diseases and on the availability to them of laboratory and other diagnostic aids. Thus a greater reported incidence of a disease today compared to a former time does not necessarily mean that the disease is more prevalent; it may merely be better recognized. In some developing countries the problem of gathering health statistics is very great. This is due to the dispersed nature of populations, the lack of trained health personnel and the lack of effective communication systems.

Mortality Mortality expresses the incidence of death in the population. Data for leading causes of death in the United States were given in Figure 1.1. It was noted that infectious diseases were once the major causes of death, whereas currently they are of much less significance. Now diseases such as heart disease, cancer, and stroke are of greater importance.

There are marked influences of age, sex, and race on mortality. Mortality is high in infants and newborns, drops sharply after the first year of life, remains low throughout childhood and early adult life, and rises again in old age. Both infant and elderly mortality have been markedly influenced by medical practices, and the rates are much lower than formerly, but these two segments of the population still show the greatest mortality. Female mortality is lower than male mortality at all ages. This difference is at least partly due to inherent differences in susceptibility of the two sexes, since even newborn males have a higher death rate than newborn females. In the United States, death rates in the black population are considerably higher than the rates in the white population. This difference appears to be due to social, economic, and environmental factors.

A genetic component may explain the fact that there is a greater concentration of disease susceptibility in certain families than in others. Interestingly, mortality is greater in unmarried than in married individuals of the same age. This difference occurs irrespective of sex or race when all causes of death are included. There is also a high correlation between the length of life of husbands and wives, and spouses also tend to die from the same or similar causes. Occupation also has a marked influence on mortality, both on age at death and cause of death.

Morbidity Morbidity refers to the incidence of disease in populations, and includes both fatal and nonfatal disease cases. Clearly, morbidity statistics more precisely define the health of the population than do mortality statistics, since many diseases that affect health in important ways have only a low

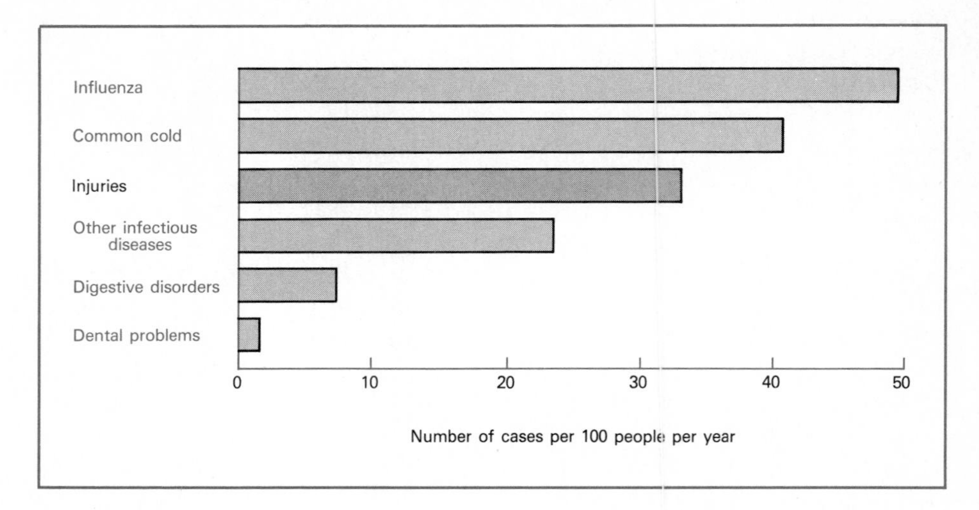

Figure 10.5 Leading causes of acute illness in the United States. The data are typical of recent years.

mortality (the common cold, for instance). Unfortunately, it is much more difficult to obtain meaningful and accurate statistics of morbidity, because there is no registration procedure such as the death certificate. A restricted number of diseases are *reportable* to public-health authorities, but, in most cases, data on morbidity must be obtained from other statistical sources, such as prepaid health-program statistics, hospital and clinic records, absenteeism records in schools and industry, and routine physical examinations. For research purposes, detailed surveys have been made of the health of populations, based on census procedures.

Leading causes of illness Figure 10.5 presents data on the leading causes of acute illness in the United States. Many other causes of illness, such as arthritis, rheumatism, heart disease, allergy, and diabetes, are considered to be long-term disorders rather than acute illnesses, and hence are not included. Although 14 percent of the American population suffers from these long-term ailments, none of these conditions are as significant as infectious diseases as leading causes of illness. As seen, the major causes of acute illness are quite different from the major causes of death given in Figure 1.1. The major illnesses are the acute respiratory diseases such as influenza and the common cold. On average, an American gets either the common cold or influenza once each year.

Reportable diseases Certain diseases are classified as *reportable diseases*. Several categories of such diseases have been established, the distinctions being based on the seriousness of the disease and the extent to which its infectivity can be controlled by public-health means. Some infectious diseases are more communicable than others, so that a series of categories of reportable diseases has been suggested by the American Public Health Association.

Class 1 diseases are diseases in which a case report is universally required by international sanitary regulations. This class includes the diseases in which quarantine (see Section 10.6) is required throughout the world, such as chol-

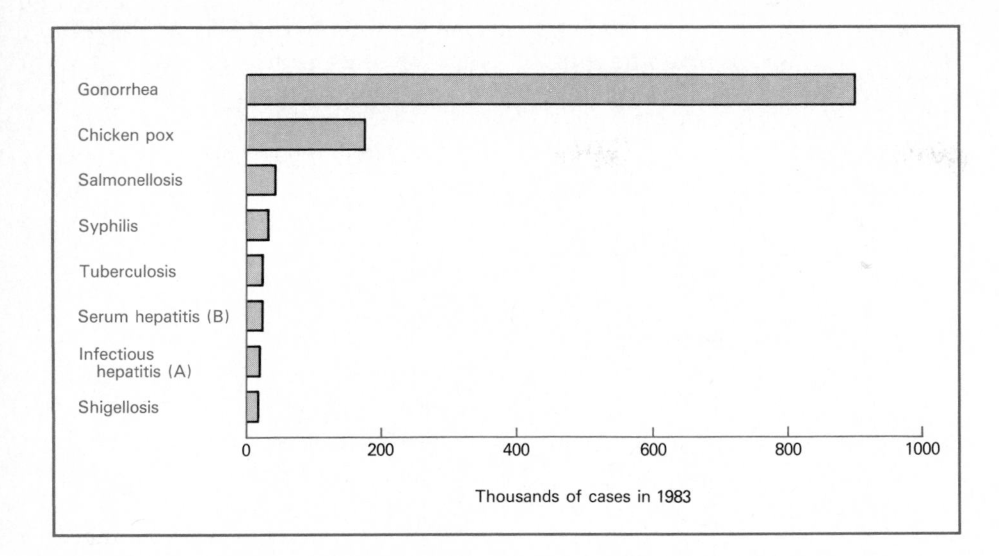

Figure 10.6 Leading reportable diseases in the United States.

era, plague, and yellow fever. *Class 2 diseases* are those in which a case report is regularly required wherever the disease occurs. Two subclasses are recognized, based on the relative urgency for investigation of outbreaks. In the first, required for the diseases typhoid fever and diphtheria, reporting is required by telephone, telegraph, or other rapid means. In the second, which includes brucellosis and leprosy, reporting is required by the most practicable means. *Class 3 diseases* are diseases that are selectively reportable in areas where the disease is known to be generally present in the population, and local jurisdictions may require reporting because of undue frequency or severity in an area. Diseases in this category include tularemia, scrub typhus, and coccidioidomycosis. *Class 4 diseases* include those diseases in which individual cases are not reportable but *outbreaks* are. These include food infections (salmonellosis) and poisonings (staphylococcal form and botulism). Pertinent data reported are the number of cases, the time period of the outbreak, the approximate size of population involved, and the apparent mode of spread. *Class 5 diseases* are diseases for which an official report is not ordinarily required, either because the diseases are not generally transmissible from person to person (blastomycosis) or are of such a nature that no practical means for control exists (common cold).

In the United States, The Centers for Disease Control (CDC) lists statistics on many reportable diseases; those with the highest incidence are shown in Figure 10.6. Gonorrhea is by far the most important reportable disease. Another sexually transmitted disease, syphilis, is also prevalent. Chickenpox is the second most common reportable disease in the U.S. In contrast to the other childhood viral diseases, such as measles, mumps, and rubella, chickenpox has not yet been effectively controlled through vaccination. Food- and water-borne diseases, such as infectious hepatitis, salmonellosis, and shigellosis, are also leading reportable diseases in the United States.

Although the incidence of reportable diseases illustrates those diseases for which health officials have great concern, such data present a biased view of morbidity among the population in general. As mentioned above, diseases like influenza and the common cold, which often are not reported, are far more prevalent causes of illness. Also, other types of sexually transmitted diseases, such as genital herpes and chlamydial nongonococcal urethritis (NGU), are probably more common than gonorrhea (see Figure 10.12). However, these diseases are also not reportable and their incidence can only be estimated.

10.4 THE STATUS OF WORLD HEALTH

So far, we have discussed the health of the population in the United States. To a considerable extent, the United States can be considered typical of those countries where public-health protection is highly developed. Other countries with similar characteristics include Japan, Australia, New Zealand, Israel, and the European countries. In quite another category as far as infectious disease is concerned are the developing countries, a category that includes most of the countries in Africa, Central and South America, and Asia. In these countries, infectious diseases are still major causes of death.

Infectious disease in developing countries As shown in Figure 10.7 and Table 10.2, there is a sharp contrast in the degree of importance of infectious diseases as causes of death in developing versus developed nations. In developing regions of the world, infectious diseases account for between 30 and 50 percent of deaths whereas infectious diseases only account for about 4–8 percent of deaths in developed regions (see Table 10.2). Diseases

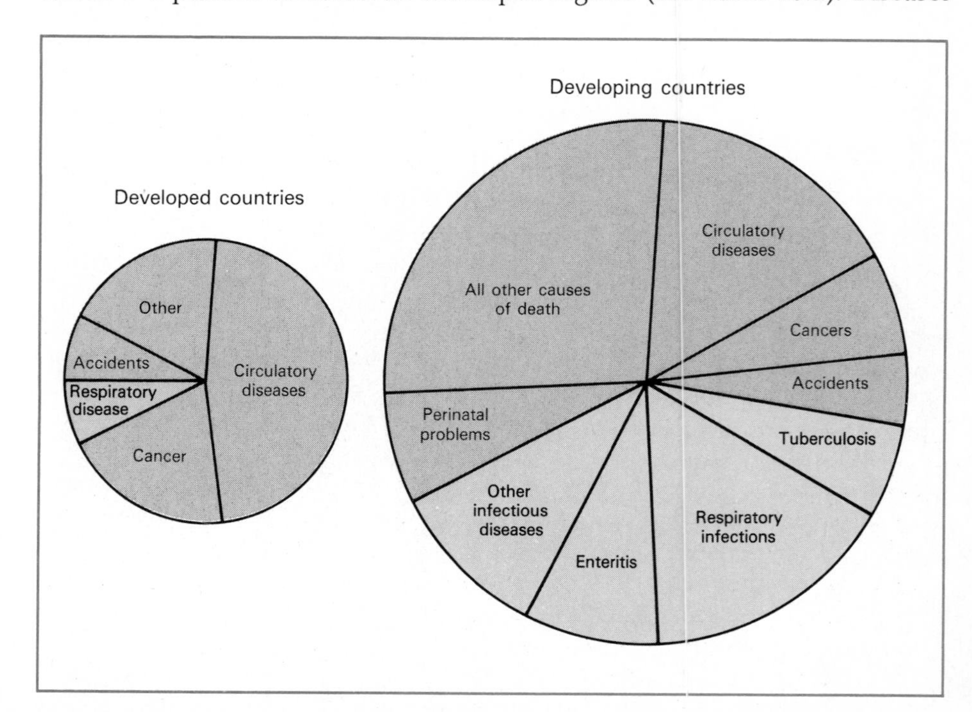

Figure 10.7 Leading causes of death in developed and developing countries. Infectious diseases are shown in color. The sizes of the circles are proportional to the relative number of deaths which occurred. The data are typical of recent years.

TABLE 10.2 Causes of death in developed and developing countries

Cause of death	Developed countries: Americas	Developed countries: Europe	Developing countries: Americas	Developing countries: Southeast Asia	Developing countries: Africa	Developing countries: Eastern Mediterranean
Infectious disease	3.6*	8.6	31.1	43.9	49.8	44.5
Cancer	21.5	18.1	9.0	4.4	2.9	4.2
Circulatory diseases	54.5	53.8	24.5	15.6	11.7	14.1
Accidents	8.4	5.6	6.3	4.3	3.8	4.1

*Percentage of deaths due to various causes in developed and developing countries in 1980. Data from World Health Statistics Annual, 1984, World Health Organization.

which were leading causes of death in the United States nearly a century ago, such as tuberculosis and gastroenteritis, are still leading causes of death in developing countries (Figure 10.7). Furthermore, the majority of deaths due to infectious disease in developing regions occur among infants and children. Thus, the age structure of deaths in developing versus developed countries is also dramatically different, as shown in Figure 10.8.

The distinct differences in the health status of people in different regions of the world is due in part to general nutritional deficiency in individuals in undeveloped countries, so that death from infection is more likely. It is also due in part to lower levels of public health protection, making infection more likely in the first place. Statistics on disease in developed countries show that control of many diseases is possible. However, statistics on the worldwide

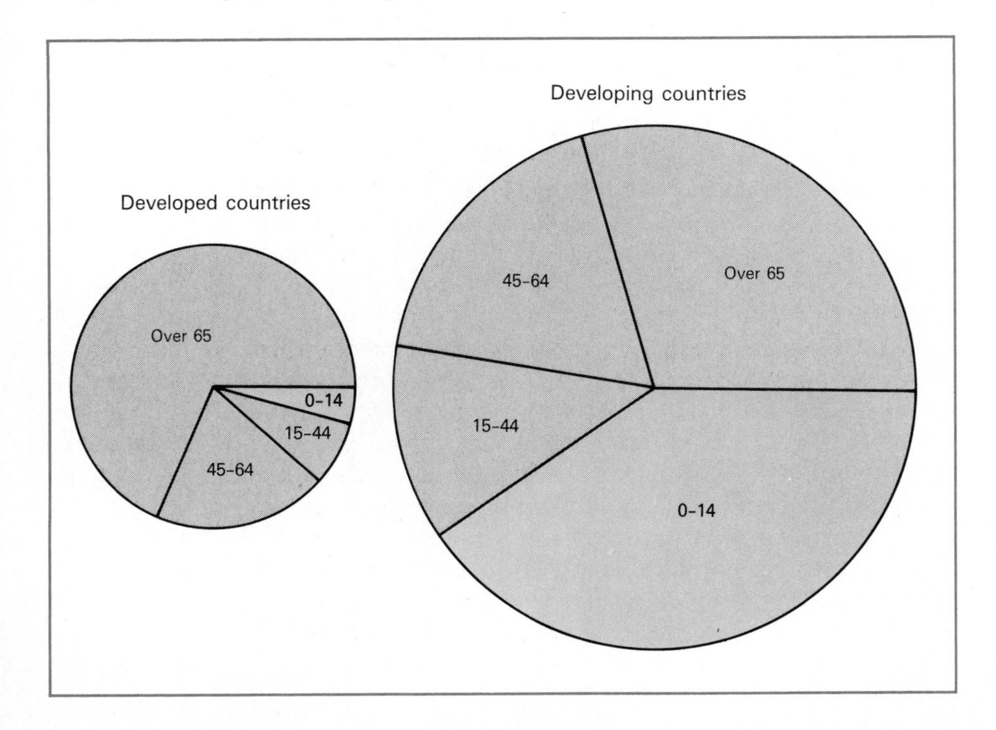

Figure 10.8 Estimated age composition of deaths in developed and developing countries. Infant and child deaths are shown in color. The sizes of the circles are proportional to the relative number of deaths which occurred. The data are typical of recent years.

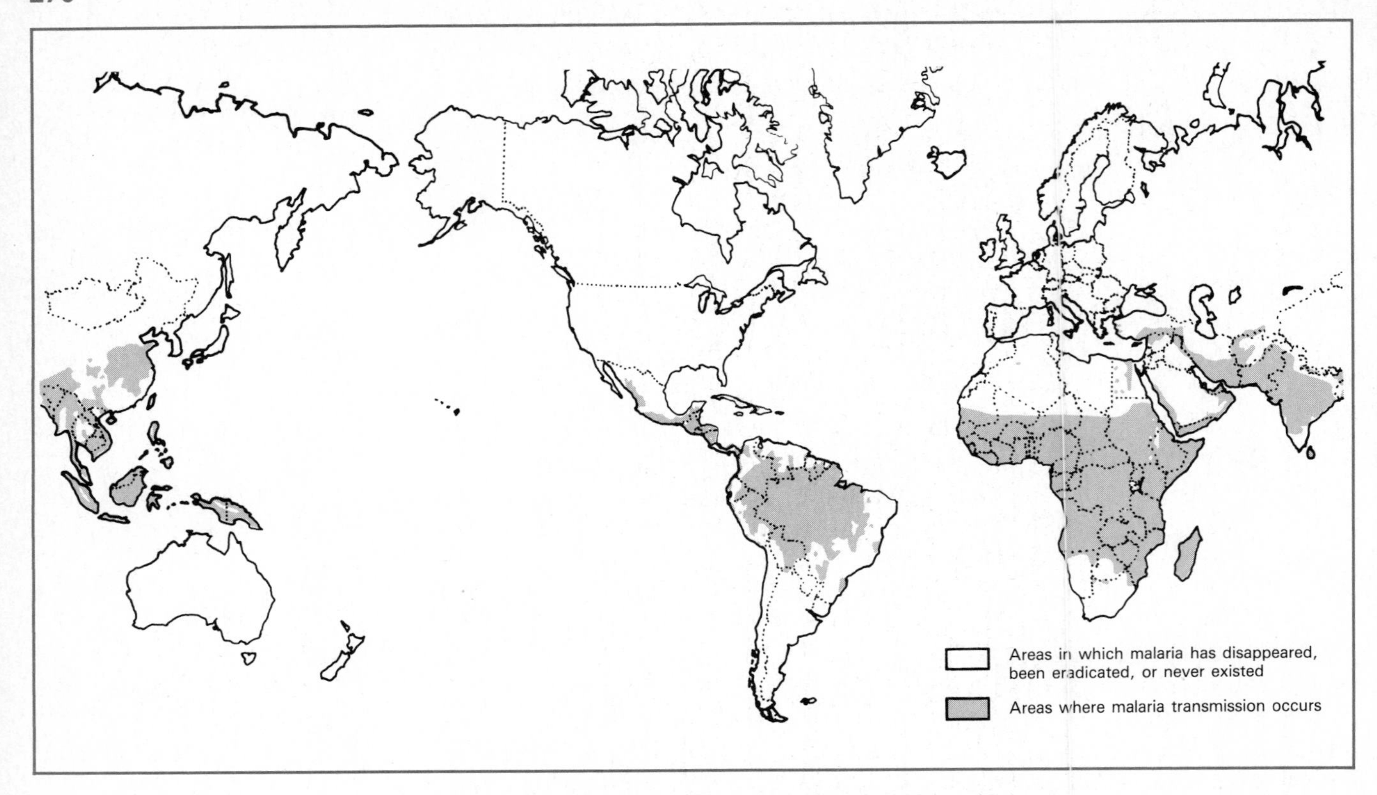

Figure 10.9 Areas in the world where malaria is endemic.

incidence of disease show that infectious disease remains an important problem about which we must all be concerned.

Global distribution The worldwide distribution of a disease of particular importance, malaria, provides an example of the difference in health of humans in different regions (Figure 10.9). As we have noted, an endemic disease is one which is constantly present in a population. The regions in which a disease is considered to be **endemic** are those locations in which it is possible to get the disease. Malaria is endemic in Africa, South and Central America, the Middle East, and Indochina, but not in the United States or Europe. This is the result of the limited distribution of the *Anopheles* mosquito, the vector required for transmission of the pathogen between humans, but also because public health measures for controlling the disease by eliminating the mosquito have only been applied in certain regions. Recent estimates suggest that in 1982 as many as 7.8 million people per year were afflicted with malaria. Even this staggering number is probably grossly underestimated because it does not include data from southern Africa, one of the most important endemic areas, due to inability to obtain reliable statistical information in that region.

Travel to endemic areas The high incidence of disease in many parts of the world is also a concern for people travelling to such areas. Figure 10.10

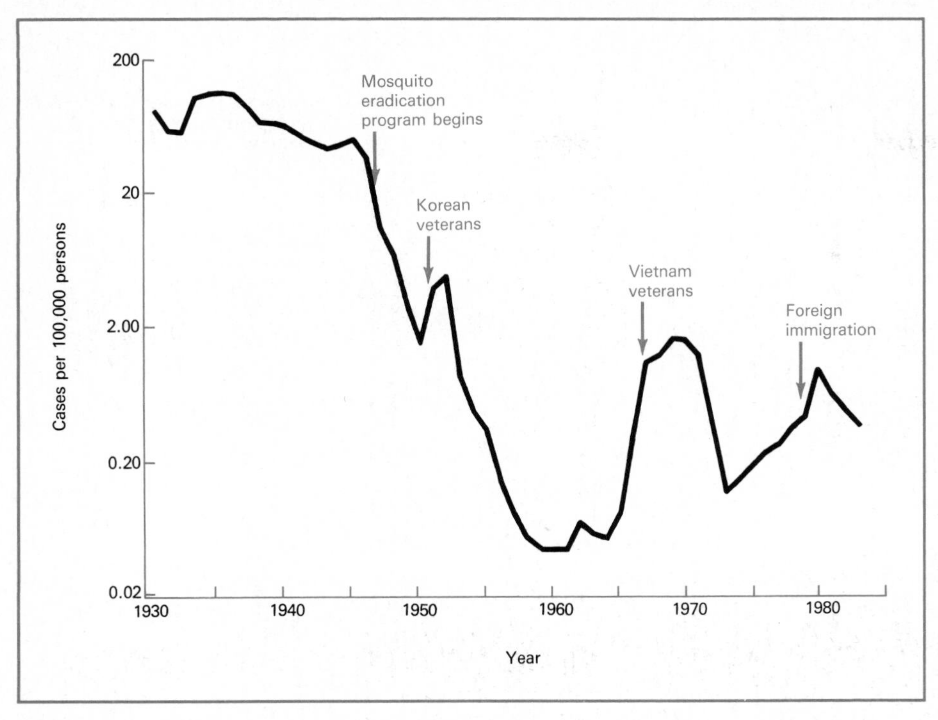

Figure 10.10 Incidence of malaria in the United States, showing the effect of insect eradication, and of travel by U.S. citizens to regions where malaria has been endemic. Note that the scale is logarithmic. Although the rises after 1950 are significant, the most important factor in control of malaria was the mosquito eradication program of the 1940s.

shows, for example, how the incidence of malaria among U.S. citizens was affected by travel into war zones in which malaria was, or still is, endemic. It is possible to be immunized against many of the diseases which are endemic in foreign countries. Recommendations for immunization of U.S. citizens travelling abroad are shown in Table 10.3. Many foreign countries currently require immunization certificates for cholera and yellow fever, but many other immunizations are only recommended for people who are expected to be at high risk. There is also risk in many parts of the world of exposure to diseases for which there is no vaccine available. These include amebiasis, dengue fever, encephalitis, giardiasis, malaria, rabies, and typhus. Travelers are advised to take reasonable precautions such as avoiding insect or animal bites, only drinking water which has been properly treated, and undergoing chemotherapeutic programs after exposure is suspected.

10.5 HOW DISEASES AFFECT HUMAN POPULATIONS

Populations are at risk of infection if a disease is endemic in a certain geographical area. Some diseases continuously affect a small percentage of the population. For other diseases, **outbreaks** occur. The outbreak may be relatively small and confined, as in the case of food poisonings. When an outbreak of disease occurs in an unusually large number of individuals in a community at the same time, the disease is said to be **epidemic**. The arrival of a new pathogen in a susceptible community is one common cause of epidemics. Figure 10.11 shows how an epidemic outbreak of the common cold occurred

TABLE 10.3 Immunizations required or recommended for U. S. travelers abroad

Disease	Destination	Recommendation
Cholera	Many central African nations, India, Pakistan, S. Korea, Albania, Malta	*Vaccination required* if entering from or continuing to endemic areas
Yellow fever	Many Central and South American, and African countries	*Vaccination often required* for entry; or if entering from or continuing to endemic areas
Plague	Mostly rural mountainous or upland areas of Africa, Asia and South America	*Vaccination recommended* if direct contact with rodents is anticipated
Infectious hepatitis (A)	Specific tropical areas and many developing countries	*Passive immunization recommended* for long stays
Serum hepatitis (B)	Africa, Indochina, eastern and southern Europe, USSR, Central and South America	*Vaccination recommended*
Typhoid fever	Many African, Asian, Central and South American countries	*Vaccination recommended*

From Health Information for International Travelers, 1984, U. S. Department of Health and Human Services.
Vaccinations are not required, but are recommended, for diphtheria, pertussis, tetanus, polio, measles, mumps, and rubella. Most U. S. citizens are already immunized through normal immunization practices (see Table 9.7).

in Spitsbergen, an isolated island community in the Arctic Ocean. Boats can reach Spitsbergen only during the summer months. In this example, the arrival of the first boat, bearing a passenger infected with a particularly virulent strain of virus, resulted in an epidemic outbreak of the common cold. Island communities are particularly susceptible to introduction of pathogens. Another example further emphasizes this point. In 1707, smallpox virus was brought to Iceland in the contaminated clothing of a student who had died from the disease while studying abroad. In the next few months, nearly all of the 50,000 Icelanders were infected, and 16,000 to 18,000 people (nearly one-third of the population) died!

When epidemics occur in many areas of the world at the same time they are called **pandemics**. Pandemic outbreaks of bubonic plague, for example, killed a quarter of the European population during the Middle Ages. In 1918–1920, about 20 million people were killed in a pandemic caused by a particularly virulent strain of influenza virus.

Factors influencing disease outbreaks Many factors may influence the development of disease outbreaks. As mentioned above, a population may be highly susceptible to a pathogen that is normally absent from the area. If the pathogen is then introduced, an explosive epidemic can occur. Eventually, however, immunity develops as infected individuals recover, and the incidence of disease declines. Likewise, when a group of susceptible individuals moves into an area where a disease is endemic, they may become infected, resulting in a small epidemic.

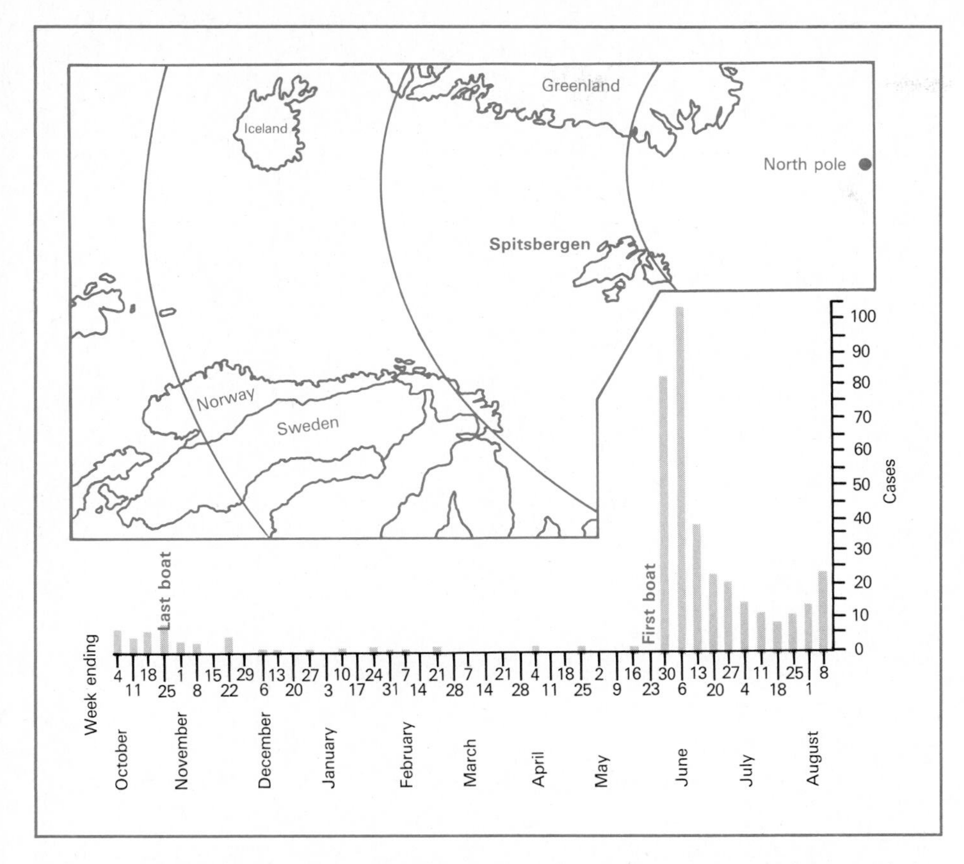

Figure 10.11 An example of an explosive epidemic caused by the introduction of a virulent pathogen into a susceptible population. The graph illustrates the incidence of colds throughout the year in an isolated community on Spitsbergen, in the Arctic Ocean. Note the sharp rise in the number of cases soon after the arrival of the first boat of the summer. The map shows the remoteness of the Spitsbergen location.

The same disease may reemerge even in an immune population, if the population is exposed to a pathogen of a different antigenic type. For example, there are over 100 antigenically different types of viruses which can cause the symptoms of common cold. Similarly, there are about 50 antigenically different types of *Streptococcus pyogenes*. This explains in part why it is possible to get diseases like the common cold and streptococcal sore throat repeatedly.

An epidemic may also develop in a previously immune population because the pathogen itself changes (by genetic mutation or recombination) to a new antigenic type. Such changes explain the emergence of new strains of influenza virus, as will be discussed in Chapter 13.

A decrease in the nonspecific resistance of a population may result in an epidemic. As we have noted, resistance to infectious disease can be influenced by the nutritional status and general well-being of an individual. Deterioration of living standards can lead to decreased resistance of a whole population and hence to increases in incidence of disease. Certain diseases arise more readily in response to such causes than others. Tuberculosis is one of the best examples of a disease whose incidence rises as living standards decline. In the past, deterioration of sanitary conditions in military camps has led to epidemic outbreaks of cholera, dysentery, and typhus, which probably were more important than weapons as causes of death among soldiers. In modern times, refugee camps are often sites of epidemics of cholera and other water-

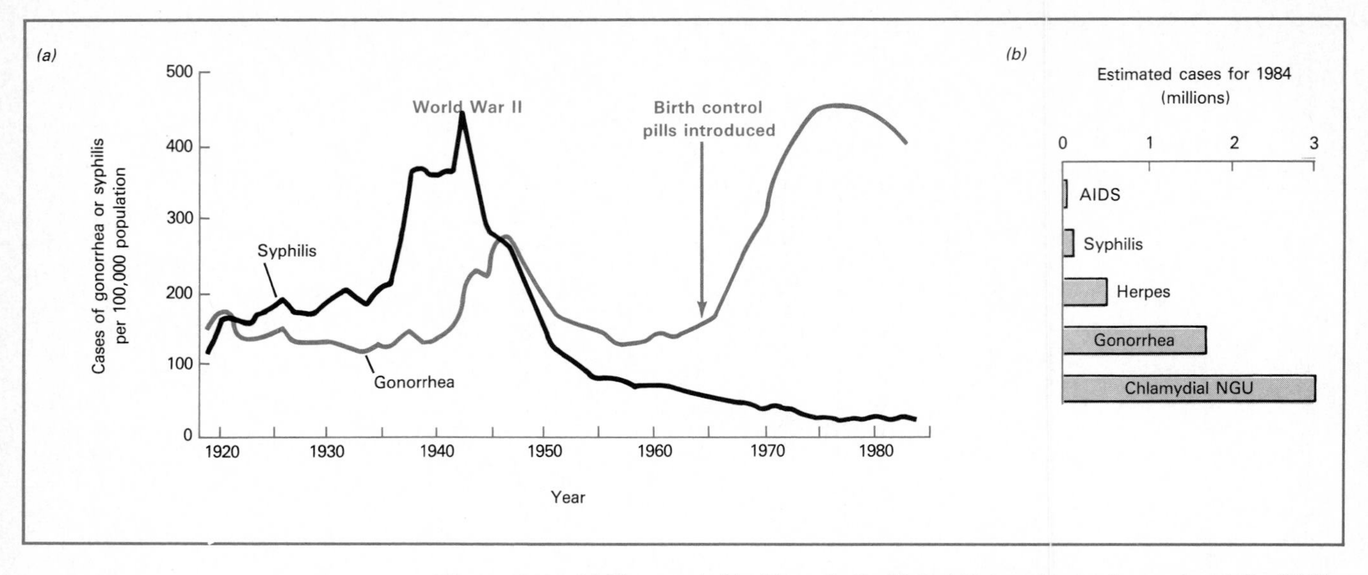

Figure 10.12 (a) The reported incidence in the United States of gonorrhea and syphilis. (b) The estimated number of actual new cases of sexually transmitted diseases during a recent year, including those which are not reportable. As seen, chlamydial nongonococcal urethritis (NGU) is probably the most common venereal disease.

borne gastrointestinal diseases. Political and military developments frequently decrease standards of living, resulting in epidemic disease.

Changes in behavior and personal habits may also lead to epidemics. Customs and habits influence the efficiency of transfer of pathogens. Incidence of sexually transmitted diseases increases drastically when sexual promiscuity increases. As shown in Figure 10.12, the incidence of gonorrhea and syphilis increased during World War II. More recent increases in sexually transmitted diseases occurred after the marketing of birth control pills.

The frequency of occurrence of intestinal infections is greatly affected by procedures for sewage disposal, water purification, hand washing, and other aspects of personal hygiene. The importance of water purification in reducing the incidence of water-borne diseases is discussed in Section 15.4. Evidence suggests that recently there has been an increase in the frequency of outbreaks of water-borne diseases (Figure 10.13). This rise may be due to increased complacency with regard to adequate water purification, especially in non-community water supplies, or to better reporting. An increase in public complacency may also be a factor in the increase in incidence of other diseases. For example, failure to follow immunization schedules predisposes a segment of the population to diseases which are otherwise effectively controlled by vaccination. A resurgence of measles during the 1970s, and on college campuses in the 1980s (see the Box on page 367), may have resulted from relaxed attitudes towards the need for immunization.

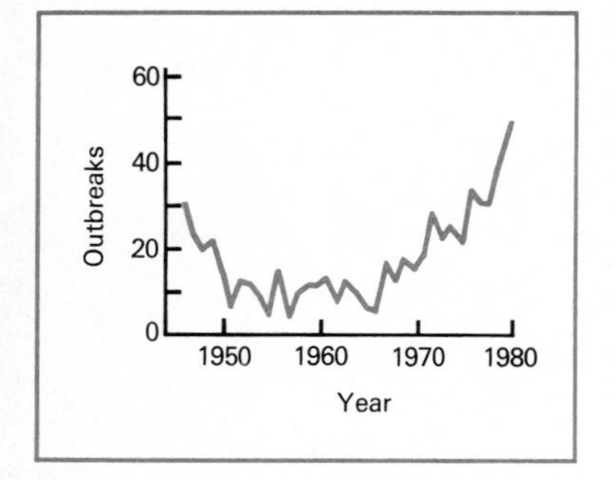

Figure 10.13 Annual occurrence of water-borne disease outbreaks in the United States, showing a rise in recent years.

Cycles of disease Certain diseases occur in cycles. For instance, acute respiratory diseases usually are more prevalent during the winter months (Figure 10.14*a*). In part the periodicity may be due to increased contact among

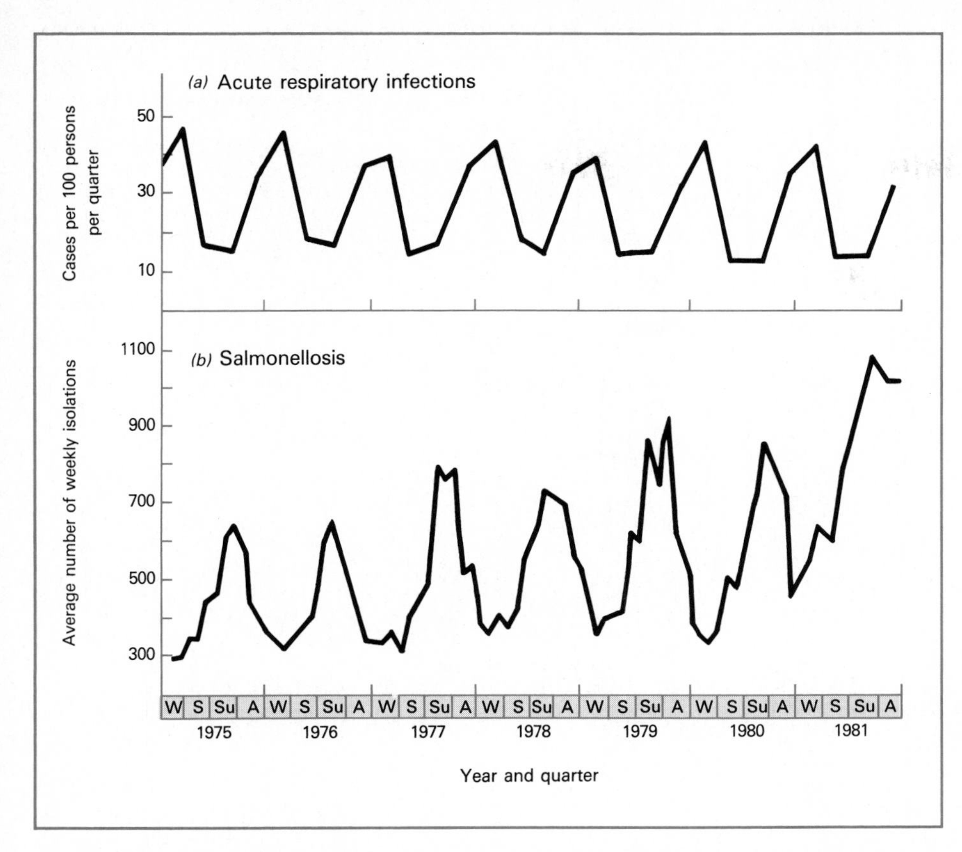

Figure 10.14 Seasonality of infectious disease. (a) Acute respiratory infections, which show a peak in late winter. (b) Salmonellosis, a water- and food-borne disease which shows a peak in late summer. (W, winter; S, spring; Su, summer; A, autumn.)

individuals during winter, since respiratory diseases are usually transmitted by airborne secretions from the respiratory tract. Diseases with a winter-spring seasonal occurrence include chickenpox and, in the past, other childhood viral diseases which have now been effectively controlled by vaccination (measles, mumps, and rubella). Epidemics of these diseases occur (or occurred) commonly among school children, in whom opportunity for respiratory transmission is high. During an epidemic, the level of immunity among school children exposed to the disease increases, so that eventually the spread of the virus is limited and the epidemic subsides. Entry of new students in lower grades who have not yet been exposed to the virus lowers the population immunity level and a new epidemic can occur in the following year.

Salmonellosis is another example of a disease which occurs in cycles (Figure 10.14*b*). This disease may be acquired from contaminated food or water. Its incidence is highest in summer months, presumably a time when environmental conditions are more favorable for the growth of *Salmonella* in foods.

10.6 PRINCIPLES OF INFECTIOUS DISEASE CONTROL

Infectious disease control can never be an individual matter but must involve participation of the entire community. The control of disease falls under the general heading of public health and preventive medicine. Because govern-

TABLE 10.4 Epidemic diseases and their control

Disease	Infective organism	Infection sources	Entry site	Method of spread	Incubation period	Disease control method
Bacillary dysentery	*Shigella*	Contaminated water and food	Gastrointestinal tract	Patients and carriers; fecal-oral route	24–48 hours	Detection and control of carriers; inspection of food handlers; decontamination of water supplies
Brucellosis	*Brucella melitensis* and related organisms	Milk or meat from infected cattle, goats and pigs	Gastrointestinal tract	Oral ingestion of infective material	6–14 days	Milk pasteurization; control of infection in animals
Diphtheria	*Coryne-bacterium diphtheriae*	Human cases and carriers; food; objects	Respiratory route	Nasal and oral secretions; respiratory droplets	2–5 days	Active immunization with diphtheria toxoid or toxin-antitoxin mixture; case quarantine; disinfection of carriers
Rubella (German measles)	Virus	Human cases	Probably respiratory	Probably respiratory droplets	10–22 days (average 18)	Patient isolation when pregnant woman is in household; rubella vaccine
Gonorrhea	*Neisseria gonorrhoeae*	Urethral and vaginal secretions	Urethral or vaginal mucosa	Sexual intercourse	3–8 days	Chemotherapy of carriers and potential contacts; case-finding and treatment of patients
Infectious hepatitis (A)	Virus	Contaminated food and water	Gastrointestinal tract	Fecal-oral route	2–6 weeks	Sanitary precautions applied to infected cases; passive immunization with antiserum
Influenza	Virus	Human cases	Respiratory tract	Respiratory	18–36 hours	Vaccine
Malaria	Protozoa, *Plasmodium vivax*	Human cases	Skin (insect blood-meal)	Mosquitoes	2 weeks	Coordinated measures for wide-scale mosquito control; prompt detection and effective treatment of cases
Measles	Virus	Human cases	Respiratory	Nasal secretions	11–14 days	Measles vaccine
Meningococcal meningitis	*Neisseria meningitidis*	Human cases and carriers	Respiratory	Respiratory droplets	Variable	Group chemotherapy with sulfadiazine (when strain is sensitive to sulfonamides)
Paratyphoid fever	*Salmonella paratyphi*	Contaminated food and water	Gastrointestinal tract	Infected urine and feces	7–24 days	Control of public water sources, food vendors, and food handlers; treatment of carriers; individual vaccination with *S. paratyphi* A and B vaccine
Pneumococcal pneumonia	*Streptococcus pneumoniae*	Human carriers; patient's own pharynx	Respiratory	Respiratory droplets	Variable	Control of upper respiratory infections; avoidance of alcoholic intoxication; communicable disease precautions applied to cases
Poliomyelitis	Poliovirus	Human cases and carriers	Gastrointestinal tract	Infected feces and pharyngeal secretions	4–7 days	Wide-scale application of poliovirus vaccine; case isolation
Rocky Mountain spotted fever	*Rickettsia rickettsii*	Infected wild rodents, dogs, wood ticks, and dog ticks	Skin	Tick bites	3–12 days	Avoidance of tick-infected areas, or wearing of protective clothing in such areas; frequent search for and prompt removal of ticks; specific vaccination

TABLE 10.4 Continued

Disease	Infective organism	Infection sources	Entry site	Method of spread	Incubation period	Disease control method
Syphilis	*Treponema pallidum*	Infected exudate or blood	External genitalia; cervix; mucosal surfaces; placenta	Sexual intercourse, contact with open lesions; blood transfusion; trans-placental inoculation	10–90 days	Case-finding by means of routine serologic testing and other methods; and adequate treatment of infected individuals
Tuberculosis	*Mycobacterium tuberculosis*	Sputum from human cases; milk from infected cows	Respiratory or gastro-intestinal mucosa	Sputum; respiratory droplets; infected milk	Variable	Early discovery and adequate treatment of active cases; milk pasteurization
Typhoid fever	*Salmonella typhi*	Contaminated food and water	Gastrointestinal tract	Infected urine and feces	5–14 days	Decontamination of water sources; milk pasteurization; individual vaccination; control of carriers
Whooping cough (pertussis)	*Hemophilus pertussis*	Human cases	Respiratory tract	Infected bronchial secretions	12–20 days	Active immunization with *H. pertussis* vaccine; case isolation

mental units vary in desire and ability to carry out public-health services, it is not possible to present a comprehensive picture of what *is* done. Instead, we provide here an overview of what *can be* done. The approaches used to control many important diseases are given in Table 10.4. It is obviously more desirable to prevent the occurrence of diseases than to respond after they occur, but both approaches are important since not all diseases can be prevented.

Responding to disease outbreaks Recognition of an outbreak begins with proper diagnosis of the disease in individuals by physicians and the reporting of the disease to some centralized body. As statistics accumulate, public-health authorities recognize that an epidemic is in progress. Notification of the public and the medical profession via newspapers, radio, and television is then carried out. In addition to treatment of those afflicted, public health officials must act to protect the population at large. The most difficult task is to locate the precise cause of the disease outbreak. By locating the source of a current outbreak, the disease can be more quickly limited, and preventive measures can be put into effect.

Importance of Epidemiology *Epidemiologists* are public-health officials who try to locate the cause of the epidemic by analyzing the way in which the disease spreads during the outbreak. Their recommendations have a significant impact on the way in which control of the disease is carried out after the epidemic has begun. In the case of a well-known disease, such as food-borne salmonellosis, the major problem may be to locate the contaminated food product, and to find out why it became contaminated. In the spring of 1985, for example, the largest recorded outbreak of food-borne salmonellosis ever to occur in the United States affected over 18,000 people in Illinois and several surrounding states. By examining case reports epidemiologists were able to pinpoint the source to milk supplied by a single dairy plant (see the Box on page 110). Removal of this milk from stores and shutdown of the plant limited the outbreak. It took much longer to determine how the milk had become contaminated.

Sometimes an epidemiologist is confronted with a new disease and must work even harder to discover its cause and way to limit its spread. About a decade ago, for example, epidemiologists had to cope with an outbreak of legionellosis for the first time (see the Box on page 84). Now, epidemiologists are trying to understand the disease AIDS, so that methods for controlling its spread can be devised.

Quarantine Quarantine involves the limitation of the freedom of movement of individuals with active infections, to prevent the spread of disease to other members of the population. The time limit of quarantine is the longest period of communicability of the disease. Quarantine must be done in such a manner that effective contact of the infected individual with healthy individuals is prevented. Quarantine is not the same as strict isolation, used for unusually infectious diseases in hospital situations (see Section 10.8).

Quarantine of ships was extensively used in attempts to control introduction of infectious diseases into the United States by immigrants. Ships arriving in New York harbor in the late nineteenth and early twentieth century were held in quarantine until it could be ascertained that no individuals suffering from cholera or typhoid fever were on board. At times, passengers were not allowed to leave their ship for many days, although the effectiveness of such measures in preventing introduction of the pathogen into the United States was uncertain. Through the 1930s, quarantine was required for a number of infectious childhood diseases, such as measles, chickenpox, and mumps: residences in which quarantined children were housed had placards affixed to the outside. Such measures were found to have little public-health significance in the control of the spread of these diseases, and quarantine is no longer required, although it is still advisable to prevent the contact of infected children with other possibly susceptible children. Currently, cholera, yellow fever, and plague are considered quarantinable diseases. These are endemic in some areas of the world (see Table 10.3), but are not a problem in the United States. Quarantine is, however, still widely used in the control of infectious disease in domestic animals (see Section 17.7).

10.7 PROCEDURES FOR PREVENTION OF INFECTIOUS DISEASE

Many diseases can be controlled before they occur through a variety of means including immunization, insect control, drug treatment, and control over the modes of transmission of the pathogen. These measures are largely the responsibility of public health officials and must be coordinated through government health agencies. However, individuals also have a significant influence on the prevention of diseases, since the transmission of many diseases is affected by individual and social behavior. Thus, education of the public is of central importance. Adults should be encouraged to maintain awareness of current public health practices. School health programs are of great value. Schools can also improve the health of individuals directly through services such as immunization programs, screening of pupils for disease, studies of physical growth, and diagnostic services.

Immunization Many diseases have been effectively controlled by vaccination of human populations. An example was shown in Figure 9.18 for the disease polio. Control of diseases through immunization requires effective vaccines (discussed in Section 9.10) as well as effective programs for administration of vaccines. In the United States, routine immunization of children (see Table 9.7) has effectively controlled diphtheria, whooping cough (pertussis), tetanus, measles, mumps, and rubella. Although the incidence of these diseases has been dramatically reduced, they are still endemic so that the population immunity level must be maintained. Outbreaks of any of these diseases can occur due to the lack of immunity in some segment of the population.

In some cases it has been possible to completely **eradicate** a disease through immunization. Rabies, for example, has been eliminated in some countries, although not in the United States. Rabies eradication in England was achieved by immunization of dogs and by strict quarantine of dogs entering the country. Eradication of rabies in England has been helped by the fact that large wild areas where rabies could remain established in rodents or other animals do not exist in England, so that infection of dogs from this source does not occur. In the United States, despite immunization of dogs, rabies is still a problem because it is endemic in wild animal populations. Eradication is considerably easier to achieve in islands (such as England) than in large continental areas because control of the movement of populations is easier.

The worldwide eradication of smallpox is perhaps the ultimate achievement in control of disease through vaccination. This disease was one of the most devastating diseases known to man, affecting hundreds of thousands of people per year, but it has now been completely eliminated due to an effective global immunization campaign sponsored by the World Health Assembly. Several features of the disease were important to the success of the eradication program. These included: (1) the lack of a significant reservoir other than humans; (2) the lack of a carrier state; (3) the recognized danger of the disease among peoples of diverse cultures; (4) the ease with which symptoms could be recognized; and, (5) the presence of a simple and effective vaccine which confers active immunity for 3–5 years. Unfortunately, most other diseases lack these features and will be much more difficult to eradicate worldwide. One important lesson learned during the eradication program was that *mass vaccination* was less effective than *surveillance and containment*, immunization

THE FALL OF SMALLPOX

Smallpox, one of the scourges of humankind, is the first major infectious disease of humans to be completely eradicated. The basis for worldwide eradication was the availability of a highly effective live virus vaccine, and the use of this vaccine throughout the world. The eradication of smallpox can be considered to be the most dramatic example of the power of immunization, and constitutes a stunning display of worldwide cooperation. The campaign began first in the developed countries, and the disease was completely eliminated from North America by the end of the 1960s. In 1966 the World Health Assembly decided to begin an intensified effort of global eradication of smallpox, coordinated by the World Health Organization. Areas such as South America, Africa, India, and Indonesia were selected as major regions in which efforts were to be concentrated. Initial attempts to *mass-vaccinate* populations were much less successful than later efforts involving a strategy called *surveillance-and-containment*. This required vaccination only of those people in contact with a diseased individual, to prevent the immediate spread of the virus. Despite interruption by cultural barriers, climate, and warfare, the areas where smallpox was endemic were reduced during the 1970s. By 1977, the countries in Africa which were the last to have smallpox cases were declared free of smallpox, and in 1980 the World Health Assembly certified that smallpox had been eradicated worldwide. Vaccinations for smallpox are now no longer required anywhere in the world.

of the people to whom the virus would next spread. The program used to eradicate smallpox is described in the Box above.

Control of vectors Insect-borne diseases and others transmitted by living vectors can be eradicated if the *vector* is eliminated. This is most effective when only a single insect species or a group of related species is responsible for the transmission of the pathogen. Eradication of the insect is most effectively achieved by the use of an insecticide, and the introduction of the insecticide DDT had great impact on eradication measures for malaria (see Figure 10.10), as well as yellow fever. Elimination of insect vectors without the use of insecticides can be achieved by draining swamps and eliminating other insect-breeding places or by the introduction of biological agents that are pathogenic for the insects themselves (*biological control*).

The *malaria* parasite is transmitted by the *Anopheles* mosquito, and elimination of this mosquito can lead to eradication of the disease. Malaria has been eliminated from the United States for many years, and because the vector is absent, infection of the population should not occur even if infected individuals enter the country (for instance, returning military personnel and immigrants). *Yellow fever* is transmitted to humans by another mosquito, *Aedes aegypti*. The elimination of this mosquito from the United States many years ago was made easier by the fact that the mosquito lives primarily in warm-climate areas so that it had a restricted habitat in this country. Complete elimination of yellow fever in jungle areas is impossible because the mosquito cannot be completely eradicated and the monkey population serves as a reservoir of infection; the virus is transmitted from monkey to monkey by mosquitoes other than *A. aegypti*.

Drug treatment In principle, drugs such as antibiotics might be used to eradicate diseases since if all infected individuals were treated, the transmission of the pathogen through the population would be quickly stopped. The problem with this approach is that the location of infected individuals is often difficult, especially if the disease is not rapidly fatal or debilitating or when the infected individual is contagious long before symptoms begin. It is also essential, of course, that a highly effective drug be available. Some success with penicillin has been achieved in eradicating *yaws*, a skin disease caused by a spirochete. Penicillin is highly effective against the pathogen, and a single dose usually leads to cure. The ease and effectiveness of penicillin treatment has made this approach feasible in the eradication of yaws. Mass treatment campaigns have been initiated in those communities where the prevalence of the disease is 10 percent or more. In such campaigns, every individual in the community is given a single large dose of penicillin. In communities where the incidence is between 5 and 10 percent, all children under 15 years of age and the contacts of infectious cases are treated. For populations with rates less than 5 percent, more selective treatment is recommended, and only immediate family and other obvious contacts of infected individuals are treated.

Eradication of yaws has been attempted in Jamaica, but it would be difficult to initiate a program in large countries, because even though the antibiotic is inexpensive, considerable expense is incurred in operating a mass treatment program. Antibiotics could not be used for eliminating virus diseases, and even with bacterial diseases there is the potential danger of the selection of antibiotic-resistant mutants. Thus, this approach to disease eradication does not have widespread utility.

Interrupting the mode of transmission The spread of water- and food-borne diseases is controlled primarily by proper sanitation. The pathogens which cause these diseases usually exit the host in fecal matter and after contaminating food or water, they enter the body through the oral cavity. Thus, these infections are often referred to as *fecal-oral route* infections. Severing of the fecal-oral connection can thus be an effective way to control these diseases. Control of water-borne disease will be covered in Chapter 15 and control of food-borne disease in Chapter 16.

Among the most difficult diseases to control are the sexually transmitted diseases. Incidence of these diseases depends greatly on the degree of sexual promiscuity in the population. Thus, these diseases are social as well as medical problems. For those sexually transmitted diseases caused by bacteria, such as gonorrhea, syphilis, and chlamydial nongonococcal urethritis, treatment is available, yet a worldwide resurgence of these diseases has occurred recently (see Figure 10.12). The responsibility for control of these diseases lies mainly with individuals who must be able to recognize symptoms and seek treatment. The only way to break the mode of transmission is to limit contact with sexual partners and alert partners that they should also seek treatment. This approach has not occurred with most sexually transmitted diseases, perhaps because treatment *is* available. The emergence of a new sexually transmitted disease, acquired immunodeficiency syndrome (AIDS), is a case in point. The high likelihood of death among those infected, and the lack of a cure for the disease, is forcing an increased appreciation of the need for personal responsibility in controlling the spread of such diseases.

10.8 HOSPITAL INFECTIONS

The hospital may not only be a place where sick people get well, it may also be a place where sick people get sicker. The fact is that cross-infection from patient to patient or from hospital personnel to patients presents a constant hazard. Hospitals are especially hazardous for the following reasons: (1) Many patients have weakened resistance to infectious disease because of their illness; (2) Hospitals must of necessity treat patients suffering from infectious disease, and these patients may be reservoirs of highly virulent pathogens; (3) The crowding of patients in rooms and wards increases the chance of cross-infection; (4) There is much movement of hospital personnel from patient to patient, increasing the probability of transfer of pathogens; (5) Many hospital procedures, such as catheterization, hypodermic injection, spinal puncture, and removal of samples of tissues or fluids for diagnosis, carry with them the risk of introducing pathogens to the patient; (6) In maternity wards of hospitals, newborn infants are unusually susceptible to certain kinds of infection; (7) Surgical procedures are a major hazard, since not only are highly susceptible parts of the body exposed to sources of contamination but the stress of surgery often diminishes the resistance of the patient to infection; (8) Many drugs used for immunosuppression (for instance, in organ transplant procedures) increase susceptibility to infection; (9) Use of antibiotics to control infection carries with it the risk of selecting antibiotic-resistant organisms, which then cannot be controlled if they cause further infection.

The broad area of hospital sanitation includes all aspects of operations within the hospital involving the cooperative efforts of many individuals at different levels of responsibility. *Asepsis*, or lack of contamination, is achieved through the interaction of several types of professional and nonprofessional personnel. Sterilization of materials of various sorts, and proper cleaning of some material for reuse, are essential. Monitoring is needed to ensure the success of efforts to achieve asepsis. Occasionally isolation of patients for their own protection, or for the protection of others, is required. Finally, proper disposal of corpses must be considered. All of these will be discussed after we consider the types of infections commonly acquired within the hospital.

Hospital pathogens Between 5 and 6 percent of the patients who enter hospitals acquire infections there, and up to 4 percent of those infections actually contribute to the death of the patient. Hospital infections are often called *nosocomial infections* (*nosocomium* is the Latin word for hospital). Hospital infections are partly due to the prevalence of diseased patients, but are largely due to the presence of pathogenic microorganisms which are selected for by the hospital environment. Hospital infections most commonly occur during surgery; newborn and pediatric services have the least problems. All parts of the body are subject to these infections, but urinary tract infections are most common. *Escherichia coli* is the most common cause of urinary tract infections in hospitals. Also, certain strains of *E. coli* are causes of epidemics of gastrointestinal infection in newborn babies in maternity wards.

One of the most important and widespread hospital pathogens is *Staphylococcus aureus*. It is most commonly associated with skin, surgical, and lower respiratory tract infections, and is a particular problem in infections acquired by newborns in the hospital. Certain strains of unusual virulence have been so widely associated with hospital infections that they are sometimes designated "hospital staphylococci." The habitat of these staphylococci is the upper respiratory tract, usually the nasal passages, and they often become established as "normal flora" in hospital personnel. In such healthy personnel the organism may cause no disease, but these symptomless carriers may be a source of infection of susceptible patients. Since staphylococci are resistant to drying, they survive for long periods on dust particles and other inanimate objects and can subsequently infect patients. Because of the potential seriousness of infection with hospital staphylococci, careful application of principles of hospital sanitation is necessary.

Pseudomonas aeruginosa is important in causing infections of the lower respiratory and urinary tracts. It is also an important cause of infections in burn patients who have lost their primary barrier to infection. *P. aeruginosa* exhibits one of the most significant features complicating the treatment of nosocomial infections, *drug resistance*. Isolates of this bacterium from patients with hospital infections are commonly resistant to many antibiotics. A somewhat lower degree of resistance has been noted among *S. aureus* isolates, whereas *E. coli* isolates are generally sensitive to antibiotics. Antibiotic-resistant pathogenic bacteria in hospitals generally contain plasmids coding for *multiple drug resistance*, as described in Chapter 7.

Hospital housekeeping practices The first level of proper hospital sanitation is the responsibility of housekeeping personnel. All surfaces including floors, walls, ceilings, windows, and toilets should be frequently cleaned with germicidal solutions to kill microorganisms. This also reduces dust, which can harbor microorganisms such as staphylococci which resist drying. Particular attention must be paid to housekeeping in high-risk areas such as isolation units, surgeries, nurseries, autopsy rooms, bathrooms, showers, and utility and treatment rooms. Bandages, dressings, surgical suturing materials, hypodermic syringes and needles, tubing, glassware, scissors and other surgical instruments, catheters, forceps, and a wide variety of other items are sterilized and used in great number. To ensure delivery of properly sterilized items to the patient's bed or surgery, hospitals usually employ a central sterile-supply department, where cleaning, packaging, and sterilization occur, although small hospitals may purchase presterilized disposable goods, which they use and discard.

There are three categories of supplies: dirty (or contaminated), clean, and sterile. Dirty items are those that have been used, generally in contact with a patient. Disposable items must be separated from reusable items and incinerated. Reusable items must first be decontaminated and cleaned, then wrapped and sterilized. An important principle is that the three categories of items must be kept carefully separated so that there is no chance of mixup and so that dirty items will not contaminate sterile ones.

Several sterilization methods are used in hospitals depending on what is to be sterilized. The most widely used method is autoclaving (see Chapter 4). It is used to sterilize forceps, scissors, and other metal items, as well as dressings and other cloth goods. There is a considerable need for *sterile water* and *sterile solutions* in the hospital. These are often purchased presterilized from pharmaceutical firms, but can also be sterilized in the hospital. Gas sterilization (also described in Chapter 4) is used to sterilize heat-sensitive items, such as rubber and plastic goods, and delicate instruments. Chemical sterilization, sometimes called cold sterilization, is used for certain items, such as thermometers, lensed instruments, polyethylene tubing, catheters, and inhalation and anaesthesia equipment. In many cases, chemical treatment is used not to *sterilize* but rather to *disinfect*, reducing the total microbial load or eliminating possible pathogenic organisms present on the material. Chemical agents used include ethyl alcohol, iodine, formaldehyde, iodophors, quaternary ammonium components, phenol, sodium hypochlorite, and glutaraldehyde. Since many of these sterilization agents are irritating or toxic, it is important to rinse treated materials that are to be used in the body in order to remove chemical residues.

During surgery, the patient is unusually susceptible to internal infection, and even organisms from the skin, not ordinarily pathogenic, may initiate infection. For this reason, surgeons and other personnel involved in surgery must carry out careful skin antisepsis. Simple washing does not kill organisms and may actually cause an increase in bacterial numbers on the surface by bringing to the surface organisms embedded in the hair follicles. The best preparations for skin antisepsis in medical practice contain hexachlorophene, iodophors, or quaternary ammonium compounds mixed with foam wetting agents. Such an antiseptic cleansing agent must act both to remove all dirt and to kill organisms. The value of handwashing procedures for reducing the microbial load on the skin is illustrated in Figure 10.15.

One means to ensure the effectiveness of hospital practices is to use culture techniques that permit the counting of the microbiological load of objects or surfaces within the hospital. Of most value is the use of microbiological sampling as an adjunct to infection-control programs, when specific hospital epidemics are being investigated. If a series of similar cases or a significant epidemic occurs in a hospital and epidemiological evidence suggests that some group of articles, such as sterile solutions, nebulizer (spray) bottles, catheters, or sterile instruments are involved in the outbreak, intensive microbiological sampling of these items would be initiated, with the aim of identifying the contaminated items responsible for the epidemic and determining how contamination is occurring. Finally, it must be emphasized that microbiological sampling is only an adjunct to an infection-control program and should never replace the establishment and implementation of careful hospital practices. A trained hospital epidemiologist can often more readily identify problem areas by visual examination than by microbiological sampling, except in very unusual situations.

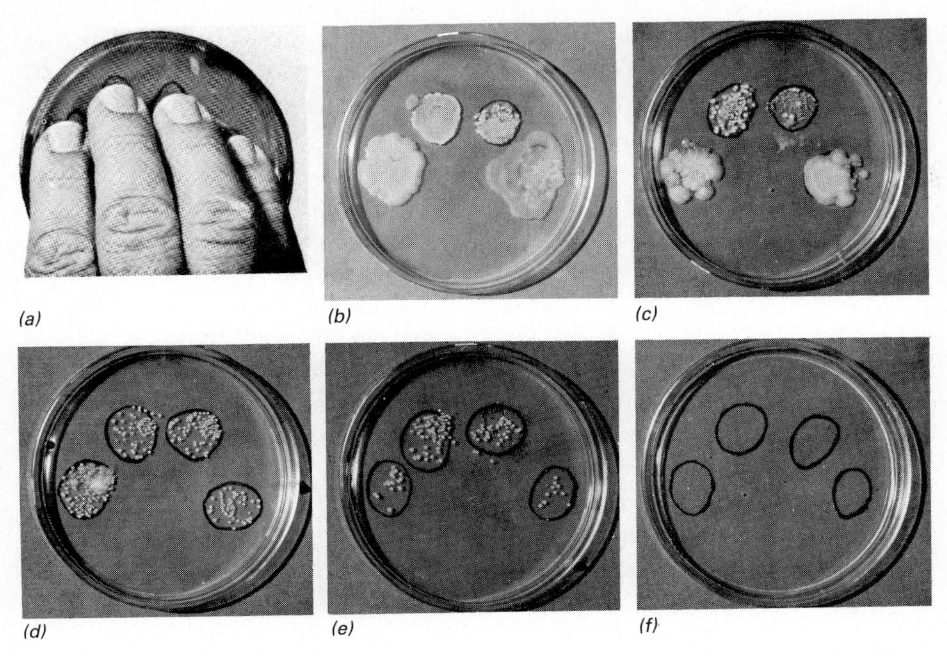

Figure 10.15 The value of handwashing procedures for reducing the microbial load on the skin. (a) Dirty fingers touch a plate of nutrient agar. (b) After 24-hours incubation. A large number of microbial colonies have grown where each finger was placed. (c) Before the fingers were placed on the agar they were given a 20-second rinse with cold water. There is little reduction in bacterial numbers. (d) After a 20-second wash with soap and water. Note considerable reduction in bacterial numbers. (e) After an additional 20-second wash with soap and water. Further reduction has occurred. (f) After use of a sanitizing solution (an iodine-containing compound). No bacterial colonies are seen. Note, however, that although there are no bacterial colonies seen in (f), the hands are not really sterile, since bacteria are embedded in the crevices of the skin. After vigorous rubbing, some of these viable bacteria would probably be brought to the surface.

Isolation of patients When there is great danger that a hospital patient may infect others, he or she may may be placed in isolation. The extent of isolation precautions depends upon the severity of the disease and its degree of communicability. The patient is placed in a private room and kept out of contact with all personnel except those specifically involved in his or her care. The room should be free of all unnecessary furniture and equipment; everything needed for the care of the patient is kept in the room, and the room has its own toilet facilities. Attendants of the patient are kept to a minimum; those present wear masks and gowns and discard them immediately after leaving the unit. All food is transferred to a tray kept permanently in the patient's room, and uneaten food or other solid waste is collected in paper bags for incineration. Bedding and towels used by the patient are placed in a special bag labeled CONTAMINATED and sterilized before routine cleaning. Feces, urine, and other body excretions may be poured directly in the toilet, but if this is not convenient they should be placed in a container with a solution of chlorinated lime and left standing for 4 hours before discarding. Books, magazines, and papers must be burned following contamination. Letters may be read or signed without contamination by placing them between sheets of newspaper or paper towels, with only the writing showing, so that the patient does not touch the letter.

There may be rare cases when even the foregoing procedures are not sufficient and the patient must be placed in a room with specially filtered air to avoid exit of pathogens into the outside air. Because of the expense of such extreme isolation, it may be preferable for the hospital to construct a special isolation unit, either in a separate building or in a separate wing, where several nursing units and the necessary laundry and cleaning facilities are available separate from the rest of the hospital. The isolation unit is entered through airlocks with positive air pressure from filtered air. Personnel change clothes within the unit and, at the end of the working day, bathe completely and discard all working clothes. Each room in the unit is designed so that it has no connecting ducts or channels to other rooms, and personnel cannot move directly from one room to another.

Reverse isolation is used to protect patients with unusual susceptibility to infection. Patients in this category include those with extensive burns, premature infants, patients with immune-deficiency diseases, patients who have received transplants and have been treated with immuno-suppressants, and patients who have received intensive radiation or corticosteroid therapy. Such patients may succumb to infections by organisms not generally considered pathogenic for normal individuals. Any source of contamination is potentially dangerous, so that the patient must be maintained in an environment where the level of asepsis is comparable to that in an operating room. All persons with suspected infection must be excluded from the room, and all personnel entering the room must wear clean caps and sterile masks and gowns. When it is necessary to touch the patient, sterile gloves must be used, and precautions are taken to ensure that all objects the patient touches are sterile. This type of protective isolation is a continuous, 24-hour responsibility and requires the use of specially trained personnel who are aware of the requirements.

Infants born prematurely are unusually susceptible to infection and may be placed in isolators similar to those used for maintaining germ-free animals (see Section 8.2). The chance of contaminating pathogens reaching the interior of such isolators is minimal, and the survival rate of such infants has been raised considerably over the past several years.

Isolation procedures vary with the nature of the infection, and specific recommendations have been made by the U.S. Public Health Service. Special self-adhesive cards are available that can be affixed to doors of patient's rooms, to call attention to the isolation procedures in force. An example is illustrated in Figure 10.16.

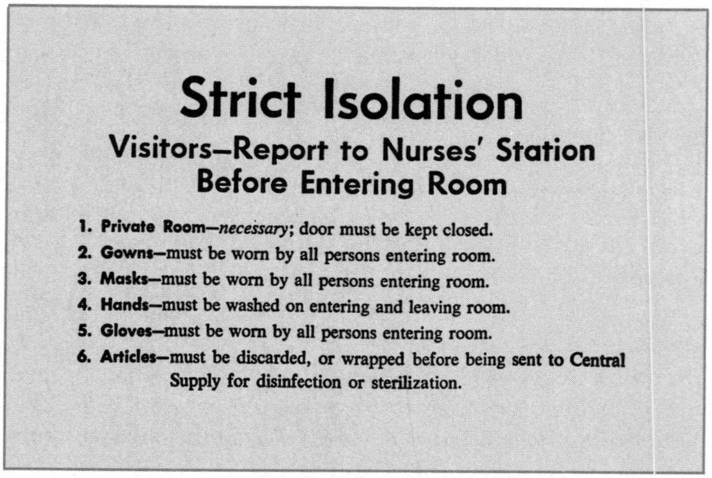

Figure 10.16 An example of an isolation notice which would be fixed to the door of a patient's room, as needed.

Strict Isolation

Visitors—Report to Nurses' Station Before Entering Room

1. **Private Room**—*necessary*; door must be kept closed.
2. **Gowns**—must be worn by all persons entering room.
3. **Masks**—must be worn by all persons entering room.
4. **Hands**—must be washed on entering and leaving room.
5. **Gloves**—must be worn by all persons entering room.
6. **Articles**—must be discarded, or wrapped before being sent to **Central** Supply for disinfection or sterilization.

10.9 EMBALMING AND DISPOSAL OF THE DEAD

Although procedures for disposing of the dead were based originally on superstitious customs or religious practices, there are good scientific reasons for safe and sanitary disposal procedures. Corpses may be important sources of infectious organisms and must be treated to make them sanitary. In addition, corpses rapidly putrefy, and if they are not buried immediately, they must be properly preserved.

The living body has tremendous powers to resist bacterial growth, but once death occurs, resistance is quickly lost, and bacteria that were not harmful in the living body multiply and spread rapidly. Even if death has been due to bacterial infection, it is not these pathogenic bacteria that attack the corpse but putrefactive bacteria found in the normal flora of the body or in the environment. These bacteria produce gases such as hydrogen sulfide, hydrogen phosphide, mercaptans, methane, carbon dioxide, ammonia, and hydrogen. Several of these gases are quite odoriferous, so that the putrefying body may exude a powerful stench.

Embalming is the process by which the dead body is treated to preserve and disinfect it. There are two reasons for embalming: (1) to preserve the body for a funeral; (2) to disinfect the body. Embalming fluid is usually injected into the body through the arterial system, generally under considerable pressure in order to force it into all parts of the body. The active ingredient of the embalming solution is formaldehyde, which is a potent antimicrobial agent and one that kills rapidly on contact. Various additives are often put in the aqueous formaldehyde solution, including anticoagulants, wetting agents, dyes, humectants (to help retain moisture in the tissues), and perfumes.

If an embalmed corpse is not desired, a quicker, less expensive, and more sanitary way of disposing of a body is by cremation. The corpse, in a wooden coffin, is placed in the cremation chamber, which has been previously heated to 600 to 700°C. The body and coffin serve as additional fuel, and the temperature rises to about 1000°C; the temperature is maintained at this level until combustion is completed, usually about 1 to 2 hours. The cremation chamber should be so designed that no smoke issues, since rising particles might carry with them live bacteria. In the fire, all microbes are destroyed rapidly, and all that remains after cremation is the mineralized body skeleton, which is broken up into pieces and crushed to a fine ash with a mortar and pestle. The resulting ash, weighing about 5 lb, is placed in a suitable urn for burial or disposal.

SUMMARY

Pathogenic microorganisms may be contracted from many different **reservoirs**, including human **carriers**. Diseases spread via many different **modes of transmission**, such as through air, water, food, on inanimate objects, through close personal contact, or via animals. Though a disease may be **endemic** within a population, it becomes of greater concern when **epidemic** or **pandemic** outbreaks occur. **Epidemiologists** respond to *outbreaks* of disease by determining the cause and mode of transmission, and by designing means to control and limit the outbreak to ensure public health. Diseases may also be controlled before they occur by immunization programs, eradication of insect **vectors**, water treatment, food surveillance, and modification of social behavior.

Partly because of effective control measures, the current health status in the United States is good with respect to infectious diseases; many infectious diseases which were once leading causes of death are now less important. **Mortality** is now mainly due to cancer and heart disease. Infectious diseases, such as influenza and the common cold, however, still account for most of the **morbidity** in the United States. **Sexually transmitted diseases** are among the leading **reportable diseases**, but because they are frequently not reported, their true significance as a leading public health problem is even higher. The health status in developed countries is far better than in developing regions of the world, where infectious diseases account for 30–50 percent of deaths, and infant mortality due to infectious disease is very high.

Control of the spread of disease within hospitals is particularly difficult. In fact, many people acquire infections while in hospitals (sometimes called *nosocomial* infections). Careful housekeeping, and well organized systems for distribution of sterile materials, collection and disposal of contaminated materials, isolation of patients requiring special attention, and monitoring of microbial contamination, are required to minimize the risk of infection. Proper embalming of the dead also reduces the risk of spread of disease.

KEY WORDS AND CONCEPTS

Reservoir
Carrier
Zoonosis
Vector
Airborne transmission
Water-borne transmission
Food-borne transmission
Sexually transmitted
Venereal
Fecal/oral route
Mortality
Morbidity
Reportable disease
Endemic
Outbreak
Epidemic
Pandemic
Epidemiology
Hospital (nosocomial) infection

STUDY QUESTIONS

1. Describe four ways in which microorganisms may leave the human body and be transferred to another person.
2. Name an inanimate and an animate *reservoir* for pathogenic microorganisms.

3. What is a *carrier*? Why are carriers important in public health?
4. Give examples of diseases transmitted via the airborne, water-borne, and food-borne routes.
5. Syphilis and gonorrhea are classic examples of *sexually transmitted* diseases. Name two other *venereal diseases* which are important in the United States today.
6. Name a disease transmitted via the *fecal/oral* route. How might the transmission of this disease be prevented?
7. Give an example of a *zoonosis* which is transmitted directly to humans. Give an example of a zoonosis which is transmitted to humans via a *vector*.
8. What are the leading causes of *mortality* today in developed countries? In developing countries?
9. What are the leading causes of *morbidity* in the United States today?
10. Why isn't the incidence of *reportable disease* necessarily a valid measure of the true incidence of disease in a population?
11. What governmental agency in the United States is responsible for compiling statistics on the incidence of disease?
12. Name a disease which is of major importance in the world today, but which is virtually absent from the United States.
13. Name three diseases which are not important in the United States, but which are *endemic* in other countries. What precautions should the foreign traveler take?
14. What is the difference between an *outbreak*, an *epidemic*, and a *pandemic*? Give an example of each.
15. Name a disease which exhibits cycles. Why does the incidence of such a disease show a peak in one season and not another?
16. List three general ways by which any infectious disease may be controlled.
17. What are the features of smallpox which made it possible to *eradicate* the disease worldwide.
18. Name a disease which has been controlled by eliminating the *vector* by which it is transmitted, rather than by elimination of the pathogen itself.
19. How can the mode of transmission be interrupted in water-borne diseases? In sexually transmitted diseases?
20. Why is the risk of infection especially great in hospitals.
21. What characteristic of the microorganisms that cause hospital infection makes treatment difficult?
22. What is the difference between *isolation* and *reverse isolation*? Give an example of a situation requiring each type of isolation.
23. What microbiological reasons are there for disposing of the dead by cremation or embalming and burial?

SUGGESTED READINGS

AMERICAN HOSPITAL ASSOCIATION. 1979. *Infection control in the hospital, 4th edition*. Chicago, Illinois. (An excellent brief textbook on the principles of control of hospital infection.)

BENENSON, A.S. (editor). 1985. *Control of communicable diseases in man, 14th edition*. American Public Health Association. Washington, D.C.

CENTERS FOR DISEASE CONTROL. *Morbidity and mortality weekly report*. United States Department of Health and Human Services. Atlanta, Georgia. (This weekly publication reports statistics on notifiable diseases.)

HOPKINS, D.R. 1983. *Princes and peasants, smallpox in history*. The University of Chicago Press. Chicago, Illinois. (An excellent book on the history of smallpox throughout the world, including its eventual eradication.)

WORLD HEALTH ORGANIZATION. 1980. *The global eradication of smallpox*. Geneva, Switzerland. (The final report of the global commission for certification of smallpox eradication.)

11

Bacterial Diseases

In the last three chapters we have considered general aspects of the infectious disease process within individuals and populations. In this and the next two chapters we will consider individual diseases caused by microorganisms. For each disease, we focus on the causal organism, mode of transmission, manner of production of disease symptoms, diagnostic procedures for the disease, chemotherapy, and epidemiology. Procedures used for culturing and identifying the pathogen are also given where appropriate. These chapters can be read straight through to give some idea of the nature and extent of infectious diseases, or they can be used as a reference when information about a specific disease is sought.

11.1 STREPTOCOCCUS

Members of the genus *Streptococcus* are Gram-positive cocci arranged in chains (Figure 11.1*a*). Streptococci are of considerable practical importance to human beings. As producers of lactic acid, certain streptococci play important roles in the production of yogurt, buttermilk, silage, and other fermented products. However, some members are pathogenic to humans and animals. Before antimicrobial agents were available, streptococcal infections constituted one of

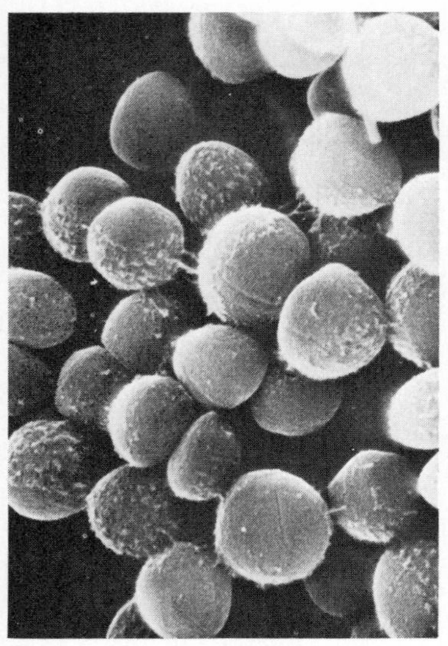

(a)

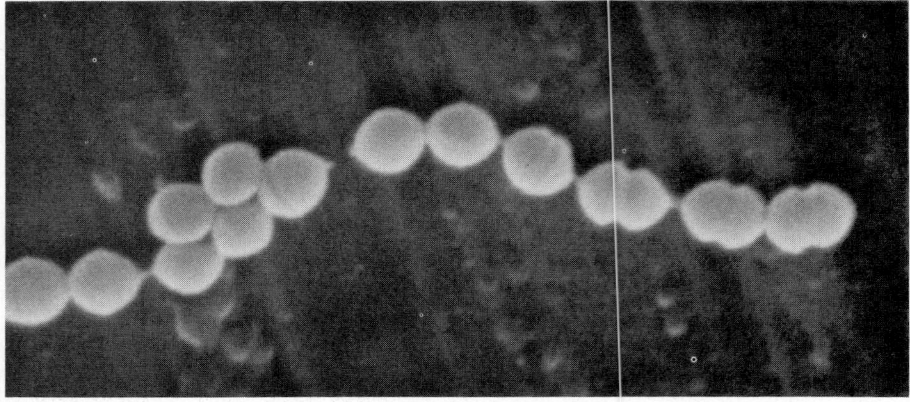

(b)

Figure 11.1 Pathogenic cocci, as viewed in the scanning electron microscope. (a) *Staphylococcus.* Magnification, 11,000 ×. (b) *Streptococcus.* Magnification, 13,500 ×.

the leading causes of death. Even today, one of the ten leading causes of death is pneumonia, which is caused primarily by *S. pneumoniae.*

Pathogenic species of the genus *Streptococcus* cause hemolysis on blood agar, owing to formation of a toxin (see Figure 8.19). Most streptococci are divided into antigenic groups based on the presence of specific cell-wall polysaccharides. These antigenic groups are designated by letters: groups A through S are recognized, and those hemolytic streptococci found in human beings usually contain the group A antigen. These group A streptococci are called the **pyogenic** streptococci and are most frequently associated with human disease. Pyogenic means "pus-forming" and refers to the characteristic symptoms induced by these organisms when infecting the skin or peripheral areas of the body. A variety of toxins and enzymes are produced that cause destruction of phagocytic and other cells which accumulate at the site of infection, leading to the formation of pus. One product, **erythrogenic toxin**, is responsible for the characteristic rash of scarlet fever. It is interesting, however, that the gene for this toxin is present in a virus which infects streptococci. In order to cause scarlet fever, the invading streptococcus must carry this lysogenic bacteriophage or prophage.

Group A streptococci are classified into a number of *antigenic types*, based on the immunological nature of a cell-surface protein called the *M protein* (see Figure 9.1). The streptococcal M protein is associated with resistance of the organism to phagocytosis. There are more than 40 M protein types, which are given numerical designations, that is, type 1, type 12, type 14, and so on. The study of M protein types is of use in following the spread of a specific strain of streptococcus through a population during an epidemic. Some types are often associated with a particular disease condition; for instance, glomerular nephritis, a disease of the kidney, can result as a secondary complication after infection with *S. pyogenes* type 12.

A wide variety of diseases are associated with streptococcal infection. These include mastitis (a disease of the mammary glands), impetigo (a superficial skin infection), peritonitis, streptococcal sore throat, pneumonia, puerperal sepsis (a disease of the uterus following childbirth), erysipelas (a generalized infection of the body, sometimes called "blood poisoning"), glomerular nephritis, and rheumatic fever. In most streptococcal infections, the causal organism is transmitted through the air, and infection is initiated in the upper respiratory tract. The interaction between *S. pyogenes* and the host during streptococcal sore throat, for example, was described in Section 8.6. If unchecked, the organism may spread from the throat to other parts of the body, where more generalized infections can begin. Individuals vary greatly in their susceptibility to streptococcal infection. Many people are symptomless carriers of group A streptococci in their upper respiratory tracts and may serve as reservoirs for the infection of more susceptible individuals. Virtually all strains of group A streptococci are highly sensitive to penicillin and other antibiotics active against Gram-positive bacteria, and resistant strains rarely develop; therefore, most acute streptococcal infections may be readily treated. Before the availability of penicillin however, group A streptococcal diseases were among the most frequent causes of death in humans (see the Box on page 212).

Streptococcal sore throat

The one streptococcal disease that is still of major concern is rheumatic fever. As described in Section 9.8, this is an autoimmune disease which results when the body's immune response to streptococci also damages the host. Although treatment of the symptoms of rheumatic fever is not possible, control of the disease can be effected by continuous administration of penicillin to susceptible individuals, which hinders the initiation of minor respiratory infections and thus prevents the onset of the autoimmune reaction. Another such autoimmune complication is glomerular nephritis.

Rheumatic fever

Streptococcus mutans and *S. sanguis* have been implicated as prime causal agents of dental caries (tooth decay), as was described in Section 8.1. Susceptibility to tooth decay varies greatly among individuals and is affected by inherent traits in the individual, dental hygiene, and diet. Ability to cause tooth decay (cariogenicity) is correlated with the ability of these streptococci to produce an extracellular slime or gum that enables them to adhere firmly to the tooth surface. One component of the slime material is called *dextran* and is produced only from sucrose (table sugar); this may be one reason that sucrose is especially cariogenic. The organisms adhere to the tooth surface, especially in crevices where they are difficult to remove by brushing, and convert sugars to acids, which cause decalcification of the hard dental tissue of the tooth.

Dental caries

The pneumococci (*Streptococcus pneumoniae*) are a distinct group of streptococci, differing from the other streptococci in that they tend to lyse readily, either spontaneously in the late exponential phase or after the addition of bile salts (hence, the pneumococci are said to be "bile-soluble"). These bacteria are the most frequent causal agents of pneumonia, although pneumonia can also be caused by other bacteria, such as *Staphylococcus aureus, Haemophilus influenzae*, and *Mycoplasma pneumoniae*, as well as by viruses. Pneumococcal pneumonia often develops as a secondary infection following other respiratory illnesses, such as influenza. The bacteria invade and spread in alveolar tissues of the lung, and the host produces an inflammatory response. The resultant increase in phagocytes leads to blockage of major sections of the lung. A capsule is important in protecting *S. pneumoniae* from phagocytosis. In later stages, macrophages are important in eliminating the pathogen and bringing the infection under control. If untreated, however, the fatality rate for strep-

Pneumonia

tococcal pneumonia is high (about 30 percent). Antibiotic therapy is important in decreasing the chance of death, and as with other streptococcal infections, penicillin is the drug of choice. However, the discovery of penicillin-resistant and multiple drug-resistant *S. pneumoniae* points out that penicillin should be used judiciously. *S. pneumoniae* is also an important causative agent for other important diseases, such as otitis media (an infection of the middle ear, common among children) and bacterial meningitis.

11.2 STAPHYLOCOCCUS

The genus *Staphylococcus* contains common parasites of human and animals, including some that are occasionally serious pathogens. Staphylococci are Gram-positive cocci which divide in several planes and hence form irregular clumps (Figure 11.1*b*). Staphylococci are relatively resistant to drying and hence can be readily dispersed in dust particles through the air. They are able to grow on media containing a high concentration (7.5 percent) of sodium chloride, and this property is exploited in their selective isolation from natural materials, since very few other bacteria from nonsaline environments show such tolerance.

Boils, pimples, and impetigo

In humans, two species are important: *S. epidermidis*, a nonpigmented, nonpathogenic form usually found on the skin or mucous membranes; and *S. aureus*, a yellow-pigmented form associated with pathological conditions, including boils, pimples, and impetigo, but also pneumonia, osteomyelitis, carditis, meningitis, and arthritis. Those strains of *S. aureus* most frequently causing human disease produce a number of extracellular enzymes or toxins. At least four different hemolysins have been recognized, a single strain often being capable of producing more than one hemolysin, and the production of these is responsible for the hemolysis seen around colonies on blood agar plates.

Another substance produced is **coagulase**, an enzymelike factor that causes fibrin to coagulate and form a clot (see Section 8.5). The production of coagulase is generally associated with pathogenicity. It seems likely that blood clotting induced by coagulase results in the accumulation of fibrin around the cocci and thus renders them resistant to phagocytosis. In addition, the formation of such fibrin clots results in the walling off of the infected area, making it difficult for host defense agents to come into contact with the bacteria (Figure 11.2). Strains are tested for coagulase production by mixing 0.5 ml of a dense suspension of bacteria with 5 ml of human plasma and incubating at 37°C. A coagulase-positive strain will usually cause clotting after a few hours incubation. Most *S. aureus* strains also produce **leukocidin**, a complex factor that causes the destruction of leukocytes, allowing the *S. aureus* cells to escape unharmed. Production of leukocidin in skin lesions such as boils and pimples results in much cell destruction and is one of the factors responsible for pus formation. Other extracellular factors produced by some strains include proteolytic enzymes, hyaluronidase, fibrinolysin, lipase, ribonuclease, and deoxyribonuclease.

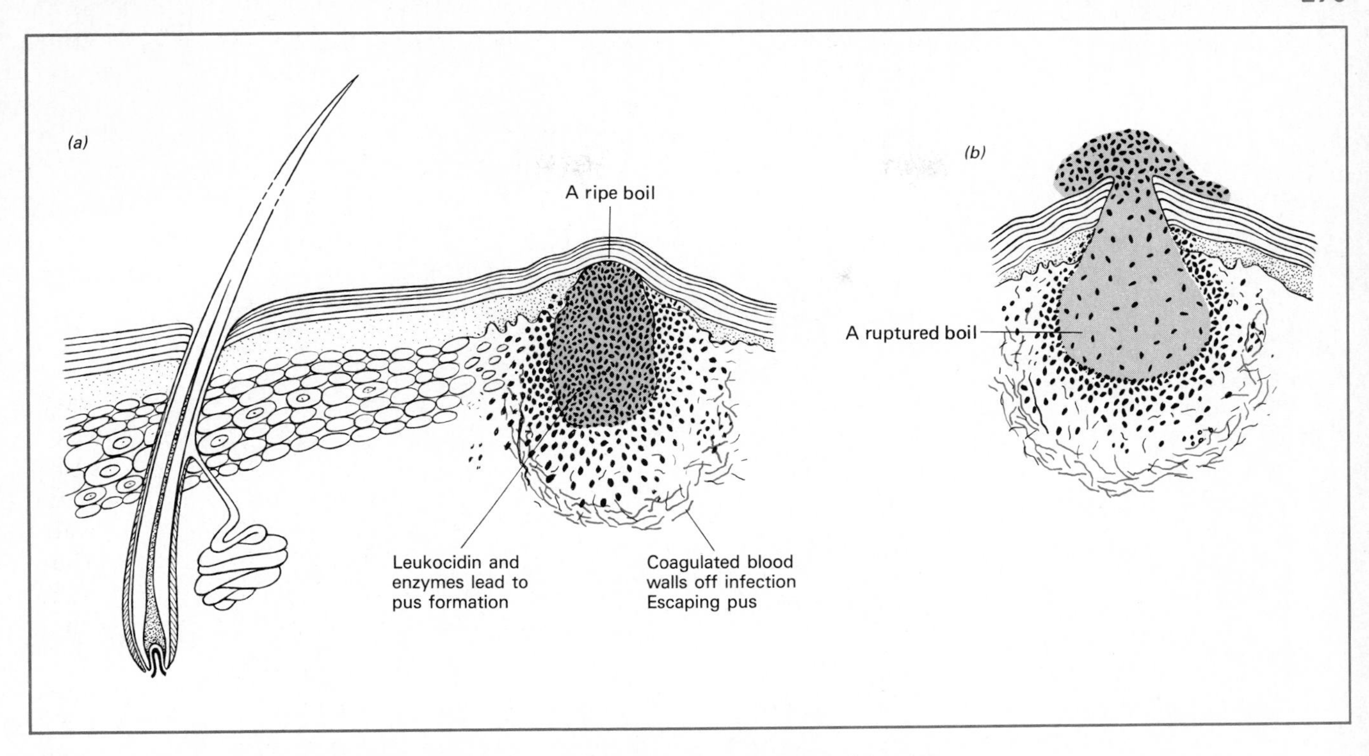

Figure 11.2 The structure of a boil. (a) Staphylococci initiate a localized infection of the skin and become walled off by coagulated blood. (b) The rupture of the boil releases pus and bacteria.

The most common habitat of *S. aureus* is the upper respiratory tract, especially the nose and throat, and many people are carriers throughout most of their lives; however, the strain being carried probably varies from time to time. Most infants become infected during the first week of life and usually acquire the strain associated with the mother or with another close human contact. In most cases, these strains do not cause pathological conditions, and serious staphylococcal infections occur only when the resistance of the host is low because of hormonal changes, debilitating illness, or treatment with steroids or other anti-inflammatory drugs. Hospital epidemics have occurred in recent years (see Section 10.8), which have usually involved antibiotic-resistant strains. Extensive use of antibiotics has often resulted in the selection of resistant strains of *S. aureus*, and hospital epidemics with antibiotic-resistant staphylococci may occur when patients whose resistance to infection is lowered (owing to other diseases, surgical procedures, or drug therapy) receive contamination from hospital personnel who are normal carriers of antibiotic-resistant staphylococci. Control of such hospital epidemics requires careful attention to the maintenance of asepsis.

Hospital infections

Toxic shock syndrome, another disease caused by *Staphylococcus aureus*, appeared suddenly in the late 1970s, peaked in 1980, and has been decreasing since (Figure 11.3). It almost exclusively affects women 15–34 years in age, and usually occurs during menstruation. In 99 percent of the cases in menstruating women, tampons were used. Specific brands of tampons

Toxic shock syndrome

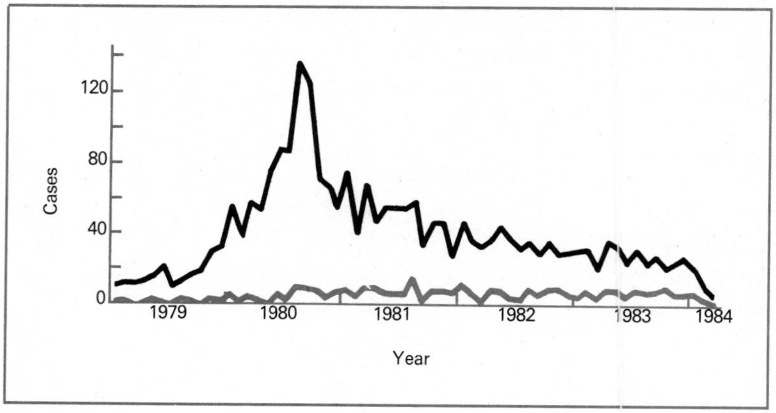

Figure 11.3 Changes in the incidence of toxic shock syndrome over a 6-year period. All cases occurred during menstruation except those shown by the colored line.

associated with the occurrence of toxic shock syndrome have been removed from the market. The development of toxicity occurred apparently because certain highly absorbent brands of tampons promoted the growth of toxin-producing staphylococci, the toxin spreading from the tampon where this staphylococci had grown. The symptoms are fever, low blood pressure, diarrhea, and a rash. Initially about 10 percent of patients with toxic shock syndrome died; now the fatality rate is about 2 percent.

Many strains of *S. aureus* produce a toxin which causes the most common type of food-borne gastrointestinal disorder (discussed in Section 16.2).

11.3 NEISSERIA

The genus *Neisseria* consists of Gram-negative aerobic cocci that live in human beings. They are nutritionally highly specialized, grow well only near body temperature, and are extremely sensitive to inhibitory materials. Two species are pathogenic for humans: *Neisseria gonorrhoeae*, the causal agent of gonorrhea; and *N. meningitidis*, a frequent cause of spinal meningitis. These two organisms are often given the colloquial names of *gonococcus* and *meningococcus*, respectively.

Gonorrhea

Gonorrhea is a sexually transmitted disease that apparently occurs only in human beings; experimental animals have not been successfully infected. Gonorrhea is one of the most widespread human diseases, and is in fact, the most common notifiable disease in the United States (see Figure 10.6). In contrast to syphilis (another important sexually transmitted disease of humans discussed later), gonococcus infection only occasionally results in serious complications or death, but frequently results in sterility in females. The disease symptoms are quite different in the male and female. In the female, the symptoms are usually a mild vaginitis that is difficult to distinguish from vaginal infections caused by other organisms; the infection may easily go unnoticed. In the male, however, the organism causes a painful infection of the urethral canal. In addition the organism can cause eye infections which may lead to blindness in the newborn or adult.

The causal agent, *N. gonorrhoeae,* is quite difficult to culture on initial isolation from pathological material. Many components of ordinary culture media are inhibitory, but starch, serum, or heated whole blood, when added to culture media, adsorbs these toxic materials and makes growth possible. On initial isolation, most strains require an atmosphere containing 2 to 10 percent CO_2, and their temperature limits are very narrow; growth does not occur below 30°C or above 38.5°C. The organism is killed quite rapidly by drying, sunlight, and ultraviolet light. This extreme sensitivity probably explains in part the sexually transmitted nature of the disease; that is, the organism is transmitted from person to person predominantly by intimate sexual contact. However, the eyes of infants may become infected during birth by mothers who carry the organism. Prophylactic treatment of the eyes of all newborns with silver nitrate or an antibiotic cream is required by law in most states and has helped to control the disease in infants. Before this, such eye infections were the leading cause of blindness in the United States.

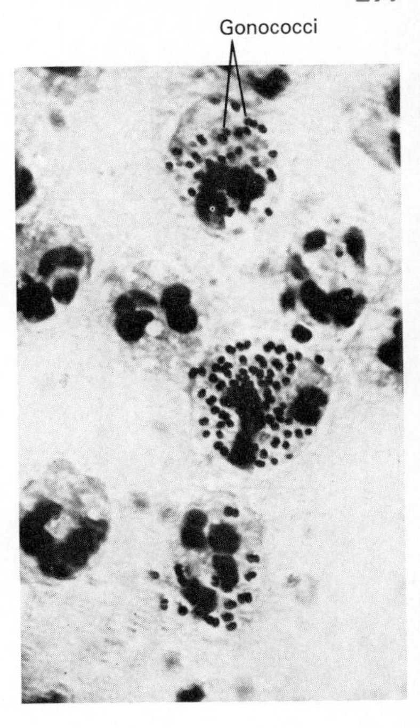

Figure 11.4 Photomicrograph of *Neisseria gonorrhoeae,* the causal agent of gonorrhea. The bacteria are present within human phagocytes. A single phagocyte is about 10 μm in diameter.

The organism usually enters the body through the epithelium of the genitourinary tract. The presence of a pilus seems to be important in attachment of the bacterium to epithelial tissue. The production of a protease which specifically degrades immunoglobulin of class IgA may also be important in helping *N. gonorrhoeae* evade this host defense and in establishing infection in the epithelium. Pili also appear to have a role in preventing phagocytosis, and in fact, *N. gonorrhoeae* are able to grow within phagocytic cells, as illustrated in Figure 11.4. The toxicity of the organism is probably due exclusively to an endotoxin and no extracellular products significant in pathogenicity seem to be produced.

Treatment of the infection with penicillin is highly successful; a single injection usually results in elimination of the organism and complete cure. Strains moderately resistant to penicillin have been known since at least 1958 but had not presented any special problem since they could be controlled with higher doses and repeated injections of the antibiotic. The recommended dosage today is 4.8 million units of penicillin, contrasted to the 200,000 units which was sufficient before resistant strains arose. More seriously, new strains of gonococci have recently been identified which are resistant to even higher doses of penicillin. For this new class of strains, no amount of penicillin is lethal; instead, the organism inactivates the drug by action of an enzyme called *penicillinase,* or *β-lactamase.* The number of cases caused by penicillinase-producing *N. gonorrhoeae* is small in comparison with the total number of cases of gonorrhea, but the rise in their incidence (see Figure 5.16*b*) is alarming and suggests that these "super-gonorrhea" strains may pose a most serious public-health threat. Another antibiotic, spectinomycin, can be used and is effective against these new strains, but recently spectinomycin-resistant penicillinase-producing *N. gonorrhoeae* strains have also been reported.

Despite the ease with which most gonorrhea can be cured with penicillin, the incidence of gonococcus infection remains relatively high (see Figure 10.12). The reasons for this are twofold: (1) acquired immunity does not prevent repeated infection; (2) symptoms are such that even at its height the disease may go unrecognized. Thus, a promiscuous infected person can serve as a reservoir for the infection of many other people. The rise in the incidence of gonorrhea in the past 15–20 years is thought to be due to an actual increase in the occurrence of the disease as well as to improvement in the reporting of the cases to public-health authorities. An increase in the number of people in the most infection-prone age group (20–24), the availability of birth control pills, and a general increase in sexual activity may have all contributed to this rise. The recent plateau in the number of reported cases (Figure 10.12) is probably due to a reduction in the percentage of individuals in the infection-prone age group, increased disease prevention activities on the part of health authorities (screening of high-risk populations, and providing sexual-partner referral services to patients), and education of high-risk individuals to prevent exposure and to encourage early medical care for themselves and their sexual partners.

Meningitis

Neisseria meningitidis is the second most frequent causal agent of **meningitis**. Other bacteria important as causative agents of meningitis are *Streptococcus pneumoniae* and *Haemophilus influenzae*. The natural habitat of *N. meningitidis* is the nose and throat. The organism possesses much the same fastidious nature as *N. gonorrhoeae*, and similar culture conditions are used in its isolation. Meningococcus is also sensitive to drying, heat, and other adverse environmental conditions, but it is sufficiently resistant so that transmittal from person to person via the respiratory tract occurs, although relatively close contact is required. The number of people who carry meningococcus in the upper respiratory tract is quite high; most carriers have infections with no noticeable symptoms.

In certain instances, the organism invades the bloodstream from the respiratory tract and sets up a generalized infection of the body. In some of these cases, the organism invades the central nervous system and becomes established in the meninges (the membranes surrounding the brain and spinal cord). The symptoms of meningitis are severe headache, muscular spasm, stiff neck, and exaggerated reflexes; if untreated, these are followed by convulsions, coma, and death. When these symptoms develop, the organism can usually be cultivated from samples of the cerebrospinal fluid. Invasion depends on the presence of an antiphagocytic capsule and pathogenicity is associated with the production of a characteristic endotoxin. The organism is extremely sensitive to sulfonamides, penicillin, and most other antibiotics, although drug-resistant strains are appearing with greater frequency. Successful treatment of meningitis requires the use of a drug that will penetrate the meningeal membrane, and the sulfonamides are often preferred because they possess this property. However, penicillin can also be used; although it does not normally penetrate to the spinal fluid in healthy persons, it does so when the meninges are acutely inflamed.

The incidence of meningococcal infections in the U.S. is shown in Figure 11.5. No major epidemics have occurred since 1946, but outbreaks have been fairly common in military camps and barracks, situations where overcrowding, fatigue, and exposure to carriers are found (note in Figure 11.5 the large increase in cases during World War II). Meningococcal infections are most common among infants and small children as also shown in Figure 11.5, and the possibility of outbreaks in child-care centers also exists. Vaccines are available which protect against several *N. meningitidis* strains, as well as *Hae-*

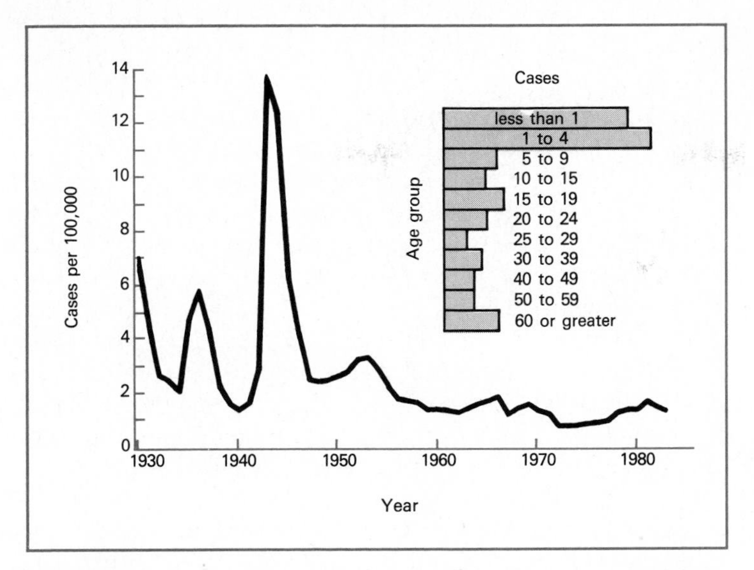

Figure 11.5 Incidence of meningococcal infection in the United States and the age distribution of cases, showing the marked prevalence of the disease in very young children.

mophilus influenzae, the most common causative agent of bacterial meningitis (see below). These may prove to be useful in controlling outbreaks.

11.4 HAEMOPHILUS

Organisms of the genus *Haemophilus* are small Gram-negative rods, characterized by a growth requirement for heme, the red pigment of blood: hence the origin of the genus name. A number of species of *Haemophilus* are known, all of which are found in association with animals or humans. The most common of these, *H. influenzae*, received its name because it was first erroneously described as the causal agent of influenza, which in reality is a viral disease. *H. influenzae* is an encapsulated organism that is invasive because the capsule prevents phagocytosis.

Meningitis

H. influenzae is the leading cause of bacterial meningitis, as mentioned in Section 11.3. A single strain causes most cases of meningitis, and a vaccine consisting of the purified capsular antigens of this strain is available. The Centers for Disease Control Immunization Practices Advisory Committee has recommended that this vaccine be used to immunize all children at age 24 months. *H. influenzae* is second in importance only to *Streptococcus pneumoniae* as a causative agent of inner ear infection (otitis media); it may also cause sore throat (pharyngitis), sinusitis, epiglottitis, and occasionally, pneumonia. The bacterium shows a high resistance to penicillin G so that the recommended drug for chemotherapy is ampicillin. However, because many strains are ampicillin resistant, chloramphenicol is often prescribed.

11.5 BORDETELLA

Pertussis (whooping cough)

An organism morphologically and physiologically similar to *Haemophilus* is *Bordetella pertussis*. *B. pertussis* is the causal agent of **pertussis** (or **whooping cough**), a disease most commonly afflicting children under the age of one

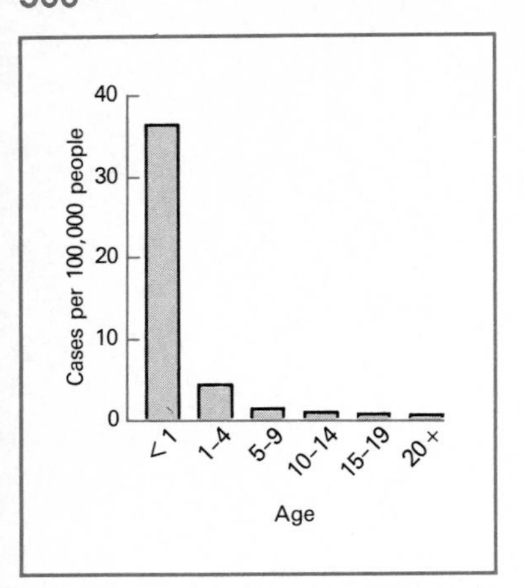

Figure 11.6 Age distribution of pertussis (whooping cough) in the United States.

year (Figure 11.6). *B. pertussis* is also encapsulated but, unlike *Haemophilus*, does not require heme as a growth factor. The organism is transmitted via the aerosol produced by violent coughing of an infected individual. The pathogen attaches to ciliated cells of the upper respiratory tract via pili. *B. pertussis* is not very invasive, but after colonizing the upper respiratory tract it produces exotoxins which bring about the major symptoms. The most characteristic symptom is the whooping sound which results from deep inhalation when the infected person is starved for air after a violent coughing episode. The microorganism is sensitive to erythromycin and other drugs so that chemotherapy is quite effective.

A vaccine against *B. pertussis* has been in use since the early 1940s. Pertussis is one of the three diseases for which protection is now provided in the DPT shot administered to children (see Table 9.7). In the United States, vaccine use has coincided with a dramatic decrease in the incidence of pertussis (Figure 11.7*b*). However, the vaccine does have side effects in a rare number of children. Concern over harmful effects associated with pertussis vaccine has resulted in ineffective immunization in Great Britain and major epidemics still occur. The important issue of risk versus benefit from vaccination is considered in the Box on page 302.

11.6 CORYNEBACTERIUM

The genus *Corynebacterium* comprises a group of aerobic, nonmotile, nonsporulating, Gram-positive rods. Corynebacterial cells often have swollen ends so that the rods appear club-shaped, hence the origin of the name (*koryne* is the Greek word for *club*). Members of the genus *Corynebacterium* are widespread: some are common inhabitants of the soil, others are causal agents of plant diseases, and still others are organisms living with or pathogenic to humans and higher animals. One species, *C. acnes*, is a common inhabitant of the human skin and has been implicated in the skin condition **acne**.

Diphtheria

The best-known and most widely studied species is *C. diphtheriae*, the causal agent of **diphtheria**. At one time, diphtheria was a major cause of death in children, but today it is quite rare. Diphtheria is of historical significance because it was the first infectious disease whose symptoms were shown to be due to an *exotoxin*, and it was also the first disease to be well controlled by immunization procedures (see Figure 11.7*a*). The organism is strictly an inhabitant of the respiratory tract, being unable to invade other parts of the body. It is transmitted from person to person by the respiratory route. Upon its establishment in the upper respiratory tract, the organism multiplies on the mucous membranes and produces a potent exotoxin that causes damage to adjacent cells and creates a favorable environment for further growth. The inflammatory response of throat tissues to infection results in the formation of a characteristic structure called a *pseudomembrane*, which consists of altered tissue cells and bacteria. This pseudomembrane can result in mechanical blockage of the throat, leading to death by suffocation. In most cases, however, death results from toxemia arising from the spread of the toxin throughout the body. Thus, even though *C. diphtheriae* is weakly invasive, its powerful

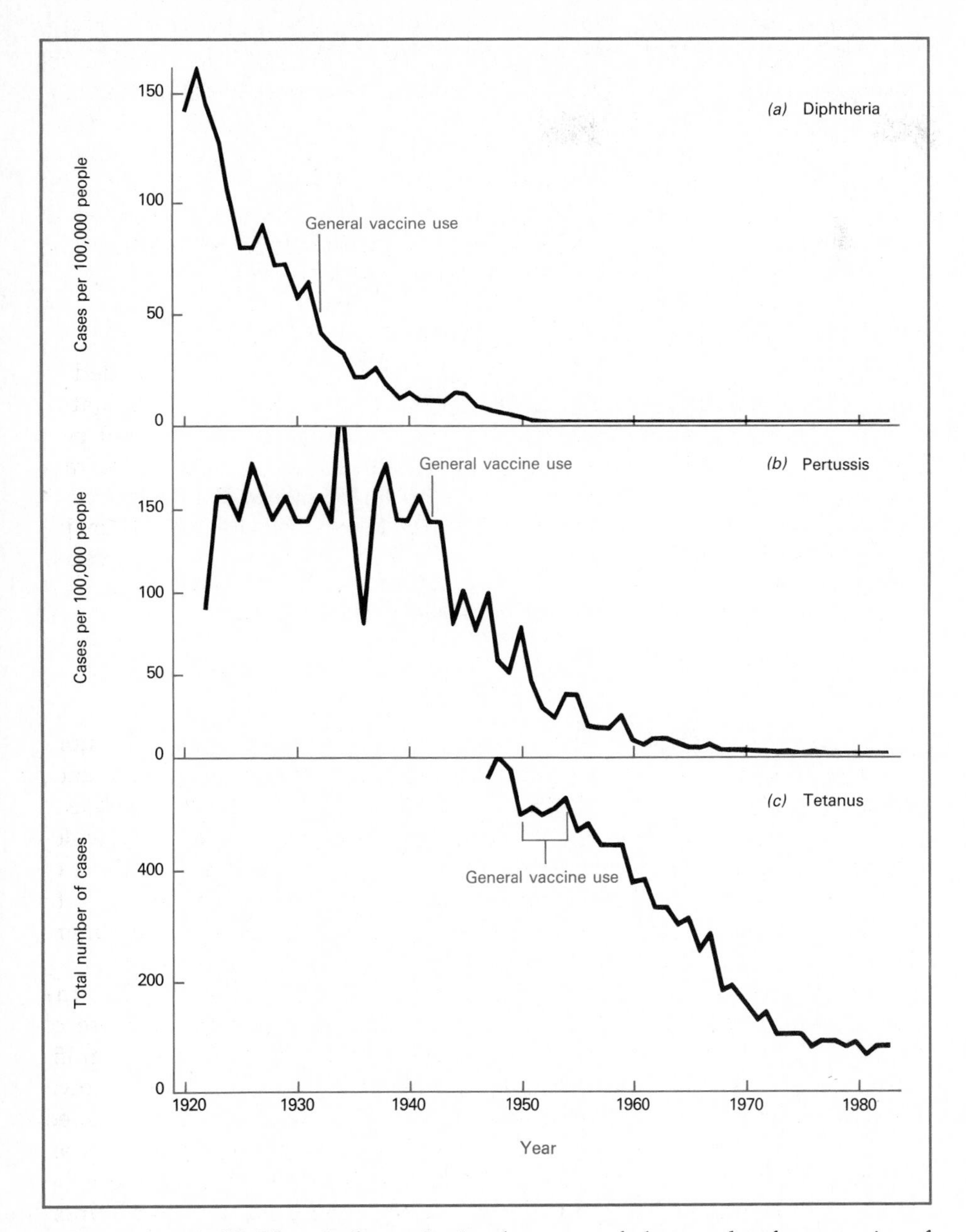

Figure 11.7 The relationship between vaccine use in the United States and disease incidence for the three diseases now controlled by the DPT vaccine. (a) Diphtheria. (b) Pertussis. (c) Tetanus.

toxin makes it highly virulent. As in the case of the scarlet fever toxin of *Streptococcus pyogenes* (see Section 11.1), the diphtheria toxin is actually carried in the genetic material of a specific prophage which can establish a lysogenic infection of *C. diphtheriae*.

Administration of formalin-treated toxin (toxoid) results in the formation of antitoxin antibodies, and infants and small children are generally given routine immunizations as a part of the DPT series (see Table 9.7). Antitoxin antibodies completely neutralize the action of the toxin but do not prevent the establishment of the organism in the upper respiratory tract. Inapparent

RISK VERSUS BENEFIT IN VACCINATION

During a recent winter a shortage of DPT vaccine in the United States caused doctors to alter their normal schedules for immunization against the important diseases, diphtheria, pertussis (whooping cough), and tetanus. This shortage resulted when two of the three major vaccine suppliers decided to discontinue production, due to fears about vaccination hazards, and when some of the vaccine supplied by the third company failed to meet manufacturer's specifications. Although the occurrence of adverse reactions associated with vaccines is rare, about one in a million, the liability expenses associated with such cases were apparently beginning to erode the profitability of vaccine production. With any vaccine there is the question of whether the benefits outweigh the risks involved in using it. In the case of pertussis vaccine, there have been debates over the frequency and severity of such adverse reactions as anaphylaxis, seizures, and encephalopathy (a neurological disorder).

What are the facts? In the United States, use of vaccines has been associated with a dramatic decline in the incidence of pertussis (Figure 11.7*b*). In 1934, for example, there were close to 300,000 cases and 8,000 deaths; in 1984 there were only about 2000 cases and very few deaths. The vaccine is reported to provide protection in 70–90 percent of those immunized. To some, proof of the vaccine's efficacy was not apparent; questions on the justification of use of the vaccine were raised, based on the severe adverse reactions which sometimes resulted. Some solid information has been obtained from observations in Great Britain. Public controversy in England and Wales had led to a decrease in vaccine use and the immunization level of English and Welsh children dropped from 76–81 percent in 1974 to 30 percent in 1978. Since pertussis affects mainly infants and small children, this left the most susceptible portion of the population without immunity. In 1977–1979 and again in 1981–1982, two whooping cough epidemics, affecting over 100,000 children, occurred in England and Wales which were almost as severe as those which had occurred before the vaccine was introduced. This unfortunate calamity showed that in the absence of vaccination many more people are seriously affected by the disease than would be injured by side-effects of vaccination.

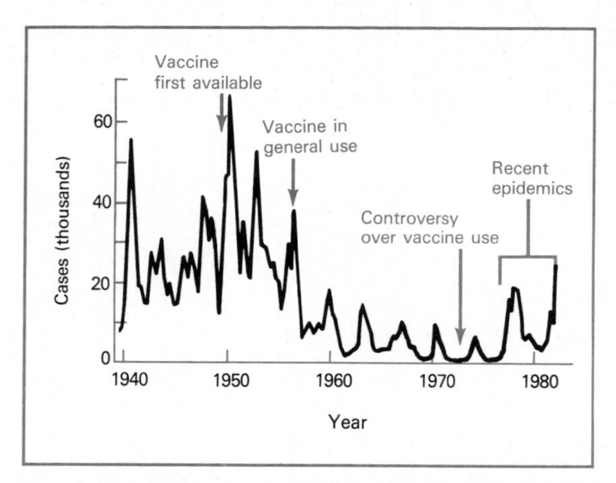

Incidence of pertussis (whooping cough) in England and Wales, showing impact of vaccine use on the disease frequency.

infections are therefore common, and carriers can be the source of infection of unimmunized individuals. Occasional small outbreaks of diphtheria can arise owing to the migration of unimmunized individuals into an area where the population is predominantly immunized but carries inapparent infections. To determine whether an individual is immune to diphtheria, the *Schick test* can be used. This involves injecting under the skin a minute amount of diphtheria toxin; if a person is immune, antibodies present will neutralize the toxin, whereas in a nonimmune individual the toxin will induce formation of an area of redness and swelling. Diphtheria therapy is best effected by passive immunization with antitoxin, since this will neutralize any circulating toxin. Antibiotic therapy is also used, but never alone, since the antibiotic will not affect toxin already circulating in the body.

11.7 CLOSTRIDIUM

The genus *Clostridium* consists of endospore-forming, Gram-positive rods (see Figure 2.16). All are obligate anaerobes but are not killed by O_2 if they exist in the endospore form. Clostridia live primarily in soil and are capable of causing disease in humans only under specialized conditions. Those causing disease include *C. tetani*, causing tetanus, *C. botulinum*, causing botulism, and *C. perfringens*, causing gas gangrene and food poisoning. These pathogenic clostridia produce specific toxins that are responsible for the disease symptoms.

Tetanus

Tetanus is a generally fatal disease, the symptoms of which are due to a potent toxin that acts on the central nervous system, causing spastic paralysis (Figure 11.8). The causal agent, *C. tetani*, frequently infects dirty wounds, especially deep puncture wounds where extensive tissue damage results in anaerobic conditions. It remains localized at this initial site of infection, but

Figure 11.8 A soldier dying of tetanus. Photograph of a painting by the famous Scottish anatomist, Sir Charles Bell. (Original painting in the Royal College of Surgeons, Edinburgh.)

the toxin can spread to distant parts of the body. Once the toxin combines with the nerves, its action cannot be reversed. Hence, treatment of tetanus is difficult, and may require passive immunization with antitoxin.

Prevention of tetanus is by routine immunization of children with toxoid. The benefits of the tetanus toxoid vaccine were demonstrated by the protection it provided to U.S. soldiers during World War II. Battle wounds are frequently infected with *C. tetani*, but the incidence of tetanus was very low in World War II. Since the incidence of tetanus in the general public is very low, routine vaccination began only when it was found that the toxoid could be combined with diphtheria and pertussis vaccines as a part of the DPT series (see Table 9.7). There has been a consistent decline in tetanus incidence correlating with the use of this vaccine (Figure 11.7*c*).

Gas gangrene

In **gas gangrene** the causal agent, *C. perfringens*, also infects wounds, but it is more invasive than *C. tetani*, primarily because of a series of toxins it produces that cause extensive destruction of tissue. (The term *gangrene* refers to dead tissue, and the gangrenous condition may result from either microbial or nonmicrobial causes.) During *C. perfringens* infection, the dead tissue frequently contains gaseous products of the bacterium's activities, and the infected region may become filled with liquid or gas. In such dead tissue, anaerobic conditions develop, permitting the organism to grow more extensively. Some of the toxins also spread to distant parts of the body and cause destruction of heart tissue and other internal organs, leading to death. Immunization against all the toxins is not possible, so that control is mainly by preventing the spread of the pathogen through use of antibiotics and surgery. If the infection is in an arm or leg, amputation of the portion of the member that is diseased may eliminate the pathogen from the body. Gas gangrene develops most frequently as a result of infection of large and very dirty wounds. Infection can be prevented by *immediate* treatment of such wounds by surgical cleaning and antibiotics.

C. perfringens is also a common cause of *food poisoning*. Although it causes a much milder food poisoning than that caused by *C. botulinum* (see below), it is much more frequently encountered, ranking second only to *Staphylococcus aureus* as a cause of food-poisoning outbreaks in the United States. Food-poisoning strains produce a gastroenteritis characterized by abdominal distress followed by diarrhea. The disease is fairly mild and self-limiting, usually of one day's duration, with onset about 12 hours following ingestion of heavily contaminated food. Food poisoning is considered in greater detail in Chapter 16.

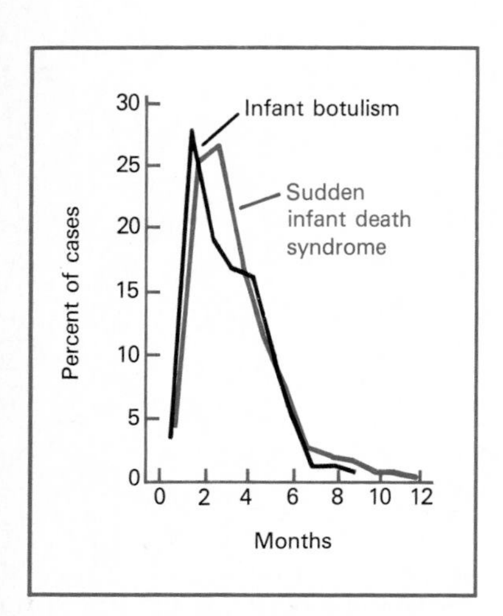

Figure 11.9 Age distribution of infant botulism and sudden infant death syndrome in the United States.

Another clostridial food poisoning is due to *Clostridium botulinum*. Although rare, botulism is much more serious than other food poisonings. As we noted in Section 8.4, the botulinum toxin is the most poisonous toxin known. Botulism is discussed in Chapter 16. In addition to food poisoning, a second form of botulism called **infant botulism** has been recognized. It occurs almost exclusively in children between the ages of 2 weeks and 6 months (Figure 11.9). Interestingly, the timing is indistinguishable from that of *sudden infant death syndrome*, although the diseases are probably unrelated. Infant botulism is an infection rather than an intoxication. As we have noted, *C. botulinum* is generally unable to grow in the body, but growth can occur in the case of infant botulism. The infant has a weakly established microflora so that *C. botulinum* is able to establish itself in some cases, producing the toxin causing paralysis. In severe cases of infant botulism, death results. The incidence of infant botulism has increased dramatically over the past decade, perhaps due to greater awareness of the disease, and there are presently more cases of this form than of botulism associated with contaminated foods.

Infant botulism

11.8 BACILLUS

The genus *Bacillus* consists of aerobic, endospore-forming, Gram-positive rods (see Figure 2.1*a*). Members of the genus *Bacillus* are easy to isolate from soil or air and are among the most common organisms to appear when soil samples are streaked on agar plates containing various nutrient media.

Anthrax

Anthrax is the most significant human disease caused by a *Bacillus* species, *B. anthracis*. Anthrax was the first disease shown conclusively to be caused by a bacterium, and its study by Koch provided one of the foundations for the development of microbiology (see Chapter 1). Primarily a disease of farm animals, anthrax is only occasionally transmitted to human beings. *Bacillus anthracis* is a very large bacillus, 1 to 1.5 μm wide by 4 to 8 μm long, and the cells usually remain attached after division; long chains are formed on agar to create colonies of characteristic morphology. The anthrax bacillus forms a capsule that protects it from phagocytosis and, probably because of this capsule, the pathogen is highly invasive, growing well throughout the body; in later stages of the disease, large numbers of bacilli are found in the blood. In the body, the organism produces an exotoxin that is responsible for most of the disease symptoms. The toxin causes shock, electrolyte imbalance, swelling, hemoconcentration, and acute kidney failure.

The organism infects human beings only when they come in contact with diseased animals or their hides or animal products (Figure 11.10). Many kinds of wild and domestic animals are susceptible, including cattle, sheep, horses, goats, pigs, mink, dogs, deer, birds, and even frogs and fish, although not all animals are equally susceptible. An animal usually becomes infected by ingesting bacilli or endospores from an infected carcass. At death, large numbers of bacilli are present in the blood and tissues, and if the carcass is opened, sporulation can occur (in warm countries, high environmental temperatures favor this process). If sporulation has occurred, the organism can remain viable in bone for long periods of time and is transmitted to uninfected animals when they eat the bones. Animals are usually attracted to the carcass because of their craving for salt or their need for bone minerals; the latter is observed especially among animals living on forage in phosphate-deficient areas. In dry soils and in areas of pasture land, anthrax endospores can remain alive for very long periods, but in most agricultural soils, they do not persist for more than a few years. The organism is not able to grow directly in soil.

Human infection is primarily an occupational hazard in the meat-packing and tanning industries. Invasion most often occurs by way of the skin, usually through a small scratch or abrasion, and the primary lesion develops as an inflamed pustule or blister. In most cases, the disease is self-limited, but the

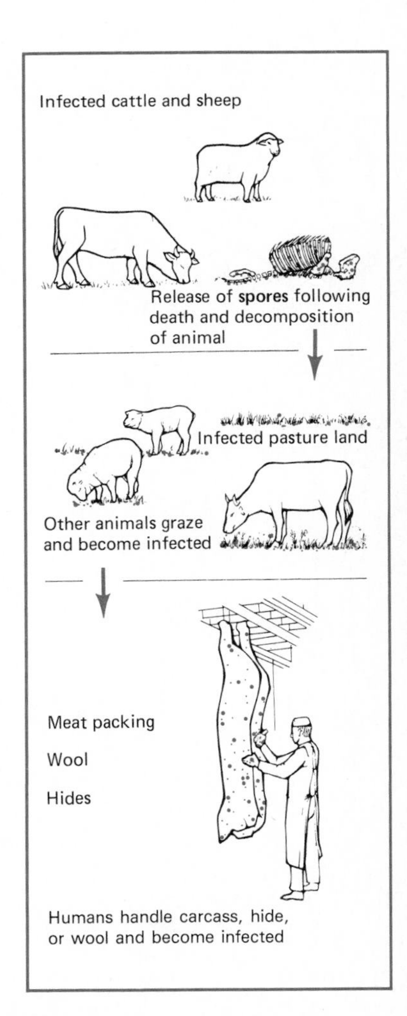

Figure 11.10 Transmission of anthrax bacteria among animals and humans.

organism may spread from the initial site of infection and initiate a fatal systemic infection. Another form of anthrax in humans, called "woolsorter's disease," is a respiratory infection resulting from the inhalation of endospores.

Immunity to anthrax develops through production of antibodies against the toxin. The first attenuated vaccine was the anthrax vaccine developed by Louis Pasteur. Despite Pasteur's early success, subsequent work has shown that proper attenuation of the pathogen is difficult, and even today, anthrax vaccine is only recommended as a preventive measure for those at high risk of exposure to anthrax, such as veterinarians, and as a means of controlling industrial outbreaks. The human disease can be effectively treated with penicillin, the tetracyclines, or erythromycin, provided the symptoms are detected before bacteremia develops.

Bacillus cereus causes a type of food poisoning which is only rarely reported in the United States, but is more common in Europe (see Chapter 16).

11.9 MYCOBACTERIUM

The genus *Mycobacterium* contains rod-shaped nonsporulating aerobic organisms that exhibit a characteristic staining property called *acid-fast*. In the acid-fast staining procedure, a mixture of the dye basic fuchsin and phenol is used, the preparation being treated by slow heating, which forces the stain into the cells. The preparation is then decolorized with acid alcohol. The fuchsin dye is removed by the acid from other organisms but the dye is retained by mycobacteria. The reason mycobacterial cells are acid-fast is because of the presence of a unique lipid fraction of the cell wall which is absent in other bacteria. Thus, the acid-fast stain is a convenient diagnostic test for mycobacteria. The most important species are those pathogenic to humans: *M. tuberculosis*, the causal agent of tuberculosis, and *M. leprae*, the causal agent of leprosy.

Tuberculosis

Tuberculosis has been one of the great scourges of society. Pulmonary tuberculosis, also called "consumption" or "phthisis," has been recognized as a disease entity for hundreds of years. Robert Koch first showed in 1882 that tuberculosis was caused by a bacterium when he successfully cultured *M. tuberculosis* in the laboratory and succeeded in establishing an experimental infection in guinea pigs. The organism, which is an obligate aerobe, has simple nutritional requirements and will grow on a synthetic medium containing acetate or glycerol as a sole carbon source and ammonium as a sole nitrogen source. Growth is stimulated by fats; egg yolk, which is high in fats, is often added to culture media to achieve more luxuriant growth. Perhaps because of the high lipid content of its cell walls, the organism is able to resist such chemical agents as alkali or phenol for considerable periods of time. This resistance is used in the selective isolation of *M. tuberculosis* from sputum, the sample first being treated with sodium hydroxide for 30 minutes before being neutralized and streaked on the isolation medium. Since avirulent and weakly virulent strains are widespread, any isolate must be tested to confirm virulence, and for this, the guinea pig is the animal of choice because it is highly susceptible to *M. tuberculosis* infection. Pathological material or a culture of virulent organisms injected subcutaneously will cause the formation of characteristic tubercle nodules at the site of injection, and the spread of the organism from the initial site may result in secondary tuberculosis nodules in the spleen or peritoneal cavity within 4 to 5 weeks. Death of the animal usually occurs in 6 weeks to 3 months, and autopsy will reveal lesions in organs throughout the body.

The interaction of the human host and *M. tuberculosis* is an extremely complex phenomenon, being determined in part by the virulence of the strain but probably more importantly by the specific and nonspecific resistance of the host. Cell-mediated immunity plays an important role in the development of disease symptoms. It is convenient to distinguish between two kinds of human infection: **primary** and **postprimary** (or reinfection). Primary infection is the first infection that an individual receives and often results from inhalation of droplets containing viable bacteria derived from an individual with an active pulmonary infection. Dust particles that have become contaminated from sputum of tubercular individuals are another source of primary infection. The bacteria settle in the lungs and grow. A delayed-type hypersensitivity reaction (see Section 9.7) results in the formation of aggregates of activated macrophages, called *tubercles*, characteristic of tuberculosis. In a few individuals with low resistance the bacteria are not effectively controlled, and an acute pulmonary infection occurs, which leads to the extensive destruction of lung tissue, the spread of the bacteria to other parts of the body, and death.

In most cases, however, acute infection does not occur, and the infection remains localized and is usually inapparent; later it subsides. But this initial infection hypersensitizes the individual to the bacteria or their products and consequently alters the response of the individual to subsequent infections. A diagnostic test, called the **tuberculin test**, can be used to measure hypersensitivity. If *tuberculin*, a protein fraction extracted from the bacteria, is injected intradermally into the hypersensitive individual, it elicits at the site of injection a localized immune reaction characterized by hardening and reddening of the site 1 to 2 days after injection. An individual exhibiting this reaction is said to be *tuberculin positive*, and many healthy adults give positive reactions as a result of previous inapparent infections. A positive tuberculin test does not indicate active disease but only that the individual has been exposed to the organism at some time.

It is in tuberculin-positive individuals that the postprimary type of tuberculosis infection can occur. When renewed pulmonary infections occur in tuberculin-positive individuals, they are usually chronic infections that involve destruction of lung tissue, followed by partial healing and a slow spread of the lesions within the lungs. Spots of destroyed tissue may be revealed by X-ray examination (Figure 11.11), but viable bacteria are found in the sputum or stomach washings only in individuals with extensive tissue destruction. In many cases, symptoms in tuberculin-positive individuals are a result of reactivation and growth of bacteria that have remained alive and dormant in the lungs for long periods of time. Malnutrition, overcrowding, stress, and hormonal imbalance often are factors predisposing to secondary infection.

Drugs such as streptomycin, isonicotinic acid hydrazide (INH), and *p*-aminosalicylic acid (PAS) are the most effective antituberculosis agents, the

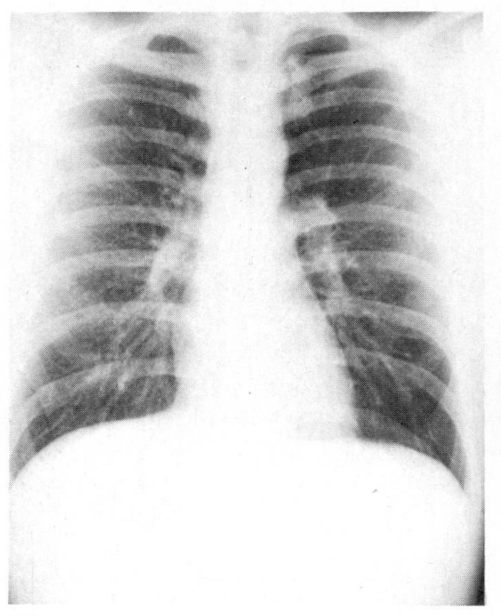

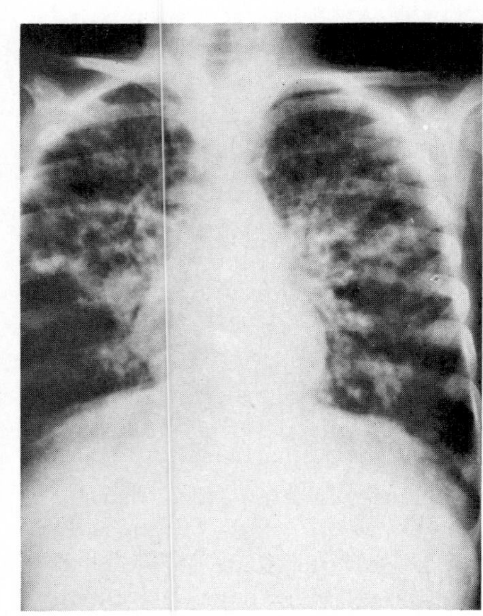

(a) (b)

Figure 11.11 X-ray photographs. (a) Normal chest X ray. The faint white lines are arteries and blood vessels. The heart is visible as a white bulge in the lower right quadrant. (b) An advanced case of pulmonary tuberculosis; white patches indicate areas of disease.

latter two being fairly specific in acting against mycobacteria. These drugs are used in combination to avoid complications due to drug resistance. To eliminate the organism from all the sites within destroyed lung tissue where it has become established, long-term therapy is necessary. Drug therapy does not lead to the direct healing of damaged tissue, but by destroying the bacteria it prevents the establishment of new sites of infection and new tissue damage.

The severity of tuberculosis in Europe and North America had begun to subside in the late nineteenth century, even in the absence of chemotherapeutic measures, owing to improved nutritional and socioeconomic conditions. Most people became tuberculin positive at an early age, and so the disease was mainly of the postprimary or reinfection type. The introduction of chemotherapeutic measures in the post-World War II era changed this situation considerably (Figure 11.12). The elimination of the bacteria from infected individuals has greatly reduced the death rates in many countries. Today, tuberculin-positive individuals comprise a smaller proportion of the population than they once did. Although tuberculosis is no longer a leading cause of death in the United States it is still among the top ten reportable diseases (see Figure 10.6). At one time, extensive surveys were made using X rays and the tuberculin test to detect infected individuals, who could then be treated, but the incidence is now low enough that the cost of such extensive surveys is hardly justified by the number of new cases detected. Tuberculosis remains a leading cause of death in developing countries (Figure 10.7). At present there may be as many as 15–20 million cases of tuberculosis worldwide. Though the incidence of tuberculosis was once high among infants and children, it is now mainly a disease of adults over 30 years of age, particularly among those with a poor socioeconomic environment.

A live vaccine, the **BCG strain** of *M. tuberculosis*, has been available for many years, but the desirability of its use has been the subject of much debate. Vaccination of tuberculin-negative individuals with BCG can lead to tuber-

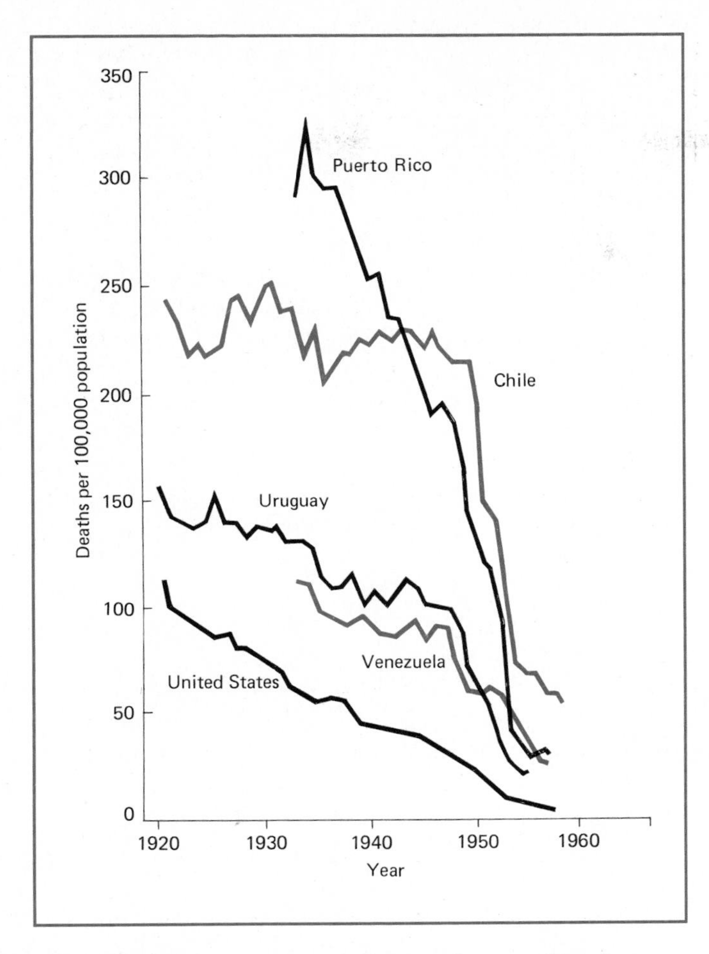

Figure 11.12 Efforts by public health authorities in many countries resulted in a marked decline in deaths due to tuberculosis by the end of the first half of the twentieth century. Note that, except for Chile, the most extensive declines took place before antimicrobial therapy became available, probably due to improved nutrition and general health of the populations.

culin-positive status, with an increase in resistance to infection, but the vaccination confers the same hypersensitive state induced by natural infection and so does not eliminate the chance of the postprimary type of infection. Since vaccination renders the individual tuberculin-positive, it also eliminates the possibility of using the tuberculin test at a later date to detect a new infection. Vaccination in many countries is now recommended only for tuberculin-negative individuals who have a high probability of infection, such as children of tubercular parents. In the United States, the tuberculin test is used to detect cases of tuberculosis in infants, but immunization is not advised.

Tuberculosis in cattle at one time was a serious disease. The bovine strains of *M. tuberculosis* are also highly virulent for human beings, and in the days before public health measures were widespread, many people became infected from contaminated dairy products. Because the organism entered the body by way of the gastrointestinal tract, the site of infection was usually not the lungs but the lymph nodes, and the organism subsequently became localized in the bones and joints. Pasteurization of milk and other dairy products and elimination of diseased cattle have virtually eradicated this type of human tuberculosis in Europe and North America.

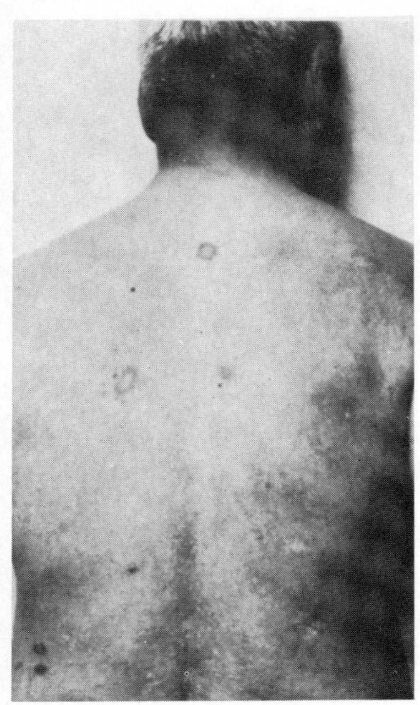

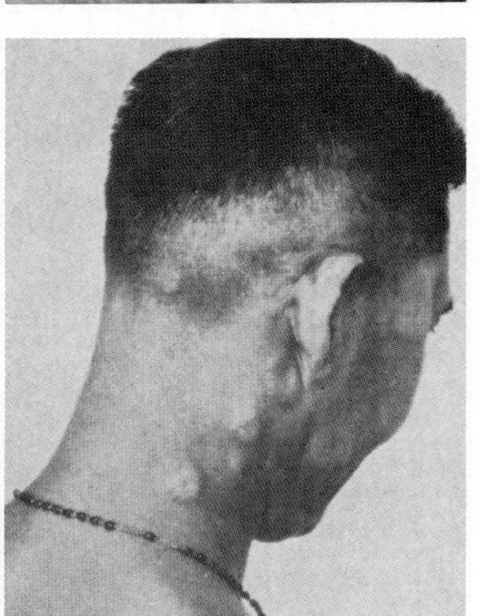

Figure 11.13

The disease **leprosy** is caused by *Mycobacterium leprae*. This bacterium, which has not been successfully cultivated, grows slowly in skin causing lesions such as those shown in Figure 11.13. In more severe cases, lesions can be extensive, coalescing and causing large folded, deformed areas in the skin. In addition to cutaneous infections, a neural form of leprosy may occur in which peripheral nerves are infected, leading to loss of sensation.

Leprosy is not a highly contagious disease. Little is known of the pathogenicity of *M. leprae* or of its mode of transmission, though it would seem likely that it is transmitted by direct contact. The bacterium grows within macrophages, causing an intracellular infection which can result in an enormous population of bacteria within the skin. In many areas of the world the incidence of leprosy is very low, although in ancient times it was apparently much more common, perhaps due to crowding and poor sanitation. In other areas, such as tropical areas, the incidence is higher; leprosy remains a problem for some 15 million people. Prolonged treatment with the drug diaminodiphenylsulfone (DDS) used to be recommended, but a rise in resistance to this drug has changed treatment so that DDS is now administered together with other drugs.

11.10 ENTERIC BACTERIA

The enteric bacteria comprise a relatively homogeneous group: they are Gram-negative, nonsporulating rods, either nonmotile or motile by peritrichous flagella (see Figure 2.21). They are facultative anaerobes with relatively simple nutrition, fermenting sugars to acids or alcohols. Among the enteric bacteria are several genera pathogenic to man or animals, as well as other genera of practical importance.

Members of the genus *Escherichia* are almost universal inhabitants of the intestinal tract of man and warm-blooded animals, although they are by no means the dominant organisms in these habitats. The best-known species is *Escherichia coli*. Though *E. coli* is often thought of as a harmless member of the normal flora, it can be involved in several different kinds of diseases, including urinary tract infections, diarrheal disease, pneumonia, meningitis, and wound infections. As mentioned in Section 10.8, *E. coli* is one of the most frequently observed causes of hospital infections, especially those of the urinary tract.

There are several different types of diarrheal disease caused by *E. coli*. This Gram-negative bacterium produces an *endotoxin* which is responsible for symptoms in certain kinds of infections. However, some strains of *E. coli* also produce an *enterotoxin*, an extracellular toxin which acts against the intestinal tract. Enterotoxin-producing strains of *E. coli* (*enterotoxigenic E. coli*) are a frequent cause of *infant diarrhea* in developing countries and of *travelers' diarrhea*, acquired by tourists visiting such countries. Studies have shown that as many as 50 percent of U.S. travelers to Mexico and Central and South America suffer diarrheal disease, usually after eating uncooked foods (such as lettuce in salads) or drinking impure water. Enterotoxigenic *E. coli* may account for 40–50 percent of these cases. The enterotoxins cause increased

Travelers' diarrhea

permeability of the intestinal lining, resulting in fluid loss and diarrhea. Some strains of *E. coli* also cause outbreaks of infant diarrhea in nurseries. Other forms of *E. coli* are more invasive and cause severe dysentery-like or bloody diarrheal diseases in adults. Most strains are susceptible to streptomycin, the tetracyclines, and chloramphenicol.

Infant diarrhea

Although *Salmonella* and *Escherichia* are very closely related, members of the genus *Salmonella* are almost always pathogenic to humans or warm-blooded animals. In humans, the most common diseases caused by salmonellas are typhoid fever and a type of gastroenteritis called salmonellosis. The pathogenicity of salmonellas is due primarily to the action of endotoxins.

The important disease **typhoid fever** is caused by the species *Salmonella typhi*. This organism is able to resist digestion after phagocytosis and can live and reproduce intracellularly in macrophages and other phagocytic cells. *Salmonella typhi* is transmitted from person to person in food or water that has been contaminated from fecal sources. Initial replication of the organism occurs in the intestinal tract, and once established there some of the organisms enter the lymphatic system draining the intestine, travel to the bloodstream, and then become disseminated throughout the body (Figure 11.14). The organism grows especially well in the gall bladder, the spleen, and the lymph nodes, but may also reproduce in other organs. After the organism is established in the tissues, the characteristic fever develops, probably as a result of the release of endotoxin (see Figure 8.20). Diarrhea is not as common a symptom of typhoid fever as it is of salmonella gastroenteritis (see below). One diagnostic sign of the disease is the transitory appearance of *rose spots* on the trunk of the body. The intestinal phase of the disease can be treated with a variety of antibiotics active against Gram-negative bacteria. Even if the or-

Typhoid fever

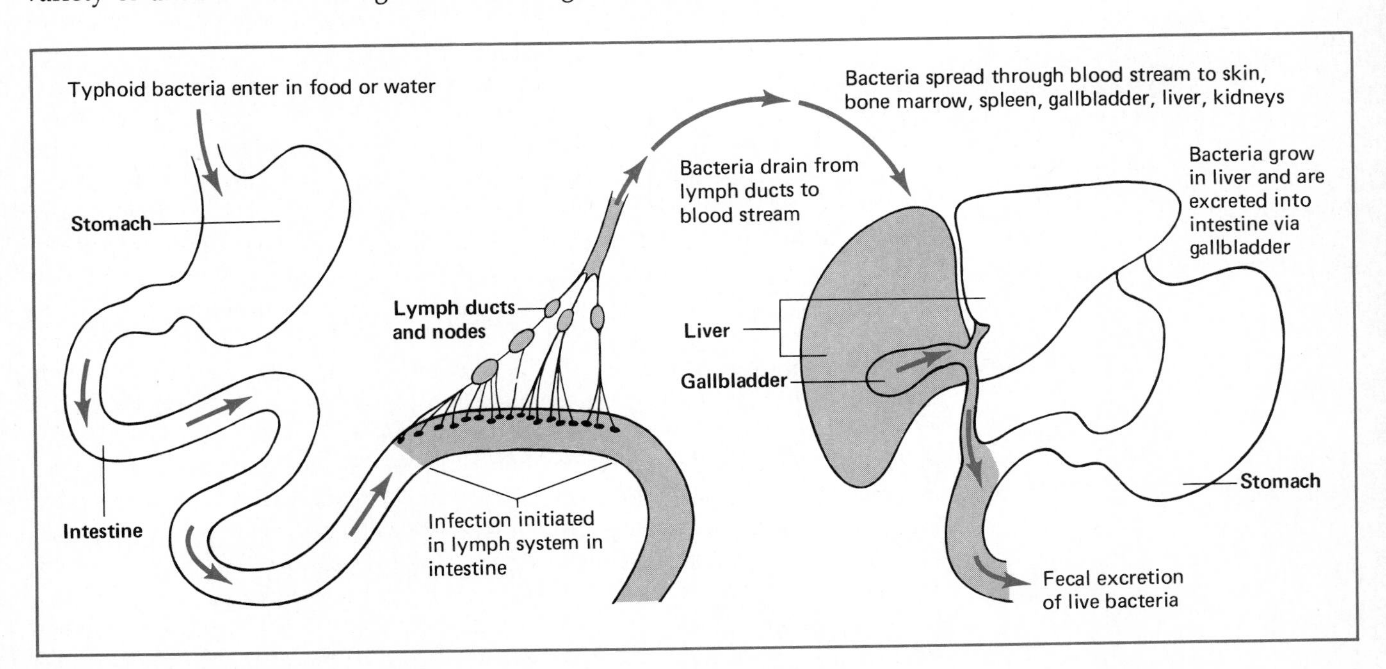

Figure 11.14 Bacterial infection in typhoid fever.

ganism is eliminated from the intestinal tract by antibiotic therapy, however, organisms derived from the biliary tract may continually appear in the feces; this is probably the situation during the chronic carrier state. The chronic intracellular phase can be treated effectively only with chloramphenicol, an antibiotic that penetrates phagocytic cells readily. Since this antibiotic is only bacteriostatic, antibiotic therapy must be continued long enough so that the host defenses can eliminate all viable bacteria, and indeed, chloramphenicol is not always able to eliminate the chronic carrier state. (Chloramphenicol is also potentially toxic to the host, so its use must be carefully monitored.)

Typhoid fever is a disease that has been controlled most dramatically by such public-health measures as pasteurization of milk, treatment of sewage, purification of water, and elimination of chronic carriers such as food handlers (discussed in the Box on page 259 and in Chapter 16). However, even today typhoid fever is not completely eradicated, and a breakdown in water-supply treatment could easily lead to a return of the disease. Sporadic cases are still reported around large cities, especially during the summer months. A killed-bacteria vaccine for *S. typhi* is available and is widely used for those parts of the world where epidemics are still common. It is also recommended for those travelling to endemic areas (see Table 10.3).

Salmonellosis

Salmonellosis, caused by salmonella species other than *S. typhi*, is the most frequently reported type of **gastroenteritis** in the United States (see Figure 10.6). Assuming that only a small portion of cases are actually reported, it is thought to be among the most significant types of contagious bacterial diseases in the United States. Infection most commonly occurs after eating contaminated food, so that it is often considered a type of "food poisoning." It is not, however, a true food poisoning but an infection derived from food; that is, the symptoms develop as a result of multiplication of the organism in the intestinal tract. In contrast to *S. typhi*, the enteritis-inducing species rarely spread out of the intestine. Between 8 to 48 hours after inoculation, onset of the symptoms occurs, and it is usually sudden with headache, chills, vomiting, and diarrhea, followed by a fever that lasts a few days. The disease is usually self-limited, but antibiotic therapy may be desirable. Diagnosis is from symptoms and by culture of the organism from the feces. The disease may be controlled by the use of public health measures, especially in monitoring food handlers and food preparation (see Chapter 16). In the Box on page 110 we considered the epidemiology of the largest epidemic of food-borne salmonellosis ever to occur in the United States. This was caused by milk contaminated with a *Salmonella* strain which showed resistance to several antibiotics.

Shigellosis (bacterial dysentery)

The genus *Shigella* is closely related to *Escherichia* and *Salmonella*. Species of *Shigella* are commonly pathogenic to man, causing a rather severe gastroenteritis, **shigellosis**, also called **bacterial dysentery** to distinguish it from amoebic dysentery (discussed in Chapter 12). High fever, chills, convulsions, abdominal cramps, and frequent bloody stools are characteristic symptoms of shigellosis. The pathogenicity of some strains of *Shigella* seems related to their ability to produce a powerful endotoxin. Since the organism apparently is incapable of invading the bloodstream as *Salmonella typhi* often does, the disease is confined to the intestine and is usually self-limited; it is rarely fatal and spontaneous cure occurs within a few days. *Shigella* is found almost exclusively in humans and primates, infection of other animals being rare. Communities with poor sanitation measures are often affected, and quite frequently, military troops in combat zones, where proper sanitation is unavailable, are stricken. The reported occurrence of shigellosis in the U.S. is nearly as high as that for salmonellosis (Figure 10.6) and it is a leading cause of infant mortality in developing countries.

11.11 YERSINIA

The causal agent of **bubonic plague**, *Yersinia pestis*, has in recent years been found to be closely related to *Escherichia*. Until the twentieth century, bubonic plague was one of the greatest scourges of the human race and was responsible for epidemics that killed millions of people. In the Middle Ages, a plague epidemic caused the death of one-quarter of the population of Europe; epidemics much less extensive have occurred in various parts of world up to the present. There is good reason to believe that the organism was considerably more virulent in the extensive epidemics of the Middle Ages than it is today, but why this should be so is not known. The disease in the Middle Ages was often associated with severe hemolysis, producing dark skin in the dying person: hence the name *Black Death* for bubonic plague. This symptom is not seen today. The disease is still endemic in parts of Asia, South America, and Africa, and in a modified form, it is found in rodents in the western United States.

Bubonic plague

Plague is now a disease which is more common in rodents than in humans. The causal organism is transmitted from one animal to another by fleabite. In areas of the world where people live in close proximity to rodents, the flea may transmit the organism from an infected rodent to a human, but rarely is the organism passed directly from one person to another. As a result of a fleabite, the bacteria are inoculated into the lymphatic system and move to the regional lymph nodes where they multiply and cause the formation of enlarged lymph glands (called *buboes*; hence the name "bubonic" plague). In the later stages of the disease, the bacteria become disseminated throughout the body. The fatality rate of bubonic plague is high in untreated cases, often approaching 100 percent. Death results from production by the bacteria of a potent endotoxin, but other virulence factors probably also are involved. Invasiveness is determined at least in part by a capsule that prevents phagocytosis. Interestingly, the optimum temperature for growth of *Y. pestis* is 28°C rather than the normal human body temperature of 37°C, but at 28°C the virulence factors are not produced. The bacteria also multiply in the flea; possibly the low temperature optimum encourages multiplication in the insect. In some cases in humans, the bacteria become established in the lungs, leading to the condition called *pneumonic plague*. This disease is highly contagious, the bacteria being transmitted from person to person by droplets and strict hospital isolation (see Figure 10.16) is necessary.

Recovery from plague resulting in lifelong immunity involves the formation of opsonizing antibodies which are active against the bacterial outer wall layer. The disease is treated with antibiotics such as the tetracyclines, chloramphenicol, and streptomycin, and if antibiotic therapy is begun early enough, recovery may be assured. Public health measures involve the elimination of rats, the chief reservoir of the organism in urban areas. Elimination of plague in wild rodents is virtually impossible, and these animals thus constitute a reservoir from which the organism might move back into urban centers if rat control were not carried out effectively. Vaccination is recommended for people travelling to endemic areas and for those who expect to be exposed to rodents (see Table 10.3).

Another *Yersinia* species, *Y. enterocolitica*, infects wild and domestic animals, and causes a form of *gastroenteritis* in humans. Disease may result from human contact with infected animals (primarily from pigs), and from contaminated foods such as milk. This disease is thus another form of food-borne gastroenteritis (see Chapter 16). After infection, the bacteria attach to and invade the gut epithelium, where enterotoxin production leads to characteristic symptoms, including diarrhea, headache, malaise, and fever associated with convulsions. Another common symptom, severe abdominal pain, has misled physicians to suspect appendicitis and to perform unneeded surgery as a result. Antibiotic treatment is possible, but the susceptibility of isolates from individual cases varies and so must be tested before therapy is begun.

11.12 FRANCISELLA

Tularemia (rabbit fever)

The causal agent of the disease **tularemia** is *Francisella tularensis*. Tularemia, a widespread disease of wild rodents such as squirrels and rabbits, was first discovered in 1911 in Tulare County, California, and it is from there that the disease receives its name. It mainly affects rodents, but many other domestic and wild animals may be infected. The organism is transmitted in rodents by blood-sucking insects. Only occasionally is the organism transmitted to humans, either by the bite of a tick, or through the handling of an infected carcass. The disease is common in hunters, who acquire the infection during the dressing of a rabbit or squirrel: hence the common name "rabbit fever." *F. tularensis* enters the body through the skin, and infection is concentrated in the lymph glands, producing symptoms of headache, body pain, and fever; it sometimes leads to death. The mechanism of pathogenicity is poorly understood. Antibiotics active against Gram-negative bacteria can be used in therapy, streptomycin being the drug of choice because of its bactericidal action. The disease is best prevented by avoiding contact with the viscera of wild rodents or by using rubber gloves when cleaning susceptible animals. A live attenuated vaccine is available for immunization of people at risk, including trappers, shepherds, and sheepshearers.

11.13 BRUCELLA

The genus *Brucella* consists of small, Gram-negative, nonsporulating rods, which are usually pathogenic to animals or humans. Several species are closely related, being differentiated primarily by certain cultural characteristics and by the host in which they are most commonly found. *Brucella abortus* has its main reservoir in cattle, *B. melitensis* in goats and sheep, and *B. suis* in pigs, but any of the species may be found in all these animals or in humans. Any disease caused by a brucella is called **brucellosis**. Various disease syndromes are seen, of which the most important are *infectious abortion* in cattle and *undulant fever* in humans. The organisms are aerobic and have a complex nutrition, and certain species require an atmosphere of 5 to 10 percent CO_2 for growth. Pathogenic isolates produce a surface antigen that enables them to grow intracellularly in phagocytes, and in the host an intracellular site is the natural habitat. In cattle, *B. abortus* shows a marked specificity for the reproductive tract, due to the presence in fetal tissues of *erythritol*, an alcohol that greatly stimulates the growth of *B. abortus*. Brucellosis is widespread in cattle, and these animals can be infected in a variety of ways, such as by the venereal route, ingestion, the skin, and inhalation. Once introduced into a herd, the organism spreads readily and may eventually infect most of the animals. Not all pregnant cows abort, but a 30 to 50 percent abortion rate is not uncommon. Many animals recover completely, but may continue to harbor virulent organisms that may infect other animals.

Brucellosis (undulant fever)

Although the classic mode of transmission to humans is through contaminated milk, brucellosis is most often an occupational hazard of meat packers, livestock producers, dairy workers, and others who have exposure to animals and animal products. The organism enters through the gastrointestinal epithelium or abraded skin, then invades the lymphatic system and the bloodstream, becoming selectively localized in the spleen, liver, lymph nodes, bone marrow, and kidneys. Erythritol is not present in humans, and the organism shows no tendency to localize in the reproductive tract. Disease symptoms are due primarily to the presence of an endotoxin, the most characteristic signs being chills, fatigue, headache, and backache. Another symptom is a fever that undulates, increasing at night and dropping in the daytime, from which is derived the name *undulant fever* (Figure 11.15). Therapy with streptomycin and the tetracyclines is effective but must be prolonged becaus the intracellular localization of the organisms results in partial protection from

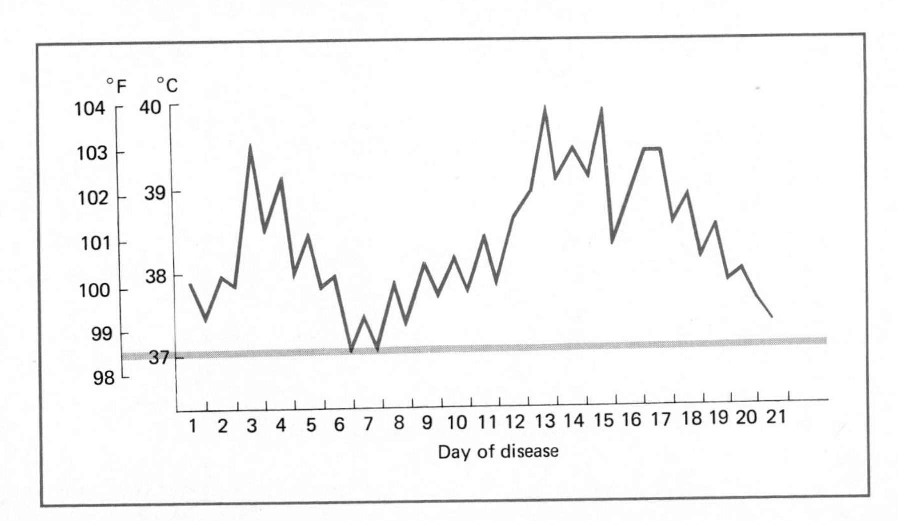

Figure 11.15 Undulating body temperature in a patient with brucellosis.

the antibiotic. Antibiotic therapy often results in an aggravation of symptoms immediately following administration, apparently because the presence of the antibiotic leads to bacterial lysis and sudden release of large amounts of endotoxin. Antibody develops readily in response to *Brucella* infection, and a rise in antibody titer is used in diagnosis, since the symptoms themselves are often too ill-defined to permit a precise identification of the disease. Recovery does not result in complete immunity, however, and reinfection may occur. The best diagnostic procedure is to culture the organism, usually from a blood sample. The brucellas are rarely transmitted from person to person, so that control of the disease can be effected through elimination of infected cattle. After immunological surveys of dairy herds to indicate the presence of *Brucella*, infected animals are segregated from the herd or slaughtered. Uninfected herds are vaccinated with a live attenuated vaccine. Pasteurization of milk is also an important procedure in preventing the spread of the organisms to human beings.

11.14 PSEUDOMONAS

The genus *Pseudomonas* comprises aerobic Gram-negative rods motile by means of polarly arranged flagella (see Figure 2.21). Most species are harmless soil organisms, but a few are pathogenic to humans or animals. Species which are pathogenic may show some adaptation to the human habitat. For instance, unlike nonpathogenic pseudomonads, pathogenic strains are able to grow well at 37–38°C, temperatures not found in most soils. *Pseudomonas aeruginosa* is frequently associated with infections of the urinary tract in humans. This species is not an obligate parasite, however, since it is also common in the soil. It thus appears to be primarily an *opportunistic pathogen*, initiating infections in individuals whose resistance is low, such as burn patients, cancer patients, immunosuppressed patients, and patients with cystic fibrosis. *P. aeruginosa* is one of the most frequent causes of hospital infections, as mentioned in Section 10.8. As we noted there, the organism is naturally resistant to many of the widely used antibiotics, so that chemotherapy is often difficult. Polymyxin, an antibiotic that is not ordinarily used in human therapy because of its toxicity, is effective against *P. aeruginosa* and can be used with caution.

Hospital infections

11.15 VIBRIO

The genus *Vibrio* consists of Gram-negative, polarly flagellated cells that are usually curved (comma-shaped) rods. Most are nonpathogenic, but some vibrios are causal agents of disease in animals and humans. The most important is *Vibrio cholerae*, which is the specific cause of the disease **cholera**. Cholera is one of the most common infectious human diseases and one that has had a long history, including pandemic outbreaks involving hundreds of thousands of people, even in recent years. The bacteria spread via the fecal-oral route; as many as 20 percent of the population in endemic regions may be carriers. The organism is transmitted almost exclusively via water, and studies

Cholera

of its distribution in the nineteenth century played a major role in demonstrating the importance of water purification in urban areas. Today, the disease is virtually absent from the Western world, although it is still common in many other areas. It is one of the few diseases for which many countries require vaccination certification before entry into the country is permitted (see Table 10.3).

Vibrio cholerae is an aerobe that has relatively simple nutritional needs. Also, it is quite insensitive to alkaline conditions, being able to grow at pH 9.0 to 9.6, at which most other intestinal bacteria are inhibited. This resistance to alkalinity makes possible the use of a high pH selective medium for primary isolation of *V. cholerae.* Although the bacteria grow readily in the intestinal tract, they do not invade the rest of the body. The organism produces two substances that affect the intestinal mucosa: an enzyme called *neuraminidase* that attacks the substance holding the intestinal cells together in a tissue; and an *enterotoxin* that affects the permeability of the intestinal wall and thus causes a profound electrolyte imbalance and water loss. The combined action of these two substances results in an enormous loss of water through the large intestine, up to 10 to 20 liters per day, causing extreme dehydration followed by shock and death. If untreated, the fatality rate may be as high as 60 percent. However, in many cases the disease can be mild and self-limited, running its course in a week or so. Antibiotic therapy with streptomycin or the tetracyclines is effective only in the very early stages of the disease. More important in treatment are measures taken to correct fluid imbalance, such as intravenous injection of isotonic fluids, which usually bring about a dramatic recovery. Control of cholera primarily depends on adequate sanitation measures, such as water purification and sewage treatment. Vaccines of killed cells are available, but they are not always successful and confer immunity of only short duration. They are used only in areas where the danger of contracting the disease is great.

In the last decade there have been a few sporadic outbreaks of disease associated with *V. cholerae* in the United States. These outbreaks have been associated with consumption of improperly cooked seafoods, suggesting that a new mode of transmission may exist. *V. cholerae* has been found to be present in estuarine waters along the Mid-Atlantic and Southeast coasts.

Another species, *Vibrio parahaemolyticus*, causes a type of food-borne gastroenteritis associated with consumption of raw seafood. Outbreaks have occurred in the United States, but this disease is far more prevalent in areas where raw seafood is a major component of the diet. In Japan, for example, 25 percent of reported cases of diarrhea are caused by this organism.

11.16 CAMPYLOBACTER

Bacteria of the genus *Campylobacter* are Gram-negative, curved rods resembling *Vibrio*, but unlike members of this latter genus, *Campylobacter* is microaerophilic, growing best at reduced oxygen tensions. *C. fetus* is economically important because it causes *infectious abortion* in cattle and sheep. In some cases, this organism has also been associated with fatal infections of fetuses. *C. jejuni* also infects animals, and sometimes humans. Outbreaks of gastroenteritis caused by this species are usually associated with contaminated foods or water (see Chapters 15 and 16).

11.17 SPIROCHETES

The spirochetes are bacteria with a unique morphology and mechanism of motility. They are widespread in aquatic environments and in the bodies of warm-blooded animals. Many of them cause diseases of animals and humans, of which the most important is **syphilis**, caused by *Treponema pallidum*. The spirochete cell is typically a slender, flexuous body in the form of a spiral, often of considerable length (see Figure 2.15*d*). Motility occurs by means of a bundle of fibers called the axial filament attached at the cell poles and wrapped around the cell in a spiral fashion. These structures are analogous to flagella, but the pattern of motility is quite different from that of flagellated bacteria. Whereas flagellated bacteria swim rapidly, the spirochetes move in the manner of a snake.

Syphilis

Of *Treponema* species, *T. pallidum*, the causal agent of **syphilis**, is the best known. *T. pallidum* cells are extremely thin, and for this reason they are almost invisible with the ordinary light microscope, whether unstained or treated by conventional dyes. Living cells are visible unstained by use of the dark-field microscope (Figure 11.16*a*), and dark-field microscopy has been extensively used to examine exudates from suspected syphilitic lesions. In nature, *T. pallidum* is restricted to humans, although artificial infections have been established in rabbits and monkeys. Although *T. pallidum* has never been grown in pure culture, it has been grown in tissue culture using rabbit

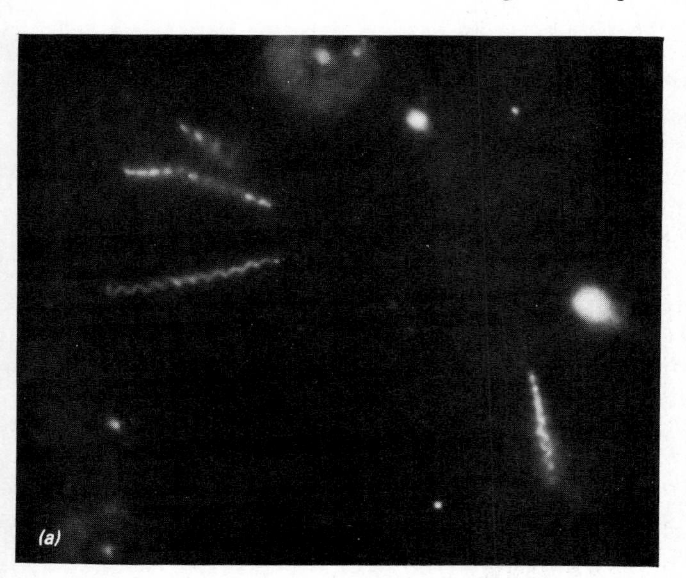

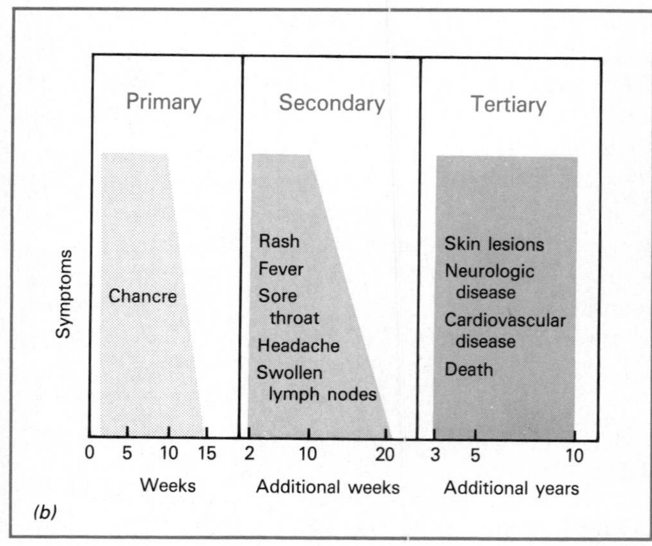

Figure 11.16 The spirochete of syphilis and its consequences. (a) Photomicrograph of the organism, *Treponema pallidum*, dark-field microscopy. A single spirochete cell is between 10 and 15 μm in length. (b) Stages in the disease.

epithelial cells. Reduced O_2 levels are required for growth. The organism is quite sensitive to increased temperature, being rapidly killed by exposure to 41.5 to 42.0°C. The heat sensitivity of *T. pallidum* is also reflected in the fact that the organism becomes most easily established in cooler sites of the body, such as the genital organs of the male, although once established in other areas of the body, it will multiply there. The extreme sensitivity to drying of the organism at least partially explains why transmittal between persons is only by direct contact, usually sexual intercourse.

The disease in humans exhibits variable symptoms. The organism does not pass through unbroken skin, and initial infection most probably takes place through tiny breaks in the epidermal layer. In the male, initial infection is usually on the penis, whereas in the female, it is most often in the vagina or cervix. In about 10 percent of the cases, infection is in the oral region. During pregnancy, the organism can be transmitted from an infected woman to the fetus; the disease acquired in this way by an infant is called *congenital syphilis*.

Syphilis is a *persistent infection* which occurs in distinct phases over a period of many years, as shown in Figure 11.16*b*. The organism multiplies at the initial site of entry and a characteristic primary lesion known as a *chancre* is formed within 2 weeks to 2 months. Dark-field microscopy of the exudate from syphilitic chancres often reveals the actively motile spirochetes. This is the *primary* stage of the disease process. In most cases, the chancre heals spontaneously, and the organisms disappear from the site. Some, however, spread from the initial site to various parts of the body, such as the mucous membranes, the eyes, joints, bones, or central nervous system, and extensive multiplication occurs. A hypersensitive reaction to the treponema takes place, which is revealed by the development of a generalized skin rash; this rash is the key symptom of the *secondary* stage of the disease. At this stage the patient's condition may be highly infectious, but eventually the organisms disappear from secondary lesions and infectiousness ceases. The subsequent course of the disease in the absence of treatment is highly variable. About one-fourth of untreated patients undergo a spontaneous cure, and another one-fourth do not exhibit any further symptoms, although the infection may persist. In about half of these patients, the disease enters the *tertiary* stage which follows a latent period lasting, in some cases, many years. Symptoms range from relatively mild infections of the skin and bone to serious infections of the cardiovascular system or central nervous system. Involvement of the nervous system is the most serious phase of the illness, since generalized paralysis or other severe neurological damage may result. In the tertiary stage, only very few organisms are present, and most of the symptoms probably result from hypersensitivity reactions to the spirochetes.

Penicillin is highly effective for therapy, and the early stages of the disease can usually be controlled by a series of injections over a period of 1 to 2 weeks. In the secondary and tertiary stages, treatment must extend for longer periods of time. Since penicillin kills only growing cells, the death rate is a function of how rapidly the pathogen is growing. Thus it is understandable that with such slow-growing organisms as *T. pallidum*, penicillin therapy must be prolonged.

Despite the relative effectiveness of penicillin in curing syphilis, the disease is still common (see Figures 10.6 and 10.12). This is due mainly to the social problems of locating and treating sexual contacts of infected individuals, as we have already mentioned in relation to gonorrhea.

The genus *Leptospira* contains a large number of diverse, widely distributed species. Some are harmless aquatic organisms, others are harmless parasites of animals, and some are pathogens, causing the disease *leptospirosis* (also called Weil's disease). Although many of the pathogenic forms can infect humans, the natural reservoirs are primarily domestic and wild animals. Rodents are the natural hosts of most leptospiras, although dogs and pigs are also important carriers of certain strains. Leptospiras ordinarily enter the body through the mucous membranes or through breaks in the skin. After a transient multiplication in various parts of the body, the organism localizes in the kidney and liver, causing nephritis and jaundice. The organism passes out of the body in the urine, and infection of another individual is most commonly by contact with water contaminated with the urine of infected animals. Sewer workers and those who swim in stagnant ponds and canals run the greatest risk of infection by this means. Therapy with penicillin, streptomycin, or the tetracyclines is possible but may require extended courses to eliminate the organism from the kidney; this is probably because of the slow growth and protected location of the leptospiras. Domestic animals are vaccinated against leptospirosis by means of a killed virulent strain; dogs are usually immunized routinely with a combined distemper-leptospira-hepatitis vaccine.

Leptospirosis

11.18 MYCOPLASMA

One bacterial group, the *mycoplasmas,* is composed of bacteria which lack cell walls. Members of this group are widespread in nature and have been frequently isolated from warm-blooded animals. Mycoplasmas are often pathogenic, causing serious diseases in cattle and birds; One species is the causal agent of pleuropneumonia in cattle. *Mycoplasma pneumoniae* causes a condition in humans known as *primary atypical pneumonia*. The disease occurs primarily in children and young adults; it is usually marked only by upper respiratory symptoms, and very rarely by pneumonia, or involvement of the lungs. Tetracyclines have been effective in the treatment of mycoplasmal pneumonia. However, currently available diagnostic tests are slow, so treatment of patients remains difficult. *Ureoplasma urealyticum* is an important cause of urethritis not associated with gonococci. Bovine contagious pleuropneumonia, caused by *M. mycoides,* has been a serious problem to the cattle industry throughout the world. Another important mycoplasmal disease is *agalactia* of sheep and goats, a disease that primarily affects the mammary glands.

Primary atypical pneumonia

11.19 RICKETTSIA

The rickettsias are small bacteria that have a strictly intracellular existence (see Figure 8.10) in vertebrates, and that are also associated at some point in their natural cycle with blood-sucking arthropods, such as fleas, lice, or ticks.

Rickettsias cause a variety of diseases in humans and animals, of which the most important are typhus fever, Rocky Mountain spotted fever, scrub typhus (tsutsugamushi disease), and Q fever. Rickettsias take their name from Howard Ricketts, a scientist of the University of Chicago, who first provided evidence for their existence and who died from typhus fever, caused by *Rickettsia prowazekii*. The rickettsias die quickly outside their hosts; they must be transmitted from animal to animal by arthropod vectors. When the arthropod obtains a blood meal from an infected vertebrate, rickettsias present in the blood are inoculated directly into the arthropod, where they penetrate to the epithelial cells of the gastrointestinal tract, multiply, and appear later in the feces. When the arthropod feeds upon an uninfected individual, it then transmits the rickettsias either directly with its mouthparts or by contaminating the bite with its feces. However, the causal agent of Q fever, *Coxiella burnetii*, can also be transmitted to the respiratory system by aerosols.

The most important and widespread rickettsial disease is **typhus fever**, also called "epidemic louse-borne typhus," caused by *R. prowazekii*. Typhus is an acute infectious disease that has a high fatality rate. It has probably afflicted human beings since ancient times. The symptoms of typhus frequently resemble those of typhoid fever, the disease caused by *Salmonella typhi*, and in earlier years, the two diseases were frequently confused; hence the origin of the name *typhoid*, which means "typhus-like." Typhus fever has occurred frequently in armies during military campaigns, often with disastrous consequences; indeed, this was one of the reasons for the downfall of Napoleon's army in 1812. Typhus is a frequent accompaniment of famine and civil disruption. In contrast to most of the other rickettsial diseases, typhus occurs naturally only in humans. The causal agent is transmitted from person to person by two species of lice, the human body louse and the human head louse. The rickettsias are released in louse feces. The fecal matter containing the agent gets rubbed into the puncture made by the louse and is carried to the bloodstream. The organism multiplies inside cells lining the small blood vessels, and clots occur in the vessels as a result of destruction of infected cells. A characteristic symptom of typhus is a rash, appearing first in the trunk and then spreading over the whole body except for the face, palms, and soles. Fever and general lethargy (torpor, sluggishness) of the patient result, with death on the ninth or tenth day of illness. The tetracyclines and chloramphenicol are highly effective against *R. prowazekii*. The control of epidemics, however, is directed primarily at the louse. Insecticides are applied to all persons known to have been in contact with infected individuals, and in widespread epidemics all persons in the community must be so treated. Typhus is still reported from a few mountainous areas of the world, but no typhus cases among U.S. travelers have been observed since 1950. Vaccination is thus neither required, nor recommended (see Table 10.3). Although typhus is at present not a problem in Western civilization, it could presumably become so again if standards of personal cleanliness and public hygiene were considerably reduced.

Typhus fever

Rocky Mountain spotted fever, caused by *Rickettsia rickettsii*, was first recognized in Idaho and Montana in the 1900s, but is actually more common in the eastern U.S. than in the Rocky Mountain region (Figure 11.17). All the

Rocky Mountain spotted fever

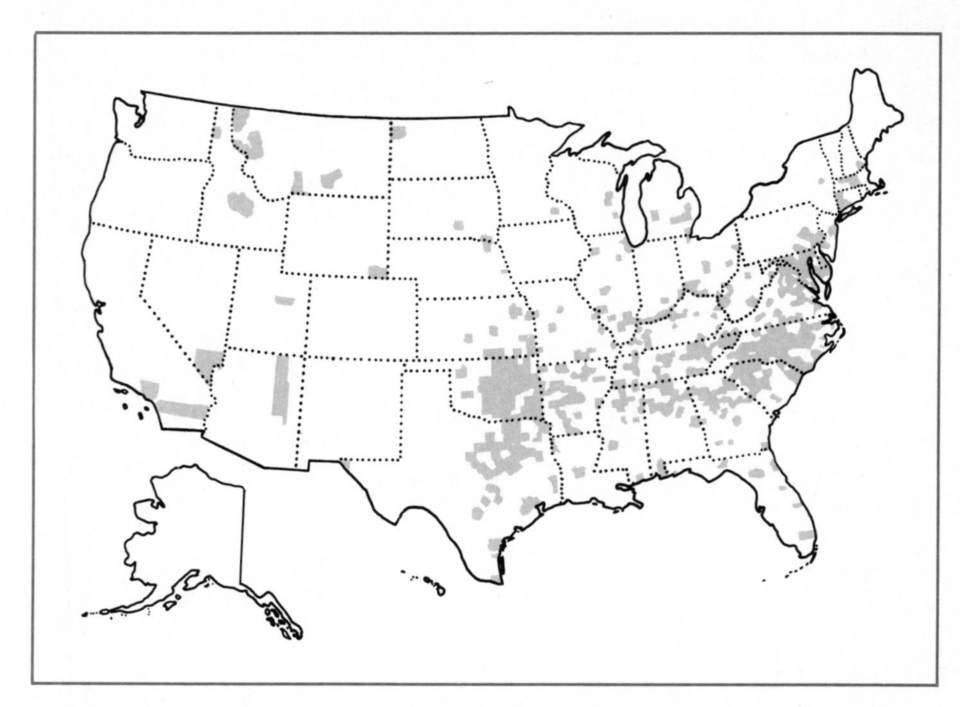

Figure 11.17 Regions of the United States where Rocky Mountain spotted fever is most common. The shaded areas indicate locations where cases of the disease have been reported in a recent year.

rickettsias that cause spotted fevers multiply in both the nucleus and the cytoplasm of host cells, whereas other rickettsias, such as that which causes typhus, grow exclusively in the cytoplasm. The spotted fevers are all transmitted by ticks; the onset of the disease occurs 1 to 2 weeks after the tick bite and is marked by chills, fever, and a rash that begins peripherally on the ankles, wrists, and forehead, and spreads to the trunk. The disease is more common in younger people, but the fatality rate is higher among older people (about 15 percent for people over 70). Tetracyclines and chloramphenicol are effective chemotherapeutic agents. Ticks are the vectors and also the primary reservoirs for the organisms; the rickettsias do not harm their arthropod host. *R. rickettsia* may also be cycled between ticks and wild animals.

11.20 CHLAMYDIA

The genus *Chlamydia* consists of obligate intracellular parasites that are small like the *rickettsia* but even more dependent on their hosts for nutritional factors required for growth and reproduction. Unlike the rickettsias, the chlamydias are not transmitted by arthropods. Some are primarily airborne invaders of the respiratory system whereas others are transmitted by direct contact.

Trachoma

Trachoma, an eye infection of humans, is caused by *Chlamydia trachomatis*. The disease, the most common cause of blindness in the world, affects over 400 million people, of whom 6 million are totally blind. Trachoma is widespread in developing regions of South America, the Middle East, Asia, and Africa; in North America, Native Americans are most commonly involved.

Conditions of poor public sanitation and personal hygiene encourage the disease, which probably is transmitted primarily by eye-to-finger-to-eye contact, and by towels and clothing. Transmission by flies is also suspected but not proved. In many communities nearly all children are infected in their early years; the infection becomes chronic in a few of these cases, and blindness occurs primarily in this group. The disease affects only the epithelial cells of the eye, and its onset is marked by inflamed conjunctiva. Development of characteristic follicles and scars in the conjunctiva is followed by vascularization and infiltration of the cornea, which produces partial or complete blindness. Various complications may ensue, and as a result secondary bacterial infections of trachomatous eyes are common. Tetracyclines and sulfonamides may be used to limit these complications.

The most prevalent chlamydial infection in the United States is probably a disease of the genitourinary tract which is sexually transmitted. Many different pathogens are responsible for causing urethritis not associated with gonococci, called *nongonococcal urethritis* (*NGU*). Included are the herpes virus (see Chapter 13), the mycoplasma *Ureoplasma urealyticum*, and the protozoan *Trichomonas vaginalis* (see Chapter 12). However, about half of the cases of NGU are caused by *C. trachomatis*. **Chlamydial NGU** occurs more frequently than gonorrhea (see Figure 10.12), and hence is the most frequent sexually transmitted disease. The clinical signs of the disease resemble those of gonorrhea, but often the disease is asymptomatic. In men, a mild urethritis occurs, accompanied by a slight discharge, but this may go unnoticed. Most women show no apparent symptoms of chlamydial infection, the organism remaining unnoticed in the cervix. In a small percentage of cases, uncontrolled chlamydial NGU can lead to serious consequences in both sexes, including testicular swelling, narrowing of the urethra, and prostate inflammation in men, and blockage of the fallopian tubes in women, causing sterility. Chronic debilitating complications may follow, including eye problems, arthritis, skin eruptions, and ulcers in the mouth area. Babies born to mothers with NGU have a high risk of eye infection as well as some risk of a pneumonia-like illness, and middle ear infection.

Chlamydial NGU

Chlamydial NGU is relatively difficult to diagnose since the organisms are not easy to culture and specialized techniques are required. Although penicillin is routinely prescribed for most sexually transmitted diseases, it is useless against chlamydial NGU. Once it is suspected, effective treatment can be undertaken, consisting of a week-long course of tetracycline or sometimes erythromycin.

Nongonococcal urethritis is frequently observed as a secondary infection to gonorrhea. After seeming to be cured, gonorrhea patients may develop symptoms again but in at least half of these cases the symptoms are due not to the recurrence of gonorrhea, but to NGU acquired at the same time as the gonorrhea. Since NGU has a longer incubation period than gonorrhea, this later development of NGU is understandable. The importance of being checked for complete recovery from symptoms is clear, since patients having post-gonococcal urethritis caused by chlamydia will require additional therapy.

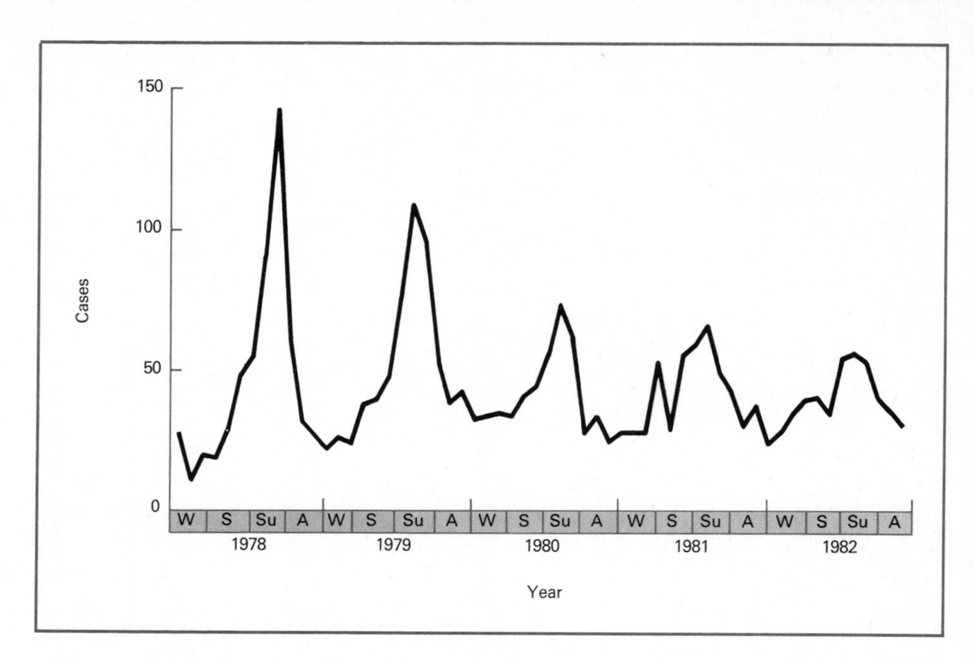

Figure 11.18 Seasonal occurrence of legionellosis (Legionnaire's disease) in the United States. (W, winter; S, spring; Su, Summer; A, autumn.)

11.21 LEGIONELLA

Legionellosis (Legionnaire's disease)

The discovery of the bacterium *Legionella pneumophila* following the 1976 outbreak of **legionellosis (Legionnaire's disease)** in Philadelphia, was described in the Box on page 84. Although *Legionella pneumophila* is an intracellular parasite in humans it can grow in natural aquatic environments; it is primarily an opportunistic invader of humans. Although the 1976 outbreak of Legionnaires' disease resulted in discovery of *L. pneumophila,* the disease was not new. Analysis of serum from earlier outbreaks of a similar nature revealed that those patients also had antibodies against this bacterium.

Legionellosis shows a seasonal occurrence with a peak in late summer months (Figure 11.18). Consistent with this seasonal distribution is the observation that a common source of the organism appears to be air conditioner cooling towers; *L. pneumophila* grows in the cooling tower waters. Airborne droplets are formed and dispersed by the rapid passage of air through the cooling towers. Once inhaled, the organism causes an intracellular infection within the lungs, leading to pneumonia. Pneumonia caused by *L. pneumophila* is an important hospital infection, accounting for 10–30 percent of hospital acquired pneumonias. The organism is sensitive to erythromycin, which is the drug of choice.

SUMMARY

Many different bacteria cause a variety of infectious diseases in humans. **Pyogenic streptococci** cause many different diseases, including streptococcal sore throat, rheumatic fever, and scarlet fever; other streptococci cause such diseases as tooth decay and pneumonia. **Staphylococci** also cause many diseases, including skin infections and toxic shock syndrome.

Sexually transmitted diseases are caused by several quite different kinds of bacteria. Spirochetes are well known as the cause of **syphilis**, and *Neisseria gonorrhoeae*, as the cause of **gonorrhea**. Despite the prevalence of these diseases, **chlamydial nongonococcal urethritis (NGU)** has become a far more frequent sexually transmitted disease. Other chlamydia are responsible for much of the blindness which occurs worldwide. Other *Neisseria* cause bacterial meningitis, though this disease is more commonly caused by *Haemophilus influenzae*.

Three important bacterial diseases, **whooping cough** (caused by *Bordetella pertussis*), **diphtheria** (caused by *Corynebacterium diphtheriae*), and **tetanus** (caused by *Clostridium tetani*) have been effectively controlled through the **DPT** immunization series. Other clostridia cause **gas gangrene** and food poisonings such as botulism.

A number of important **food- and water-borne diseases** of the gastrointestinal tract, including diarrheal diseases, salmonellosis, typhoid fever, and shigellosis are caused by **enteric bacteria**. Another water-borne disease, cholera (caused by *Vibrio cholerae*) remains an important health problem in many areas of the world.

Many bacterial diseases are transmitted from animals to man. **Anthrax** (caused by *Bacillus anthracis*), **brucellosis** (caused by *Brucella* species), and **tularemia** (caused by *Francisella tularensis*), for example, are hazards for those who come into contact with infected animals and their products. Other diseases, including **bubonic plague** (caused by *Yersinia pestis*) and **rickettsial diseases** (such as **typhus** and **Rocky Mountain spotted fever**) are transmitted via arthropod vectors.

The incidence of **tuberculosis** (caused by *Mycobacterium tuberculosis*) has been effectively reduced through antibiotic use, but tuberculosis and the related mycobacterial disease **leprosy** remain important in developing regions. Other diseases of the lung, such as pneumonia, can be caused by a number of different bacteria. For example, *Mycoplasma* are unique bacteria which lack a cell wall; some cause **primary atypical pneumonia**. Another form of pneumonia, **legionellosis**, is caused by the newly discovered bacterium *Legionella pneumophila*.

KEY WORDS AND CONCEPTS

Streptococci
Pyogenic
Erythrogenic toxin
Pneumonia
Staphylococci
Coagulase
Leukocidin
Toxic shock syndrome
Gonorrhea
Meningitis
Whooping cough
Pertussis
Diphtheria
Tetanus
DPT (immunization series)
Gas gangrene
Botulism
Anthrax
Tuberculosis
Leprosy
Enterotoxigenic *E. Coli*
Typhoid fever
Salmonellosis
Shigellosis

Bubonic plague
Tularemia
Brucellosis
Undulant fever
Cholera
Spirochete
Syphilis
Mycoplasma
Rickettsia
Typhus fever
Rocky Mountain spotted fever
Chlamydia
Nongonococcal urethritis (NGU)
Legionellosis

STUDY QUESTIONS

1. Name five diseases caused by bacteria and the name of the causal organism.
2. Name a disease caused by *pyogenic streptococci*. What symptom is characteristic of pyogenic infection? How would you test a clinical isolate to see if it were a Group A streptococcus?
3. Explain how *streptococcal sore throat* can lead to the autoimmune disease *rheumatic fever*.
4. Name two bacterial diseases that would not occur were it not for lysogenic bacteriophages. Explain how the virus is involved in the disease state.
5. How do streptococci cause *tooth decay*?
6. How are *pneumococci* and *gonococci* alike? How are they different?
7. Explain how the enzyme *coagulase* helps staphylococci protect themselves from host defenses. What other antiphagocytic factors do some staphylococci possess?
8. What are the symptoms of *toxic shock syndrome*? Who is most at risk of getting this disease?
9. Name three sexually transmitted diseases caused by bacteria. Which bacteria cause them? Rank these diseases in terms of the frequency with which they occur. Which one is the most debilitating?
10. Penicillin has been the drug of choice for treatment of *gonorrhea*. Why is it no longer always effective?
11. Name three bacterial diseases whose incidences have been dramatically reduced by immunization programs.
12. Name two diseases which can be caused by a number of different types of bacteria. Would it be simple or difficult to vaccinate against these diseases? Why?
13. Why are antibiotics alone insufficient for the treatment of *diphtheria*?
14. Name three diseases caused by *clostridia*. Why are clostridial diseases of particular concern in wound infections?
15. How is *infant botulism* different from the classical form of botulism?
16. Name two bacterial diseases which are occupational hazards of those who work with animals or animal products.
17. What factors have contributed to the decline in deaths due to *tuberculosis* in this century? Does a positive *tuberculin test* indicate an active case of tuberculosis? Explain.

18. Name three bacterial diseases which are still significant health problems in developing countries.
19. Name three water-borne diseases caused by bacteria. Which are caused by *enteric bacteria*?
20. Name three types of bacterial food poisoning, and one bacterial food-borne infection.
21. Why is rat control in urban areas still needed to ensure public protection against the disease *bubonic plague*?
22. Name a bacterial disease which is transmitted directly from wild animals to humans. Name a bacterial disease transmitted from wild animals to humans via arthropod vectors.
23. Name three types of bacteria which commonly cause hospital infections.
24. What is the most unique property of *mycoplasma*?
25. Explain how *rickettsia* and *chlamydia* are different from ordinary bacteria.
26. Name three bacteria which are so sensitive to the environment that they must be transmitted between hosts directly, or via an animate vector.
27. What different organisms (bacteria or other microorganisms) can cause the disease *nongonococcal urethritis* (*NGU*)?
28. Although *Legionella pneumophila* was only recently discovered, the disease *legionellosis* is not new. What type of evidence shows this to be true?

SUGGESTED READINGS

BENENSON, A.S. (ed.). 1985. *Control of communicable diseases in man, 14th edition*. American Public Health Association, Washington, D.C.

CENTERS FOR DISEASE CONTROL. *Morbidity and mortality weekly report*. United States Department of Health and Human Services, Atlanta, Georgia.

DAVIS, B.D., R. DULBECCO, H.N. EISEN, and H.S. GINSBERG. 1980. *Microbiology, 3rd edition*. Harper & Row, Publishers, Philadelphia, Pennsylvania.

HOEPRICH, P.D. (ed.). 1983. *Infectious diseases, a modern treatise of infectious processes, 3rd edition*. Harper & Row, Publishers, Philadelphia, Pennsylvania.

JAWETZ, E., J.L. MELNIK and E.A. ADELBERG. 1982. *Review of medical microbiology, 15th edition*. Lange Medical Publications, Los Altos, California.

JOKLIK, W.K., H.P. WILLETT, and D.B. AMOS. 1984. *Zinsser microbiology, 18th edition*. Appleton-Century-Crofts, East Norwalk, Connecticut.

LENNETTE, E.H., A. BALOWS, W.J. HAUSLER, JR., and H.J. SHADOMY. 1985. *Manual of clinical microbiology, 4th edition*. American Society for Microbiology, Washington, D.C. (Presents detailed methods for the isolation and characterization of pathogenic bacteria from clinical specimens. Also presents common pathogenic bacteria and their properties.)

12

Fungal and Protozoal Diseases

Some of the diseases which occur most commonly among humans are caused by eucaryotic microorganisms, either fungi or protozoa. Allergies to fungal spores, athlete's foot, and urethritis caused by protozoa may, for instance, occur within a large percentage of the human population. Though many diseases caused by fungi and protozoa are not especially serious, some of the protozoan diseases have been and remain scourges of the human population, particularly in developing countries in tropical regions. Malaria and African sleeping sickness, for example, affect millions of humans each year, and kill a significant portion of those affected. Also, some of the diseases caused by opportunistic fungal or protozoan pathogens can kill patients whose immune systems are suppressed. In this chapter we will describe the diseases caused by some of these eucaryotic microorganisms.

Treatment of diseases caused by eucaryotes is difficult because antifungal and antiprotozoal agents are also often toxic to humans, so that toxic side effects of drug therapy often occur. There are, in fact, only a few useful antibiotics known which are toxic to fungi or protozoa but nontoxic to humans. The chronic nature of many of the fungal and protozoal diseases also complicates their treatment. Also, in contrast to the bacterial and viral diseases, control of diseases caused by fungi and protozoa through vaccination has not yet been accomplished. Thus, it is important to try to understand the epidemiology of these diseases so that their transmission may be prevented.

12.1 FUNGAL DISEASES

Most fungi are harmless to humans, occurring as saprophytes in nature. The molds we see most commonly are those growing on bread or fruits, or those decomposing organic matter in the soil; these are relatively harmless. Only a few fungi are pathogenic to animals; it has been estimated that only about 50 species cause human disease.

It should be noted, however, that many fungi cause plant diseases, and fungi have probably caused more suffering to humans via their roles as plant pathogens than as human pathogens. Plant diseases are discussed further in Chapter 17.

Most fungi that cause disease in humans are not obligate pathogens. Their primary habitat is the soil, but some exist as part of the flora of the human body. Transmission of the organism from person to person is rare, so that fungal diseases are not notably contagious.

Infectious diseases caused by fungi are called **mycoses** (Figure 12.1). In many of the common mycoses, infection is *superficial* and only involves the skin. When the protective barrier of the skin is broken, however, some fungi may cause *subcutaneous* infections. A few fungi cause diseases called *systemic mycoses* which involve internal organs of the body; and such diseases are often debilitating. Many systemic and subcutaneous mycoses are due to infections caused by opportunistic fungi which are ordinarily harmless. Examples of superficial, subcutaneous, and systemic mycoses are given in Table 12.1. Fungi may also cause disease without causing infection. For example, inhalation of mold spores may lead to allergy. Finally, ingestion of toxins produced by certain fungi may lead to poisoning or hallucinogenic reactions (for example, mushroom poisoning, see the Box on page 38).

12.2 SUPERFICIAL MYCOSES

As we noted, fungi primarily cause skin infections. In these case, infections by relatively noninvasive fungi involve mainly the dead keratinized layer of the skin (see Section 8.1 and Figure 8.2), and usually do not provoke a significant host response. One disease, *pityriasis versicolor*, for example, involves discoloration of skin without irritation, and is often mainly a cosmetic problem, because the discolored skin fails to tan evenly with uninfected skin (Figure 12.2*a*). A group of mycelial fungi, called the *dermatophytes*, cause a number of superficial skin infections which are very common and which lead to somewhat more serious symptoms (Figure 12.2*b* and *c*).

Pityriasis versicolor

Since Roman times, the superficial mycoses caused by dermatophytes have been given the medical name "tinea," which actually means *small insect larvae*. The common name used for many of these conditions is *ringworm*. These names are obviously misnomers, since fungi rather than worms or

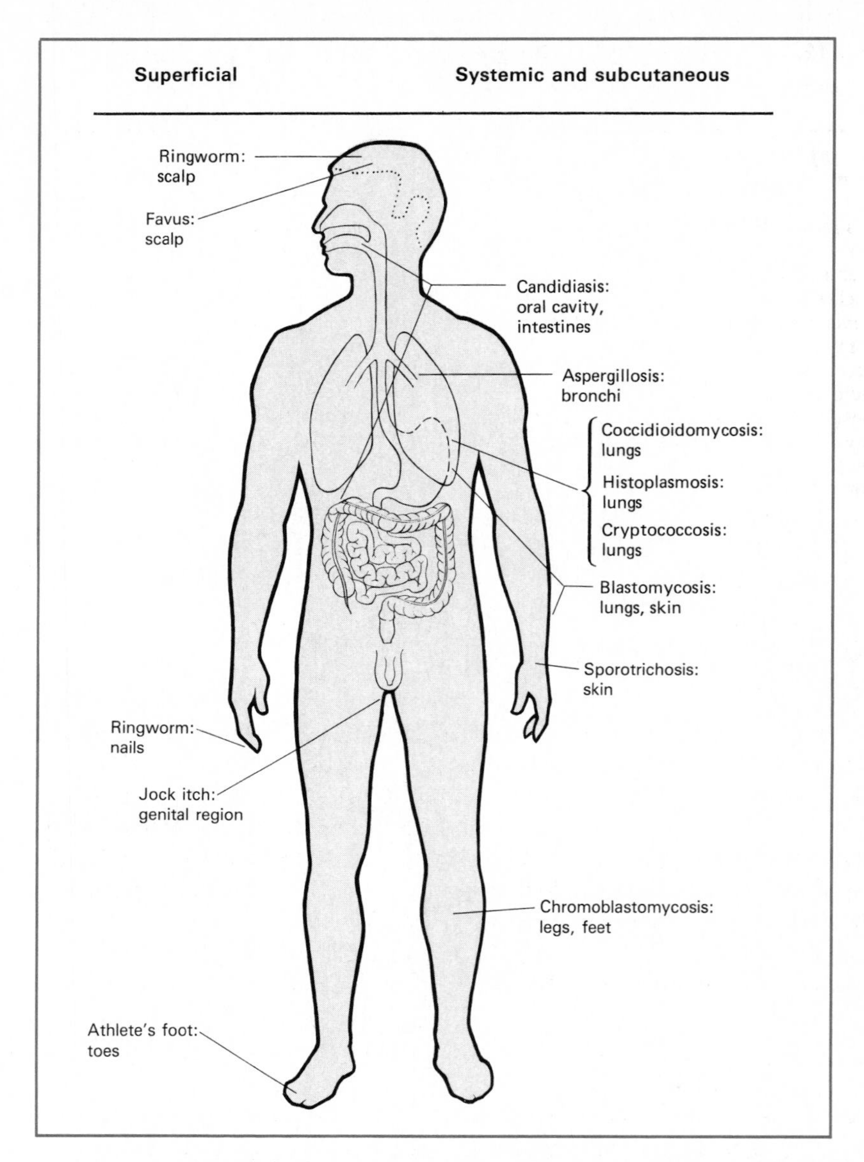

Figure 12.1 Infectious diseases caused by fungi.

larvae cause the disease; nevertheless, most superficial mycoses are still referred to by these names. Species of three genera of dermatophytic fungi may be involved, including *Trichophyton, Microsporum,* and *Epidermophyton.* Different species of these genera are commonly found in humans, animals, and soil, so that superficial mycoses may result from contact with any of these reservoirs.

Dermatophytic infections (ringworm, athlete's foot, jock itch)

Different dermatophyte species are found associated predominantly with different tissues. The most common fungal infection in humans is probably *tinea pedis,* more commonly called *athlete's foot,* which affects between 30 and 70 percent of the population. *Tinea capitis* (ringworm of the scalp), and *tinea favosa* (favus), involve dermatophyte infection of the scalp. Another condition, *tinea cruris,* results when the pathogenic fungus grows in the skin

TABLE 12.1 Some pathogenic fungi and the diseases they cause

Disease	Causal organism	Main disease foci
Superficial mycoses (dermatomycoses)		
Ringworm	*Microsporum*	Scalp of children
Favus	*Trichophyton*	Scalp
Athlete's foot	*Epidermophyton*	Between toes, skin
Jock itch		Genital region
Subcutaneous mycoses		
Sporotrichosis	*Sporothrix schenckii*	Arms, hands
Chromoblastomycosis	Several genera	Legs, feet
Systemic mycoses		
Cryptococcosis	*Cryptococcus neoformans*	Lungs, meninges
Coccidioidomycosis	*Coccidioides immitis*	Lungs
Histoplasmosis	*Histoplasma capsulatum*	Lungs
Blastomycosis	*Blastomyces dermatitidis*	Lungs, skin
Candidiasis (opportunistic)	*Candida albicans*	Oral cavity, intestinal tract

around the genital region, leading to a condition called "jock itch." Other regions which may be infected by dermatophytes include the nails, the beard, the torso, the penis, and the vagina.

The invasion caused by the fungus (Figure 12.2*c* and *d*) in each of these cases is similar; growth within the superficial tissues causes itching, reddening, cracking, and scaling of the skin. Some fungi can grow on the surface of the hair, while others can grow within the hair fiber itself. The initiation of a dermatophytic infection generally depends on host factors such as hormonal balance and various other physiological influences. Age is a factor in susceptibility to athlete's foot, which is common in adults and rare in children; and ringworm of the scalp, which is frequent in children but rare in adults. Adults may be resistant to ringworm because of increased secretions by the sebaceous glands, especially the secretion of fatty substances with antifungal activity. One of these antifungal substances, *undecylenic acid,* is used as a component of many ointments marketed for the control of fungal infections of the skin. Moisture is an important factor in the development of many surface infections. Athlete's foot usually occurs between the toes because moisture accumulates in this region, promoting fungal growth.

There are only a few antibiotics effective for treatment of fungal infections, as discussed in Chapter 5. For superficial infections, topical applications of some antifungal drugs, such as tolnaftate, micronazole, or clotrimazole are usually of some value. Griseofulvin has been used extensively for superficial mycoses. Although it is ineffective if applied directly to the skin, it is effective if given orally because the drug can accumulate in the newly synthesized keratin-containing tissues and render them resistant to fungal infection. Because such tissues gradually rejuvenate, griseofulvin therapy must be continued for prolonged periods until the newly synthesized tissues have replaced those sloughed off.

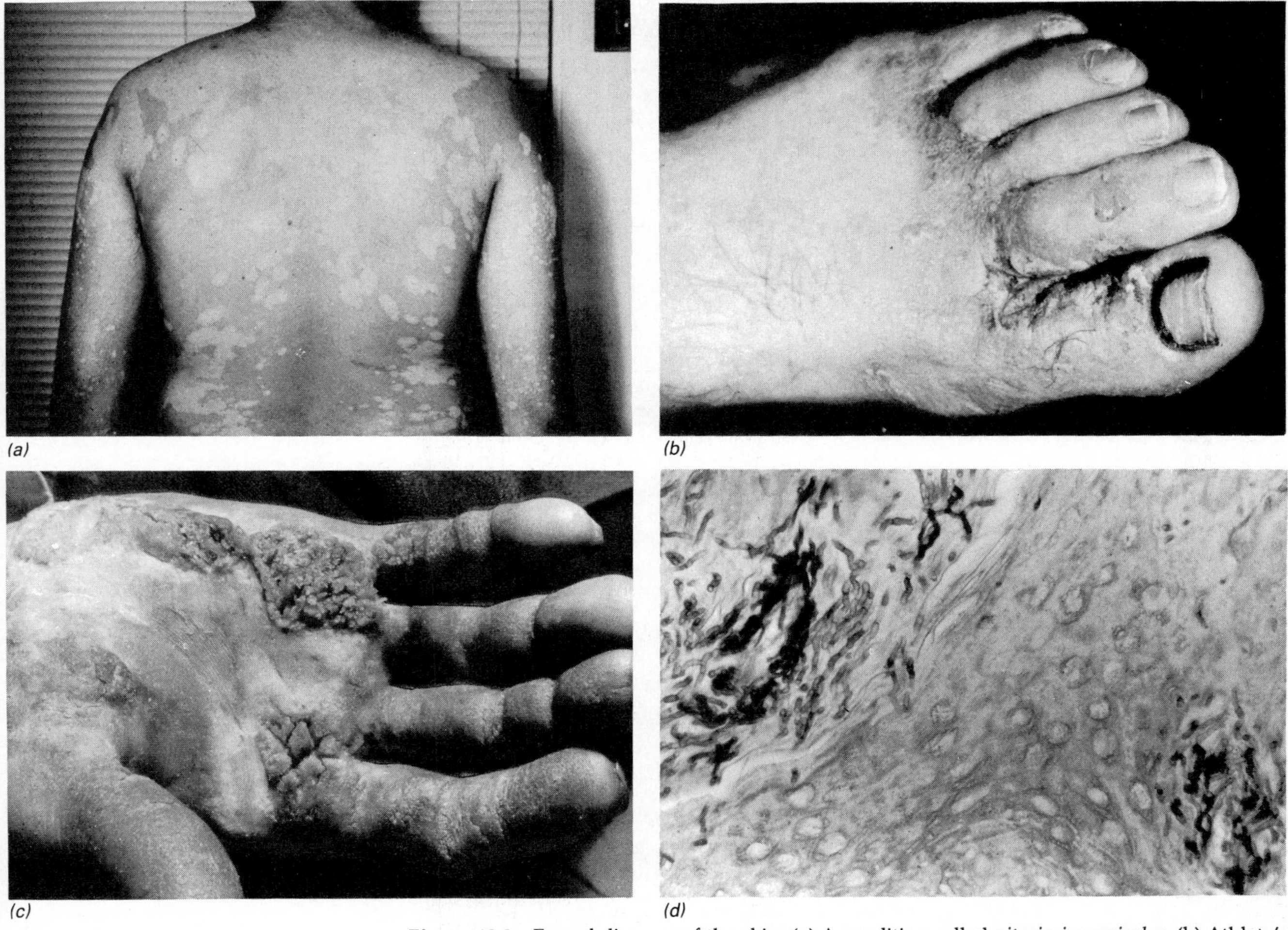

Figure 12.2 Fungal diseases of the skin. (a) A condition called *pityriasis versicolor*. (b) Athlete's foot (*tinea pedis*). (c) Infection of the hand caused by the fungus *Epidermophyton*. (d) Photomicrograph of *Epidermophyton* in infected tissue. The fungal hyphae are the dark structures in the top of the photo.

12.3 SUBCUTANEOUS AND SYSTEMIC MYCOSES

Subcutaneous mycoses may result when the skin is broken and some fungi penetrate and cause more extensive damage to subcutaneous tissues. One example is the disease *sporotrichosis*, sometimes called "rose-gardener syndrome" as it may occur after a prick by a rose thorn, resulting in the inoculation of the fungus *Sporothrix schenckii* beneath the skin. Sporotrichosis is frequently a problem among people in Central and South America, often resulting in systemic infection of the lymphatic system draining the infected subcutaneous tissue.

In the disease *chromoblastomycosis*, subcutaneous infection in the legs and feet leads to the slow development of elevated growths on the skin which have an irregular appearance similar to the surface of cauliflower (Figure 12.3). Another subcutaneous fungal disease, *rhinosporidiosis*, leads to the formation

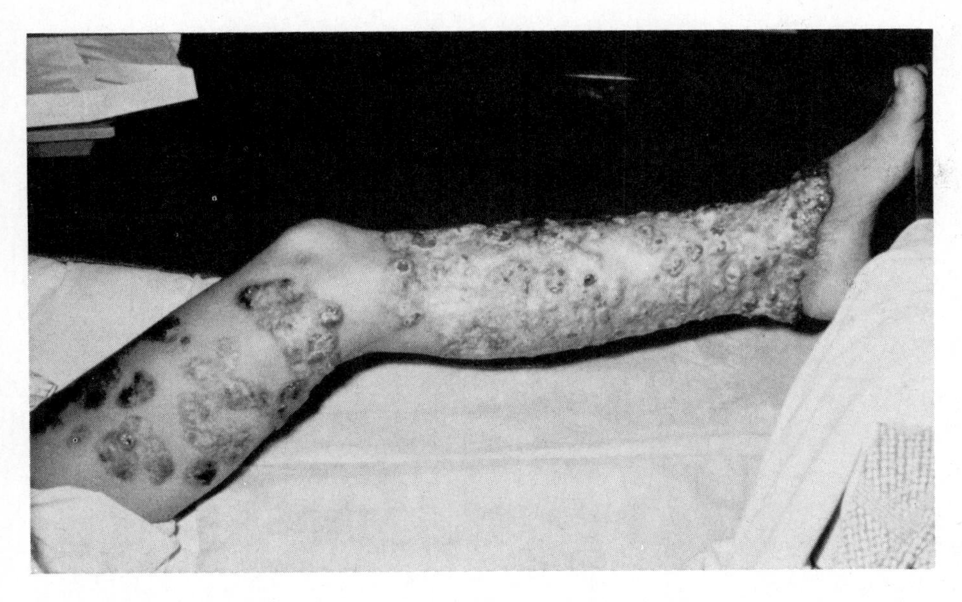

Figure 12.3 Extensive subcutaneous fungal infection, a condition called *chromoblastomycosis.*

of similar growths inside the nose on the nasal epithelium. Even more extensive deformation of tissue can occur in the disease *mycetoma,* an infection usually involving the feet. Although sporotrichosis occurs among gardeners and greenhouse employees, most subcutaneous fungal infections are rare in temperate regions (including the United States), and are mostly a problem in tropical or subtropical regions (especially in South and Central America, Africa, and India).

Systemic mycoses are diseases caused by yeasts or by filamentous fungi which can grow in a yeastlike manner at 37°C or as filaments at lower temperatures. Such fungi are said to be *dimorphic,* as they can exist in two forms (Figure 12.4). Usually, the infecting organisms reside in the soil and enter the body by inhalation. A mild respiratory disease ensues that has no characteristic symptoms and often resembles a cold. Subsequently, the organisms may become disseminated throughout the body and a chronic infection occurs, of which the symptoms are highly variable; general debilitation is common, but fatal infections are generally rare.

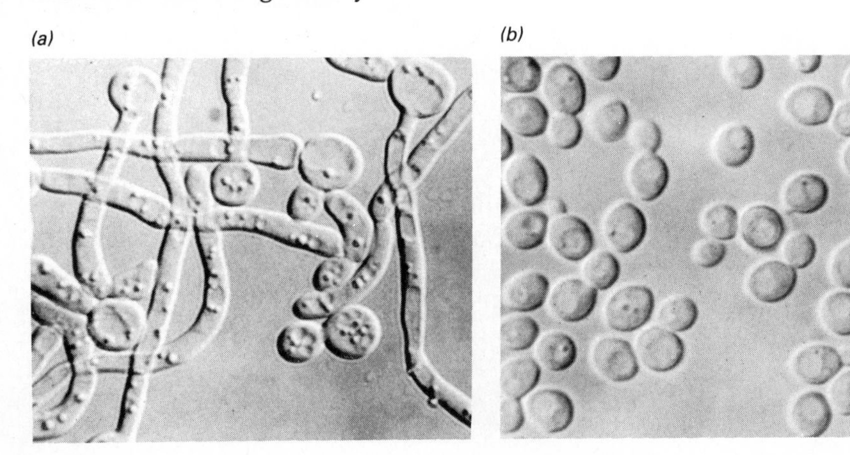

Figure 12.4 A dimorphic fungus can exist in either (a) mycelial (mold) or (b) yeast form. These are photomicrographs of the pathogenic fungus *Candida albicans.* A single yeast cell is about 7 μm in diameter.

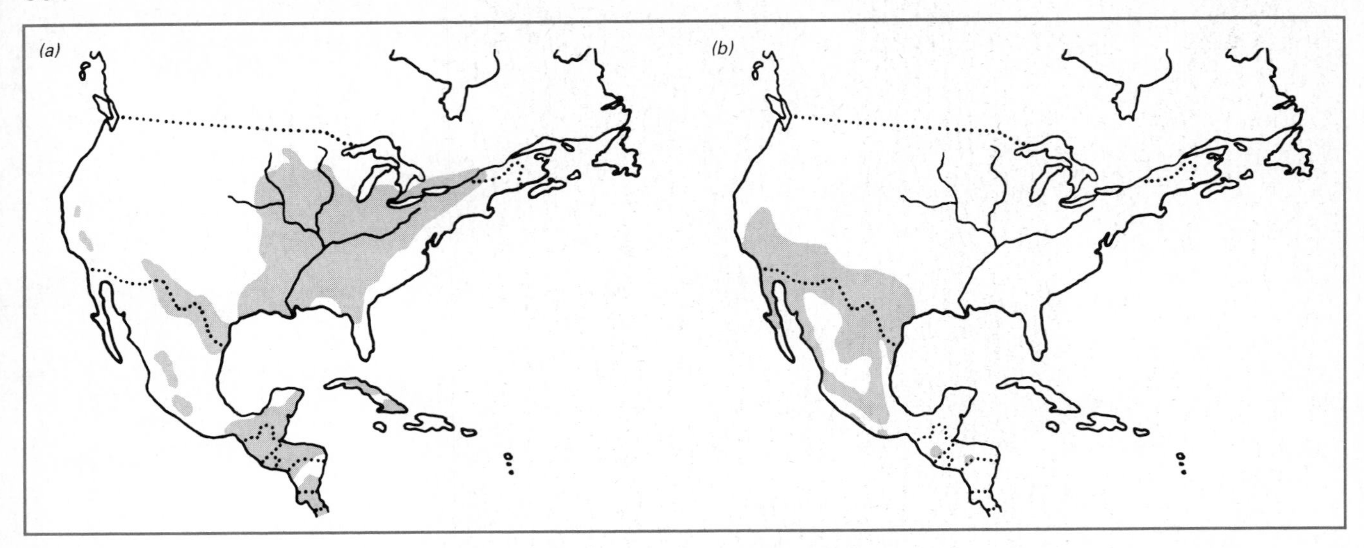

Figure 12.5 Geographic incidence of two systemic fungal diseases in North and Central America. (a) Histoplasmosis. (b) Coccidioidomycosis.

Histoplasmosis

Certain systemic mycoses are found only in restricted geographical areas. *Histoplasmosis* is found in high incidence in certain parts of the midwestern United States, especially in rural areas of the Ohio and Mississippi River valleys (Figure 12.5*a*). It is also found in some regions of South America and Africa. The causal agent, *Histoplasma capsulatum*, is found in large numbers in soils receiving droppings of chickens and other birds. Infection begins after inhalation of airborne spores from such contaminated soils. *H. capsulatum* grows initially in the yeast form within macrophages. After a cell-mediated immune response occurs, activated macrophages destroy the pathogen and limit the infection. Many human beings may become infected and develop a pulmonary disease which is usually mild and may even go unnoticed. Serologic tests have shown that as many as 90 percent of people over the age of 20 living in endemic areas have been exposed to *H. capsulatum*, indicating the widespread nature of this disease. In patients whose cell-mediated immune system is not totally competent, the pulmonary disease may become more involved, and dissemination to many other organs may result. Most cases of histoplasmosis are mild and do not require treatment. In more serious cases, the antibiotic amphotericin B is the drug of choice.

Coccidioidomycosis (San Joaquin Valley fever)

Coccidioidomycosis (also called "San Joaquin Valley fever") is generally restricted to desert areas of southwestern United States (Figure 12.5*b*); it also occurs in some regions of South America. The causal agent, *Coccidioides immitis*, grows in desert soils, and the spores are disseminated on dry windblown particles and inhaled. In endemic areas, as many as 80 percent of the inhabitants may be infected. In most people, a cell-mediated immune response causes macrophage activation, and the result is a mild lung infection. Allergies may, however, result from mild *C. immitis* infections. In a small number of cases, often involving people with an impaired cell-mediated immune response, more serious symptoms result, including chronic lung disease and infection of the meninges, bone, or skin. Prolonged treatment with amphotericin B is usually required, and harmful side effects from chemotherapy may occur.

The incidence of other systemic fungal diseases is not so well known because skin tests do not exist with which to survey populations. *Blastomycosis,* also common in the eastern United States, is another pulmonary disease which can occasionally spread to the skin and other organs. It is caused by *Blastomyces dermatitidis.*

Infection by *Cryptococcus neoformans* leads to the disease *cryptococcosis.* The distribution of this fungus is worldwide. Again, infections first occur in the lungs and are usually mild. Rarely, and usually associated with some problem in immunocompetence, the fungus may spread through the body, with involvement of the central nervous system as well as other organs. Amphotericin B is used in the treatment of both diseases.

Opportunistic fungi Some fungi are not normally pathogenic but may become so under certain conditions of lowered host resistance. These opportunistic fungi may become virulent pathogens for those suffering from diabetes, certain cancers, and in those treated with immuno-suppressive agents, such as radiation. They may also appear in those treated with massive doses of orally administered broad-spectrum antibiotics as a consequence of the elimination of the normal intestinal bacteria. Also, although most hospital infections (see Chapter 10) are caused by bacteria, fungi account for 7 percent of hospital-derived infections. Species of the yeast genus *Candida,* perhaps the most common opportunistic fungal pathogen, account for about 5 percent of these infections. Opportunistic fungal infections are also found in AIDS (acquired immunodeficiency syndrome) patients, whose immune system is impaired.

Candidiasis (thrush, vaginitis)

In addition to its role in hospital infections, *Candida* is often present in the normal flora of the mouth, vagina, and intestinal tract. When it becomes invasive, a variety of lesions may appear, depending on location. One of the most common syndromes is *thrush,* in which the organism grows in white patches on the mucous membranes of the mouth and pharynx (Figure 12.6). Thrush is most common in the first days of life in newborns and in the terminal stages of a wasting disease. Another common problem is vaginitis caused by *Candida* species; it occurs in 10–17 percent of all women, and is more common during pregnancy. When *Candida* becomes more invasive in immunocompromised patients, many different and more severe complications may result.

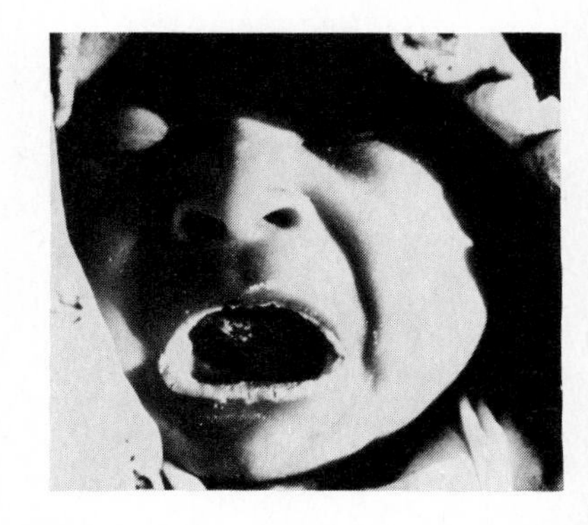

Figure 12.6 The disease *thrush,* caused by the dimorphic fungus *Candida albicans,* an opportunistic pathogen.

Many other ubiquitous fungi are ordinarily harmless but cause opportunistic infection in immunocompromised patients. These include members of the genus *Aspergillus,* which commonly cause pulmonary disease, and several fungi related to the bread mold, *Rhizopus,* which can invade the nose and throat and spread rapidly throughout the sinuses, eyes, and brain, killing 90 percent of those infected. Although opportunistic fungal infections are rare and usually involve people whose immune system is not normal, these infections can be serious, and are thus important when they occur.

TABLE 12.2 Some types of allergy caused by fungi

Type of allergy	Fungal allergens
Asthma	*Alternaria, Helminthosporium, Drechslera, Cladosporium, Penicillium, Aspergillus*
Grain dust allergies	Many fungi
Occupational allergies	
Mushroom picker's lung	*Pleurotus ostreatus*
Farmer's lung	*Aspergillus, Penicillium* (also certain actinomycete bacteria)
Maple bark stripper's lung	*Cryptostroma corticale*
Maltster's lung	*Aspergillus clavatus*
Wood pulp worker's lung	*Alternaria* species
Cheese worker's lung	*Penicillium casei*

12.4 FUNGAL ALLERGIES

Mold spores are commonly found in the air in amounts up to 100,000 spores per cubic meter. Thus, fungal spores are dispersed widely, and are inevitably inhaled by humans. Many allergies result from inhalation of mold spores (Table 12.2). Such allergies may occur in many individuals within a population, or in a particular group exposed to unusually large numbers of fungal spores because of their occupation.

Asthma

Asthma, a condition affecting many individuals, may follow inhalation of spores of many different types of fungi. After a sensitizing exposure, subsequent exposures may lead to an immediate-type hypersensitivity reaction in the upper respiratory tract, resulting in difficulty in breathing. The mechanism of immediate-type hypersensitivity was described in Section 9.7. Sometimes these allergy symptoms can be severe and may lead to death. Those people most commonly affected are adults over the age of 40 or children. In addition to air-borne distribution, fungal allergies can arise from the spores of fungi which are commonly found on cereal grains.

Alternatively, because of their occupation some individuals may be exposed to unusually large numbers of mold spores. When moldy materials are disturbed under confined conditions, as in a grain-storage bin or silo, the density of spores in the air may reach 10^9 per cubic meter! People exposed to such high concentrations may develop allergic reactions of either the immediate- or delayed-type, but the condition is usually chronic. A good example is the disease called "Farmer's Lung," common among farmers who become exposed to large numbers of spores released during threshing or harvesting. Although antibody-mediated allergic responses may occur in Farmer's Lung, delayed-type hypersensitivity probably occurs more frequently, leading to chronic and debilitating lung disease. Other examples of fungal allergies which occur in association with certain occupational hazards are listed in Table 12.2.

12.5 FUNGAL TOXINS

Some fungi produce chemicals which are toxic to humans. There are two categories of fungal toxins: those which are produced by mushrooms con-

TABLE 12.3 Some toxin-producing fungi

Fungus	Toxin	Disease
Mushroom poisons		
Amanita phalloides (death cap)	Amanitin, phallotoxins	Violent diarrhea, cramps, abdominal pain, liver and kidney failure, coma, death
Most *Gyromitra* species	Gyromitrin	Damage to liver, gastrointestinal tract, central nervous system; sometimes coma and death
Cortinarius orellanus	Orellanines	Liver and kidney damage, death
Galerina sulciceps	—	Spasms, nausea, death
Coprinus atrementarius (inky cap)	Coprine	Hot flushed feeling, palpitation, chest pain, nausea, vomiting (following alcohol consumption)
Inocybe and *Clitocybe* species	Muscarine	Perspiration, salivation, cramps, diarrhea, headache
Many other mushrooms	Gastrointestinal irritants	Nausea, vomiting, cramps, diarrhea
Mushroom hallucinogens		
Amanita muscaria	Ibotenic acid, muscimol, muscazone	Incoordination, dizziness, jerking, expanded perception and understanding
Psilocybe cubensis (flesh of the gods) and other *Psilocybe*, *Paneolus*, and *Conocybe* species	Psilocybin	Hallucinations
Mycotoxins		
Aspergillus flavus	Aflatoxin	Liver damage liver cancer
Aspergillus ochraceus	Ochratoxin	Fatty liver
Aspergillus versicolor	Sterigmatocystin	Liver damage, liver cancer
Aspergillus clavatus	Patulin	General cell damage, cancer
Claviceps purpurea	Ergot poisons	Hallucination, smooth muscle contraction, loss of limbs
Fusarium species	Trichothecenes	Blistering, internal bleeding
Fusarium graminearum	Zearalenone	Estrogenic hormone
Penicillium citrinum	Citrinin	Kidney damage
Stachybotrys species	Stachybotrytoxin	General cell damage, hemorrhage

sumed directly as food, and those which are produced when certain molds grow on other food products. Mushroom poisoning is called *mycetismus*, whereas the toxins produced by filamentous fungi (molds) are referred to as *mycotoxins*. The fungi that produce these toxins, the types of toxins, and their actions are given in Table 12.3.

Mushroom poisoning (mycetismus)

The danger of poisonous mushrooms comes from the fact that mushrooms are considered such a delicacy that many people hunt their own. Though some mushrooms are very flavorful, others are deadly. For example, a single cap of the mushroom *Amanita phalloides*, appropriately named the "death cap," contains more than enough poison to kill a normal human. There is no known antidote. This points out how careful one must be in the identification of mushrooms (see the Box on page 38).

Numerous mushrooms produce toxins (Table 12.3). Some cause mild gastrointestinal irritation, but many cause severe gastrointestinal complications, liver and kidney failure, and neurologic disorders which can lead to coma and death. In the United States there are several hundred cases of mushroom poisoning each year; about two-thirds of these cases involve children less than five years old.

Some mushrooms produce chemicals which elicit psychological effects. Members of three genera, for example, produce *psilocybin*, which is structurally similar to lysergic acid diethylamide (LSD). *Psilocybe cubensis*, called "flesh of the gods," has been gathered and used by people from some tribal American cultures as a part of their religious experience. Another species, *Amanita muscaria*, produces a series of components which elicit a more controlled psychological effect. This fungus has been used by some Old World priests to provide "religious visions." Because this fungus is rare and the active ingredient is excreted in active form in the urine, groups involved in *muscarine* culture often collect and drink the urine of those who have already "seen heaven."

Some fungi produce chemical substances called **mycotoxins**. Some of these toxins are extremely active against humans, causing a wide variety of diseases: blood diseases, disturbances of nerve function, liver damage, kidney malfunction, limb deformation during embryonic growth, hormonal disturbances resulting in changed sexual behavior and decreased fertility, and cancer. Because of the extreme toxicity of some of the toxins, the diseases are often fatal. The fungi that produce mycotoxins (Table 12.3) can grow in foods, feeds, and grain, producing microscopic colonies invisible to the consumer. If the food is not obviously moldy, it could be eaten, the toxin ingested, and a disease elicited. The most dramatic examples of mycotoxins are the **aflatoxins**, a group of products of the fungus *Aspergillus flavus*. In some areas of Africa where climatic conditions are very favorable for the growth of *A. flavus* on peanuts and grains, the incidence of liver cancer in the population is markedly higher than in areas where aflatoxin is not so commonly present in foods. In developed countries, such products are now carefully monitored before being distributed to consumers. In some regions, another fungal toxin can be a significant problem. The fungus *Claviceps purpurea* grows in the heads of cereal grains, replacing the grain with bodies containing *ergot poisons*, which are structurally related to LSD (see the Box on the next page). In addition to causing hallucinogenic effects and abortion, ergot poisons cause smooth muscle contraction which may limit blood flow to extremities and cause loss of limbs. A mixture of toxins produced by *Fusarium* species causes blistering and eventually internal bleeding when absorbed through the skin.

Farm animals may also be affected by eating moldy feed, and serious economic effects on animal production can occur. Thus, an understanding of the factors influencing fungal growth and mycotoxin production in foods and feeds is of great importance. The agricultural importance of mycotoxins is considered further in Chapter 17.

12.6 PROTOZOAL DISEASES

The overall nature and cellular structure of protozoa was discussed in Section 2.9. There are over 30,000 species of protozoa, most of which are harmless.

WITCHCRAFT AND ST. ANTHONY'S FIRE: THE FUNGAL CONNECTION

In 1951 some of the inhabitants of Pont-St. Esprit, a small town in France, were visited by a horrible torment. Convinced that an evil spirit had entered their bodies, they ran in panic through the streets. Several people even died. Eventually, this baffling affliction was recognized as a modern form of *St. Anthony's Fire*, a disease which was widespread in northern Europe in the Middle Ages. The name of the disease was derived from the fact that the Roman Catholic Order of St. Anthony then operated hospitals to treat patients with the disease, and from the symptoms, which included severe itching and burning. The disease is now known medically as *ergotism*, and is caused by the fungus *Claviceps purpurea*, a plant pathogen which grows in rye kernels and produces the ergot poisons. Apparently, the people of Pont-St. Esprit had eaten rye bread made from grain contaminated with the ergot fungus. Among the ergot toxins are several mind-altering compounds related to the psychedelic drug lysergic acid diethylamide (LSD). The 1951 affliction in France was an aberration, since the disease in the twentieth century has been mostly eliminated by careful cleaning of the rye grain.

Perhaps the most fascinating visitation to which ergotism has been attributed was the witchcraft terror that took place in Salem, Massachusetts in 1692. Salem, one of New England's oldest cities, was the site of the famous Salem witchcraft trials. Although it is difficult at this late date to be certain of an ergot connection, the Salem case is an interesting example of how a microbe might cause an extreme alteration in human behavior. In Salem about 30 people, chiefly women, were executed, at least one by the horrible medieval procedure of being pressed to death under heavy weights. These piteous creatures were charged with practicing the occult arts, but one hypothesis is that their minds had been deranged by eating ergot-infested grain. Cotton Mather, the infamous Congregational minister who typified the most extreme form of American Puritanism, was fascinated by the presence of the supernatural in daily life and is generally credited with stimulating the frenzy which culminated in the Salem witchcraft trials. Superstition and fantasy are still common mind-sets of primitive cultures, and in more refined forms are even in vogue in so-called advanced civilizations. One of the more fascinating aspects of microbiology is that extreme examples of grotesque delusions may have a fungal connection.

Protozoa commonly occur in aquatic environments where they participate in decomposition of organic matter and are an important part of the aquatic food chain. They also are essential in the digestion process in ruminant animals (see Chapter 17). It is interesting that ciliated protozoa rarely cause human disease. In contrast, many flagellated, amoeboid, and sporozoan protozoa cause important diseases of humans.

12.7 DISEASES CAUSED BY FLAGELLATED PROTOZOA

Most protozoa that are motile by means of flagella are harmless aquatic or soil organisms, but a few are pathogenic to animals and humans. The most important pathogenic flagellates are the **trypanosomes**. These organisms cause a number of serious diseases of humans and vertebrate animals, including the feared disease **African sleeping sickness**. At any one time as many as one million Africans may suffer from this disease.

African sleeping sickness

The trypanosomes are highly variable in morphology, both from species to species and within a single species at various stages of the life cycle. In

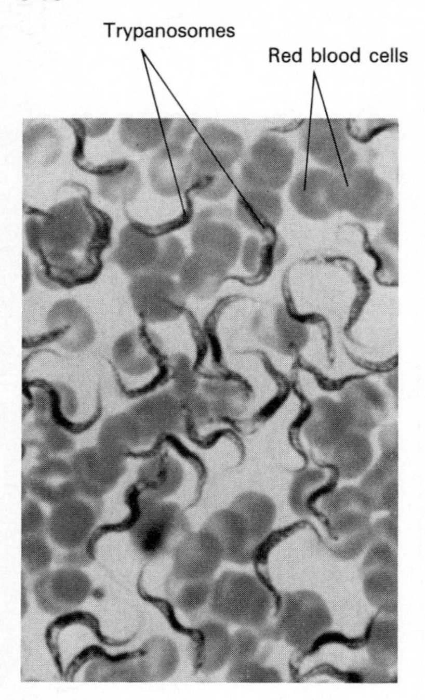
Trypanosomes
Red blood cells

Figure 12.7 Photomicrograph of the causal agent of African sleeping sickness, *Trypanosoma gambiense*, in blood. The flagellate cells are about 10 μm in length.

Trypanosoma, the genus infecting humans, the protozoa are rather small, around 20 μm in length, and are thin, crescent-shaped organisms with a single flagellum that originates in a basal body and folds back laterally across the cell, where it is enclosed by a flap of surface membrane. Both the flagellum and the membrane aid in propelling the organism, making possible an effective movement even in blood, which is rather viscous. Organisms of several sizes and shapes are seen in the blood of infected individuals, with short and broad forms, long and slender forms, and intermediate forms all seen in the same blood smear (Figure 12.7). *Trypanosoma gambiense* is the species that causes the chronic and eventually fatal African sleeping sickness. In humans, the parasite lives and grows primarily in the bloodstream, but in the later stages of the disease, invasion of the central nervous system may occur, causing an inflammation of the brain and spinal cord that is responsible for the characteristic neurological symptoms of the disease. In late stages, weakness, mental lethargy, and sleepiness, eventually leading to continuous sleep, are characteristic symptoms. Trypanosomes undergo change of their surface antigens, resulting in evasion of the host immune response. Trypanosomes may have hundreds of different genes coding for hundreds of different antigens. By selectively expressing these genes via gene transposition (see Chapter 7), the organism presents unrecognized antigens to the immune system, leading to waves of infection in the bloodstream.

Trypanosoma gambiense is transmitted from person to person by the tsetse fly, a blood-sucking fly of the genus *Glossina*, found only in Africa (Figure 12.8). The organism divides in the intestinal tract of the fly and then invades the insect intestinal cells, where further multiplication is accompanied by an

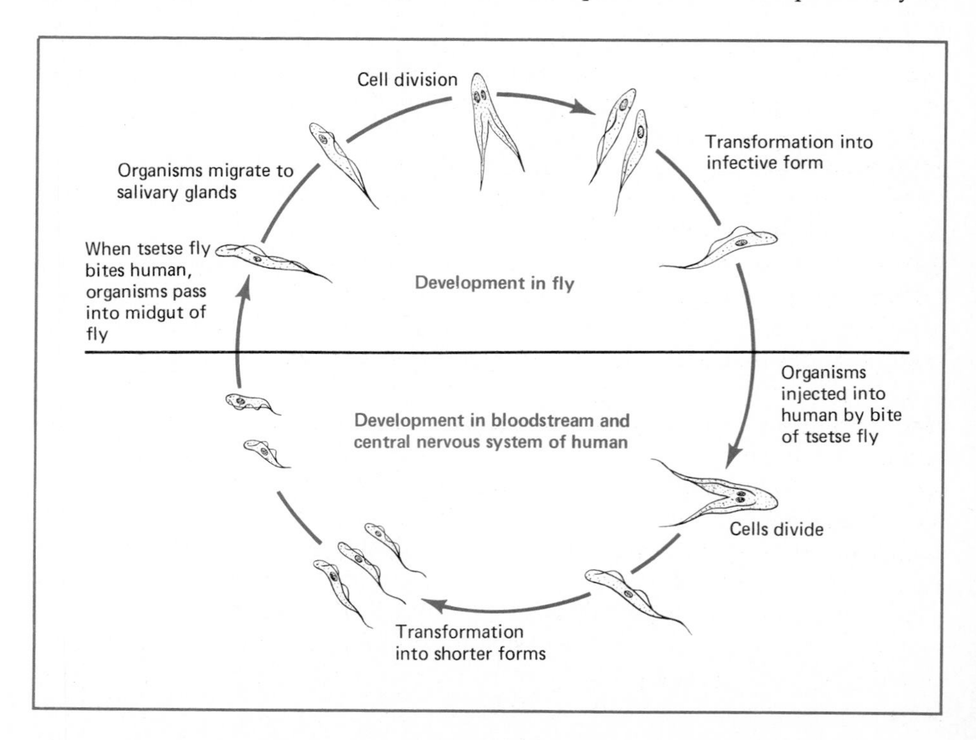
Cell division
Transformation into infective form
Organisms migrate to salivary glands
When tsetse fly bites human, organisms pass into midgut of fly
Development in fly
Organisms injected into human by bite of tsetse fly
Development in bloodstream and central nervous system of human
Cells divide
Transformation into shorter forms

Figure 12.8 Life cycle of *Trypanosoma gambiense*.

alteration in morphology. The organism then invades the salivary glands and mouthparts of the fly, from which sites it may be transferred to a new human host. The spread of the disease can be prevented by destruction of the tsetse fly. Control of the fly requires the use of rather potent insecticides, such as DDT. By use of such insecticides and by clearing of underbrush in which the flies congregate, the tsetse fly has been eliminated from many parts of West Africa that previously had been heavily infested. Several new drugs have also been developed for treating the disease in humans, and these have also reduced the incidence of the disease by reducing the number of sources from which the flies could become infected. Another trypanosome, *T. brucei*, infects domestic animals such as cattle and pigs, causing symptoms similar to African sleeping sickness and seriously restricting the ability to raise cattle in certain parts of Africa. This parasite, which is probably a variant of the form that attacks humans, is also transmitted by a species of tsetse fly, whose control is effected at the same time that the vector of *T. gambiense* is eliminated. Control of the tsetse fly has made possible the establishment of a cattle industry in many parts of Africa previously unsuitable for this purpose. Another trypanosome species, *T. cruzi*, causes a zoonosis called Chagas' disease, common in Central and South America.

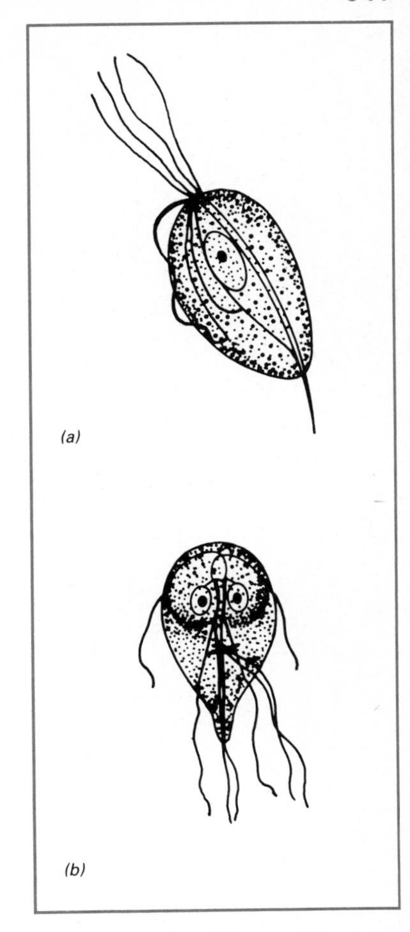

Figure 12.9 Two flagellated protozoa which are pathogenic to humans. (a) *Trichomonas vaginalis*, a common cause of nongonococcal urethritis (NGU). (b) *Giardia lamblia*, a common water-borne pathogen which causes giardiasis.

The *trichomonads* are protozoa which live in of the vertebrate alimentary and genitourinary tracts. These flagellates have three to six flagella, one of which typically is trailed behind the cell during movement (Figure 12.9*a*). Most of the pathogenic forms produce cysts, which are the infective stages by which the species is transferred from host to host. There are three species which grow in humans: *Trichomonas tenax* (or *T. buccalis*), found in the mouth, is nonpathogenic and most likely is transmitted by kissing; *T. hominis*, in the colon; and *T. vaginalis*, in the genitourinary tract. The last species, despite its name, is also found in the genitourinary tract of males. A surprisingly high proportion of the human population harbors one or more of these three species; *T. hominis* is found in 2 percent and *T. tenax* in 10 percent of the population; and *T. vaginalis*, in 25 percent of females and 4 percent of males.

Trichomoniasis

Infections of the urogenital tract by *T. vaginalis* are very common. Since the spread from one infected individual to another may occur during sexual intercourse, it has been estimated that **trichomoniasis** may be one of the most common sexually transmitted diseases. It is one of the diseases described collectively as *nongonococcal urethritis* (*NGU*) (see Section 11.20). Cells of *T. vaginalis* do not form cysts, but are able to survive for a few hours outside the body if conditions are warm and moist; thus, contamination via towels, underwear, or toilets may also be possible. The symptoms are usually mild, sometimes being confused with those of gonorrhea, and include irritation, itching, and purulent discharge. More severe symptoms may result if there is a secondary bacterial infection. The organism can be recognized microscopically in exudates, thus permitting diagnosis. Trichomoniasis can be treated with metronidazole, but sexual partners as well as others in the household of a person known to be infected should receive treatment.

Giardiasis

The flagellate *Giardia lamblia* (Figure 12.9*b*) causes an acute form of gastroenteritis called **giardiasis**. This disease appears to be increasing in incidence and has been associated with many water-borne disease outbreaks in recent years. Between 1946 and 1980, *G. lamblia* accounted for 5.6 percent of the outbreaks of water-borne illness in the United States, and these outbreaks involved more individuals than the common water-borne outbreaks due to bacteria: salmonellosis and shigellosis (see Table 15.2). Giardiasis is associated with drinking unfiltered water in wilderness areas, and because of this is sometimes called "backpacker's disease." As many as 20 percent of people in the United States may be carriers of *G. lamblia*. The organism spreads by the fecal-oral route, but wild animals, as well as humans, harbor the pathogen. Cysts are ingested and the organism multiplies in the gastrointestinal tract. Symptoms include pain, cramps, foul-smelling diarrhea (sometimes alternating with constipation), flatulence, malaise, and weight loss. The drugs quinacrine hydrochloride and metronidazole are useful in treatment. The prevention of giardiasis involves proper water treatment, but this can be a problem as the cysts are not killed by common disinfectants such as chlorine. Filtration is required to remove the cysts from the drinking water supply (see Chapter 15). The organism is also killed by boiling, so that for sanitation in wilderness areas either filtration or boiling are recommended.

12.8 DISEASES CAUSED BY AMOEBOID PROTOZOA

A wide variety of amoebas live in tissues of humans and other vertebrates, the usual habitat being the oral cavity or the intestinal tract. *Entamoeba histolytica* can serve as an example of a pathogenic amoeba. This organism is found in the intestinal tract of a high percentage of individuals living in regions where sanitation is poor. In Egypt, for example, a large percentage of the population live within a few miles of the Nile River. Poor sanitation and common exposure to water probably explain the fact that nearly three-fourths of the people in some Egyptian communities carry *E. histolytica* in their intestinal tract. In the United States only about 1–3 percent of the population are carriers. In many cases, infection causes no obvious symptoms, but in some individuals it produces ulceration of the intestinal tract, which results in a diarrheal condition called **amoebic dysentery**. *E. histolytica* may also spread to the liver and cause abscesses. The organism is transmitted from person to person in the cyst form. Cyst germination occurs in the intestinal tract, and the amoebas divide by binary fission, establishing a population which invades and feeds on cells of the intestinal epithelium, resulting in ulcer formation (Figure 12.10). Subsequently the amoebas form cysts, in which form they leave the intestinal tract in the feces. Amoebic dysentery is diagnosed by microscopic examination of the stool for cysts, which can be separated and concentrated by centrifugation in a solution of zinc sulfate. The latter material is of high specific gravity, and the cysts float to the surface; a loopful of the surface film can be stained with iodine and examined under the microscope. The cysts are resistant to drying and hence can survive for long periods of time outside the body. *Entamoeba histolytica* is transmitted by fecal contamination of water supplies and by houseflies. Treatment is by the use of the drugs metronidazole and diiodohyroxyquin.

Amoebic dysentery

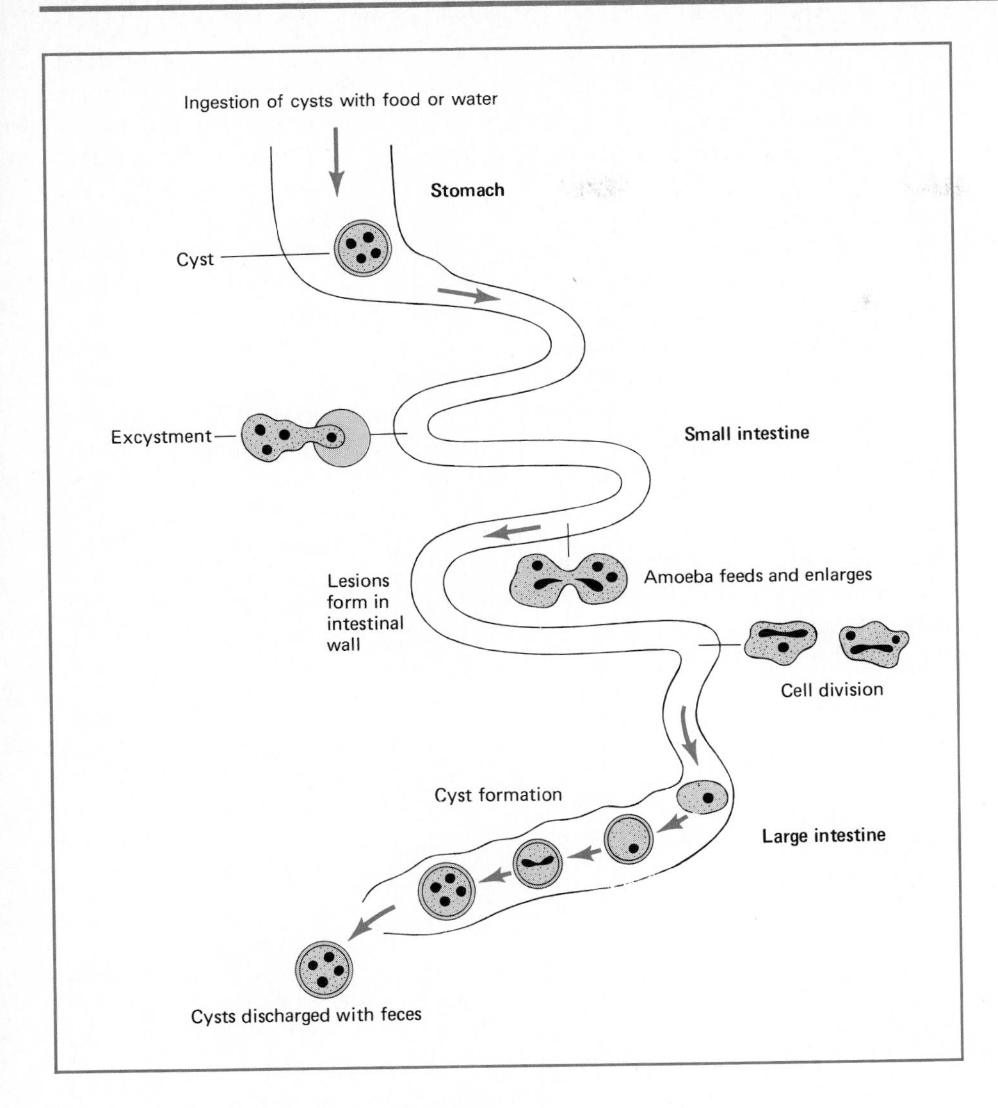

Figure 12.10 Life cycle of a parasitic amoeba, *Entamoeba histolytica*, in the human intestine.

12.9 SPOROZOA AND MALARIA

The *Sporozoa* constitute a large class of protozoa, all of which are obligate parasites. They are characterized by a lack of motile adult stages and by a nutritional mode of life in which food is generally not ingested but is absorbed in soluble form through the outer wall, such as occurs in bacteria and fungi. Although the name "sporozoa" implies the formation of spores, members of this group do not form true resting spores, as do some bacteria, algae, and fungi, but instead, produce structures called *sporozoites* which are involved in transmission to new hosts. Numerous kinds of vertebrates and invertebrates serve as hosts for sporozoa, and in some cases, an alternation of hosts occurs, with some stages of the life cycle occurring in one host and some in another. The most important members of the class *Sporozoa* are the coccidia, parasites of birds, and the plasmodia (malaria parasites), which infect birds and mammals, including humans. Our discussion will be restricted to the plasmodia.

Malaria

The *malaria parasite* is one of the most important human pathogens and has played an extremely significant role in the development and spread of human culture. Malaria is still endemic in a large area of the world (see Figure 10.9) and millions of people are infected annually. Four species infect humans: *Plasmodium vivax, P. flaciparum, P. malariae,* and *P. ovale*. These differ in the degree of severity of symptoms they cause and in certain aspects of their life cycles. *P. falciparum* causes the most severe form of malaria. However, we shall consider *P. vivax* in greatest detail, as it is the most widespread species and the one about which the most information is available.

This organism carries out part of its life cycle in humans and part in the mosquito, by which vector it is transmitted from person to person. Only mosquitoes of the genus *Anopheles* are involved, and since these inhabit primarily the warmer parts of the world, malaria occurs predominantly in the tropics and subtropics. The life cycle of *P. vivax* is complex (Figure 12.11). The human host is infected by plasmodial sporozoites, small elongated cells produced in the mosquito, which localize in the salivary gland of the insect. When biting, the female mosquito inserts her proboscis into a capillary, thereby inoculating the sporozoites directly into the bloodstream. The sporozoites are carried throughout the body and are removed from the blood by phagocytic cells of the lymph nodes, spleen, and liver. Replication occurs in the liver,

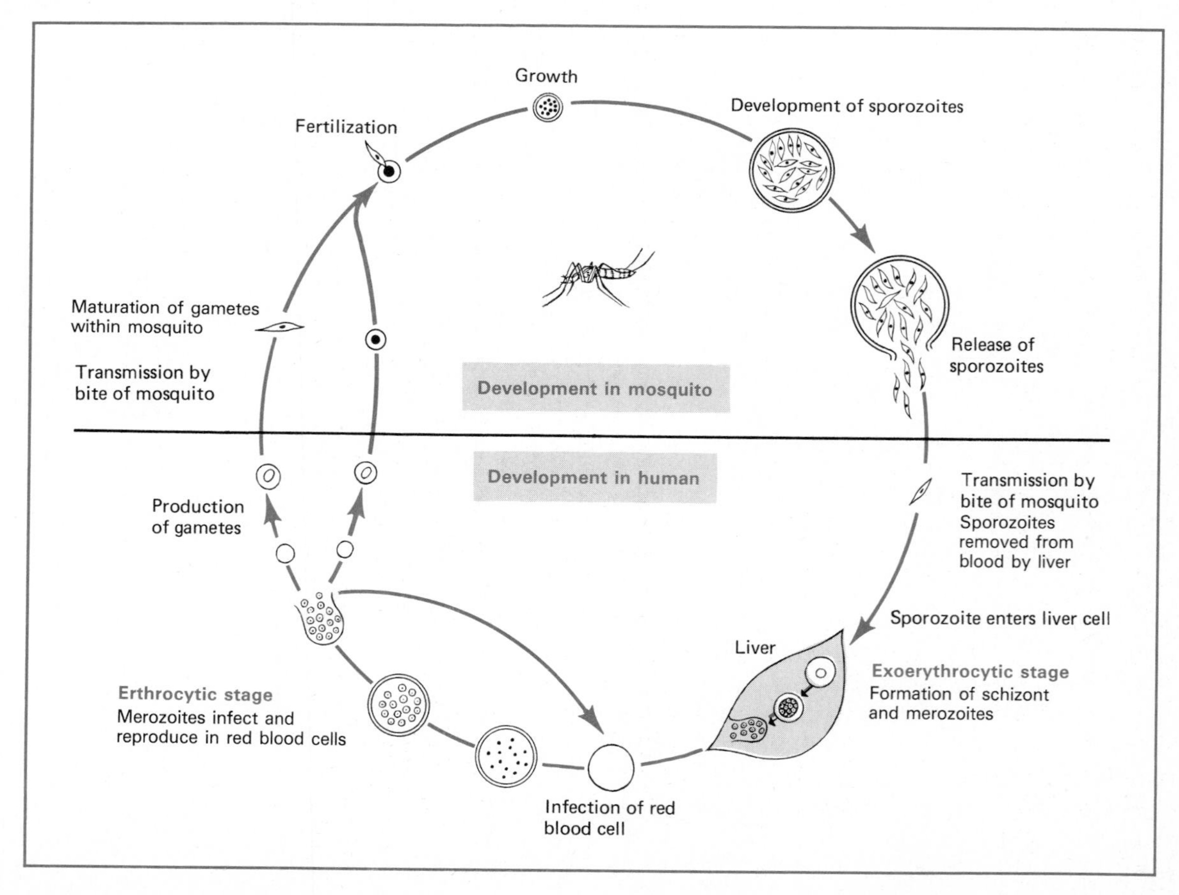

Figure 12.11 Life cycle of the malaria pathogen, *Plasmodium vivax*.

and during this stage (termed the *exoerythrocytic stage*), the parasites are absent from the blood. The sporozoite becomes transformed to a *schizont*, which replicates by enlarging and segmenting into a number of small cells called *merozoites*; these cells are liberated from the liver into the blood stream. Some of the merozoites infect the red blood cells, initiating the *erythrocytic stage* (Figure 12.12). The cycle in the red blood cells proceeds as in the liver and usually occupies a definite period of time, 48 hours in the case of *P. vivax*. It is during the erythrocytic stage that the characteristic symptoms of fever and chills occur, the chills occurring when a new brood of cells is liberated from the erythrocytes. Anemia and enlargement of the spleen are also common symptoms. Not all the cells liberated from red cells are able to infect other erythrocytes; those which cannot, called *gametocytes*, are infective only for the mosquito. If these gametocytes happen to be ingested when another insect of the proper species of *Anopheles* bites the infected person, they mature within the mosquito into *gametes*. Two gametes fuse, and a zygote is formed; the zygote migrates by amoeboid motility to beneath the outer wall of the insect's intestine, where it enlarges and forms a number of sporozoites. These are liberated, some of them reaching the salivary gland of the mosquito, from where they can be inoculated into another person; and the cycle begins again.

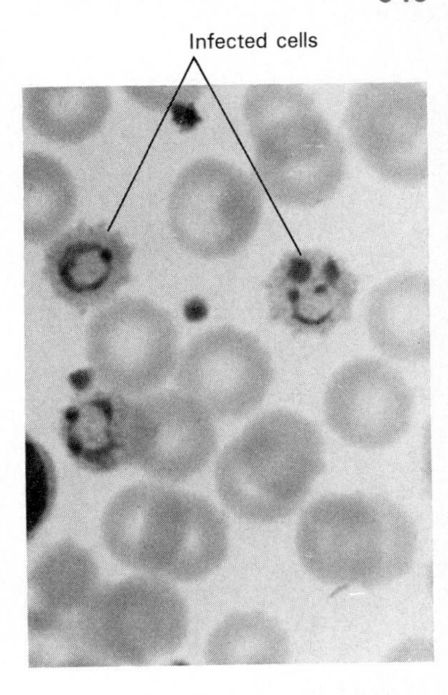

Figure 12.12 Photomicrograph of red blood cells infected with the malaria pathogen, *Plasmodium vivax*.

Conclusive evidence for the diagnosis of malaria in humans is obtained by examining blood smears for the presence of infected cells; the various stages of replication in the erythrocytic phase are observed microscopically. Diagnosis of the disease in the tissues (exoerythrocytic stage) is much more difficult.

The complex cycle shown in Figure 12.11 can be broken by eradication of the *Anopheles* mosquito; destruction of the vector has essentially eliminated malaria from the United States and Europe (see Figure 10.10). Even so, some infected persons may continue to serve as reservoirs for the plasmodium, and a renewed spread of the disease may result if the mosquito returns to these areas. Chloroquine is the drug of choice for treating the erythrocytic phase, but it does not cause a complete cure of the disease. The stages outside the bloodstream (the exoerythrocytic stages) can be eliminated by use of the drug primaquine, and in combination these drugs can effect a complete cure. Reinfection, of course, can still occur. Since 1978, chloroquine-resistant *P. falciparum* strains have been found in many areas; cases involving these strains must be treated with quinine sulfate plus pyrimethamine. As mentioned in Section 10.7, malaria in the United States is commonly associated with travellers returning from endemic areas. A mixture of chloroquine and fansidar (sulfadiazine plus pyrimethamine) are recommended as prophylactic measures for those traveling to such regions. Recently, however, complications following use of fansidar have led to more cautious administration of this drug as a chemoprophylactic.

Plasmodium vivax has been very difficult to grow in culture, and because of its specificity for humans, experimental infection has also been difficult. For these reasons, most of the experimental work on malarial parasites has been done with species of *Plasmodium* that infect birds or rats, and it is with these forms that most of the studies on the development of new drugs have been carried out. Recently, scientists have been able to grow *P. falciparum* in laboratory cultures, and this has permitted the investigation of the basis for natural resistance (see below).

Natural resistance to malaria Red blood cells are the targets of the four species of the protozoan *Plasmodium* which can cause the disease malaria. During the erythrocytic stage of the disease, *Plasmodium* must locate and invade red blood cells, within which they multiply (see Figure 12.12). During the course of evolution, some humans have developed differences in the makeup of their red blood cells which increase their resistance to malaria. These changes in human red cells decrease the effectiveness with which *Plasmodium* cells can either enter or grow within the red blood cells. As mentioned in Section 9.9, red blood cells possess chemicals on their surface which make them antigenically unique. The major blood groups are determined by the A and B antigens, but there are many other antigens on red blood cells, including one type called the Duffy antigens. About 90 percent of black people in West Africa, or blacks in North America who are descendants of West Africans, lack the Duffy blood group antigens. Most West African blacks are also resistant to *Plasmodium vivax,* the species which is endemic in this region. It has been shown that red blood cells which lack the Duffy antigen cannot be invaded by *Plasmodium vivax* cells, apparently because the parasite binds to chemical groups of the Duffy antigen in the initial penetration of the red blood cell. Resistance to malaria is thus linked to a change in the human blood groups so that the binding site for the infectious agent is no longer present.

An interesting correlation exists between the distribution of malaria and the distribution of the genetic disease *sickle cell anemia* (see Figure 12.13). Red blood cells contain the protein pigment hemoglobin which is used to

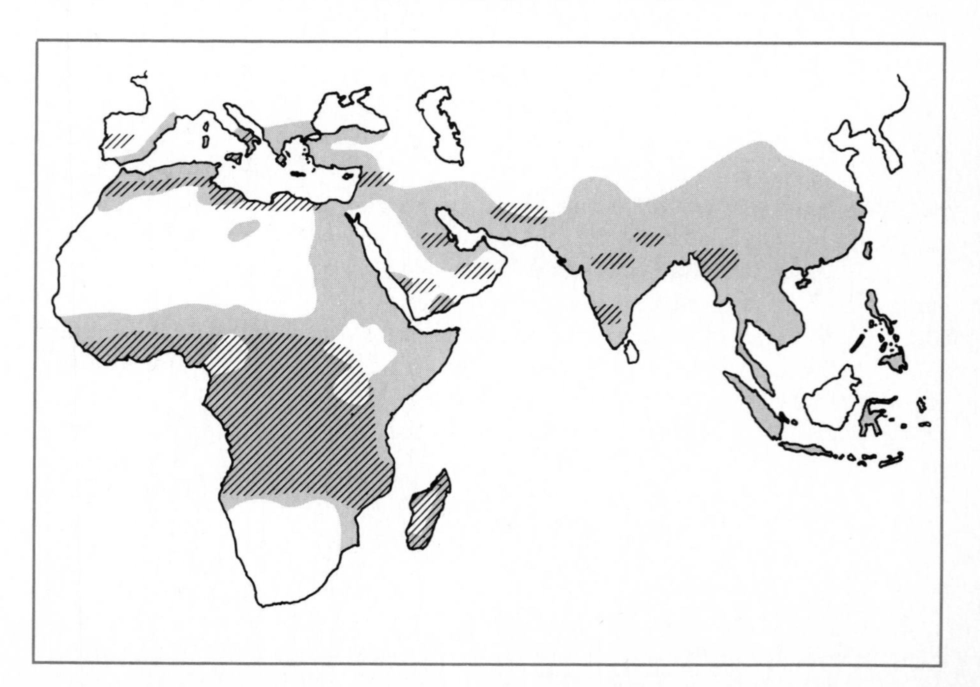

Figure 12.13 Correlation between the distribution of malaria (color) and sickle cell anemia (lines).

transport oxygen throughout the body. Sickle cell anemia results from a mutation of a single nucleotide in the gene coding for hemoglobin; one amino acid change results in the formation of faulty hemoglobin. Because humans receive a copy of the hemoglobin gene from each parent, they may possess two normal hemoglobin genes, two faulty genes, or one normal and one faulty gene. People who possess two faulty hemoglobin genes develop sickle cell anemia and usually survive poorly. Individuals who possess one copy of the normal and one copy of the faulty hemoglobin gene, a condition called "sickle cell trait," have in their red cells both types of hemoglobin. Unless the individual with sickle cell trait is deprived of oxygen (such a condition might occur at high elevation where air does not contain as much oxygen) sickle cell disease does not occur and the individual is healthy. Because most of West Africa is at low elevations, oxygen deprivation has not been a problem to those with sickle cell trait. Although *Plasmodium falciparum* cells are able to invade the red blood cells of people with sickle cell trait, they are unable to grow within these cells. Thus, by carrying a defective gene, individuals become more resistant to an infectious disease to which they are at high risk.

SUMMARY

Mycoses are infectious diseases caused by fungi. Although most fungi live in soil, many can invade the body and cause disease. **Dermatophytic** fungi invade only the keratinized surface cells of skin, causing **superficial mycosis**. Examples include **athlete's foot**, one of the most common infectious diseases, and **ringworm**. When the physical barrier of skin is broken, **subcutaneous mycosis** can develop. One example is the disease **sporotrichosis**, which is common among gardeners, and may occur after the prick of a rose thorn inoculates the fungus beneath the skin. A few fungi, notably **dimorphic** fungi, can cause **systemic mycosis**. **Histoplasmosis** and **coccidioidomycosis**, two examples of systemic mycosis, are very common lung diseases which can occasionally disseminate and become debilitating. **Opportunistic mycosis** can threaten the life of an immune-compromised patient.

Fungi also produce chemicals which can be harmful. Many mushrooms produce poisons and/or hallucinogens. **Mycotoxins**, such as **aflatoxin** and **ergot poison** cause significant food contamination problems. The spores of molds are also common causes of **allergies**.

Many protozoa cause disease, and some protozoan diseases are among the most important diseases affecting humankind. **African sleeping sickness**, caused by a **trypanosome**, affects millions of Africans and their livestock each year. Another flagellated protozoan causes **trichomoniasis**, one of the most common causes of nongonococcal urethritis. Other protozoa are well known as the causal agents of such water-borne diseases as **giardiasis** and **amoebic dysentery**. **Malaria**, one of the most significant diseases in the world, is also caused by a protozoan. Interestingly, **natural resistance** to malaria has coevolved with diseases like **sickle-cell anemia**, to protect those in endemic regions.

KEY WORDS AND CONCEPTS

Mycosis
Superficial mycosis
Dermatophyte
Ringworm
Subcutaneous mycosis
Systemic mycosis
Dimorphic fungi
Opportunistic infection
Mycotoxin
Aflatoxin
Ergot poison
Trypanosome
African sleeping sickness
Trichomoniasis
Giardiasis
Amoebic dysentery
Sporozoan
Malaria
Sickle-cell anemia

STUDY QUESTIONS

1. Explain the difference between *superficial, subcutaneous,* and *systemic mycosis.* Give an example of each type.
2. What are *dermatophytic* fungi? What specific region of the human body are they able to colonize? How can their growth in the human body be controlled?
3. Name a *systemic mycosis* endemic to the eastern United States. Name a systemic mycosis endemic to the western United States.
4. Draw a diagram showing the major features of *dimorphic* fungi.
5. What type of evidence suggests that *histoplasmosis* is extremely common among inhabitants of endemic regions?
6. Give an example of an *opportunistic mycosis.* Why do individuals with AIDS (acquired immunodeficiency syndrome) often have them?
7. Name two types of fungal allergies which can be considered occupational hazards.
8. Why is the mushroom *Amanita phalloides* called the "death cap"?
9. What do some mushrooms produce that causes them to be referred to as "flesh of the gods"?
10. Name two types of *mycotoxins,* and explain how each is significant to humans.
11. Explain how the causative agent of African sleeping sickness is able to evade the host's immune system.
12. How have *trypanosome* diseases, such as African sleeping sickness, been controlled?
13. What are the possible reservoirs for the protozoan that causes *giardiasis*? Why is chlorination insufficient to remove it from contaminated water? What methods are used to eliminate it?
14. Describe the life cycle of the protozoan which causes *amoebic dysentery.* How do *cysts* aid in the transmission of the pathogen?
15. Where does *malaria* occur? Why is malaria no longer a problem within the United States?

16. Describe two ways in which *natural resistance* to malaria has developed.

SUGGESTED READINGS

BENENSON, A.S. (ed.). 1985. *Control of communicable diseases in man, 14th edition.* American Public Health Association, Washington, D.C.

DONELSON, J.E. and M.J. TURNER. 1985. *How the trypanosome changes its coat.* Scientific American, 252: (No. 2), 44–51.

FRIEDMAN, M.J. and W. TRAGER. 1981. *The biochemistry of resistance to malaria.* Scientific American 244: (No. 3), 154–164. (A review of the relation between resistance to malaria and the genetic disease sickle cell anemia.)

JOKLIK, W.K., H.P. WILLETT, and D.B. AMOS. 1984. *Zinsser microbiology, 18th edition.* Appleton-Century-Crofts, Norwalk, Connecticut.

MARKELL, E.K. and M. VOGE. 1981. *Medical parasitology, 5th edition.* W.B. Saunders Company, Philadelphia, Pennsylvania.

RIPPON, J.W. 1982. *Medical mycology, the pathogenic fungi and the pathogenic actinomycetes.* W.B. Saunders Company, Philadelphia, Pennsylvania.

13

Viruses and Viral Diseases

Viruses are infectious particles that replicate only in living cells. We discussed the structure of viruses in Section 2.10 and the reproduction of viruses in Section 6.5. Here we discuss viruses as disease entities. All viruses contain nucleic acid as genetic material, although some may contain RNA rather than DNA. The genetic material is surrounded by a protein coat which protects the nucleic acid when it is outside the host cell; some viruses also have a membrane surrounding the protein coat. Since viruses have no metabolism of their own, they can be thought of as obligate parasites which depend on the metabolism of the host cells they infect for their reproduction.

Viruses are known which infect animals, plants, bacteria, algae, and fungi. In this chapter we will consider those viruses which infect animal cells and cause diseases in humans. Usually the host range of a virus is limited to one or a few different animals. Many viruses can attack only the cells of certain tissues within an animal, and this determines the course of infection in a host. Viral diseases of other animals and plants will be considered in Chapter 17.

The major groups of animal viruses are shown in Table 13.1. The variety of shapes and arrangements of some of these viral groups was illustrated in Figure 2.30. One of the major features used in classifying animal viruses is the type of nucleic acid the virus contains. As shown in Table 13.1, there are

TABLE 13.1 The major groups of animal viruses and diseases they cause in humans

Virus family	Type of genetic material	Human diseases
DNA viruses		
Poxvirus	Double stranded	Smallpox, cowpox
Herpesvirus	Double stranded	Fever blisters, genital herpes, chickenpox, shingles, mononucleosis, some cancers
Adenovirus	Double stranded	Sore throat, diarrheal disease
Papovavirus	Double stranded	Warts, some cancers?
Iridovirus	Double stranded	—
Parvovirus	Single stranded	—
RNA viruses		
Picornavirus	Single stranded	Polio, gastroenteritis, common cold
Orthomyxovirus	Single stranded	Influenza
Paramyxovirus	Single stranded	Measles, mumps, common cold, pneumonia
Coronavirus	Single stranded	Diarrheal disease
Reovirus	Double stranded	Infant diarrhea
Rhabdovirus	Single stranded	Rabies
Togavirus	Single stranded	Encephalitis, rubella, yellow fever, dengue
Bunyavirus	Single stranded	Encephalitis
Orbivirus	Single stranded	Colorado tick fever
Arenavirus	Single stranded	Lassa fever
Retrovirus	Single stranded	AIDS, human T-cell leukemia

six groups of DNA viruses and eleven groups of RNA viruses. Moreover, the DNA or RNA may exist either as a single or as a double-stranded molecule. The DNA of all cellular organisms is always double stranded (see Section 6.1).

Viruses exhibit a wide range of sizes, but all are too small to be seen without an electron microscope. In fact, it was in the field of virology that the electron microscope made its first great contributions to biology. The electron microscope reveals the structural variety among different animal viruses (Figure 13.1). The **naked viruses** have only a protein coat surrounding the genetic material. Because of the regular assemblage of proteins in the coat, naked viruses often have a crystalline appearance and are said to be **icosahedral** (Figure 13.1*a* and *b*). **Enveloped viruses** possess a membrane surrounding the protein coat and are often irregular in shape (Figure 13.1*c* and *d*). Some enveloped viruses have distinctive appearances and their names reflect their unique morphology. *Coronaviruses* have radial projections resembling the corona of the sun (Figure 13.1*c*); *rhabdoviruses* are bullet shaped (Figure 13.1*d*). Some viruses are named for the types of infections they cause, for example *poxviruses* include the smallpox virus, and *herpesviruses* include the virus which causes genital herpes. Within virus families, individual viruses have names of various origins, some even named after the site of an outbreak (for example, Coxsackievirus is named after Coxsackie, New York). Obviously, the rules for naming viruses are not the same as those used to name cellular organisms in the binomial nomenclature system (see Section 2.4).

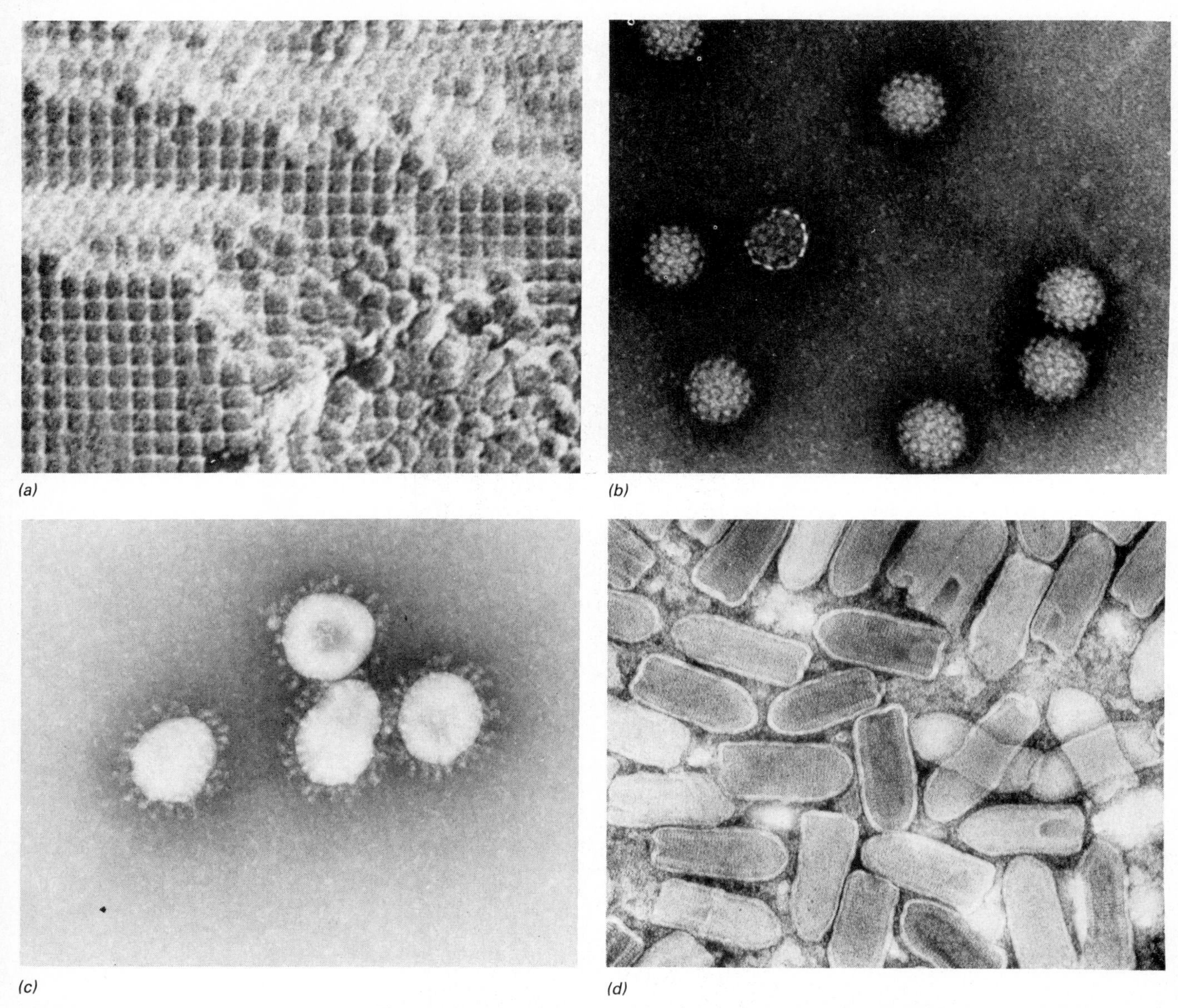

Figure 13.1 Electron micrographs of animal viruses. **Naked icosahedral viruses:** (a) Poliovirus particles, arranged in a crystalline sheet. (b) Human wart virus. **Enveloped viruses:** (c) Coronavirus. (d) Rhabdovirus. The magnifications of all the micrographs are about the same. A single particle of rhabdovirus is about 100 nm across.

13.1 TECHNIQUES FOR VIRUS CULTIVATION

Because viruses only grow in living hosts, their laboratory study cannot be done by the conventional culture methods of microbiology. Experimental animals are used for some studies, but the most important technique for culturing animal viruses is the use of **animal-cell cultures** (sometimes called *tissue* cultures). Animal cells used to support virus growth may be either 1) nongrowing primary tissues isolated from various organs of an animal or 2)

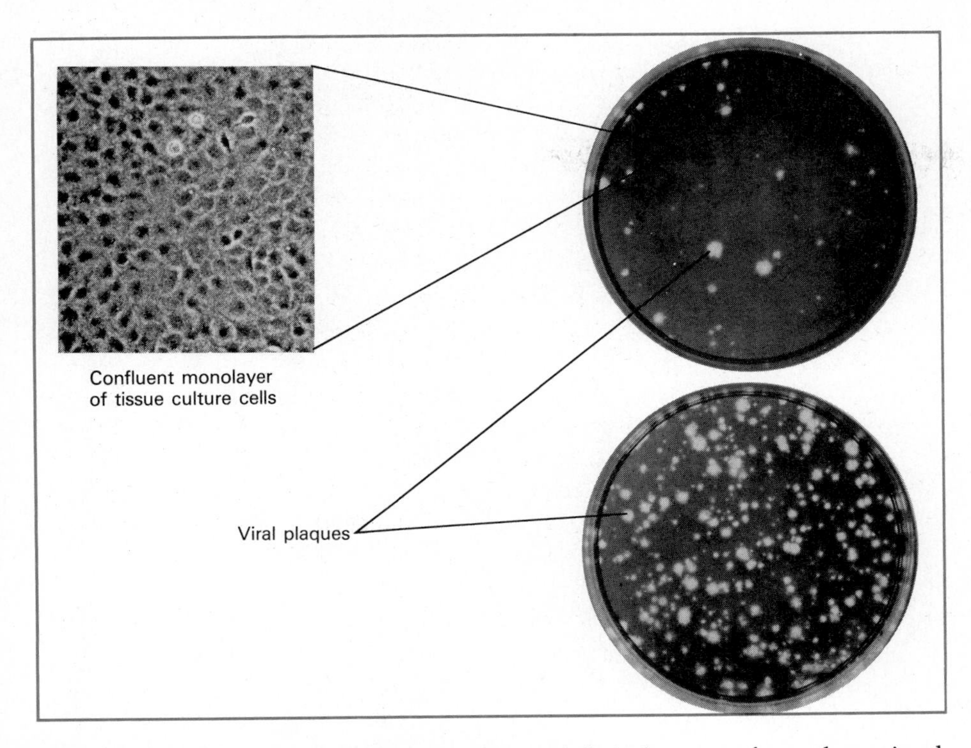

Figure 13.2 Cell cultures in monolayers within petri plates. Note the presence of plaques where virus lysis has occurred. Also shown is a photomicrograph of a cell culture.

cell cultures that are capable of growing indefinitely away from the animal host. Cell cultures are maintained in complex media containing salts, glucose, amino acids, vitamins, other growth factors, usually some natural materials such as serum or embryo extract, and antibiotics (to reduce bacterial contamination). In such media, animal cells may grow or subsist for considerable lengths of time. Cell cultures may be grown in suspension or in thin sheets or monolayers on glass, agar, or plasma or fibrin clots. Monolayers are especially useful for observing pathological changes that may be induced by viruses, since the cells can be observed directly under the microscope. For example, cell monolayers are used to obtain *plaques* suitable for virus quantification (Figure 13.2). Dispersed cells may be useful for mass virus propagation and for biochemical studies of virus reproduction. Cell cultures offer advantages over laboratory animals in being cheaper, more reproducible, and simpler; most important, they make it possible to use human cells for propagating viruses that will not grow in nonhuman hosts.

13.2 VIRUS REPRODUCTION

The reproduction of a virus within a host cell is a complex process which can be broken down into several stages. First, the virus must *adsorb* to the cell which it will infect. This is mediated by specific receptors on the surface of the host cell, and in part determines which type of host cells a virus may infect. Animal cells may have thousands of receptors coating their surface so that many virus particles may attach to a single cell.

Once the virus has attached to the animal cell, it must *penetrate* the cell. In Section 6.5 we discussed the reproduction of viruses that infect bacteria and showed that they inject their genetic material into the cell, leaving the protein coat on the exterior. Animal viruses, on the other hand, often enter a host cell intact, usually being taken up by invagination of the cell membrane so that they become enclosed within a vesicle inside the cell (Figure 13.3*a*). Once inside, the virus protein coat is lost and the DNA or RNA which is the infectious part of the virus particle is released.

Multiplication of the virus is directed by the DNA or RNA of the virus. In Section 6.5 we discussed the concepts involved in reproduction of infectious virus particles within the host cell. In essence, a virus must *replicate* its nucleic acid, and carry out *protein synthesis* to produce the essential virus proteins (protein coat, enzymes, proteins associated with the envelope). The specific steps differ depending on the type of nucleic acid the virus possesses. As outlined in Figure 6.15, when DNA viruses infect cells, the viral DNA replaces host DNA as a template for both replication and transcription in the production of messenger RNA (mRNA) for synthesis of viral protein. With RNA viruses, multiplication is often less complex. In some RNA viruses, the viral RNA serves directly as the mRNA. This is the case for poliovirus, a single stranded RNA virus. As shown in Figure 13.4, the entire poliovirus RNA is translated as a single message into one very large protein which is subsequently cleaved into a few functional viral proteins. Cleavage leads to the production of the poliovirus protein coat, and two enzymes, a protease and an RNA-directed RNA polymerase. The RNA polymerase can use the viral RNA as a template from which more copies of poliovirus RNA are produced.

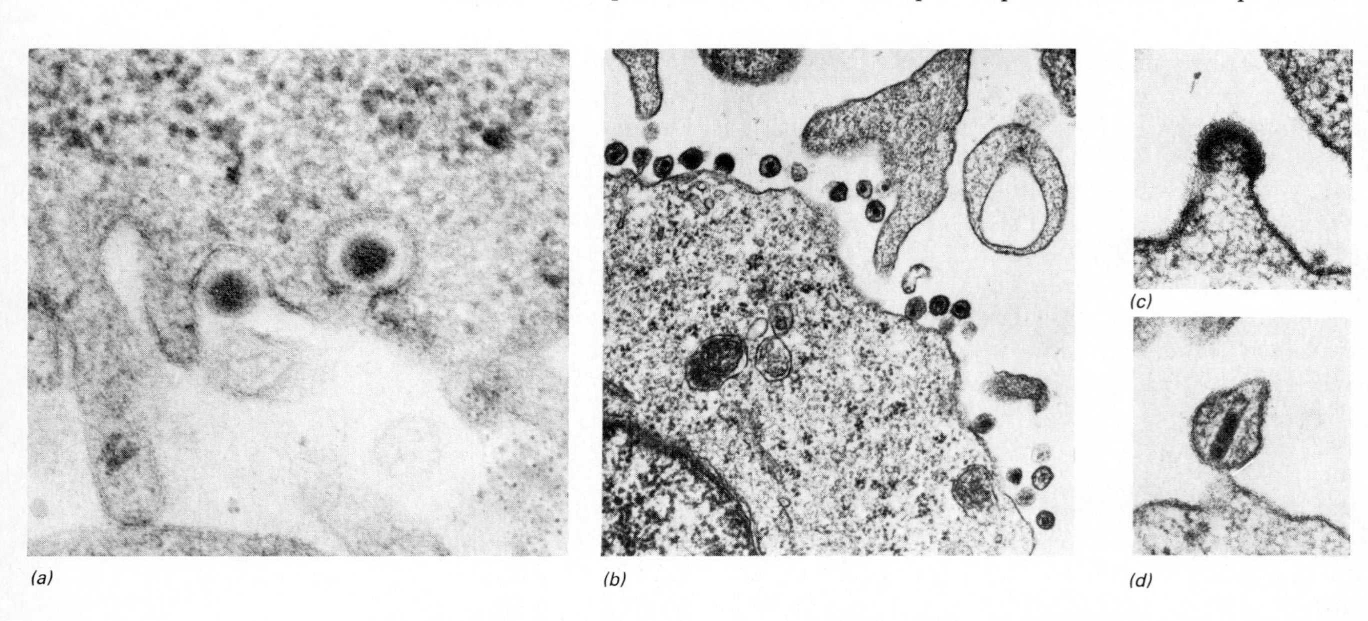

Figure 13.3 The penetration (a) and release (b-d) of virus particles from an animal cell. Electron micrographs. (a) Two adenovirus particles, each about 50 nm in diameter, at the periphery of a cell. (b-d) AIDS virus particles, each about 100 nm in diameter, budding from the host cell membrane.

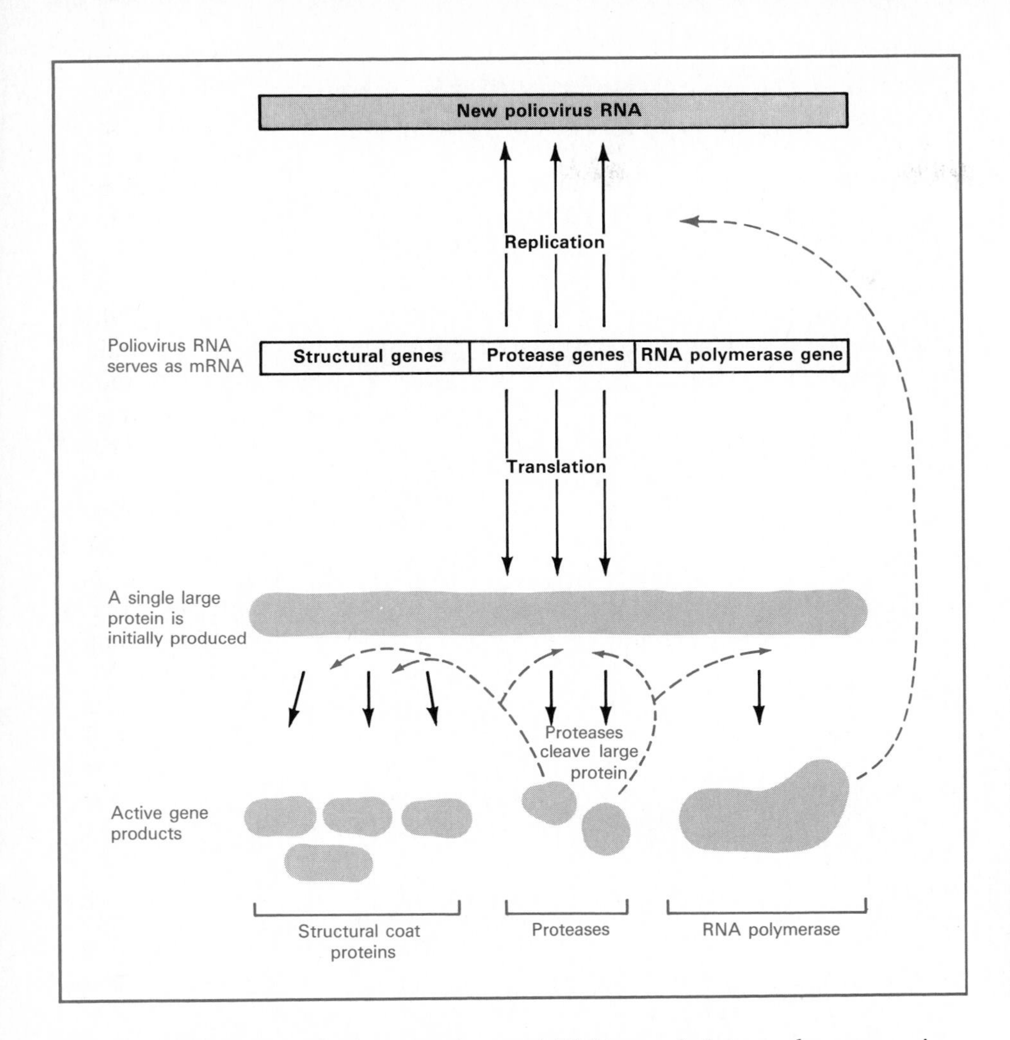

Figure 13.4 The reproduction of poliovirus. The single-stranded RNA of the virus is translated directly as a messenger RNA, with the production of one large protein molecule. This protein is cleaved, leading to the production of the active viral proteins, including the structural coat protein and the RNA polymerase which brings about the replication of the poliovirus RNA. The assembly of intact poliovirus from coat protein molecules and RNA then follows.

The result is that both the proteins and RNA needed to make more virus particles are produced; assembly into intact poliovirus particles follows.

In Section 6.5 (see Figure 6.16), we also discussed an alternate type of infection of the host cell called a *lysogenic infection*. In this case, viral DNA becomes integrated into the DNA of the host bacterial cell and only at a later time does the virus undergo reproduction. Similarly, some animal viruses may integrate their nucleic acid within the DNA of the host cell. It may be easy to understand how a DNA virus can integrate its DNA with host DNA, but how can an RNA virus accomplish this? In Section 6.5 we discussed *retroviruses* and showed that they have the enzyme *reverse transcriptase* which produces DNA from RNA; the DNA is then integrated into the host DNA. Retroviruses have been shown to be able to cause certain forms of cancer in animals, and in some cases in humans (see Section 13.12).

After intact virus particles have been formed within the host cell, they are *released* from the cell. Naked virus particles may be released all at once upon lysis of the host cell. Enveloped viruses derive their envelope from the membrane of the host cell. This occurs by a *budding* process through patches of membrane which encircle the virus particles (Figures 13.3*b*, *c*, and *d*).

Effect of viral infection on the host cell Viruses can have varied effects on cells. **Lytic infection** results in the destruction of the host cell. However, there are several other possible effects following viral infection of animal cells (Figure 13.5). When enveloped viruses are formed by budding, the release of virus particles may be slow and the host cell may not be lysed. Such infections may occur over long time periods and are thus referred to as **persistent infections**. Viruses may also cause **latent infection** of a host. The effect of a latent infection is that there is a delay between infection by the virus and the appearance of symptoms. Fever blisters, caused by the herpes simplex type 1 virus, result from a latent viral infection; they reappear sporadically as the virus emerges from latency. The latent stage in viral infection of animal

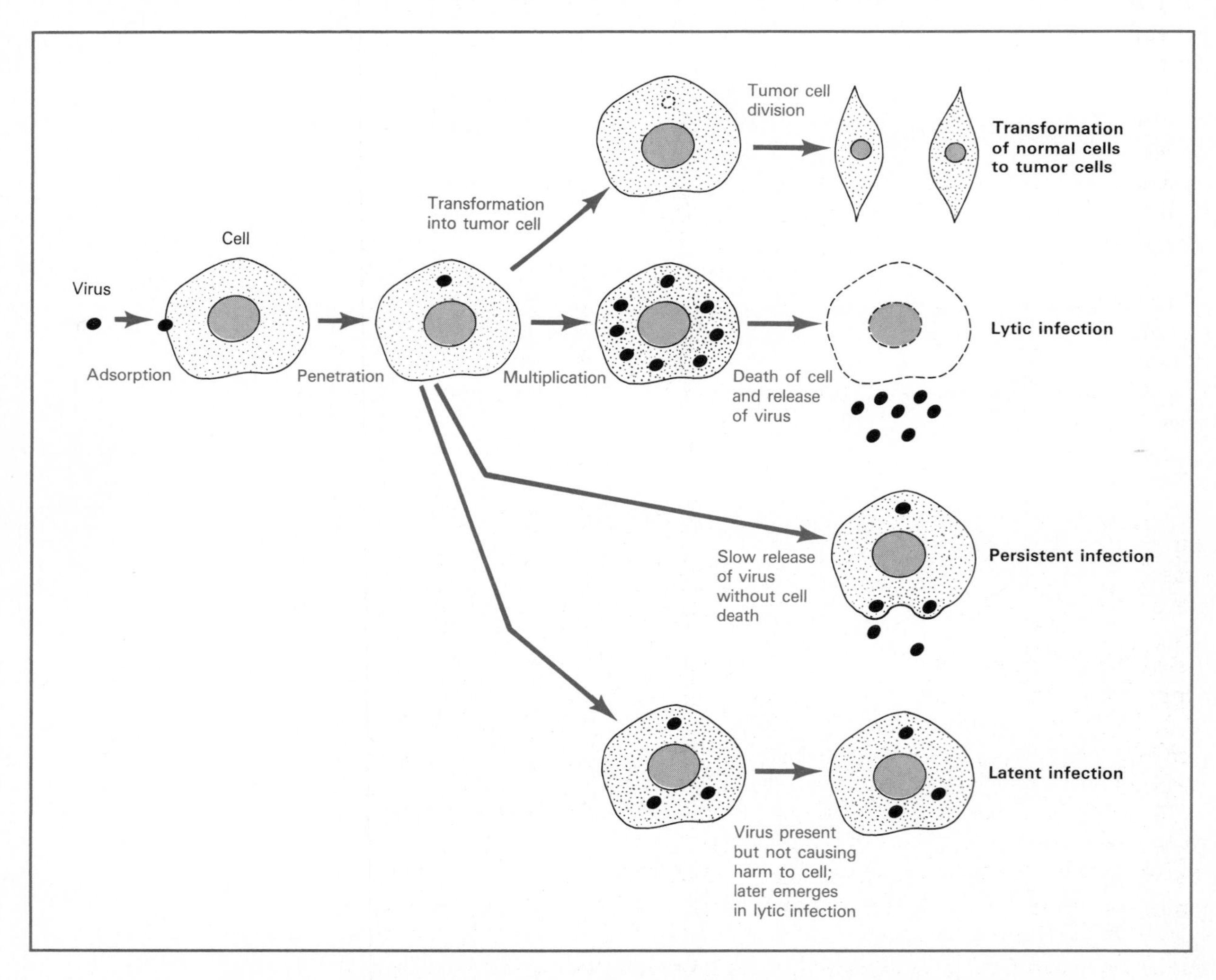

Figure 13.5 Possible effects that animal viruses may have on cells they infect.

cells may not, however, be due to integration of viral DNA into host DNA as is generally the case with bacteriophage latent infections.

Many animal viruses have the potential to change a cell from a normal cell into a cancer cell. This process, called **transformation**, can be induced by infections of cells with viruses capable of causing the formation of tumors. One of the key differences is that cells which have been transformed by tumor viruses do not cease growing when they contact adjacent cells. Rapidly growing cells pile up into accumulations of cells as would occur in a tumor. The role of viruses in causing cancer will be considered later in this chapter.

13.3 CONTROL OF VIRUS GROWTH

Control of virus disease can be accomplished either by preventive measures such as vaccination or by use of chemotherapeutic agents. We discuss the general principles of virus control in this section and give specific examples later in this chapter.

Antiviral drugs Since viruses depend on their host cells for many processes, it is difficult to inhibit virus multiplication without at the same time affecting the host cell itself. Because of this, the spectacular successes achieved with antibacterial agents (see Chapters 5 and 11) have not been followed by similar successes in the search for specific antiviral agents. Through the use of cell culture systems, many chemicals have been tested which might inhibit either replication of viral nucleic acid (often carried out by virus-encoded enzymes) or virus protein synthesis. Several compounds have been found to inhibit viral multiplication in cell culture, but these chemicals are either less effective or toxic when administered to the animal. The few compounds which have been successfully used to restrict some viral infections in humans will be considered as we discuss specific diseases.

Vaccines As we will show in the following sections, vaccines have been important in the control and even elimination of many important viral diseases. Because of the lack of effective virus chemotherapy (see above), vaccines are even more important in the control of virus diseases than they are for bacterial diseases. We discussed vaccination in Section 9.10. However, in some cases, vaccine development has not been possible, either because the virus cannot be grown in tissue culture, or because of a problem with the purity of the vaccine. The development of new techniques in genetic engineering (see Section 7.7) has made possible novel approaches to the production of vaccines.

Two possibilities for engineering virus vaccines are outlined in Figure 13.6. Since the viral proteins, rather than the entire virus particle, serve as the antigens which stimulate the immune system, it is possible to transfer the viral genes encoding such proteins into bacteria or yeast, which then produce the viral proteins in large quantity (Figure 13.6*a*). A vaccine for foot-and-mouth disease, an important viral disease in animals, has been developed in this way. In addition, a protein antigen of the virus which causes serum

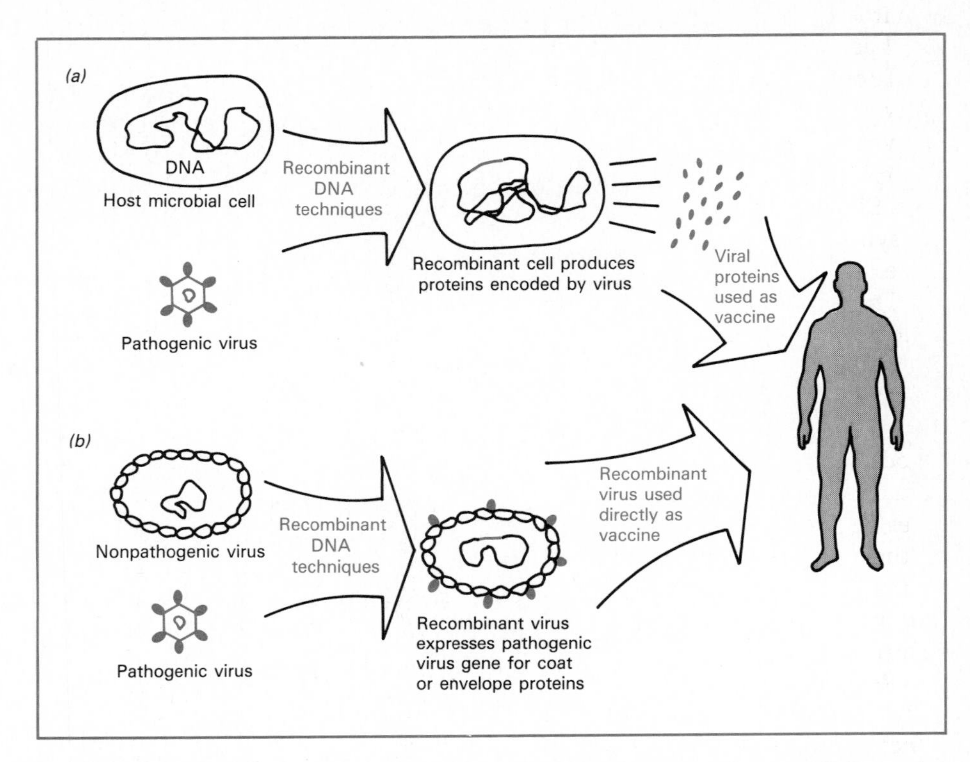

Figure 13.6 Two strategies for use of recombinant DNA techniques in the creation of virus vaccines. (a) Recombinant microorganism produces viral protein which is used as a vaccine. (b) Nonpathogenic recombinant virus producing proteins of a pathogenic virus is used as a live-virus vaccine.

hepatitis in humans can now be produced by bacteria and yeast. Often, the purified proteins of viruses are less effective than live viruses containing the proteins in stimulating the immune response. Thus, another promising approach is to transfer the genes encoding the antigenic proteins of a disease-causing virus into a virus which is nonpathogenic in humans; the recombinant virus is a type of live-virus vaccine (Figure 13.6*b*). During a mild infection, the harmless virus produces the antigens of the disease-causing virus, and the immune system responds. Foot-and-mouth virus genes have, for example, been transferred into vaccinia, a poxvirus of cattle, and infection with the genetically engineered poxvirus causes active immunity to foot-and-mouth disease. It is hoped that genetic engineering will permit the control of many diseases which still significantly affect human lives.

13.4 SMALLPOX

Smallpox

We begin our discussion of viral diseases with a disease that no longer exists! Although smallpox has been known for centuries as the cause of major epidemics killing millions, it has been completely eradicated everywhere in the world due to an effective vaccination program. The campaign to eradicate smallpox was discussed in the Box on page 280. The vaccination program made use of a virus related to smallpox called *vaccinia,* which infects cows. Although smallpox has been completely eradicated, vaccinia virus is still available and provides a useful research tool to study the molecular biology of poxvirus reproduction. As noted in the previous section, vaccinia virus has also been used in genetic engineering of virus vaccines.

Though smallpox infection does not occur anywhere in the world, it is interesting to review the course of the disease as it once occurred. The smallpox virus, one of the poxviruses, is a double-stranded DNA virus. The virus multiplies in the cytoplasm at four or five localities in the cell at the same time. Since DNA synthesis in eucaryotes normally occurs only in the nucleus, the synthesis of smallpox virus DNA in the cytoplasm is an unusual occurrence.

The virus is present in large numbers in pustules on the skin and spreads through the air, entering a new host through the upper respiratory tract. Growth is first initiated on the mucous membranes of the upper respiratory tract, followed by multiplication in lymphoid tissues that drain the respiratory tract. Just before the development of symptoms, the virus is found in the bloodstream, and from there it passes to the skin, mucous membranes, and organs such as the heart, liver, spleen, and kidney. During the initial phase, fever, headache, and backache occur, followed by prostration. A rash appears on the skin, and soon after this the temperature falls to normal and recovery occurs. Vesicles teeming with virus form crusts on the skin, which develop into the characteristic pockmarks that disfigure the skin. In severe cases, death occurs, usually due to hemorrhage and generalized toxemia.

Recovery from smallpox infection confers complete immunity. Historically, smallpox is important because immunization with vaccinia was the first effective procedure for control of a human disease. (The term *vaccination* comes from the use of the cowpox virus, *vacca* meaning "cow.") The immunization procedure was developed by Edward Jenner in the early part of the nineteenth century. Until the early 1970s smallpox was still endemic in many areas of the world, but efforts by the World Health Organization resulted in total eradication of the disease.

13.5 INFLUENZA

A common route of transmission of viruses is through airborne droplets; infection then begins in the upper respiratory tract. Included among these viral diseases are the most important causes of infectious disease in the United States, influenza and the common cold, as well as measles, mumps, rubella, and chickenpox. We discuss influenza first.

Frequently in the winter people suffer from a short-lived fever associated with soreness and redness of the respiratory passages and a dry cough. Such a condition is commonly called "flu" or *influenza*. Influenza is among the leading causes of illness in the United States (see Figure 10.5) and, because of secondary complications such as pneumonia, it remains as the only infectious disease among the leading causes of death in the United States (see Figure 1.1).

Flu (influenza)

Influenza is caused by an *orthomyxovirus*, a single-stranded enveloped RNA virus (Figure 13.7). The RNA is present as eight separate fragments rather than as one continuous molecule, and these fragments are surrounded

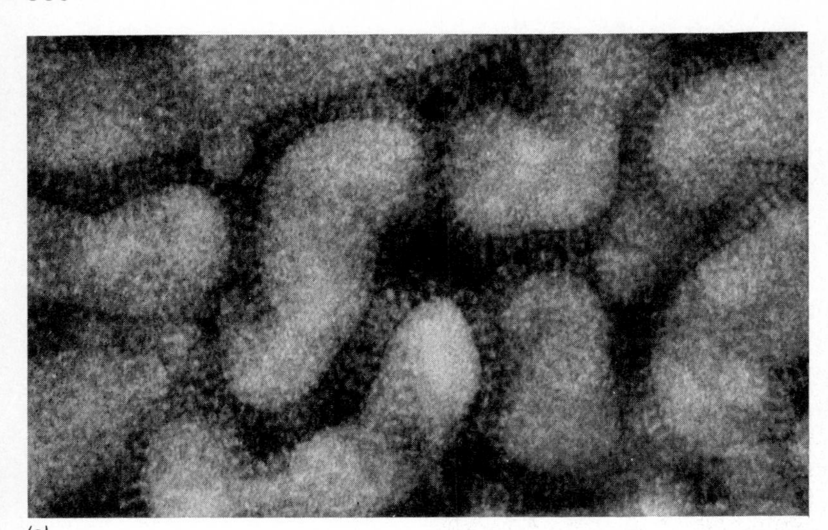

(a)

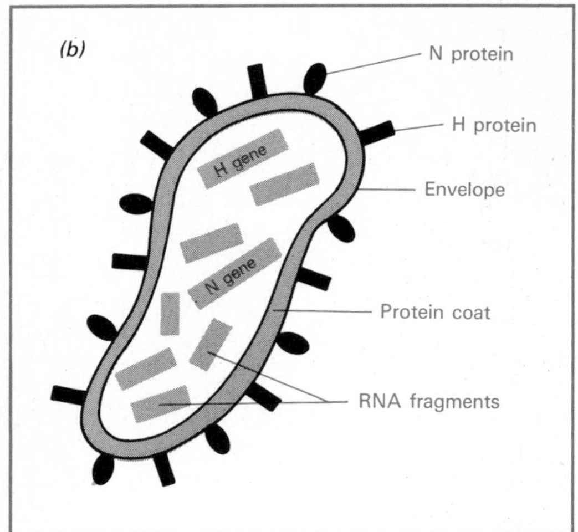

Figure 13.7 Influenza virus. (a) Electron micrograph of virus particles. An individual particle is about 80 nm in diameter. (b) Diagram of an influenza virus particle, showing the various constituents of the virus particle.

by a thick protein coat. An envelope surrounds the coat and embedded in it are two surface proteins which are important in infection and immunity. One, called *hemagglutinin* or *H protein* is responsible for attachment to the host cell; the other, called *neuraminidase* or *N protein,* is an enzyme which cleaves a chemical component of the virus receptor on the cell surface, allowing release of the virus from the cell. These antigens are important to the epidemiology of influenza virus infection, as will be described below.

Influenza virus exists in nature not only in humans but also in birds, horses, and pigs. It is transmitted from person to person through the air, primarily in droplets expelled during coughing and sneezing. The virus infects the ciliated epithelium of the upper respiratory tract and occasionally invades the lungs. Localized symptoms such as sore throat and respiratory difficulty are due to destruction of the ciliated epithelium and lung tissue by the virus. Systemic symptoms include an abrupt fever from 3 to 7 days, chills, fatigue, headache, and general aching. Recovery is usually spontaneous and rapid. Most of the serious consequences of influenza infection occur not because of the viral infection but because bacterial invaders may be able to cause severe infections in persons whose resistance has been lowered. Especially in infants and elderly people, influenza is often followed by bacterial pneumonia; death, if it occurs, is usually due to the bacterial infection. A contributing factor is the destruction of ciliated epithelium by the virus, which may decrease the effectiveness of mucociliary flushing as a mechanism to remove bacteria from the upper respiratory tract.

There are a number of recognized strains of influenza virus, each antigenically distinct from others. Since the immune system responds separately to each strain, an individual is not protected against all strains after infection by a single strain, so that influenza infection can occur many times. The types of influenza virus are given the designations A, B, or C. Within each type, differences in the H and N protein antigens determine the various infectious strains; there are five major H antigens and two major N antigens. Changes in the H and N antigens occur frequently and account for the different strains of influenza virus which cause annual epidemic outbreaks (Figure 13.8).

How does the influenza virus change its antigens? One mechanism, called **antigenic drift**, involves mutational change. Since new and different antigens will increase the chance for influenza virus to infect a population which is immune to the former antigenic type, the new strain resulting from antigenic drift may be more fit to survive. Antigenic drift accounts for many of the new strains of influenza virus that cause mild epidemics year after year, such as the A/Victoria, A/Port Chalmers, and A/England strains shown in Figure 13.8.

Influenza viruses may also change their antigens in a more dramatic and sudden way. For example, an epidemic of Hong Kong flu, which occurred in 1968, was caused by a strain possessing H and N proteins which were completely unlike those of strains which had caused earlier epidemics. The mechanism by which sudden antigenic change, called **antigenic shift**, occurs may involve reassortment of genomes rather than mutation. As noted, influenza virus RNA does not exist as a single molecule, but rather as eight different RNA molecules. The genes coding for the H and N proteins are located on different molecules. It has been shown that two different influenza viruses can infect a single host cell. For example, a human influenza virus strain and an animal influenza virus strain might infect an animal at the same time. As multiplication of both viruses occurs within a single cell, RNA fragments of both viruses are produced. During assembly of intact virus particles, some virus particles will contain RNA derived from each of the two strains. This results in the formation of a new hybrid influenza virus strain, infectious for humans, containing either an H or an N protein which had been common in animal strains, but not found among human influenza virus strains. The Hong

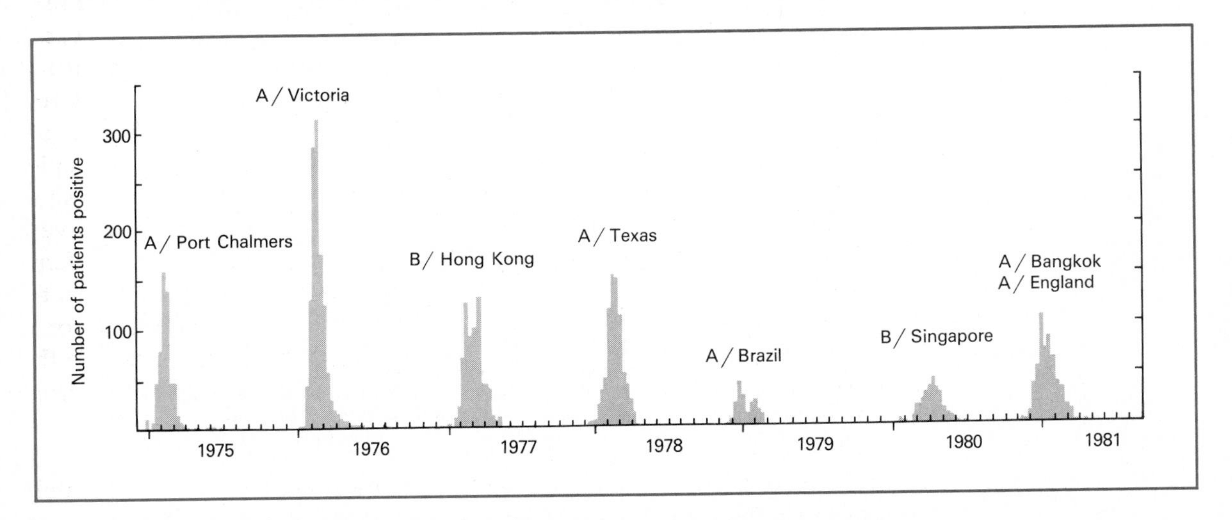

Figure 13.8 Incidence of influenza virus cases in a single city (Houston, Texas), and the virus strain which was cultured. Note the marked seasonality of the disease and the dominance in each year of a different virus strain.

Kong flu strain, in fact, possesses an H protein which resembles the H protein of an influenza virus strain isolated from ducks, and is totally unlike the H protein of the strain which caused epidemic outbreaks in the years prior to the emergence of the Hong Kong strain in 1968.

Antigenic shift has been associated with the pandemic influenza outbreaks which have occurred during this century, including the 1957 emergence of Asian flu, the 1968 outbreak of Hong Kong flu, and presumably earlier pandemics such as the one in 1918–1920. One of the best studied pandemics was the 1957 Asian flu pandemic which provided an opportunity for a careful study of how a worldwide epidemic develops. The epidemic began with the development of a new form of the virus of marked virulence and differing from all previous strains in antigenicity. Since immunity to this strain was not present in the population, the virus was able to advance rapidly throughout the world (Figure 13.9). It first appeared in the interior of China in late February, 1957, and by early April had been brought to Hong Kong. It spread from Hong Kong along air and naval routes and was apparently transferred to San Diego, California, by naval ships. An outbreak occurred in Newport, Rhode Island, on a naval vessel in May. Other outbreaks occurred in various parts of the United States. Peak incidence occurred in the last two weeks of

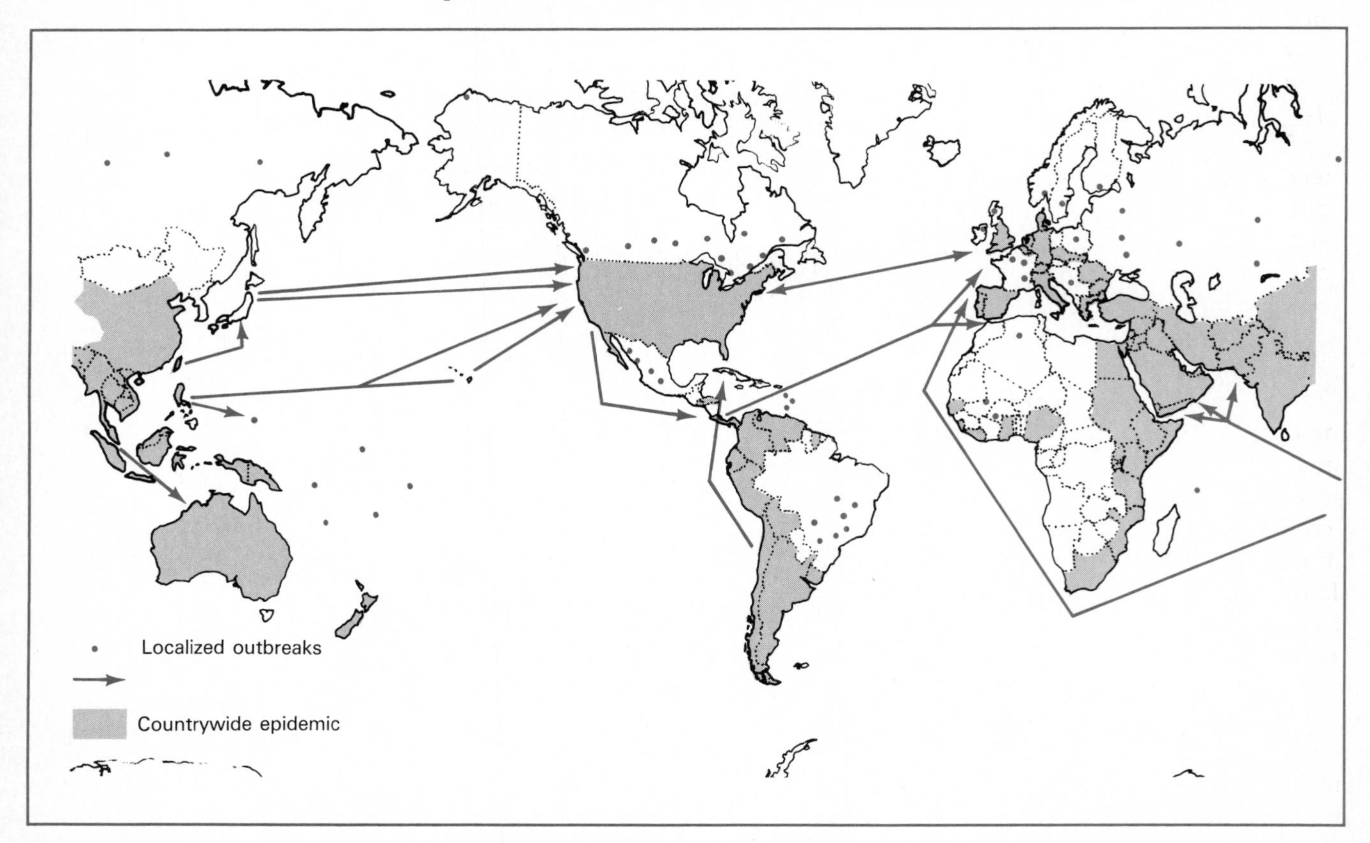

Figure 13.9 Route of spread of a major influenza epidemic, the Asian flu pandemic of 1957. The epidemic had its origin in China and spread to the rest of the world via Hong Kong.

October, during which time 22 million new cases developed. From this period on, there was a progressive decline.

Control of influenza epidemics can be carried out by vaccination, but this is complicated by the fact that so many different virus strains may cause the disease. Obviously, the vaccine must be derived from the strain causing the epidemic. Vaccines prepared from several different virus strains can be mixed, producing what is called a *polyvalent* vaccine. If a new strain has arisen, vaccine for it will of course not be available, but by careful worldwide surveillance, it is sometimes possible to obtain cultures of a new strain before it has reached epidemic proportions and thus to produce vaccine ahead of the epidemic. This is costly, however, and may result in production of large amounts of vaccine that are never needed. The "swine flu" vaccination program in the United States in 1976 was an example of an attempt at mass vaccination against a particularly virulent influenza virus strain, but the epidemic in question never developed. One complication of mass vaccination for control of an infectious disease may be the harmful side effects which result from the use of the vaccine. A more common procedure is to restrict vaccination to those most likely to succumb to severe or fatal illness, such as the aged and those suffering from chronic debilitating diseases. The families of these persons should also be vaccinated. Duration of immunity for influenza is not long, usually only for a few years, so that revaccination is necessary if a new epidemic occurs.

Influenza may also be controlled by the use of the chemical *amantadine*. This drug has been used as a chemoprophylactic agent to prevent the spread of influenza to those at high risk. It may also help to shorten the course of infection somewhat. The alleviation of symptoms of influenza by the use of aspirin is not recommended, as there is evidence of a link between aspirin treatment of influenza and *Reye's Syndrome* (a rare, but occasionally fatal, illness involving the central nervous system).

13.6 THE COMMON COLD

The common cold is almost as important as influenza as a leading cause of illness in the U.S. (see Figure 10.5). The symptoms are much milder than those of influenza, commonly including sore or scratchy throat, a mild cough, increased nasal secretions, and malaise; fever and other systemic manifestations, such as lung involvement, may occur in children but are rare in adults. Many different viruses can elicit the symptoms of the common cold and this, of course, complicates the control of the disease through vaccination. For example, *rhinovirus*, a type of *picornavirus*, is a frequent cause of the common cold, but there are are over 100 antigenically different types of rhinoviruses.

Common cold

Cold viruses are not highly infectious: even when nasal secretions are inoculated directly into volunteers, the percentage of detectable infections is only 30 to 40 percent. When a volunteer is placed in an isolated room and inoculated with nasal secretions from a person with active cold infection, the disease (when it appears) runs a quick course (Figure 13.10). No experimental animal except the chimpanzee has been found to be susceptible to cold viruses,

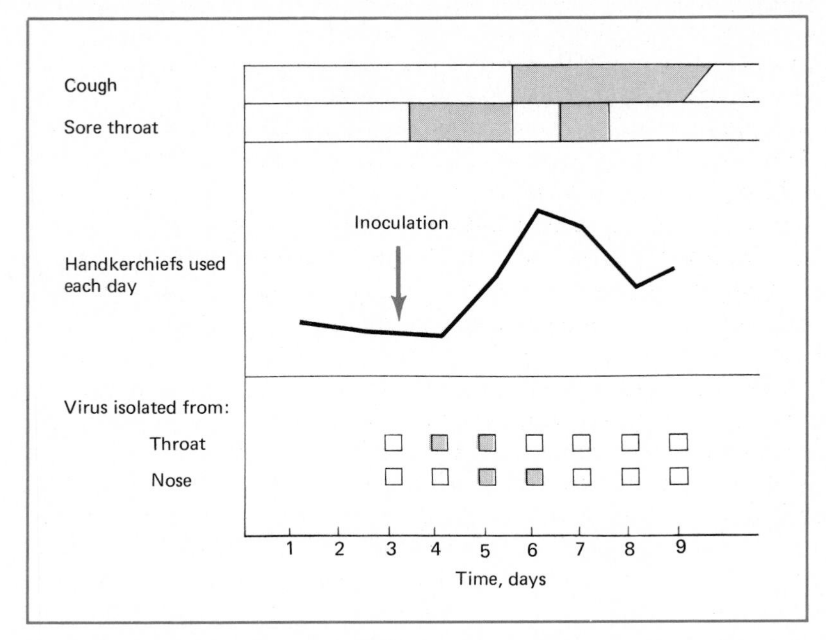

Figure 13.10 The course of a cold in a human volunteer inoculated intranasally with a cold virus. The severity of the symptoms were roughly quantified by the number of handkerchiefs used each day. In the lower section, the filled squares indicate positive virus culture, the open squares negative culture.

and this animal is too expensive to keep and too difficult to work with for routine laboratory use. Techniques for growing the viruses in human cell cultures are available, and many strains have been isolated in tissue culture from persons with typical colds.

Studies have shown that direct contact, such as hand to hand contact, may be more important than airborne droplets in the spread of the viruses that cause the common cold. Although there is no way of treating the disease, or of preventing its occurrence through vaccination, it may be possible to interrupt the transmission of the viruses. Impregnation of tissue paper with iodine, which kills rhinoviruses, apparently limits the spread from infected to uninfected individuals.

13.7 CHILDHOOD VIRUS DISEASES

Measles, mumps, rubella, and chickenpox are virus diseases which in the past have been extremely common infections among children. They are caused by different types of viruses and exhibit different symptoms, sometimes with severe complications. Due to the effectiveness of immunization programs, the incidence of three of these diseases has been dramatically reduced (Figure 13.11). The immunization of all children in the United States with the MMR (measles, mumps, rubella) series is recommended (see Table 9.7). Chickenpox remains one of the most common reported diseases (see Figure 10.6).

Measles

Measles *Measles* virus (sometimes called *rubeola* virus) is a member of the *paramyxovirus* group of enveloped single-stranded RNA viruses. It grows in nature only in humans, although monkeys can be infected experimentally

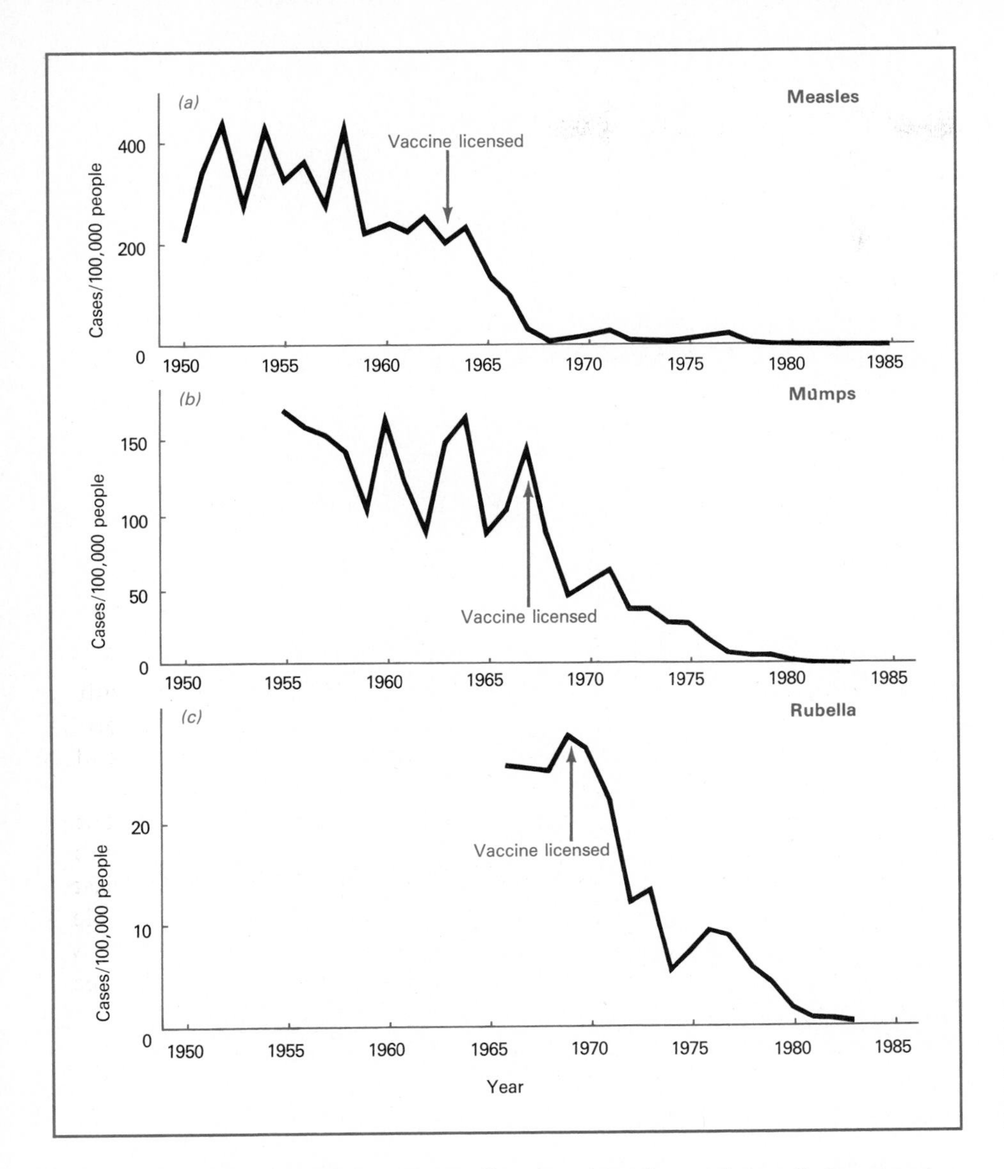

Figure 13.11 The effect of vaccines on the incidence in the United States of the major childhood virus diseases now controlled by the MMR (measles, mumps, rubella) vaccine. (a) Measles. (b) Mumps. (c) Rubella.

and may also become infected naturally when kept in captivity. Only primates have been successfully infected, but tissue-culture methods have permitted extensive research studies.

Measles is highly contagious and the virus is transmitted from person to person by the respiratory route. The virus infects cells of the nose and throat and then spreads via the lymph system throughout the body. Virus can be found in the blood and in the secretions of the nose and throat. Initial symptoms are chills, followed by sneezing, running nose, redness of eyes, cough, and fever. Characteristic spots, called *Koplik's spots,* appear on the mucous membranes and around the salivary glands and are an early diagnostic symptom of the disease. Fever and cough become worse and a rash appears, first on the forehead and behind the ears, then on face, neck, limbs, and trunk.

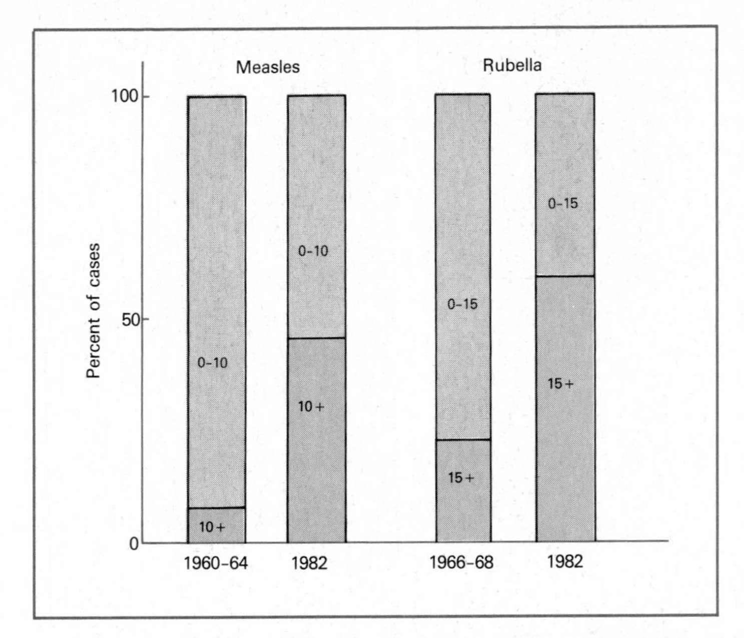

Figure 13.12 Age shift in susceptibility to measles and rubella. Percent of cases is shown for younger (gray) and older (color) population segments.

There may be inflammation of the eyes and sensitivity to light. Pustules are not formed; the rash is probably an allergic response to the presence of virus products in the body rather than a result of direct viral multiplication in the skin cells. Cytotoxic T cells appear to be important in the elimination of virus-infected cells.

A variety of complications may occur as a consequence of measles, including bronchopneumonia, inner ear infection, and encephalomyelitis. Recovery usually confers long-term immunity. Measles occurs in nonimmunized populations in epidemics at about 3-year intervals, and in many countries nearly 90 percent of the people over 20 years of age have had the disease; it remains an important disease in developing countries. In the U.S., vaccination of young children has resulted in a shift in the age group most often affected (Figure 13.12). Measles often occurs now in limited epidemics on college campuses among students who were never immunized as children (see the Box on the next page).

Rubella (German measles)

Rubella *Rubella*, or *German measles*, is caused by an enveloped single-stranded RNA virus of the *togavirus* group. The upper respiratory disease caused by this virus is similar to measles but is milder, is of shorter duration, and has fewer complications; infection may even be inapparent. Although rubella is generally a mild disease, rubella infection of mothers during the first trimester (3 months) of pregnancy allows placental transmission of the virus and induces serious damage in the fetus. The infant may be stillborn; if it survives, severe deformities such as deafness, heart disease, eye defects, mental retardation, and other complications are common. A rubella epidemic in the United States in 1964 caused disabilities in nearly 20,000 infants. Immunization of all susceptible individuals is required, not so much to prevent what is a rather mild disease in children, but rather to prevent infection of

MEASLES AND RUBELLA EPIDEMICS ON COLLEGE CAMPUSES

Vaccination programs begun in the mid- to late-1960s have dramatically reduced the incidence of measles and rubella (see Figure 13.11). Because vaccination is routinely done between 15–16 months of age, there has been an age shift in those susceptible to these diseases (Figure 13.12). More than half of the cases of both diseases now occur among adolescents and young adults, 5–20 percent of whom have somehow missed being vaccinated.

In the early to mid-1980s, up to 38 percent of the reported measles cases in the United States have been college-associated. The usual scenario is that a student who lacks immunity travels to a foreign country where measles (or rubella) is more common. Having acquired the disease, the student returns and becomes the source of infection for the many nonimmunized students on today's college campuses. In such nonimmune populations, the virus spreads rapidly and the cost to the student is usually a few days of missed classes and a visit to the doctor. However, some college students have developed respiratory complications and died, and pregnant female students who acquired rubella have terminated their pregnancies to avoid the risk of having malformed babies. The cost to a university of implementing the necessary health measures during such an epidemic may be high, up to a quarter of a million dollars on a large campus. Students may soon find themselves forced to show evidence of immunization against these diseases before admission to college is allowed. Imagine having to show an immunization record in order to attend a college athletic event! This actually occurred recently at an eastern university, where such extreme measures were taken to limit transmission of the disease during an epidemic of measles.

pregnant women by eliminating the virus from children.

As with measles, rubella now commonly affects older people (see Figure 13.12) and often occurs in epidemics on college campuses and army camps, where nonimmunized individuals congregate.

Mumps This disease was once widespread, with as many as 90 percent of the individuals of a given population exposed to the virus by the age of fifteen, but is now less common due to effective immunization. Mumps is caused by a *paramyxovirus*, an enveloped, single-stranded RNA virus. Like orthomyxoviruses, paramyxoviruses possess H and N antigens, but also another protein, called the *F protein*, which is involved in the spread of the virus from cell to cell. Mumps virus is spread only among humans by airborne droplets. After adhering to cells of the upper respiratory tract, infection of the salivary glands with resultant inflammatory host responses leads to swelling around the jaws, the most characteristic symptom. The virus spreads through the bloodstream and may infect other organs, sometimes leading to more severe symptoms. A fairly frequent complication is meningitis. More rare infections involving the central nervous system are also more debilitating; encephalitis is an example. About one-fourth of males who have passed the age of puberty experience a painful swelling of the testes during mumps infection, but this seldom leads to any serious problem. Eventually, the host produces antibodies which neutralize the H, N, and F surface proteins and decrease the virulence of the mumps virus. Cytotoxic T cells are probably involved in attacking virus-infected cells. Because of the effectiveness of the

Mumps

immune system in limiting infection, live attenuated virus vaccine has been extremely useful in reducing the incidence of mumps in the last two decades (Figure 13.11*b*).

Chickenpox Despite the name of this disease, the virus which causes it is not a poxvirus, but rather a type of *herpesvirus* which possesses double-stranded DNA and an envelope surrounding the protein coat. The virus itself is called Herpes Zoster or Varicella-Zoster virus as it causes two different syndromes: varicella or *chickenpox*, a disease which usually afflicts children; and zoster, or *shingles*, a disease which mainly afflicts people over the age of 40 who have once had chickenpox. Chickenpox is still such a common childhood disease that most children become infected. It is very contagious and is probably transmitted via airborne droplets from infected to uninfected people. The virus is thought to invade through the upper respiratory tract and, like many upper respiratory diseases (see Figure 10.14), it occurs primarily during the winter months when human contact is increased. However, there are no upper respiratory symptoms associated with chickenpox. The virus presumably enters the bloodstream and spreads, causing a characteristic pox-like rash over the entire body, beginning at the trunk and head, then spreading to the extremities. In children there are rarely other symptoms and, except for a mild fever in some cases, the afflicted child may otherwise feel normal. In adults other symptoms are usually present, including fever, malaise, loss of appetite, headache, and more severe rash; viral pneumonia may even develop. There is no treatment for chickenpox, but a vaccine is currently being developed and tested. Drugs may be prescribed to alleviate fever or the itchiness of the rash, but aspirin is not advised because of a possible association with Reye's Syndrome.

Chickenpox

Varicella-Zoster virus also has the ability to cause latent infection. After chickenpox infection, the virus apparently relocates to nerve cells where it may be dormant for many years. Exactly how the virus exists during this time is unknown, but in later years the virus may become activated and cause infection again. Virus particles apparently travel via sensory nerves to the skin and cause the localized lesions of *shingles*. The cell-mediated immune system is apparently important in preventing shingles, as immunosuppressed patients show a high incidence of the disease. Complications such as pneumonitis, hepatitis, and encephalitis may occur, and death can even result. Passive immunization with serum containing immunoglobulin against the Varicella-Zoster virus can be effective in preventing the disease in high risk individuals, but there is otherwise no treatment.

Shingles

13.8 HERPES INFECTIONS

Several diseases are caused by viruses which apparently cannot survive long outside the host. These are transmitted by direct contact in saliva, during sexual intercourse, or through intravenous injections. Many of these diseases are caused by members of the herpesvirus family.

Several different types of herpes viruses may cause infections which are transmitted by direct contact. *Herpes simplex* virus causes either *fever blisters* (herpes simplex type 1) or the sexually transmitted disease *genital herpes* (herpes simplex type 2). The former is a disease which is common among young people, with one third or more individuals in a given population showing exposure to the herpes type 1 virus by adulthood. Genital herpes is mainly found in adults, most commonly of age 14–29, the most sexually active segment of the population. As pointed out in Figure 10.12, genital herpes is one of the most important sexually transmitted diseases. The marked increase in incidence of genital herpes is shown in Figure 13.13.

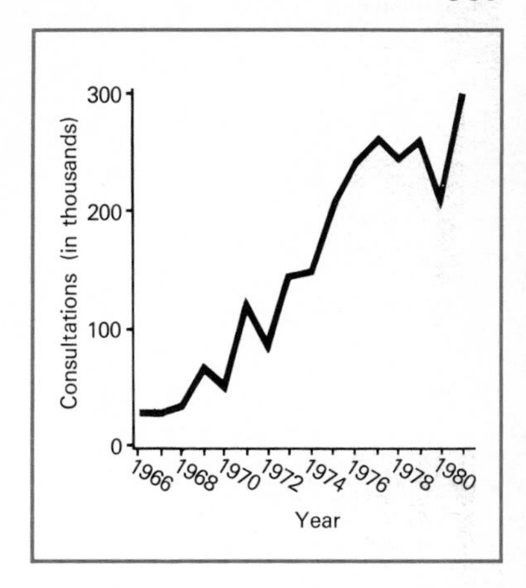

Figure 13.13 Rise in incidence in genital herpes in the United States, as shown by frequency of consultation of patients with physicians.

In both herpes diseases, the virus invades through the skin near the site of entry into the body and causes localized lesions as a result of lytic infection and the host inflammatory response. Herpes type 1 virus causes lesions in the oral cavity, or fever blisters which occur mainly on the lips. After the lesions heal, the virus enters a latent stage within nerve cells. At a later time, and following as yet undefined stimuli, the virus may emerge from latency and cause recurrence of the skin lesions. Herpes type 2 virus causes similar lesions in the vagina or urethra, leading to painful itchy ulceration and vaginal or urethral discharge. Systemic symptoms such as fever and malaise may accompany the more acute symptoms. Sometimes, however, symptoms may be so mild as to go unnoticed.

Genital herpes

Genital herpes is also a potentially latent viral infection. Primary symptoms usually last two to three weeks, but symptoms may reappear as many as eight times per year. Genital herpes may spread to the newborn as the baby passes through the birth canal of an infected mother. Infants born with congenital herpes may suffer a variety of serious systemic complications and death is a common result. Caesarean section is recommended for delivery if the mother is known to have genital herpes. A second possible complication of infection with herpes type 2 virus, cervical cancer, has been suggested but has not yet been proven (see Section 13.12). Although no antiviral drug can yet provide an effective cure, *acyclovir* has shown some promise in reducing the effects of viral infection. The best measures against genital herpes are preventive. As with other sexually transmitted diseases, this means that one should limit sexual contact with infected individuals, and advise sexual partners when infection is suspected.

Infectious mononucleosis

Two other herpes viruses can cause infections associated with direct contact modes of transmission. Serologic tests have shown that both of these viruses cause widespread infections. One, called *Epstein-Barr virus*, causes the disease **infectious mononucleosis**. The virus is spread by intimate oral contact and is not otherwise contagious. Infection begins in the mouth and throat, but spreads to the bloodstream and lymphatic system where the virus infects B cells of the immune system. Common symptoms include chills, sweats, severe sore throat, fever, swollen lymph nodes, and enlarged spleen or liver. The infection is self limited in most patients and complications are rare. Another herpesvirus, called *cytomegalovirus*, can also be a cause of infectious mononucleosis. It appears to be transmitted via several different routes, including direct oral contact, sexual intercourse, infected blood, and airborne particles. It is most dangerous when transmitted congenitally from mother to newborn, as the newborn may suffer severe complications leading to central nervous system disorders, deafness, or other difficulties which are often fatal. The role that Epstein-Barr virus plays in causing cancer will be discussed in Section 13.12.

13.9 ACQUIRED IMMUNODEFICIENCY SYNDROME (AIDS)

AIDS

Since it was first reported around 1980, AIDS has become an increasing concern, especially among homosexual men who are at particularly high risk, but now also to the general public since the disease may also spread through infected blood products and heterosexual contact. There has been a dramatic rise in the incidence of AIDS (Figure 13.14) with tens of thousands of individuals affected, about half of whom have died. AIDS is a disease of the immune system, as noted in Section 9.6. A type of retrovirus, called *human T-lymphotrophic virus* type III (or HTLV-III), has been identified as its probable cause (see cover photo and Figure 13.3*b*, *c*, and *d*). This virus has been isolated

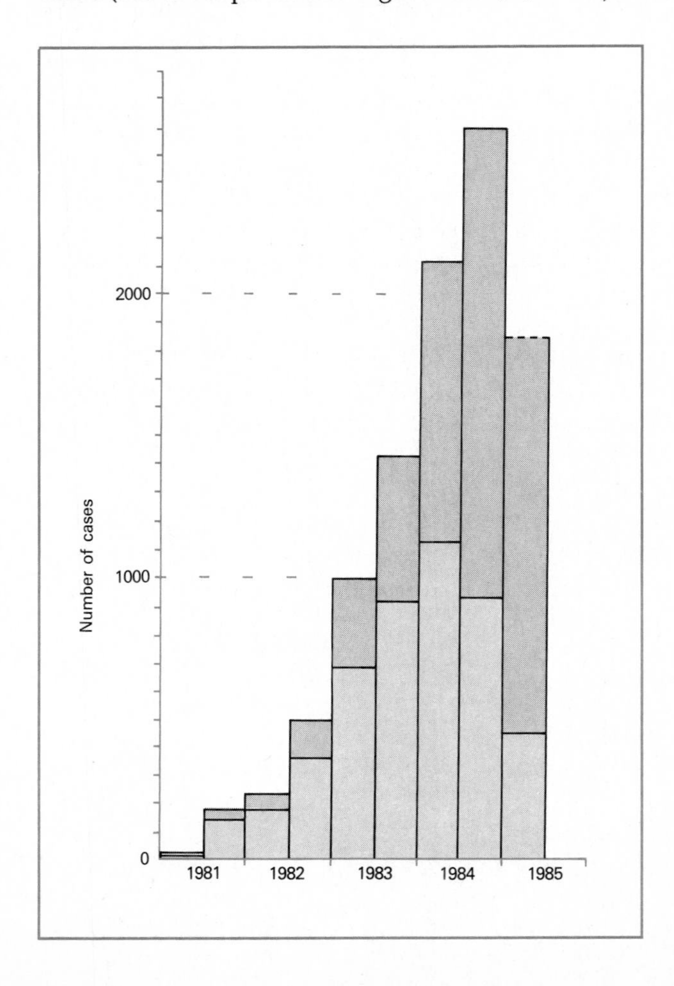

Figure 13.14 Incidence of acquired immunodeficiency syndrome (AIDS) in the United States, with known deaths shown in color. The number of cases for 1985 are only for the first several months of the year.

from a high percentage of patients who have the disease; also, antibodies against this virus are more common in AIDS patients than in people who are not in high risk groups. The fact that many individuals in high risk groups also possess antibodies against this virus suggests that exposure to AIDS is far more frequent than indicated by the actual number of known clinical cases (probably over one million people have been exposed to HTLV-III virus). The virus infects helper T cells leading to severe impairment of the immune system (see Section 9.6). Patients with AIDS are extremely prone to opportunistic infections by other microorganisms or to a rare form of cancer, called Kaposi's sarcoma. Three-fourths of AIDS patients have either Kaposi's sarcoma or pneumonia caused by the protozoan *Pneumocystis carinii*. Other frequent opportunistic pathogens include protozoa (*Toxoplasma* and *Cryptosporidium*), fungi (*Candida* and *Cryptococcus neoformans*), bacteria (*Mycobacterium avium-intracellulare*), and viruses (cytomegalovirus and herpes simplex virus). It is infection with an opportunistic pathogen which results in death of the patient.

AIDS has affected mainly adult male homosexuals, who account for almost 75 percent of the total number of cases. This may also explain the unusual geographic distribution of the disease; nearly half of the cases have occurred in either New York City or San Francisco, where there are large homosexual communities. However, AIDS is not just restricted to homosexuals. The virus may also spread through contaminated needles shared by intravenous drug users, in contaminated blood or blood products to those who receive transfusions (hemophiliacs receive frequent injections of blood products and are especially at risk), or congenitally from an infected parent to a newborn. Frequency of cases among the groups at risk of acquiring AIDS are listed in Table 13.2. Although the data do not suggest that heterosexual contact is as important as other routes of transmission, there is great concern that heterosexual promiscuity may result in spread of the disease to the general public.

There is currently no known cure for AIDS. Since it is such a debilitating disease, research is ongoing to try to improve the detection, treatment, and prevention of the disease. Since the discovery of the virus that causes AIDS,

TABLE 13.2 Distribution of acquired immunodeficiency syndrome (AIDS) among high risk groups

Risk group	Percent of cases
Adult cases (99% of total cases)	
Homosexuals/bisexuals	73.4
Intravenous drug users	17.0
Hemophiliacs	0.7
Transfusion recipients	1.4
Heterosexual contacts	0.8
Child cases (1% of total cases)	
Parents with or at risk of AIDS	71.7
Transfusion recipients	13.3
Hemophiliacs	5.3

From Morbidity and Mortality Weekly Report 34 (no. 18):245, published May, 1985.

serologic techniques have been developed which permit detection, including a radioimmunoassay to test for one of the major HTLV-III viral proteins, and an ELISA test (see Section 9.12) to test for antibodies in serum or donated blood or blood products which react with HTLV-III virus. The presence of antibodies against the AIDS virus as measured by this test correlates well with the ability to isolate living AIDS virus from members of high risk groups, so that the test is useful in screening for those who might have AIDS, or who once had it. Because of the possibility of false positive test results, serum is only considered positive if the reaction occurs repeatedly. The ELISA test has also been valuable in screening the blood supply of the United States. About 0.25 percent of the blood donated in the United States has been donated by people who have been exposed to the AIDS virus, as evidenced by reaction in the ELISA test. For the moment, prevention of the disease is being attempted through several recommended practices. These include education of those at high risk, elimination of sexual contact with infected individuals, elimination of high risk individuals or their sexual contacts as blood donors, screening of blood and blood products for evidence of AIDS exposure, heat treatment of blood products used by hemophiliacs to inactivate HTLV-III virus, and use of separate needles by intravenous drug users.

13.10 VIRAL DISEASES TRANSMITTED BY THE FECAL-ORAL ROUTE

A number of important diseases caused by viruses are transmitted mainly through contaminated food and water. Some of these, such as polio and hepatitis, are well known diseases, caused by members of the picornavirus family, called *enteroviruses*. Others are less well known but probably cause a major portion of the gastrointestinal illness that occurs in the United States and other countries.

Polio Poliovirus is one of the best known and studied enteroviruses. The paralytic disease it causes has been known for centuries, but is now relatively rare due to the development and use of effective vaccines (see Figure 9.18).

Poliovirus reproduces primarily in the cells of the intestinal tract of humans. It usually causes only mild symptoms that cannot be distinguished from many other trivial illnesses, and the disease may be called "summer cold" or "intestinal flu." The virus may occasionally pass into the blood or lymph system with no ill effects. In rare instances, it invades the central nervous system and causes the paralytic disease known as *poliomyelitis* or *infantile paralysis*. Although infection with the virus may be very common, paralytic polio disease is relatively rare.

Polio

The development of cell-culture methods made it possible to grow the virus in large numbers, opening the way for the production of vaccines. A *live-virus vaccine*, first introduced during the period 1955 to 1957, consists of virus that has been attenuated so that it no longer causes paralysis, although it still reproduces in the intestine. It may be given orally rather than by injection, and a single dose is sufficient since the virus reproduces in the body and can continue to induce immunity for some period of time. A possible disadvantage of the live-virus vaccine is that the virus might revert to a virulent form and cause paralytic infection, but this has not occurred even after vaccination of millions of individuals. Vaccination with oral poliovirus vaccine is now routine in the immunization of infants (see Table 9.7).

Hepatitis *Infectious hepatitis* is a disease of the liver caused by another enterovirus (hepatitis A virus) that is transmitted by the oral route and enters humans primarily through fecal contamination of water, food, or milk. The disease is endemic throughout the inhabited areas of the world, and outbreaks of epidemics occur in many regions. Explosive water-borne epidemics have occurred in areas where proper control of water supplies is not maintained, and food-borne epidemics have recently become of considerable concern. Hepatitis is among the leading reportable diseases in the U.S. (see Figure 10.6).

Hepatitis

Among foods transmitting the virus, the most significant are shellfish (oysters and clams) taken from polluted waters. These animals live on the sea bottom in estuaries and shallow marine bays and feed by passing through their bodies large amounts of silt-laden seawater. Particles in the seawater are filtered out in the animal and serve as its source of food. Pathogenic microorganisms, including infectious hepatitis virus derived from domestic raw sewage in the water, will also be filtered out and concentrated by the animal. No problem with infectious hepatitis is experienced in eating such shellfish if they are cooked, since heating destroys the virus, but oysters in particular are often eaten raw, and many epidemics have occurred in areas where shellfish waters are subject to sewage pollution.

Serum hepatitis (caused by hepatitis B virus) resembles infectious hepatitis in symptoms, but the virus is transmitted by direct contact with infected blood and blood products during transfusion, or through contaminated needles. A major difference from infectious hepatitis is that serum hepatitis may be associated with severe complications, such as nephritis, cirrhosis, and liver cancer (see Section 13.12).

There is no specific therapy for those with hepatitis. The best means of controlling infectious hepatitis is to prevent its spread by improving the general sanitation of the region through use of proper sewage-treatment systems, elimination of flies, and prevention of fecal contamination of food and milk supplies by infected food handlers. (General practices of food sanitation are further discussed in Chapter 16, and sewage-treatment systems are considered in Chapter 15.) The best control of serum hepatitis is by constant vigilance in the use of all materials used for blood sampling and transfusions and by avoiding the use of blood donors having a history of hepatitis. A vaccine is now available so that immunization against hepatitis B virus is possible.

Diarrheal diseases caused by viruses Diarrheal diseases of the gastrointestinal tract are among the leading causes of illness in the United States (see Figure 10.5), and are second only to respiratory infections as a leading cause of death due to infectious diseases in developing countries (see Figure

Diarrheal diseases

10.7). One of the most important viruses causing diarrhea in infants aged 6–24 months is *rotavirus*, a member of the reovirus family. This is a naked double-stranded RNA virus which causes a lytic infection of intestinal epithelial cells, leading to watery diarrhea. Sometimes vomiting and fever also occur. Protection is provided by antibodies reaching the infant from mother's milk, and decreases in the popularity of breast feeding may partly account for the high incidence of diarrheal disease among infants in developing countries. The disease is usually self limited, but if severe dehydration occurs it can be fatal.

Norwalk virus is an RNA virus whose relation to other viruses is not yet certain. It is a significant cause of epidemic diarrheal outbreaks, commonly involving schools, camps, and other places where people congregate (see Table 15.2). Symptoms include nausea, vomiting, cramps, often diarrhea, and sometimes fever and chills. Certain coronaviruses may also cause diarrheal disease in older children and adults.

13.11 RABIES

Rabies

Some viral diseases are transmitted to humans through direct contact with vertebrates; rabies is a classic example. The rabies virus is an enveloped single-stranded RNA virus of the *rhabdovirus* family (see Figure 13.1*d*). It attacks the nervous tissue of all warm-blooded animals. Rabies has been known since ancient times as a disease of dogs, but it is also common in wild animals such as rodents, foxes, skunks, and bats. The virus multiplies in the salivary gland and appears in the saliva; because of this it is readily transmitted by biting. Rabies virus is highly neurotropic; after the virus has entered the bloodstream following the bite, it passes to the nervous tissue where multiplication takes place, and the nerve cells degenerate (Figure 13.15). Characteristic inclusion bodies in the cytoplasm of nerve cells, called *Negri bodies*, are formed during virus multiplication and are demonstrated by a special staining method. The length of time before symptoms appear is highly variable, ranging from 2 to 40 weeks, depending on the size, location, and depth of the wound and on the amount of virulent saliva introduced.

When the virus has been passed for many generations in chick embryos, it becomes attenuated and is no longer capable of invading the central nervous system. This modified virus can then be used in the production of rabies vaccine for dogs. It can also be used to protect humans who have been bitten by rabid animals. The rabies vaccine was first developed by Louis Pasteur by attenuation of the virus through inoculation in rabbits, and Pasteur's procedure was of historical importance in demonstrating the effectiveness of vaccination.

The human mortality rate is considered close to 100 percent. If it is known that a person has been bitten by a rabid animal, that individual must immediately receive treatment to inactivate the virus while it still remains localized. Because of the relatively long incubation period, vaccination of a person who has been bitten may be successful. If the incubation period is less than 30 days, however, the development of immunity may not be rapid

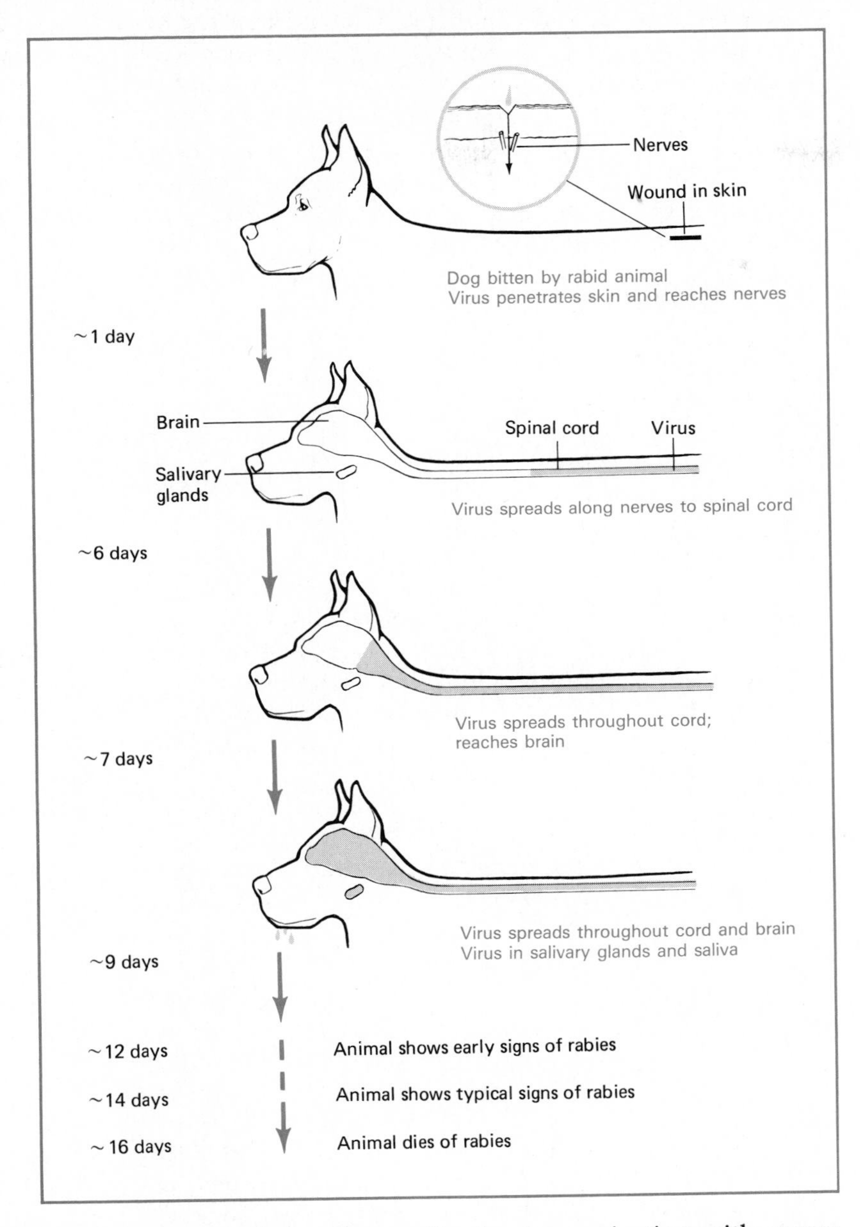
Nerves
Wound in skin
Dog bitten by rabid animal
Virus penetrates skin and reaches nerves
~1 day
Brain
Salivary glands
Spinal cord
Virus
Virus spreads along nerves to spinal cord
~6 days
Virus spreads throughout cord; reaches brain
~7 days
Virus spreads throughout cord and brain
Virus in salivary glands and saliva
~9 days
~12 days
Animal shows early signs of rabies
~14 days
Animal shows typical signs of rabies
~16 days
Animal dies of rabies

Figure 13.15 Progress of the rabies virus through an infected dog.

enough to prevent onset of the disease. Passive immunization with serum from an immune individual may be used in some cases. Vaccine for humans is prepared from brain tissue of infected rabbits, usually by killing the virus with phenol or ultraviolet light. Since an allergic reaction to rabbit brain tissue may occur, vaccine is rarely used unless there is good evidence of exposure to rabies. More recently, the attenuated live-virus vaccine used for dogs has been used in humans, since it does not cause sensitization.

The spread of rabies infection is quite complex, since the virus can maintain itself in nature only if infected animals can transmit the virus before they die. As we mentioned above, rabies virus is also found in nature in many wild animals, usually carnivores, including foxes, skunks, mink, weasels, wolves, and bats. The fact that bats carry rabies is important, since these animals may be symptomless and serve as a reservoir for the virus in nature. Bats have attacked persons, and some of these bats have been shown to be infected. In Central and South America the vampire bat, a blood-sucking animal, frequently attacks cattle and transmits the virus to them. Rabies infection may be a major factor in controlling the population sizes of wolves, foxes, wild dogs, and other caninelike species: when the population of these animals builds up, an increase in incidence of rabies can bring about a sharp drop in their numbers.

13.12 VIRUSES AND CANCER

The word *cancer* describes a condition rather than a specific disease. There are many different types of cancers, and there can be many different causes for them. The common feature of all cancers is that they are conditions in which normal cells of the body become abnormal. Cancerous cells are different from normal cells in many ways, but one key difference is the lack of **contact inhibition**. This is the process in which cells stop growing when they come in contact with other cells. Cancerous cells, lacking contact inhibition, grow profusely, leading to the accumulation of large masses of cells, called *tumors*. The scientific term for tumor, **neoplasm**, is descriptive of this condition, as it literally means "new shape," referring to the development of new and different growth from old tissue.

Cancer

Not all tumors are seriously harmful. The body is able to wall off some tumors so that they do not spread and invade other normal tissues; noninvasive tumors are said to be **benign**. Human warts, caused by papilloma viruses (a type of papovavirus), are good examples of benign tumors. Other tumors are **malignant** and can invade and destroy normal body tissues and organs. In advanced stages of cancer, malignant tumors may develop the ability to spread to other parts of the body and initiate new tumors, a process called **metastasis**.

How does a normal cell become cancerous? The process of tumor formation can be broken down into several steps, as outlined in Figure 13.16. The first step leading to a cancerous condition, *initiation*, involves a genetic change within a cell. This step may be induced by many chemicals, called *carcinogens*, or by physical stimuli, such as ultraviolet light or X rays, all of which are mutagenic (see Figure 7.4). Cancer is a complex phenomenon and often many factors other than the cause of initiation are necessary for the potentially cancerous cell to actually begin tumor formation. Thus, factors such as age, sex, genetic predisposition, diet, sunlight, smoking, the state of the immune system, and other chemicals (such as nitrosamines, aflatoxins) may be involved in the *promotion* of the initiated cell into a cancer cell which develops into a tumor. Tumor *progression* leads to increased invasiveness, malignancy, and possibly metastasis.

Viruses that cause animal cancers Although tumor cells arise in various ways, some tumors clearly have been shown to be induced by viruses.

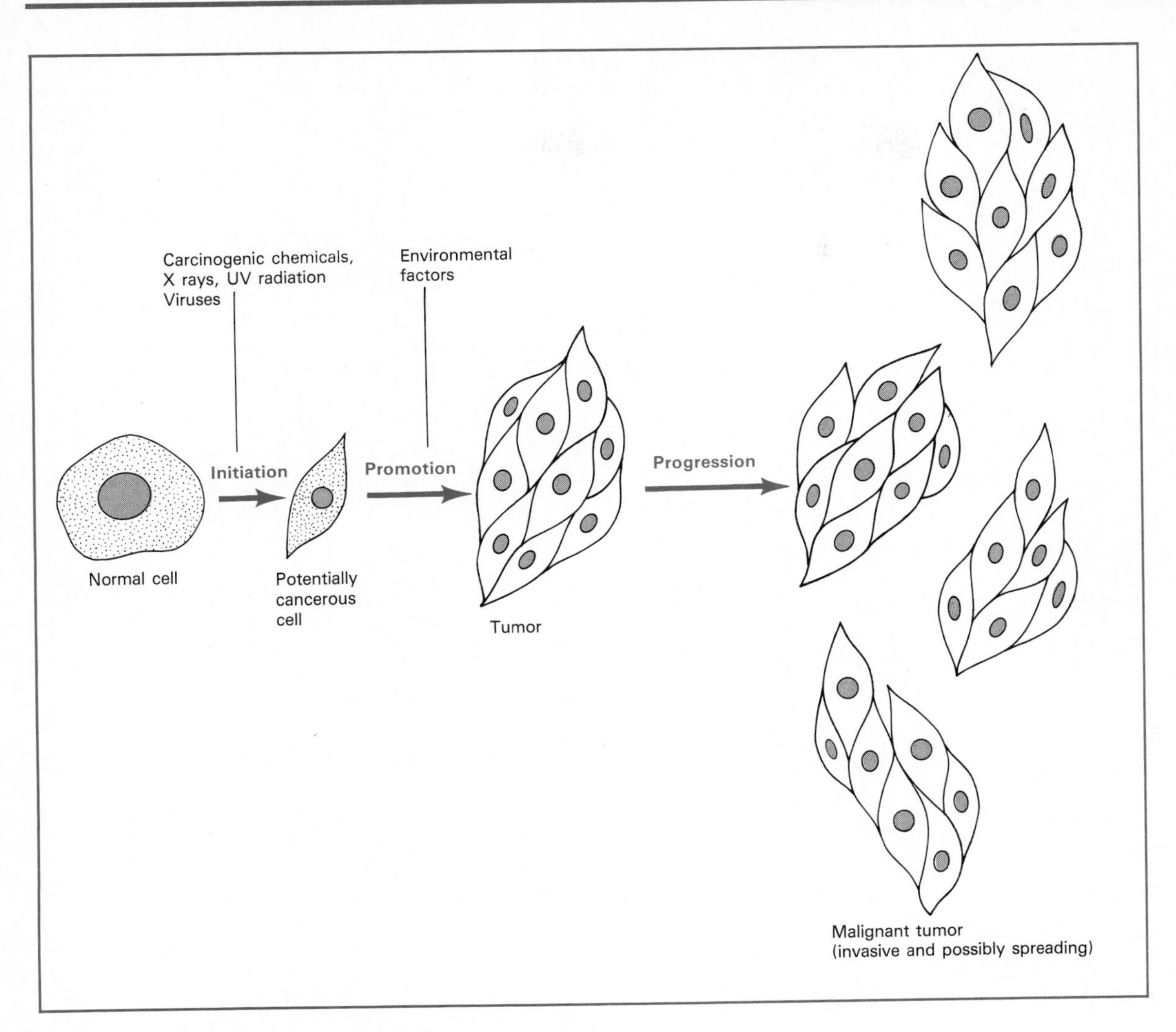

Figure 13.16 Stages in the development of a tumor from a normal cell.

Both DNA and RNA viruses can cause tumors in animals. A tumor of chickens, known as *Rous sarcoma,* was the first tumor shown to be induced by a virus. A number of other tumors of birds were subsequently found to have a viral origin. Shope papilloma and Shope fibroma are virus-induced tumors of rabbits. A mammary-gland tumor of mice has been discovered to be induced by a virus transmitted in the mother's milk. The polyoma virus (from *poly,* "many," and *oma,* "tumor") causes a wide variety of tumors in mice and other animals.

Research on viruses that cause animal cancers has led to increased understanding about how viruses are involved in tumorigenesis. When a tumor virus enters a cell, it does not cause a lytic infection, but rather it **transforms** the cell into a cancer cell. There are many ways to recognize whether cells have been transformed. In tissue culture, transformed cells grow profusely and collect into aggregates (Figure 13.17); transformed cells may also cause tumors when injected into animals.

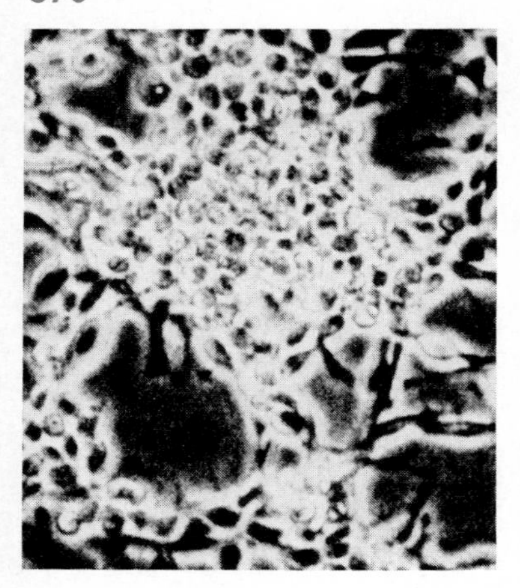

Figure 13.17 Microscopic appearance of a cell culture in which some of the cells have been transformed into tumor cells. The normal cells are elongated because they spread out on the glass culture dish. The tumor cells have lost contact inhibition and have piled up to make a small clump.

Cancer viruses enter cells and either integrate into the host cell chromosome, or exist as an extrachromosomal element. RNA tumor viruses, the retroviruses, synthesize DNA from RNA using the enzyme *reverse transcriptase* (see Figure 6.17), before integrating into the DNA of the host chromosome. Virus integration can lead to genetic change in the host cell. Alternatively, a viral gene may be expressed, and the gene products cause the cell to transform into a cancer cell. Transformation of cells by polyoma virus and retroviruses seem to occur in this way. Genes that code for proteins which can bring about cell transformation are called **oncogenes**. A surprising fact is that most normal animal cells contain genes which are nearly identical in structure to oncogenes of retroviruses, and these cellular genes also have the potential to transform cells. During integration of viral DNA into host chromosomes, these host *proto-oncogenes* may be switched on, leading to the transformation of the cell.

Viruses and human cancers It is difficult to prove the viral origin of human cancers, as it is unethical to inject such viruses into humans. Thus, the suggestion of a causal relationship is made on the basis of indirect evidence of several different types. For several types of human cancers, there is growing evidence of the involvement of viruses (Table 13.3).

TABLE 13.3 Some human cancers which may be caused by viruses

Cancer	Virus	Evidence
Adult T-cell leukemia	Human T-cell leukemia virus (type 1)	Virus isolated from patients; distributions of virus and cancer correlate; high incidence of antibody against virus in exposed individuals
Burkitt's lymphoma	Epstein-Barr virus	Virus isolated from cells of patients; virus transforms B cells in culture; virus is tumorigenic
Nasopharyngeal carcinoma	Epstein-Barr virus	Viral DNA present in tumor cells; virus transforms epithelial cells in culture; virus is tumorigenic
Hepatocellular carcinoma (liver cancer)	Hepatitis B virus	Correlation between chronic hepatitis and liver cancer; virus and cancer distributions correlate
Cervical cancers	Herpes simplex type 2 virus	Higher incidence of cancer in women with genital herpes; higher incidence of antibody against virus in women with cervical cancer
Skin and cervical cancers	Papilloma virus	Viruses cause benign human tumors (warts); malignant tumors may develop from warts in patients with a rare hereditary disease; genital warts may convert to epithelial cell carcinoma in the genitourinary tract

1. *Direct isolation* of a virus from cancer patients. This has been observed for Burkitt's lymphoma, a type of cancer found in East Africa and New Guinea, and for adult T-cell leukemia.
2. *Demonstration* that an isolated virus is able to transform tissue culture cells or to cause experimental tumors in animals. This is true of Epstein-Barr virus, associated with Burkitt's lymphoma and nasopharyngeal carcinoma.
3. *Observation* of viral DNA in tumor cells. This is true for Epstein-Barr virus in nasopharyngeal carcinoma.
4. *Increased incidence* of antibody against a virus in individuals who live in an area where a certain cancer is found. This is true for adult T-cell leukemia.
5. *Correlation* between the geographic distribution of a virus and of a specific type of cancer. Such correlation exists between liver cancer and hepatitis B virus, and between adult T-cell leukemia and human T-cell leukemia virus (type 1).
6. *Correlation* between a certain known virus infection (or antibodies against this virus) and a certain type of cancer. For example, women with genital herpes are more likely to also have cervical cancer, and people with chronic hepatitis are more likely to develop liver cancer.

The most convincing evidence of a causal relationship in humans is in the case of adult T-cell leukemia, a form of cancer which occurs mainly in Southwest Japan and the Caribbean. Human retroviruses have been suspected for a long time, but have only recently been isolated. One reason for their suspected existence was the presence of proto-oncogenes in human DNA structurally like those carried by retroviruses of animals. Human T-cell leukemia virus (type 1) has been isolated from these cancer patients. There is a correlation between the geographic distributions of the cancer and the virus, and there is a high incidence of antibody against the virus in people living in regions where the cancer is found. Definitive proof that any or all of the viruses listed in Table 13.3 actually cause the cancers with which they are associated cannot be obtained by the experimental infection necessary to fulfill Koch's postulates. However, with the future development of vaccines, a reduction of incidence of these cancers resulting from vaccination against the suspected virus, may show that the virus is the etiologic agent.

SUMMARY

Because viruses are obligate intracellular parasites, they commonly cause disease. Our ability to culture viruses has improved through the development of **animal-cell cultures**, so that we can now better understand the molecular nature of virus infection. Animal viruses must *penetrate* host cells before *multiplication* can occur. New virus particles then *bud* through the cell membrane and are *released* into the extracellular environment. Animal viruses, like bacteriophages, may cause **lytic** or **latent** infections. Other possible effects of animal viruses include **persistent infection** of the host cell, or **transformation** of the host cell into a cancer cell.

Because there are very few effective antiviral drugs, control of viral diseases through immunization is important. **Smallpox**, the first disease for which vaccination was developed, has been completely eradicated through vaccination. Three of the major childhood viral diseases, **measles**, **mumps**, and **rubella**, have also been effectively controlled by vaccination. Genetic engineering methods are providing new vaccines for viral diseases.

Some viral diseases which are among the leading causes of disease cannot be controlled with vaccines. Because of **antigenic shift** and **antigenic drift**, the virus that causes **influenza** constantly changes resulting in annual epidemics and occasional pandemics. Only *polyvalent* vaccines provide complete protection. The **common cold** is caused by over 100 antigenically different viruses, so that a single vaccine would be useless.

Sexually transmitted viral diseases are among the most prevalent and worrisome of modern diseases. **Genital herpes**, caused by a virus related to the one which causes **fever blisters**, is one of the most common venereal diseases. **AIDS** is rarer, but more debilitating as it destroys much of the immune system, leaving patients prone to opportunistic infections.

Though some important water-borne viral diseases, such as **polio**, have been controlled with vaccines, others such as **hepatitis**, and **diarrheal diseases** remain significant public health problems.

We have learned much about the events of **cancer** from studying the viruses that cause **tumors** in experimental animals. They **initiate** the change of the cell from normal to cancerous, so that other factors may **promote** the loss in the transformed cell of **contact inhibition** and produce a **tumor**. Not all tumors are dangerous; some are **benign** and some are **malignant**, with the ability to invade surrounding tissue and possibly **metastasize**. **Oncogenes** are viral genes which code for products that can transform cells, and normal cells themselves often contain **proto-oncogenes**. There is mounting evidence that some human cancers, such as cervical cancer, liver cancer, and more rare cancers, may be caused by viruses.

KEY WORDS AND CONCEPTS

Enveloped virus
Icosahedral virus
Animal-cell culture
Latent infection
Persistent infection
Transformation
Smallpox
Influenza
Antigenic shift
Antigenic drift
Common cold
Measles
Mumps
Rubella
Chickenpox
Genital herpes
Infectious mononucleosis
Acquired immunodeficiency syndrome (AIDS)
Polio
Infectious hepatitis
Serum hepatitis
Rabies
Cancer
Contact inhibition
Neoplasm
Benign tumor
Malignant tumor
Metastasis
Oncogenes

STUDY QUESTIONS

1. What features are used to classify animal viruses?
2. What are the advantages of *animal-cell cultures* over experimental animals for the culture of viruses?
3. List the steps which occur during the infection of a host cell by an animal virus. Which steps are different from a bacteriophage infection of a bacterial cell?
4. Describe how the *poliovirus* genome codes for reproduction in its RNA genome. What are the events involved in poliovirus reproduction within the host cell?
5. How is a *persistent infection* of a cell by a virus different from a *lytic infection*?
6. Compare the importance of antiviral drugs in the control of viral diseases with the importance of antibacterial drugs in the control of bacterial diseases.
7. Describe two ways in which genetic engineering techniques may be used to produce viral vaccines.
8. Name four viral diseases which have been effectively controlled through immunization. What viral disease has been *eradicated* using a vaccine?
9. What is the difference between *antigenic shift* and *antigenic drift*? Explain how these two processes initiate epidemic and pandemic outbreaks of *influenza*.
10. Why is it not possible to effectively control the *common cold* with vaccines?
11. What is a *polyvalent* vaccine?
12. Which viral diseases are associated with *Reye's syndrome*? What medication is not recommended for relief of the symptoms of these diseases?
13. Explain why *measles* and *rubella* now are often seen in epidemics on college campuses.
14. Explain the relationship between the diseases *chickenpox* and *shingles*. Which is a *latent infection*?
15. Name three viral diseases which may be transmitted *congenitally*. In each case, what complications may result?
16. Most of those afflicted with *acquired immunodeficiency syndrome (AIDS)* die. What is the usual cause of death?
17. List the groups of individuals who are at high risk of getting AIDS. What is the best means of avoiding AIDS for members of each of these groups?
18. Name three water-borne viral diseases. Which water-borne viral disease has been effectively controlled by vaccination?
19. Name two viral diseases which may be contracted through the use of dirty needles.
20. Why is it difficult to eradicate the disease *rabies*?
21. What steps are involved in the conversion of a normal cell to a cancer cell? Which step is caused by tumor viruses? What property of normal cells have cancer cells lost?

22. Give an example of a *benign* tumor and an example of a *malignant* tumor. What is the difference between these types of tumors?

23. Give two examples of human cancers which might be caused by viruses. What kinds of evidence suggest that viruses are the causal agents?

SUGGESTED READINGS

BENENSON, A.S. (ed.). 1985. *Control of communicable diseases in man, 14th edition.* American Public Health Association, Washington, D.C.

FRAENKEL-CONRAT, H. and P.C. KIMBALL. 1982. *Virology.* Prentice-Hall, Englewood Cliffs, New Jersey.

JOKLIK, W.K., H.P. WILLETT, and D.B. AMOS. 1984. *Zinsser microbiology, 18th edition.* Appleton-Century-Crofts, East Norwalk, Connecticut.

PALMER, E.L. and M.L. MARTIN. 1982. *An atlas of mammalian viruses.* CRC Press, Boca Raton, Florida.

WEINBERG, R.A. 1983. *A molecular basis for cancer.* Scientific American 249: 126–142. (A review of oncogenes and the role they play in causing cancer.)

WEISS, R.A. 1985. *Unraveling the complexities of cancer.* Pages 1–21 *in* P.W.J. Rigby, and N.M. Wilkie, eds., *Viruses and Cancer.* Cambridge University Press. (A review of past and present ideas on the role of viruses as causes of cancer.)

14

Environmental and Global Microbiology

We often think of three states of matter: solid, liquid, gas. The earth itself can be divided into three major regions which correspond to these three states of matter: (1) The *solid* portion of the earth, composed of rocks and soil, called the **lithosphere**. (2) The *aquatic* regions of the earth, the lakes, rivers, and oceans, called the **hydrosphere**. (3) The *gaseous* regions of the earth, called the **atmosphere**. Biologists frequently use another term, the **biosphere**, to refer to the living organisms present on or near the surface of the earth. Microorganisms are a major, indeed vital, part of the biosphere.

When we think of the biosphere, we think of life in a global context (Figure 14.1). Although many of the processes which take place in and on the earth are strictly physical or chemical, biological processes are also of major global significance. The activities of organisms alter the chemistry and physics of the lithosphere, hydrosphere, and atmosphere in many important ways. As an example, the oxygen of the atmosphere, so vital for aerobic (including human) life, is a product of the photosynthetic activities of green plants, algae, and cyanobacteria. In one way, we can think of living organisms of the biosphere as *catalysts* of chemical reactions; many of these reactions would not occur at all, or would occur only slowly, in the absence of life. Microorganisms carry out a vast number of chemical transformations that

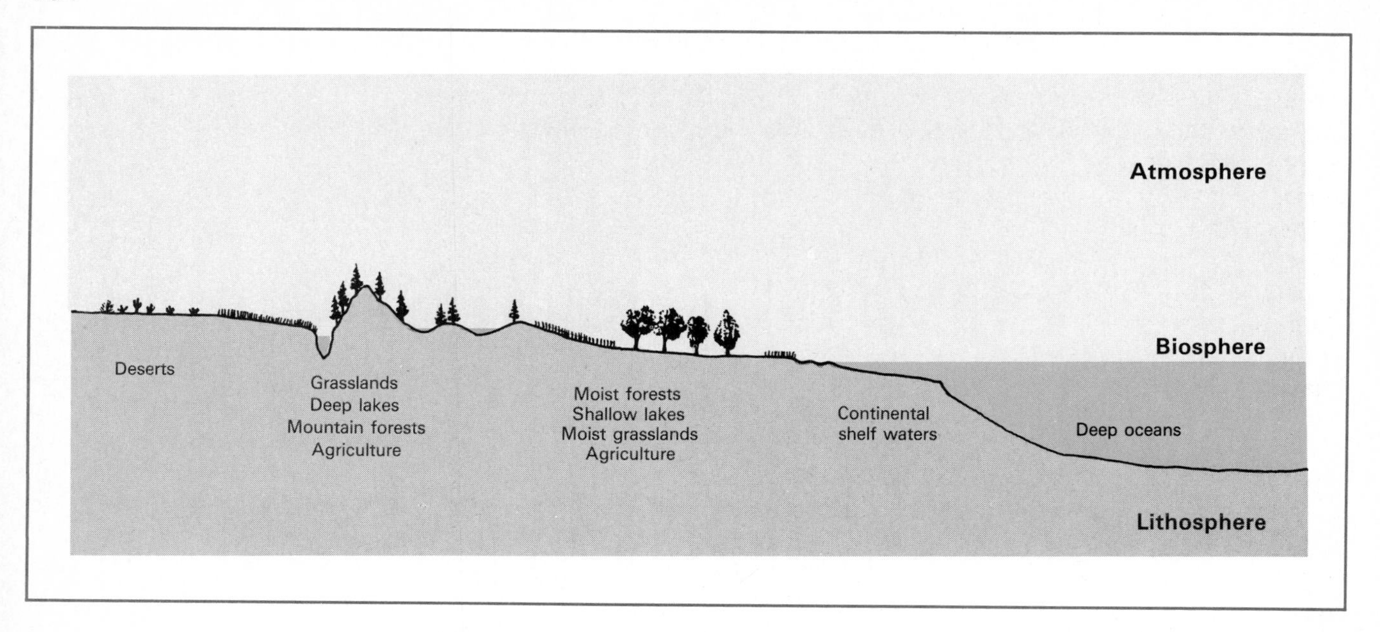

Figure 14.1 The global ecosystem, showing relationships between atmosphere, biosphere, and lithosphere.

affect the earth. It is the purpose of the present chapter to consider such microbial activities in a global context.

Why should we be concerned about the global activities of microorganisms? Microorganisms play far more important roles in nature than their small sizes suggest. Their activities are so extensive and diverse that even when they are present in relatively low numbers they may have significant effects on the environment. We already discussed earlier in this book some of the most dramatic effects of microorganisms: their roles as causal agents of infectious disease. But pathogenic microorganisms constitute only a small fraction of the kinds of organisms present in the world. The vast majority of microorganisms do not cause disease, but that fact does not make them less important or less interesting; they carry out a wide variety of other processes of importance to the well-being of humans. Because microorganisms are usually invisible, their existence in an environment may go unsuspected; yet without them, higher organisms would quickly disappear from the earth.

When we think of the roles of life forms in the world at large, the word *environment* frequently comes into use. **Environment** refers to everything surrounding a living organism: the chemical, physical, and biological factors and forces which act on a living organism. Although the earth provides suitable environments for microbial growth, we do not find the same organisms everywhere. In fact, virtually every environment, no matter how slightly it differs from others, probably has its own particular complement of microorganisms that differ in major or minor ways from the organisms of other environments. Because microorganisms are small, their environments are also small. Within a single handful of soil, many microbial environments exist, each providing conditions suitable for the growth of certain microorganisms.

When we think of microorganisms living in nature, we must learn to "think small."

In the present chapter, we will be discussing the roles of microorganisms in the global system. We will be considering both the fundamental microbial processes that alter the physics and chemistry of the globe as well as some global activities of microorganisms that have major practical significance. Some of these practical applications are beneficial, whereas others are harmful. The present chapter will also introduce some of the important concepts of environmental microbiology which will find application in following chapters on water, food, and agricultural microbiology. We will also study briefly some of the ways in which microorganisms cooperate with higher plants and animals in ways which are mutually beneficial.

14.1 MICROORGANISMS AND ECOSYSTEMS

An **ecosystem** is usually defined as the total community of organisms living together in a particular place, and includes the physical and chemical environment in which the organisms live. Each organism interacts with its physical and chemical environment and with the other organisms in the system so that the ecosystem can be viewed as a kind of *superorganism* with the ability to respond to and modify its environment. A good example of an ecosystem is a lake (Figure 14.2). The sides of the lake define the boundaries of the ecosystem, and within these boundaries the organisms live and carry on their activities, greatly modifying the characteristics of the lake as well as each other.

Energy enters an ecosystem mainly in the form of sunlight which is used by photosynthetic organisms in the synthesis of organic matter. Since the step involves the initial synthesis of organic matter, it is often referred to as *primary*

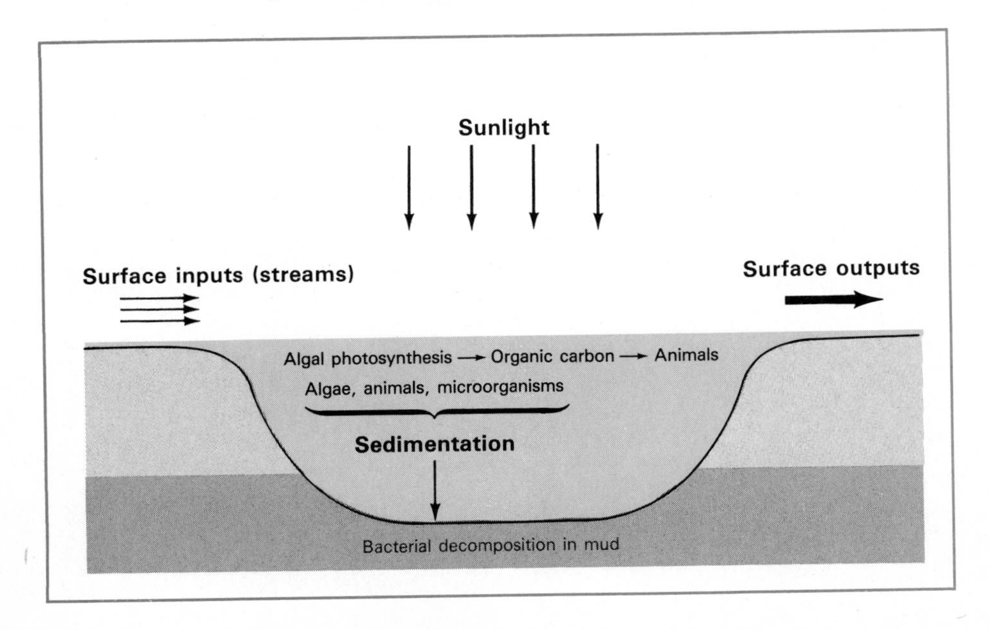

Figure 14.2 The lake, an example of an ecosystem.

production. Some of the energy contained in this organic matter is dissipated by the photosynthesizers themselves during respiration, and the rest is available to *herbivores*, which are animals that consume the photosynthesizers. Of the energy entering the herbivores, one portion is dissipated by them during respiration, and the rest is used in synthesizing the organic matter of the herbivore bodies. Herbivores are themselves consumed by carnivorous animals, and these *carnivores* are eaten by other carnivores, and so on. At each step in this chain of events, a portion of the energy is dissipated as heat. Any plants or animals that die, whether from natural causes, injury, or disease, are attacked by microorganisms and small animals, collectively called *decomposers*. The decomposers also utilize energy released by plants or animals in the form of excretory products. All these reactions constitute a *food chain* or *food web*. Because there is a loss of energy at each stage of the food chain, ultimately practically all of the biologically useful energy that is used to convert materials to organic matter by the photosynthesizers is dissipated; usually only very small amounts are stored. Because of this, energy is said to *flow through* the ecosystem. The quantitative relations can be expressed by an energy-flow diagram, such as the one shown in Figure 14.3. Not all ecosystems

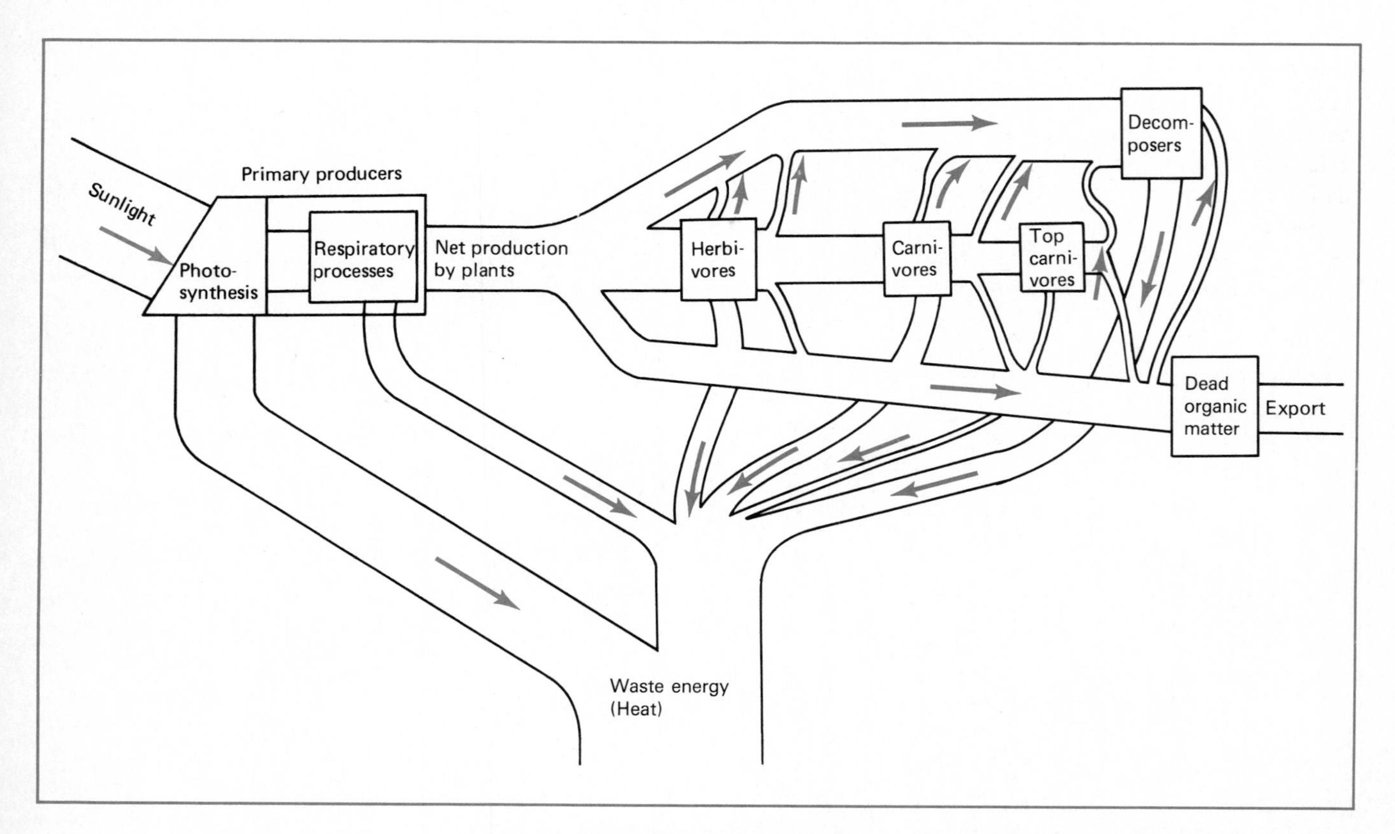

Figure 14.3 Energy flow through an ecosystem. Energy enters the ecosystem as sunlight, is converted into chemical energy by photosynthetic organisms (primary producers), and is then consumed by heterotrophs.

derive their energy from sunlight through photosynthesis. Primary production driven by chemical energy via lithotrophic bacteria occurs in a few ecosystems, notably the deep-sea thermal vents described in the Box on page 76.

Biogeochemical cycles Although the energy fixed by photosynthesizers is ultimately dissipated as heat, the chemical elements that serve as nutrients usually are not lost from the ecosystem. For instance, carbon from CO_2 fixed by plants in photosynthesis is released during respiration by various organisms of the food chain and becomes available for further utilization by the plants. Nitrogen, sulfur, phosphorus, iron, and other elements taken up by plants are also released through the activity of the decomposers and are thus made available for reassimilation by other plants. Therefore, although energy flows through the ecosystem, chemical elements move in *cycles* within the system. In some parts of the cycle the element is oxidized, whereas in other parts it is reduced; for many elements a **biogeochemical cycle** can thus be defined, in which the element undergoes change in oxidation state as it is acted upon by one organism after another. In addition to this *redox* or *oxidation-reduction* cycle, it is also possible to define a transport cycle, which describes the movement of an element from one place to another on earth, as for instance, from land to air or from air to water (Figure 14.4). Such a transport cycle may or may not also involve a redox cycle. For instance, when oxidation or reduction leads to conversion of a nonvolatile substance to one that is volatile, the latter can then be transported to the air, so that the transport cycle is coupled to the redox cycle. In other cases, oxidation or reduction does not lead to a change in state and has no influence on transport.

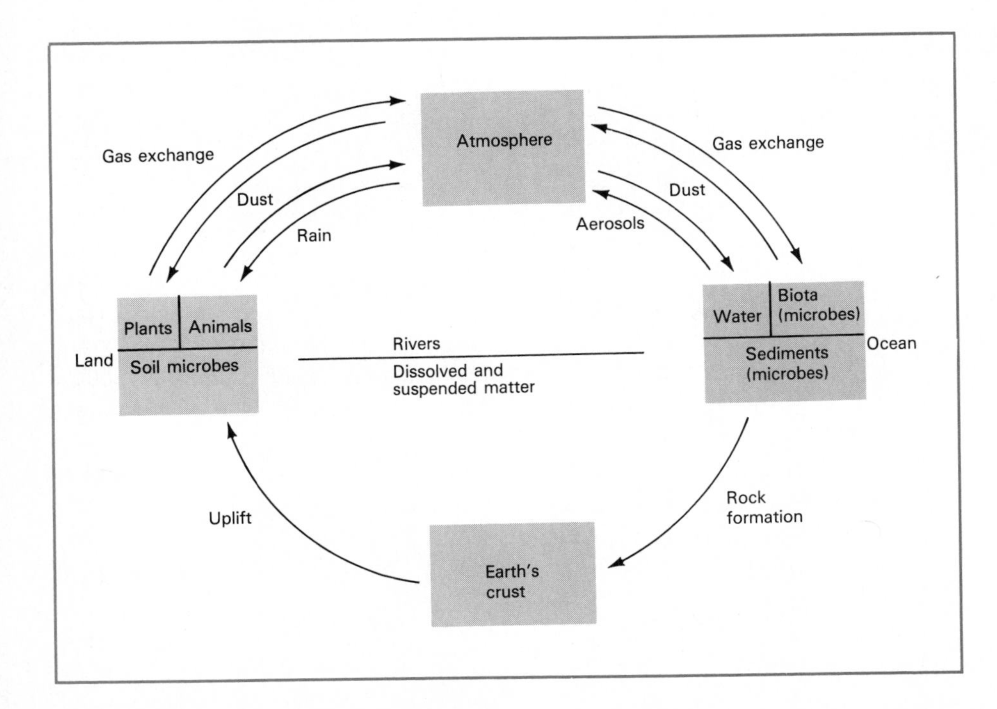

Figure 14.4 The biogeochemical cycle, showing the transport of an element from one location to another on earth. Living organisms are responsible for chemical changes that lead to mobilization of elements.

14.2 MICROORGANISMS IN NATURE

The natural habitats of microorganisms are exceedingly diverse. Any habitat suitable for the growth of higher organisms will also permit microbial growth, but in addition, there are many habitats unfavorable for higher organisms where microorganisms exist and even flourish. In fact, some microorganisms are probably the hardiest living things, as they thrive in environments considered extreme from the human point of view (see the Box on the next page). Because microorganisms are usually invisible, their existence in an ecosystem is often unsuspected, yet their activities are almost always of central importance to the function of that ecosystem.

What is the environment of a microbe? The microbial environment is very small, and within a single soil crumb or upon a single root surface there may be a variety of distinct environments. Each of these tiny environments is called a **microenvironment**; it is suitable for the growth of certain kinds of organisms but not for others. Of most importance is the fact that microbes become concentrated on surfaces. **Surfaces** are solid or liquid interfaces to which microbes attach and grow. On such surfaces, microbial numbers and activity will usually be much higher than away from the surface.

Because microorganisms live in microenvironments, it is difficult to measure their activities in nature. We may be able to count the *numbers* of microbes in a given environment, but we find it difficult to determine how fast they are growing. Because microorganisms form spores and other structures which are dormant, a viable count often tells us little about whether the microbe that was counted was really doing something at the time of sampling. The microbial ecologist has developed procedures for measuring the activities of microorganisms in nature which are based on measurements of microbial metabolism. Microbes carrying out respiration cause changes in oxygen and carbon dioxide concentrations of the habitat, and such changes can be measured by sensitive chemical methods. In addition, when microbes grow, they take up nutrients from their surroundings, and expel waste products into their surroundings, and changes in such products can be measured. Very often, the microbial ecologist uses radioactive tracers to measure metabolic changes in the environment, because the assay of radioactive tracers is exceedingly powerful and sensitive. For instance, if the process of microbial photosynthesis were to be measured, this would result in the use of light energy for the uptake of carbon dioxide. A radioactive form of carbon dioxide, $^{14}CO_2$, would be used, and the incorporation of the radioactive carbon would provide a measure of photosynthesis (Figure 14.5). One of the surprises when such measurements are made is that the *activity* of an organism is often *not* correlated with its numbers. For instance, photosynthesis by algae in lakes or the oceans may be slow or fast at different times, even when approximately the same number of algal cells are present. For instance, features of the environment such as temperature or light may control how rapidly microbial activity occurs.

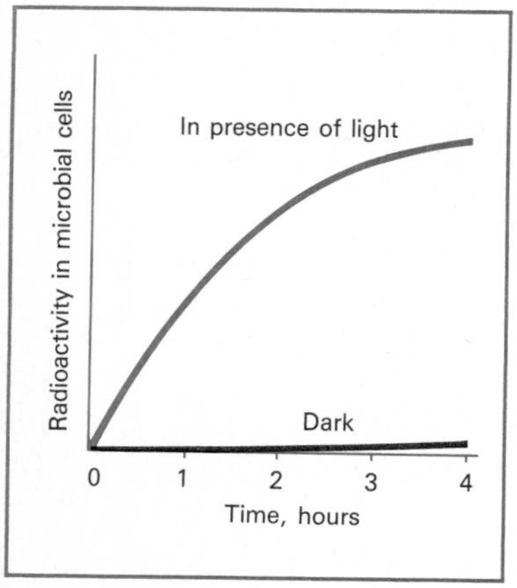

Figure 14.5 Measurement of microbial activity in an ecosystem by means of a radioactive tracer, carbon-14 (^{14}C). The process of photosynthesis is measured by assessing the uptake of radioactive carbon dioxide ($^{14}CO_2$) into the photosynthetic microorganisms in a lake water sample.

Soil as a microbial habitat In terrestrial environments the most extensive microbial growth takes place on the surfaces of soil particles (Figure

THE LIMITS OF LIFE

Though animals and plants grow only in moderate environments, some microorganisms can grow under extreme conditions. For instance, the upper temperature limit for plants or animals is about 50°C, but some bacteria living in hot springs can grow at temperatures up to the boiling point (100°C). Though the upper temperature limit for life has not yet been determined, we may learn more about it by studying deep-sea hot springs (see the Box on page 76), where extreme pressures keep water in a liquid state to over 350°C. There is good evidence that some deep-sea vent bacteria can grow at over 110°C.

Other environmental parameters may also be extreme. For example, halophilic bacteria grow so profusely that pink colors are seen in some very salty lakes and evaporation ponds. Microorganisms can grow in the extremely cold and dry valleys of Antarctica where water is very scarce. Acidophilic algae, fungi, and bacteria thrive in environments as acid as pH 0. Some of the most interesting microorganisms are those which live where more than one parameter is extreme. *Sulfolobus acidocaldarius*, for example, can grow in nearly boiling hot springs which are at pH 1.

Algae are visible in the run-off channel from a hot spring in a thermal area of Yellowstone National Park.

The knowledge gained by studying microorganisms inhabiting extreme environments may be applied to such practical problems as pollution caused by thermal wastes or acid rain, and spoilage of salted foods, as well as to the use of some of these microorganisms in industrial processes. Perhaps the best reason to study microbial life in extreme environments is to learn the limits of life itself. In the quest to discover life elsewhere in the universe, we have already learned that extraterrestrial environments are extreme when viewed from the human point of view, but other forms of life may be able to survive under such conditions.

14.6). Soils form as a result of the combined action of physical, chemical, and biological forces, which cause the breakdown of the rocks of the lithosphere and their conversion into the soil fabric. We discuss soils and soil formation further in Chapter 17. Near the surface of the earth, a soil will generally be rich in organic matter, and microbial numbers will be high, whereas deeper in the earth the soil will be poorer in organic matter and microbial numbers will be low.

One of the major factors affecting microbial activity in soil is the availability of water, whose presence depends on rainfall and on the moisture-holding capacity of the soil particles. In marshy soils or under conditions of high rainfall, the space between the soil particles may be completely saturated

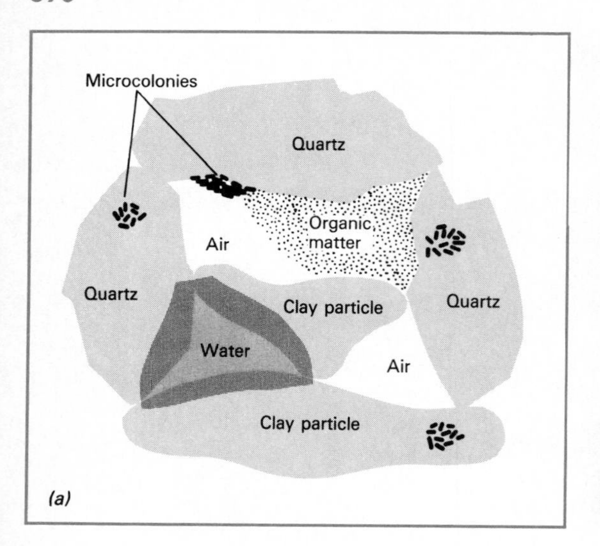

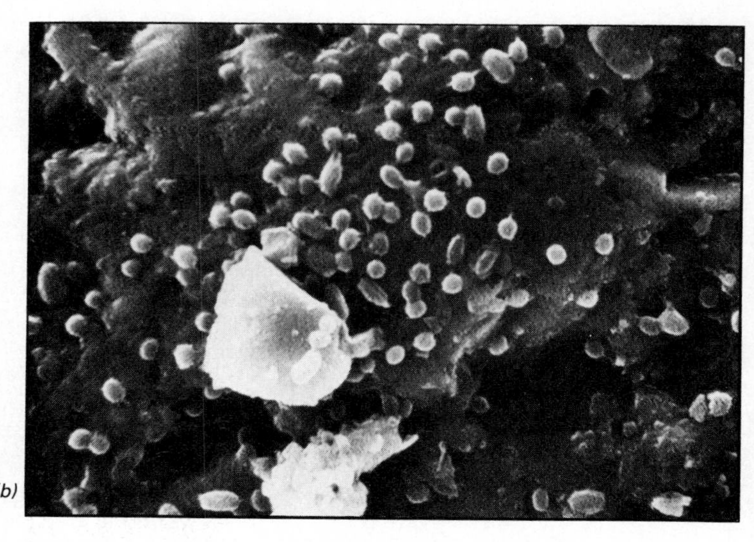

Figure 14.6 Soil as a microbial habitat. (a) Diagram of a soil particle, showing microcolonies of microorganisms that develop in association with the organic and mineral material of the soil. Note that water is present in small pockets within the soil. (b) Scanning electron micrograph of a soil particle, showing the presence of microcolonies of bacteria.

with water. Even when a soil lacks such a free-water phase, each soil crumb is a surface to which moisture can adsorb; often the major part of the moisture of a soil available to microbes is that portion adsorbed to soil particles. Many soil microorganisms can grow quite well using merely this bound water, whereas others only grow when pools of liquid water are present.

In addition to water, microorganisms need nutrients for growth (see Chapter 4). Microorganisms obtain their nutrients both from those substances dissolved in the liquid water and from the soil particles themselves. The dominant soil microorganisms are heterotrophs and require organic matter to grow. Because soils in general are fairly low in organic matter, the greatest microbial activity is found in the organic-rich surface layers and in the regions adjacent to plant roots. Dead plant and animal materials that become incorporated into soil become rich sources of microbial nutrients.

14.3 THE CARBON CYCLE

The biogeochemical cycle of carbon is shown in Figure 14.7. Energy from the sun is used by higher plants and algae to convert carbon dioxide into organic matter. When the plants die, this organic matter becomes decomposed and eventually carbon dioxide is formed again. The organisms involved in the decomposition process and the biochemical steps that occur depend to a considerable extent upon whether decomposition occurs aerobically or anaerobically. In most soils, decomposition is generally aerobic, whereas in lakes it is aerobic in the top part and anaerobic in the mud and in the deeper water.

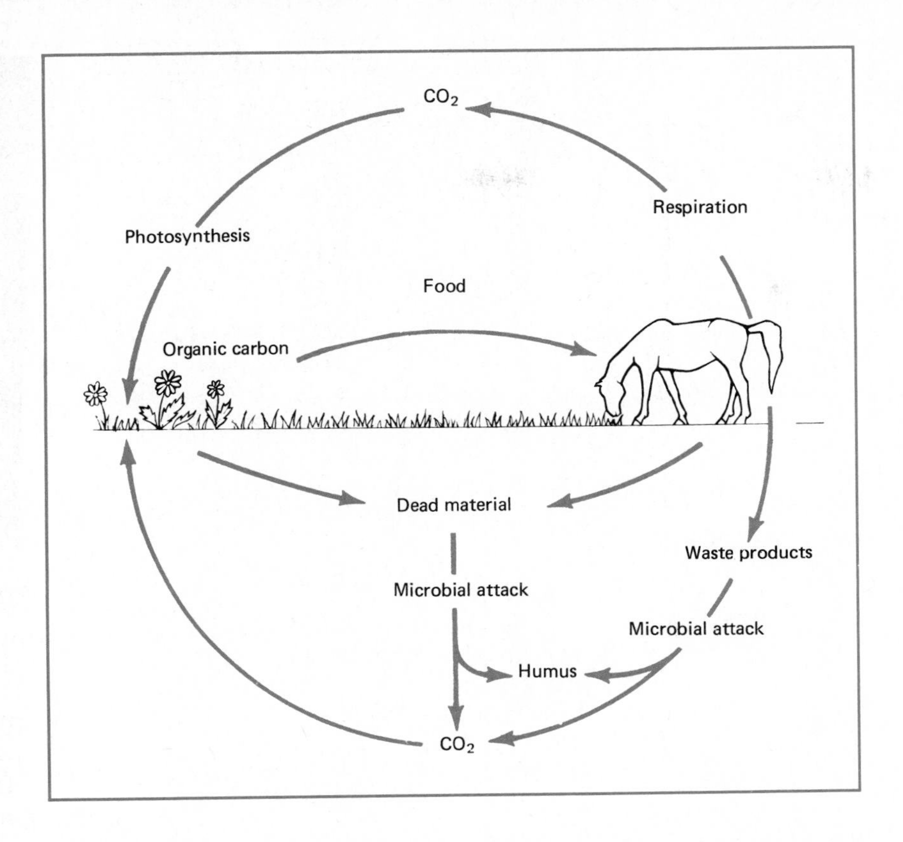

Figure 14.7 The carbon cycle.

Organic matter decomposed anaerobically is acted upon almost exclusively by bacteria. The initial steps involve fermentation to organic acids, followed by the conversion of these acids to methane, hydrogen, and CO_2. If organic acids accumulate, this often leads to the inhibition of further decomposition processes, and the organic material may then slowly accumulate. Coal is thought to have originated when enormous accumulations of organic matter in the anaerobic zones of shallow marshes and lakes of the past were subsequently modified by purely geological processes. In many coals the remains of plants are still easily visible in fossil form. Petroleum is thought to have been created in a similar way from organic deposits in marine environments. Organic matter deposited as coal or oil represents carbon and energy lost from the ecosystem. Much of this material is now being converted back to CO_2 as these fuels are combusted (see the Box on the next page).

In aerobic situations, decomposition is a combined effort of animals and microorganisms. The animals are involved to some extent directly in decomposition, by eating and digesting plant remains, but probably their most important role is to fragment large plant parts into smaller pieces, making them much more susceptible to microbial attack. The shredding and chewing action of animals produces a rich array of small particles that are quickly colonized by bacteria and fungi. In aerobic environments, decomposition usually goes essentially to completion, with almost all the carbon returned to CO_2. A small

MICROORGANISMS AND THE GREENHOUSE EFFECT

Over the past 50–100 years there has been a significant increase in the carbon dioxide (CO_2) concentration of the atmosphere. Why do we care about this increase in carbon dioxide in the atmosphere? Atmospheric scientists tell us that a carbon dioxide build-up will trap energy from the sun near the earth's surface, in the same way that glass does in a greenhouse. If this energy cannot escape, it is converted to heat, and the temperature of the earth will gradually increase. The increased temperature of the earth could cause the melting of polar ice and the destruction by flooding of such major cities as New York, Philadelphia, and Boston (mile-high Denver should be safe!).

Where is this carbon dioxide coming from, and do microorganisms play any role? One possibility is that the increase in carbon dioxide is a result of the burning of fossil fuels, those coals and oils that we are taking out of the earth and using to drive our civilization. There is no part for microorganisms in this model.

But another explanation has an important microbial component. In this model, the increased carbon dioxide is a result of the massive deforestation which has taken place in the vast tropical rain forests in recent decades. An area of tropical forest the size of Great Britain is lost each year. When forest material is harvested, only a part of the wood is removed, the rest being left on the ground to rot. This rotting process is primarily microbial.

Microorganisms of the tropical forest floor thus oxidize the organic carbon of the wood to carbon dioxide, thereby increasing the atmospheric concentration. How important are microorganisms in this process? We do not know, but because of their immense numbers, we would not be surprised if they played a crucial role in bringing about the greenhouse effect.

amount of carbon remains in an undegradable or very slowly degradable fraction called *humus*.

14.4 PETROLEUM MICROBIOLOGY AND THE CARBON CYCLE

Petroleum contains a large array of organic carbon compounds that are potential nutrients for microorganisms. As long as petroleum remains buried in the earth, away from oxygen, it is resistant to microbial attack because there are no microorganisms which can attack it anaerobically. However, as soon as it is brought into contact with oxygen during petroleum-recovery activities, it is rapidly colonized by a variety of aerobic microorganisms. Many organic compounds present in petroleum are *hydrocarbons*, containing only carbon and hydrogen. During refining operations, hydrocarbons are separated from the crude petroleum and are converted into the major fractions used in gasoline and oil. Large numbers of microorganisms, both bacteria and fungi, attack and degrade hydrocarbons. Because of the wide variety of microbial activities possible with crude petroleum and its derivatives, petroleum microbiology is a large and important field.

Petroleum prospecting Prospecting is one of the most interesting ways in which microorganisms have been of aid to the petroleum industry. Associated with liquid and solid petroleum is a gaseous fraction consisting of methane, ethane, and propane. In petroleum-producing regions these gases may seep to the surface and provide nutrients for the growth of specific hydrocarbon-utilizing bacteria. Where one finds bacteria capable of oxidizing these gases, there is a strong suggestion that a petroleum deposit is nearby. Looking for methane-utilizing organisms in searching for petroleum is not practical, since methane is produced biologically in many systems that are not related to petroleum. Ethane, however, is not produced biologically in significant amounts and is almost always associated with petroleum, so that detection of ethane-utilizing organisms has been used to discover petroleum reserves. Under some conditions, it may be easier to detect the ethane-utilizers than the ethane gas itself. Since geological methods of locating petroleum deposits have thus far been adequate, microbiological petroleum prospecting has not found wide use. However, it may become more important in the future, as petroleum reserves become depleted.

Petroleum decomposition The decomposition by microorganisms of petroleum and petroleum products is of considerable importance. It is virtually impossible to keep moisture from bulk storage tanks; it accumulates as a layer of water beneath the petroleum. At the petroleum-water interface, bacteria develop in large numbers (Figure 14.8), and molds and yeasts may also grow. Microbial growth has been an especially serious problem in the kerosene-based fuels used in jet airplanes. When such microbe-containing fuels are burned, fuel strainers rapidly become clogged, leading to power loss or stalling. In addition, microbial growth on the inside surfaces of the fuel tanks of aircraft can lead to corrosion of the tanks. Several control methods can be used: (1) The fuel can be filtered through membrane filters, which remove microorganisms. (2) Inhibitors of microbial growth can be added. (3) Corrosion can be minimized by coating the inside surfaces of the fuel tanks with more resistant substances, such as polyurethane. (4) Aircraft fuel tanks may be washed out at regular intervals with an antimicrobial agent. (5) Some hydrocarbon fractions of the fuel are more readily attacked than are others; if these fractions could be removed from the fuel, its storage life may be lengthened.

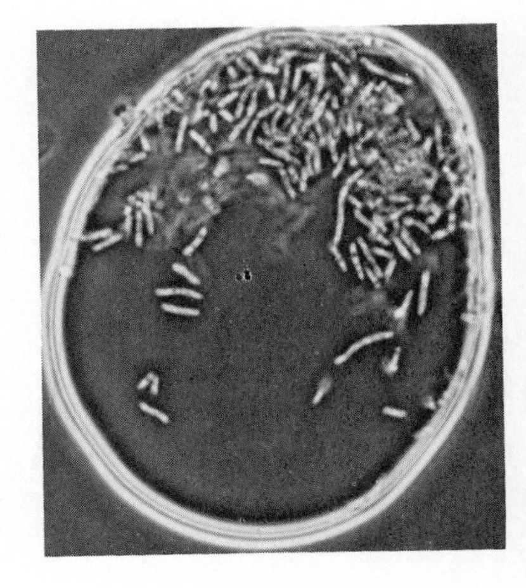

Figure 14.8 Hydrocarbon-oxidizing bacteria associated with an oil droplet. The bacteria are concentrated in large numbers at the oil-water interface.

Microorganisms that degrade petroleum hydrocarbons can also be useful to humans. Catastrophic tanker wrecks and oil-well blowouts can cause massive oil spills which pollute shorelines and kill aquatic animals and plants. The chronic leakage of oil from automobiles, street drains, and other sources may contribute just as much oil to soils and water. Such spills encourage the development of oil-degrading microorganisms and their action can be a significant factor in the natural cleanup process. Microbiologists are even trying to use genetic engineering to develop "super-oil-eating bacteria" which could be seeded directly onto an oil spill to accelerate the biodegradation of the oil. The term **biodegradation** indicates that decomposition of the oil is a biological process.

Petroleum recovery Microorganisms play both beneficial and harmful roles in the petroleum recovery process itself. One detrimental effect of microorganisms is observed during the drilling of an oil well. Large amounts of water are used at the site of the drill bit, the function of the water being to cool and lubricate the drill bit and to bring the rock cuttings to the surface. When the water is brought back to the surface, it is pumped into a pit where the rock particles settle out, and the liquid is then recycled. The water used, called a "drilling fluid" or "drilling mud," usually contains organic substances that are added to thicken it and make it flow better. Some of the additives used are starch, modified cellulose, lignin, and natural gums, most of which are excellent substrates for the growth of bacteria. Bacterial growth in the drilling fluid can so greatly change its properties that it no longer functions properly, so that inhibitors are nearly always added. The chemical inhibitor used must be inexpensive, effective, and noncorrosive. Representative types of inhibitors include quaternary ammonium compounds, organic amines, phenols, and formaldehyde.

Microbial activity in the oil well itself can have detrimental effects that may seriously affect the recovery of petroleum. The most important is the growth of microorganisms in the pores of the rock within which the petroleum is trapped, thus plugging the pores and preventing the movement of the petroleum out of the rock reservoir and into the well. Microbial growth in petroleum will occur whenever water and air are present. In the original (undrilled) petroleum reservoir, neither of these substances is present, but when the well is opened they are admitted to the reservoir. Water is often injected into the reservoir intentionally to force out the petroleum and thus increase recovery. This injection water naturally carries air and bacteria with it, and with the rich source of organic matter present in the petroleum deposit, conditions are good for bacterial growth (Figure 14.9). Many of the bacteria that grow produce capsules, and the slimy bacterial masses effectively block the pores.

Since it is virtually impossible to maintain a sterile petroleum reservoir, inhibitors such as those used in drilling are added to the injection waters to prevent plugging caused by bacteria. The inhibitor used must of course be active against the bacteria involved, but other important properties are low cost, ease of handling, solubility in both water and oil, lack of corrosion of pipes and pumps, and lack of toxicity to humans. If there is any danger of the injection water entering the ground-water supply used for humans, the inhibitor must also be nonpolluting to natural waters and should be biodegradable.

Gum-producing bacteria also play *beneficial* roles in the recovery of petroleum by water injection. Such gum-producing bacteria are used to seal up the larger pores of oil-bearing formations so that the injection water is forced through the smaller pores. The bacterium most commonly used has been *Xanthomonas campestris*, which produces **xanthan gum**. This gum has the property of swelling when hydrated so that it can serve as an effective sealer of the larger pores. Interestingly, *Xanthomonas campestris* is a plant-pathogenic bacterium, capable of causing a wilt of cabbages and some other crops by growing within the water-conducting tubes of the plant, thus blocking, by virtue of its gum, the flow of water. When used in the petroleum recovery process, whole bacterial cultures are used, the dried preparations of gum-rich

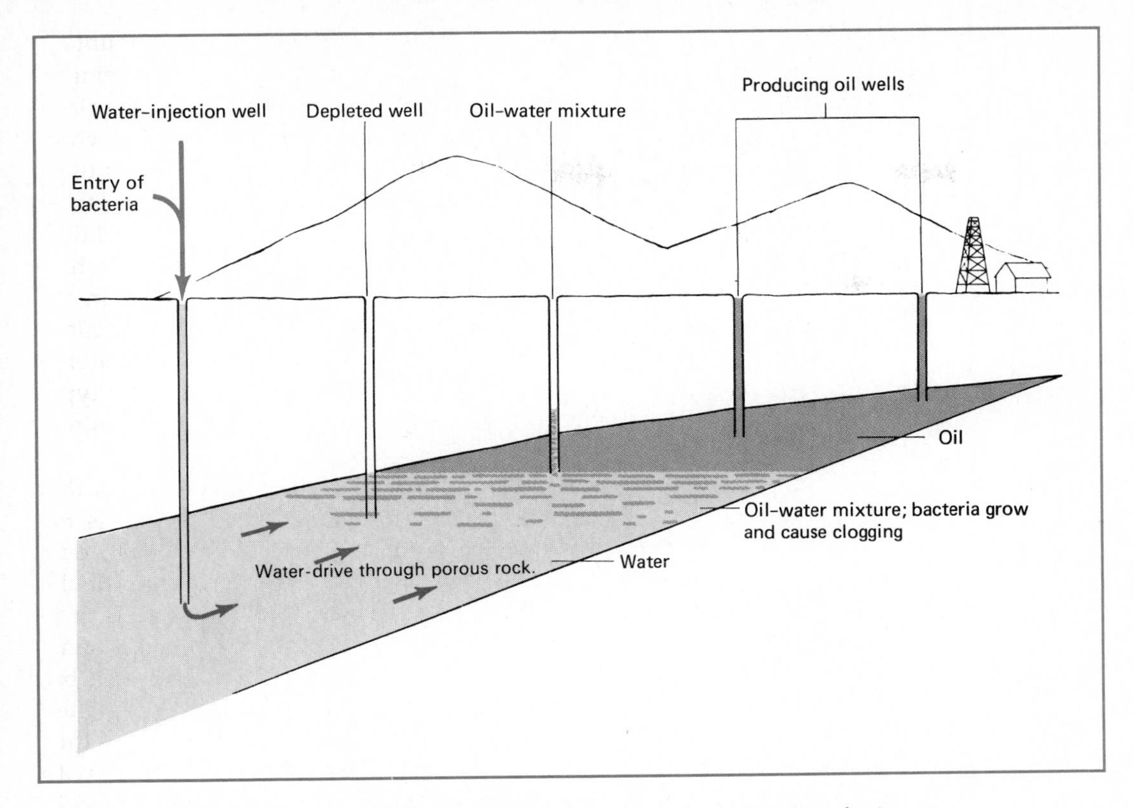

Figure 14.9 Injection of water into deep wells that are becoming depleted aids in forcing out the remaining oil, but bacterial growth in the injection water may cause clogging. To prevent bacterial growth, microbial inhibitors are added to the injection water.

bacteria being added directly to the injection fluid. Several commercial firms provide the dried preparations of *Xanthomonas campestris* to the petroleum industry, and research is underway to discover even more favorable gum-producing bacteria. It is interesting that, depending on the situation, gum-producing bacteria can be either beneficial or harmful in the petroleum recovery process.

14.5 THE NITROGEN CYCLE

The biogeochemical cycle of nitrogen is shown in Figure 14.10. Nitrogen is found on earth at several different oxidation states: nitrate (NO_3^-), the most oxidized form, nitrite (NO_2^-), nitrous oxide (N_2O), nitrogen gas (N_2), and ammonia (NH_3) and organic nitrogen (primarily in the form of amino acids), the most reduced forms of nitrogen.

Nitrogen fixation The largest amount of nitrogen on earth is present in the form of nitrogen gas (N_2) in the atmosphere. Nitrogen gas has the important chemical property that it is very stable and is not directly available for most organisms. Some conversion of nitrogen gas to ammonia and nitrate

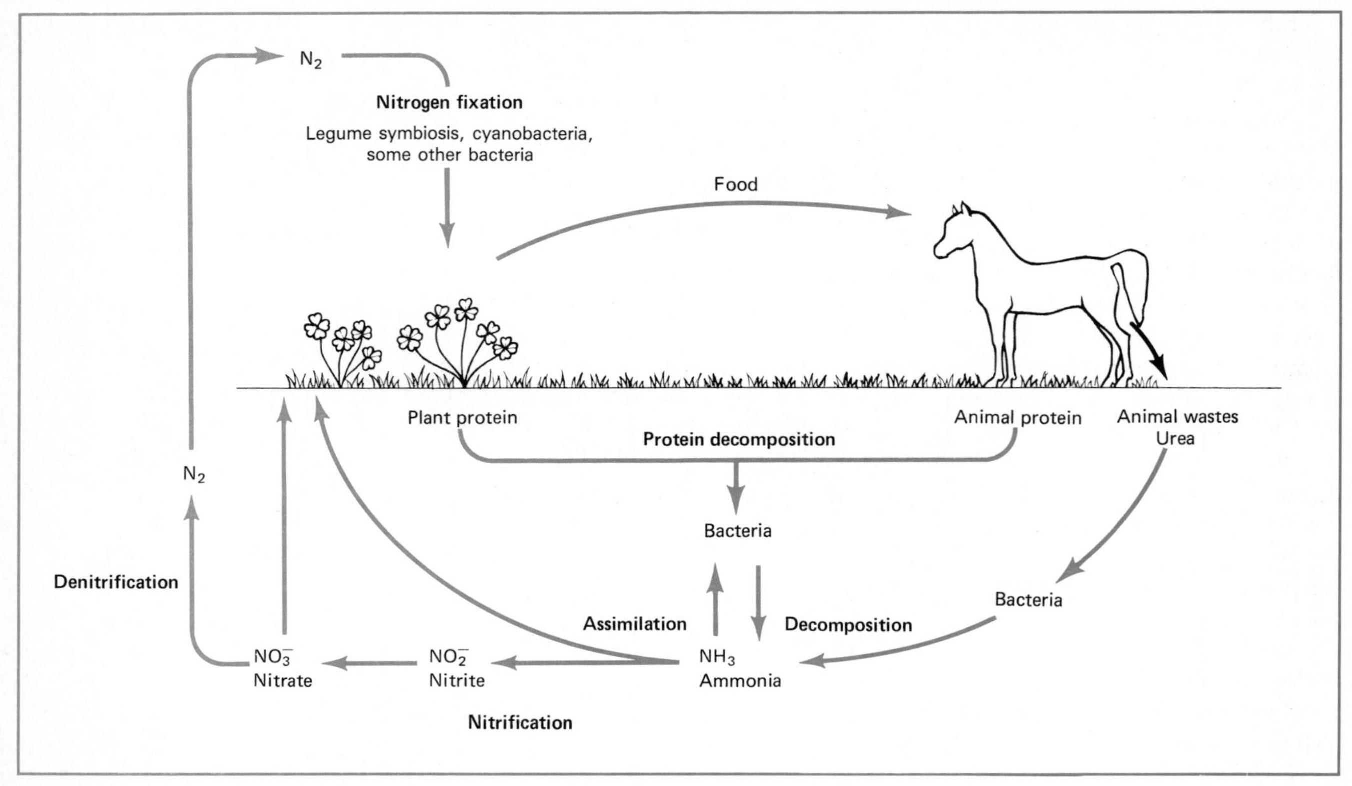

Figure 14.10 The biological nitrogen cycle.

occurs chemically in the atmosphere via action of lightning discharges, and some nitrogen gas is converted industrially into combined nitrogen, either through combustion processes (for example, the burning of coal or gasoline results in oxidation of nitrogen gas to nitrogen oxides) or in the manufacture of nitrogen fertilizers. But the most important process by which atmospheric nitrogen is rendered available to organisms is through the process of biological nitrogen fixation. On a global basis, about 85 percent of nitrogen fixation is biological, the rest occurring due to spontaneous and industrial processes.

Biological nitrogen fixation is one of the most important microbiological processes on earth. All the microorganisms that carry out nitrogen fixation are procaryotes, and nitrogen-fixing organisms include a variety of different bacteria. We can distinguish two kinds of nitrogen-fixing organisms: free-living and symbiotic nitrogen fixers. A symbiotic nitrogen-fixing organism carries out this process only when living in association with a plant, whereas a free-living organism is able to fix nitrogen when living directly in soil or water with no other organism involved.

Free-living nitrogen-fixing organisms include two genera of bacteria that are widespread in soil, *Azotobacter* and *Clostridium*. *Azotobacter* species are aerobic, whereas *Clostridium* species are anaerobic. Studies of nitrogen fixation in soils of various types have shown that fixation occurs most readily under anaerobic conditions, suggesting that *Clostridium* is most important. Nitrogen fixation by these free-living bacteria requires the presence of considerable amounts of organic matter that serves as energy source for the bacteria. Nitrogen fixation is therefore more common in soils rich in organic matter than in organically poor soils.

Attempts to increase nitrogen fixation in soils by inoculation with free-living nitrogen fixing bacteria have generally been unsuccessful. Failure to stimulate nitrogen fixation by inoculation is probably due to the fact that these free-living microorganisms are already widely distributed on earth; whether or not they act depends primarily on the presence in the soil of sufficient organic matter to permit their growth. In fact, some increase in the combined nitrogen content of a soil may be brought about simply by enriching the soil with organic matter (for instance, plant residues, sewage sludge). When such organic additions are carried out, the nitrogen-fixing bacteria already present can grow and flourish.

The other major free-living nitrogen-fixing organisms are photosynthetic, the *cyanobacteria*. Not all cyanobacteria fix nitrogen, but those that do so may be responsible for considerable addition of nitrogen to the habitats in which they live. Since cyanobacteria grow using light as sole energy source and CO_2 as sole carbon source, nitrogen fixation by these organisms does not require the presence of organic matter. These procaryotes are widespread in lakes, streams, and the ocean, and are also found frequently in the surface crusts of desert and other arid soils. Another important habitat of cyanobacteria is in rice paddies (Figure 14.11). When a paddy is flooded, the cyanobacteria develop rapidly in the warm shallow water and are probably responsible for addition of a significant amount of combined nitrogen.

Figure 14.11 A rice paddy in southern India. Nitrogen-fixing cyanobacteria often grow extensively in such partially aquatic systems.

The most important **symbiotic nitrogen-fixing organisms** are bacteria of the genus *Rhizobium*, which live in association with plants called legumes. Because of its agricultural importance, this type of nitrogen fixation is discussed in Chapter 17. A number of nonleguminous plants also enter into relationships with nitrogen-fixing bacteria. Among the most important are alder trees (genus *Alnus*), which form a root symbiosis with actinomycete bacteria of the genus *Frankia*. The cultivation of *Frankia* presented difficult problems for microbiologists, primarily because the organism grows exceedingly slowly, but patience and careful work has finally resulted in pure cultures of these interesting nitrogen-fixing bacteria. Alder trees are often pioneers on bare or disturbed soil, and it is thought that the ability of these trees to use atmospheric nitrogen makes it possible for them to succeed on soils that are too poor in nitrogen for other tree species.

When nitrogen is fixed, it is reduced to ammonia (NH_3), and the ammonia produced is then converted into organic nitrogen, primarily in the form of amino acids in the proteins of the cells. Thus, initially the nitrogen that is fixed is held within the protoplasm of nitrogen-fixing organisms and is not available to plants. Later, when the nitrogen-fixing organisms die, their cells are decomposed by other bacteria, and the organic nitrogen is converted back to NH_3, a process called *mineralization*. The ammonia is then available directly to plants as a nutrient, or as described below, it can be converted further to nitrate, which is also available as a plant nutrient.

Nitrification The conversion of ammonia to nitrate, called **nitrification** is an important microbial process that occurs commonly in soils where ammonia is present. Ammonia is often added to soils as a fertilizer in the form of anhydrous (gaseous) ammonia and it is also produced by the microbial decomposition of organic nitrogen. It is also the first product of nitrogen fixation, just discussed. Nitrification is brought about by a special group of bacteria called the **nitrifying bacteria**, which obtain their energy for growth from the oxidation of ammonia or nitrite (see Section 3.7). These lithotrophic bacteria are autotrophs, being able to grow in completely inorganic media, using CO_2 as sole source of carbon.

Nitrification occurs in two stages: (1) the oxidation of ammonia (NH_3) to nitrite (NO_2^-), brought about by one group of bacteria; and (2) the oxidation of nitrite to nitrate (NO_3^-), brought about by another group of bacteria. The combined action of both these groups leads to the conversion of ammonia to nitrate. Nitrification requires the presence of oxygen and, hence, occurs only in aerated, well-drained soils. Under anaerobic conditions, such as in water-logged soils, nitrification is inhibited. Nitrate is taken up by plants more efficiently than is ammonia. But nitrification may not necessarily be beneficial in agricultural practice; since nitrate is very soluble in water, it is rapidly leached from the soil and is then unavailable in any form to the plants.

Nitrification also occurs commonly in aerobic sewage-treatment plants from the ammonia released by the mineralization of organic nitrogen. Excessive fertilization with anhydrous ammonia may lead to large accumulations of nitrate in soils. The leaching of nitrate from such soils into the groundwater can lead to nitrate pollution of wells, a potentially serious condition. Nitrate is detrimental to humans if consumed in large amounts, and government regulations require that the nitrate concentration of drinking water remain below 10 $\mu g/ml$, a value often exceeded in agricultural areas where excessive nitrogen fertilization occurs. In regions where excessive nitrate concentrations occur in well water, alternate sources of water must be found.

Denitrification Denitrification is an anaerobic process that results in the conversion of nitrate to gaseous products, usually N_2 and N_2O, which then escape into the air. Denitrification is thus a harmful process in that it leads to a loss of combined nitrogen. A number of bacteria are able to denitrify. Denitrifying bacteria are facultative anaerobes, and denitrification occurs only under anaerobic conditions, being inhibited by the presence of O_2. Denitrification is a type of anaerobic respiration (see Chapter 3). Most denitrifiers are heterotrophs and therefore organic compounds are necessary for it to occur. The process occurs commonly in the anaerobic sediments of lakes and estuaries and in soils that become flooded.

14.6 THE SULFUR CYCLE

The biogeochemical cycle of sulfur is shown in Figure 14.12. Sulfur is an important plant nutrient, and several forms of this element exist in nature. There are three forms that are of practical significance: *sulfate* (SO_4^{2-}), which is the most oxidized form; *sulfide*, in the form of either hydrogen sulfide (H_2S) or metal sulfides such as iron sulfide (FeS), which is the most reduced form;

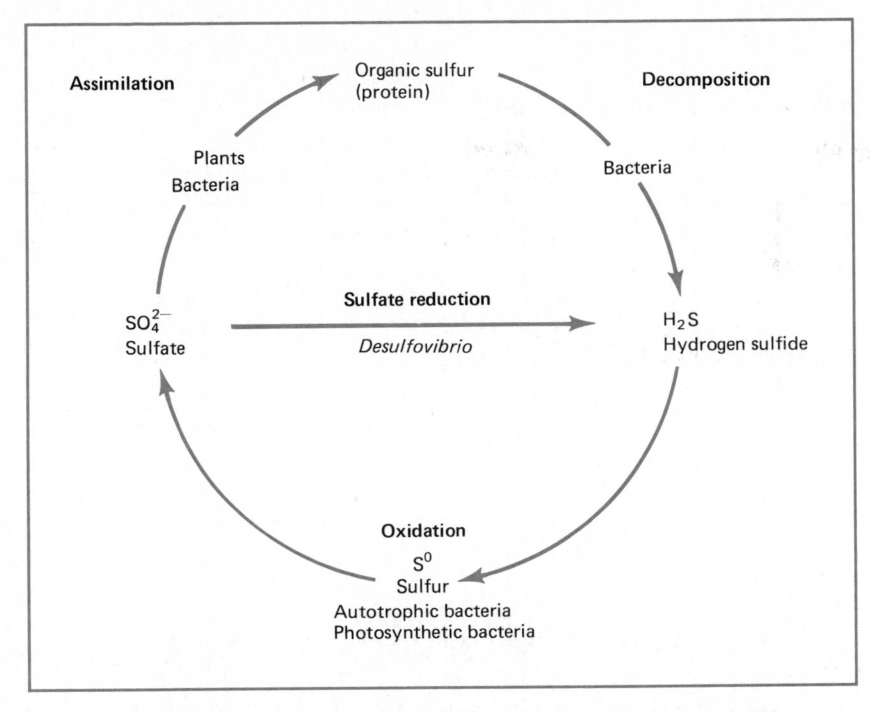

Figure 14.12 The sulfur cycle.

and *elemental sulfur* (S^0), which is intermediate in oxidation level. The conversion of sulfur from one oxidation state to another is carried out in nature most commonly by microorganisms.

Sulfate reduction Under anaerobic conditions, sulfate may serve as an electron acceptor in anaerobic respiration and be reduced to hydrogen sulfide, a toxic product. This conversion is carried out by a special group of bacteria called the **sulfate-reducing bacteria**; for example, bacteria of the genus *Desulfovibrio*. If a soil becomes waterlogged and hence anaerobic, sulfate reduction can occur; the toxic hydrogen sulfide that accumulates may cause damage to plants and animals. Sulfate reduction and the resultant hydrogen sulfide production may also cause serious corrosion of iron and steel pipes buried in soils, the hydrogen sulfide attacking the metal, making holes and pits in the pipe.

Sulfur and sulfide oxidation The oxidation of *elemental sulfur* to sulfate is a microbial process of considerable significance. When elemental sulfur is oxidized; sulfuric acid (H_2SO_4) is produced, reducing the pH of the soil. Soils too alkaline for normal plant growth can thus be acidified by adding elemental sulfur to them. Also, elemental sulfur is a common form of sulfur fertilizer, but it is not utilizable directly by plants. Upon its conversion by microbes to sulfate, it becomes available as a plant nutrient. *Sulfur bacteria* are lithotrophs; the most common are members of the genus *Thiobacillus*, and representatives of this genus are widespread in soils.

A third sulfur transformation, the oxidation of *sulfides* to sulfuric acid, is usually harmful because it leads to development of excessive acidity in soil and water. Sulfides are fairly common in low-lying soils along seacoast areas, especially in warmer parts of the world. As long as the soils are flooded they are anaerobic, and the sulfides remain reduced. If the soils are drained for agricultural purposes, however, the aerobic conditions that develop lead to rapid oxidation of sulfide and the formation of excessive sulfuric acid, resulting in a soil pH too low for the growth of plants. Another kind of soil showing high acidity is that which forms in the waste banks of many surface-mining areas where sulfides are present in large amounts. The acidity makes the growth of plants and the successful revegetation of such areas impossible. This problem is discussed in more detail in the next section.

14.7 MINING MICROBIOLOGY

In the mining of coal and minerals (copper, lead, and zinc), microbial activities can lead to effects that are either beneficial or harmful. A serious form of water pollution, **acid mine drainage**, occurs frequently in coal mining operations and occasionally in mineral mining. In addition, the same bacteria that cause acid mine drainage can also play a beneficial role in the extraction of minerals from some kinds of low-grade copper, lead, and zinc ores, a process called **microbial leaching**. In both cases, the bacteria act by attacking sulfide minerals present in the coal or ore deposit. A mineral frequently present in association with coal is an iron sulfide known as *pyrite*; this substance has a goldlike glitter, a factor responsible for its common name, "fool's gold." In the case of copper, lead, and zinc, the minerals themselves are in the form of sulfides, usually associated with iron sulfide. The attack of bacteria on these sulfides results in both formation of sulfuric acid and solubilization of the metal.

Acid mine drainage Not all coals contain iron sulfide, so that acid mine drainage does not occur in all coal-mining regions; where acid mine drainage does occur, however, it is often a very serious problem. Mixing of acid mine drainage with natural waters in rivers and lakes causes a serious degradation in the quality of the natural water, since both the acid and the dissolved metals are toxic to aquatic life. In addition, such polluted waters are unsuitable for human consumption and industrial use. An understanding of the factors involved in acid mine drainage may help us control this situation. Since bacteria are a prime factor in acid mine drainage, an understanding of their properties and activities may aid in preventing its occurrence.

Attack on the sulfide minerals involves the breakdown of iron sulfide into sulfuric acid and ferrous iron and is done by the lithotrophic bacterium *Thiobacillus ferrooxidans*. The sulfuric acid is not changed further, but the ferrous iron liberated can also be oxidized by this bacterium, producing ferric iron. This forms an insoluble yellow precipitate, called "yellow boy" by miners, which coats polluted streams and rivers, making these waters unsightly. *Thiobacillus ferrooxidans* is an acidophilic bacterium; therefore it is able to remain active under the acidic conditions that it produces. Oxygen is required for growth and for the oxidation of sulfides, and the organism develops only where aerobic conditions prevail. *T. ferrooxidans* is autotrophic and is therefore able to grow on a completely inorganic medium; in the coal and ore-bearing materials that are uncovered, there is almost always a plentiful supply of the nitrogen, phosphorus, potassium, sulfur, magnesium, and other nutrients needed for growth.

The properties of this organism help to explain how acid mine drainage develops. As long as the coal is unmined, oxidation of its sulfide minerals cannot occur, since neither air nor the bacteria can reach it. When the coal seam is exposed, it quickly becomes contaminated with the bacteria, and the presence of oxygen makes oxidation of the sulfide minerals possible. If conditions are appropriate, the acid produced can leach out of the mine into a nearby stream.

The most dramatic examples of conditions promoting acid mine drainage are found in **strip mines**. A *strip mine,* also called an *open cast,* or *surface, mine,* is one in which the coal is so near the surface that it can be mined by removing (stripping) the overlying rock and soil (called *overburden*) (Figure 14.13). Coal seams within 30 to 50 meters of the surface can be mined by this method. The coal is removed from the seam, and then the overburden

(a)

(b)

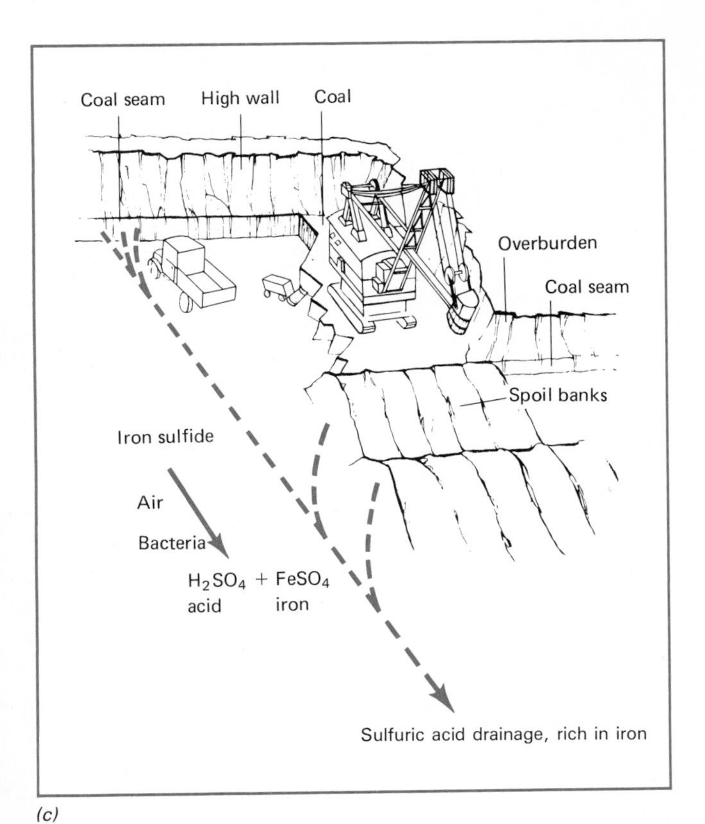

(c)

Figure 14.13 Coal-mining operations. (a) A large shovel is used to remove overburden during strip-mining operations. (b) Spoil bank with acid lake is formed as a result of strip mining. Note the absence of vegetation on this very acidic soil. (c) Origin of acid mine drainage in strip mine operations.

is replaced. The overburden from one pit is deposited in the pit formed by the previous stripping. Because the rock and soil is considerably loosened, it can never be completely redeposited in the pit, and a series of low hills, called *spoil banks*, are formed (Figure 14.13*b*). These spoil banks often contain pyritic materials oxidized by *Thiobacillus ferrooxidans*, so that the lakes and streams become acidic. Modern practice requires that any acid-producing spoil be buried deep in the pit so that it cannot be a source of acidity. Unfortunately, in many older strip mine areas this practice was not followed, and such older areas may be sources of acid drainage for many years. Further, the soil of such spoil banks becomes acid, making it difficult or impossible to establish vegetation. Such acid spoil banks are veritable deserts (see Figure 14.13*b*), even if they are in regions where plentiful rainfall occurs.

When the coal is removed from the mine, it is always intermixed with rocks and other nonburnable material which must be removed before the coal can be marketed. This material, called *coal refuse*, is removed in the coal preparation plant and is usually discarded in nearby coal refuse piles. This refuse is often very high in iron sulfides and is a major source of acid pollution. Within the coal refuse pile, high populations of *T. ferrooxidans* quickly develop, and extremely acidic conditions result. Rain falling on the coal refuse pile leaches acid out, and streams draining such piles often become very acidic. Since many refuse piles are abandoned, the responsibility for controlling acid mine drainage from them may rest on no one legally and they are serious sources of pollution. Eventually, all the sulfide becomes oxidized and all the acid leaches out, but this takes many years.

Since acid mine drainage would not develop in the absence of bacterial activity, it might seem logical to prevent its occurrence by using chemicals or other agents that kill or inhibit the growth of bacteria. A number of such agents are available, including antibiotics, antiseptics, and organic acids. None of these agents have yet proved of practical value in controlling acid mine drainage. They are either too expensive to use in the enormous quantities necessary to have significant impact, or they may be difficult or impossible to deliver at appropriate concentrations to the active sites of acid production deep within the ore materials, or they may not be sufficiently active in inhibiting or killing the bacteria. To date, the most effective means of controlling acid mine drainage is the use of mining practices that keep air and bacteria from the acid-producing materials.

The acid produced from coal-mining operations can be neutralized by use of lime and neutralization is the most widely employed method of eliminating this water-pollution nuisance. However, one problem with use of lime as a neutralizing agent is that the lime particles become coated with a layer of ferric hydroxide, so that fresh lime must be added periodically.

Microbial leaching Microbial leaching is the process by which mineral ores are solubilized using microbial action so that the minerals contained in the ores can be recovered. The organism primarily used in leaching is *Thiobacillus ferrooxidans*. The ores from which copper, lead, and zinc are obtained are often sulfide ores, and *T. ferrooxidans* is able to oxidize these sulfide minerals in a manner similar to the way it oxidizes iron sulfide. In addition, iron sulfide is usually present in association with these other sulfide minerals in the ore body and is also oxidized. As a result of the oxidation of the sulfide minerals, not only is sulfuric acid produced, but also the metal of the sulfide mineral is solubilized. Leaching is used most extensively in copper mining, and as copper becomes scarcer and low-grade ores therefore become more valuable as copper sources, leaching will become even more important.

In the leaching process the ore, usually containing 0.5 percent or less of copper, is placed in huge piles, and acid water is sprayed over the top of the pile (Figure 14.14). This water slowly percolates down through the pile and creates the appropriate conditions for the growth of *T. ferrooxidans* within the pile. This organism oxidizes the sulfide minerals, creating more sulfuric acid and liberating copper. At the bottom of the leach dump, the copper-laden acid water exits and is carried to a processing plant where the copper is removed. After the copper is removed, the water is still very acidic and is high in ferrous iron. This water is usually carried to an oxidation pond where it is further acted upon by *T. ferrooxidans*, which oxidizes the ferrous iron to the ferric form. The water, now laden with ferric iron, is recycled through the leach dump, and the ferric iron itself reacts with further sulfide minerals to solubilize more copper. The bacteria thus play two roles: in the leach dump itself, and in the oxidation pond (Figure 14.15).

Figure 14.14 A typical dump used in microbial leaching. The low-grade ore has been piled on the side of the mountain and acidic water is allowed to percolate over the surface of the pile. The leach water leaving the bottom of the dump is collected in a small pond and pumped to the copper-recovery site.

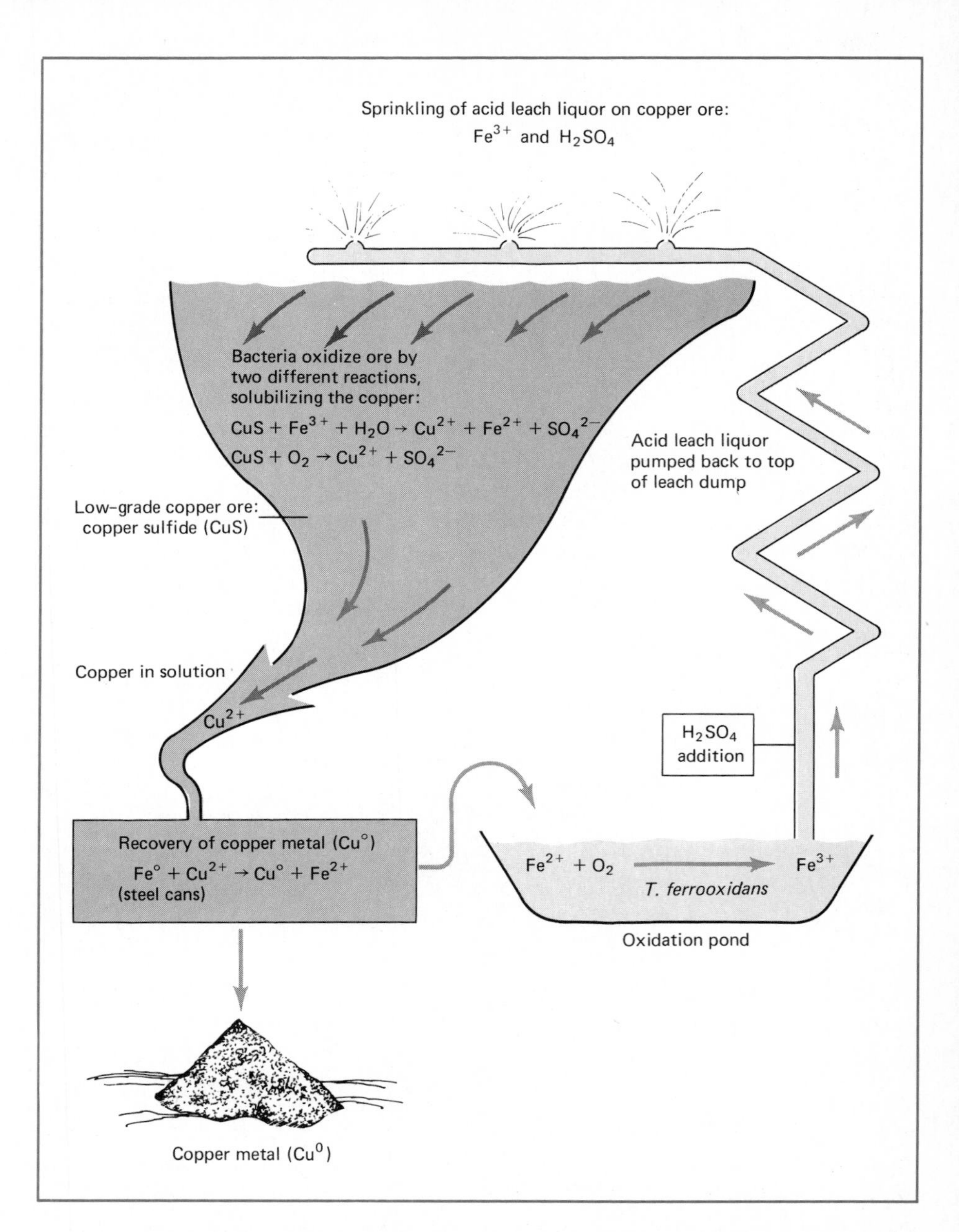

Figure 14.15 Microbial activity in leach dumps is responsible for recovery of large amounts of metals from low-grade ores.

The actual operation of the leach dump is carried out without any careful attempt to promote bacterial action. The process of leaching was actually developed by copper companies without any real knowledge that bacteria were involved and is being continued today mostly by using empirical methods. Although less widely used, leaching is also carried out in some uranium-mining operations. The importance of microbial leaching is shown by the fact that in the United States, about 25 percent of the copper recovered from ores has come from leaching operations.

SUMMARY

In this chapter we have presented only a few of the many ways in which microorganisms play roles in the world at large. Some of these microbial activities are beneficial to humans, whereas others are harmful. An **ecosystem** is defined as the total physical and chemical environment in which a set of microbes live. Though their **microenvironments** are not large in scale, microorganisms play many important roles in ecosystems. They function as **primary producers**, channeling energy into the ecosystem, and as **decomposers**, converting dead plants and animals back into their basic constituents. They carry out oxidations and reductions on various elements, such as carbon, nitrogen, and sulfur, and thus are the prime biological agents in the **biogeochemical cycles**. A major form of organic carbon on earth is **petroleum**, present in vast underground deposits. Microorganisms can carry out a variety of processes related to petroleum, both harmful and beneficial, and petroleum microbiology is a large and important field of study.

The ability of microorganisms to convert nitrogen gas (N_2) from the air into combined nitrogen, a process called **nitrogen fixation**, is one of the most important contributions that microorganisms make to ecosystem function. Nitrogen fixation makes available to higher organisms nitrogen from the air, an inexhaustible nitrogen source.

The oxidation and reduction of sulfur compounds in nature is carried out primarily by microorganisms. Such transformations are important in the **sulfur cycle**, but there are also significant practical implications of microbial activities on sulfur. Coal and metal ores often contain or are associated with sulfide minerals, and the oxidation of these sulfide minerals by microorganisms can have either harmful or beneficial effects. In coal-mining operations, the oxidation of sulfide minerals results in the formation of sulfuric acid, an important pollutant, leading to the formation of **acid mine drainage**. In mineral mining, especially of copper, the ability of *Thiobacillus ferrooxidans* to oxidize copper sulfide minerals is used to recover useful ore from low-grade deposits in a process called **leaching**.

KEY WORDS AND CONCEPTS

Lithosphere
Hydrosphere
Atmosphere
Biosphere
Environment
Ecosystem
Biogeochemical cycle
Microenvironment
Carbon cycle
Primary production
Biodegradation
Nitrogen cycle
Nitrogen fixation
Nitrification
Denitrification
Sulfur cycle
Leaching
Acid mine drainage

STUDY QUESTIONS

1. What is the *biosphere*?
2. Define *ecosystem*. Give two examples of ecosystems.
3. What is a *food chain*?

4. Discuss the energy relations of ecosystems. How do decomposing microorganisms affect the energy relations of ecosystems?
5. Discuss the roles that microorganisms play in the *carbon cycle*. Contrast the processes of the carbon cycle under aerobic and anaerobic conditions.
6. Describe two detrimental effects that microorganisms can have in petroleum production. In what way can microorganisms be of benefit in petroleum prospecting?
7. Why is the process of *nitrogen fixation* so important?
8. Discuss some of the organisms involved in nitrogen fixation.
9. What is *nitrification*? Contrast with *denitrification*.
10. Diagram the *sulfur cycle*, and indicate the kinds of microorganisms involved in different steps.
11. How do sulfur bacteria cause the formation of *acid mine drainage*? How can acid mine drainage be prevented?
12. Describe the strip-mining process, and indicate the places where sulfur bacteria can develop.
13. How do the same sulfur bacteria that cause acid mine drainage carry out a beneficial effect in the recovery of minerals from low-grade ores?

SUGGESTED READINGS

ATLAS, R.M. (editor). 1985. *Petroleum microbiology*. Macmillan, New York.

ATLAS, R.M. and R. BARTHA. 1981. *Microbial ecology. Fundamentals and applications*. Addison-Wesley, Reading, MA.

BRIERLEY, C.L. 1982. *Microbiological mining*. Scientific American 247: 44–53.

CHIRAS, D.D. 1985. *Environmental science. A framework for decision making*. Benjamin-Cummings Publishing Co., Menlo Park, CA.

KRUMBEIN, W.E. (editor). 1983. *Microbial geochemistry*. Blackwell Scientific Publications. Oxford, England.

REVELLE, R. 1982. *Carbon dioxide and world climate*. Scientific American 247: 35–43.

15

Microbiology of Water and Wastewater

Water is essential for life, but it can also be a hazard, as it can carry pathogenic microorganisms and toxic chemicals. Probably the greatest contribution that urbanization has made to the advance of human civilization has been the provision of pure and reliable water for its inhabitants. In the United States, water use in urban areas averages about 160 gallons per person per day and is distributed as follows: residential, 40 percent; industrial, 25 percent; commercial, 15 percent; public, 5 percent; other, 15 percent. Of the 60 gallons per person per day consumed for interior residential purposes, 40 percent is used for toilet flushing, 30 percent for bathing, 15 percent for laundry, and 15 percent for drinking and miscellaneous (irrigation, auto washing, etc.).

The need for reliable sources of water for domestic use is obvious. However, we must also be concerned about treating the wastewater which is generated by human water use, since the *water supply* for one community may contain the *wastewater* of another community. This is especially true along major river systems, where all of the water in the river may be *reused* by humans several times as it flows from its source to the sea. In this chapter we will consider treatment of water for human consumption, and treatment of the wastewater generated by humans; both are essential in the control of water-borne diseases.

15.1 WATER SUPPLY

Water originates initially as precipitation (rain or snow), which falls to the earth, percolates through the soil, and enters rivers and streams. These water courses carry the water to the lakes and the sea, from which it evaporates and returns to the atmosphere. It can again fall to the earth as precipitation, completing what is called the **hydrologic cycle** (Figure 15.1). The microbial content of natural waters varies greatly with location and with time. Rain and snow contain only small numbers of microorganisms, mainly spores and dormant cells of bacteria, fungi, and algae. When water percolates through the soil it picks up significant numbers of microorganisms, but many of these are gradually removed as the water penetrates deeper. Groundwater generally has the lowest microbial content, and deep groundwaters are often almost devoid of organisms. Surface waters, on the other hand, often have large numbers of microorganisms, derived both from the soil (as runoff) and from microorganisms growing in lakes and streams. The microorganisms in the sea differ from those in fresh water or soil, being adapted to the higher salt content of ocean waters.

Most of the microorganisms which are native to natural waters are nonpathogenic and hence are harmless when consumed in small numbers. Pathogenic microbes found in surface or ground waters are usually not capable of growth in these environments, but have reached these waters as a result of contamination of the water by human activity.

Drinking water Not all water is fit for human consumption. Drinking water must be free of contaminants harmful to the human body. The average

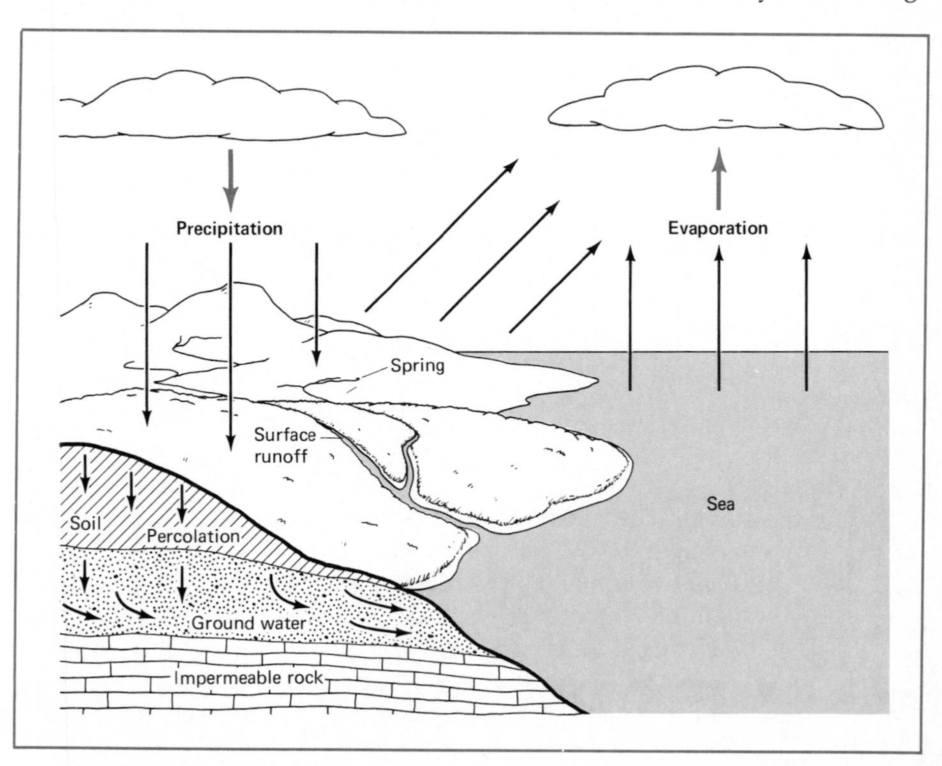

Figure 15.1 The hydrologic cycle.

TABLE 15.1 Sources of water for domestic use

Source	Advantages	Disadvantages
Ground water	Reliable supply, constant temperature	Limited quantity, water often hard
Dug well—shallow	Inexpensive	Undependable, often contaminated
Drilled well—deep	More dependable, less contaminated	Expensive
Spring	Inexpensive	Often contaminated, water often hard, very limited supply
Surface water	Inexpensive	Unreliable supply, temperature varies, often contaminated
Rivers	Readily available	Often turbid
Lakes	Less turbid	Few good lakes
Reservoirs	Reliable supply	Expensive, excess algal growth
Rain water	Soft water, uncontaminated	Unreliable supply, lack of storage
Cistern	Easily polluted by runoff	Storage expensive

individual drinks between 1 and 2 liters of water a day, depending on the climate and season. Not all of this drinking water comes out of the tap in the home. The water in beverages comes from the bottling or canning plant, milk (which is mostly water) comes from the dairy farm, and so forth. Thus, even if one is certain of the purity of one's own drinking-water supply, there is no guarantee that all of the water consumed is free of contaminants.

Urban water supply The water system of a city is an essential ingredient that makes the city possible. It is virtually impossible to have a high-density population without some sort of central water-supply facility. Sources of water vary widely (Table 15.1). The most widely used sources are surface waters, such as lakes and reservoirs. Surface waters are often subject to pollution so that they must be treated before use. Ground waters provide excellent water sources, but are less widely available than surface waters. After treatment, the water may be pumped to a reservoir for storage and then distributed to individual homes and businesses via an extensive network of underground pipes. Great care must be taken to ensure that water pipelines remain intact, since sewer pipes are usually also present underground, and a real possibility exists for leakage of sewage into the water supply (called a *cross-connection*), resulting in a serious health hazard.

15.2 PATHOGENIC ORGANISMS TRANSMITTED BY WATER

We discussed the general nature of disease epidemics in Chapter 10. Organisms pathogenic to humans that are transmitted by water include bacteria, viruses, and protozoa (Table 15.2). Note, however, that for over half of the

TABLE 15.2 Water-borne disease outbreaks due to microorganisms[a]

Disease	Causal agent	Outbreaks, percent[b]	Cases, percent[c]
Bacteria			
Typhoid fever	*Salmonella typhi*	10	0.5
Shigellosis	*Shigella* species	9	9
Salmonellosis	*Salmonella paratyphi*, etc.	3	12
Gastroenteritis	*Escherichia coli*	0.3	2.5
	Campylobacter species	0.3	2.5
Viruses			
Infectious hepatitis	Hepatitis A virus	11	1.6
Poliomyelitis	Polio virus	0.2	0.01
Diarrhea	Norwalk virus	1.5	2
Protozoa			
Dysentery	*Entamoeba histolytica*	0.1	0.05
Giardiasis	*Giardia lamblia*	7	13
Unknown etiology			
Gastroenteritis		57	58

[a]Compiled from data provided by the Centers for Disease Control.
[b]Of 650 outbreaks in recent decades.
[c]Of 150,000 cases over the same period.

cases of water-borne illness in the United States, no cause in known. Organisms transmitted by water usually grow in the intestinal tract and leave the body in the feces. Fecal pollution of water supplies may then occur, and if the water is not properly treated, the pathogens enter the new host when the water is consumed. Because water is consumed in large quantities, it may be infectious even if it contains only a small number of pathogenic organisms. The pathogens lodge in the intestine, grow, and cause infection and disease.

Bacteria Probably the most important pathogenic bacteria transmitted by the water route are *Salmonella typhi*, the organism causing typhoid fever, and *Vibrio cholerae*, the organism causing cholera (see Sections 11.10 and 11.15). Although the causal agent of typhoid fever may also be transmitted by contaminated food and by direct contact from infected people, the most common means of transmission is the water route. *Typhoid fever* has been virtually eliminated in many parts of the world, primarily as a result of the development of effective water treatment methods (see Section 15.4). However, typhoid fever does still occur occasionally, either due to a breakdown in water-purification methods, contamination of water during floods, or cross-contamination of water pipes by leaking sewer lines. The causal agent of *cholera* is transmitted almost exclusively by the water route. At one time cholera was common in Europe and North America, but the disease has been almost entirely eliminated from these areas by effective water purification (see Section 10.7). Both *V. cholerae* and *S. typhi* are eliminated from sewage during sewage treatment and hence do not enter water courses receiving treated sewage effluent. More frequent than typhoid, but less serious are salmonellosis caused by species of *Salmonella* other than *S. typhi*, and shigellosis caused by *Shigella* species. As shown in Table 15.2, the largest number of cases of water-borne bacterial disease in the United States during the last several decades have been due to these two diseases, which are among the most frequently reported diseases (see Figure 10.6).

It may seem surprising that drinking water can be a vehicle for bacterial disease agents even when the water looks clear and uncontaminated. However, the number of bacteria necessary to start an infection is not especially large. Feeding trials using human volunteers have shown that with some bacteria only a few cells are necessary to cause infection. Since people drink fairly large amounts of water, even if bacterial numbers are low in the drinking water, sufficient bacteria may be ingested to initiate an infection.

Viruses Viruses transmitted by the water route include poliovirus, the Norwalk virus (see Section 13.10), and the virus causing infectious hepatitis. *Poliovirus* has several modes of transmittal, and transmission by water may be of serious concern in some areas. Before introduction of polio vaccine, the disease was commonly encountered during the summer months in polluted swimming areas. Infectious hepatitis (Section 13.10) is caused by hepatitis A virus, which should be distinguished from hepatitis B virus that causes serum hepatitis. Hepatitis A virus is a serious water-borne virus disease agent at present (see Table 15.2); however, the virus is also transmitted in foods, and most of the infectious hepatitis cases probably arise by food-borne rather than water-borne means.

It should be recalled that viruses are not cells and lack cellular structure. Because of this, they are more stable in the environment and are not as easily destroyed as cells. However, both polio and infectious hepatitis virus are eliminated from water by purification, as discussed in Section 15.4.

Protozoa A pathogenic protozoan transmitted via the water route is *Entamoeba histolytica*, the causal agent of amoebic dysentery (see Section 12.8). This amoeba lives in the intestine and forms resistant structures called *cysts*, which are excreted with the feces and are able to survive in the environment for long periods of time. If they contaminate the water, it can then serve as a source of infection. Amoebic dysentery is most common in those parts of the world where sanitation standards are low.

Another protozoal agent that is responsible for many outbreaks and individual cases of water-borne disease is *Giardia lamblia*, the causal agent of giardiasis (see Figure 12.9). The symptoms of giardiasis are generally mild and include diarrhea, flatulence, and discomfort. There has been a noticeable increase in incidence of giardiasis in the United States over the past decade. As discussed in Section 12.7, giardiasis is most common when unfiltered water is consumed.

15.3 MICROBIOLOGICAL ASSAY OF WATER

Even water that looks clear and pure may be sufficiently contaminated with pathogenic microorganisms to be a health hazard. One of the main tasks in water microbiology is the application of laboratory methods that can be used to detect the microbiological contaminants which might be present in drinking water. It usually is not practical to examine drinking water directly for the various pathogenic organisms that might be present. As discussed in Section 15.2, a wide variety of organisms may be present, including bacteria, viruses, and protozoa. To check each drinking-water supply for each of these agents would be a difficult and time-consuming job. In practice, **indicator organisms** are used instead. These are organisms associated with the intestinal tract whose presence in water indicates that the water has received fecal contamination. The presumption is that the presence of the indicator warns of the possible presence of pathogens.

Coliforms as indicators The most widely used microbial indicator of water quality is the coliform group of organisms. The **coliform group** is defined in water bacteriology as all the aerobic and facultatively anaerobic, Gram-negative, non-spore-forming, rod-shaped bacteria that ferment lactose with gas formation within 48 hours at 35°C. This is an operational rather than a taxonomic definition and includes a variety of organisms, mostly of intestinal origin. In practice, the coliform organisms are almost always members of the enteric bacteria (see Section 11.10). The coliform group includes the organism *Escherichia coli*, a common intestinal organism, plus the organism *Klebsiella pneumoniae*, which lives not only in the intestine but also in soil and water. The definition also currently includes organisms of the species *Enterobacter aerogenes*. Some of the organisms in the coliform group may not even be derived from the intestine and therefore may not be of direct sanitary significance. For evaluating drinking-water quality, this deficiency is not a problem, since we are concerned with having a certain margin of safety in our drinking-water analyses and can accept a test that counts coliforms in excess of those naturally present.

The coliforms are suitable as indicators because they are common inhabitants of the intestinal tract, both of humans and warm-blooded animals, and are generally present in the intestinal tract in large numbers. When excreted into the water environment, most coliform organisms eventually die, but they do not die at a faster rate than the pathogenic bacteria *Salmonella* and *Shigella*, and both the coliforms and the pathogens behave similarly during water-purification processes. Thus, it is likely that if coliforms are found in water, the water has received fecal contamination and may be unsafe. There are very few organisms in nature meeting the definition of the coliform group that are not associated with the intestinal tract. It should be emphasized that the coliform group includes organisms derived not only from humans but from other warm-blooded animals as well. Since many of the pathogens (e.g., *Salmonella, Leptospira*) found in warm-blooded animals also infect humans, an indicator that signals both human and animal pollution is desirable.

Standard methods All sampling and testing should be done using standard methods, following carefully controlled procedures. Such procedures are described in considerable detail in a publication of the American Public Health Association, Washington, D.C., entitled *Standard Methods for the Examination of Water and Wastewater*. The Drinking Water Standards established by the

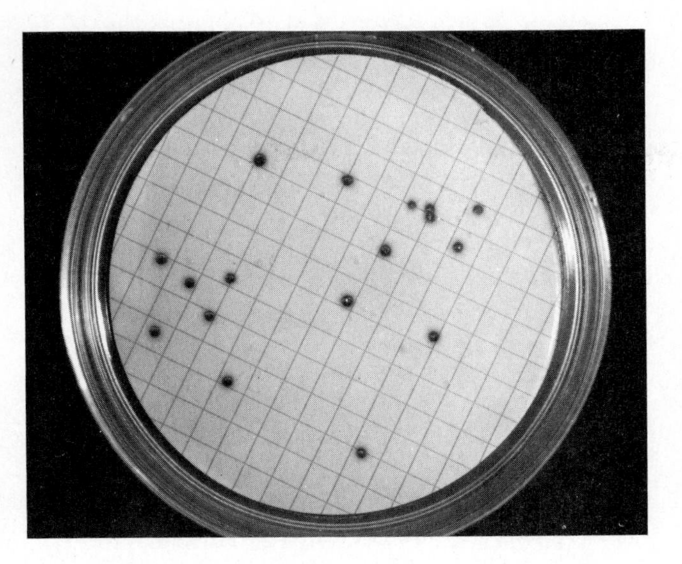

Figure 15.2 Coliform colonies growing on a membrane filter. A drinking water sample has been passed through the filter and the filter placed on a culture medium which is both selective and differential for lactose-fermenting bacteria (coliforms). The dark color of the colonies is characteristic of coliforms. A count of the number of colonies gives a measure of the coliform count of the original water sample.

U.S. Environmental Protection Agency specify that the methods prescribed by *Standard Methods* must be used.

Two types of procedures are used for the coliform test, the most-probable-number (MPN) procedure, and the membrane filter (MF) procedure (see Section 5.2). The most-probable-number (MPN) procedure employs liquid culture medium in test tubes (lactose broth with provision for detecting gas formation), samples of drinking water being added to replicate tubes of medium.

In the membrane filtration (MF) procedure, the sample of drinking water is passed through a sterile membrane filter that removes the bacteria; the filter is placed on a culture medium for incubation. This is the procedure most commonly used in the assessment of drinking water. The general MF procedure was discussed in Section 4.2. For drinking water, at least 100 ml of water is filtered, although in clean water systems, even larger volumes can be filtered (thus increasing the margin of safety). After filtration of a known volume of water, the filter is placed on one of several special culture media which are selective and which will indicate the presence of coliforms (see Figure 4.7). The coliform colonies (Figure 15.2) are counted, and from this value the number of coliforms in the original water sample can be determined. The MF procedure permits the determination of coliform numbers in 1 day while the MPN procedure requires 3 to 4 days. However, MF does not work well on water samples containing large amounts of suspended matter, silt, algae, or bacteria, since these interfere with both filtration and colony growth. Therefore, such water samples must be evaluated by the MPN procedure.

15.4 WATER PURIFICATION

It is a rare instance when available water is of such clarity and purity that no treatment is necessary before use. Water treatment is carried out both to make the water safe microbiologically and to improve its chemical characteristics for domestic and industrial purposes. Treatments are performed to remove pathogenic and potentially pathogenic microorganisms and also to decrease turbidity, eliminate taste and odor, reduce or eliminate nuisance chemicals such as iron or manganese, and soften the water to make it more useful for the laundry. The kind of treatment that water is given before use depends on the quality of the water supply.

Domestic water purification The necessary degree of water purification depends on the source water. If waters are used from rivers heavily impacted by industrial or domestic pollution, extensive purification is necessary (Figure 15.3). In such cases, the water is first pumped to **sedimentation**

Figure 15.3 Aerial view of a water-treatment plant of a large city, showing the flow of water from the river through the plant.

basins, where sand, gravel, and other large particles settle out. Most water supplies are subjected to **chemical coagulation**. Chemicals containing aluminum and iron are added, which under proper control of pH form a flocculent, insoluble precipitate that traps organisms and adsorbs organic matter and sediment, carrying them out of the water. After the chemicals are added in a mixing basin, the water containing the coagulated material is transferred to a settling basin where it remains for about 6 hours, during which time the coagulum separates out. About 80 percent of the turbid material, color, and bacteria are removed by this treatment.

After coagulation, the clarified water is usually *filtered* to remove the remaining suspended particles and microbes. Filters can be of the slow sand, rapid sand, or pressure type. **Slow sand filters** are suitable for small installations such as resorts or rural locations. The water is simply allowed to pass through a layer of sand 2 to 4 feet deep. Eventually, the top of the sand filter will become clogged, and the top layer must be removed and replaced with fresh sand. **Rapid sand filters** are used in large installations. The rate of water flow is kept high by maintaining a controlled height of water over the filter. When the filter becomes clogged, it is clarified by backwashing, which involves pumping water up through the filter from the bottom. **Pressure filtration** is similar to rapid sand filtration except that the water is pumped through the filter under pressure. Pressure filters are used mainly in small installations such as swimming pools and industrial plants. From 98 to 99.5 percent of the total bacteria in raw water can be removed by proper settling and filtration.

Chlorination is the most common method of ensuring microbiological safety in a water supply. Chlorine is an oxidizing agent that reacts with any organic matter present in the water. In sufficient doses, it causes the death of most microorganisms within 30 minutes. In addition, since most taste- and odor-producing compounds are organic, reaction with chlorine reduces or eliminates them. Chlorine can be added to water either from a concentrated solution of sodium or calcium hypochlorite or as a gas from pressure tanks. The latter method is used most commonly in large water-treatment plants, as it is most amenable to automatic control.

Since chlorine is used up during reaction with organic materials, if a water supply is high in organic materials, sufficient chlorine must be added so that there is a residual amount left to kill the microorganisms after all reactions with organic materials have occurred.

The significance of filtration and chlorination for ensuring the safety of drinking water cannot be overemphasized. Figure 15.4 illustrates the dramatic drop in incidence of typhoid fever in a major American city after these two water purification procedures were introduced. Similar results were obtained in other major cities. The dramatic improvement in the health of the American people in the early decades of the twentieth century was due to a very large extent to the establishment of satisfactory water purification procedures (see the Box on page 417).

Emergency water-supply treatment We are rarely called upon to treat our own water, but in emergencies and in remote parts of the world, such

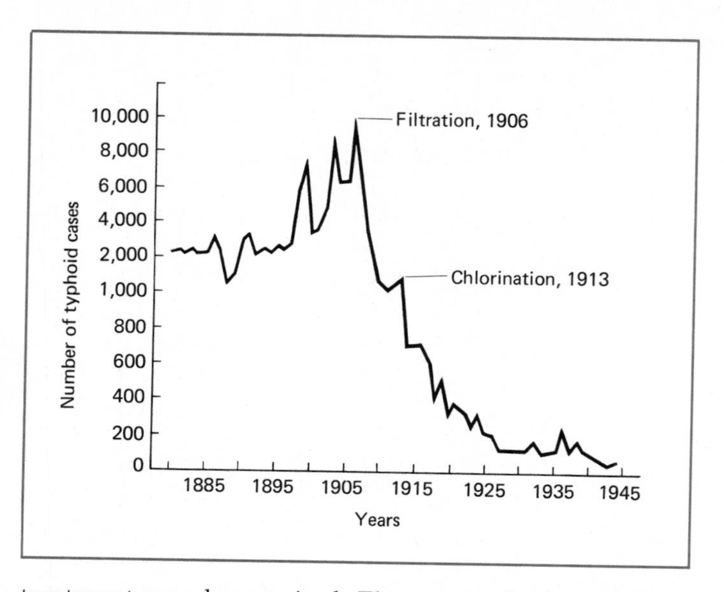

Figure 15.4 The dramatic effect of water purification on incidence of water-borne disease. The graph shows the incidence of typhoid fever in the city of Philadelphia during the early part of the twentieth century. Note the marked reduction in incidence of the disease after the introduction of filtration and chlorination.

treatment may be required. The most reliable and simplest method of purifying small quantities of water is boiling, a 10-minute treatment usually being sufficient. Water from a hot water tank, which has usually been kept at a temperature of 55 to 60°C (about 130 to 140°F) for many hours, is probably microbiologically safe, since virtually all pathogens are killed at such temperatures. *Filtration* can be accomplished on a provisional basis by passing the water through a barrel or drum filled with sand. The rate of filtration should not exceed 50 gallons per day per square foot of filter surface area. The filtered water should then be chlorinated.

Chlorination alone is satisfactory for water that is not grossly polluted. Suitable preparations, both powders and liquids, can be purchased at grocery and drug stores. The active ingredient in powders is calcium hypochlorite and in liquids, sodium hypochlorite. The strength is given on the label as percent of available chlorine. Since these materials deteriorate with age, fresh supplies should be obtained from time to time. Common household bleach provides 5.25 percent available chlorine; used at full strength, six drops are sufficient to treat 1 gallon of water. It is important to allow the water to stand for 30 minutes after the chlorine is added to ensure that all organisms are killed. Chlorination tablets suitable for use on hiking and camping trips can also be obtained, but will not be effective for *Giardia lamblia*, a common problem in wilderness areas. To reduce the risk of getting giardiasis, water must be filtered or boiled.

15.5 DRINKING WATER STANDARDS

Drinking water standards in the United States originated in 1914, when standards were prescribed by the federal government for the drinking water provided by common carriers such as railroads. However, it was not until 1975 that these standards were applied to drinking waters throughout the country.

WATER PURIFICATION: ROBERT KOCH SHOWED THE WAY

A rational approach to water purification required an understanding of the microbial role in infectious disease, as well as careful observation of water-borne epidemics. Among those who contributed importantly to this development was Robert Koch (Figure 1.7), who discovered the causal agent of cholera, *Vibrio cholerae,* and who showed the importance of water filtration, through observation of a specific cholera epidemic in Hamburg, Germany.

When the Hamburg epidemic broke out in 1892, Koch made the striking observation that the city of Altona, immediately adjacent to Hamburg, had a much lower incidence of cholera than Hamburg itself. Both cities obtained their water from the Elbe River, but Hamburg had the better water source, obtaining its supply from a point on the river above the city, whereas Altona obtained its water from below the city. Because its source was inferior, the city of Altona had to treat its water by running it through sand filters, whereas the city of Hamburg used its water unfiltered. Koch showed clearly that cholera was associated with only the Hamburg water supply, and even on streets which formed the border of the two cities, the Hamburg side was attacked by cholera while the Altona side was almost free of it. In Koch's words:

> "Here then we have a kind of experiment, which performed itself on more than a hundred thousand human beings, but which, despite its vast dimensions, fulfilled all the conditions one requires of an exact and absolutely conclusive laboratory experiment . . . The group supplied with unfiltered Elbe water suffers severely from cholera, that supplied with filtered water very slightly . . . For the bacteriologist nothing is easier than to give an explanation of the restriction of the cholera to the sphere of the Hamburg water supply. He need only point out that cholera bacteria got into the Hamburg water from the outlets of the Hamburg sewers . . . Altona received water which was originally much worse than that of Hamburg, but which was wholly or almost wholly freed of cholera bacteria by careful filtration."

Within the next decade after Koch's studies, water filtration became accepted practice in all water-purification systems using river water, leading to a dramatic drop in the incidence of water-borne disease (see Figure 15.4).

The Safe Drinking Water Act provides a framework for the development by the Environmental Protection Agency of drinking water standards for the whole country. Although these standards are still under development, there seems to be general agreement that the measurement of coliforms provides the greatest assurance of the microbiological safety of drinking water. The standards specify the frequency of coliform analysis and the allowable limits for numbers of coliforms present.

The drinking water standards apply to both community and noncommunity water systems. A *community water system* is defined as one that supplies at least 15 service connections used by year-round residents or regularly supplies at least 25 year-round residents. *Noncommunity systems* serve primarily transient-population areas, such as camp grounds, gasoline service stations, restaurants, and so forth, which may have their own water systems.

The sampling frequency specified in United States Drinking Water Regulations is related to the size of the population served by the water system. Noncommunity water systems are required to sample only once each calendar quarter during the period when the system is in operation. Community water systems are required to sample for coliforms at regular time intervals, the frequency of sampling being related to the size of the population being served. For the smallest community water systems, serving populations of 25 to 1,000 people, only one sample per month is required. At the other extreme, large cities are required to sample several hundred times a month.

Samples must be taken at points throughout the distribution system that are representative of the conditions within the system. This is extremely important, since water that leaves the water-purification plant free of contamination may become contaminated within the distribution system. Pipes may develop leaks and become contaminated from outside; cross-connections may result if there are breaks or defects in both the water pipes and the sewer pipes serving an area. Under these conditions, there is a real possibility of movement of intestinal pathogens from the sewer system to the water system, leading to significant contamination. By requiring that samples be taken throughout the distribution system, health hazards resulting from cross-connections should be avoided.

The Drinking Water Regulations specify that laboratory procedures for coliform counts be performed in accordance with the recommendations outlined in the book, *Standard Methods for the Examination of Water and Wastewater*. When the membrane filtration technique is used, the coliform count shall not exceed one colony per 100 ml of water filtered. When the coliform count is greater than the regulations permit, the supplier must obtain at least two additional check samples collected at the same point. If these check samples are positive, additional check samples must be analysed until at least two consecutive check samples are not positive.

If the water continues to fail to meet the standard, corrective action must be taken. Further, the water supplier is required to report within 48 hours any failure to comply with the regulations. If a community water system fails to comply with the coliform standards, water users must be notified through the news media. The intent of these requirements is to ensure that the public is totally aware of any problems that the water system might have and the measures that are being taken to rectify them.

15.6 MICROBIOLOGY OF WATER PIPELINES

The water pipeline, usually hidden from sight, is a crucial link between the individual and the source of purified water. Water lines are subject to a number of microbiological onslaughts, and a knowledge of these can help prevent any serious problems from developing. Water pipes may be constructed of steel, galvanized iron, brass, copper, wood, reinforced concrete, asbestos cement, or plastic. Galvanized iron, brass, copper, and plastic are used for small water systems and within buildings, and asbestos cement, iron, and steel are used in cities and other places where large amounts of water are distributed. Corrosion and clogging of pipelines are always potential problems (Figure 15.5).

Pipe corrosion by hydrogen sulfide is brought about by *sulfate-reducing bacteria*, which convert sulfate (SO_4^{2-}) to hydrogen sulfide (H_2S) when ad-

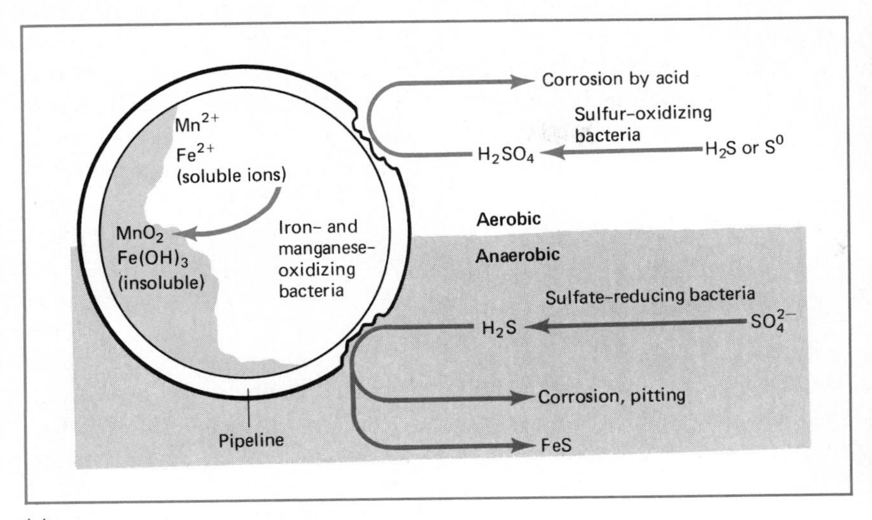

(a)

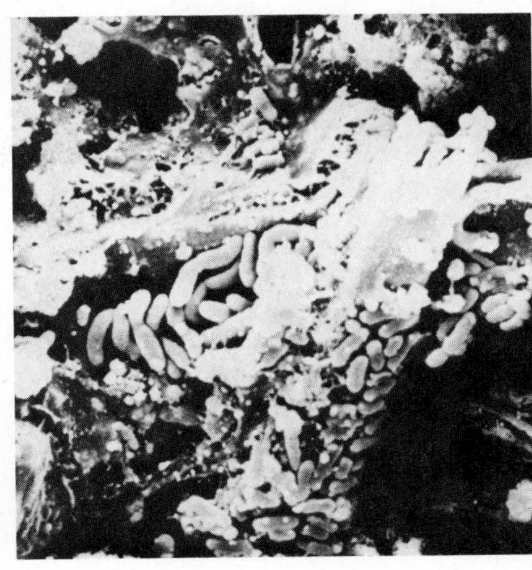

(b)

Figure 15.5 Corrosion and clogging of pipelines as a result of microbial action. (a) Various microbial processes that can occur in a pipe. (b) Scanning electron micrograph of a large bacterial colony found within the water distribution pipe for a large American city.

equate organic matter is available. The hydrogen sulfide attacks the iron of pipes and converts it into iron sulfide, causing pitting and deterioration of the pipe. These bacteria are obligate anaerobes and are especially of significance in the corrosion of pipes buried in areas in which oxygen is absent, such as waterlogged soils and marine and freshwater muds. Such corrosion is especially serious in areas where seawater has infiltrated the soil, since seawater is very high in sulfate. An unprotected pipe buried under anaerobic conditions in such an area can be destroyed in a few years. Pipes can be protected by coating them with asphalt or other impervious material, or it may be preferable to use pipe made of noncorrodible material, such as plastic.

Iron- and *manganese-oxidizing bacteria* are lithotrophs which can cause clogging of water-supply pipelines. In many water supplies, especially wells, iron and manganese are present in soluble form. When this water is brought to the surface and aerated, iron and manganese bacteria oxidize these substances and large amounts of insoluble precipitate can form. Difficulties arise both because of the clogged pipeline and because water containing these precipitates is unsightly, has a metallic flavor, and causes the staining of laundry. Iron causes a rusty red stain, and manganese, a black stain. Iron and manganese can be removed from a water supply before distribution by aeration, followed by settling or filtration.

Microorganisms that have not been removed in the water-purification process may be retained in viable form within the water pipe (Figure 15.5*b*). Drinking water derived from surface water supplies often has small amounts of organic matter present, derived from the lake or stream providing the water. This organic matter, if not removed during flocculation and filtration, can serve as a food source for slime-producing bacteria, which attach and form coatings on the pipe surface. These *bioforms* can then develop into large deposits within the pipes. Pathogenic organisms may be retained within these slimes, and certain of them may even grow. Subsequent release of these organisms in a *slug*, or clump, may cause infection in individuals receiving that portion of water. Slime build-up and growth of pathogenic bacteria within the pipes can be prevented by maintenance of a residual-chlorine concentration in the water. Periodic flushing of pipes, preferably annually, is also used to eliminate build-up of slimes and precipitates.

15.7 WATER POLLUTION

Water pollution occurs when undesirable effluents enter water courses and so change water quality that the water is unfit for human use without treatment. The major source of water pollution is *sewage* produced in urban areas. Sewage is rich in organic matter and contains vast numbers of microorganisms, some of which are pathogenic. Addition of sewage to a natural water often causes serious degradation of the quality of the water, making it unfit as a source of drinking water.

Biochemical oxygen demand (B.O.D.) A major result of the addition of sewage to a water course is the depletion of the dissolved oxygen of the water. This is due to the fact that the sewage contains organic matter; aerobic bacteria in the water oxidize this organic matter in respiration processes, thus consuming the oxygen. The decrease in oxygen content makes the water course partially or completely anaerobic, and this leads to the development of objectionable odors, flavors, and toxic materials in the water. When the water becomes anaerobic, many animals such as fish die, and their remains putrefy and add further foul odors and organic matter to the water (Figure 15.6). Also, decomposition of organic materials takes place much more slowly in the absence of oxygen; the purification processes in the water course are therefore slowed, and a thick, unsightly, organic-rich sediment may accumulate on the bottom of the water course.

The consumption of oxygen by bacteria is called *biochemical oxygen demand*, usually abbreviated B.O.D., and the extent of oxygen consumption is a function of the amount of oxidizable organic matter present in the water. The B.O.D. is commonly used as a measure of the degree of organic pollution of waters. The B.O.D. is determined by taking a sample of water, aerating it well, placing it in a sealed bottle, incubating for a standard period of time (usually 5 days at 20°C), and then determining the residual oxygen in the water at the end of incubation. During the 5-day incubation period, microorganisms present in the water grow, oxidize the organic matter, and consume oxygen. The amount of oxygen consumed is roughly proportional to the amount of biodegradable organic matter present. (B.O.D. is not a measure of the total amount of organic matter as some organic compounds can not be oxidized by microorganisms.)

Sewage pollution may be followed by purification, the processes in which the quality of the water is returned toward normal. When purification occurs without human intervention, it is called *self-purification* and occurs as a result

Figure 15.6 A fish kill resulting from the development of anaerobic conditions in a river, caused by excessive organic matter in the water.

of microbiological, chemical, and physical processes. Microbiological changes include death of many intestinal microorganisms present in the sewage and growth of normal aquatic microorganisms able to oxidize organic matter entering the system. Chemical changes include oxidation of organic matter, release of nitrate and phosphate, and reoxygenation of the water by solution of oxygen into the water from the air. The most important physical changes involve sedimentation, in which particulate matter settles out of the water onto the bottom of the water course.

It is the purpose of sewage-treatment systems (see Section 15.8) to replace these relatively slow, uncontrolled self-purification processes with rapid, well-controlled purification processes. The sewage-treatment process, carried out under careful control before the material enters the watercourse, attempts to ensure that the quality of the water course is not degraded.

15.8 SEWAGE-TREATMENT SYSTEMS

Sewage is collected into *sanitary sewers* from homes, businesses, and many industries and is carried to the sewage plant for treatment before discharge into streams. Water from rain and snow is also often collected in sewers called *storm sewers*; this water is not usually polluted and is often discharged directly into streams without treatment. The separation between sanitary sewers, which contain human wastes, and storm sewers, which contain surface runoff, is an important one, since the volume of storm sewer water is often high and would overtax the sewage-treatment process.

Sewage treatment consists of a series of processes in which undesirable materials in the water are removed or rendered harmless. Oxygen-consuming organic matter (B.O.D.) is destroyed, silt, clay, and other debris (called *suspended solids*) are removed, pathogenic microorganisms are killed, and the total number of microorganisms is reduced. There are many designs for sewage-treatment systems; the best design to be used for a specific system generally depends upon local factors. Figure 15.7 shows the sequence of steps in a typical urban sewage-treatment system. Both biological and nonbiological treatments are used. Nonbiological treatments include coarse and fine screening, sedimentation, sand filtration, chemical treatment to induce flocculation, and incineration. Biological treatments include various digestion and oxidation processes in which organic matter and inorganic materials are removed from sewage through the action of living organisms. The key organisms in biological treatment are bacteria, although fungi, algae, protozoa, and even higher animals are involved in some treatment processes. Biological treatment processes can be divided into two groups: digestion processes, which are anaerobic; and oxidation processes, which are aerobic. The principal digestion process is *anaerobic sludge digestion*, which functions in the treatment of insoluble organic-rich wastes; these are high in fiber and cellulose. Installations for oxidation processes include *trickling filters, activated-sludge systems*, and *lagoons* (oxidation ponds). In the following pages, we discuss the basic features of these various microbiological treatment processes.

Anaerobic digestion processes in sewage treatment The insoluble material that settles in any part of a sewage-treatment system is called *sludge*. The major components of sludge are proteins and organic fibers rich in cellulose, and anaerobic sludge digestors are used to decompose these substances (Figure 15.8). The process is carried out in large tanks from which oxygen is completely excluded. Under the anaerobic conditions that prevail, fermentation occurs, and the sugars released in the digestion process are converted to methane and carbon dioxide. Methane is a burnable gas, and in many

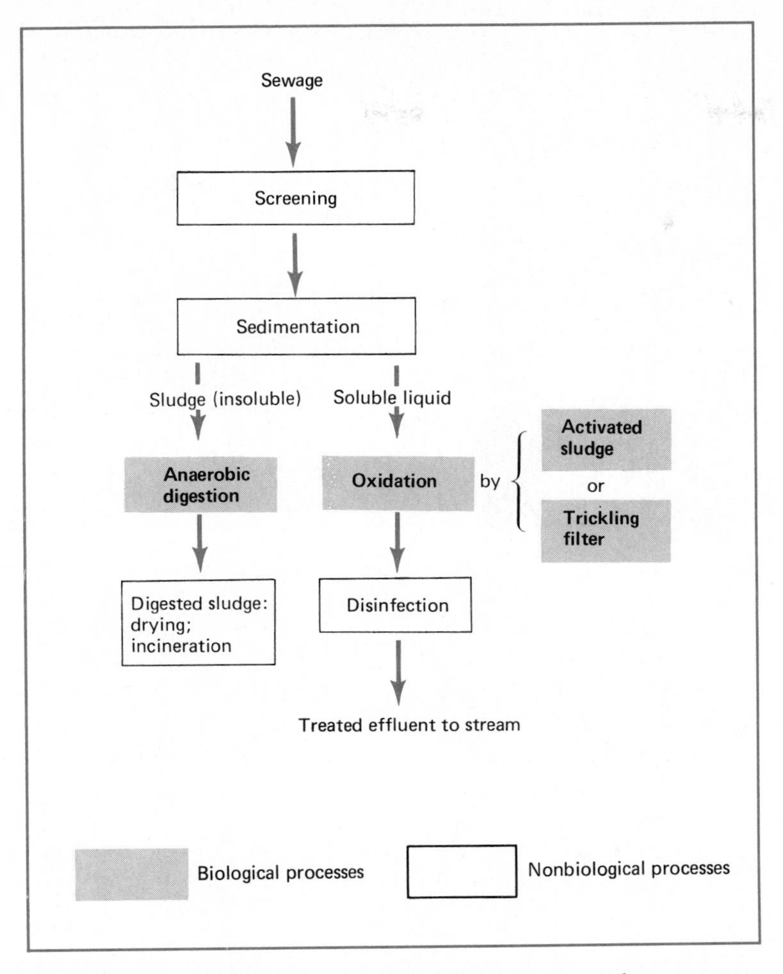

Figure 15.7 Steps in sewage treatment.

sewage plants, the methane produced in the digestion tank is used as a source of heat or power for the whole sewage-plant complex.

Methane is formed by a special group of bacteria, the methanogenic bacteria, that are members of the archaebacteria group (see Figure 2.31). These bacteria form methane only from a restricted range of compounds, of which the most important are hydrogen gas, H_2, and acetic acid, CH_3COOH. The formation of methane from H_2 is a type of anaerobic respiration (see Section 3.7), in which CO_2 is used as the electron acceptor. The whole process of methane formation is thus dependent on the effective functioning of this special group of bacteria which have only a restricted range of compounds that they can work on. The H_2 and acetic acid used by these methanogens are formed by fermentative bacteria which break down the organic matter of the sewage. As we discussed in Section 3.6, fermentation is an anaerobic process in which internal electron acceptors generated from the organic substrate itself are used. In the fermentation process, reduced end products are released. In anaerobic sewage digestion, the two major reduced end products of the fermentative bacteria are H_2 and acetic acid, which are converted to methane by the methanogens.

(a)

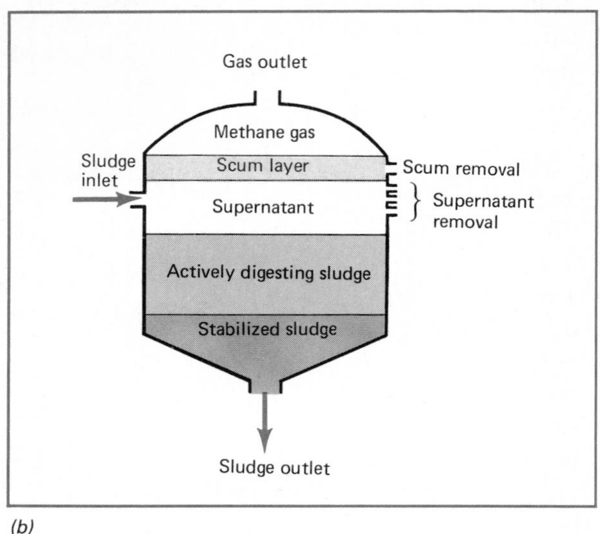

(b)

Figure 15.8 The anaerobic sewage digestion process. (a) An anaerobic sludge digestor. Only the top of the tank is shown; the rest is underground. (b) Inner workings of an anaerobic sludge digestor.

The whole process of sludge digestion occurs in four stages:

1. *The digestion phase*: initial digestion of fiber and other insoluble materials to sugars, amino acids, and other soluble products.
2. *The acid-production stage*: fermentation of the soluble materials to organic acids and alcohols by bacteria.
3. *The acid-regression stage*: conversion of organic acids and alcohols to acetic acid, hydrogen (H_2) plus carbon dioxide (CO_2) carried out by a different bacterial group.
4. *The methane-production stage*: conversion of acetic acid and H_2 plus CO_2 to methane (CH_4) by methane-producing bacteria.

The time taken for sludge to become digested is from 2 weeks to a month, and in a well-operated plant, 90 percent or more of the B.O.D. of the sludge can be removed; that is, 90 percent of the biologically oxidizable organic matter is converted to CO_2, and H_2O, and methane (which may itself be burned to CO_2 and H_2O).

A serious problem in the management of many sludge digestors is the development of acid conditions. If stages 3 and 4, the acid-regression and methane formation stages, do not immediately follow stage 2, the acid-production stage, the pH may drop to a value so low that the bacteria are killed or inhibited. When this happens, the whole process stops, gas production ceases, and B.O.D. removal does not occur. This condition often results from overload of the digestor with too much sludge. The management of a sludge digestor is not easy, and a knowledge of the microbiology of the process is essential.

The undigestible materials remaining in the sludge are removed at intervals and dewatered to reduce their bulk. Dewatering is often difficult with certain kinds of sludges, and the material may have to be pumped to a large field for settling and drying. If this is the case, considerable care must be taken to ensure that the drying sludge does not pollute nearby streams or lands. The dried sludge may be useful as a soil builder, but it is important that the sludge does not contain any harmful materials, such as live viruses or heavy metals. Many sludges contain undesirably large amounts of heavy metals such as copper, chromium, lead, and zinc derived from the water pipes or from industrial processing. Because of the potential toxicity of these heavy metals, it must be ascertained that they remain firmly tied up in the sludge particles and do not leach into the soil solution where they could be taken up by the plants being cultivated. If it is predominantly organic in nature, the dried sludge can be burned in an incinerator, care being taken to ensure that no air pollution results from the incineration process. After incineration, the ash residue is buried. If the sludge is mainly inorganic and hence not burnable, it is usually buried.

Aerobic sewage-treatment systems Aerobic systems are used to treat the soluble components of sewage that remain after the sludge has been removed. Tremendous volumes of liquid that must be treated in this way are generated by urban populations. The liquids contain a large variety of organic compounds, none present in very large amounts. Before these organic compounds can be acted upon by the living organisms of the treatment system, they must be concentrated in some way. Concentration involves the adsorption of the organic materials to some type of surface upon which the treatment organisms can grow. After concentration, the organic compounds are oxidized by aerobic bacteria in respiratory processes. As we noted in Section 3.6, organic compounds oxidized in respiration are converted to carbon dioxide, a harmless inorganic compound. It is the goal of aerobic sewage treatment systems to convert all readily oxidizable organic compounds to CO_2.

Two major types of aerobic treatment processes are used, called *trickling filter* and *activated sludge*. In the *trickling filter process*, the sewage is passed over a bed of rocks. As the liquid slowly trickles through the bed, the organic matter adsorbs to the rocks, and microbial growth occurs. A film of microorganisms, called a *biofilm*, develops on the rocks, and within the biofilm, oxidation of organic matter takes place. The effluent of a trickling filter is low in organic matter, the B.O.D. being reduced about 80 to 95 percent.

The **activated-sludge process** (Figure 15.9) is the most common type of oxidation process used in large urban installations. The wastewater to be treated is mixed and aerated in a large tank, and slime-forming bacteria grow and form flocs. These clumps constitute what is called the activated sludge; they are composed of several kinds of bacteria and in addition contain protozoa and other animals that live by feeding on the bacteria. One of the principal bacteria in the floc is the organism *Zoogloea ramigera*, a small, Gram-negative rod that forms an extensive extracellular slime. This organism forms complex, fingerlike colonies that provide the basis of the floc (Figure 15.10). The other microorganisms attach to the floc and build up the complex, active particle.

(a)

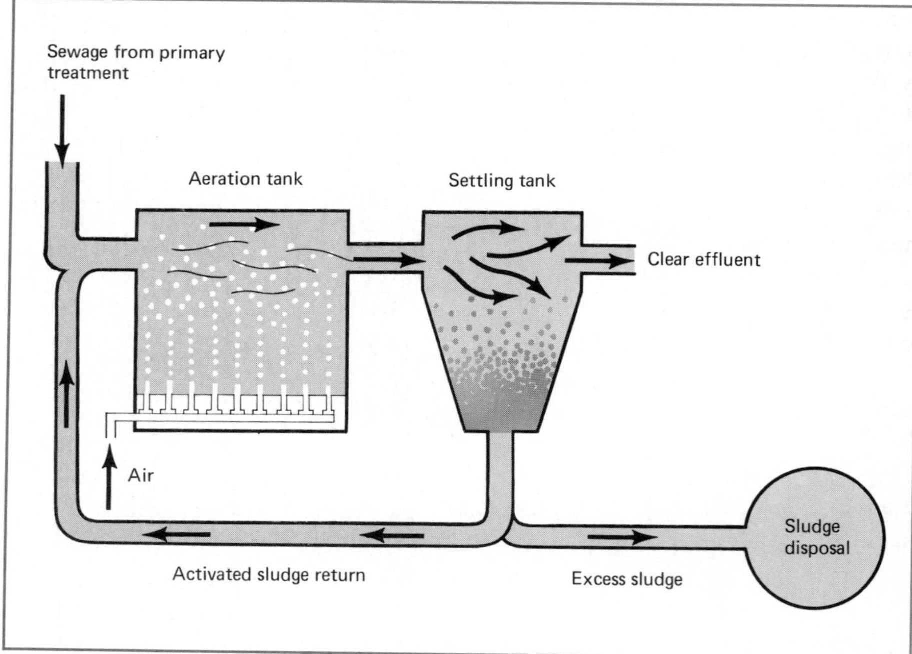

(b)

Figure 15.9 The activated-sludge process. (a) An aeration tank of an activated-sludge installation. (b) Diagram of the flow of liquid through an activated-sludge system.

Oxidation of organic matter by the activated-sludge flocs occurs in a manner similar to that described for trickling filters. The treatment operates continuously, the wastewater being pumped into the tank and the effluent containing flocs passing out at another location. The holding time in the activated-sludge tank is generally only a few hours. The effluent containing the flocs is pumped into a holding tank or clarifier where the flocs settle. Some of the floc material is then returned to the aerator as inoculum for the continuation of the oxidation process, and the rest is either dried and discarded or is pumped to the anaerobic sludge digestor for treatment.

A serious problem in the operation of some activated-sludge systems is *bulking;* the flocs do not settle sufficiently rapidly, and the clarification process is inhibited. Bulking usually occurs because of unusual growth of filamentous bacteria, which form a loose floc rather than the tight floc that is formed by *Zoogloea ramigera*. Excessive growth of filamentous bacteria is probably due to the presence in the wastewater of certain kinds of organic compounds that favor their growth over the *Z. ramigera* population. If bulking occurs, the sewage-plant operator is faced with an unclarified effluent that is high in organic materials. No simple remedy for bulking is available, but if it is caused by the presence in the wastewater of some unusual effluents, perhaps from industrial or food-processing plants, it may be possible to remove these effluents from the main treatment process and handle them separately. About 75 to 90 percent decrease in B.O.D. occurs in the activated-sludge process. The efficiencies of various sewage-treatment processes are compared in Table 15.3.

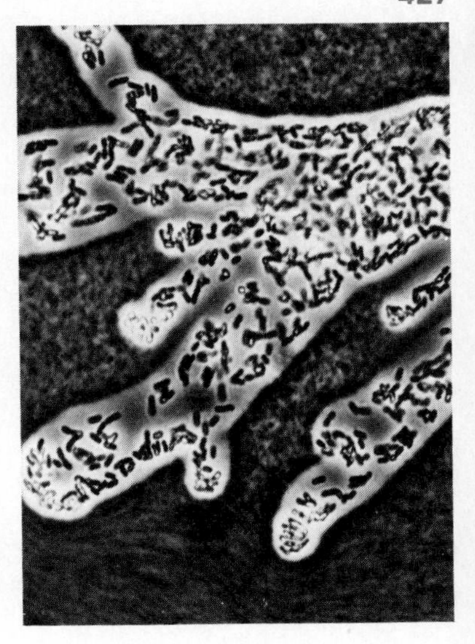

Figure 15.10 Photomicrograph of a floc formed by *Zoogloea ramigera*, the characteristic organism of the activated-sludge process. Note the large number of small, rod-shaped bacteria and the characteristic fingerlike projections of the floc. To visualize the floc under the microscope, the sample was treated with india ink, a process called *negative staining*.

Substances resistant to decomposition Most organic substances in domestic sewage decompose readily during sewage treatment. In contrast, many chemicals present in industrial wastes are resistant to decomposition by the microorganisms present in normal sewage-treatment systems. Some of these compounds are toxic; if so, they can sometimes be effectively handled merely by diluting them to a point where they no longer constitute a hazard. Other compounds are not toxic but are merely resistant to decomposition. Sometimes these materials cannot be treated biologically at all, and chemical means of treatment must be used. In some cases, it may be possible to develop a microbial population that will degrade the compound. If a compound can be decomposed by microorganisms, it is **biodegradable**; if it cannot be decomposed, even after attempts to obtain a population capable of degrading it, it is considered *nonbiodegradable*. It is often observed that when a new substance is first introduced into a sewage-treatment system, it decomposes slowly, but after a period of time, decomposition is rapid as a result of development of a suitable microbial population. Therefore, when beginning the treatment of a new material, it is important to ensure that the proper population is present before introducing material into the treatment plant.

TABLE 15.3 Efficiency of sewage-treatment processes

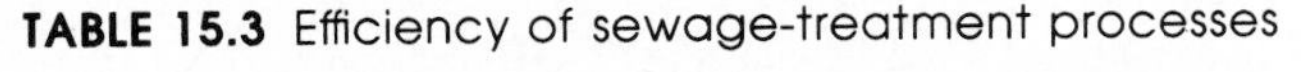

Treatment	Percent Reduction		
	Suspended solids	B.O.D.	Bacteria
Sedimentation	40–70	25–40	25–75
Sedimentation plus trickling filter	75–90	80–95	90–95
Activated sludge	85–95	85–95	90–98

If a material is not decomposed, it will pass through the system unchanged and will be discarded into the water body receiving the treated effluent. For instance, nonbiodegradable detergents will pass through a system unchanged and enter receiving waters where they can cause foaming. Since microbial populations able to degrade these compounds do not readily develop, the problem of foaming has been reduced by legislation banning the sale of nonbiodegradable detergents.

Discharge of treated sewage Although the treatment processes just discussed bring about marked reduction in the harmful materials present in raw sewage, a potential for the deterioration of the body of water receiving the effluent still exists because of the large volumes of treated sewage that are discharged, even when the concentrations of harmful materials are quite low. Also, since many sewage-treatment systems do not operate properly 100 percent of the time, there are situations in which essentially untreated wastes might be discharged. It is vital that the sewage-plant operator monitor the effluent by performing analytical tests and procedures. Total coliform counts, B.O.D. determinations, and chemical analyses for harmful materials such as ammonia are essential. There should be no or few coliforms in the effluent. The level of B.O.D. permitted depends on state and federal regulations but should be maintained at a low, constant level.

The potentially pathogenic microorganisms present in raw sewage are virtually all eliminated during sewage treatment, either by sedimentation and removal or because they die during the treatment process. However, to ensure that the final effluent is free of harmful microorganisms, it is often *chlorinated* before it is discharged into the receiving water body. Although chlorination is used primarily to destroy any intestinal microorganisms that might have survived passage through the plant and would contaminate the receiving water, it also leads to the decomposition of materials in the effluent that would otherwise cause unpleasant odors. However, chlorination of treated sewage presents problems of its own, because potentially harmful chlorinated organic compounds are often formed. To avoid chlorination, other possible agents for disinfecting treated sewage are *ozone* and *ultraviolet radiation,* both of which are effective disinfecting agents.

Viruses present special problems in sewage treatment. Because of their small size and noncellular structure, they do not behave like bacteria and other pathogenic organisms during the treatment process. Many viruses are not killed by either anaerobic or aerobic treatment processes. Viruses attached to particles settle out in the sedimentation processes, and some adsorb to the flocs in activated sludge, but this does not lead to their destruction, and they may remain viable in the final dried sludge. Virus particles that remain in the liquid may pass through the whole treatment process and end up in the effluent, presenting a possible health danger if they are discharged into a stream. Viruses in the effluent can be eliminated by disinfection, although the doses required may be higher than those required to eliminate bacteria. Thus, an effluent that shows a negative coliform count is not necessarily free of viruses. The greatest hazard arises if the stream that receives the virus-laden effluent is used by a downstream community as its water supply, since viruses may also survive many water-purification processes.

The proper functioning of urban society depends upon the proper activity of microorganisms growing in these large sewage treatment systems. Every urban citizen is part owner of a large and costly microbial cultivation system, the sewage treatment plant.

Lagoons, septic tanks, and privies All the sewage-treatment systems just discussed involve large installations, complicated engineering, and skilled operation by large staffs of people. They are suitable for urban and industrial use but are beyond the means or resources of individual homeowners or small communities. It is vital that all sewage be treated before it is discharged in order to avoid public-health problems and environmental degradation. In situations where conventional systems are not feasible, certain types of treatment processes are available that can provide at least some measure of public protection. None of these methods is desirable in areas with dense population distribution; rather, they can be used in rural areas and in small, remote communities.

The *sewage lagoon,* sometimes called an *oxidation pond* or *stabilization pond,* is simply a large shallow pond into which sewage clarified by sedimentation is pumped. In the lagoon, bacteria oxidize organic substances to inorganic matter. These inorganic materials serve as nutrients for the growth of algae, which use sunlight and carry out photosynthesis. In a well-balanced lagoon, 80 to 95 percent of the B.O.D. may be removed. The oxygen produced by the algae during photosynthesis keeps the surface of the lagoon aerobic, thus preventing the development of anaerobic conditions, which generally lead to foul odors. A lagoon is usually odor-free. Pathogenic bacteria are usually destroyed in the lagoon.

A *septic tank* is a water-tight tank used at rural and some suburban residences to provide minimal treatment of sewage where municipal sewer lines are not available. The tank is a concrete, metal or masonry structure through which the raw sewage passes (Figure 15.11). During the time the sewage is in the tank, settling and minimal sludge digestion occur; the effluent liquid is then distributed into the soil. Eventually the sludge builds up in the bottom of the tank to a point where operation is impaired, and the tank must then be pumped out or otherwise cleaned. Some microbiological digestion of sludge does occur in septic tanks, although under less efficient and less well controlled conditions than in modern treatment plants. In septic-tank effluent, the total suspended solids are reduced by about 65 percent and B.O.D. is reduced by 57 percent. These values should be compared to those already given for standard treatment processes in Table 15.3; the low efficiency of the septic tank is clearly indicated. The level of fecal coliforms in effluent is also quite high (over one million per 100 ml) compared to that found after conventional sewage treatment. Finally both sludge and effluent from septic-tank systems may contain high levels of viruses.

The extent to which a given soil can accept septic-tank effluent will depend on its permeability. In a porous, highly permeable soil, the effluent spreads readily and no difficulties arise; in a hard clay soil of low permeability,

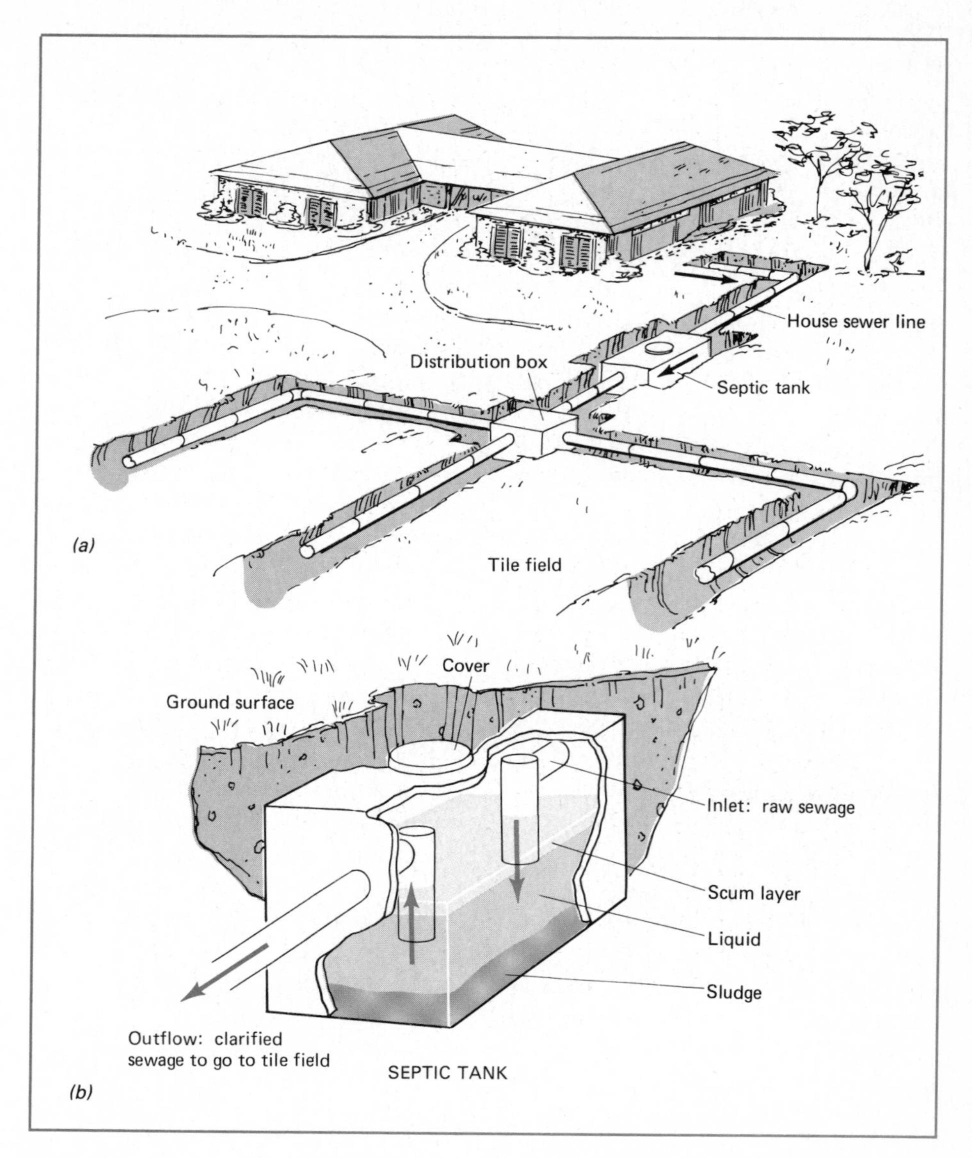

Figure 15.11 Septic tank installation, as used in a small-scale sewage treatment system in a rural area. (a) Distribution system. (b) Inner workings of the septic tank.

the effluent cannot spread well but will only follow cracks and crevices in the soil, resulting in poor purification. The soil serves as a crude sand filter for septic-tank effluent, and much of the purification process occurs in the soil rather than in the tank. To determine the capacity of soil to accept septic-tank effluent, a soil percolation test is run. Such a test is required by law in many areas before a septic tank may be installed.

A *privy* (outhouse) is merely a hole or pit in the ground above which a toilet seat and some kind of shelter are mounted. Privies are used mainly in very rural areas, such as farms or primitive campgrounds, where population density is quite low, but they are also used in temporary installations, such as military outposts or construction sites. Decomposition of waste matter occurs within the pit, both by digestion and oxidation, but the process is slow and inefficient. Odor is frequently a problem, and disease transmission may be common and serious. Transmission of pathogens occurs often by flies, which may pick up pathogens on their legs and transport them to nearby dwellings. Screened shelters and covered toilet seats hinder the entry of flies and hence reduce transmission of pathogens. Pathogens can also be carried out of the pit into the surrounding soil by leaching and may be carried through crevices and cracks in the soil to the water table.

15.9 ALGAE AND WATER POLLUTION

A serious problem in many lakes and reservoirs used as sources of water is the growth of algae. Algae are undesirable because they cause bad odors and flavors in water and may produce toxic materials of potential danger to humans, pets and livestock. Algal growth is favored by warm water temperatures, high sunlight, and adequate sources of nutrients, especially nitrates, phosphates, and carbon dioxide. The deleterious enrichment of water courses by nutrients is called **eutrophication**. Eutrophication is a consequence of agricultural and urban development in a watershed, and is a very difficult problem to control. Algal growth in temperate lakes and reservoirs is most common in summer and is rare in winter. Occasionally, in late summer and early fall, algal growth may be so heavy that the water resembles pea soup. This condition, called an *algal bloom* (Figure 15.12), occurs when algae float to the surface and drift into backwaters where they become concentrated. Bacteria attack and decompose the algae, causing a reduction in oxygen, which in turn leads to the death of fish and other animals and the development of foul and putrefying odors.

Reservoirs for domestic water supplies are often good habitats for algal growth, because they are generally relatively shallow and receive large amounts of algal nutrients from the watershed. In addition to the odor and flavor problems that may develop, heavy algal growth in a reservoir makes filtration and disinfection of the water difficult, thus markedly increasing the cost of water purification. Algal growth in reservoirs is generally controlled by application of copper sulfate. In many water supplies, copper sulfate is applied routinely at 2- to 4-week intervals throughout the period from April to October. Because copper sulfate is toxic to fish (and in high doses to humans also), its use in reservoirs must be carefully controlled. The recommended amounts are well below the known toxic levels. It is most desirable to carry out routine microscopic examinations of the reservoir for algae throughout the warm season. When microscopy reveals an increase in algal numbers, copper sulfate should then be applied, before the algal bloom begins. In this way, uncontrolled increases in algal numbers can be eliminated completely.

Eutrophication is a serious economic problem because algal growth adds organic matter back into the water, thus increasing the B.O.D. and causing a deterioration of water quality. The algal nutrients are inorganic and are released by mineralization during the decomposition of the organic matter in the nutrient source (such as sewage) by bacteria. Organic nitrogen is converted to ammonia; the ammonia is then oxidized to nitrate. Organic sulfur is con-

Figure 15.12 An algal bloom results from eutrophication. The thick algal scum decomposes, causing oxygen depletion and foul odors.

verted to hydrogen sulfide, which is then oxidized to sulfate. Organic phosphorus is converted to inorganic phosphate. Nitrate and phosphate are especially important in water pollution because they are effective nutrient sources for algae.

SUMMARY

In this chapter, we learned some of the microbiological principles involved in the handling and purification of water and wastewater. Vast amounts of pure fresh water are needed for domestic purposes, and care must be taken to ensure that this water is free of harmful microorganisms. Water can be the vehicle for the transmittal of a number of pathogenic organisms, including bacteria, viruses, and protozoa. The most serious pathogens transmitted by water are *Salmonella typhi,* the causal agent of typhoid fever, and *Vibrio cholerae,* the causal agent of cholera. Other agents include *Shigella, Salmonella* species other than *S. typhi,* the viruses causing polio, infectious hepatitis, and diarrheal diseases, and the protozoa causing amoebic dysentery and giardiasis.

Purification of water is carried out by sedimentation, filtration through sand, and coagulation. **Chlorination** is the most common method of ensuring the microbiological safety of a water supply. Microorganisms are good indicators of water pollution, and the quality of water is usually defined in terms of its microbial load. The **coliform count**, the most commonly used indicator of water pollution, measures the concentration of fecal bacteria in the water supply. The federal Drinking Water Regulations specify the coliform count for drinking water.

Pollution of water by domestic sewage and industrial wastes can result in serious degradation of water quality. One measure of organic pollution of water is the **biochemical oxygen demand (B.O.D.)**, which measures the oxygen consumed by the aquatic system after addition of organic wastes. To remove B.O.D. and harmful microorganisms from sewage, a variety of treatment methods are employed, both biological and nonbiological. One biological method is the **anaerobic digestion process**, in which cellulose and other fibrous organic materials are digested and fermented by anaerobic bacteria with the production of **methane**. Other processes involve oxidation by aerobic bacteria of soluble organic materials present in sewage. Two aerobic systems are the *trickling filter,* in which the liquid is allowed to trickle over a bed of rocks upon which the microorganisms grow, and the **activated-sludge process**, in which the liquid is aerated in large tanks. The end product of sewage treatment systems is a liquid low in B.O.D. and lacking pathogenic microorganisms; it can be discharged into a river or lake without causing harm. An insoluble undigestible residue, the *sludge,* is also produced, which is usually disposed of by burial. We thus see that in all stages of the handling and disposal of sewage and solid waste, microorganisms play important roles.

KEY WORDS AND CONCEPTS

Hydrologic cycle
Indicator organism
Membrane filtration procedure
Filtration
Chlorination
Coliform group
Pollution
Biochemical oxygen demand (B.O.D.)
Anaerobic digestion process
Methanogenic bacteria
Activated-sludge process
Biodegradable
Eutrophication

STUDY QUESTIONS

1. Both surface and ground waters are used as sources of drinking water. Contrast these two types of sources from the point of view of availability, likelihood of pollution, and degree and type of treatment necessary to make them potable.

2. What are the most important water-borne diseases? Identify the pathogen causing each. Group these as bacterial, viral, and protozoal.
3. Define coliforms. How does the coliform group differ from the organism *Escherichia coli*?
4. Describe briefly the *membrane filtration* procedure for analysis of coliforms in drinking water.
5. Outline briefly the steps in the purification of water intended for domestic use. Discuss briefly the function of each step.
6. As an aftermath of floods, earthquakes, or other disasters, a city's water supply may be interrupted. What means of purification would you use to ensure a supply of safe water for drinking?
7. The presence of coliforms is generally taken as evidence of fecal contamination. Why? If coliforms are present, does this necessarily indicate *human* fecal contamination?
8. How may microorganisms be involved in the corrosion or clogging of pipelines?
9. Pathogenic organisms may be harbored in pipelines under certain conditions. Why? Is this of any health significance?
10. Discuss the nature, measurement, and significance of *B.O.D.*
11. Describe the events that occur downstream from a point where untreated sewage enters a river.
12. What is meant by self-purification of a stream? Why is this not a reliable way of handling domestic or industrial wastes?
13. Domestic sewage in populated areas is usually treated by a multistep process. Outline a typical treatment process, describing at each step what is occurring microbiologically and in what ways the entering sewage is changed.
14. What is sludge digestion? Is it aerobic or anaerobic? Describe the steps in sludge digestion.
15. Describe the function of the *activated-sludge* plant used for the aerobic treatment of sewage.
16. What types of problems may arise if *nonbiodegradable* compounds are present in sewage?
17. Compare the efficiency of sewage treatment by activated-sludge and septic-tank systems. Based on this comparison, what is your opinion of the value of septic tanks in built-up (suburban) areas?
18. Try to find out what sewage-treatment methods are used in the area where you live. Where is the treated sewage in your area discharged?
19. Many sewage-treatment plants are grossly overloaded and do not operate properly, thus discharging poorly treated sewage. Try to find out how well the sewage-treatment plant in your area is operating. Is it meeting the standards of your state?
20. What causes an *algal bloom*? How can algal growth be controlled in large bodies of water?
21. What is *eutrophication*? How can it be prevented?

SUGGESTED READINGS

AMERICAN PUBLIC HEALTH ASSOCIATION. 1985. *Standard methods for the examination of water and wastewater, 16th edition.* American Public Health Association, Washington, D.C.

BITTON, G. 1980. *Introduction to environmental virology.* John Wiley and Sons, New York.

CENTERS FOR DISEASE CONTROL. *Morbidity and Mortality Weekly Report.* Atlanta, Ga.

GELDREICH, E.E. 1966. *Sanitary significance of fecal coliforms in the environment.* U.S. Department of Interior, Federal Water Pollution Control Administration, Publication WP-20-3.

MITCHELL, R. (editor). 1978. *Water pollution microbiology, Vol. 2.* John Wiley & Sons, New York.

NATIONAL RESEARCH COUNCIL, NATIONAL ACADEMY OF SCIENCES. 1977. *Drinking water and health.* Washington, D.C.

"U.S. DRINKING WATER REGULATIONS." *Federal register, 40,* 248.

16

Food Microbiology

Microorganisms have both beneficial and harmful effects on foods, and the study of microbiology encompasses both these aspects. Beneficial roles of microorganisms include their use in preparing foods such as sausage, cheese, and pickles. On the other hand, microorganisms are responsible for some of the most serious kinds of food poisonings and toxicities and also cause spoilage of a wide variety of food and dairy products. We discussed briefly in Chapter 1 the economic importance of food-related activities involving microbiology (see Table 1.3). It is convenient to divide our present discussion into separate sections dealing with the general principles of food spoilage, including what pathogens are likely to be found and how they are detected, and the various methods of food preservation. There are also special sections on dairy and meat products because of the unique problems relating to preservation and microbial control in these highly perishable foods.

16.1 FOOD SPOILAGE

Food spoilage is probably the most serious economic consideration of the food-processing industry. Foods are subject to attack by microorganisms in a variety of ways, and such attack is usually harmful to the quality of the food.

The physical and chemical characteristics of the food determine its degree of susceptibility to microbial attack, the kinds of microorganisms that will affect it, and the kinds of spoilage that will result. Food-processing techniques have been developed that make it possible to reduce or prevent microbial spoilage and thus ensure long shelf life for the food product.

A food is considered spoiled when it is in such a state that a discriminating consumer will not eat it. Spoilage is not an absolute characteristic of a food, since custom plays an important role in whether a food is considered desirable to eat. For instance, although soured milk may be considered spoiled, buttermilk is not considered spoiled even though it is sour. An important distinction must be made, therefore, between foods that are not eaten because they taste bad and foods that are not eaten because they are poisonous or harmful.

Microbiologists are involved in food-spoilage problems in the following ways: (1) *research*; identifying types of microorganisms causing food spoilage, assessing the conditions under which they develop, evaluating the potential harm to humans of different food spoilage organisms, and predicting the circumstances under which spoilage will develop; (2) *development*; developing and perfecting methods for processing foods so that they can be stored safely; (3) *quality control*; checking stored foods for microbial spoilage and making recommendations for discarding unsuitable foods; (4) *consultation*; evaluating food-processing plants from a microbiological viewpoint and recommending modifications in operation to reduce microbial contamination and spoilage.

Foods vary in the ease with which they undergo microbial spoilage:

1. *Highly perishable foods* are those that must be processed and stored carefully if they are to be preserved. These include meats, fish, poultry, eggs, milk, and most fruits and vegetables; in short, the basic daily foods in many parts of the world.
2. *Semiperishable foods* can be kept for fairly long periods of time if they are stored well. Examples include potatoes, some apples, and nuts.
3. *Stable* or *nonperishable foods* can be stored for long periods of time without difficulty. These foods, which are stable because they are low in moisture content, include sugar, flour, rice, and dry beans. In addition to these naturally stable foods, many perishable foods (for example, meats, fruits, and vegetables) can be converted to stable ones by drying.

At any step along the route from farmer to consumer, foods can become contaminated with microorganisms. At each step, the kinds of contaminating organisms may be different, and consequences of such contamination may also be quite different. At the farm, foods can become contaminated by soil organisms, many of which are endospore formers and hence resistant to heat, or by organisms from other sources in the farm environment. Microbial growth may actually occur in the product during storage or handling, so that the product is altered even before it reaches the processing stage. In the food-processing plant, contamination may come from the water used in processing, from contaminated equipment, from the packages in which the final product is placed, or from the employees handling the food. After processing, the food may become contaminated during storage in warehouse, store, or home. Each link in the chain from farmer to consumer must be carefully controlled if the consumer is to be provided with a product that is safe, clean, and palatable.

Microbial growth in foods Many factors are involved in determining the keeping qualities of foods—that is, how long a food will retain its quality and flavor and resist spoilage. Properties of the foods themselves as well as the conditions under which they are stored influence this. The most important factors influencing microbial growth in foods are: (1) moisture and water status; (2) temperature; (3) pH; (4) oxygen availability; (5) physical state of the food (for instance, whether it is whole or ground); and (6) chemical constitution, including the presence of microbial agents.

A factor of foremost significance in food preservation is the **moisture content** of the food. If food is sufficiently dry, microbial growth is, of course, impossible since all organisms require water for life. Indeed, the drying of foods is one of the oldest forms of food preservation. The amount of water in foods varies greatly and different microorganisms are able to grow at different moisture levels. However, water availability does not depend solely on water content; water absorbed to surfaces may or may not be available for microbial growth, depending on how tightly it is adsorbed and how effective the microorganism is in removing it. Also, when solutes are dissolved in water, the extent to which they are hydrated also affects the availability of the water.

A common way of expressing water availability is in terms of **water activity**. Water activity (a_w) is related to the vapor pressure of water in the air over a solution or substance and is estimated by measuring the relative humidity of the vapor phase. Relative humidity (R.H.) expresses the ratio of the amount of water in air at a given temperature to that which air could hold if it were saturated with water. In normal usage, R.H. is given as a percent while a_w is expressed as a fraction. Thus an a_w of 0.75 equals 75 percent relative humidity. The a_w of most fresh foods is above 0.98. Table 16.1 shows the a_w values for a variety of foods and also lists some microorganisms, showing the lower limit of a_w which permits their growth. Most can grow at an a_w of almost 1.00 down to the value given in the table. Most bacteria and many yeasts are unable to grow at an a_w below 0.9, while molds can often grow at 0.9 to 0.8. At lower a_w values only a few specialized fungi and yeasts, called *xerophilic organisms*, are able to grow. Thus, in dried foods, spoilage would be expected to be caused primarily by the xerophilic forms.

Although water activity really determines the propensity for microbial growth, to a first approximation water content can be used as a measure of the likelihood of microbial susceptibility. Different foods vary in the minimum moisture content above which spoilage will occur. The minimum level in dry milk is 8 percent water; in dried whole eggs 10 to 11 percent; in flour 13 to 15 percent; in nonfat dry milk 15 percent; in dehydrated fat-free meat 15 percent; in dry beans 15 percent; in other dehydrated vegetables 15 to 20 percent; and in dehydrated fruits 18 to 25 percent. At moisture levels below those values, growth of even xerophilic organisms is not generally possible.

TABLE 16.1 Water activities of common foods and limits for growth of microorganisms

Water activity, a_w	Typical foods	Microorganisms*
1.00	Distilled water	
0.99	Fresh foods: meats, fish, fruits, and vegetables	Many Gram-negative bacteria
0.95	Bread	Most Gram-positive bacterial rods
0.90	Ham	Most Gram-positive cocci, *Lactobacillus*, *Bacillus*, most yeasts, the fungi *Fusarium* and *Mucor*
0.85	Salami, salted meat	*Staphylococcus*, *Saccharomyces rouxii* (in salt), *Debaryomyces* (in salt)
0.80	Jams, jellies, fruitcakes	*Saccharomyces bailii* (in sugar), *Penicillium*
0.75	Salted fish	*Halobacterium*, the fungi *Wallemia*, *Aspergillus*, and *Chrysosporum*
0.70	Dried foods: cereals, grains, dried fruits, candy	
0.65		The fungus *Eurotium*
0.60		*Saccharomyces rouxii* (in sugar), *Xeromyces bisporus*

*Typical microorganisms growing at this water activity or above.

A knowledge of these moisture levels permits selection of the appropriate moisture conditions for storage. Interestingly, a liquid food high in sugar or salt (for example, a syrup or brine) resembles a dry food even though it is high in water, because it has a relatively low a_w. The sugar or salt causes water to pass out of the microbial cell, which then becomes dehydrated and dies. Thus spoilage of syrups is caused by the same kinds of xerophilic fungi that spoil partially dry foods.

One of the most crucial factors for microbial growth in food is **temperature** (Figure 16.1). In general, the lower the temperature of storage, the less rapid the spoilage rate. In cold climates, natural temperatures may be low enough so that foods can be safely stored without refrigeration, at least during the cold part of the year, but in tropical areas, unrefrigerated food may spoil within days or even hours, and only if artificial refrigeration is available can lengthy storage be ensured. In foods stored nonfrozen at refrigeration temperatures, 0 to 7°C, microbial growth can occur, although slowly. The organisms in such foods are called psychrotrophs (able to grow at low temperatures) and include various genera of bacteria and fungi. Some of these organisms cause changes in the flavor and odor of foods, and are therefore considered spoilage organisms. Foods most susceptible to such spoilage are the highly perishable ones, such as meat and milk. Because of the presence and activity of psychrotrophs, meat and milk usually cannot be stored at refrigeration temperatures for longer than a week or so.

Another major factor affecting food microbial growth is the **acidity** of the food, its pH, which we discussed previously in Chapter 4 (see Figure 4.1). Foods vary widely in pH, although most are neutral or acidic. Microorganisms

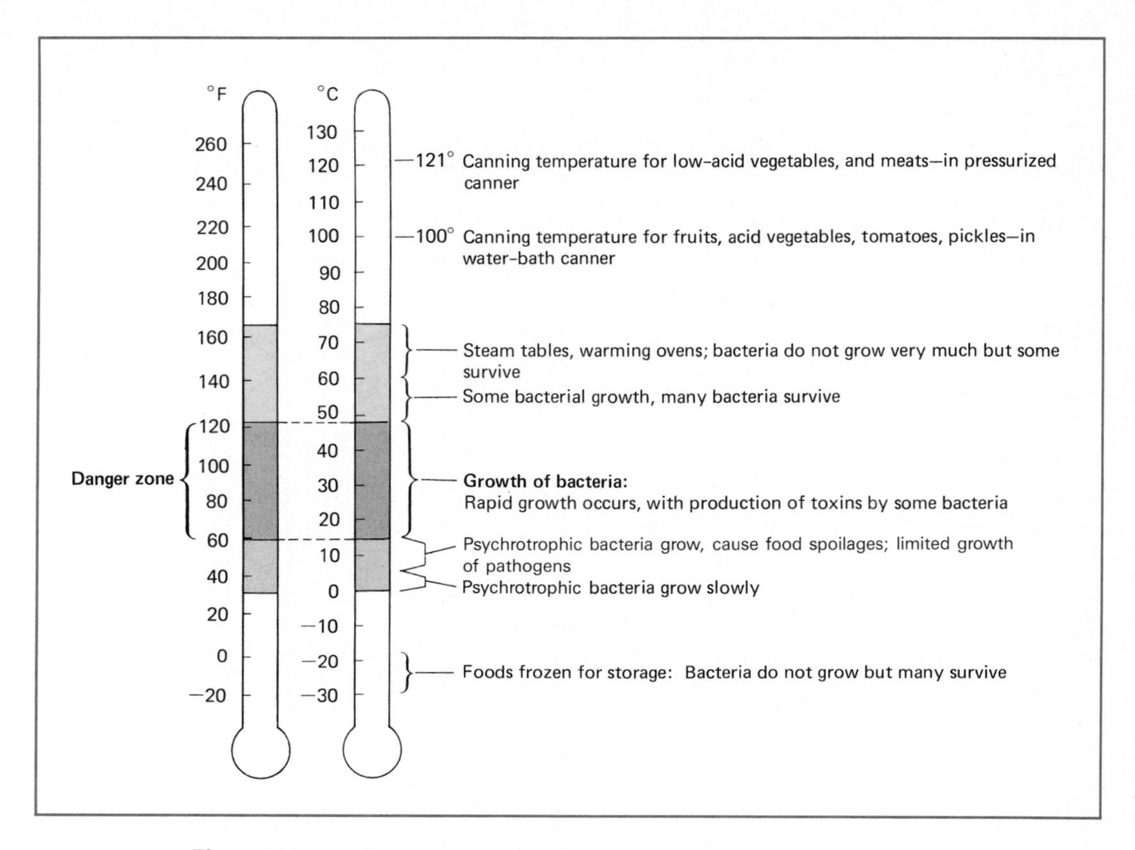

Figure 16.1 Important temperatures in food microbiology.

differ in their ability to grow under acidic conditions. Many bacteria are unable to grow at pH values below 5, whereas a few, (for example, the lactic acid bacteria) can grow at pH values of 4. Below pH 4, only a few rare bacteria are able to grow, and these are not often found in foods. However, at these acid values, many molds and yeasts grow quite well, and it is these organisms which are usually responsible for spoilage of acidic foods.

Oxygen availability influences microbial growth directly, since it is required by aerobes, and indirectly, since it influences the oxidation-reduction potential. The **oxidation-reduction potential** of food influences the kinds of microorganisms that are able to grow. Oxidation and reduction reactions were discussed in Chapter 3. When electrons are transferred from one compound to another, an electrical potential is produced which can be measured, and which can be referred to as the oxidation-reduction potential (O/R potential or E_h). Aerobic microorganisms require positive E_h (an oxidizing environment) for growth, while anaerobes require a negative E_h (a reducing environment). Foods with positive oxidation-reduction potentials include fruit juices and fresh and ground meats. Cheese has an approximately neutral oxidation-reduction potential, whereas aged meat has a negative potential. As would be expected, generally organisms growing in foods with positive E_h are aerobes, those in foods with E_h near zero may be microaerophilic (requiring a small amount of oxygen but being inhibited by larger amounts) and those in foods with negative E_h are anaerobes.

The presence or absence of oxygen in the stored food affects the extent and type of microbial growth. All filamentous fungi require oxygen for growth, as do most yeasts and many bacteria. On the other hand, some bacteria are obligate anaerobes, growing only when oxygen is absent. Canned foods are usually completely anaerobic and hence are usually spoiled only by anaerobes. Foods such as jams and jellies are kept from developing mold growth by covering them with a layer of paraffin, which keeps out oxygen, or by sealing them tightly to avoid entrance of oxygen.

Still another major factor affecting microbial growth in a food is its **physical structure**. A food that consists of small pieces is much more susceptible to spoilage than a food that consists of large pieces, since in the former much more surface is exposed and available for microbial growth. However, some foods have protective coverings such as shells, waxy coatings, or scales that prevent or decrease the extent of microbial invasion. When foods are cut, diced, chopped, or ground, moisture is released and microorganisms are spread more widely in this moisture, greatly increasing the susceptibility of the food to spoilage. For instance, ground beef spoils much more rapidly than the whole beef chunks from which it is prepared.

A number of **chemical properties** of foods affect microbial growth. Foods, of course, provide the nutrients for microbial growth, but foods vary widely in nutritional qualities for both human and microbe. Meats, for instance, are high in protein and fat, while many vegetables are high in starch and cellulose, and fruits may be high in sugars. In the decay of meat, microorganisms that produce enzymes capable of breaking down proteins predominate; such organisms are called *proteolytic*. In the softening or rotting of vegetables and fruits, *pectinolytic* microorganisms, able to break down the pectin materials that hold the plant cells together, predominate.

In addition to providing energy for microbial growth, foods must provide the other essential nutrients such as nitrogen, sulfur, and phosphorus. Nitrogen and sulfur come mainly from amino acids of the food protein, and phosphorus comes from the nucleic acids of the food. Foods are also often rich in vitamins and other microbial growth factors.

16.2 FOOD-BORNE DISEASES

There are many microorganisms which can be identified as causes of food-borne disease. In a recent year the Centers for Disease Control reported 568 outbreaks of food-borne disease in the United States. For 250 of these outbreaks the responsible microorganism could be identified (confirmed outbreaks) while no cause could be identified for the remaining 318 outbreaks. Food-borne disease is usually classified as either *food poisoning*, caused by toxins produced by the microorganism before the food is eaten, or *food infection*, caused by growth of the microorganism in the human body after the contaminated food has been eaten. Table 16.2 shows the incidence of major

TABLE 16.2 Food-borne disease outbreaks due to microorganisms[a]

Responsible microorganism	Outbreaks, percent[b]	Cases, percent[c]
Bacteria		
Salmonella sp.	39	31
Staphylococcus aureus	25	38
Clostridium perfringens	13.5	14
Clostridium botulinum	10.5	0.3
Shigella sp.	5	4.5
Campylobacter jejuni	2.5	7
Vibrio parahaemolyticus	2	0.2
Yersinia enterocolitica	0.5	4
Virus		
Hepatitis A	2	1

[a]Compiled from data provided by the Centers for Disease Control.
[b]Of 600 outbreaks reported in recent years.
[c]Of 8000 cases in a single year.

food-borne disease outbreaks in the United States in recent years. We discuss only the most serious food-borne diseases in this section.

Food poisoning and toxins The growth of microbes in foods, although usually rendering the food unsightly and possibly bad tasting, does not necessarily render it unfit for human consumption. However, several microorganisms produce powerful toxins that can cause serious illness or death, and when one of these toxins is present, the food is highly dangerous. Because of the severity of the effects of these microbial toxins, any food showing obvious microbial growth should be discarded, even if there is no direct evidence that the food is toxic. Unfortunately, not all foods that contain microbial toxins show obvious spoilage, so that food poisoning can develop even from seemingly unspoiled foods.

Botulism is the most severe type of food poisoning; it is often fatal and occurs following the consumption of food containing the toxin produced by the anaerobic endospore-forming bacterium *Clostridium botulinum*. Although botulism is a rare disease, it remains a perennial threat in the United States. During a 70-year period in this century, 659 outbreaks comprising 1,696 cases were reported, resulting in 959 fatalities. The fatality ratio was very high during the first half of this century (60–70 percent), but has fallen to less than 10 percent in recent years. This decline is undoubtedly a result of improvements in diagnosis, availability of *Clostridium botulinum* antitoxins, and increased skill in caring for patients, especially the maintenance of respiratory function by mechanical means if necessary.

Clostridium botulinum normally lives in soil or water, but its endospores may contaminate raw foods before harvest or slaughter. If the foods are properly processed so that the endospores are killed, no problem arises; but if viable spores are present, they may begin to grow and produce toxin. Even a small amount of the resultant toxin (see Section 8.4) can render the food poisonous. Outbreaks of botulism have most frequently been related to home-processed foods, since the processing is sometimes not well controlled, and only a small percentage of cases are caused by commercially processed food. The majority of outbreaks (60 percent) were caused by contaminated vegetables, 13 percent were traced to preserved fruits, and 12 percent to contaminated fish or fish products.

The toxin itself is destroyed by heat, so that properly cooked food should be harmless upon ingestion, even if it did originally contain the toxin. Most cases of botulism occur as a result of eating foods that are not cooked after processing. Canned vegetables and beans are often used without cooking in making cold salads. Similarly, smoked fish and meat and most of the vacuum-packed sliced meats are often eaten without heating. If these products contain the botulinum toxin, ingestion of even a small amount will result in this severe and highly dangerous type of food poisoning.

In addition to causing food poisoning, *C. botulinum* can also produce food infection in infants. *Infant botulism* is caused by growth of the bacteria after ingestion and has only been recognized as a distinct disease since 1976. However, in the United States it is now more prevalent than botulism caused by food poisoning (see Figure 11.9). In 1984, the Centers for Disease Control reported 93 cases of infant botulism and only 20 cases of food-poisoning botulism.

Another *Clostridium* species, *C. perfringens,* causes a milder form of food poisoning, which is not usually fatal. This organism produces an enterotoxin during sporulation in the intestine; the enterotoxin acts on the gastrointestinal tract, causing abdominal cramps and diarrhea. A transient carrier state follows recovery and lasts for some weeks. The organism normally is found in soil and sewage and also occurs in the intestinal tract of humans and animals. There are some strains which have especially heat-resistant endospores, and it is these that most commonly cause this type of food poisoning. The sequence of events that can lead to *C. perfringens* food poisoning is shown in Figure 16.2. The food may be infected before it is processed, or noncontaminated food may come into contact with contaminated dishes, utensils, or food workers; the organisms thus introduced multiply before the food is eaten. The organism is quite commonly found in cooked and uncooked foods, especially raw meat, poultry, and fish, in most dust samples, and in many water supplies. However, ingestion of a large dose of *C. perfringens* (10^8 to 10^9 organisms) is required for food poisoning to occur. Thus, although small numbers of the organism may frequently be consumed, with proper food processing and handling the organisms do not multiply to the point where they constitute an infective dose.

Staphylococcal food poisoning is the most common type of food poisoning and is caused by certain strains of *Staphylococcus aureus.* These strains produce an enterotoxin that is released into the surrounding food; if food containing the toxin is ingested, severe reactions are observed within a few hours, including nausea with vomiting and diarrhea. The kinds of foods most commonly involved in this type of food poisoning are custard- and cream-filled desserts, poultry, meat and meat products, gravies, egg and meat salads, puddings, and creamy salad dressings. If such foods are kept refrigerated after preparation, they remain relatively safe, as *Staphylococcus* is unable to grow at low temperatures. In many cases, however, foods of this type are kept

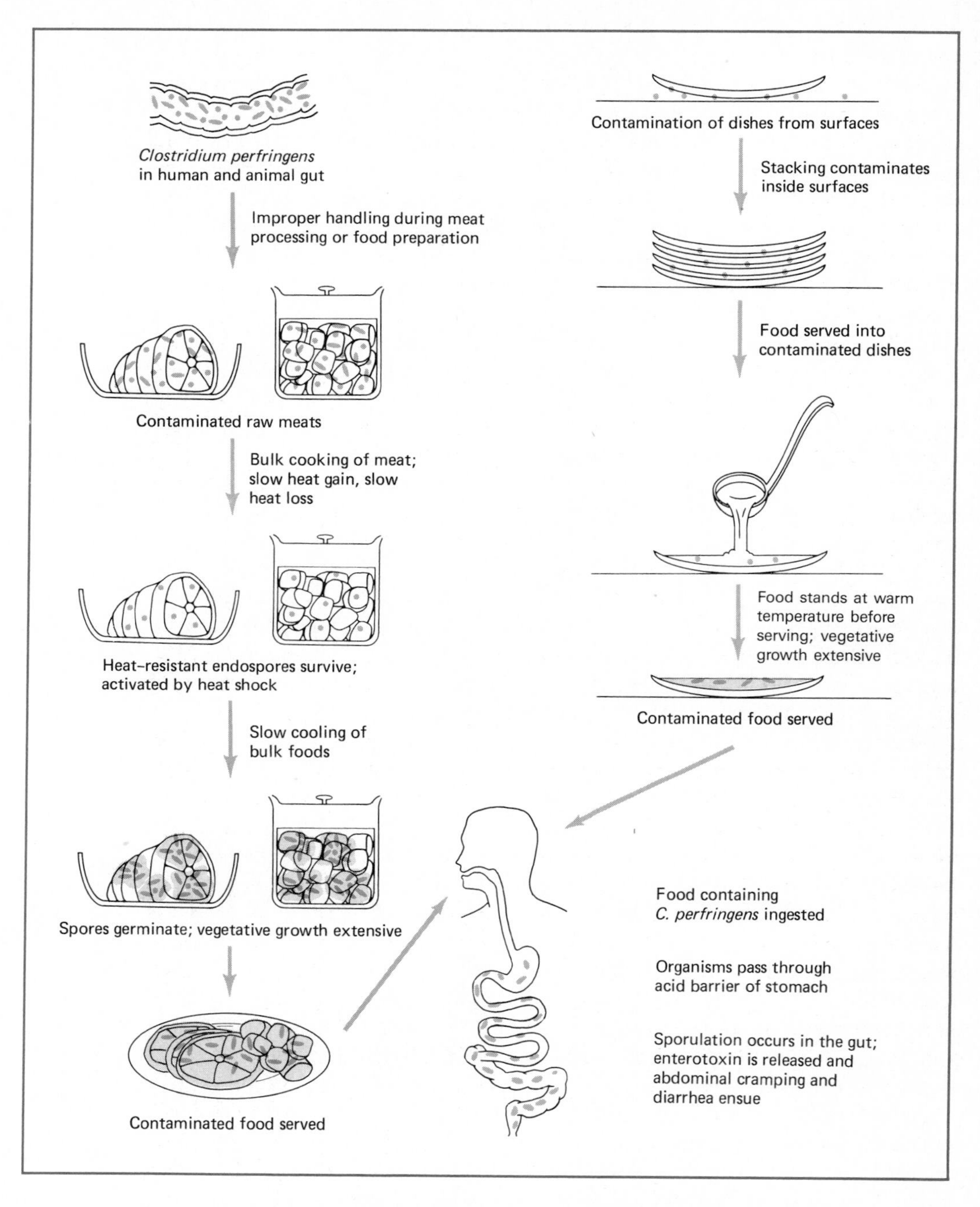

Figure 16.2 Sequence of events that may lead to *Clostridium perfringens* food poisoning.

warm for a period of hours after preparation, such as in warm kitchens or outdoors at summer picnics. Under these conditions, *Staphylococcus*, which might have entered the food from an infected food handler (a carrier) during preparation, grows and produces enterotoxin. Many of the foods involved in staphylococcal food poisoning are not cooked again before eating, but even if they are, this toxin is relatively stable to heat and may remain active.

Staphylococcal food poisoning often occurs in small epidemics among people eating at a common location, such as a banquet, dormitory, or large picnic. When foods are prepared in large quantities, facilities are often not available to keep the cold foods cold and the hot foods hot, so that foods are often left at warm indoor or outdoor temperatures. Such warm temperatures are appropriate for staphylococcal growth, which ensues if the inoculum is present. Staphylococcal food poisoning can be prevented by careful sanitation methods so that the food does not become inoculated, by storage of the food at low temperatures to prevent staphylococcal growth, and by the disposal of foods stored for more than a short time at warm temperatures. It should be emphasized that the source of inoculum in staphylococcal food poisoning is most likely to be the person who prepared the food. As we have discussed in Section 11.2, *S. aureus* is a common inhabitant of the upper respiratory tract. In this habitat, the enterotoxin involved in food poisoning is not a factor. The food handler is generally a symptomless carrier, and it is only when the bacterium is transferred into the food during the food preparation process that the events are initiated that lead to staphylococcal enterotoxin production.

Salmonella food infection Although sometimes called a food poisoning, gastrointestinal disease caused by food-borne *Salmonella* is in reality a **food infection** because symptoms arise only after the pathogen has grown in the intestine. Virtually all species of *Salmonella* are pathogenic for humans: one *S. typhi*, causes the serious human disease typhoid fever, and a number of other species cause food-borne gastroenteritis (see Section 11.10). The organism reaches food by contamination from food handlers; or in the case of such foods as eggs or meat, the animal that is the source of the food may be contaminated (Figure 16.3). The foods most commonly involved are meats and meat products (such as meat pies and sausage), poultry, eggs, and milk and milk products. If the food is properly cooked, the organism is killed and no problem arises, but many of these products are eaten uncooked or partially cooked. *Salmonella* causing food-borne gastroenteritis is often traced to products made with uncooked or slightly-cooked eggs, such as custards, cream cakes, meringues, pies, and eggnog. Previously cooked foods that have been warmed and held without refrigeration or canned foods held for a while after opening often support the growth of *Salmonella* if they have become contaminated by an infected food handler. Cutting boards and utensils used to prepare food such as poultry which may contain *Salmonella* may be improperly washed after use and thus contaminate new foods.

In order for intestinal infection to occur, a significant number of viable cells must be ingested. The number of organisms required for infection depends on the *Salmonella* species and the host's susceptibility, but usually ranges from 100,000 to 50 million. It is likely that mildly contaminated foods will not often cause the disease, and this therefore usually means that the organism must multiply in the food before it is eaten in order for an infection to occur. However, a smaller number of bacteria may cause the disease in infants and small children as well as in certain elderly adults. For this reason,

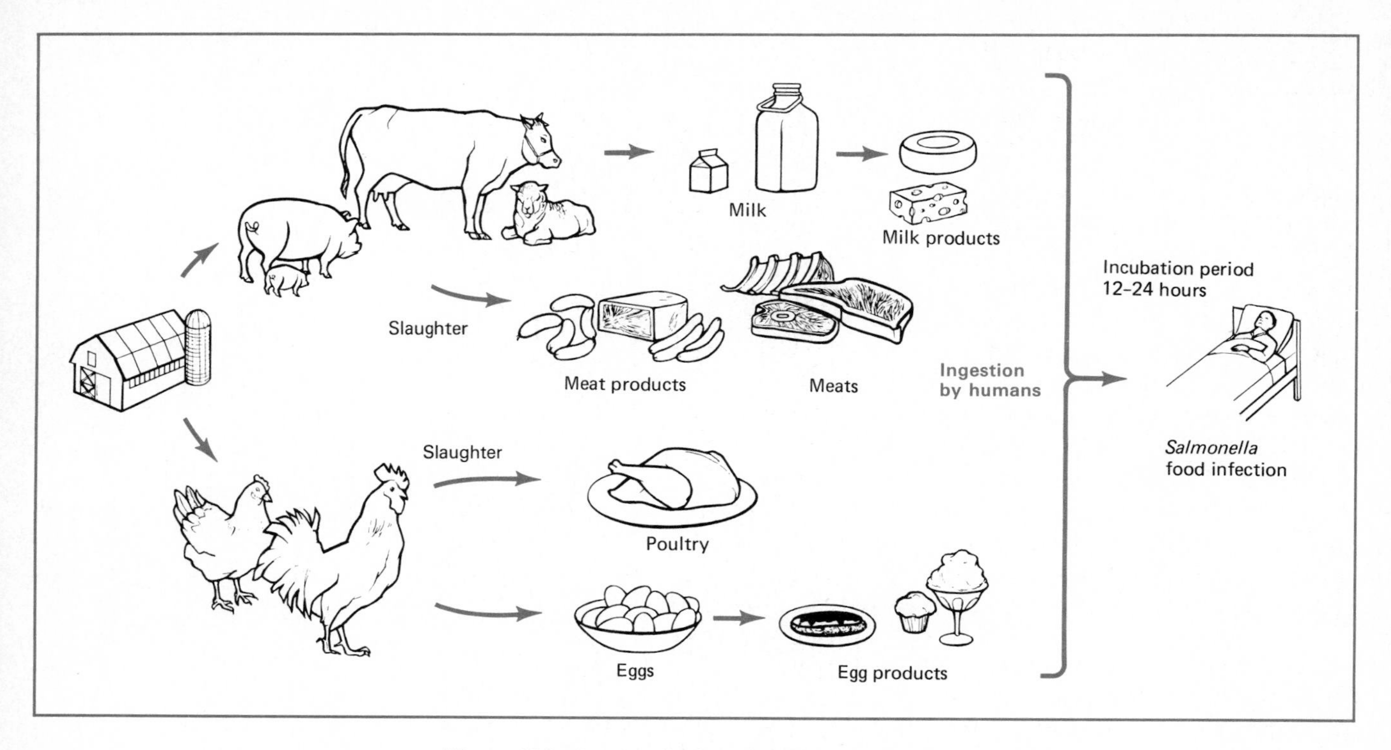

Figure 16.3 Spread of *Salmonella* from food animals to humans.

the presence of salmonellas at any level in food should be considered unacceptable and should be regarded as a health hazard.

Growth occurs most rapidly if the food has been stored under warm conditions. The length of time following ingestion before symptoms appear is usually 12 to 24 hours, which is longer than the time required for *Staphylococcus* food poisoning to appear. This longer incubation time is due to the fact that *Salmonella* must multiply before symptoms develop. The principal symptoms are nausea, vomiting, abdominal pain, and diarrhea, usually preceded by headache or chills. The disease is rarely fatal, and recovery usually occurs, even without treatment, after 2 to 3 days. Proper food sanitation practices minimize the incidence of this food-borne disease.

In summary, two kinds of illness arise from eating food, food poisoning and food infection. With a *food poisoning*, the bacteria grow in the food before it is cooked and produce a toxin, and it is the toxin which causes the harm. In the case of a *food infection*, the food is just the source of the bacteria that then grow in the body. Food poisoning is controlled by preparing and storing food under conditions that prevent microbial growth. Food infection is controlled by preventing the food from becoming contaminated in the first place.

Investigating an outbreak Once a diagnosis or tentative diagnosis of food-borne disease has been made, the cause of the outbreak must be determined. The purpose of an investigation is to determine what conditions led to the outbreak and, more importantly, to gather information that will

TABLE 16.3 Food-specific attack rate table

Foods	No. of persons who ate specific foods				No. of persons who did not eat specific foods				Difference in percent
	Sick	Well	Total	Percent sick	Sick	Well	Total	Percent sick	
Braised beef	74	17	91	81	2	9	11	18	+63
Peas	48	20	68	71	28	6	34	82	−11
Cabbage salad	36	12	48	75	40	14	54	74	+1
Buttered biscuits	46	12	58	79	30	14	44	68	+11
Peaches	62	22	84	73	14	4	18	78	−5
Milk	60	16	76	79	16	10	26	62	+17

The data are for a single outbreak of food poisoning. The analysis suggests that braised beef is the responsible food (see text).

help prevent a recurrence. All facts and samples must be collected and studied as soon as possible after the outbreak occurs. If the outbreak is severe, a trained epidemiologist from the state or local health department should head the investigation:

1. *Case histories* should be obtained. The suspect food may have been widely distributed in homes and restaurants via a common supplier, or it may have been quite localized, as at a banquet or picnic. If a particular meal is suspected, case histories should be taken from both those who became ill and those who did not.
2. *Careful inspection* of all food preparation methods should be made, and full information as to food storage, health of food workers, menu, and so on should be obtained.
3. *Microbiological analysis* of leftover foods should be taken to corroborate the findings.

After all the information has been obtained, an attack rate table is prepared (Table 16.3). From such a table, the food responsible for the outbreak may be identified, as the attack rate would be much higher for persons who ate the contaminated food than for those who did not. In the case shown in Table 16.3, the attack rate for those eating braised beef was 81 percent whereas for those not eating it, the rate was 18 percent. The difference in attack rates between eaters and noneaters is the significant factor. For other noncontaminated foods, eaters and noneaters had about the same rates.

16.3 ASSESSING MICROBIAL CONTENT OF FOODS

All fresh foods have some viable microbes present. The purpose of assay methods is to detect evidence of abnormal microbial growth in foods or to detect the presence of specific organisms of public health concern, such as *Salmonella*, *Staphylococcus*, or *Clostridium botulinum*. For solid foods, preliminary treatment is usually required to suspend in a fluid medium those mi-

TABLE 16.4 Suggested microbiological guidelines for selected foods*

Food	Plate count no./gram	Coliforms no./gram
Precooked frozen foods	100,000	10
Frozen pies	50,000	50
Gelatin	3,000	10
Raw meat, fresh or frozen	5,000,000	50
Cooked meat	1,000,000	10
Ice cream and frozen desserts	50,000	20

*The guidelines vary widely from one state to another; there are no federal standards.

croorganisms embedded or entrapped within the food. The most suitable method for treatment is high-speed blending. Examination of the food should be done as soon after sampling as possible, but if it cannot begin within 1 hour of sampling, the food should be refrigerated. A frozen food should be thawed in its original container in a refrigerator and examined as soon as possible after thawing is complete.

A plate count is frequently done to determine the total viable count of organisms in the food. This will give some indication of the overall quality of the food. Appropriate dilutions are made and samples are plated onto a standard culture medium. The plates are incubated at appropriate temperatures: 0 to 5°C for psychrotrophs, 30 to 35°C for mesophiles, and 55°C for thermophiles. Table 16.4 presents some suggested standards in effect in various states. It should be noted that these figures are recommended only; compliance is voluntary. For assay of specific food-borne pathogens, more specialized tests are made. In some cases the pathogen must first be enriched in a special medium which will enhance its growth and increase its numbers to the point where there are enough cells to be identified. Then a selective medium is usually employed so that the suspected pathogen will be easily identified. The particular media employed of course depend on which organism is being assayed.

A number of serologic methods are available for the detection of microbial products in foods. These include methods employing antibody specific for the toxin, for instance *staphylococcal enterotoxins* or *botulinum toxin*. The liquid containing the suspected toxin is allowed to react with known antibody and precipitation will occur if toxin is present. Even more sensitive than precipitation are fluorescent antibody (FA) and radioimmunoassay (RIA) techniques (see Chapter 9), which have been applied to the assay of staphylococcal enterotoxins. These procedures are extremely sensitive and quite rapid, requiring only 3 to 5 hours. Special procedures used in the enumeration of microorganisms in milk and meat are discussed in Sections 16.6 and 16.7.

16.4 FOOD PRESERVATION

From a knowledge of the factors involved in food spoilage, as outlined in Section 16.1, effective measures for food preservation can be devised. It has been in part through the development and perfection of effective methods for food preservation that human civilization has been able to develop into the highly industrialized and urbanized system found today in most parts of the world. The major methods of preserving foods are drying, refrigeration and freezing, pickling, preserving with sugar or salt, canning, and addition of chemical preservatives. Each of these methods has certain advantages and certain disadvantages when used with various types of foods.

Drying The simplest form of food preservation is drying. The method works because of the simple fact that microorganisms need water to grow and metabolize. If a food is dried to a point where it contains less water than the level needed by an organism for growth, then there can be no growth of the organism. As we saw in Table 16.1, there are some organisms which can grow at relatively low moisture levels, but between an a_w of 0.7 and one of 0.6 only a very few organisms can grow, and below 0.6 none can grow. Dried and freeze-dried foods have an a_w in these low ranges or below, and contain less than 25 percent water; their shelf life is very long and no microbial growth can occur as long as the product remains dry. The fact that no microorganisms can grow in a dried food does not mean that there are no organisms present, however. Although some organisms are destroyed during a drying process, many types can be recovered from dried foods. If the foods are stored under conditions of high relative humidity, or if the foods are rehydrated and allowed to stand at favorable temperatures, then microbial growth may certainly occur, with attendant loss in food quality.

Sun-drying is the oldest form of food preservation, and is still used for drying many fruits such as apples, plums, apricots, currants, and raisin grapes (Figure 16.4). The product is spread out in a thin layer on a cloth or plastic screen in the sun. Fish is sun-dried all over the world, generally in conjunction with salting, brining, or smoking. Artificial drying, or *dehydration*, has widely replaced sun-drying where fuel energy is available. Foods are placed on racks and dehydrated in cabinet or tunnel dryers, in which dry warm air is circulated around the food.

Spray-drying is extensively used for the large-scale drying of liquid foods, such as milk, eggs, fruit juices, coffee, and tea. The product is sprayed into the top of a cone-shaped chamber; heated air in the chamber dries the food, which is collected at the bottom (Figure 16.5*a*).

Drum-drying is commonly used to dry milk, pulped fruit, potato flour, and various prepared foods. The food is fed onto the outside of a large, rotating, heated drum; the food dries as the drum rotates, and is scraped off (Figure 16.5*b*). More rapid drying at lower temperatures can be attained if the drum drying is done in a vacuum chamber (Figure 16.5*c*). *Freeze-drying*, or *lyophilization*, is a process in which the product is frozen and the moisture removed under vacuum while the food product remains in the frozen state. The advantage of freeze-drying is that changes in flavor and texture are kept to a minimum. The disadvantage is that the process is expensive.

Once a product is dried, it must be wrapped and sealed in such a way that it remains dry. Plastic and cellophane film are the cheapest wrapping materials, but they are easily broken and hence must be packed in a protective cover. A dried product kept dry should remain free of microbial growth in-

Figure 16.4 Sun-drying of apricots.

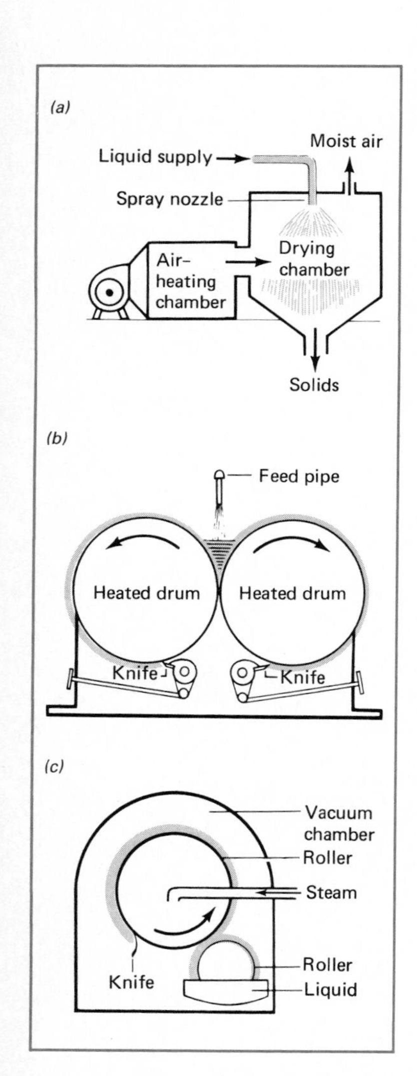

Figure 16.5 Drying methods for liquid food products. (a) Spray-drying. (b) Drum-drying. (c) Vacuum-drying.

definitely. With time, its flavor and nutritional qualities will probably deteriorate because of chemical changes, usually oxidation by air, but a dried product stored in the complete absence of air may remain unchanged for an indefinite period.

Low-temperature storage Storage at temperatures just above freezing (2 to 5°C) greatly prolongs the shelf life of foods. Such temperatures are easily achieved with mechanical refrigeration powered by electricity or gas. The availability of mechanical refrigeration has considerably changed eating habits, as it has made possible the storage of many potentially perishable foods. Also, refrigerated transport by truck, rail, and air has made possible the distribution of fresh foods to areas quite distant from those where they are raised. Storage at these temperatures for extended periods is not possible, as spoilage will eventually occur, usually as a result of growth of psychrotrophic microorganisms.

Long-term storage can be carried out, however, if temperatures below freezing are used. Foods freeze at temperatures below 0°C, but complete and solid freezing occurs only at temperatures below about −20°C. It should be emphasized that even if a food looks solidly frozen, it may have pockets of liquid where microbial growth can occur. Storage of foods in the frozen state is one of the most effective means of ensuring long-term storage, although freezing may alter the physical and chemical characteristics of foods. Quick freezing is usually more effective than slow freezing in preserving the quality of the food. Note that freezing does not kill all the microorganisms present in the food but may actually maintain their viability. Thus, foods should be frozen soon after harvest, before microbes have had a chance to grow. During long-term frozen storage, care should be taken to ensure that the food does not thaw and refreeze, since during the thaw cycle microbial growth may occur.

Pickled or fermented foods Acid is often used to prevent the growth of microorganisms in perishable foods; this process is usually called *pickling*. Foods commonly pickled are cucumbers (sweet, sour, and dill pickles), cabbage (sauerkraut), and some meats and fruits. The food can be made acid either by addition of vinegar or by allowing acidity to develop directly in the food through microbial action, in which case the product is called a *fermented* food. In addition to the increased keeping qualities of pickled foods, their flavors are often distinctive, and they are delicacies in many cuisines.

The microorganisms most commonly involved in food fermentations are the lactic acid bacteria, the acetic acid bacteria, and the propionic acid bacteria. In most cases, the acids formed are the result of fermentation of the sugars occurring in the foods. The microorganisms involved in the fermentation may be present in the raw food when it is harvested from the field, or they may be added as cultures called *starter cultures*.

When green vegetables such as cabbage, lettuce, and spinach are shredded or chopped and allowed to stand, a lactic acid fermentation usually develops as the naturally occurring lactic acid bacteria act upon the sugars in the liberated plant juices. Salt is usually added to such plant materials at the start of the process to prevent the growth of other bacteria, the lactic acid bacteria being more resistant to inhibition by salt than are most spoilage organisms. The amount of acid that develops depends on the amount of sugar in the food. If the natural sugar content is low, sugar can be added to increase the acid output. The most commonly fermented green vegetable is cabbage, and the product obtained is sauerkraut. Another common food fermentation is that in which cucumbers are converted to pickles. The cucumbers are placed in a salt solution in large vats and allowed to ferment for several weeks or longer, during which time the lactic acid and flavor produced by the microorganisms becomes absorbed by the cucumbers, greatly altering their flavor, color, and texture.

Pickling of foods can also be carried out by mixing the product directly with vinegar, in which case the process is not really a fermentation. However, vinegar itself is a microbial product, made by the action of acetic acid bacteria on alcoholic juices such as wine or alcoholic apple juice (cider). The industrial manufacture of vinegar will be discussed in Chapter 18.

Preserves, jellies, and salted products Because of the dehydrating effect of syrups (discussed earlier in this chapter), foods soaked in these solutions can be preserved for long periods of time. Foods preserved in this way are mainly fruits. Jellies are made from fruit juices by adding sugar and a gelling agent such as pectin. Preserves are made by cooking fruit pieces in a sugar syrup, during which process the natural pectins of the fruit are extracted and cause gelling when the product is cooled. Both preserves and jellies are subject to spoilage by the growth of fungi (molds) tolerant of high sugar concentrations. Since all these molds are aerobes, they grow only on the surface of the product, and such surface growth can be prevented by sealing the container either with a tightly fitting lid or with a layer of paraffin.

Canning Canning is a process in which foods sealed in cans are heated so as to kill all living organisms or at least assure that there will be no growth of residual organisms in the can. When the can is properly sealed and heated, the food should remain stable and unspoiled for a prolonged period, and can be stored without refrigeration. However, foods stored for extremely long periods sometimes show loss in quality due to chemical changes or growth of residual organisms. Home-canned foods are usually prepared in glass containers, whereas commercial products are most often in tin-coated steel cans. In any canning process, it is important to select containers that do not leak and to seal them properly. Glass containers for home-canned foods are usually sealed with lids containing rubber gaskets. The filled jars are heated for the proper length of time at the proper temperature, then cooled and sealed. After sealing, the jars should be checked for leaks before being stored in a cool, dry place. Commercial cans must be constructed of tin plate sufficiently heavy so that leaks do not occur, and the cans must be carefully inspected for holes, pits, or signs of corrosion. During canning operations, frequent inspection of can seams (Figure 16.6) must be made to be sure that the can-sealing equipment is working properly.

For processing, all nonacid foods must be heated under pressure, whereas acid foods can be processed in a boiling water bath. Time and temperature

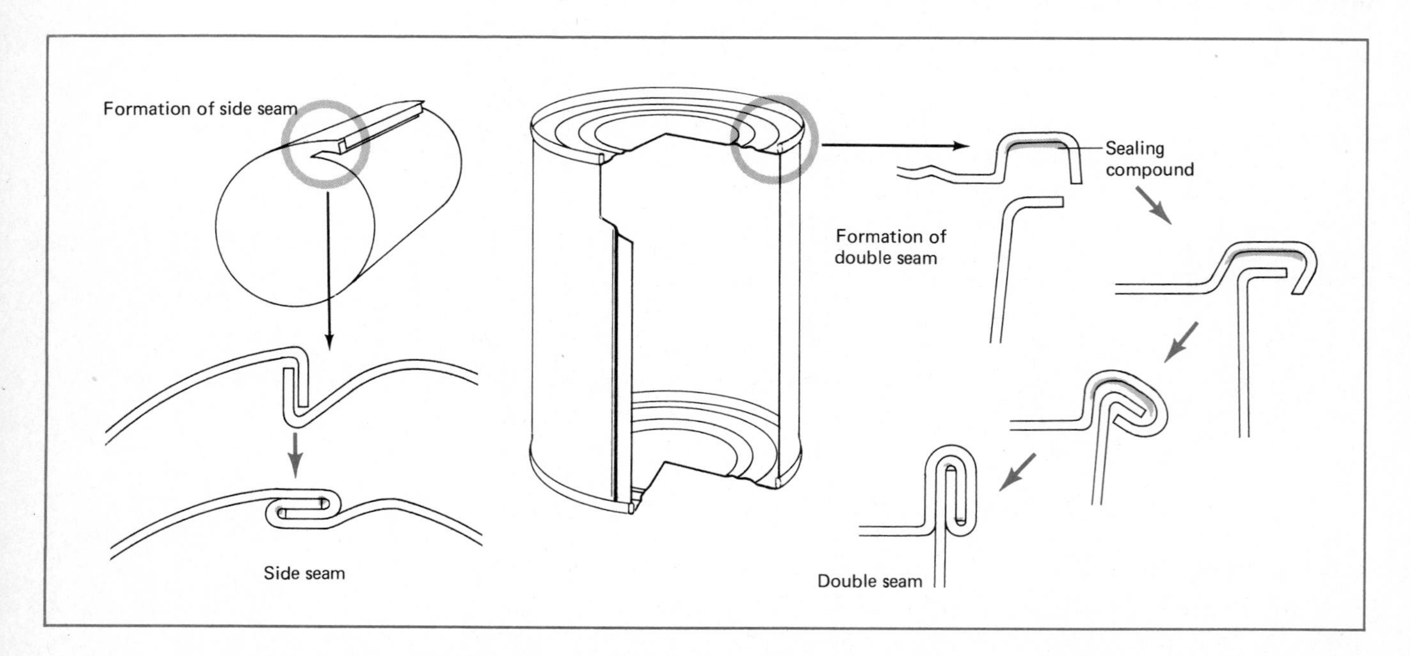

Figure 16.6 Construction and sealing of a metal can.

for heating vary depending on ease with which the organisms can be eliminated. Acid foods can be preserved by heating at 100°C for 15–20 minutes. Nonacid foods must be heated at autoclave temperatures (116–121°C) and for prolonged periods of time, 50–90 minutes. For preparation in the home, instruction for specific foods can be found in home-canning guides. It is not desirable to heat foods much longer than necessary, since prolonged heating affects the nutritional and eating qualities of the food. It should be noted that "sterilize" is not an entirely accurate term to use in relation to canning. The numbers of organisms are reduced greatly during the heating process, and in most cases the product probably is sterile, but in fact if the initial load of organisms in the food is high, then not every cell may be killed. Heating times long enough to guarantee the absolute sterility of every can would alter the food so greatly that it would be unpalatable. Thus a fine balance is kept among reduction in microbial numbers, eradication of all pathogens, and palatability.

Commercial canning operations (Figure 16.7) consist of the following stages: (1) *Blanching,* a preliminary treatment of the product with hot water or live steam. It is done with all vegetables, the purpose being to soften the tissues so that the can may be filled properly, to eliminate air from the product so that there is less air in the can, and to destroy enzymes of the vegetable that might cause undesirable changes in the flavor, texture, or color of the product before canning. (2) *Cleaning* the cans. (3) *Filling* the cans. It is important that each can be as full as practicable, to eliminate the dead space at the top. (4) *Exhausting* the can to remove air. This can be done either by heating the filled can for a few minutes or sucking out the air with a vacuum pump. It is important to have a vacuum in the final sealed can, since this ensures absence of oxygen and, hence, a reduction in corrosion and spoilage.

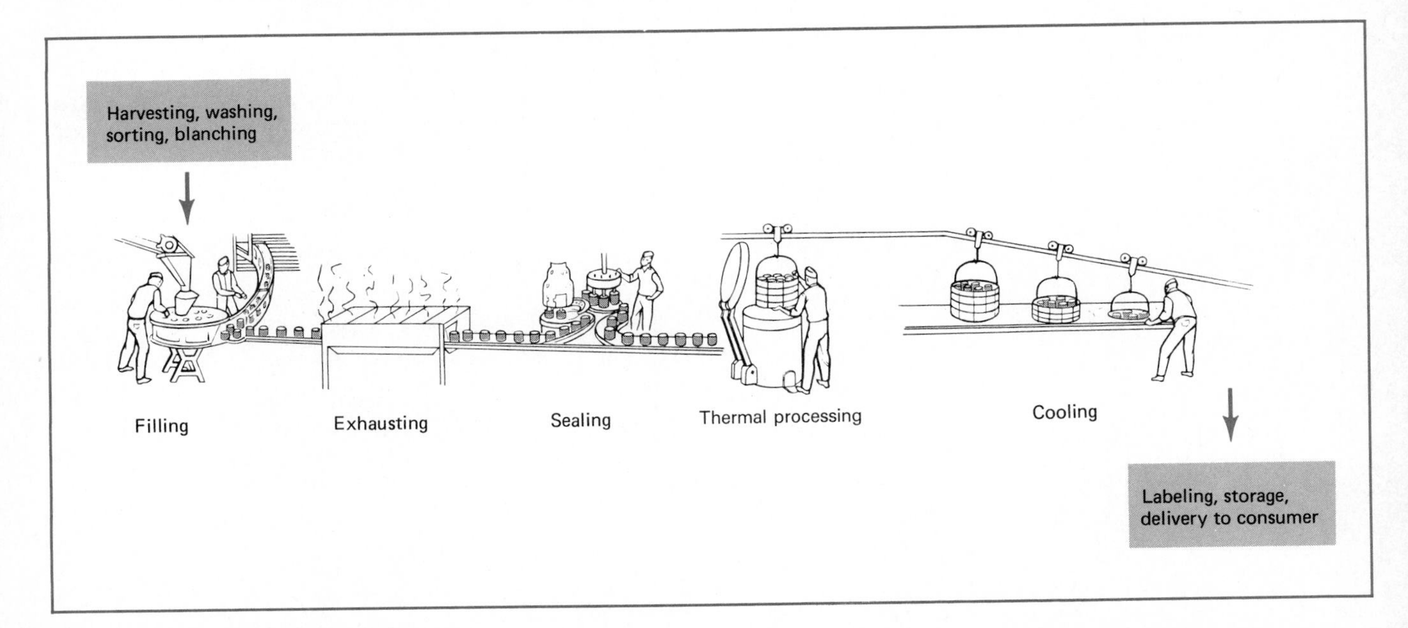

Figure 16.7 Commercial canning operations.

(5) *Closing* the can. This is usually done by machine. It requires construction of a good seal on the seams (see Figure 16.6), and the proper operation of the sealing machine must be checked from time to time. (6) *Marking* the can with a code number to indicate the product, date of packing, and the grade. This ensures identification of the can even in the absence of a label and, in the case of a processing defect, permits determination of when and where the product was canned so that other cans of the same processing batch can be identified and eliminated from market. (7) *Thermal processing* of the cans by heating in a steam bath or autoclave. (8) *Cooling* the cans, preferably under controlled decrease of pressure to prevent buckling. Cooling by immersion in water can be a serious source of contamination, since during cooling, some water may be sucked into the can through any tiny holes present, and if this water has a microbial load, some organisms could enter the can. All cooling waters should be chlorinated to maintain at least 1 ppm residual chlorine at the discharge end of the can cooler. (9) *Handling and storage* of the cans under such conditions that contamination does not occur. Dropping, bumping, or rattling of cans should be avoided, since any impact to the can may lead to the development of leaks. The outsides of the cans should be dried immediately after leaving the cooling system, and the cans should be stored dry, since bacteria then cannot develop on the outsides of the cans. If completely dry conditions cannot be maintained, water containing 3 to 5 ppm residual chlorine should be continuously sprayed.

Bacterial contamination can be greatly reduced by proper design and operation of a canning factory. Cleanliness is the prime factor, but elimination of equipment that could permit bacterial growth and hence be a source of contamination is also important. Microbial spoilage may be due either to improper heating or to post-processing leakage. By determining the kinds of bacteria responsible for the spoilage, these two causes can be distinguished, since spoilage due to improper heating will virtually always be caused by a single heat-resistant, endospore-forming species; whereas if leakage occurs, a mixture of nonspore-forming bacteria will usually be found. A major role of the microbiologist in the canning industry is in the investigation of spoilage problems so that recommendations can be made to rectify deficiencies in the canning process.

When spoilage problems are being investigated, the can should first be examined externally for abnormal conditions (Figure 16.8). The canning industry has evolved terms for different abnormalities. *Flippers* are cans that appear flat but have one end that becomes forced out when the can is knocked against a flat surface. The flipper represents an early stage of contamination by a gas-producing microorganism, enough gas having been produced to relieve the vacuum in the can. The *springer* represents a later stage of gas accumulation in which one end of the can is permanently bulged, and if pressure is applied to this end, it will slip in but the other end will flip out. A *soft swell* is a can with both ends bulged, but not so tightly that the ends cannot be pushed in with the thumbs, and a *hard swell* is a can bulged at both ends so tightly that no indentation can be made with thumb pressure. If sufficient gas pressure builds up, *bursting* may occur at one of the seams.

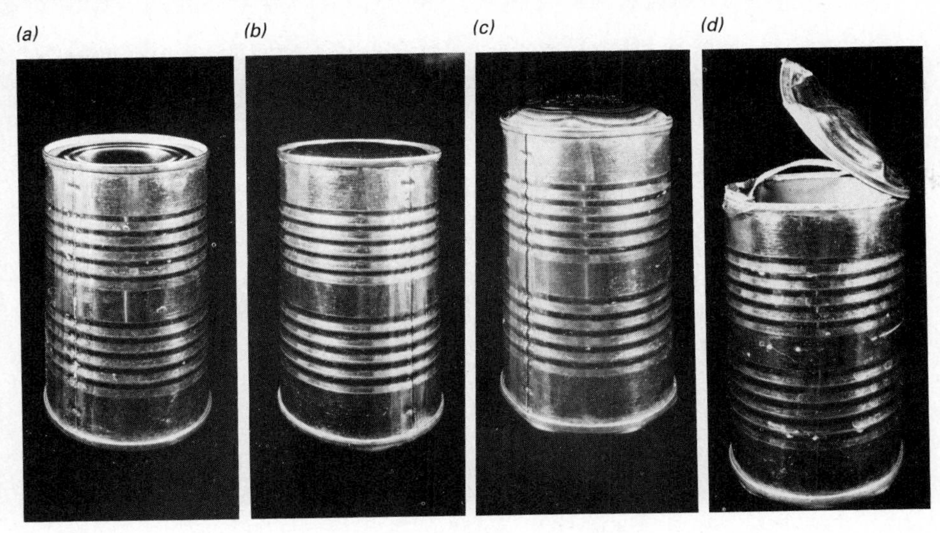

Figure 16.8 Changes in cans as a result of microbial spoilage. (a) Normal can; note that the top of the can is indented due to negative pressure (vacuum) inside. (b) Slight swell resulting from minimal gas production. Note that the lid is slightly raised. (c) Severe swell due to extensive gas production. Note the great deformation of the can. This can is potentially dangerous, and could explode if dropped or hit! (d) The can shown in (c) was dropped and the gas pressure resulted in a violent explosion. Note that the lid has been torn apart.

A wide variety of spoilage problems can develop, and the canner should be alert to all of them. If spoilage does occur, a careful analysis of the canning operation is necessary to identify the source of difficulty. Even if spoilage is not occurring, routine bacteriological checks on cans removed at random from the processing line should be carried out, especially for those items that are difficult to sterilize. Many canners produce very large numbers of cans in central plants for shipping long distances. Quality control is essential if consumers are to be confident that they are buying a reliable, entirely safe product. If it should happen that the deadly toxigenic bacterium *Clostridium botulinum* develops in the product, the whole output from the same processing operation must be destroyed, a quantity that could amount to many thousands of cans. Alert bacteriological control of the canning process is a proper and inexpensive way to avoid such difficulties.

Chemical food preservation Although chemicals should never be used in place of careful food sanitation, there are a number of antimicrobial chemicals that find wide use in the control of microbial growth in foods. These are summarized in Table 16.5. The chemicals listed in Table 16.5 are all in a category called by the Food and Drug Administration, "Generally Regarded as Safe" (GRAS). These chemicals have been used for many years with no

TABLE 16.5 Chemical food preservatives

Chemical	Foods
Sodium or calcium propionate	Bread
Sodium benzoate	Carbonated beverages, fruit juices, pickles, margarine, preserves
Sorbic acid	Citrus products, cheese, pickles, salads
Sulfur dioxide	Dried fruits and vegetables
Formaldehyde (from food-smoking process)	Meat, fish
Sodium nitrite	Smoked ham, bacon

evidence of human toxicity. However, a newly available chemical could not be used in food without lengthy toxicity testing to determine whether its use involved any risks.

16.5 FOOD SANITATION

In restaurants, cafeterias, and other public food services, large numbers of people are served food prepared in central kitchens. Because of the dangers of food-borne poisonings and infections, good sanitation practices are necessary. Effective sanitation begins with cleanliness but extends beyond this to the employment of practices that prevent excessive microbial contamination of foods, utensils, and preparation equipment (pots, kettles, grinders, fryers, and the like). The word *sanitize* is often used in the food industry to describe treatments of equipment and utensils so as to destroy microorganisms. Such treatments rarely sterilize but do reduce the microbial load and virtually always kill all potential pathogens present.

Any establishment offering food to the public should have an effective food-sanitation program. Such a program begins with proper control and training of the food handlers. Food handlers must be free of infectious disease, especially diseases such as typhoid fever, diphtheria, and tuberculosis, the causal agents of which can easily be transferred from the body to food. Discharging wounds or lesions are potentially dangerous since they may be a source of contamination of food with food-poisoning staphylococci. Routine medical examinations of food handlers may be of value in detecting infection, but of even greater importance is the education of the food handlers to make them aware of the importance of their health and of proper personal hygiene. Infected persons should not handle food, utensils, or equipment, and managers of food establishments should ensure that employees are aware of their responsibilities in this regard.

The best quality food available should be used, and it should be handled and stored in such a manner as to prevent spoilage. Prepared foods present the biggest problem: to be available for quick service, they must be stored in large quantities considerably ahead of time, thereby presenting ample opportunity for spoilage. For refrigeration, such prepared foods should be placed in shallow pans, not over 3 inches deep, so that they will cool quickly. Salads and other foods should not be mixed with the hands, but with a large spoon or wooden paddle. Vegetables to be eaten raw should be of high quality and should be thoroughly scrubbed under running water before serving. Steam tables and other devices for keeping food hot before serving should be kept clean and should be drained and cleaned completely once a day. The temperature of the foods on a steam table should always be maintained above 65°C.

In most urban areas, routine inspection of eating establishments is carried out by public-health officials. The importance of frequent, meaningful inspections should be obvious. A standard reporting form is generally used, and any infringements of prescribed procedures are reported to the management and to the appropriate government agency.

16.6 DAIRY MICROBIOLOGY

Milk and milk products are excellent, high-quality foods, providing both nutritional and culinary values. However, milk is extremely susceptible to spoilage by microorganisms, and the microbiologist plays a major role in the dairy industry in the control of milk quality. Further, microorganisms are used in a beneficial way in the manufacture of many dairy products, including buttermilk, yogurt, cheese, and sour cream; but proper management of microorganisms in the production of these items is required to ensure quality and dependability of the product. Finally, milk may be a major agent in transfer of disease-causing microorganisms from cows to humans, and great care must be taken to ensure that milk and milk products are free of such pathogens.

In modern dairy operations (Figure 16.9), cows are milked by machine, and the milk is conveyed to holding tanks where it is immediately cooled and stored until it can be transported to the dairy plant. At the plant, the milk is pasteurized, bottled, and held for distribution to the consumer. Some dairy plants also process milk for use in making various products such as cheese, dry milk, or butter. At each stage on the way from cow to consumer, the milk must be carefully handled and checked to be sure that its microbiological quality is maintained.

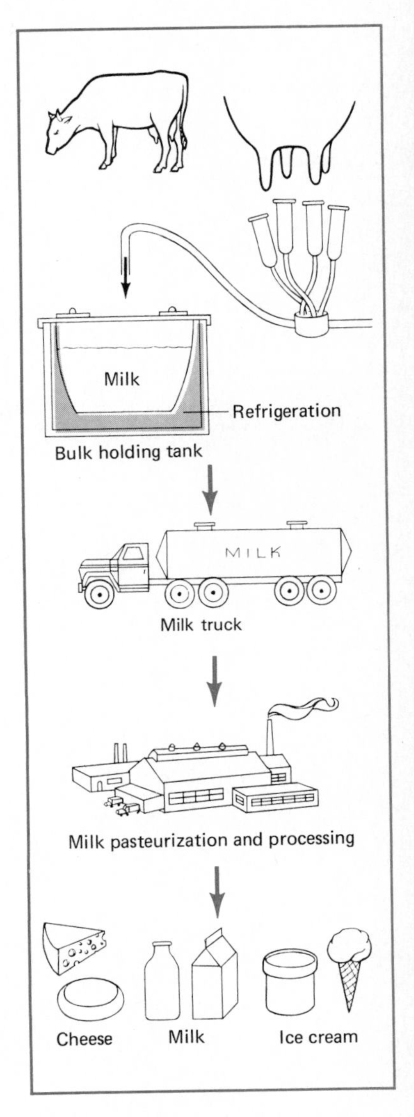

Figure 16.9 Steps in milk processing.

The nature of milk Cow's milk consists of a complex mixture of constituents including fat, protein, and carbohydrate; the most plentiful component is water, which comprises 87 percent of the weight of milk (Table 16.6).

Milk is an ideal medium for the growth of microorganisms because it is rich in microbial nutrients, and even nutritionally exacting (*fastidious*) microorganisms are able to grow successfully in milk. However, there are some peculiarities about the composition of milk that should be noted. The sugar of milk is lactose, and many microorganisms are unable to utilize it as a source of carbon and energy. Those organisms that can utilize lactose possess an enzyme called β-galactosidase which catalyzes the splitting of lactose into its constituent sugars, glucose and galactose, which then are utilized further. Another peculiarity is that the protein of milk is present primarily in the form of casein; microorganisms using the casein as a source of nitrogen, possess an enzyme (protease) that will break the casein down into its constituent amino acids.

Microorganisms in milk Milk from the udder of a healthy cow contains very few microorganisms. Almost immediately after removal from the cow, however, milk can become contaminated, and from the time of milking to the time of consumption, contamination continues to be possible. Most microorganisms that enter milk have the potential for growth if the temperature is appropriate. In clean milk stored at low temperature very little microbial growth will occur, since the only organisms present come from the udder of the cow and these have temperature optima of 35 to 40°C. However, if the milk has become contaminated with organisms from the environment, microbial growth may occur even if the milk is kept cool, since at least some of the contaminants will be psychrotrophs able to grow at low temperatures.

TABLE 16.6 Composition of cow's milk

Constituent	Percent by weight
Water	87
Fat	3.8
Lactose (milk sugar)	4.7
Protein	
Casein	3.0
Lactalbumin	0.4
Ash (minerals)	0.75
Others (vitamins, flavor ingredients, etc.)	0.15

Microorganisms in milk can be classified in two groups: pathogenic and nonpathogenic. The pathogenic organisms are derived either from the animal or from persons handling the animal and its milk. A number of microorganisms that cause diseases of cattle also infect man, of which the most important are *Mycobacterium tuberculosis*, causing bovine and human tuberculosis, and *Brucella abortus*, causing abortion in cows and undulant fever (brucellosis) in humans. Microorganisms that are derived from milk handlers include the causal agents of typhoid fever (*Salmonella typhi*), bacterial dysentery (*Shigella dysenteriae*), gastroenteritis (various *Salmonella* species), and Q fever (*Coxiella burnetii*). Even if milk is produced under the most careful conditions and has a low bacterial count, it may still be dangerous if it contains pathogenic organisms.

The nonpathogenic organisms in milk are usually derived from the environment and, although of little or no danger to humans, may still be responsible for undesirable effects such as souring, curdling, or production of off-flavors. The organisms entering milk after it leaves the cow are derived from the utensils, dust, manure, and water present around the dairy farm or from persons handling milk or equipment. These organisms include the lactic acid bacteria (*Streptococcus lactis, Lactobacillus casei*, and *L. acidophilus*), the coli-aerogenes group (*Escherichia coli, Enterobacter aerogenes*), spore formers (*Bacillus and Clostridium*), and the pseudomonads (*Pseudomonas* species). Another bacterium, *Streptococcus agalactiae*, is one of the causes of bovine mastitis, a disease of the udder of the cow (see Chapter 17), and this bacterium may be present in milk of untreated infected cows.

Pasteurization The pasteurization of milk (see Section 5.3) has primarily been responsible for the production of milk that is safe and reliable for human consumption. In addition to being pathogen-free, pasteurized milk has much better keeping qualities than does unpasteurized milk, making it possible to distribute milk safely from dairy farm to urban center. It should be emphasized that pasteurization is not the same as sterilization, since in pasteurization, the heat treatment is not sufficient to guarantee the killing of all organisms. Although pasteurization alters the flavor and reduces the nutritional value of milk slightly, these disadvantages are minor when compared to the advantages. Indeed, it is virtually impossible in many parts of the world today to purchase nonpasteurized (so-called raw) milk.

The times and temperatures chosen for pasteurization are those needed to destroy the most heat-resistant nonsporeforming pathogens found in milk—*Mycobacterium tuberculosis* and the rickettsia which causes Q fever, *Coxiella burnetii* (Table 16.7). Many other bacteria are of course also killed; in fact, at least 99 percent of the nonpathogenic bacteria are destroyed. All yeasts, molds, Gram-negative bacteria, and most of the Gram-positive bacteria are killed. The bacterial count of pasteurized milk generally ranges from 2,000 to 20,000 bacteria/ml, whereas the bacterial count of raw milk will generally range from 500,000 to 3,000,000. In terms of spoilage during low-temperature storage, most of the bacteria remaining will present little problem; it is the psychrotrophic bacteria (able to grow at low temperature) that enter the milk after pasteurization that cause the main problems. By close attention to san-

TABLE 16.7 Sensitivity of bacterial pathogens to the pasteurization process

Organism	Disease	Time (min) required for destruction of pathogens at 60°C
Coxiella burnetii	Q fever (respiratory infection)	30
Mycobacterium tuberculosis	Tuberculosis	20
Corynebacterium diphtheriae	Diphtheria	1
Salmonella typhi	Typhoid fever	2
Shigella dysenteriae	Bacterial dysentery	10
Brucella abortus	Undulant fever (brucellosis)	10 to 15
Streptococcus agalactiae	Mastitis (in cows)	Under 30
Streptococcus pyogenes	Streptococcal infections (in humans)	Under 30

itary conditions in the dairy plant, contamination with psychrotrophic bacteria can be minimized or eliminated, thus greatly extending the keeping qualities of the milk.

There are three major pasteurization methods: the low-temperature method, the high-temperature short-time (HTST) process (so-called *flash pasteurization*), and the ultra-high-temperature method. The *low-temperature method*, or "batch-holding process," involves heating the milk in bulk at 63–66°C (145–151°F) for 30 minutes (Figure 16.10). The heating can be done after the milk is bottled but is more commonly done in bulk in tanks of 500 to 2,000 liters. The walls of the tank within which the milk is placed are heated with either hot water or steam to the desired temperature, and the milk is agitated with a motor-driven stirrer to ensure that the heat is transferred throughout the milk. The milk must be uniformly heated to the desired temperature, and care must be taken to ensure that there are no pockets or

(a)

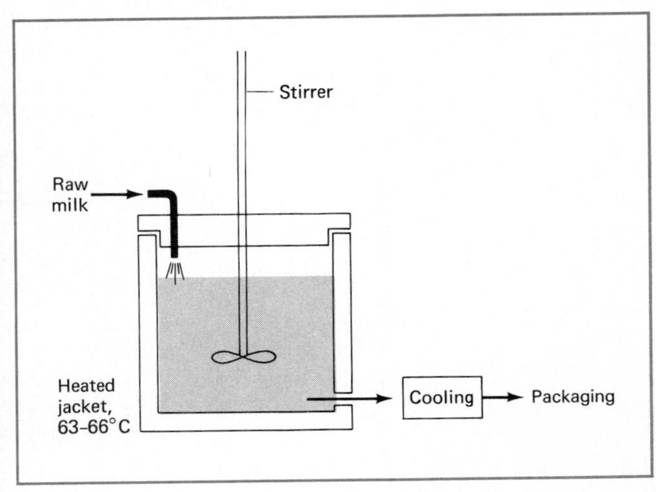

(b)

Figure 16.10 Pasteurization of milk by the batch pasteurization process. (a) Photograph of a typical batch pasteurizer. (b) Diagram of the process.

dead spaces in the tank where milk might escape the heat treatment. After heating, the milk is cooled quickly to avoid growth of thermophilic bacteria and the development of disagreeable flavors. This method is mainly used now for processing milk to be used in making dairy products such as cheese and ice cream.

Liquid milk for drinking is now pasteurized primarily by the *high-temperature short-time pasteurization* (HTST) process (Figure 16.11). The HTST process is a continuous-flow method in which the milk is passed through a tube where it is heated to 76–78°C (170–172°F) for at least 15 seconds and then quickly cooled. This method requires more elaborate equipment than

(a)

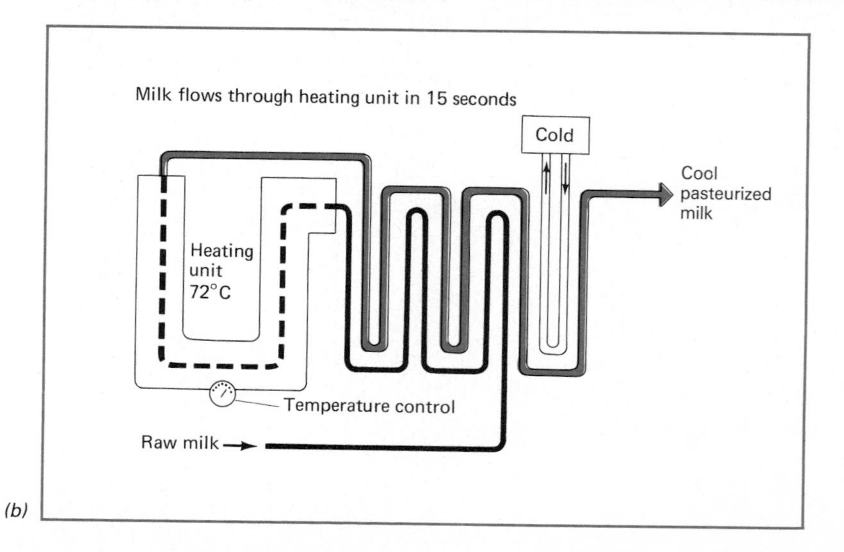

(b)

Figure 16.11 Pasteurization of milk by the continuous flow *high-temperature short-time* process. (a) Photograph of a typical installation. (b) Diagram of the process.

does the batch pasteurization method; however, it is far better suited to large dairy operations, where milk is bottled continuously. Much more milk can be pasteurized in the same space because there is no necessity for a large area for bulky holding tanks. However, although pathogens are destroyed just as effectively by both methods, heat-tolerant bacteria are less readily destroyed by flash pasteurization, so that if raw milk is high in such bacteria, the bacterial count of the pasteurized product may be high. Since the milk is heated only very briefly, undesirable flavors are lower than with batch pasteurized milk, and on balance, the advantages of this method far exceed the disadvantages in most dairy operations (for an exception, see the Box, page 110). This method is by far the one most commonly used for the pasteurization of milk in the United States and in many European countries, as well.

Even higher temperatures and shorter treatment times can used to pasteurize milk with the *ultra-high-temperature* (UHT) *treatment*. Steam is used to heat the milk, either by direct injection or indirectly through a heat-exchange system, the product reaching temperatures over 130°C for less than 1 second. The milk is virtually sterile and has long keeping qualities, although flavor changes may occur, with the milk having a boiled taste or a flavor sometimes called "cabbagey." Some reduction in vitamin content also occurs.

Pasteurization by the normal methods gives milk and its other fluid products an average shelf life of 7 to 10 days. Since the turnover time of milk is short, this shelf life usually is adequate, and the more extreme heating treatments to increase shelf life further are not worth the decrease in flavor and quality. However, for such products as cream that have a relatively slow turnover time on the shelf, a longer shelf life is desirable. Storage times of 30 or more days can be attained by more vigorous heating followed by aseptic packaging, to give a product that may not be absolutely sterile but has so few organisms that it can be termed "commercially sterile." The UHT method is widely used for the production of individually packaged portions of cream for coffee, for example.

Bacteriological examination of milk The determination of the number of bacteria present in milk is part of routine examination procedures. Two types of determinations are used: the direct microscopic count and the viable count. A number of methods are available for performing these tests, each with its advantages and disadvantages.

The *direct microscopic count* (see Section 5.2) is sometimes called the Breed test (Figure 16.12). With a special pipette, 0.01 ml of milk is spread uniformly over an area of 1 sq cm. The slide is dried, treated to remove fat, and stained. The slide is then examined with an oil-immersion lens, and the number of organisms in 30 fields is counted. From this count and a knowledge of the microscope field size, the number of organisms per milliliter of milk can be calculated. The direct microscopic count is a rapid method requiring relatively little apparatus or time. It has the disadvantages that both dead and living organisms are counted and that it is relatively insensitive, since only a small amount of milk is examined (0.01 ml), and if the milk is low in bacterial numbers, there are so few organisms per slide that the count is not very accurate. With milks of high bacterial count (over 500,000/ml), the method

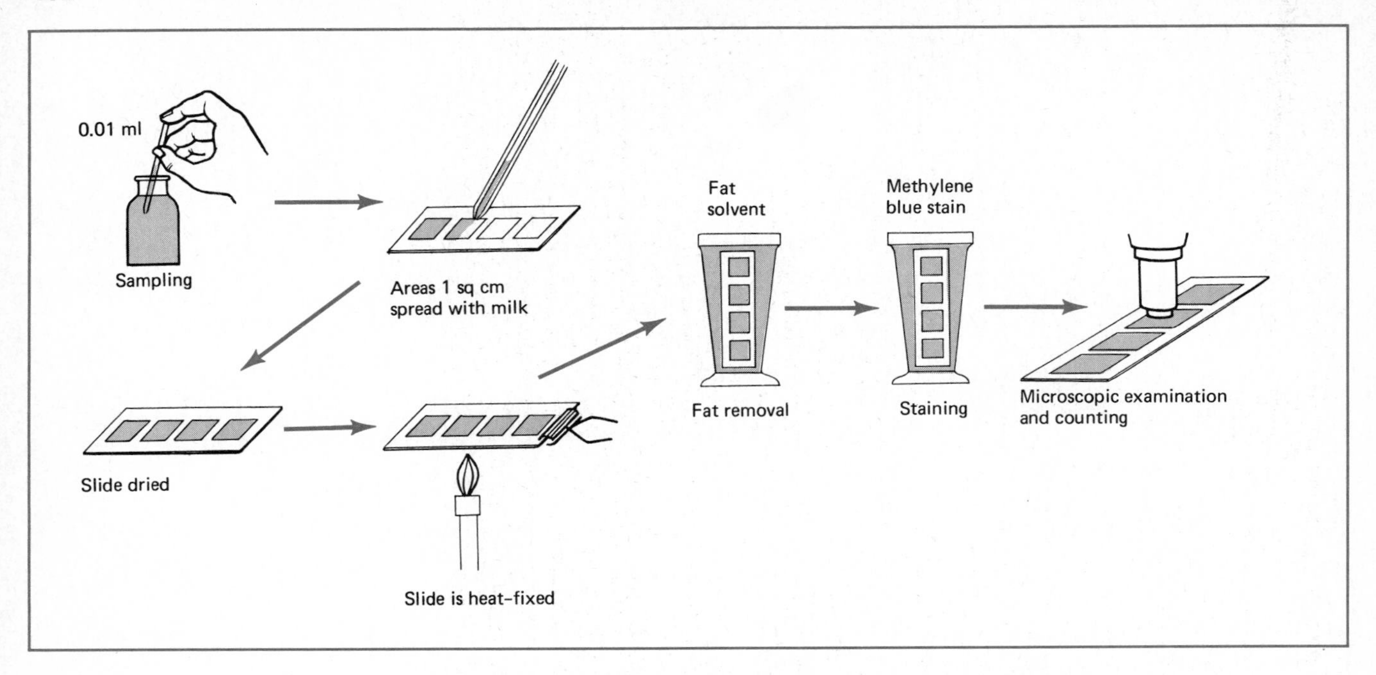

Figure 16.12 The Breed test, a direct microscopic count of milk.

is just as accurate as viable count methods, and the results are available much sooner.

Viable counts (see Section 5.2) are more difficult and time-consuming than direct microscopic counts, but they are more sensitive, indicating the presence of a smaller number of bacteria and, of course, only measuring those organisms that are viable. A viable-count procedure also offers the option of identification of the kinds of organisms present, since colonies can be picked and studied.

The *standard plate count* is a carefully standardized procedure designed to permit an estimate of the total number of viable bacteria present (Figure 16.13). A general-purpose agar culture medium is used, and for government-regulated milk supplies, the medium is specified by the regulating agency. *Coliform counts* are performed on pasteurized milk samples and other dairy products to detect recontamination after treatment, especially to indicate possible fecal contamination by food handlers. The *Escherichia coli*, or coliform, test for dairy products follows the same principle as that for water, as described in Section 15.3.

It should be emphasized that it is only because of the application of careful and precise microbiological controls on the quality of milk that milk has become a reliable and safe product for human consumption. When most communities were generally rural, milk was available locally, and careful quality control was of less importance; but in the present urban world, milk is produced at considerable distances from its place of consumption. It is produced in large volumes, is transported in bulk to central dairies, and may be stored for many days before consumption. The microbiologist plays a vital role in ensuring that the product remains safe and satisfying to consume.

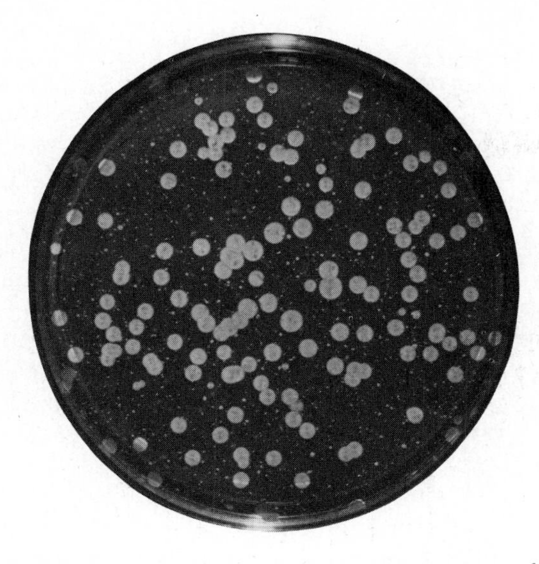

Figure 16.13 Typical pour plate from a viable count of milk.

Government regulation of milk The extremely perishable nature of milk makes it essential to control microbial multiplication during handling, transportation, and storage. Milk regulations are designed to ensure that the milk delivered to the public is safe to drink. These regulations are therefore public-health regulations and are usually administered by an agency of the government responsible for public-health matters. For pasteurized milk, two bacteriological criteria are used: the standard plate count and the coliform count. Standards, of course, vary with the country, climate, and conditions of manufacture, but standards for pasteurized milk widely accepted in developed countries are as follows: standard plate count not to exceed 30,000 bacteria/ml and coliform count not to exceed 1/ml. Other regulations may specify the standard and coliform counts of raw milk, the absence of tuberculosis and brucellosis in the dairy herd, and other matters. Milk products are also regulated for chemical quality, such as butter fat and milk solids, but these aspects do not involve the microbiologist.

Fermented milk products Not all microbial action in milk is considered harmful. In the production of yogurt, buttermilk, and other fermented milk products, microbial action cause beneficial changes. The microorganisms involved are specific kinds of lactic acid bacteria of the genera *Streptococcus*, *Lactobacillus*, and *Leuconostoc* and their function is twofold: (1) to produce lactic acid, which leads to an acidification of the milk; and (2) to produce flavor ingredients that give the final product its distinctive taste. Because of its acidic character, a fermented milk has quite good keeping qualities in comparison to unfermented milk, and historically this was the reason for manufacturing these products. However, at present the keeping qualities are not as important as the flavor that develops, since modern pasteurized fresh milk also has excellent keeping qualities.

In noncommercial manufacture of fermented milks, the bacteria for a new batch are obtained by adding a small amount of product from a previous batch, but in commercial processes, carefully selected pure cultures, called **starter cultures**, are used. By the inoculation of appropriate starter cultures into pasteurized milk, a fermented product can be obtained that is uniform and of high quality. In the manufacture of some products, two bacteria are used. One, usually a variety of *Streptococcus*, is responsible for most of the *lactic acid* produced. The second, usually a variety of *Leuconostoc*, is mainly responsible for the production of flavor and aroma. This second organism is usually a minority member of the starter culture mixture but is nonetheless important. The flavor and aroma are due primarily to the production of *diacetyl*, and this substance is produced by the bacterium through its action on the citric acid that is always present in small amounts in milk.

In many modern dairy operations, starter cultures are not maintained but are purchased in dried or frozen form from commercial firms. In this way, a culture of optimum quality is always available, ensuring uniformity in the product. The milk used for production of starter cultures is pasteurized and cooled before inoculation. The starter cultures are usually not used directly for inoculating the bulk milk, but rather as the inoculum for an appropriate volume of the milk which will be finally used as the bulk milk inoculum. This means that the dairy-plant operator must have some understanding of the manner of handling of cultures and should be familiar with microbial growth curves. Many problems can develop when starter cultures are not properly maintained. If the starter culture has two organisms, they can become imbalanced during the buildup of the inoculum. Antibiotics which have passed into the milk from the cow can inhibit the growth of the bacteria. Viruses (bacteriophages) able to attack the starter bacteria may enter the culture and destroy it or reduce its activity. The quality of the starter culture is evaluated by determining the rate of development of acid and the degree of acidity obtained, and by checking the odor and flavor of a test fermentation. If difficulty arises, it is best to begin anew with a fresh starter culture.

Fermented milks are made in most parts of the world, and local products are given special names. *Yogurt* is the most widely consumed form of sour milk in Europe and North America. For the commercial production of yogurt, the milk is heated to 90°C for 15 to 30 minutes, then cooled to 43°C; it is then inoculated with 2 percent starter culture and held at 43°C until the desired acidity is reached, after which it is cooled to 5°C and held chilled until sold. The acidity actually causes thickening by altering the milk protein casein. There are minor differences among particular manufacturers: temperatures, times of incubation, final pH, fat content, and concentrations of additives such as skim milk powder and sucrose. The organisms involved in production of yogurt are *Streptococcus thermophilus* and *Lactobacillus bulgaricus*; in commercial production, the organisms are generally added in a balanced mixture as a starter culture. *S. thermophilus* grows best at first, removing oxygen by respiration and producing the weakly acidic conditions that favor the growth of *L. bulgaricus*, which in turn produces most of the acid. The organisms together convert nearly all the milk sugar (lactose) to lactic acid, producing only trace amounts of by-products. These are, however, important by-products, since they give yogurt its characteristic flavor. *S. thermophilus* produces diacetyl, and *L. bulgaricus* produces acetaldehyde.

Cheese Cheese is a dairy product formed by precipitating the casein of milk in the form of a *curd*. The curd holds most of the fat and other suspended materials of the milk, and the water and dissolved constituents, called the *whey*, are allowed to drain away. The curd is then usually allowed to ripen, and it is during the ripening process that most of the distinctive characteristics of cheese develop. Microorganisms play two roles in cheese manufacture: (1) they are usually responsible for the souring of milk, which aids in the production of curd; (2) they play key roles in the ripening process, and hence, microorganisms can be considered responsible for the distinctive characteristics of cheese.

There are literally hundreds of different kinds of cheeses; many are of strictly local production, whereas others are manufactured and marketed over wide areas. Many cheeses can be stored for long periods of time without spoilage and hence can be transported for long distances. It is of course this resistance to spoilage that was originally the main reason for the manufacture of cheese, although in most parts of the world today, cheese is valued because of its flavor rather than because of its storage qualities.

The first step in cheese making is the *curdling* of the milk. The milk used for cheese making should be pasteurized to destroy any pathogenic organisms present if the cheese is not to be aged for a long time. It is usually considered that if the cheese is aged more than 60 days, it need not be made with pasteurized milk, since pathogens will be destroyed during the aging process. A starter culture is then added. The first function of the bacteria is to produce acid, which is necessary for the development of a satisfactory curd. However, the main agent in the development of the curd is not bacteria, but the enzyme *rennin*, which is added to the milk at the time of the starter culture inoculation. Rennin is an enzyme (prepared commercially by fungal culture or by extraction from the lining of the calf stomach) which coagulates casein. Curd can be formed through the action of rennin and other enzymes alone or through the action of bacterial acids alone, but curds for making most cheeses are achieved through the combined action of rennin and bacteria.

The curd is processed so as to lose water and shrink, and this leads to the production of a curd of desirable consistency (Figure 16.14). The moisture content is determined by the consistency of the curd, which affects how well it retains water, and by the length of time the curd is allowed to drain. The curd is shaped into forms of a desirable size and then salted. Salt not only adds flavor but also promotes the further extraction of water from the curd. Salt is also inhibitory to the growth of many spoilage microorganisms but not to the growth of the ripening organisms. The moisture content of the curd will vary with the cheese variety, hard cheeses having much less moisture than soft ones. Unripened cheeses such as cottage cheese are packaged at this stage, either directly or after mixing with additives, and no more microbial processing is involved.

The whey may be used to make ricotta or other whey cheeses, which are manufactured by nonmicrobiological processes. The whey is also used as a starting material for several industrial microbial processes, such as in the manufacture of lactic acid and alcohol. The whey may also be discarded, but since it contains many organic compounds it is a potential source of serious

Figure 16.14 Cheese manufacture. Pouring the curd into forms to allow drainage.

water pollution and must be treated in much the same way as domestic sewage.

For the manufacture of most cheeses, the next step is ripening (Figure 16.15), and during this process the cheese undergoes changes in flavor, texture, and consistency. Some of the ripening is due to continued action of rennin or of enzymes originally in the milk; however, microorganisms produce the most distinctive changes. The organisms responsible for ripening are either added with the starter or spread on the surface of the curd at the initiation of ripening, depending on the type of cheese to be made.

For the hard cheeses such as Swiss, Cheddar, or Provolone, the curd is formed into large blocks which are carefully wrapped with paraffin or plastic film to prevent surface growth of microorganisms. In general, hard cheeses ripen slowly and are not consumed for months or even years after manufac-

(a)

(b)

Figure 16.15 Curing of soft-ripened cheese. (a) Blocks are smeared with appropriate ripening microorganisms. (b) A "smear" of microorganisms on the outside of a block of semisoft cheese.

ture. During ripening, excess moisture loss and undesirable surface spoilage are prevented by the airtight coating.

For the soft and semisoft cheeses such as Camembert, Brie, and Liederkranz, ripening takes place through the action of microorganisms growing on the surface of the block. Enzymes and flavor components produced by the organisms spread through the cheese and gradually convert the whole block into the desired product. Soft-ripened cheeses are always produced in small blocks so that the microbial products can spread through the cheese rapidly. Organisms involved in the ripening are smeared on the surface, and blocks are not wrapped during ripening but are stored in curing rooms (see Figure 16.15*a*) of fairly high humidity to encourage microbial growth. The surface growth, usually called *smear* (see Figure 16.15*b*), is composed of a characteristic mixture of organisms in which yeasts, *Micrococcus* species, and *Brevibacterium linens* are the main components.

In the production of Roquefort, Blue, and Gorgonzola cheeses, which are semisoft, a fungus, *Penicillium roqueforti*, is the main organism involved. Spores of the mold are mixed with either the milk or the curd. Before the cheese is wrapped for ripening, tiny holes are drilled into the curd. During the 3 to 6 months of ripening the mold grows both in the small natural openings in the curd and in these tiny machine-made holes. The fungus pervades the cheese and sporulates, giving the product its characteristic blue-streaked appearance. The flavor is caused mainly by fungal products that accumulate in the cheese.

In Swiss-style cheese, gas-forming organisms develop, and the gas formed causes the production of the characteristic "eyes" of the cheese. The gas is CO_2, which is produced by *Propionibacterium*, an organism that ferments lactic acid with the production of propionic acid, acetic acid, and CO_2. The flavor of Swiss cheese is due partly to the propionic acid produced by this bacterium. *Propionibacterium* is thus responsible for the two main characteristics of Swiss cheese: its flavor and the characteristic holes.

During ripening, microbes do several things. They ferment the sugar and lactic acid remaining in the curd and produce flavor constituents, especially propionic acid and acetic acid. They digest some of the protein of the curd, converting it to amino acids. In soft-ripened cheeses, protein digestion proceeds quite extensively, thus causing the softening of the cheese. The amino acids liberated by the protein digestion contribute to the flavor of the product, but in addition, their microbial decomposition leads to the production of ammonia, hydrogen sulfide, and other constituents that impart distinctive flavors to many cheeses.

In a process as complicated and lengthy as cheese manufacturing, it is understandable that defects may develop in the product. Many of these defects are caused by the action of undesirable microorganisms. Of these, the most important are fungi (molds), which grow readily on surfaces of cheese exposed to oxygen. Rigid sanitation precautions in the cheese factory and careful wrapping of the product to keep air from it will ensure the absence of mold contamination. In some cases, gas formation is an undesirable change in cheese induced by microorganisms. Sugar-fermenting microbes entrapped in the curd can produce gas, usually CO_2, and cause the formation of holes, cracks, and fissures in the cheese. Microorganisms may also produce undesirable flavors and colors in cheese that will reduce its market value. The cheeses most susceptible to spoilage are the unripened ones, since these are high in moisture content. Such cheeses must be eaten within a short time after manufacture. At the other extreme, the hard-ripened cheeses are of such low moisture content that they can be kept for months or years without serious spoilage.

16.7 MEAT MICROBIOLOGY

Meats are among the most nutritious of foods, but they are also among the most susceptible to microbial spoilage. In addition, they are agents of transmission of several serious infectious diseases from animals to humans. Thus, the microbiology of meat and meat products is of considerable practical importance. Most meat is processed and prepared in large central installations called *packing houses,* where slaughtering, cutting, processing, and in some cases packaging of meat is done. Inspection and disease control are under the jurisdiction of qualified veterinarians who have received extensive training in microbiology.

When an animal is killed for meat production, a number of changes occur which permit microbial growth: circulation stops, the oxidation-reduction potential drops, the supply of nutrients to the tissue stops, fermentation begins (with resultant pH drop), protein denaturation begins, immune defenses cease, and microorganisms begin to multiply. In the whole carcass, the organisms which are involved are those of the animal's normal flora as well as those added during slaughter from the knives, storage room, etc. If the carcass is not cooled internally, organisms of the normal flora, such as *Clostridium perfringens* and various enteric bacteria, will cause spoilage. On the other hand, these organisms, adapted to the body temperature of the animal, will not grow at refrigeration temperatures; during cold storage, spoilage is primarily a surface phenomenon due to external sources of organisms. In addition to bacteria, there may be some molds and yeasts, especially when the surface is dry.

Meat cuts develop "surface shines," due to growth of slime-forming bacteria on the surface of the meat. Consequences of microbial growth in cuts of meat include color changes, from the normal red to shades of green, brown, or gray, due to microbial production of peroxides or hydrogen sulfide which react with the red pigment in meat; rancidity, through attack by microbes on the fats; taint, due to production of unpleasant-tasting compounds during microbial metabolism; and putrefaction, caused by breakdown of the protein of meat and the release of disagreeable-smelling nitrogen compounds such as ammonia, amines, indole, and skatole.

Slaughterhouse hygiene The quality of the meat is closely related to the condition of the live animal at the time of slaughter and the degree of contamination to which the carcass is subjected. Animals to be slaughtered should be carefully inspected while still alive to determine that they are free of infectious diseases (see Chapter 17). Within the slaughterhouse, care should be taken to avoid contamination of the animal with material laden with microorganisms. Soil and dung should be removed from the animal by preliminary washing. Cleanliness in the slaughterhouse is important, and dirt and debris should be cleansed from the floor and walls of the rooms, preferably with a disinfectant such as sodium hypochlorite. For the kill, the animal is first stunned with a heavy blow to the head, and then bled to death. Much of the contamination of meat comes from the hide of the animal during the skinning process. Removal of the hide in a manner that restricts the spread of contamination is very important.

After slaughter, every animal should be subjected to postmortem examination (Figure 16.16), with special attention to the lymph nodes, visceral organs, and the exposed parts of the carcass. If evidence of disease or other abnormal condition is found during the routine examination, the carcass and its parts are given a more extensive examination. If the abnormal condition is localized in one portion of the carcass, this part may be disposed of and the rest passed on for processing; however, if the whole carcass is to be condemned, it is passed to a holding area for conversion into fertilizer and other nonfood items. It should emphasized that the inspection of meat is purely visual, so that only gross evidence of infection is obtained.

The carcass is normally refrigerated immediately after slaughter in order to prevent spoilage, although because of its bulk, it does not cool instantly. The carcass is aged in the cold (always below 3°C), and the meat gradually becomes more tender, usually reaching an optimum tenderness after 2 weeks of aging. During the aging process, there is always the possibility of bacterial spoilage, especially by psychrotrophic bacteria that may contaminate the sur-

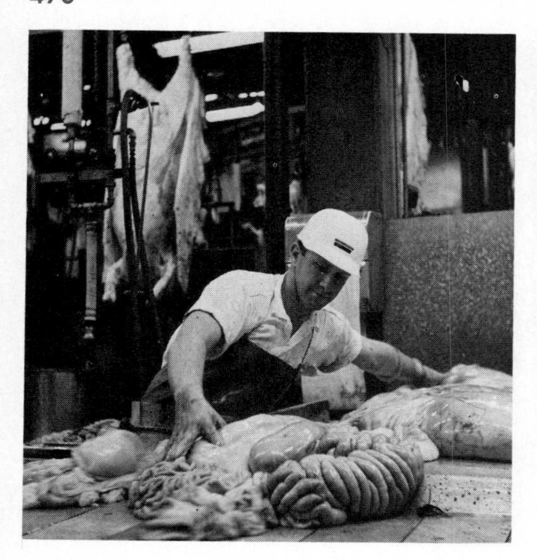

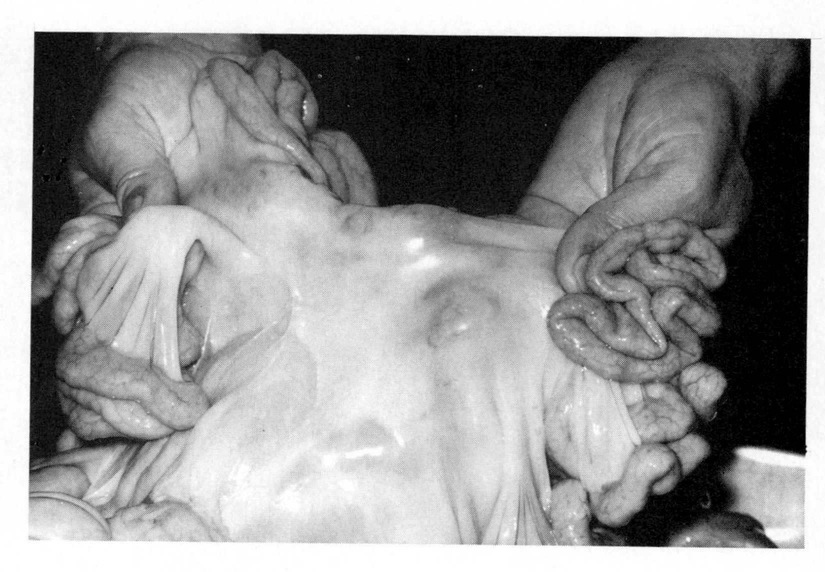

Figure 16.16 Inspection of the carcass in a meat-packing plant. The organs of the animal are carefully examined for evidence of infectious disease.

face of the meat. After aging, the carcass is cut and marketed. Cutting exposes the meat to increased contamination by providing greater surface area for microbial growth. Once cut, the meat should be quickly consumed, and it is generally considered that the retail establishment should not display meat for more than 3 days. In the home, even if the meat is properly refrigerated, fresh meat should be consumed within 1 to 4 days of purchase unless it is frozen. The quality of meat is best retained if it is frozen rapidly and stored carefully wrapped to prevent drying. Some suggested standards for allowable microbial counts in meats are included in Table 16.4. For a possible role of meat in spreading antibiotic-resistant bacteria, see the Box on the next page.

Meat preservation and curing Since fresh meat is one of the more perishable food items, a major task of the meat industry is to develop and employ methods for adequate meat storage. Two general approaches are possible: preservation and processing. *Preservation* involves such treatments as refrigeration, freezing, canning, and drying. *Processing* of meat involves the use of some method by which the fresh meat is modified, such as grinding, curing, addition of seasonings or other chemicals, heat treatment, smoking, or microbial fermentation. Such processing leads to significant changes in the flavor and texture of the product and also results in an increased shelf life. Typical processed meat products include ham, bacon, corned beef, and a wide variety of sausages.

Refrigeration is the most widely used method of meat preservation. Although the shelf life of the meat is extended, spoilage will eventually occur. As with other refrigerated foods, psychrotrophic microorganisms, able to grow at refrigeration temperature, are responsible for the spoilage. Slowing the growth of psychrotrophs by using temperatures close to 0°C will extend the shelf-life of meat. For frozen storage of meat, temperatures of about −20°C

ANTIBIOTIC-RESISTANT *SALMONELLA*: FROM CATTLE TO CONSUMER

A recent salmonellosis outbreak in the upper Midwest illustrates the complexity of the human food chain and underscores the pervasive hazard of uncritical use of antibiotics. Beef cattle in South Dakota which had been fed low doses of chlortetracycline to stimulate growth were slaughtered in Minnesota and sent to Nebraska for processing into boxed beef. Some of the boxes of beef were sent to meat brokers in Minnesota and Iowa, from which they went to a number of supermarkets. (Such boxed beef is usually sold to supermarket chains for grinding into hamburger.) Shortly thereafter in South Dakota, Minnesota, and Iowa, outbreaks occurred caused by an unusual bacterium, *Salmonella newport* (identified by a characteristic plasmid), which was resistant to ampicillin and tetracycline. Some of the patients were food handlers and in several cases these handlers had tasted the raw hamburger. Despite their separate locations, *all* of the patients were infected by the *same* bacterial strain, firm evidence for a common source. Although only 18 cases were detected, the number of undetected cases may have been much larger, since approximately 70,000 pounds (32,000 kg) of potentially contaminated meat was distributed throughout a four-state area. It was not possible to obtain bacterial cultures from the original beef cattle, since all the animals had been slaughtered, but dairy cattle on an adjacent farm harbored the same *Salmonella newport* strain.

This study clearly shows that antibiotic-resistant bacteria of animal origin can enter the human food chain and cause serious human disease. The emergence and selection of such antibiotic-resistant bacteria may be a consequence of their use as growth stimulants in animal feed (see Chapter 17). Thus, prudent use of antimicrobial agents is clearly indicated.

are generally used in commercial practice. Even colder temperatures are better, although more expensive to maintain. Freezing has no noticeable effect upon the color, flavor, odor, or juiciness of meat as judged after cooking, although during long-term freezer storage there is a gradual decrease in odor and flavor.

Canning of meat involves either moderate heating (about 70°C), which, although insufficient to completely sterilize, is used on cured meats such as hams to extend their shelf life, or more severe heating (121°C), which is applied to uncured meats to produce a sterile product that can be stored indefinitely without refrigeration. Most canned meat products are heated in the can, and because of the bulk of the product, the slow rate at which heat penetrates to the center, and the high pH of meat, fairly long heating times are necessary. Some flavor changes will inevitably take place, but these disadvantages are offset by the greatly lengthened shelf life.

Dehydration of meat is a centuries-old process, still carried out under primitive conditions in many parts of the world using the sun or wood fires as a source of heat. In commercial practice, meat is occasionally dried with hot-air dryers, but recently freeze-drying of meat has become widespread, since the product retains more of the flavor and texture of fresh meat. Dehydrated meat can be stored unrefrigerated for long periods of time without spoilage and is ideal when a lightweight product is desired. It is used widely for military rations and on safaris, expeditions, and wilderness trips.

Chemical preservatives (see Table 16.5) have been widely used in the past, but in modern practice addition of chemicals to meats has been greatly curtailed. To be useful, a chemical preservative not only must extend the shelf life of meat but also should not impart an undesirable flavor, odor, or color and should be nontoxic to humans. It is the latter requirement that has curtailed use of chemical preservatives, since the toxicity of many agents upon long-term human consumption has not been established. Agents added to meats in other countries (although not permitted in the U.S.) include sodium benzoate, formaldehyde, salicylic acid, sulfite, and boric acid. The antibiotics chlortetracycline and oxytetracycline were formerly approved for addition to raw poultry at levels no greater than 7 ppm, since these antibiotics are destroyed by heat and no significant residue is left after proper cooking, but this permission was revoked by the Food and Drug Administration in 1966. It should be emphasized that no chemical preservative is a substitute for proper sanitation.

Meat curing involves the addition of salt, color-stabilizing ingredients, and seasonings to meat in order to give specific unique characteristics to the end product. Although the exact ingredients used in the cure vary according to the product, two substances are always included: salt and sodium nitrite. The prime function of the salt is as a flavoring agent; it is usually not present in high enough concentration to act as the only preservative. Nitrite has a twofold role in meat curing. It becomes reduced during processing to nitric oxide and reacts with myoglobin in the meat to form a stable pink pigment which is typical and desirable in cured meats. More importantly from a microbiologist's viewpoint, nitrite inhibits the germination of *Clostridium botulinum* spores and prevents toxin production during storage in the refrigerator for such cured meats as hot dogs, bologna, canned hams, and bacon. Without nitrite, these foods could not be prepared as they are now because of the danger of botulism. Other substances often added are: sugar, used as a flavoring ingredient and antioxidant; nitrate, added to provide a reserve supply of nitrite; ascorbic acid, which promotes the cure by aiding in the reduction of the nitrite; sodium phosphate, which retards shrinkage in smoked products; and monosodium glutamate, which is a flavor enhancer. The curing mixture is incorporated into the meat product in a variety of ways, depending on the particular item. For sausages, the ingredients are added dry or in a concentrated solution during the mixing or grinding processes. For hams, the curing mixture can be injected into the vascular system of the carcass as a brine, and for other meats it can be injected through needles directly into the meat. Alternatively, the meat cuts may be dipped into brines or a dry curing mixture can be rubbed onto the surface. The treated product is then stored at refrigeration temperature to retard bacterial growth until salt penetration is complete. Preservation from spoilage of cured meat occurs because of the dehydrating effects of the salt on microorganisms present in the meat, as discussed earlier.

Smoking of meat is carried out both to preserve the meat and to impart to it a characteristic flavor. Smoke is generated by controlled combustion of moistened hardwoods (hickory and oak are most commonly used) in a smoke generator. The chemical composition of wood smoke is quite complex, but among the chemicals identified are fatty acids, phenols, waxes, resins, and formaldehyde. Formaldehyde has been identified as the chief bacteriostatic and bactericidal substance in wood smoke. The amount of formaldehyde added to meat during smoking is too small to either impart a bad flavor or cause human toxicity. Curiously, although formaldehyde can be added legally to meat indirectly through smoking, government regulations forbid the direct addition of the purified chemical. Although at one time smoking of meat was done mainly to preserve it, today the main function of smoking is to add characteristic flavor.

Sausage is prepared from chopped and seasoned meat that is placed in a cylindrical casing derived either from the intestinal tract of an animal (natural casing) or from modified cellulose (artificial casing). There is an enormous variety of sausages, and countries and regions often have specialty sausages of distinctive flavor and character. The meat ingredients in sausage most commonly consist of those parts of the slaughtered animal not readily utilizable in other ways, such as cheek and head meat, trimmings, tripe, and belly, but higher-quality meats may also be used. Sausage may be sold fresh, cooked, or dried. Any meat can be used for making sausage, including pork, beef, lamb, and veal, but the most commonly used meat is pork. The most popular products are fresh pork sausage, which is neither precooked nor smoked, frankfurters and Braunschweiger (liver sausage), which are precooked and smoked, and meat loaves, which are cooked but not smoked. In many sausage types, fermentation is carried out by chance contaminants, although starter cultures are increasingly used in many plants. The microorganisms involved in the sausage fermentation are mainly lactic acid bacteria; commonly used starter-culture organisms include *Lactobacillus plantarum* and *Pediococcus pentosaceus* or *P. acidilactici*. Their role is to produce lactic acid from the sugar that is added to the sausage mix. The advantage of using a starter culture is that it makes possible the manufacture of a product of uniform quality in far less time (only 20 to 40 hours instead of 150 or more hours). The ground meats, salts, spices, and some sugar are mixed with the starter culture, and the mixture is stuffed directly into appropriate casings and moved to a warm (27°C) humid area for 12 to 16 hours to permit the organisms to rehydrate and return to the vegetative state. The sausage is then moved to the curing house and held at about 40°C with 90 percent humidity until the desired acid production is reached (about 15 to 20 hours, depending on the variety of sausage). A final heating at 55 to 60°C for 4 to 5 hours kills the starter bacteria so that subsequent changes during storage are eliminated. Smoke may be applied during any part of the time in the smokehouse. Sausage fermentation is a much less highly developed process than is the manufacture of fermented dairy products, and it is still mainly carried out by traditional empirical methods, although the increasing use of starter cultures seems likely to make it a much more predictable operation.

SUMMARY

In this chapter, we have surveyed the many ways in which microorganisms affect the foods we eat. A few microorganisms cause harmful changes in foods, and food spoilage and food poisoning are serious economic problems. One

TABLE 16.8 The major food-poisoning bacteria

Organism	Products involved
Salmonella spp.	Chicken, other meats; milk and cream, eggs
Staphylococcus aureus	Meat dishes, desserts
Clostridium perfringens	Cooked and reheated meats and meat products
Vibrio parahaemolyticus	Sea foods
Bacillus cereus	Rice and other starchy foods
Clostridium botulinum	Home-canned vegetables, smoked fish
Campylobacter jejuni	Chicken, milk
Yersinia enterocolitica	Pork, milk

of the tasks of the microbiologist to identify and remove the source of contamination of foods or to control the growth of microorganisms already present. A knowledge of the kinds, locations, and growth requirements of food spoilage and food-poisoning organisms is critical for the performance of this job. A summary of the major food-poisoning bacteria is given in Table 16.8.

We should emphasize, however, that not all microorganisms are bad for foods. A number of food products are made with the use of microorganisms. These include sauerkraut, pickles, buttermilk, yogurt, cheese, and sausage. In the manufacture of many of these products, selected microbial cultures are used, to permit careful control of the process.

Food sanitation is important in the *food processing industry*, in the *food market*, in the *restaurant*, and in the *home*. Employers, employees, and consumers must all be aware of potential problems from improper preparation and storage of foods. As seen in Table 1.3, there are billions of dollars in the food industry spent because of microbial considerations.

KEY WORDS AND CONCEPTS

Food spoilage
Water activity
Oxidation-reduction potential
Food poisoning
Botulism
Enterotoxin
Food infection
Food preservation
Fermented foods
Pasteurization
Starter culture
Food sanitation

STUDY QUESTIONS

1. Contrast highly perishable, semiperishable, and stable (or nonperishable) foods. List several foods in each category.
2. List five factors that affect microbial growth in foods and discuss briefly how each factor acts.

3. Discuss the steps you would take to prevent the development of *staphylococcal food poisoning*. Of *botulism*.
4. Why is it incorrect to call *Salmonella food infection* a food poisoning?
5. List four ways in which foods can be preserved, and discuss briefly the principle of each procedure.
6. Describe briefly the steps used in the bacteriological examination of a canned food suspected of being spoiled.
7. How is milk *pasteurized*? What is the main reason for pasteurization?
8. Describe briefly two methods for assessing the bacteriological quality of milk.
9. What is a *starter culture*? How are starter cultures used in the dairy industry?
10. What roles do microorganisms play in the production of cheeses?
11. Discuss the microbiological problems in the slaughter of animals and the preservation of fresh meat.

SUGGESTED READINGS

BOARD, R.G. 1983. *A modern introduction to food microbiology*. Blackwell Scientific Publications, Oxford.

JAY, J. 1984. *Modern food microbiology, 3rd edition*. Van Nostrand, New York.

SKINNER, F.A. and T.A. ROBERTS (eds.). 1982. *A symposium on food microbiology*. Academic Press, London.

17

Agricultural Microbiology

Microorganisms play many important roles in agriculture, both beneficial and harmful. In this chapter, we discuss some of these effects, including the roles of microbes in soil formation and fertility, in compost formation, in beneficial associations with plants, as causal agents of diseases of crop plants and farm animals, in the control of insect diseases of plants, and as beneficial agents in the digestive processes in cows and other ruminants. A knowledge of the harmful and beneficial effects of microorganisms in agricultural processes can be used to increase crop yields and farm animal productivity. The economic impact of microorganisms on the agricultural industry was presented in Table 1.2. The material discussed in this chapter builds to some extent on the concepts of ecology and biogeochemistry covered in Chapter 14.

17.1 THE SOIL

Soil is the basis of agriculture, and microorganisms contribute importantly to both its formation and its fertility. Soils are made up of three components: solid, liquid, and gaseous. The solid materials, both mineral and organic, constitute the soil proper; the liquid and gaseous components, water and air, vary markedly within a single soil. The mineral constituents provide the basic fabric of the soil and vary in size, shape, and chemical composition. The larger particles are sand, the finer particles silt, and the finest clay. The proportion of these different particles greatly affects agricultural utility of the soil. If finer particles predominate, water is retained well but the soil tends to become clogged with water, whereas sandy soils are well drained but often become deficient in water. The nonliving organic matter in soil, called **humus**, is important, since it increases the water-holding capacity without causing waterlogging and also improves the soil texture. The living organisms of the soil are numerous and diverse. Bacteria are most numerous, followed by fungi and protozoa, but many other organisms are also present, including algae and invertebrate animals such as earthworms, insects, mites, and millipedes. Important in the soil are the roots of plants, which penetrate extensively and greatly modify soil texture and fertility.

Soil formation The formation of soil begins with the breakdown of rock, a process called *weathering*. Weathering is a result of three kinds of processes: physical, chemical, and biological. Physical weathering involves the fragmentation of rock owing to freeze and thaw, movement of the earth (as in earthquakes), and other mechanical processes. Chemical weathering involves reactions between substances such as oxygen or water and the minerals of rocks. Biological weathering is due to the action of living organisms, the most important of which are microbes.

The first organisms to become established on a bare rock are photosynthetic: algae, mosses, and lichens (see Section 2.8). The photosynthetic organisms convert carbon dioxide into organic matter, and some of the organic matter formed is excreted and supports the growth of bacteria and fungi. These organisms produce carbon dioxide as a result of respiration, and this CO_2 combines with water to form carbonic acid, a weak acid but one capable over a long period of time of dissolving rock. Many of these bacteria and fungi produce organic acids as well as CO_2, and these also contribute to the dissolution of rock. In time a raw soil forms consisting primarily of coarse rock fragments, live and dead organisms, and debris. In this raw soil, small plants can develop, and their roots penetrate into crevices in the rock, causing further breakdown. Organic excretions from the plant roots promote further growth of bacteria and fungi, leading to increased weathering. During the breakdown of rock, minerals are solubilized, further promoting the growth of plants. Organic matter added to soil from plants is broken down by microorganisms and converted into humus. Humus, an important factor in soil fertility, binds moisture and minerals, thus improving soil texture and making plant growth more vigorous.

As weathering proceeds, the soil increases in depth, thus permitting the development of larger plants and trees. Soil animals such as worms become established and contribute to weathering by keeping the soil mixed and aerated. As water percolates down through the soil, minerals are dissolved and carried deeper, thus modifying the lower region. Eventually the movement of materials downward results in the formation of layers, leading to what is called a *soil profile* (Figure 17.1). The time involved in the formation of a soil

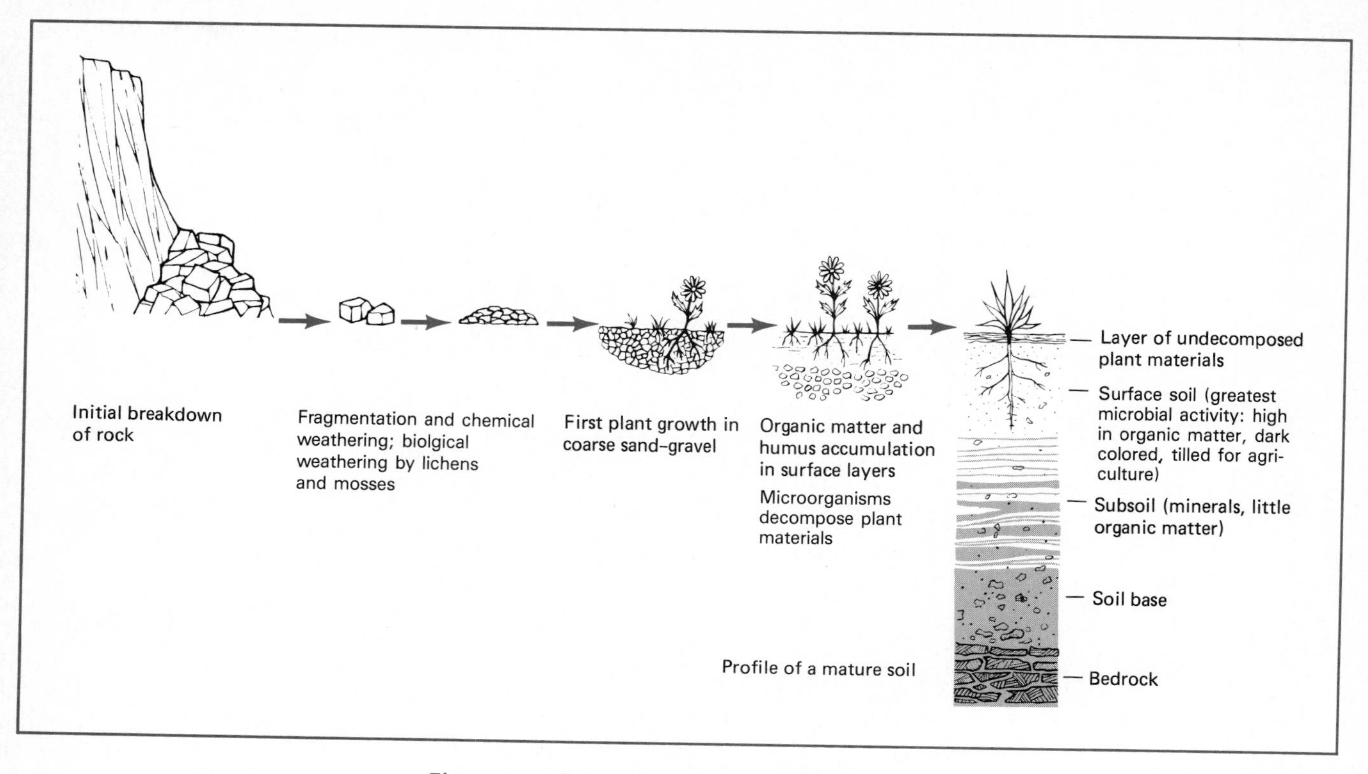

Figure 17.1 Steps in the formation of soil.

varies with climate and topography but may be on the order of hundreds of years.

Microorganisms and soil fertility Microbes contribute to soil fertility in a number of ways: (1) Through their action in decomposing organic matter they participate in humus formation. (2) Microbes can cause the release from soil particles of certain minerals that plants need for growth. Such minerals are bound to clay and humus particles; by producing acids, microbes bring about chemical reactions that result in release of these minerals. (3) Microbes can cause release of significant amounts of mineral nutrients that are bound to organic structures of dead plants or animals. This process, called *mineralization*, supplies plants with inorganic nutrients. (4) Microorganisms play important roles in the transformation of nitrogen compounds (see Section 14.5).

Compost Compost is a complex mixture of organic materials that is used for fertilization and for improving the texture of soils. It is prepared by piling leaves, straw, or other plant materials in a heap in combination with some source of nitrogen and phosphorus and allowing them to undergo decomposition (Figure 17.2). Within the pile, moist conditions are maintained and microorganisms grow well, decomposing some of the organic matter. Heat, a byproduct of microbial metabolism, is produced, causing the pile to

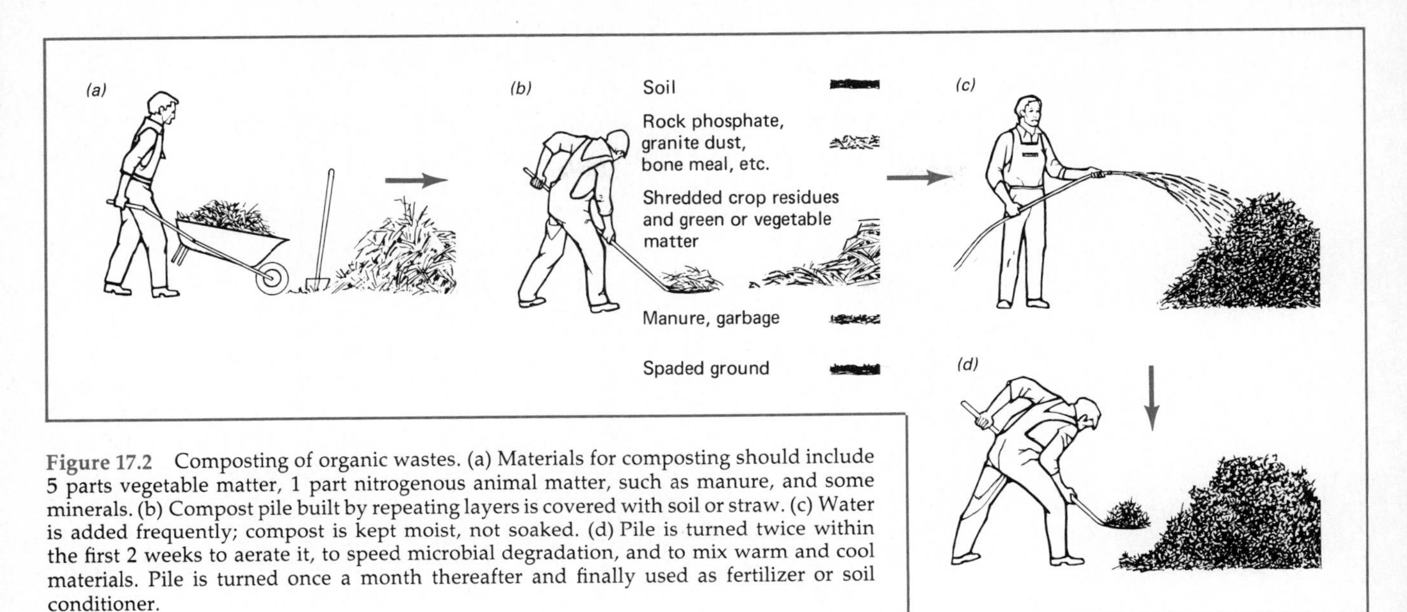

Figure 17.2 Composting of organic wastes. (a) Materials for composting should include 5 parts vegetable matter, 1 part nitrogenous animal matter, such as manure, and some minerals. (b) Compost pile built by repeating layers is covered with soil or straw. (c) Water is added frequently; compost is kept moist, not soaked. (d) Pile is turned twice within the first 2 weeks to aerate it, to speed microbial degradation, and to mix warm and cool materials. Pile is turned once a month thereafter and finally used as fertilizer or soil conditioner.

increase in temperature. This process is called *self-heating* and occurs not only in compost but in any situation in which materials rich in organic matter are piled up, such as manure, sawdust, and even coal. During the heating process in compost, thermophilic organisms can grow and some of these organisms are able to degrade cellulose and other organic materials in the plant residues. The carbohydrate and fat components of the compost material are rapidly degraded, whereas the fiber, which contains predominantly woody materials such as lignin, is less readily decomposed. The final product is reduced in weight and volume from the original by 25 to 50 percent. The compact mass of humuslike material that is produced can be returned to the soil to increase its fertility.

17.2 PESTICIDES

Pesticides are chemicals used to control weeds, insects, fungi, or other pests. Pesticides are widely used in agriculture, and some of them are applied year after year to crop land. Although some of these chemicals decompose rapidly in soil, others are highly persistent; after a number of years of use, residues toxic to animals or plants may build up in the soil and can accumulate in the food chain. Compounds that persist in the soil are resistant to microbial degradation, whereas those that disappear do so because they can be degraded either biologically or chemically.

Microorganisms can use pesticides as sources of energy, of carbon, or of nitrogen. Some organisms are able to completely mineralize these compounds, whereas others are able to oxidize them only partially. However, some pesticides synthesized by organic chemists are quite different chemically from the natural compounds in plants and animals. Since these synthetic compounds are entirely new in nature, it is not very surprising that microorganisms capable of degrading these compounds may be absent from the soil.

There are marked differences in the susceptibility of different pesticides to degradation, and this is reflected in their relative persistence in soil, as shown in Table 17.1. The decomposition of organic chemicals in soil involves both biological and nonbiological aspects. Some compounds are so unstable that they decompose rapidly in soil without the necessity of microbial action. This is true of the organophosphate insecticides listed in the table, all of which disappear quite rapidly. At the other extreme, the chlorinated insecticides, such as DDT (dichlorodiphenyltrichloroethane), are extremely stable in soil and are only slowly decomposed, even in the presence of microbial action; this is reflected in the long persistence of these compounds. Those compounds which are broken down by living organisms are called **biodegradable**.

The chemical makeup of a compound can markedly influence its susceptibility to microbial attack. This is seen in the relative persistences of two closely related herbicides: 2,4-D (2,4-dichlorophenoxyacetic acid) and 2,4,5-T (2,4,5-trichlorophenoxyacetic acid). The second compound differs from the first only by the addition of a single chlorine atom, but this markedly influences its persistence. Compound 2,4-D is readily broken down by microorganisms in soil, but the additional chlorine atom inhibits the action of microorganisms on 2,4,5-T. An example of the difference in rate of breakdown of a biodegradable and a persistent chemical is shown in Figure 17.3.

The breakdown of pesticides in soil through the activity of microorganisms requires that all conditions necessary for good microbial growth in soil be available. Since pesticides are not complete microbial foods, this means that adequate moisture and nutrients necessary for microbial growth (for example, nitrogen, phosphorus, sulfur, and other minerals) must be present in sufficient amounts. The presence of other organic compounds in soil often

TABLE 17.1 Persistence of pesticides of various types in soils

Pesticide	Time for 75 to 100% disappearance
Chlorinated insecticides	
DDT	4 years
Aldrin	3 years
Chlordane	5 years
Heptachlor	2 years
Lindane	3 years
Organophosphate insecticides	
Diazinon	12 weeks
Malathion	1 week
Parathion	1 week
Herbicides	
2,4–D	4 weeks
2,4,5–T	20 weeks
Dalapon	8 weeks
Atrazine	40 weeks
Simazine	48 weeks
Propazine	1.5 years

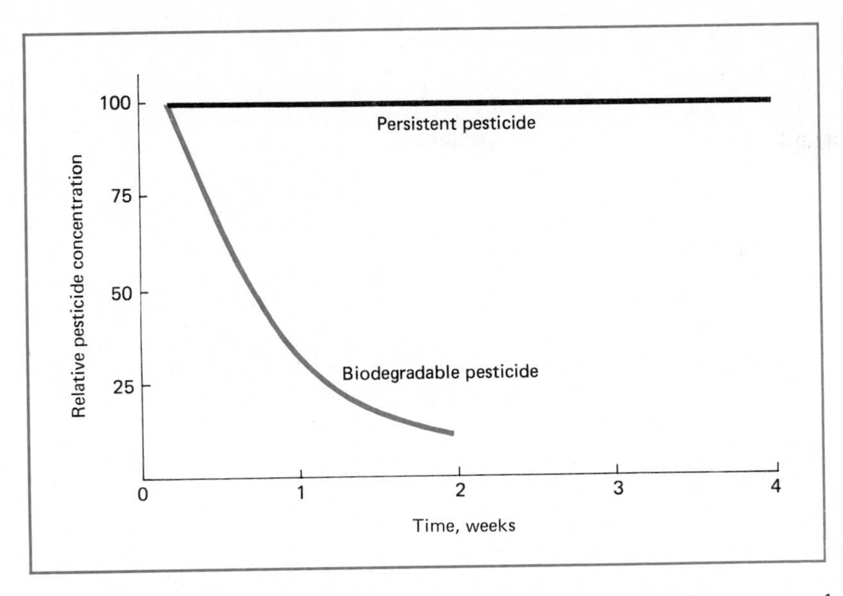

Figure 17.3 Contrast in the rate of degradation in soil of a persistent pesticide and one that is biodegradable.

promotes microbial action on pesticides by providing additional sources of food for microbial growth. Finally, it should be emphasized that only a restricted range of microbial species are able to break down a given pesticide. Thus, the proper microbial inoculum must be present. When a pesticide is first used in a particular soil, there may be very few organisms present that can degrade it. Gradually, however, a population adapted to the compound may build up in the soil, so that subsequent biodegradation proceeds much more rapidly.

When using pesticides in agricultural situations, it is essential to know their persistence times, since applications over a period of years can lead to the build-up to toxic levels of a very persistent pesticide. Thus, it is desirable to assay new compounds for biodegradability before they are made commercially available. Such assays can best be done by adding the compound to a variety of soils and incubating under favorable conditions in the laboratory, then assaying for residual pesticide. Attempts should also be made to develop microbial populations which are specifically adapted to the compound in order to study the possibility for long-term degradation.

The use of certain pesticides has been discontinued in the United States because of the toxicity of the compounds. For example, after DDT had been used for about 20 years, it was found that persistent DDT residues became concentrated along the food chain. Small amounts of nondegradable DDT finding their way into water sources were incorporated into plankton; in turn, the DDT from plankton was eventually concentrated in fish and in turn again into fish-eating birds such as the brown pelican. DDT toxicity was manifested in these birds by an unusual thinness in egg shells; eggs were too easily broken, thus preventing normal reproduction and decimating the population. Since the ban on DDT use was imposed, the brown pelican population has stabilized. Other toxic effects of DDT are possible. However, much of the DDT which has been used in the past remains with us, and the DDT still used in other countries is a constant threat to the environment. The persistence of each new compound should be studied in field trials *before* it is used widely in general agricultural practice. If the compound is unusually stable, it may be necessary to forgo its use in agriculture in order to avoid the build-up of toxic levels in soils, water sources, and animal or human populations.

By understanding the principles of biodegradation, the organic chemist may be able to synthesize pesticides that have low persistence rates in soil. We now know that certain types of molecules present more stable structures to microorganisms than others. Chemists can use this information to decide on the appropriate type of compound to synthesize for a specific agricultural task. In this way, the information on microbial degradation of pesticides can be most successfully applied to the problem of pesticide persistence.

17.3 NITROGEN IN AGRICULTURE

In Section 14.5, we discussed the nitrogen cycle and the general importance of microorganisms in nitrogen transformations in nature. Nitrogen is the element most often limiting for plant growth and a number of microbial processes influence nitrogen availability to plants. In the present section, we discuss the microbial nitrogen transformations most important agriculturally.

Root-nodule bacteria and nitrogen fixation One of the most important microbial processes in the nitrogen cycle is nitrogen fixation, the conversion of nitrogen gas, N_2, into combined nitrogen. This process is carried out only by microorganisms, and higher plants are thus dependent on microorganisms for any nitrogen they obtain via nitrogen fixation. There are two main types of nitrogen-fixing microorganisms: free-living and symbiotic. *Free-living* nitrogen fixers include the cyanobacteria and soil bacteria of the genera *Azotobacter* and *Clostridium*. Although these organisms may make significant contributions to the nitrogen budget of soils, by far the most important nitrogen-fixing organisms in soils are those which are **symbiotic**, living in association with the roots of higher plants.

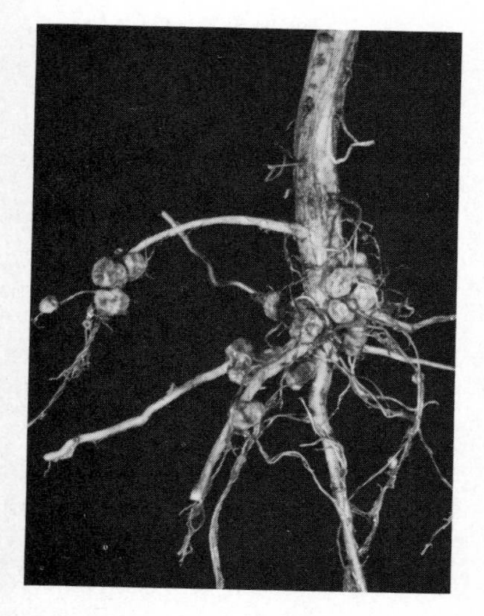

Figure 17.4 Soybean root nodules.

The most important *symbiotic nitrogen-fixing organisms* are bacteria of the genus *Rhizobium,* which live in association with legumes. The main legumes in agriculture are forage crops such as clover, alfalfa, and soybeans, and vegetable crops such as peas, beans, and soybeans. Other legumes of less importance agriculturally include many tropical trees.

Rhizobium cells are Gram-negative motile rods. They are able to infect the roots of legumes and cause the formation of **root nodules** along portions of the root system. Root nodules are tumorlike growths (Figure 17.4) in which nitrogen fixation occurs. Neither the *Rhizobium* alone nor the legume alone is able to fix nitrogen, yet the interaction between the two organisms leads to the acquisition of this property.

Nodule formation is a complex process (Figure 17.5). The roots of the legume excrete nutrients that encourage the growth of *Rhizobium* in the zone immediately adjacent to the root. The *Rhizobium* cells specifically adhere to the hairlike projections from the surface of the root (root hairs) and some of

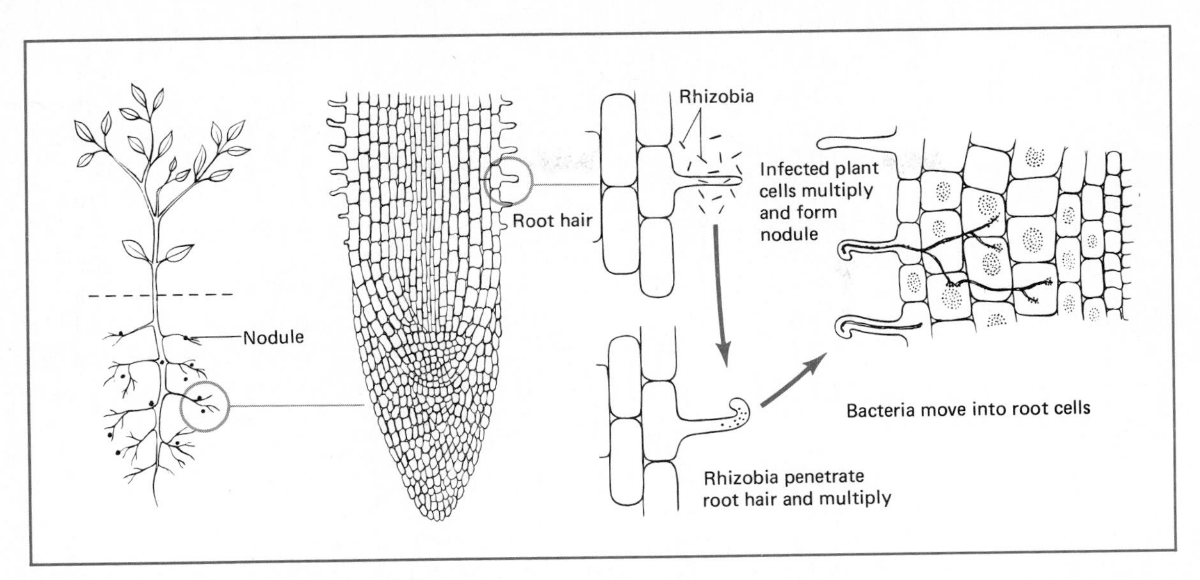

Figure 17.5 Steps in the formation of a root nodule in a legume infected by *Rhizobium*.

the bacteria then penetrate the root, passing down the root hairs and entering root cells. Infection of root cells by bacteria leads to multiple division of the root cells and the formation of the enlarged root-nodule structure. The cells of the root nodules are heavily packed with bacteria. The *Rhizobium* cells in the nodule are considerably altered in shape from the same bacteria outside the root. Within the root cells, the bacteria are swollen and misshapen and are called *bacteroids* (Figure 17.6).

The mature nitrogen-fixing nodule is pink; this color results from the production of a red hemoglobin-like protein (called *leghemoglobin*) within the nodule. Although hemoglobin is very common as a blood protein in the animal world, the legume nodule is one of the few places in the plant world where it is found. The leghemoglobin in the nodule binds O_2 and releases it at a controlled rate for the aerobic nitrogen-fixing bacteria. It is well-estab-

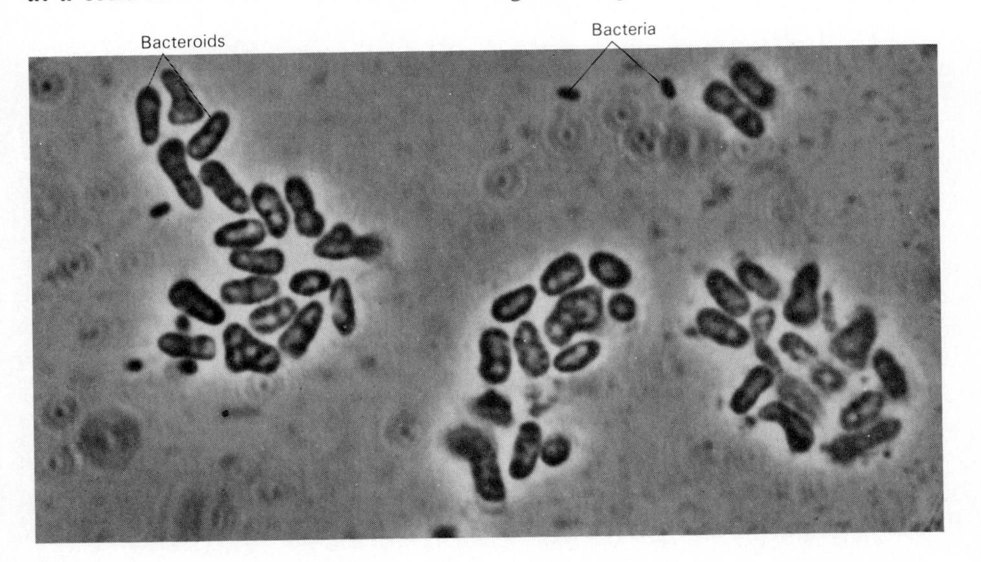

Figure 17.6 Photomicrograph of *Rhizobium* bacteroids and vegetative bacteria from a root nodule. The bacteroids are between 1 and 2 μm in diameter.

lished that nitrogen fixation is an anaerobic process and thus very sensitive to O_2, yet the bacteria need O_2 for metabolism. Since the soil surrounding the plant root is aerobic, O_2 readily penetrates to the nodule. This O_2 is firmly bound by the leghemoglobin, rendering the interior of the nodule microaerophilic, so that the nitrogen-fixation process occurs at the maximal rate.

Energy for nitrogen fixation in the nodule comes from the plant. Sugars produced in the plant leaves are transported to the nodules and promote nitrogen fixation by serving as nutrients for *Rhizobium* growth. Nitrogen (N_2) passes into the nodule and is converted into amino acids by the *Rhizobium* bacteroids. The amino acids then are transported out of the nodule to the roots, stems, and leaves and are used as a nitrogen source for plant growth.

As the root ages, the nodules eventually die and break down, releasing the bacteroids back into the soil. The bacteroids do not multiply there, but there is always a small number of normal *Rhizobium* cells present in the nodule, and these are released also. It is these normal *Rhizobium* cells that multiply in the soil to keep a reservoir of infective cells available. They can serve as the source of inoculation for a new legume.

There is a marked specificity in the *Rhizobium*-legume interaction. A single *Rhizobium* strain can grow on certain species of legumes but not others; similarly, a given legume can interact with only one or a very few of the many known strains of *Rhizobium*. A knowledge of which *Rhizobium* strains are infective for a legume is important in agriculture, since it is common practice to inoculate the seed with the appropriate strain before sowing; naturally, the correct strain of bacterium must be used. Cultures suitable for inoculation can be purchased commercially. Inoculation of seed is especially important if the legume crop has not been previously raised in the area or has not been raised for some time, since a *Rhizobium* suitable for this legume will probably not be already present in the soil.

In nitrogen-deficient soils, nodulated legumes grow better than unnodulated plants (Figure 17.7); thus, nodulation is a distinct advantage to the

Figure 17.7 Nodulated and unnodulated alfalfa plants growing in nitrogen-poor soil.

plant. It is also beneficial agriculturally, since a leguminous crop can be grown for a season and then plowed into the soil, thus adding considerable combined nitrogen to the soil. However, it is debatable whether such practice is preferable over using chemically synthesized nitrogen fertilizers. It is usually considered that in regions of high-intensity, high-cost agriculture, such as in North America and Europe, where land costs are high, chemical fertilizer may be cheaper, whereas in low-intensity agricultural areas, such as Australia and New Zealand, use of legumes may be better.

Microbial transformations of nitrogen fertilizers Nitrogen fertilizers are widely used in agricultural practice to increase plant production. Three types of nitrogen fertilizer are used: nitrate (NO_3^-), ammonia (NH_3), and urea ($H_2N{-}CO{-}NH_2$). All three of these forms are transformed in various ways by microorganisms, and the microbial transformations may be important in determining the persistence and availability of the fertilizer after its application to the field. Some of these transformations have been discussed in Chapter 14.

Potassium or sodium *nitrate* is a convenient form of nitrogen fertilizer to add to soils. The nitrate ion is very water soluble, and the nitrogen is thus rapidly available to plants. In the absence of oxygen, however, several bacteria are able to convert nitrate into nitrogen gas, N_2, thus changing the material into a form unavailable for higher plants. This process, called **denitrification**, is very undesirable agriculturally since the nitrogen is lost from the soil. Bacteria that denitrify are carrying out a type of anaerobic respiration (see Section 3.7), using the nitrate ion instead of O_2 as an electron acceptor. Thus, for denitrification to occur, soil conditions must be anaerobic, and a source of organic matter must be available for bacterial metabolism.

Since denitrification is undesirable agriculturally, many farming practices are designed to minimize it. Draining of soils to ensure that they do not become waterlogged is of value, since flooded or waterlogged soils quickly become anaerobic and thus favor denitrification. The practice of tilling soil probably inhibits denitrification, since it breaks up large soil clumps within which anaerobic conditions generally develop. However, even in well-drained soils of good texture, some denitrification may occur if large amounts of organic matter are added (as from manure or other organic fertilizer); the organic matter promotes the growth of facultatively anaerobic bacteria that use up the oxygen present and create temporary or partially anaerobic conditions. Therefore, a proper carbon-to-nitrogen balance is desirable in agricultural soils.

In large farm practice, the most commonly used nitrogen fertilizer is *ammonia* (NH_3). Ammonia is preferable to nitrate because on a weight basis it has more nitrogen (the H atoms of ammonia weigh less than the O atoms of nitrate), and it can be more conveniently applied to large fields. Ammonia is normally a gas, but it can be liquefied under pressure, and it is applied to fields in this form. A high-pressure tank is pulled across the field and the ammonia is injected from a nozzle directly into the soil. As soon as the gaseous ammonia contacts the soil particles, it is adsorbed and converted in the soil into the ammonium ion (NH_4^+), which is not gaseous. The water-soluble

NH_4^+ can be used directly by higher plants, but most commonly it is converted into nitrate through the action of the *nitrifying bacteria*, a process called **nitrification**. We discussed nitrification briefly when we considered the overall nitrogen cycle in Section 14.5 (see also Section 3.7). The nitrate formed is taken up by plants.

Nitrification is an aerobic process and is thus favored by aeration of the soil. The nitrifying bacteria are widespread in soils and water and are commonly found in fields where ammonia fertilizer is used. Nitrification occurs most readily in soils with neutral to alkaline pH, and is inhibited in acidic soils.

Urea ($H_2N{-}CO{-}NH_2$) is an organic form of nitrogen that is sometimes used as a fertilizer. It is hydrolyzed in water by the enzyme urease, to yield ammonia and carbon dioxide:

$$H_2N{-}CO{-}NH_2 + H_2O \rightarrow 2NH_3 + CO_2$$

Urease is a common enzyme in many soil bacteria, and the conversion of urea to ammonia in soil is primarily a bacterial process. The ammonia liberated from urea may then be converted to nitrate by the nitrifying bacteria.

17.4 PLANT DISEASES

Plant diseases caused by microorganisms are of major economic concern and a knowledge of methods of control of plant diseases is crucial to ensure high crop yields. Under some conditions, a plant disease can affect the well-being of a whole population or country. The classic case of this was the Irish potato blight (see Chapter 1), a fungal disease which caused a mass migration of people from Ireland.

Although the principles of host-parasite relations discussed in Chapter 8 are generally applicable to plant diseases, there are so many differences between plants and animals that a separate discussion of plant diseases is desirable. Bacteria, fungi, and viruses all cause diseases of plants, although fungi cause the most serious and widespread diseases of crop plants.

Plants, like animals, have a *normal flora* of microorganisms that are not pathogenic. Although microbes can grow on all parts of a plant (flowers, fruits, seeds, leaves, stems, and roots), it is on the roots that the most extensive normal microbial flora develops. The region in the soil immediately around the roots is called the *rhizosphere*, and here large numbers of harmless bacteria and fungi are present, living on nutrients excreted from the roots. Under some conditions, microorganisms are able to penetrate into the living root tissue itself without causing special damage. Trees often have fungi attached to their roots; this fungus-root association is called a *mycorrhiza*, and mycorrhizae are of considerable benefit to the tree.

Plant pathogens may cause disease in plants in a variety of ways: (1) by digesting the contents of host cells; (2) by producing toxins, enzymes, and other substances that kill or damage the function of host cells; (3) by absorbing food materials from the plant and thus depleting the plant's own food supply;

(4) by blocking the transportation of food, minerals, and water through the conductive vessels of the plant as a result of microbial growth in the vessels.

Plant diseases are given names based on the organ affected and on the appearance of the diseased organ (Figure 17.8). For example, there are root rots, leaf spots, fruit rots, wilts, cankers, leaf blights, galls, mildews, and rusts.

Although microorganisms are the primary causes of plant disease, *environmental factors* greatly influence the microbe's ability to attack the plant. Temperature, moisture, light, soil acidity, and availability of plant nutrients are all factors that influence the presence and severity of disease. Certain diseases may be common in summers that are cool and damp, whereas others may be more common in warm dry summers. Some plant varieties are more susceptible to specific diseases than are others, and disease incidence is thus affected by the variety of plant used.

Control of plant diseases is achieved differently than control of animal and human diseases, since rarely is an individual plant of such value that therapeutic measures are justified. Control of plant diseases is therefore primarily by prevention rather than by treatment.

Chemical methods are employed primarily to protect plants against future disease. Leaf sprays and dusts, seed treatments, soil fumigation, and soil sterilization are some of the treatments used. Bordeaux mixture, a combination of copper sulfate and lime, was first used as a leaf dust near Bordeaux, France, for controlling powdery mildew of grapes. A number of sulfur preparations are used, including both inorganic (elemental sulfur, S^0) and organic forms of sulfur. Inorganic and organic mercury compounds are also highly effective,

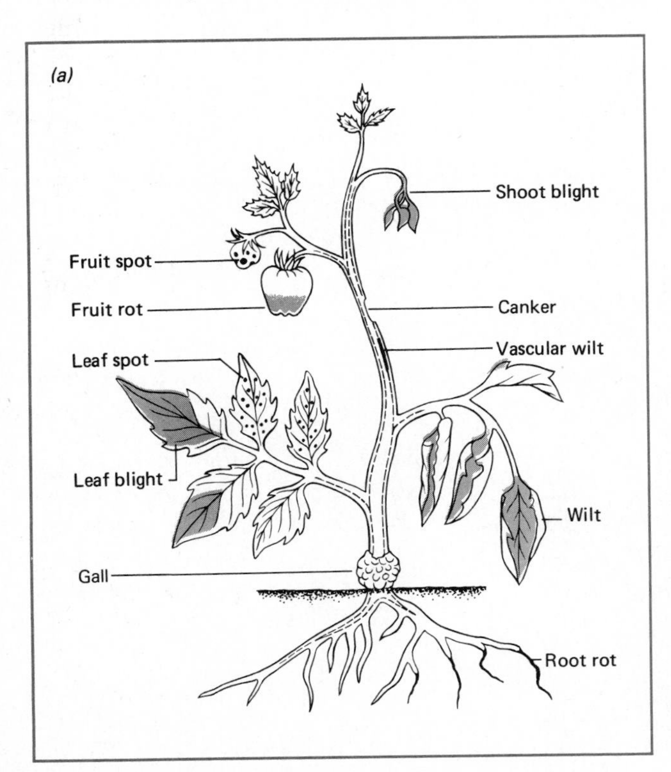

Figure 17.8 (a) A composite drawing showing typical plant diseases caused by microorganisms. (b) Tomato with bacterial spot.

(b)

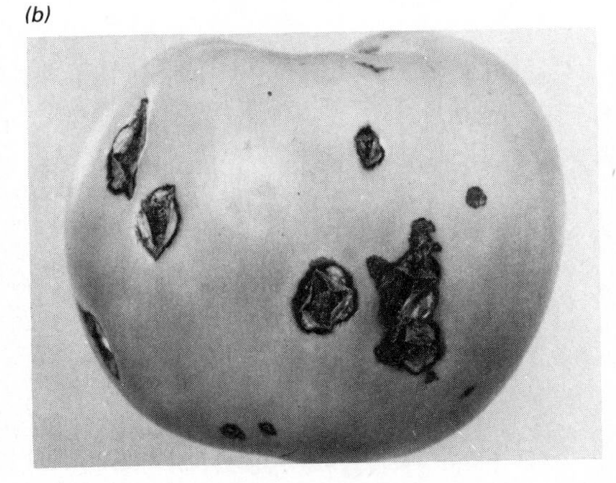

but because of the high toxicity of mercury to humans and animals, these compounds are used primarily in the treatment of seeds before planting; in that way, none of the mercury reaches the final harvested crop. A number of organic compounds are used, including both synthetic chemicals and antibiotics. Of the antibiotics, the most widely used are streptomycin and tetracycline, which control certain bacterial diseases, and cycloheximide and griseofulvin, which are used to control some fungal diseases (see Section 5.6).

An important factor in the use of chemicals to control plant diseases is the cost of treatment. Usually, when chemicals and antibiotics are used to control human diseases, cost is not the prime factor, but in agriculture, with the large areas to be treated with the chemical and the low value of the plant, the cost of treatment can be quite high and so must be balanced against the anticipated economic gain from its use. Thus, some effective chemicals are not used in practice because they are too expensive to manufacture or apply.

Other control methods employed in some cases include: (1) mild heat treatment of seeds to eliminate pathogens; (2) crop rotation, in which crops not susceptible to the pathogen are planted for a year or two so that the pathogen finds no host to grow on and dies out; (3) quarantine and inspection, to prevent the entry of a pathogen into an area where it has not yet become established; (4) eradication of alternate hosts for a pathogen (these are noncrop plants that serve as reservoirs of infection); (5) appropriate fertilization, tilling, and weed control, which promote appropriate growth of the crop to increase its disease resistance; (6) control of agents such as insects that may disperse the pathogen; (7) use of resistant varieties of the crop.

The field of plant pathology is highly developed, and tremendous advances have been made in the control of plant diseases. These controls have led to marked increases in crop yields and have to a great extent contributed to the agricultural revolution that has taken place in the world in the past 50 years. However, continual vigilance is necessary as new pathogens arise, and research and development aimed at controlling plant diseases must continue. Also, more effective and less toxic chemical agents are needed, so that the use of mercury and copper compounds potentially harmful to humans can be eliminated. In this research effort, the microbiologist and plant pathologist play major roles.

17.5 MICROBIAL INSECTICIDES

Insects cause a wide variety of diseases in plants as well as act as vectors of animal pathogens and control of insect infestation usually can greatly increase crop yield and animal health. Because a number of undesirable effects can arise from the use of chemical pesticides, methods for controlling insects by nonchemical means have been sought. One potentially useful method is the use of microbial insecticides. These are microbial toxins or viruses which kill insects.

Insects are susceptible to a wide variety of microbial pathogens (Table 17.2). In virtually every case, these insect pathogens are quite specific and have no effects on humans or higher animals, which makes them very de-

TABLE 17.2 Examples of microbial insecticides

Agent	Uses
Viruses	
Nuclear polyhedrosis virus of the corn earworm	Caterpillar pests of cotton, soybeans, tomatoes, corn, and tobacco (cotton bollworm, corn earworm, budworm, tomato fruit worm, soybean podworm, tobacco budworm)
Nuclear polyhedrosis virus of the tussock moth	Caterpillar pest of forests and shade trees
Nuclear polyhedrosis virus of the gypsy moth	Caterpillar pest of forests especially hardwoods
Nuclear polyhedrosis virus of *Autographa californica*	Caterpillar pest of vegetables and field crops
Nuclear polyhedrosis virus of sawfly larvae	Caterpillar pest of forests
Bacteria	
Bacillus popilliae	Japanese beetle grubs of lawns and pastures
Bacillus thuringiensis	Caterpillar pests of vegetables, forests, and ornamentals
Bacillus thuringiensis subsp. *aizawai*	Wax moth
Bacillus thuringiensis subsp. *israeliensis*	Mosquito larvae
Bacillus sphaericus	Mosquito larvae
Fungi	
Verticillium (Cephalosporium) lecanii	White fly, aphids
Hirsutella thompsonii	Mites on citrus
Phytophthora citropthora	Milkweed vine in citrus
Nomuraea rileyi	Caterpillar pests of vegetables, field crops, forests and forage
Protozoa	
Nosema locusteae	Grasshoppers in rangeland

sirable in comparison to chemical insecticides. One major group of insect pathogens is comprised of certain members of the genus *Bacillus*. During sporulation, these spore-forming bacteria produce a protein that is toxic to specific insects. In actual practice, the toxin-producing organisms are cultured in the laboratory and a preparation containing toxin and spores is obtained. This is then formulated into a dust or spray and distributed in the same manner as a chemical insecticide. Because of the high specificity of the toxin, there is no danger to the health of human beings or livestock. The most widely used organism is *Bacillus thuringiensis*; this toxin-forming bacillus grows in various moth larvae and causes a fatal disease. Since some moth larvae (for example, cabbage worm, tent caterpillar, gypsy moth) destroy plants, the toxin can be used to control these insect infestations.

Another species, *Bacillus popilliae*, has a much more specific action, affecting just the Japanese beetle. This agent is thus less widely useful but has proved effective for the single important insect pest that it attacks.

Several viruses have also been found effective in controlled trials. Nuclear polyhedrosis viruses and granulosis viruses affect some of the same insects that are affected by *B. thuringiensis* (see Table 17.2) but have the additional advantage that they should be able to reproduce in the larvae under field conditions after application, giving some degree of long-term control. However, in practice, repeated applications of the viruses are usually necessary because they become inactivated on foliage as a result of exposure to sunlight. Additionally, continued growth of the crop plant leads to the production of new leaves that do not have any virus.

The microbial agents that have been used so far are essentially insecticides, attacking the insects present at the time of application but not multiplying in nature so that they have little or no long-term effect. They do not differ in principle from chemical insecticides, except that they are much more specific and have minimal or no harmful effects on the environment. However, the ideal type of agent for control of insect pests is an agent that multiplies after application and is transmitted from one insect to another. Such an agent would spread through the insect population and would, if sufficiently effective, eventually wipe out the insects in the target area. To date, no microbial pathogens with these properties have been discovered.

17.6 MYCOTOXINS AND GRAIN STORAGE

Mycotoxins are chemical toxins produced by fungi; they were discussed briefly in Section 12.5. The best-known mycotoxins are those called *aflatoxins*, produced by the fungus *Aspergillus flavus*. *Aspergillus flavus* is widespread in the soil and generally causes no harm there. However, grain becomes infected with the fungus in the field very easily; under improper storage conditions, the fungus can grow in the harvested grain and produce aflatoxins. Grain is usually stored in large storage bins or silos called *grain elevators*. The most important factor determining whether the aflatoxin-producing fungus will grow is the moisture content of the grain. If the moisture content is low (less than 14 percent moisture) and the bins are kept dry, the fungus will not grow. However, if grain is harvested at too high a moisture content, as might occur at the end of an unusually wet season, the fungus may grow in the grain and produce aflatoxins (Figure 17.9*a*). Susceptible crops include wheat, corn, oats, peanuts, and rice. Species of *Fusarium*, another fungus which produces mycotoxins, are also a significant contaminant of cereal grains (Figure 17.9*b*).

Aflatoxins act by affecting the liver. In low doses, they may induce the formation of liver cancer (hepatoma), and in higher doses they cause a general liver toxicity that can lead to rapid death. Animals affected include cattle, pigs, chickens, turkeys, and horses. Also affected are fish, and extensive damage to rainbow trout in fish hatcheries has resulted from the feeding of moldy grain. In addition, humans may be affected if they eat foods such as peanut butter or flours prepared from moldy raw materials.

The variety of aflatoxins so far identified is large; the major ones include aflatoxin B_1, G_1, M_1, and M_2. All of these can be produced in grain, but of further interest is the fact that aflatoxins M_1 and M_2 can also be produced in the animal body from aflatoxins B_1 and G_1. Aflatoxins ingested by the dairy cow can pass to the milk, hence finding their way into milk products (Figure 17.10).

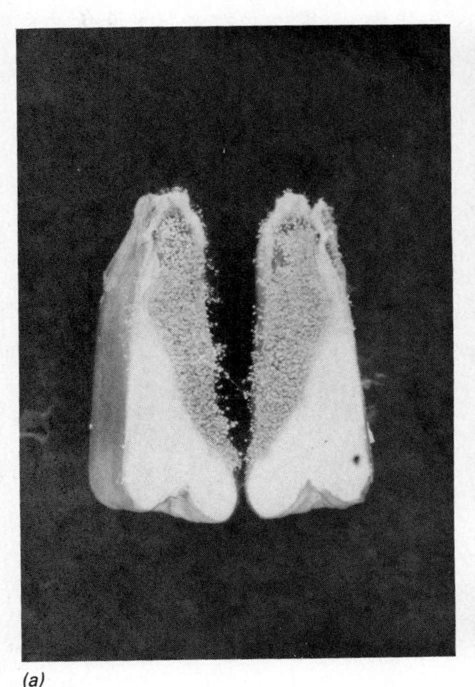
(a)

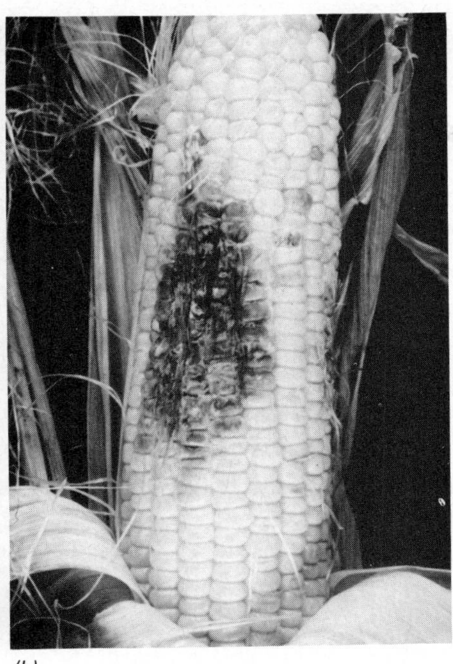
(b)

Figure 17.9 (a) Corn kernel split open to show *Aspergillus flavus* infection. (b) Ear of corn infected with *Fusarium*.

Because of the serious nature of the disease, considerable caution must be exercised to be sure that grain is stored properly. Since grain is often stored in large quantities, mycotoxin development is a worldwide problem and is of considerable economic importance. Fortunately, we have a good knowledge of the proper methods of control of these important infestations.

17.7 ANIMAL DISEASES

There are a number of ways in which a knowledge of microbiology can be applied to the care and feeding of livestock. Microbes play beneficial roles in many aspects of animal husbandry—for example, in ruminant digestion and in the preparation of silage for use as animal feed. These subjects will be discussed later in this chapter. On the other hand, many microbes present very great problems for livestock, causing sickness, loss of production, or even death. In this respect, a knowledge of some aspects of veterinary medicine can be of great use to the farmer. The principles of host-parasite relations and immunology in relation to humans which were presented in Chapters 8–10 apply directly to animal diseases as well.

Just as with humans, there are a great many diseases to which farm and domestic animals are susceptible (Table 17.3). It is incumbent upon the modern farmer to know what the possible dangers are and to combat them appropriately, whether by *preventing* diseases by immunization or by *identifying* sick animals and having their diseases *diagnosed* and *treated* by a veterinarian when necessary (see the Box on page 494).

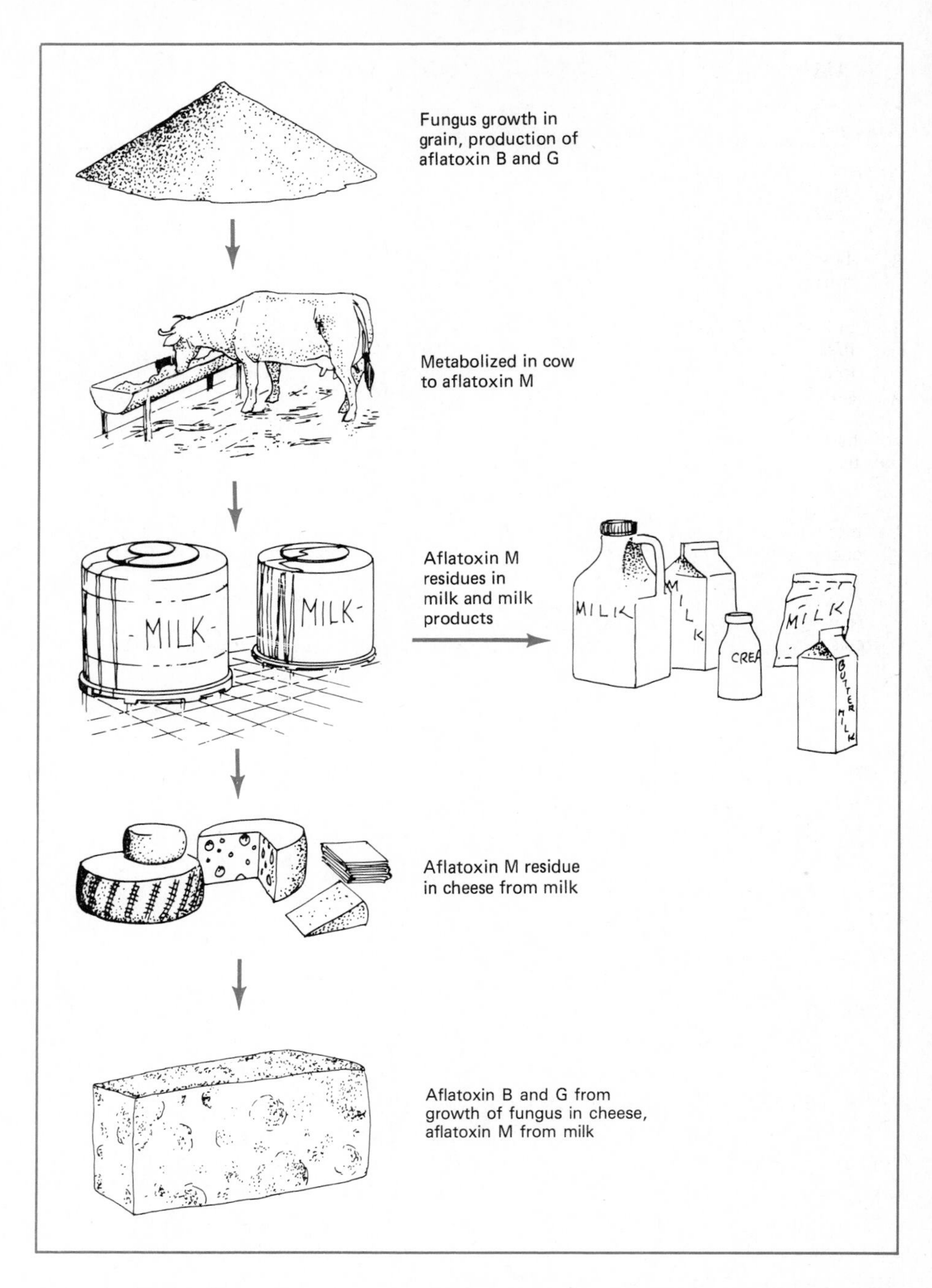

Figure 17.10 Aflatoxin production and transfer in dairy products.

Diseases under control Several diseases were once responsible for destroying large numbers of farm animals but are now controlled by effective immunization or testing programs. These include tuberculosis, brucellosis, anthrax, leptospirosis, hog cholera, rhinotracheitis, and black leg. The first two diseases, tuberculosis and brucellosis, have historically been among the most serious diseases of livestock, not only because of their effect upon the

TABLE 17.3 Infectious diseases of livestock

Disease	Pathogen	Prevention
Anthrax	*Bacillus anthracis*	Vaccination every season
Blackleg	*Clostridium chauvoei*	Vaccination
Brucellosis	*Brucella abortus*	Vaccinate young; test and slaughter
BVD (Bovine virus diarrhea)	Myxovirus	Vaccination
Calf scours	*Escherichia coli, Salmonella,* viruses	Good sanitation, colostrum, vaccination for *E. coli*
Foot-and-mouth disease	Virus	Quarantine and slaughter
IBR (Infectious bovine rhinotracheitis)	Herpes bovis virus	Vaccination
Leptospirosis	*Leptospira* sp.	Vaccination every 6–12 months
Mastitis	*Streptococcus agalactiae*	Sanitation, germicides
Metritis (uterine infection)	*Streptococcus, Corynebacterium*	Sanitation at calving, antibiotics, artificial insemination
Pneumonia	Bacteria, viruses	Good ventilation, clean, dry
Salmonellosis	*Salmonella typhimurium*	—
Shipping fever	*Pasteurella haemolytica*	Antibiotics
Trichomoniasis	Protozoa	Artificial insemination
Tuberculosis	*Mycobacterium bovis*	Test and slaughter
Vibriosis	*Campylobacter fetus*	Artificial insemination, antibiotics in semen

animals, but also because of the danger of transmission of these diseases from animals to humans. By government regulation today, however, these diseases have for the most part been brought under close control. *Tuberculosis,* caused by *Mycobacterium tuberculosis,* is an important disease of humans (see Section 11.9), which was once frequently transmitted to humans from contaminated milk taken from tuberculous cows. Milk is almost never the source of human tuberculosis today. Cows are given tuberculin tests at yearly intervals to ensure that they are free of the disease; any animals exhibiting a positive test are slaughtered. Such test and slaughter programs have essentially eradicated the disease in dairy herds. All milk must come from certified (tuberculin negative) cows. Pasteurization of milk also kills this pathogen (see Table 16.7), and would of course also ensure its safety.

Tuberculosis also occurs in swine and poultry, but a tuberculin testing program is too costly for these animals (and often inaccurate in swine). Instead, reduction in the incidence of the disease has been accomplished by careful attention to cleanliness, by removal of all poultry from large swine farms to prevent cross-contamination, and by slaughter of poultry flocks in which tuberculosis occurs.

Brucellosis, caused by several species of *Brucella,* occurs in cattle, swine, sheep, and goats, and can be transmitted to humans, generally by handling of infected animals or by ingesting nonpasteurized milk or dairy products from infected animals. Brucellosis is important in animals (see Section 11.13) primarily as a cause of abortion; in humans, the disease is called undulant fever, and is a serious, generalized disease characterized by undulating fever

THE POULTRY SLAUGHTER: "IT COULD HAVE BEEN WORSE."

When can the expenditure of $60 million to kill 17 million birds be a bargain? When it saves over *$5 billion*. This is the story of an outbreak of avian influenza, one of the most contagious diseases of chickens and turkeys. Causing an acute infection of the respiratory system, avian flu is up to 90 percent fatal in poultry, although the virus does not harm humans. Because of the extreme contagiousness of the virus, even if only a few chickens in a flock are infected there is no escape; complete eradication by slaughter is the only solution. The $60 million mentioned earlier was spent to control the outbreak of avian influenza that occurred in the eastern United States in 1983–84. Before the epidemic was over, 16 million chickens and 1 million turkeys had been intentionally killed. Poultry producers incurred direct losses of $55 million, plus additional costs for cleanup, disinfection, and transportation. Some of the poultry producers' losses were reimbursed by the federal government, which contributed $20 million in administrative costs and $40 million in direct payments to producers. But an economic analysis showed that the $60 million spent in the eradication program saved the industry $450 million, since the losses were estimated to be less than 10 percent of what they might have been without eradication. Further, consumers saved over $5 billion from the price increases that the shortages would otherwise have brought. If the epidemic had not been checked, within 6 months it would have blanketed the eastern United States, put a halt on exports, and threatened the very survival of a major portion of the poultry industry.

Some recommendations to avoid future epidemics: earlier disease detection followed by quarantine and flock eradication, use of smaller poultry houses so that fewer birds become contaminated, segregation of breeding stock in a separate area, and prevention of cross-contamination between farms. From an analysis of the 1983–84 epidemic, farmers and producers have learned that the best way to control an epidemic is "lightning-fast reaction with containment tactics." As one observer commented: "17 million birds, but it could have been worse."

and joint pain. Diagnosis of the disease in animals is accomplished by serological test. Animals which recover from the disease generally continue to harbor the organism, and can spread it through a herd or to humans. By state and federal law, all animals with a positive brucellosis test are slaughtered; this has greatly reduced the incidence of the disease. Transmission of the disease is lessened by careful sanitation, isolation of animals that have aborted, and sanitary disposal of any aborted fetal materials.

Anthrax, caused by *Bacillus anthracis,* usually occurs as a bacteremia arising within 1 to 2 days after animals ingest feed contaminated with anthrax spores. Onset may be very rapid and is usually followed by a violent death. Bloody discharges from all body openings just prior to death are common. The organisms in the blood sporulate readily, and typical endospores are released, contaminating the premises or pastureland land almost indefinitely. Humans acquire the disease most frequently by handling of infected materials or inhalation of spores (see Section 11.8). Areas of known soil contamination in the United States include some of the Central and Western states as well as a few other areas. In such areas, vaccination is effective in preventing major stock losses, but must be repeated each pasture season. Treatment of anthrax with penicillin or other antibiotics is often not effective since the disease progresses so rapidly, but, if begun early, some animals may be saved. To avoid spread of the pathogen, dead animals are either burned or buried.

Anthrax still causes serious losses in many parts of the world, but vaccination and quarantine procedures in the United States have made it less important here.

Diseases not yet controlled There are a number of diseases of farm animals which are not well controlled. For most of these, either there is no effective vaccine or the cost of individual immunizations is greater than the value of the animal or the risk of the disease. The overall cost to the farmer of many of these diseases is formidable, however, and great care must be exercised to lessen their frequency and financial impact.

Mastitis is the most common and costly disease of dairy cows. In subclinical form, this disease can occur in up to 50 percent of all dairy cows. Although there are no visible exterior changes, subclinical infection results in lowered milk production. Serious clinical infection occurs in only a small percentage of dairy cows. Clinical signs include abnormal milk, inflammation of the udder, lack of appetite, and a greatly reduced milk flow. The causative organisms are most commonly *Streptococcus agalactiae, Staphylococcus aureus,* or *Escherichia coli,* although various others may also be implicated. The disease is spread from animal to animal via the milker's hands or by improperly cleaned or malfunctioning milking equipment, with the organisms entering by way of the teat canal. Diagnosis of acute mastitis is made by visual observation of the milk for abnormalities (done by squirting milk directly onto a black strip pan to reveal milk that is watery or contains clots or flakes). Acute mastitis is also accompanied by the classic signs of inflammation of the udder and decrease in milk production. Although resistant strains may occur, treatment is often successful, with penicillin, streptomycin, and tetracycline being most effective. Prevention is preferable to treatment, however, and can be accomplished by proper sanitation and the isolation of infected animals. Since *Streptococcus agalactiae* does not live in nature outside the animal, it is possible to eradicate the disease from individual dairy herds.

Calf scours, or colibacillosis, is another disease of great economic significance on the farm. It is observed primarily in young animals—calves, piglets, and lambs. Two forms occur: enteric and bacteremic. The more common is the enteric form, in which the intestinal tract is involved and diarrhea, or "scours" occurs (Figure 17.11). The causative organisms are enterotoxin-producing strains of *Escherichia coli,* which are always present in manure. Infection is generally by the oral route during the first week after birth. Symptoms include a slight fever, abdominal pain, and diarrhea accompanied by severe fluid loss. The resulting dehydration and electrolyte loss can cause death in calves within 3 to 5 days if not treated. Piglets may succumb in as little as a day. Treatment consists primarily of fluid replacement by electrolyte solution, and good results are observed if treatment is begun soon enough. Antibiotics are routinely given as well, although they have shown only variable results. Indeed, it is common practice on some farms to allow continuous feeding of antibiotics in the water or feed as a preventive measure. Although effective, there is a very real danger with this method that selection of antibiotic-resistant strains of microorganisms may occur (see Chapter 5 and the Box on page 471). Prevention of calf scours is best achieved by being sure

Figure 17.11 A case of calf scours at an early stage.

that newborn animals receive colostrum from the mother immediately after birth, since this early milk contains protective antibodies, and also by maintaining proper sanitation, avoiding crowding, keeping calves in separate quarters, and dipping the navels of newborns in iodine to disinfect them. Calves can also be vaccinated.

"Eradicated" diseases Considerable success has been achieved in eradicating from livestock and fowl a number of serious animal diseases that had previously caused significant economic loss. Research continues on many of these eradicated diseases because they continue to be widespread in other parts of the world and can possibly be reintroduced if quarantine measures are faulty.

Foot-and-mouth disease is a highly communicable disease caused by a virus. The virus causes lesions in the mucosa of the mouth, in the skin around the hoofs, and on the teats and udder. The virus spreads throughout the whole body, and all secretions and excretions contain the virus. Foot-and-mouth disease has been introduced into the United States at least six times in the twentieth century and has been eradicated each time. The last outbreak was in California in 1929, when the virus was introduced in ships' garbage. There have been more recent outbreaks in neighboring countries: Canada in 1952 and Mexico in 1953 and 1954. In the United States, a strict quarantine is imposed immediately around any affected area, and all infected animals and exposed susceptible livestock are killed and buried on the premises. After 30 days, a few susceptible animals are introduced onto the farm by the federal government and are observed for disease. If no symptoms develop after another 30 days, permission is granted for introduction of new stock, but the farm is maintained under surveillance for an additional 90-day period. To ensure immediate reporting of possible problems, owners receive payments from the government to compensate for their livestock losses. In countries where the disease is endemic, control is by vaccination, but there is no need for vaccination in the United States, since the disease does not exist here, and the use of eradication measures when necessary are less costly than universal vaccination. A genetically engineered vaccine is now available.

17.8 RUMINANTS AND MICROORGANISMS

Ruminants are animals that possess a special organ, the *rumen,* in which the initial digestion of feeds takes place. Some of the most important domestic animals, the cow, sheep, and goat, are ruminants. In addition, a number of wild animals are ruminants, including deer, elk, moose, buffalo (bison), antelope, and camel.

Hay and grass, the main feed of ruminants, contain as their main food constituent cellulose, a material that is generally indigestible by higher animals. In humans and other meat-eating animals, cellulose passes through the intestinal tract virtually undigested. In ruminants, on the other hand, cellulose digestion does occur, through the activities of microorganisms living in the rumen.

The rumen is a large saclike organ, about 100 liters in volume in the cow and 6 liters in volume in the sheep, into which feed first enters (Figure 17.12). In the rumen, the feed is acted upon by anaerobic cellulose-digesting bacteria and protozoa. Cellulose is broken down first to glucose, and the glucose is then converted to organic acids, mainly acetic acid. These acids pass out of the rumen into the bloodstream where they are transported to the tissues and serve as the animal's main source of energy for respiration. Rumen bacteria use the simple inorganic compound ammonia as a source of nitrogen, which is present in the rumen fluid, and convert this into the organic nitrogen of protein. The growth of bacteria can be increased by feeding an additional source of ammonia along with the hay or grass. Cellulose-digesting bacteria are present in large numbers in the rumen. They pass out of the rumen into the stomach and intestine, where they are in turn digested (Figure 17.12). The main source of protein for the animal is thus the rumen bacteria. Through the mediation of the bacteria, the animal is able to obtain some nitrogen from inorganic forms; all other animals require at least part of their nitrogen in organic form, as preformed amino acids. The rumen bacteria also synthesize all the vitamins the animal needs for growth and survival. Because of the activity of these rumen bacteria, the animal is able to subsist on food materials low in protein, which are quite insufficient for the growth of nonruminant animals.

The microorganisms of the rumen are obligate anaerobic types that are specifically adapted to this habitat and are found nowhere else. Both bacteria and protozoa are found in the rumen. The bacteria are mainly nonsporulating rods and vibrios; the protozoa are ciliates. Although the protozoa are not essential for rumen function, when present they definitely contribute to the process. Their numbers are usually larger in animals fed a good ration than

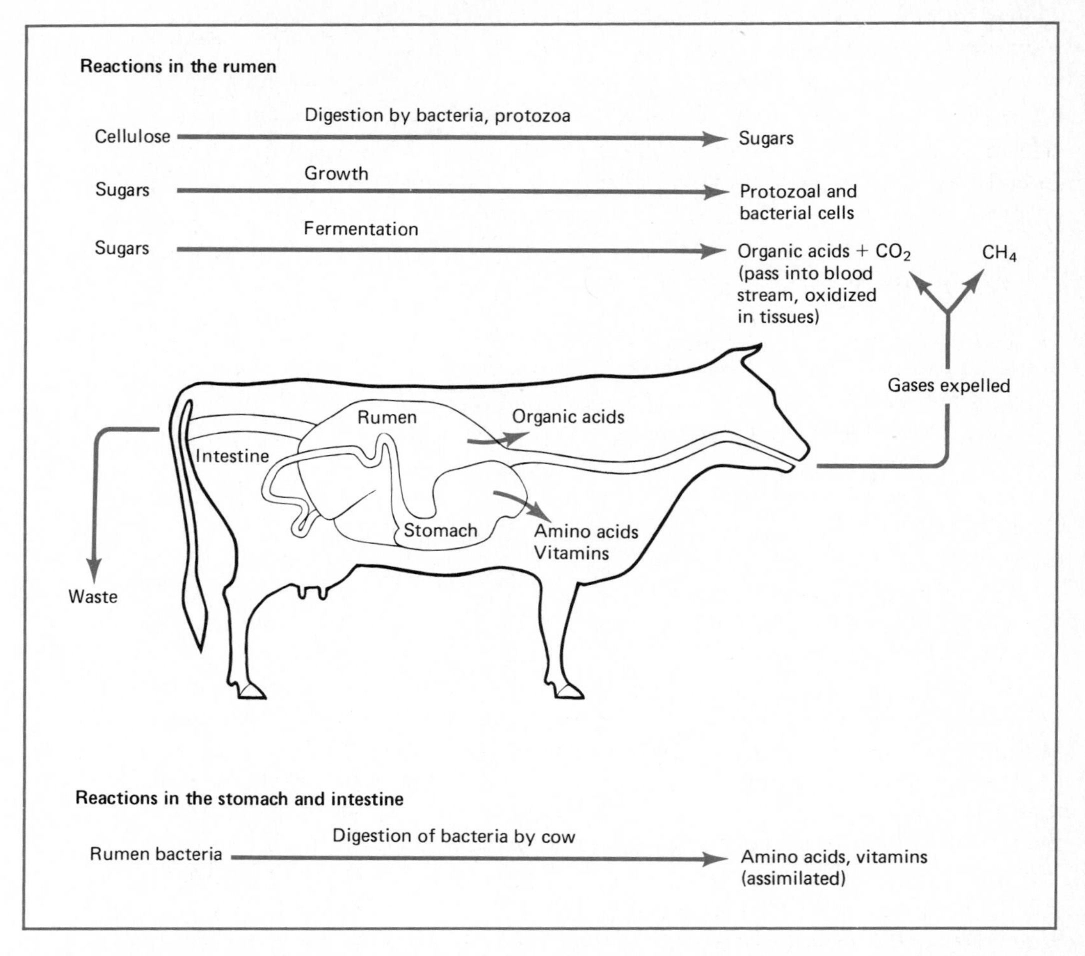

Figure 17.12 Microbial processes in the ruminant animal.

in those fed a poor diet, and their presence may possibly serve as an indication of the well-being of the animal. The protozoa digest cellulose, but they also eat and digest some of the rumen bacteria. Thus the protozoa may serve to keep bacterial numbers in check.

A knowledge of microbial processes in the rumen can be put to practical use in the intelligent management of ruminant animals. As we have seen, the protein value of cattle feeds may be increased by addition of inorganic nitrogen, since the rumen bacteria can convert inorganic nitrogen into protein. The usual nitrogen source fed to ruminants is urea, which is broken down in the rumen to carbon dioxide and ammonia, the ammonia then being converted into bacterial protein. Feeding cereal grains to cattle is a common practice, but this is wasteful and inefficient since grains are already high in protein and contain carbohydrate in the form of starch, which is readily digestible by nonruminants. Thus, there is a disadvantage in feeding grains to ruminants, since ruminants are less efficient in handling these feeds than are nonruminants such as pigs and chickens. However, some protein nitrogen must be fed, as if too much nitrogen comes from urea, the palatability of the meat is reduced. In practice, feeding grain to cattle is done primarily to fatten them for slaughter and to ensure that the flavor of meat cuts is superior. If grains are in short supply, they should not be fed to ruminants.

One unsolved problem which influences the economics of raising ruminant animals is the production of methane gas by rumen bacteria. The mechanism of anaerobic digestion and methane production was described in Section 15.8. In the rumen, cellulose fermentation to organic acids also results in production of some hydrogen (H_2) and carbon dioxide (CO_2). Methanogenic bacteria convert H_2 and CO_2 to methane (CH_4). The methane is then belched out by the animal, and since methane is an energy-rich compound, this results in an energy loss to the animal. In fact, as much as 11 percent of the energy in the feed consumed by ruminants is not used by the animal for meat or milk production, but is simply lost. Antibiotics to inhibit rumen methanogenesis have been developed, but there are conflicting results as to whether or not their use really leads to improved animal production.

17.9 ANTIMICROBIAL AGENTS IN ANIMAL FEEDS

It has been well established that the growth rate of nonruminant livestock, such as pigs and chickens, is increased if they are fed small amounts of antimicrobial agents such as antibiotics in their feed. In fact, antibiotic supplementation of feeds for these animals has been a widespread practice, since the more efficient feed use and more rapid growth permits the farmer to take the animals to market earlier. Some antibiotics and other antimicrobial agents that have been used in animal feeds are listed in Table 17.4. Antibiotics used most frequently as feed supplements are penicillin and tetracycline.

One way in which antibiotics stimulate growth is by reducing or eliminating bacteria from the normal flora that produce small amounts of toxins which decrease growth rate. One organism affected by antibiotic feeds is *Clostridium perfringens*, a bacterium present as a minor member of the normal flora of the intestinal tract, which produces several toxins. With the levels of antibiotic used in feeds, most of the normal flora of the body is unaffected, while *C. perfringens*, which is fairly antibiotic sensitive, is eliminated. With the organism gone, the toxins are also absent and growth is stimulated.

Another mechanism for growth stimulation by antibiotics relates more specifically to pathogenic bacteria. Livestock are frequently infected with path-

TABLE 17.4 Antimicrobial substances approved for use in swine, cattle, or poultry feed

Bacitracin	Novobiocin
Bambermycin	Oleandomycin
Carbadox	Oxtetracycline
Chlortetracycline	Penicillin
Erythromycin	Streptomycin
Furazolidone	Sulfamethoxypyridazine
Lincomycin	Sulfamethazine
Neomycin	Tylosin

ogens that cause minor, inapparent infections (see Section 17.7). Such infections may not be sufficiently severe to appear as frank disease but may reduce growth rate. Since livestock are frequently kept in crowded quarters, the chance of transmission of pathogens from one animal to another is high. Antibiotic feed supplements probably keep down the extent and severity of infection by these minor pathogens, thus stimulating growth of the animals.

In addition, water or feed for young animals, including ruminants, may be supplemented at certain times with an antibiotic as a *prophylactic*: that is, to prevent infection by a pathogen known to be present in the environment or in other animals. Such feeding of antibiotics is routine in prevention of calf scours (see Section 17.7) and shipping fever (a respiratory disease caused by *Pasteurella multocida* and parainfluenza virus in animals under stress such as that induced during shipping) and seems quite effective in lessening the incidence of disease. Antibiotics cannot be fed to milking animals because some of the antibiotic would appear in the milk. To avoid antibiotic residues in meat, antibiotic feeding must be discontinued well before slaughter.

It is now well established that there are possible harmful effects of routine antibiotic feed supplements. This has arisen primarily because of the development of antibiotic resistance in bacteria mediated by *resistance transfer factors*. Resistance transfer factors are plasmids that confer antibiotic resistance and are transferable from one cell to another (see Section 7.5). In the late 1960s, there was an explosive increase in antibiotic-resistance plasmids, and one explanation for their origin was that the use of antibiotic feed supplements had led to the selection of bacteria containing such plasmids. There is some indirect evidence for the transfer of such antibiotic-resistant bacteria to humans (see the Boxes on pages 110 and 471). It thus becomes essential to consider whether the obvious benefits of antibiotic feed supplementation may be overshadowed by the possible harmful effects arising from antibiotic-resistance plasmids. One alternative is to use as feed supplements antibiotics which are not used in medicine, thus avoiding the buildup of strains resistant to medically important antibiotics.

17.10 SILAGE AS AN ANIMAL FEED

Silage is a product made by allowing green hay, grass, cereal crops, or mixtures of these to undergo a bacterial fermentation, thus preserving the feed value of the crop so that it can be fed to animals at seasons of the year when fresh feed is not available. The aim of making silage is to obtain through bacterial action the production of sufficient lactic acid to inhibit the growth and activity of spoilage microorganisms. The production of silage is an old practice and is widely done in temperate climates, especially where an intensive dairy industry has developed. Silage is particularly important in dairy cattle operations, since the animals must remain on good feed year-round. The crudest method of silage production involves packing the crop into a pit dug into the ground; if properly constructed and managed, satisfactory silage can be made in such a pit silo, although the structure is not permanent. The most widely used silo in regions of intensive dairy farming is the tower silo, constructed either of concrete or metal. Silage fermentation is an anaerobic process; if air is present, the growth of spoilage microorganisms may occur. For making proper silage, the moisture content of the crop is also important. If the crop is too wet the consistency of the silage will be unsatisfactory for feeding, whereas if it is too dry the growth of microorganisms will not be satisfactory and insufficient acid will develop. In some cases, molasses may be mixed with the crop to provide an additional source of sugar for the acid-producing bacteria in the silage.

The organisms involved in the silage fermentation are mainly lactic acid bacteria of the genera *Leuconostoc, Streptococcus,* and *Lactobacillus.* These bacteria are widespread in the fields and naturally contaminate the crop, although in small numbers. Silage inoculants are also available commercially. During the early stages of the fermentation, facultative anaerobes develop and consume oxygen, as does the plant tissue itself, converting the plant mass to anaerobic conditions. Then the lactic acid bacteria take over and produce sufficient acid so that the pH is reduced to about 4. Within 2 to 5 days, the number of lactic acid bacteria in silage juice may reach 10^9 per ml. Small amounts of acetic acid and alcohol are also produced, giving the silage a characteristic aroma.

At the acid pH values common for good silage, spoilage bacteria cannot develop, and the anaerobic conditions prevent growth of spoilage fungi; thus a stable product is obtained. Under some conditions, however, such as poor packing of the chopped material in the silo, there may be some areas in the silage which are aerobic, thus allowing aerobic spoilage organisms such as fungi to grow. In other cases, undesirable anaerobic spoilage bacteria such as the butyric acid-producing clostridia or proteolytic bacteria can grow, resulting in ruined silage. This may occur if the lactic acid bacteria cannot grow well enough to produce sufficient acid to preserve the silage. Factors which might cause this are poor quality of the forage used or too high or too low a moisture content in the chopped crop.

Another potential hazard is the formation of the toxic silage gas nitrogen dioxide (NO_2) during the first days of fermentation of the ensiled material. This gas is produced by the bacterial reduction of nitrates in the plant tissue. Crops with higher than normal amounts of nitrate are most likely to favor production of NO_2. High nitrate content may result because of high levels of nitrate fertilizers used, dry growing conditions followed by rain just before harvest, unfavorable growing conditions, or plant damage. The gas is heavy, brown, and toxic. If produced, it can seep out of the silo and kill nearby animals or people. As a precaution, the silo blower should always be run to ensure adequate ventilation before entering the silo, especially during the early period of fermentation.

SUMMARY

We have seen in this chapter that microorganisms play a variety of beneficial and harmful roles in agriculture. They are essential agents in the processes involved in **soil formation** and also play major roles in promoting soil fertility.

Microorganisms are involved in the breakdown of organic matter and the production of **humus,** a stable organic fraction that increases the water- and nutrient-holding capacity of the soil. Microbes are also responsible for the decomposition of many **pesticide** residues in the soil. Microorganisms play major roles in the nitrogen economy of plants, including **nitrogen fixation** (both free-living and symbiotic), **nitrification,** and **denitrification.** Microorganisms cause many serious plant diseases. Control of plant diseases by use of chemical agents has greatly increased crop yields. Insect damage to plants can also be controlled by chemicals, but a new and potentially significant means is the use of microbial pathogens of insects. The toxins produced by these pathogens can be formulated into **microbial insecticides** of high specificity and low human toxicity. Cellulose-digesting microorganisms are key agents in the digestive process in **ruminants** (cows, sheep, goats) and permit these animals to live on feeds high in cellulose, such as hay and grass, which are not normally digestible by mammals. In addition to digestion of cellulose, the rumen microorganisms produce proteins and vitamins for the ruminant, thus permitting these animals to live on diets deficient in these vital materials. Knowledge of the beneficial and harmful effects of microorganisms has permitted the development of many agricultural practices that have greatly increased crop and animal production.

KEY WORDS AND CONCEPTS

Humus
Compost
Pesticide
Biodegradable
Symbiotic nitrogen fixation
Root nodule
Rhizobium
Denitrification
Nitrification
Microbial insecticide
Mycotoxin
Aflatoxin
Rumen
Silage

STUDY QUESTIONS

1. Discuss the sequence of steps involved in the transformation of rock to *soil*. Which organisms are significant at each step?
2. In what ways do microorganisms contribute to *soil fertility*?
3. Describe briefly the chemical and microbial changes that take place during the formation of a *compost* pile. Of what use is compost in agriculture?
4. What is the role of microorganisms in the decomposition of *pesticides* in soil? Why is it important to know if a pesticide is susceptible to microbial attack?
5. Why is *nitrogen fixation* so important to soil fertility?
6. What are the *root nodules* present on leguminous plants? How are they formed? What is their function?

7. In some regions, farmers increase the nitrogen content of their fields by planting legume crops, whereas in other regions, they add combined nitrogen directly in the form of fertilizers. What are the advantages and disadvantages of each method? When might one be preferable to the other?
8. *Nitrogen fixation* and *nitrification* tend to offset nitrogen loss due to denitrification. Discuss.
9. List three of the ways in which *plant pathogens* cause disease in plants.
10. What plant disease controls are most commonly used? What criteria must be met in selection of the control method?
11. What properties of *microbial insecticides* make them desirable for use in agriculture?
12. What are *aflatoxins*, and why are they a danger to livestock? What control measures are desirable?
13. What role has immunization played in the eradication of certain *animal diseases*? Discuss.
14. Explain the rationale behind the addition of small amounts of *antibiotics* to animal feeds.
15. Describe two ways in which *ruminant animals* depend upon microorganisms.

SUGGESTED READINGS

AGRIOS, G.N. 1980. *Plant pathology, 2nd edition*. Academic Press, New York.

ALEXANDER, M. 1976. *Introduction to soil microbiology, 2nd edition*. John Wiley & Sons, New York.

BLOOD, D.C., O.M. RADOSTITS, and J.A. HENDERSON. 1983. *Veterinary Medicine, 6th edition*. Lea and Febiger, Philadelphia.

HAYNES, N.B. 1976. *Keeping animals healthy*. Garden Way Publishing, Charlotte, VT. (A practical guide to caring for healthy and sick livestock. Many diseases discussed.)

HUNGATE, R.E. 1966. *The rumen and its microbes*. Academic Press, New York.

18

Industrial Microbiology

Industrial microbiology deals with processes involving the large-scale growth of microorganisms for the production of food, feed, vitamins, antibiotics, and industrial chemicals. It is a very diverse field and one of great economic importance. It is a well-established but ever-changing area of activity, responsible in the United States alone for an annual production of tens of billions of dollars (see Chapter 1).

With the advent of genetic engineering, which we discussed in Chapter 7, the fields of industrial microbiology and biotechnology have entered a new and exciting phase (see Tables 1.5 and 1.6). However, these fields are by no means *new* fields, since the use of bacteria and yeasts for the production of wine, beer, and vinegar, and the use of fungi to produce various food products have been traditional for centuries in many areas of the world. These traditional processes were developed empirically, without any microbiological knowledge. Indeed, the very existence of microbes and their role in these processes were not even suspected. Real growth of the field of industrial microbiology could not occur until the twentieth century, after the discipline of microbiology had become established. With the methodology and information which became available as the field developed, the intentional production of microbial metabolites could be accomplished. Lactic acid, ethanol, acetone, butanol, and riboflavin, as well as various enzymes such as proteases, amylases, and invertase could then be made by large-scale processes. However, the most significant breakthrough in industrial microbiology took place during and after World War II, when the process for the large-scale production of penicillin was developed (see the Box on page 212). A multi-national cooperative effort on penicillin production methods resulted in an understanding of key processes: aseptic manipulation of microbes, large-scale sterilization of culture media, aeration and stirring methods to ensure adequate oxygen supply, and strain improvement by genetic manipulation. These techniques were then valuable in the many other processes that were developed, both concurrently and subsequently (Table 18.1). Methods became available for the production of a variety of antibiotics as well as several amino acids and

TABLE 18.1 Examples of fermentation products developed since 1940

Time period	Chemicals	Enzymes	Therapeutic agents	Bioconversion products	Other substances
1940–1950	Itaconic acid 2-Keto-D-gluconic acid	Cellulases Pectinases	Vitamin B_{12} Penicillin G Bacitracin Streptomycin Chlortetracycline Polymyxin Neomycin		
1950–1960	Kojic acid Glutamic acid Lysine	Glucose oxidase Catalase	Amphotericin B Cycloheximide Erythromycin Griseofulvin Kanamycin Nystatin Semisynthetic penicillins Tetracycline	Steroid-oxidation products	Gibberellins Dextran Single-cell protein
1960–1970	Valine	Glucose isomerase Glucose amylase Lipase Lactases	Candicidin Cephalosporins Fusidic acid Gentamicin Lincomycin Mikamycins Monensin Pimaricin Rifamycin Spectinomycin Vancomycin	Dihydroxyacetone	Xanthan 5′-Nucleosides Insecticides
1970–1977		Rennin Melibiase Dextranase	Adriamycin Bambermycin Candidin Macarbomycin Mepartracin Medicamydin Mocimycin Myxin Sisomicin Thiopeptin Tobramycin Virginiamycin	Sterol-splitting processes Xylitol Aspartic acid	Ribose Zearalenone

nucleotides. Medically important compounds such as the steroid hormones could be made more easily and economically when it was discovered that certain of the more difficult steps in their chemical synthesis could be carried out efficiently by microorganisms. In the decades following the development of penicillin production methods, a vast array of processes for the production of enzymes used for industrial, analytical, and medical purposes have also been developed.

Of great promise at the present time is the use of genetically engineered organisms for the manufacture of nonmicrobiological products. Such products made so far include insulin, interferon, human growth hormone, and viral vaccines (see Sections 7.8, 9.10 and 13.3). These processes are for the most part possible only at a laboratory or pilot plant scale. Further scale-up to large-volume production requires careful application of the principles of industrial microbiology.

The factories and installations for growing microorganisms on a large scale are extensive and expensive, and many new problems arise that are unique to operations on such a scale. Engineers, chemists, and microbiologists work together in industrial biotechnology, the role of the microbiologist being to select and prepare suitable cultures, to monitor the process for proper growth and production, and to monitor and prevent contamination.

18.1 INDUSTRIAL FERMENTATION

Industrial products are of three basic types. In the first, the *microbial cells* themselves are the desired product, such as bakers or food yeast. In the second, the desired material is a *microbial product*, such as an acid (for example, citric acid), an alcohol (for example, ethyl alcohol), an antibiotic, a vitamin, an amino acid, or an enzyme. In the third, called a *bioconversion*, the microorganism converts a chemical compound added to the medium to another compound of economic importance.

In the usual language of industrial microbiology, any large-scale microbial process is called a **fermentation**, and the large-scale vessel in which the process is carried out is called a **fermenter**.* The size of an industrial fermenter is large, from 5,000 to 50,000 gallons (about 20,000 to 200,000 liters); a 50,000 gallon-fermenter is as tall as a three-story building (Figure 18.1). In such large vessels, the cost of the culture medium is a significant part of the total cost of the operation, and consequently, inexpensive materials are used, such as soybean meal, molasses, whey, cornsteep liquor (a by-product of the manufacture of corn starch), and distillers' solubles (a by-product of the whiskey-distilling industry). One of the main tasks of the industrial microbiologist is to find an inexpensive culture medium in which the process can be carried out efficiently. In most cases, this requires an understanding of the principles involved in the nutrition and growth of microorganisms.

*Note that the words "fermenter" and "fermentation" have different meaning in industrial microbiology than they do in biochemistry. In Section 3.6 we discussed fermentation as an *anaerobic* energy generation process. In industrial microbiology the term is used to refer to any large-scale industrial process, whether aerobic or anaerobic.

Figure 18.1 Large fermenter installation.

Sterilization With the large volumes involved in industrial fermenters, sterilization of the medium can be difficult, especially when done by batch sterilization. In such a procedure, the medium is sterilized after the fermenting tank has been filled by passing steam under pressure through a metal jacket that surrounds the outside of the vessel. Difficulties arise because the penetration of heat to the center of such a large vessel is slow, and consequently, long heating times are necessary. Alternatively, the medium can be sterilized by a process called **continuous sterilization**. In this process, which has some resemblance to the HTST pasteurization of milk (see Section 16.6), the medium is pumped through a heater where it quickly rises to a temperature sufficiently high so that sterilization occurs in a few seconds, and it is then pumped into the empty but previously sterilized fermenter. Such sterilization methods, as well as the use of heat exchangers where feasible, have greatly reduced energy costs by increasing the efficiency of heating.

In some fermentations, sterilization is not necessary, either because the growth of the desired microbe is so rapid that contaminants cannot take over or because the medium or culture conditions are unfavorable for the growth of other microbes. Sterilization is not used, for instance, in the production by yeast of alcohol for distillation.

Great care must be taken to keep the fermenter scrupulously clean. Any culture medium left within the tank or its pipes and plumbing is a potential source of nutrients for the growth of contaminants that may be difficult to eliminate.

Aeration Many industrial fermentations involve aerobic respiring organisms, and it is important that sufficient air be provided during the process. This is often difficult in large fermenters because of the great bulk of the liquid to be aerated. In order to get as much air as possible into the culture, two procedures are used simultaneously: (1) liquid in the tank is stirred rapidly with a large motor-driven blade; (2) air is forced into the bottom of the fermenter through small holes so that it rises through the liquid and quickly disperses. The air entering the tank must usually be sterilized, and this can be done by passing it through a filter made of steel or glass wool packed into a long tube. The air is often also passed through a copper tube heated to a high temperature, where incineration of contaminants will occur.

Running the fermentation Inoculation of a large fermenter is made using a fairly dense, rapidly growing culture, added in a volume equal to 5 to 10 percent of the total volume of fermenter medium. Thus, for a 15,000-gallon fermenter, an inoculum of about 1,500 gallons is required. The inoculum is usually built up in a series of stages, starting with the initial stock culture and increasing step by step in volume (see Figure 18.4). At each stage, it must be certain that the culture is behaving normally and that it is not contaminated. One of the most important roles of the industrial microbiologist is to ensure proper quality of the culture used at all stages of the process.

During fermentation, the process must be carefully monitored and controlled. Heat is produced as a result of microbial metabolism, and to avoid excessive increase in temperature the fermenter must be cooled, usually by circulating cold water through its jacket. Vigorous aeration causes foaming. This foam may be controlled by periodic additions of an antifoam agent such as lard oil or silicone, usually through an automatic foam-control system. Control of pH is often desirable, and this can also be done automatically. Experience will indicate when the fermentation is at the point where the maximum level of the desired product is obtained. At this stage, harvest begins, and the contents of the fermenter are processed in such a way as to remove the desired products.

The fermentation can be carried out as either a batch or a continuous process (Figure 18.2). In a **batch process**, the course of the fermentation is

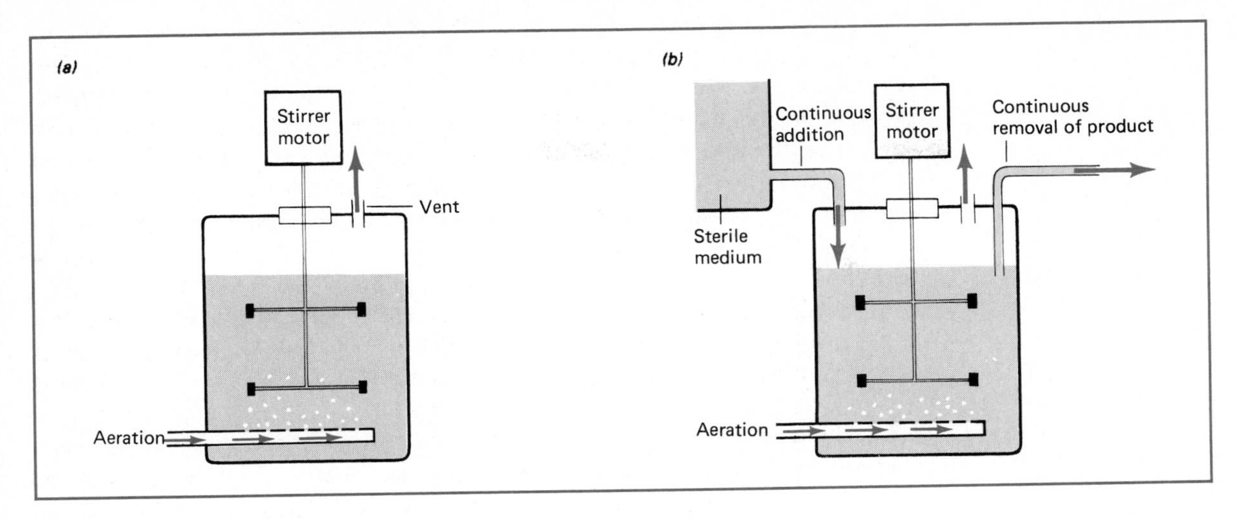

Figure 18.2 Types of fermenters. (a) Batch fermenter. (b) Continous fermenter.

similar to a typical microbial growth cycle. During exponential growth the number of cells increases rapidly, and at nutrient exhaustion the stationary phase is reached. It is quite common for the desired product to appear in the culture only after onset of this stationary phase. Products formed in the stationary phase of growth are called **secondary metabolites** to distinguish them from *primary metabolites* that are synthesized during the growth phase itself.

A milestone in biotechnological development occurred when the **continuous fermentation** process was perfected. In this method, fresh medium is pumped into the filled fermenter slowly and continuously, and overflow containing the desired product is removed continuously. The process may be carried out for days or weeks. The advantage of continuous fermentation is that the process can be kept at the stage of maximum production for long periods of time, and harvesting can be done continuously. In a batch process, there is only a small amount of time at which product yield is at maximum, so that the expensive fermenter equipment is tied up for relatively long times in unproductive phases of the cycle. However, disadvantages of continuous fermentation are that contamination may ruin the process or that the culture may degenerate during the long course of the process. In spite of potential problems, the continuous process finds ever increasing applications. It was first used for production of yeast and bacteria for use in food and feed. Other continuous fermentation processes of potential application include the production of ethanol for fuel to supplement increasingly scarce supplies of petroleum and the production of other fuels from cellulosic waste materials. Microbial production of acetone and butanol from starch (a process much used in the 1920s and 1930s) had become economically unfeasible and had thus been replaced by chemical synthesis. However, with the development of a continuous fermentation process, microbial production may become feasible.

Microbial cells are removed from the liquid by filtration or centrifugation. Filtration is usually cheaper and easier, but may not be effective with unicellular microorganisms. With fungi and actinomycetes, which are filamentous, the filter material is formed on a large rotary drum (Figure 18.3). The filter itself is usually a sheet of cotton or synthetic cloth of a tight weave, although a filter aid, such as diatomaceous earth, is often used to prevent clogging of the filter. With smaller unicellular microorganisms, the initial separation of the cells is usually done with a continuous centrifuge. Either the clear filtrate or the cells themselves may be the desired product, and processing details must be designed according to the goal of the fermentation. Purification of compounds from the filtrate follows if the desired product is a cellular metabolite.

A fermentation process that works well in the laboratory or in a small fermentation vessel may work poorly or not at all under similar conditions in large tanks. The reasons for this are often unknown or difficult to determine, and the industrial microbiologist is required to spend considerable research time developing proper conditions for large-scale fermentation. This is called **scale-up** and is of great economic importance. For research on scale-up problems, the industrial company may have a **pilot plant** in which fermenters are used and tested that are intermediate in size between small laboratory equipment and full-size industrial equipment.

The goal of an industrial fermentation is to convert a cheap raw material into a useful end product. The microbiological process must compete with synthetic chemical processes that may be able to accomplish the same task. Where microbial processes excel is in the production of complex organic chemicals such as antibiotics, vitamins, and hormones that are too difficult for the chemist to synthesize inexpensively. The successful management of industrial fermentation requires the combined efforts of chemical engineers and of microbiologists well trained in microbial physiology and genetics.

Figure 18.3 Separation of microorganisms from a fermentation broth by passage through a filter on a large rotary vacuum dryer.

18.2 YEASTS IN INDUSTRY

Yeasts are the most important and the most extensively used microorganisms in industry (Table 18.2). They are cultured for the cells themselves, for cell components, or for the end products that they produce. Yeast cells are used in the manufacture of bread, and also as sources of food, vitamins, and other growth factors. Large-scale fermentation by yeast is responsible for the production of alcohol for industrial purposes but is better known for its role in the manufacture of alcoholic beverages: beer, wine, and liquors. Production of yeast cells and production of alcohol by yeast are two quite different processes industrially, in that the first process requires the presence of oxygen for maximum production of cell material and hence is an aerobic process, whereas the alcoholic fermentation process is anaerobic and takes place only in the absence of oxygen. However, the same or similar species of yeasts are used in virtually all industrial processes. The yeast *Saccharomyces cerevisiae* was derived from wild yeast used in ancient times for the manufacture of wine and beer. The yeasts currently used are descendants of the early *S. cerevisiae*. Since they have been cultivated in laboratories for such a long time, there has been ample opportunity for selection of strains according to particular desirable properties. It is possible to breed yeasts in the laboratory, using genetic hybridization and cloning methods, to produce new strains that contain desirable qualities from two separate parent strains. By the techniques of genetic engineering, it is now also possible to transfer a desired gene from one organism into another, thus improving strains by direct intervention.

TABLE 18.2 Industrial uses of yeast and yeast products

Production of yeast cells
Baker's yeast, for bread making
Dried food yeast, for food supplements
Dried feed yeast, for animal feeds
Yeast products
Yeast extract, for culture media
B vitamins, Vitamin D
Enzymes for food industry; invertase, galactosidase
Biochemicals for research; ATP, NAD, RNA
Fermentation products from yeast
Ethanol, for industrial alcohol
Glycerol
Beverage alcohol
Beer
Wine
Whiskey
Brandy
Vodka
Rum

Baker's yeast The baker uses yeast as a leavening agent in the rising of the dough prior to baking. A secondary contribution of yeast to bread making is its flavor. In the leavening process, the yeast is mixed with the moist dough in the presence of a small amount of sugar. The yeast converts the sugar to alcohol and carbon dioxide, and the carbon dioxide gas expands, causing the dough to rise. When the bread is baked, the heat drives off the carbon dioxide (and incidentally, the alcohol) and holes are left within the bread mass, thus giving bread its characteristic light texture. That yeast contributes more to bread than carbon dioxide is shown by the fact that dough raised with baking powder, a chemical source of carbon dioxide, produces a quite different product than dough raised by yeast. Only the latter bears the name *bread*.

In early times, the bread maker obtained yeast from a nearby brewery, since yeast is a by-product of the brewing of beer. Today, however, baker's yeast is specifically produced for bread making. The yeast is cultured in large aerated fermenters in a medium containing molasses as a major ingredient. Molasses, which is a by-product of the refining of sugar from sugar beets or sugar cane, still contains large amounts of sugar that serves as the source of carbon and energy. Molasses also contains minerals, vitamins, and amino acids used by the yeast. To make a complete medium for yeast growth, phosphoric acid (a phosphorus source) and ammonium sulfate (a source of nitrogen and sulfur) are added.

Fermentation vessels for yeast production range from 10,000 to 50,000 gallons (40,000 to 200,000 liters). Beginning with the pure stock culture, several intermediate stages are needed to build up the inoculum to a size sufficient to inoculate the final stage (Figure 18.4). Fermenters and accessory equipment are made of stainless steel and are sterilized by high-pressure steam. The actual operation of the fermenter requires special control to obtain the maximum amount of yeast. It has been found that it is undesirable to add all the molasses to the tank at once, since this results in a sugar excess, and the yeast converts some of this surplus sugar to alcohol rather than turning it into yeast cells. Therefore, only a small amount of the molasses is added initially, and then as the yeast grows and consumes this sugar, more is added. Because yeast grows exponentially, the molasses is added at an exponential rate. Careful attention must be paid to aeration, since if insufficient air is present, anaerobic conditions develop, and alcohol is produced instead of yeast.

At the end of the growth period, the yeast cells are recovered from the broth by centrifugation. The cells are usually washed by dilution with water

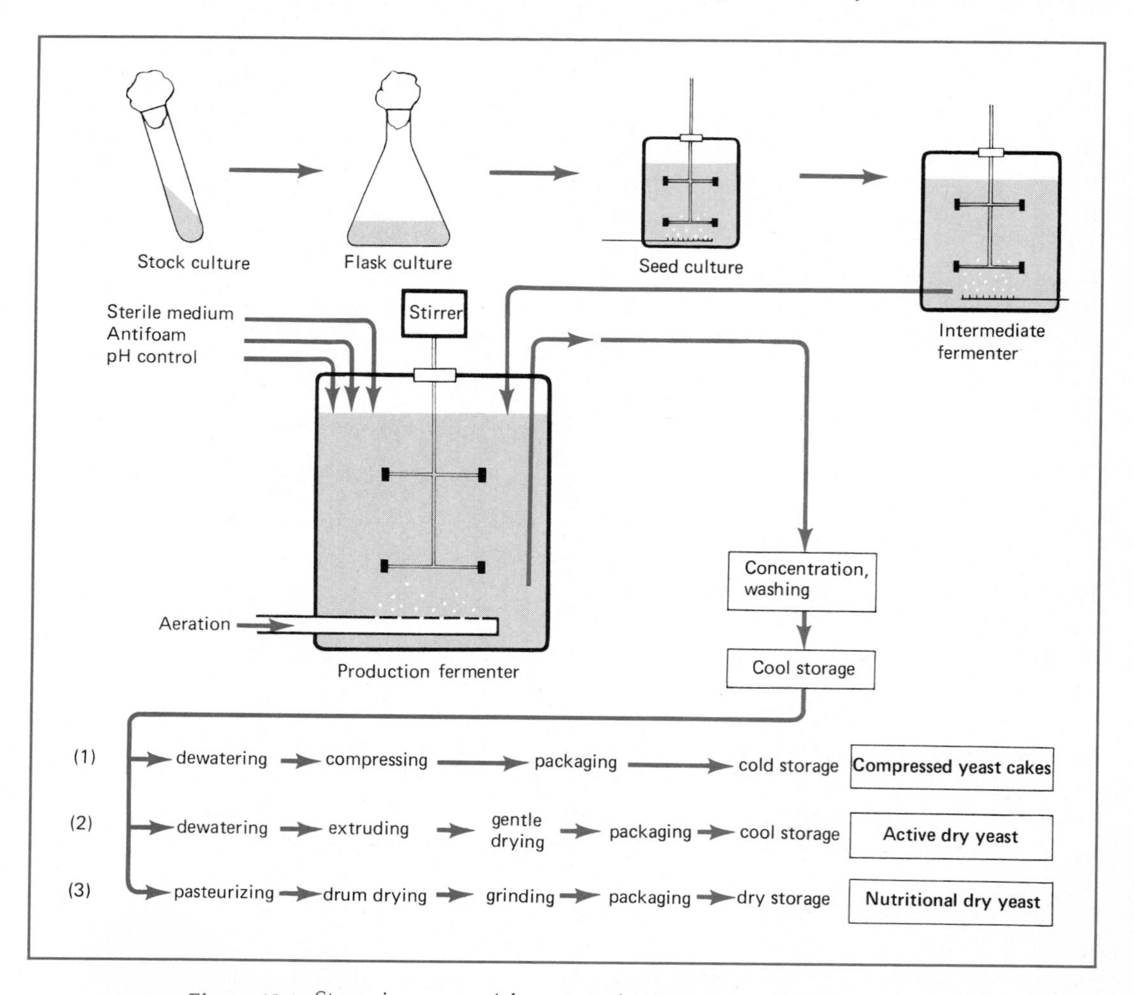

Figure 18.4 Stages in commercial yeast production.

and recentrifuged until they are light in color. Baker's yeast is marketed in two ways, either as compressed cakes or as a dry powder. Compressed yeast cakes are made by mixing the centrifuged yeast with emulsifying agents, starch, and other additives that give the yeast a suitable consistency and reasonable shelf life and then forming the product into cubes or blocks of various sizes for domestic or commercial use. A yeast cake will contain about 70 percent moisture and about 2×10^{10} cells per gram. Compressed yeast must be stored in the refrigerator so that its activity is maintained. Yeast marketed in the dry state for baking is usually called *active dry yeast*. The washed yeast is mixed with additives and dried under vacuum at 25 to 45°C for a 6-hour period, until its moisture is reduced to about 8 percent. It is then packed in airtight containers, such as fiber drums, cartons, or multiwall bags, sometimes under a nitrogen atmosphere to promote long shelf life. Active dry yeast does not exhibit as great a leavening action as compressed fresh yeast but has a much longer shelf life.

Food and feed yeast Yeast produced as food for humans or as feed for animals can be manufactured in much the same way as described for baker's yeast, or it may be a by-product of brewing or distilling. The yeast is heat-killed and usually dried. To be acceptable as food, dried yeast must be of proper flavor, color, and nutritional composition and must be free of contamination. The nutritional value of yeast to humans or animals is lower than meat or milk. However, yeast is one of the richest sources of vitamins of the B vitamin group. Although yeast is high in protein, this protein is deficient in several amino acids essential for humans, most especially the sulfur-containing amino acids. Animals fed on a diet in which all their protein comes from yeast do not grow as well as animals fed on milk or meat protein. Because of these deficiencies, yeast is used primarily as a food supplement for humans, being added to wheat or corn flour to increase the nutritional value of these foods.

18.3 ALCOHOL AND ALCOHOLIC BEVERAGES

The use of yeast for the production of alcoholic beverages is an ancient process. Most fruit juices undergo a natural fermentation caused by the wild yeasts which are present on the fruit. From these natural fermentations, yeasts have been selected for more controlled production, and today, alcoholic-beverage production is a large industry (see Table 1.3). The most important alcoholic beverages are wine, produced by the fermentation of fruit juice; beer, produced by the fermentation of malted grains; and distilled beverages, produced by concentrating alcohol from a fermentation by distillation. The biochemistry of alcohol fermentation by yeast was shown in Figure 3.19.

Wine Wine is a product of the alcoholic fermentation by yeast of fruit juices or other materials that are high in sugar. Most wine is made from grapes, and unless otherwise specified, the word *wine* refers to the product resulting from the fermentation of grape juice. Wine manufacture occurs in parts of

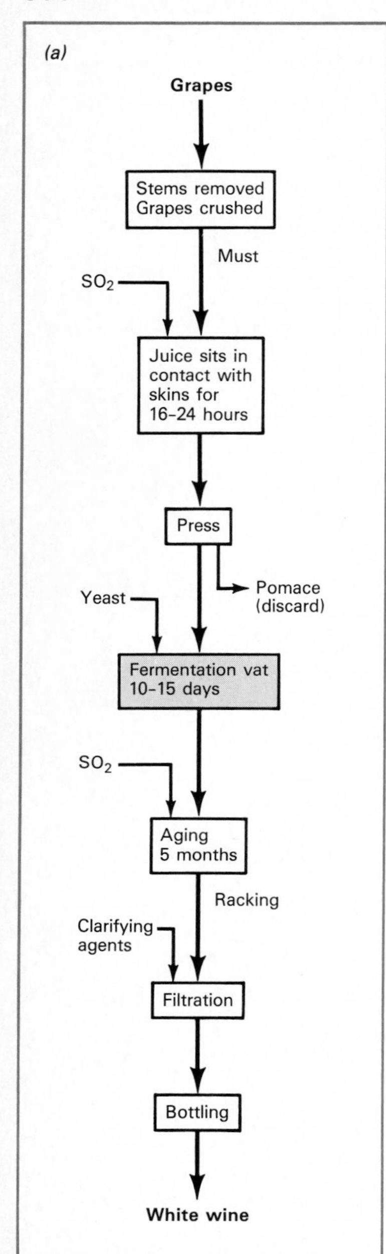

Figure 18.5 Procedure in making wine. The microbial processes are shown in color. (a) White wine. (b) Red wine. See next page.

the world where grapes can be most economically grown. The greatest wine-producing countries, in order of decreasing volume of production, are Italy, France, Spain, Algeria, Argentina, Portugal, and the United States. Wine manufacture originated in Egypt and Mesopotamia well before 2000 B.C. and spread from there throughout the Mediterranean region, which is still the largest wine-producing area in the world. Other parts of the world where wine is extensively produced often have a climate similar to that of the Mediterranean, for example, California, Chile, South Africa, and Australia. There are a great number of different wines, and their quality and character vary considerably. Dry wines are wines in which the sugars of the juice are practically all fermented, whereas in sweet wines, some of the sugar is left or additional sugar is added after the fermentation. A fortified wine is one to which brandy or some other alcoholic spirit is added after the fermentation; sherry and port are the best-known fortified wines. A sparkling wine is one in which considerable carbon dioxide is present, arising from a final fermentation by the yeast directly in the bottle.

The grapes are crushed by machine, and the juice, called *must*, is squeezed out. Depending on the grapes used and on how the must is prepared, either white or red wine may be produced (Figure 18.5). A white wine is made either from white grapes or from the juice of red grapes from which the skins, containing the red coloring matter, have been removed. In the making of red wine, the *pomace* (skins, seeds and pieces of stem) are left in during the fermentation. In addition to the color difference, red wine has a stronger flavor than white because of the presence of larger amounts of chemicals called *tannins*, which are extracted into the juice from the grape skins during the fermentation.

The yeasts involved in wine fermentation are of two types: the so-called wild yeasts, which are present on the grapes as they are taken from the field and are transferred to the juice, and the cultivated wine yeast, *Saccharomyces ellipsoideus*, which is added to the juice to begin the fermentation. One important distinction between wild yeasts and the cultivated wine yeast is their alcohol tolerance. Most wild yeasts can tolerate only about 4 percent alcohol, and when the alcohol concentration reaches this point, the fermentation stops. The wine yeast can tolerate up to 12 to 14 percent alcohol before it stops growing. In unfortified wine, the final alcoholic content reached is determined partly by the alcohol tolerance of the yeast and partly by the amount of sugar present in the juice. The alcohol content of most unfortified wines ranges from 10 to 12 percent. Fortified wines such as sherry have an alcohol content as high as 20 percent, but this is achieved by adding distilled spirits such as brandy. In addition to the lower alcohol content produced, wild yeasts do not produce many of the flavor components considered desirable in the final product, and hence their presence and growth during fermentation is unwanted.

It is the practice in many wine-producing areas to kill the wild yeasts present in the must by adding sulfur dioxide at a level of about 100 ppm. The cultivated wine yeast is resistant to this concentration of sulfur dioxide and is added as a starter from a pure culture grown on sterilized or pasteurized grape juice. During the initial stages, air is present in the liquid and rapid aerobic growth of the yeast occurs; then, as the air is used up, anaerobic conditions develop and alcohol production begins. The fermentation may be carried out in vats of various sizes, from 50-gallon casks to 55,000-gallon tanks, made of oak, cement, stone, or glass-lined metal. Temperature control during the fermentation is important, since heat produced during metabolism potentially raises the temperature above the point where yeast can function. Temperatures must be kept below 29°C, and the finest wines are produced at lower temperatures, from 21 to 24°C. Temperature control is best achieved by using jacketed tanks through which cold water is circulated. The fermenter must be constructed so that the large amount of carbon dioxide produced during the fermentation can escape but air cannot enter; this is often accomplished by fitting the tank with a special one-way valve.

After 3 to 5 days of fermentation, sufficient tannin and color have been extracted from the pomace, and the wine is drawn off for further fermentation in a new tank, usually for another week or two. The next step is called *racking*; the wine is separated from the sediment (called *lees*), which contains the yeast and organic precipitate, and is then stored at lower temperature for aging, flavor development, and further clarification. The final clarification may be hastened by addition of materials called fining agents, such as casein, tannin, or bentonite clay, or the wine may be filtered through diatomaceous earth, asbestos, or membrane filters. The wine is then bottled and either stored for further aging or sold. Red wine is usually aged for several years or more after bottling, but white wine is sold without much aging. During the aging process, complex chemical changes occur, resulting in improvement in flavor and odor, or *bouquet*.

A number of defects arise in wine as a result of microbial action, although the tannins and acids present do tend to retard the growth of many spoilage microorganisms. Pasteur discovered that when wine is heated to 63°C for 30 minutes, all spoilage organisms are destroyed and defects due to microbial growth are eliminated. This discovery was actually the origin of the process of pasteurization, and pasteurization is of great importance in the wine industry as well as the dairy industry (see Section 16.6).

Brewing The manufacture of alcoholic beverages made from malted grains is called *brewing*. Typical malt beverages include beer, ale, porter, and stout. *Malt* is prepared from germinated barley seeds, and it contains natural enzymes that digest the starch of grains and convert it into sugar. Since yeasts are unable to digest starch, the malting process is essential for the preparation of a fermentable material from cereal grains. Malted beverages are made in many parts of the world but are most common in areas with cooler climates where cereal grains grow well and where wine grapes grow poorly.

The fermentable liquid from which beer and ale are made is prepared by a process called *mashing*. The grain of the mash may consist only of malt, or other grains such as corn, rice, or wheat may be added. The mixture of ingredients in the mash is cooked and allowed to steep in a large mash tub at warm temperatures. There are a number of different methods of mashing, involving heating at different temperatures for various lengths of time; the particular combination of temperature and time used will considerably influ-

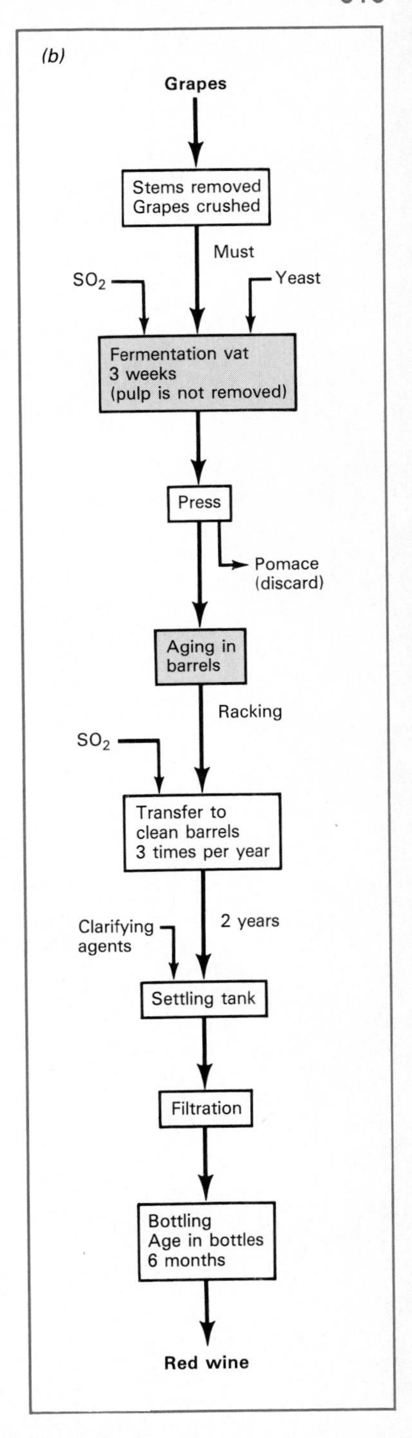

Figure 18.5(b) Procedure for making red wine.

ence the character of the final product. During the heating period, enzymes from the malt cause digestion of the starches and liberate sugars and dextrins, which are fermented by the yeast. Proteins and amino acids are also liberated into the liquid, as are other nutrient ingredients necessary for the growth of yeast.

After cooking, the aqueous extract, which is called *wort*, is separated by filtration from the husks and other grain residues of the mash. *Hops*, an herb that is derived from the female flowers of the hops plant, is added to the wort at this stage. Hops is a flavoring ingredient, but it also has antimicrobial properties, which probably help to prevent contamination in the subsequent fermentation. The wort is then boiled for several hours (Figure 18.6*a*), during which time desired ingredients are extracted from the hops, proteins present in the wort that are undesirable from the point of view of beer stability are coagulated and removed, and the wort is sterilized. Heating is accomplished either by passing steam through a jacketed kettle or by direct heating of the kettle from below by fire. Then the wort is filtered again and cooled and then transferred to the fermentation vessel.

Brewery yeast strains are of two major types: the top-fermenting and the bottom-fermenting yeasts. The main distinction between the two is that top-fermenting yeasts remain uniformly distributed in the fermenting wort and are carried to the top by the CO_2 gas generated during the fermentation, whereas bottom yeasts settle to the bottom. Top yeasts are used in the brewing of ales, and bottom yeasts are used to make the lager beers. The bottom yeasts are usually given the species designation *Saccharomyces carlsbergensis*, and

(a)

(b)

Figure 18.6 Brewing beer. (a) The brew kettle is being filled with wort. (b) The aging process is carried out in these large tanks.

the top yeasts are called *S. cerevisiae*. Fermentation by top yeasts usually occurs at higher temperatures (14–23°C) than does that by bottom yeasts (6–12°C) and is accomplished in a shorter period of time (5 to 7 days for top fermentation versus 8 to 14 days for bottom fermentation). After completion of lager beer fermentation by bottom yeast (Figure 18.6*b*), the beer is pumped off into large tanks where it is stored at a cold temperature (about −1°C) for several weeks (in German, *lager* means "to store"). Lager beer is the most widely manufactured type of beer and is made by large breweries in the United States, Germany, Scandinavia, the Netherlands, and Czechoslovakia. Top-fermented ale is almost exclusively a product of England and certain former British colonies. After its fermentation, the clarified ale is stored at a higher temperature (4 to 8°C), which assists in the development of the characteristic ale flavor.

Prior to filling bottles, cans, or kegs for the trade, beer and ale undergo a final "polishing" that may involve filtration, chill-proofing (the addition of enzymes to digest proteins that might precipitate out when the beer is cooled), carbonation, and pasteurization.

The yeast strain used for fermentation is usually carefully selected by the brewery, and at least some of the distinctive character of the product of a specific brewery is due to the yeast that is used. Some breweries have carefully selected improved strains over the years and continue to search for ways to improve their products. The genetic manipulation of some yeast strains has resulted in improved ability to ferment dextrins, the more resistant portions of starch molecules, with the result that more total carbohydrate is digested during fermentation, resulting in a product called "light" beer which is lower in carbohydrate than ordinary beer.

Some breweries pay close attention to the purity of their yeast strain and begin each fermentation with an inoculum built up to a volume of suitable size from a pure culture. However, in many breweries, the inoculum for one fermentation is derived from yeast saved from the previous fermentation, and a return to a pure culture occurs only if a defect occurs in the product. Bacterial contamination may occur during a fermentation but is not usually a problem since the pH of wort is fairly low (around pH 5), hops are present as an antimicrobial agent, and the wort has been boiled. When bacterial contamination does occur, it is usually due to acid-tolerant bacteria such as *Lactobacillus* and *Acetobacter*. Rarely, wild yeasts may develop during the fermentation and reduce the quality of the product. One role of the microbiologist in the brewery is to regulate the quality of the yeast used in the fermentation.

Distilled alcoholic beverages Distilled alcoholic beverages are made by heating a fermented liquid at a high temperature which volatilizes most of the alcohol. The alcohol is then condensed and collected, a process called *distilling*. A product much higher in alcohol content can be obtained than is possible by direct fermentation. Virtually any alcoholic liquid can be distilled, and each yields a characteristic distilled beverage. The distillation of malt brews yields *whiskey*, distilled wine yields *brandy*, and distilled molasses yields *rum*. The distillate contains not only alcohol but also other volatile products arising either from the yeast fermentation or from the mash itself. Some of these other products are desirable flavor ingredients, whereas others are undesirable substances called *fusel oils*. To eliminate the latter, the distilled product is almost always aged, usually in wood barrels. During the aging process, fusel oils are removed, and desirable new flavor ingredients develop. The fresh distillate is usually colorless, whereas the aged product is brown or yellow. The character of the final product is partly determined by the manner and length of aging, and the whole process of manufacturing distilled alcoholic beverages is highly complex. To a great extent, the process is carried out by traditional methods that have been found to yield an adequate product rather than by scientifically proven methods. Whiskey was originally almost exclusively an Anglo-Saxon (or Gaelic) product. A number of distinct whiskeys exist, usually associated with a country or region. Each of these has a characteristic flavor, owing to the local practices of fermenting, distilling, and aging. Even the word has local spellings, "whisky" being the Scottish, English, and Canadian spelling, and "whiskey" the Irish and United States spelling.

It should be noted that it is possible to distill a product in such a way that only alcohol and water are present in the distillate, and all flavor components and fusel oils are absent. This is the way in which *vodka* is manufactured, and to ensure the purity of this product, it is usually filtered through charcoal. Such a product need not be aged to be consumed, but of course, it has only the minimal flavor contributed by the alcohol itself. Vodka can be made from any starchy product, but the most usual starting materials are potatoes or rye malt. A simple solution of alcohol and water is usually called *neutral spirits*, and in addition to being consumed directly as vodka, it is added to other distilled beverages during the blending process, or in the manufacture of gin, cordials, and liqueurs. *Gin* is made by adding a mixture of flavorings to neutral spirits. The principal flavoring ingredient is juniper oil, but there are a variety of others as well. The exact blend of these ingredients is often a guarded trade secret but may include coriander, orange and lemon peel, cassia, angelica, licorice, orris, cardamon, caraway, and cinnamon.

Industrial and fuel alcohol A large amount of alcohol is needed for industrial purposes, and this can be prepared by distilling fermented mash. Because such alcohol is not being made for human consumption, many legal restrictions do not apply. Virtually any fermentable carbohydrate source can be used, the main requirement being that it is inexpensive and plentiful. Alcohol produced by yeast fermentation must compete economically with alcohol produced by chemical synthesis. The chemical process uses petroleum products as starting material, and if such products are plentiful and cheap, the chemical process is usually preferable to the fermentation process. Because of this, the production of alcohol by fermentation no longer has the prominence in industrial microbiology that it once had, although the price of petroleum can influence the economic viability of fermentation alcohol.

Microorganisms may also be used for production of alcohol for use as a fuel. In Brazil, for example, sugar cane and cassava root provide rich sources of sugars for microbial fermentations. The ethanol produced is then mixed with gasoline and burned in combustion engines. In the United States, surplus grains have been used to produce alcohol for similar purposes, but at present the cost of petroleum is low enough that microbial production of alcohol is not economically feasible. Alcohol-producing plants have been developed with government subsidies, perhaps to ensure that industries providing *alternative energy sources* will be available when petroleum supplies become more scarce.

18.4 ANTIBIOTIC FERMENTATION

We have discussed previously the laboratory study and medical significance of antibiotics (Section 5.5). As we noted, antibiotics are chemical substances that are produced by microorganisms and kill or inhibit the growth of other microorganisms. The chemical structures of most medically useful antibiotics are known, and in many cases, chemical synthesis has been achieved. However, because of the chemical complexity only rarely is it possible to synthesize an antibiotic as cheaply and as easily as it is to produce it by microbial fermentation. For this reason, almost all antibiotics for commercial use are produced by microorganisms, and antibiotic production is one of the most important areas of activity for the industrial microbiologist (Table 18.3, also see Table 18.1). Antibiotic production constitutes an enormous commercial enterprise on a worldwide scale. The annual world sales of the four most important groups of antibiotics—the penicillins, the cephalosporins, the tetracyclines, and erythromycin—is in excess of $4 billion.

Antibiotics differ from other microbial fermentation products such as alcohols, acetone, or lactic acid in that they are structurally quite complex and are usually produced in much lower concentrations during the fermentation. However, because of their high therapeutic activity and resultant great medical value, antibiotics can be economically produced even if yields are relatively low. More than 5000 antibiotics have been discovered, and about 300 new

TABLE 18.3 Some antibiotics produced commercially

Antibiotic	Producing microorganism
Bacitracin	*Bacillus subtilis*
Cephalosporin	*Cephalosporium* sp.
Chloramphenicol	Chemical synthesis (formerly *Streptomyces venezuelae*)
Cycloheximide	*Streptomyces griseus*
Cycloserine	*S. orchidaceus*
Erythromycin	*S. erythreus*
Griseofulvin	*Penicillium griseofulvin*
Kanamycin	*S. kanamyceticus*
Lincomycin	*S. lincolnensis*
Neomycin	*S. fradiae*
Nystatin	*S. noursei*
Penicillin	*Penicillium chrysogenum*
Polymyxin B	*Bacillus polymyxa*
Streptomycin	*S. griseus*
Tetracycline	*S. rimosus*

ones are found each year. Most of these are not useful, however, either because they are toxic to animals or because they are inactive when used clinically. The great majority of antibiotics are produced by the actinomycetes, especially members of the genus *Streptomyces*. Other common antibiotic producers are found in the genus *Bacillus* and among various genera of the fungi, notably *Penicillium* and *Cephalosporium*.

Because of the small amounts of antibiotic present in the fermentation liquid, elaborate methods for the extraction and purification of the antibiotic are necessary (Figure 18.7). If the antibiotic is soluble in an organic solvent which is immiscible in water, it is relatively simple to purify because then it is possible to extract and concentrate the antibiotic easily (Figure 18.8). If the antibiotic is not solvent soluble, then it must be removed from the fermentation liquid by adsorption or chemical precipitation. In all cases, the goal is to obtain a crystalline product of high purity, although some antibiotics do

Figure 18.7 Methods for extraction and purification of an antibiotic.

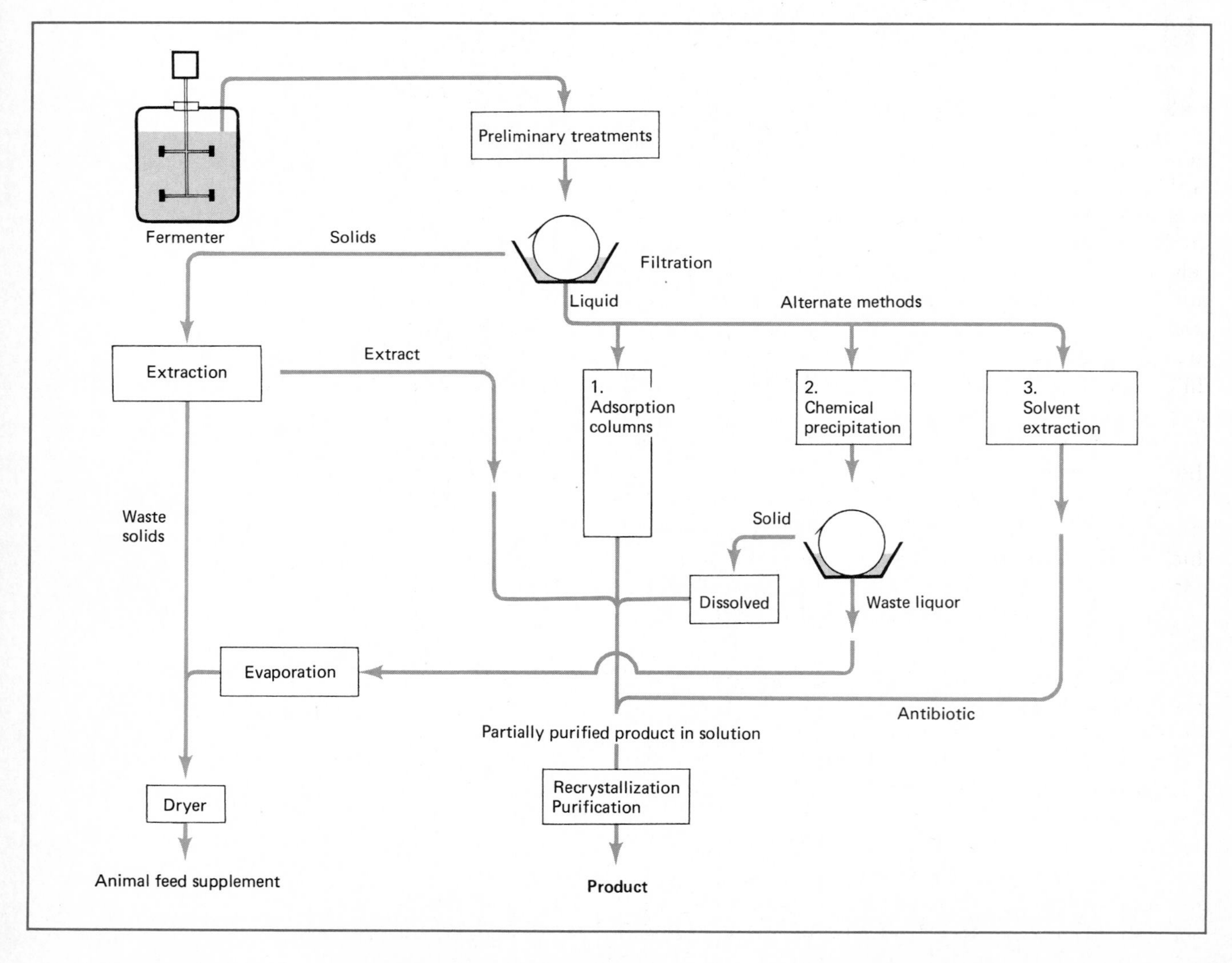

Figure 18.8 Installation for the solvent extraction of an antibiotic from fermentation broth.

not crystallize readily and are difficult to purify. A problem of some magnitude is that cultures produce other end products, including other antibiotics, and it is essential to make a commercial product with only the single desired antibiotic. One of the goals of strain development programs is to produce and select strains which make reduced amounts of undesirable byproducts. It has not been uncommon for drug companies to unknowingly produce antibiotics from which contaminants have not all been removed. If the contaminants are toxic or have other undesirable side effects, serious problems can result. Thus, diligent and careful attention to purification methods for antibiotics are essential.

The industrial microbiologist has made a very significant contribution to the antibiotic industry by developing high-yielding processes. For instance, the original laboratory strains of *Penicillium chrysogenum* produce only a few milligrams of penicillin per liter of culture—not enough for commercial production. However, by systematic exposure to a variety of mutagens, such as ultraviolet radiation, X rays, and nitrogen mustards, and by selection of appropriate spontaneous mutants, strains with greatly improved penicillin yield were developed. Such strain development in combination with improvements in fermentation media and in other aspects of fermentation technology itself have raised the yield of penicillin to 20 grams per liter, an improvement of 10,000-fold over the yields with the original strains!

Selection of high-yielding strains has been of high priority to industrial microbiologists for many years. Selection by use of mutagens was the best method available for many years, and was an effective procedure, although a rather slow and painstaking one. In recent years, the development of genetic engineering techniques has greatly improved methods for developing high-producing strains. The technique of *gene amplification* makes it possible to place additional copies of genetic information into a cell by means of a vector such as a plasmid. For example, chromosomal genes for production of a particular antibiotic can be transferred to a plasmid which will replicate rapidly in the recipient cell and thereby *amplify* the genes, increasing the production of the antibiotic for which the genes code to very high levels.

Antibiotic fermentations are all aerobic and are usually carried out in large fermenters with volumes of 50,000 to 100,000 liters. Therefore, careful attention to sterility is necessary to avoid contamination, and the procedures described earlier in this chapter for sterilizing media and air are used. Another problem that can arise during antibiotic production by bacteria of the genus *Streptomyces* is infection of the culture with virus (bacteriophage). If this happens, lysis of the culture occurs, and antibiotic yields are drastically reduced. Virus infection can be eliminated by selecting strains of the antibiotic-producing organisms that are resistant to virus. Fortunately, fermentations involving fungi and yeasts do not become infected with virus.

18.5 VITAMINS AND AMINO ACIDS

Vitamins and amino acids are growth factors that are often used pharmaceutically or are added to foods. Several important vitamins and amino acids are produced commercially by microbial processes.

Vitamins Vitamins are used as supplements for human food and animal feeds. Production of vitamins is second only to antibiotics in terms of total sales of pharmaceuticals—more than $700 million per year. Most vitamins are made commercially by chemical synthesis. However, a few are too complicated to be synthesized inexpensively but fortunately they can be made by microbial fermentation. Vitamin B_{12} and riboflavin are the most important of this class of vitamins.

Vitamin B_{12} is synthesized in nature exclusively by microorganisms. The requirements of animals for this vitamin are satisfied by food intake or by absorption of the vitamin produced in the gut of the animal by intestinal microorganisms. Humans, however, must obtain vitamin B_{12} from food or as a vitamin supplement, since even if it is synthesized by microorganisms in the large intestine, it cannot be assimilated. Microbial strains are used that have been specifically selected for their high yields of the vitamin. Members of the genus *Propionibacterium* give yields of the vitamin ranging from 19 to 23 mg/liter in a two-stage process, while *Pseudomonas denitrificans* produces 60 mg/liter in a one-stage process which uses sugar-beet molasses as the carbon source. Vitamin B_{12} contains cobalt as an essential part of its structure, and yields of the vitamin are greatly increased by addition of cobalt to the culture medium.

Riboflavin is synthesized by many microorganisms, including bacteria, yeasts, and fungi. The fungus *Ashbya gossypii* produces a huge amount of this vitamin (up to 7 grams per liter) and is therefore used for most of the microbial production processes. In spite of this good yield, there is great economic competition between this microbiological process and chemical synthesis.

Amino acids Although amino acids are present in the proteins of foods, many plant proteins are deficient in certain amino acids that are essential for human nutrition. If these plant foods are supplemented with the amino acids in which they are deficient, their food value can be considerably increased. Some of these amino acids can be made in high yields by microbial fermentation and are commercially produced. The essential amino acids that have been made by fermentation include lysine, threonine, methionine, and tryptophan. In addition, glutamic acid, a nonessential amino acid, is made in large amounts by microbial synthesis; glutamic acid is added to foods primarily as the flavor-enhancing agent *monosodium glutamate* (MSG). A variety of other amino acids are used in the food industry either alone or in combination (Table 18.4). Many amino acids are used in medicine, especially as ingredients of infusion solutions for post-operative treatment.

In the chemical industry, amino acids are often used as starting materials for the manufacture of certain polymers, for example polyalanine fibers, lysine isocyanate resins, and polymethyl glutamate coatings (all products of the plastics industry). The N-acyl derivatives of some amino acids are used in the manufacture of cosmetics and as surface-active agents.

A variety of bacteria can be used to produce amino acids. Most high-yielding strains are mutants which lack the mechanisms that normally regulate the rate of amino acid synthesis. In normal microorganisms, amino acid synthesis is carefully regulated so that amino acids are synthesized at a rate just sufficient to balance the rate of protein synthesis. Thus, free amino acids do not accumulate, since they are immediately used for protein synthesis. The mechanisms that regulate amino acid synthesis are genetically controlled (see Figure 6.12), and it is possible to isolate mutants in which a regulatory mechanism is no longer effective. In such mutants, overproduction of an amino acid can occur, and the amino acid accumulates inside the organism. Organisms used commercially also have an altered permeability, generally due to a defect in their cell membrane, a consequence being that the amino acid leaks out into the medium, from which it can be readily purified. In other fermentations, where a substance is produced in unusually high yields (for instance, vitamins or antibiotics), similar alteration of regulatory mechanisms may also occur.

TABLE 18.4 Amino acids used in the food industry

Amino acid	Foods	Purpose
Glutamate (MSG)	Various foods	Flavor enhancer
Aspartate and alanine	Fruit juices	"Round off" taste
Glycine	Sweetened foods	Improve flavor
Cysteine	Bread	Improves quality
	Fruit juices	Antioxidant
Tryptophan + histidine	Various foods, dried milk	Antioxidant, prevents rancidity
Aspartame (made from phenylalanine + aspartic acid)	Soft drinks, etc.	Low-calorie sweetener
Lysine	Bread (Japan)	Nutritive additive
Methionine	Soy products	Nutritive additive

18.6 MICROBIAL BIOCONVERSION

One of the most far-reaching discoveries in industrial microbiology is the understanding that microorganisms can be used to carry out specific chemical reactions that are beyond the capabilities of organic chemistry. The use of microorganisms for this purpose is called **bioconversion** and involves the growth of the organism in large containers, followed by the addition at an appropriate time of the chemical to be converted. Following a further incubation period during which the chemical is acted upon by the organism, the fermentation broth is extracted, and the desired product purified. Although in principle bioconversion may be used for a wide variety of processes, its major use has been in the production of certain steroid hormones.

Steroids are important hormones in animals which regulate various metabolic processes. Some steroids are also used as drugs in human medicine. One group, the adrenal cortical steroids, reduce inflammation and hence are effective in controlling the symptoms of arthritis and allergy. Another group, the estrogens and androgenic steroids, are involved in human fertility, and some of these can be used in the control of fertility. Steroids can be obtained by complete chemical synthesis, but this is a complicated and expensive process. Certain key steps in chemical synthesis can be carried out more efficiently by microorganisms, and commercial production of steroids usually has at least one microbial step.

A large number of fungi and bacteria have been discovered that can perform steroid transformations. Organisms used commercially include the fungi *Rhizopus nigrificans* and *Curvularia lunata*, and the bacteria *Streptomyces roseochromogenes* and *Corynebacterium simplex*.

18.7 ENZYME PRODUCTION BY MICROORGANISMS

Enzymes are found in all organisms, and each organism produces a large variety of enzymes, most of which, produced only in small amounts, are involved in cellular processes. However, certain enzymes are produced in much larger amounts by some organisms, and instead of being held within the cell, they are excreted into the medium. Extracellular enzymes are usually capable of digesting insoluble nutrient materials such as cellulose, protein, and starch, the products of digestion then being transported into the cell, where they are used as nutrients for growth. Some of these extracellular enzymes are used in the food, dairy, pharmaceutical, and textile industries and are produced in large amounts by microbial synthesis (Table 18.5). They are especially useful because of their specificity and efficiency when catalyzing reactions of interest at moderate temperature and pH. Similar reactions achieved by chemical means would generally require extreme conditions of temperature or pH and be less efficient and less specific.

Enzymes are produced commercially from both fungi and bacteria. The production process is usually aerobic, and culture media similar to those used in antibiotic fermentations are employed. The enzyme itself is generally formed in only small amounts during the active growth phase but accumulates in

TABLE 18.5 Microbial enzymes and their application

Enzyme	Source	Application	Industry
Amylase (starch digesting)	Fungi	Bread	Baking
	Bacteria	Starch coatings	Paper
	Fungi	Syrup and glucose manufacture	Food
	Bacteria	Cold-swelling laundry starch	Starch
	Fungi	Digestive aid	Pharmaceutical
	Bacteria	Removal of coatings (desizing)	Textile
Protease (protease-digesting)	Fungi	Bread	Baking
	Bacteria	Spot removal	Dry cleaning
	Bacteria	Meat tenderizing	Meat
	Bacteria	Wound cleansing	Medicine
	Bacteria	Desizing	Textile
	Bacteria	Household detergent	Laundry
Invertase (sucrose-digesting)	Yeast	Soft-center candies	Candy
Glucose oxidase	Fungi	Glucose removal, oxygen removal	Food
		Test paper for diabetes	Pharmaceutical
Pectinase	Fungi	Pressing, clarification	Wine, fruit juice
Rennin	Fungi	Coagulation of milk	Cheese

large amounts during the stationary phase of growth. As we have seen (Section 6.3), induced enzymes are produced only when the substrate they attack is present in the medium, and thus either the substrate must be present as one of the main medium ingredients or it must be added as a supplement at an appropriate time during the growth cycle. The potential for the production of useful enzymes has improved markedly in recent years because of the increased ease with which we can manipulate the genes of the organisms.

The microbial enzymes produced in the largest amounts on an industrial basis are the bacterial proteases, used as additives to laundry detergents. By 1969, 80 percent of all laundry detergents contained enzymes, chiefly proteases, but also amylases, lipases, reductases, and other enzymes. However, because of the recognition of allergies to these proteins in production personnel and consumers, the use of proteases was drastically reduced in 1971, with annual world sales falling from $150 million to $50 million. Recently, special processing techniques such as microencapsulation were developed to ensure dust-free preparations, and today once again 80 to 85 percent of the detergents used in Europe contain proteases, although in the United States their use is still much less widespread.

Other important enzymes manufactured commercially are amylases and glucoamylases, which are used in the production of glucose from starch. This glucose can then be acted upon by glucose isomerase to produce fructose (which is sweeter than either glucose or sucrose) resulting in the final production of a high-fructose sweetener produced from corn, wheat, or potato starch. The use of this process in the food industry has been increasing, especially in the production of soft drinks (see the Box on the next page).

Another enzyme of commercial significance is microbial rennin. It has been used in place of calf's rennin for cheese production since 1965. It is

SWEETENING A SOFT DRINK: BIOTECHNOLOGY SHOWS THE WAY

The craving for sweets is a primal urge and the soft drink is one of many devices for getting sugar into the blood stream. Annual sugar consumption in western societies is about 50 kg (110 pounds) per year. But not just any sugar will satisfy our taste buds. *Glucose,* the most widely available sugar, is not sweet enough and finds little use by soft drink makers. Historically, *sucrose* has been the sugar of commercial choice. The United States has produced some of its sucrose from sugar beets but in addition has had to import sucrose from the sugar-cane producing regions of the world. Yet *fructose,* a chemical relative of glucose, is twice as sweet as sucrose. Until the advent of biotechnology, there was no economical way of making fructose in the vast amounts needed by the soft drink industry. But by diligent application of industrial microbiology, a virtually unlimited amount of fructose can now be obtained from corn starch, a polymer of glucose, which is the dominant polysaccharide of the maize plant.

Three reactions, each catalyzed by a separate microbial enzyme, operate in sequence in the conversion of corn starch into the product called *high-fructose corn syrup.*

1. The enzyme *α-amylase* brings about the initial attack on the starch polysaccharide, shortening the chain, and reducing the viscosity of the polymer. This is called the *thinning reaction.*
2. The enzyme *glucoamylase* produces glucose monomers from the shortened polysaccharides, a process called *saccharification.*
3. The enzyme *glucose isomerase* brings about the final conversion of glucose to fructose, a process called *isomerization.*

All three enzymes are produced industrially by microbial fermentation. The end product of this series of reactions is a syrup containing about equal amounts of glucose and fructose which can be added directly to soft drinks and other food products. The savings in the United States from using domestic corn instead of imported sucrose has been over $1 billion a year.

The demand for high-fructose corn syrup is likely to increase. To date, the enzyme processes in use have been developed without the use of genetic engineering, but it is likely that recombinant-DNA technology will not only increase production of the present enzymes, but also will permit the development of completely new enzyme processes. The soft drink may continue to appear in its familiar container, but its contents will represent a refined product of the biotechnologist's art.

much simpler and less expensive to produce than calf's rennin, and seems to be equally effective in cheese production.

18.8 VINEGAR

Vinegar is the product resulting from the conversion of ethyl alcohol to acetic acid by bacteria of the genus *Acetobacter.* Vinegar can be produced from any alcoholic substance, although the usual starting materials are wine or alcoholic apple juice (cider). Vinegar can also be produced from a mixture of pure alcohol in water, in which case it is called *distilled* vinegar, the term "distilled" referring to the alcohol from which the product is made rather than the vinegar itself. Vinegar is used as a flavoring ingredient in salads and other foods, and because of its acidity, it is also used in the pickling of foods, (see Section 16.4). Meats and vegetables properly pickled in vinegar can be stored unrefrigerated for years.

The acetobacters are an interesting group of bacteria. Although aerobic, they differ from most other aerobes in that they do not oxidize their energy sources completely to CO_2 and water. Thus, when provided with ethyl alcohol, they oxidize it only to acetic acid, which accumulates in the medium. Acetobacters are fairly acid tolerant and are not killed by the acidity that they produce. There is a high oxygen demand during growth, and the main problem in the production of vinegar is to ensure sufficient aeration of the medium.

There are three different processes for the production of vinegar. The *open-vat* or *Orleans method* was the first one used and is still used in France where it was developed. Wine is placed in shallow vats with considerable exposure to the air, and *Acetobacter* develops as a slimy layer on the top of the liquid. This process is not very efficient, since the only place that the bacteria come in contact both the air and substrate is at the surface. In the *trickle method*, the contact between the bacteria, air, and substrate is increased by trickling the alcoholic liquid over beechwood twigs or wood shavings that are packed loosely in a vat or column while a stream of air enters at the bottom and passes upwards. The bacteria grow upon the surface of the wood so that they are maximally exposed both to air and liquid. The vat is called a vinegar generator (Figure 18.9), and the whole process is operated in a continuous fashion. The life of the wood shavings in a vinegar generator is long, from 5 to 30 years, depending on the kind of alcoholic liquid used in the process. The third process is the *bubble method*. This is basically a submerged fermentation process, such as already described for antibiotic production. Efficient aeration is even more important with vinegar than with antibiotics, and special highly efficient aeration systems have been devised. The process is operated in a continuous fashion: alcoholic liquid is added at a rate just sufficient to balance removal of vinegar, while most of the alcohol is converted to acetic acid. The efficiency of the process is high, and 90 to 98

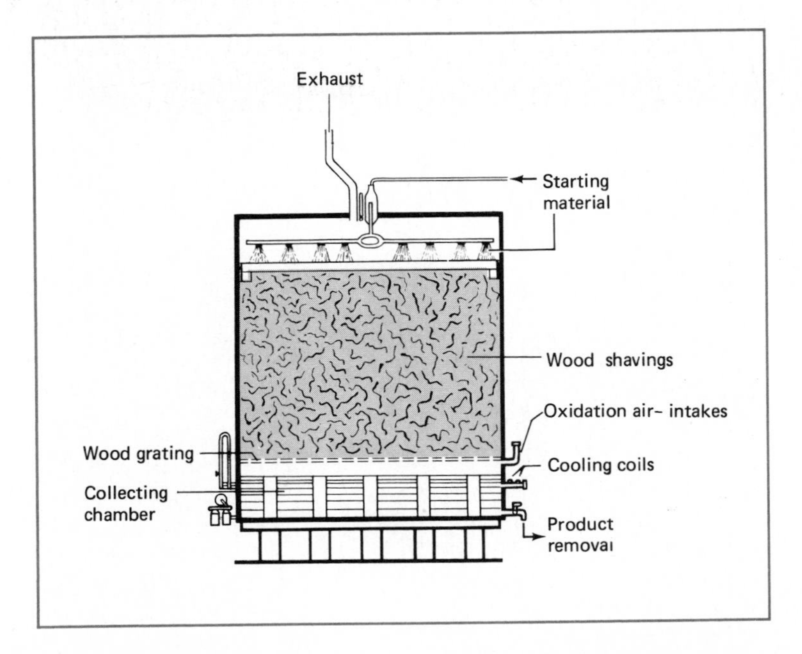

Figure 18.9 Diagram of one kind of vinegar generator. The alcoholic juice is allowed to trickle through the wood shavings and air is passed up through the shavings from the bottom. Acetic acid bacteria develop on the wood shavings and convert alcohol to acetic acid. The acetic acid solution accumulates in the collecting chamber and is removed periodically. The process can be run semicontinuously.

percent of the alcohol is converted to acid. One disadvantage of the bubble method is that the product must be filtered to remove the bacteria, whereas in the open-vat and trickle methods the product is virtually clear of bacteria since the cells are bound in the slimy layer in the former and adhere to the wood chips in the latter.

Although acetic acid can be easily made chemically from alcohol, vinegar itself is a distinctive product, the flavor being due in part to other substances present in the starting material. For this reason, the fermentation process has not been supplanted by a chemical process.

18.9 CITRIC ACID AND OTHER ORGANIC COMPOUNDS

Many chemicals are produced by microorganisms in sufficient yields so that they can be manufactured commercially by fermentation. *Citric acid,* used widely in foods and beverages, *itaconic acid,* used in the manufacture of acrylic resins, and *gluconic acid,* used in the form of calcium gluconate to treat calcium deficiencies in humans and industrially as a washing and softening agent, are produced by fungi. *Sorbose,* which is produced when *Acetobacter* oxidizes sorbitol, is used in the manufacture of *ascorbic acid,* vitamin C. (In fact, this sorbitol-sorbose reaction is the only biological step in the otherwise entirely nonbiological chemical synthesis of ascorbic acid.) *Gibberellin,* a plant growth hormone used to stimulate growth of plants, is produced by a fungus. *Dihydroxyacetone,* produced by allowing *Acetobacter* to oxidize glycerol, is used as a suntanning agent. *Dextran,* a gum used as a blood-plasma extender and as a biochemical reagent, and *lactic acid,* used in the food industry to acidify foods and beverages, are produced by lactic acid bacteria. *Acetone* and *butanol* can be produced in fermentations by *Clostridium acetobutylicum* but are now produced mainly from petroleum by strictly chemical synthesis.

Of the foregoing, **citric acid** is perhaps the most interesting product to consider here since it was one of the earliest successful aerobic fermentation products. Citric acid was formerly made commercially in Italy and Sicily by chemical purification from citrus fruits, and for many years, Italy held a world monopoly on citric acid, which resulted in relatively high prices. This monopoly was broken when the microbiological process using the fungus *Aspergillus niger* was developed, and the price of citric acid fell drastically. Today, virtually all citric acid is produced by fermentation. The process is carried out in large aerated fermenters, using a molasses-ammonium salt medium. One of the key requirements for high citric acid yields is that the medium must be low in iron since the citric acid is produced by the fungus specifically to scavenge iron from an iron-poor environment; therefore, most of the iron is removed from the medium before it is used. There has been considerable fundamental research on the citric-acid fermentation process, and some of this work has led to great improvements in the efficiency of the industrial process.

18.10 FOOD FROM MICROORGANISMS

Microorganisms can be grown to produce food for human beings, and we have discussed the production of food and feed yeast. In recent years, there has been considerable interest in the expanded production of microorganisms as food, especially in parts of the world where conventional sources of food are in short supply. Perhaps the most important potential use of microorganisms is not as a complete diet for humans but as a protein supplement. It is usually protein that is in shortest supply in food, and it is in the production of protein that microorganisms are perhaps the most successful. In many cases, microbial cells contain greater than 50 percent protein, and in at least some species, this is complete protein, that is, it contains all of the amino acids essential to humans. The protein produced by microbes as food has been called **single-cell protein** to distinguish it from the protein produced by multicellular animals and plants.

The only organism presently used as a source of single-cell protein is yeast, as already mentioned, but algae, bacteria, and fungi have also been considered. The following are desirable properties that an organism should possess to be most useful as a source of single-cell protein: (1) rapid growth; (2) simple and inexpensive medium; (3) efficient utilization of energy source; (4) simple fermentation system; (5) simple processing and separation of cells; (6) nonpathogenic; (7) harmless when eaten; (8) good flavor; (9) high digestibility; (10) high nutrient content.

Algae have been studied as sources of single-cell protein because in the presence of light they can grow on a completely inorganic medium and yield a product high in protein. For their production, a large outdoor tank is used, completely open to light and the air, into which a nutrient medium is pumped. One of the best nutrient sources for algae is domestic sewage, and of course if sewage is used it becomes more or less purified as the algae grow. (Presence of pathogens may make algae grown on domestic sewage unsuitable as food, however.) The harvesting of algae from the system makes use of a large centrifuge, and the pellet from this centrifugation is then dried. Both of these steps are expensive when carried out on a large scale. Also, the algal cells are fairly indigestible because of the presence of a thick cellulose cell wall, although they are more digestible to animals such as cattle and sheep, which have a special cellulose fermentation in the rumen.

Bacteria have many potential advantages as sources of single-cell protein. They grow rapidly, can give high yields, are high in protein, and are nutritionally versatile. Many bacteria use petroleum products as sources of energy and carbon, and because large supplies of petroleum are found in many parts of the world, bacterial single-cell protein made from this source is potentially quite valuable. Another virtue of the bacterial product is that bacteria are fairly readily digested by humans, in contrast to yeast and algae, which are rather poorly digestible. The bacterium that has been most widely studied as a source of single-cell protein is *Pseudomonas aeruginosa*, a common soil organism. However, some strains of this organism are pathogenic for humans, so that extensive testing is necessary before it can be utilized in large-scale commercial operations. The economic potential of bacterial single-cell protein depends on the price of petroleum.

Mushrooms Several kinds of **fungi** are sources of human food, of which the most important are the mushrooms. Mushrooms have been used as food by humans for many years. Both wild mushrooms and those grown commercially in special mushroom beds are used, although only the latter are produced and eaten extensively. The manner of formation of the mushroom fruiting body was illustrated in Figure 2.25.

The mushroom commercially available in most parts of the world is *Agaricus bisporus,* and it is generally cultivated in mushroom farms. The organism is grown in special beds, usually in buildings where temperature and humidity are carefully controlled. Since light is not necessary, mushrooms may even be grown in basements of homes. Another favored spot for mushroom culture is a cave. Beds are prepared by mixing soil with a material very rich in organic matter, such as horse manure, and these beds are then inoculated with mushroom "spawn." The spawn is actually a pure culture of the mushroom fungus that has been grown in large bottles on an organic-rich medium. In the bed, the mycelium grows and spreads through the substrate, and after several weeks it is ready for the next step, the induction of mushroom formation. This is accomplished by adding to the surface of the bed a layer of soil called "casing soil." The appearance of mushrooms on the surface of the bed is called a "flush" (see Figure 18.10), and when flushing occurs the mushrooms must be collected immediately while still fresh. After collection they are packaged and kept cool until brought to market. Several flushes will take place on a single bed, and after the last flush the bed must be cleaned out and the process begun again.

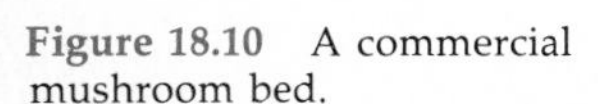

Figure 18.10 A commercial mushroom bed.

There have been some attempts to raise the mushroom fungus in aerated fermentation vats. In this case, the characteristic mushroom structure does not develop, but the mushroom flavor does, and the goal would be to use such material as a flavor ingredient in soups and sauces. However, no commercial process for deep-vat mushroom culture is yet available. Filamentous fungi other than mushrooms have also been tested for commercial deep-vat culture but have not been found satisfactory.

Although mushrooms make flavorful food, their digestibility and nutritional value are not very high. They are low in protein content and deficient in certain essential amino acids; they are also not exceptionally rich in vitamins. The mushrooms and the filamentous fungi are definitely inferior to yeast as food sources, although they serve as valuable flavoring ingredients (see also the Box on page 38).

SUMMARY

We have seen in this chapter that microorganisms play many significant roles in industrial activities. The economic benefit from the proper control of microorganisms for industrial purposes is great, and future important developments can also be anticipated. A large-scale microbial process is usually called a **fermentation**, and the vessel in which it is carried out is a **fermenter**. The size of an industrial fermenter is large, and many complications arise in its operation that are not found in laboratory culture of microorganisms, such as cost of medium ingredients, difficulties of sterilization, avoidance of contamination, and harvesting of product. **Yeasts** are the most important and most extensively used microorganisms in industry. Yeast cells are produced in large amounts for the baking and animal-feed markets and for the production of such alcoholic beverages as beer, wine, and distilled liquors. The second largest industrial use of microorganisms is in the production of **antibiotics**; the most common organisms involved here are members of the bacterial genera *Bacillus* and *Streptomyces* and the fungal genus *Penicillium*. Other products manufactured industrially with the aid of microorganisms include vitamins, amino acids, steroid hormones, enzymes, and organic acids. Some microorganisms other than yeasts that have been used as sources of food include algae, certain bacteria, and mushroom fungi.

KEY WORDS AND CONCEPTS

Fermentation
Fermenter
Secondary metabolite
Antibiotic
Bioconversion
Single-cell protein

STUDY QUESTIONS

1. What is an industrial *fermenter*? How is it used in industrial microbiology? Describe briefly how it is constructed.
2. Discuss the problems involved in sterilizing industrial fermenters.

3. Contrast the ways in which yeast is used in the baking and in the brewing industries.
4. Name four different products manufactured with the use of yeast. Discuss briefly the process involved in the manufacture of each product.
5. From what you know about the two substances, which do you think would have historically been the first product to have been discovered: wine or beer?
6. How do wild yeasts and wine yeasts differ? How are wild yeasts eliminated from the wine-making process?
7. What is the role of malt in the brewing process? Of hops?
8. Contrast top and bottom fermentation as they apply to beer. By which process is lager beer made?
9. How are microorganisms used in the manufacture of *antibiotics*? What problems arise in the manufacture of antibiotics that do not arise during the manufacture by fermentation of alcohol?
10. How are microorganisms used in the manufacture of vitamins? Amino acids? Enzymes? Give an example of each.
11. What is meant by microbial *bioconversion*? Describe one product that is made by this procedure.
12. Vinegar production is a highly aerobic process. Why? Describe several methods used to ensure that O_2 is available during the vinegar-making process.
13. What is meant by *single-cell protein*? How might it be made using algae? Bacteria?
14. What is the difference between a *primary* and *secondary metabolite*?

SUGGESTED READINGS

Advances in applied microbiology. Academic Press, New York. (This book appears annually and contains reviews on various topics of interest to industrial microbiology.)

CRUEGER, W. and A. CRUEGER. 1984. *Biotechnology, A textbook of industrial microbiology*. Sinauer Associates, Sunderland, MA. (A useful introductory textbook to large-scale industrial processes.)

DEMAIN, A.L. and N.A. SOLOMON. 1985. *Biology of industrial microorganisms*. Benjamin Cummings Publishing Co. Menlo Park, CA. (An advanced reference covering the nutrition, biochemistry, physiology, ecology, taxonomy, and genetics of the organisms which are important in industrial microbiology.)

Progress in industrial microbiology. Churchill Livingstone, Edinburgh. (This book appears annually and contains reviews on various topics of interest on industrial microbiology.)

REED, GERALD. 1981. *Prescott and Dunn's Industrial microbiology*. AVI Publishing, Westport, CT.

SCIENTIFIC AMERICAN, Volume 245: (no. 3), 1981. (An entire issue devoted to topics in industrial microbiology.)

Glossary

ABO system The most important classification for designating blood groups.
Acidophile Any organism that grows best under acidic conditions (low pH).
Acne An eruptive skin disease associated with the bacterium *Corynebacterium acnes.*
Acquired immunity Immunity which develops in response to foreign antigens in the body, involving either the production of antibodies or activated T cells which react specifically with the foreign antigen.
Acquired immunodeficiency syndrome (AIDS) A disease caused by viral infection of helper T cells resulting in inability to produce antibodies; usually transmitted by sexual intercourse, by direct injection of contaminated blood or blood products, or by intravenous injections using contaminated needles.
Actinomycete Gram-positive bacterium which usually grows as irregular, often branched filaments and forms a fungus-like mycelium.
Activated-sludge process An oxidative treatment of sewage; waste is mixed and aerated in tanks, and microbes degrade the organic materials present.
Activation energy Energy needed to make substrate molecules more reactive.
Adjuvant A substance that improves the effectiveness of an antigen without being an antigen itself.
Aerobe An organism that can use O_2 as an electron acceptor in metabolism.
Aerophobic Harmed by oxygen; obligately anaerobic.
Aerosol A suspension of particles in air.
Aerotolerant Able to tolerate exposure to oxygen, but oxygen not used in metabolism.
Aflatoxin A mycotoxin produced by *Aspergillus flavus.*
Agammaglobulinemia An immune system disorder in which antibodies cannot be produced; cell-mediated immunity is still present; treatment is by passive immunization.
Agglutination The reaction between antibody and cell-bound antigen that results in clumping of the cells.
AIDS See *Acquired immunodeficiency syndrome.*
Algae Eucaryotic photosynthetic organisms, including unicellular and multicellular forms. All contain chlorophyll.
Algal bloom Extremely heavy algal growth in a body of water; algae float to the surface and the water resembles pea soup.
Algicidal Able to kill algae; usually a type of antimicrobial agent used to poison algae to prevent their excessive accumulation.
Alkalophile Any organism that grows best under basic conditions (high pH).
Allergy An unusual sensitivity to a substance normally harmless to others. The host mounts an immunologic response against an allergen and reacts whenever exposed again to the allergen.
Aminoglycoside A type of antibiotic which contains a sugar bonded to an amino (NH_3) group; examples are streptomycin and tobramycin, which inhibit protein synthesis in procaryotic microorganisms.

Ampicillin A semisynthetic antibiotic produced by chemical modification of penicillin, produced by the mold *Penicillum*; inhibits formation of the bacterial cell wall.
Anabolism The biochemical processes involved in the synthesis of cell constituents from simpler molecules; usually requires energy.
Anaerobe An organism that grows or metabolizes without oxygen (air). *Obligate anaerobes* grow only in the absence of oxygen; *facultative anaerobes* grow either with or without oxygen.
Anaerobic respiration The oxidation of a substrate using an electron acceptor other than oxygen, such as nitrate, sulfate, and CO_2.
Anaphylactic shock An extreme antigen-antibody reaction in allergy that produces acute asthma and sometimes results in death; of most concern in drug allergies.
Anion A chemical species bearing negative ionic charge; they interact with cations in ionic bonding reactions.
Anoxygenic Not producing oxygen; a type of photosynthesis carried out by purple and green bacteria in which compounds other than water serve as electron donors, and oxygen is not produced.
Antagonism A relationship between two organisms in which one harms the other.
Antibacterial Having action against one or more kinds of bacteria; as a result bacteria may be either killed or inhibited.
Antibiotic A chemical agent produced by one organism that is harmful to other organisms.
Antibiotic resistance The ability of an organism to tolerate antibiotics which formerly killed or inhibited it; often associated with excessive use of antibiotics.
Antibody A specific protein (*immunoglobulin*) found mainly in blood serum, formed against a specific antigen and reactive with it.
Antibody-mediated immunity The portion of acquired immunity due to the production of antibodies.
Anticodon A sequence of three nucleic acid bases on transfer RNA molecules which recognizes and binds to three corresponding bases (called a *codon*) of messenger RNA; during protein synthesis this interaction ensures that the amino acid encoded by the codon is added to the growing protein.
Antifungal Acting to inhibit or kill fungi; usually a type of chemical which acts against fungi.
Antigen A substance that induces a specific antibody- or cell-mediated immune response; usually a protein or polysaccharide.
Anti-inflammatory Acting to reduce the inflammatory response of the body; antihistamines, for example, block the action of histamines which promote inflammation.
Antimicrobial Acting to inhibit or kill microorganisms, possibly of several different types.
Antioxidant A chemical used to prevent oxidation from occurring; usually a compound which is a good reducing agent.
Antiseptic An agent that kills or inhibits growth, for use on skin or *mucous membranes* but not to be used internally.
Antiserum Purified or unpurified blood serum which contains antibodies against a specific antigen or collection of antigens.
Antitoxin An antibody active against a toxin.
Antiviral Able to inhibit or kill viruses; usually a chemical which acts against viruses.
Archaebacteria One of the three major evolutionary lines of cellular life forms including procaryotic microorganisms distinct from *eubacteria*; the group includes methanogenic, halophilic, and some sulfur-metabolizing bacteria.
Asepsis The absence of microorganisms.
Aseptic technique The procedures used in handling cultures, media, and equipment so that only the desired organisms (if any) are present, with no contaminants.
ATP Adenosine triphosphate; a nucleotide used to store chemical energy within cells; it is produced in catabolic reactions and provides the energy needed in biosynthetic reactions.
Attenuation of virulence Loss of a pathogen's ability to cause disease, generally due to overgrowth of nonvirulent strains in laboratory cultures. Attenuated strains are often used as vaccines.
Autoantibody An antibody which acts against antigenic determinants present on cells or components of the body which produced it.
Autoclave An apparatus for sterilizing by heat under steam pressure.
Autoimmune Having action against the body (self); a type of disease in which autoantibodies attack components of the body which produced them.
Autotrophic An organism able to utilize CO_2 as sole source of carbon.
Avirulent Not able to cause disease; nonpathogenic.
Bacteremia The presence of bacteria in the blood.
Bacteria Small procaryotic organisms, which commonly have a spherical, rod, or spiral shape but are sometimes more complex.
Bactericidal Capable of killing bacteria.
Bacteriochlorophyll A type of pigment similar to *chlorophyll*, used by purple and green bacteria to convert light energy into chemical energy during photosynthesis.
Bacteriophage A virus that infects bacteria.
Bacteriostatic Capable of inhibiting bacterial growth without killing.
Bacteroids Irregular-shaped bodies of the bacterial genus *Rhizobium*, which carry out nitrogen fixation within root nodules on the roots of leguminous plants.
Basidiospores Sexual spores of mushrooms; formed on gills on the mushroom cap and spread by the wind.
B cell A type of lymphocyte which responds to foreign antigens to become an antibody-producing cell (see *plasma cell*).
Benign Unable to invade surrounding tissue; contained; contrast with *malignant*.
Biochemical oxygen demand (B.O.D.) The amount of oxygen consumed by microorganisms in a quantity of water, determined by the amount of oxidizable organic matter in the water.

Bioconversion The microbial conversion of a chemical to one of economic importance.
Biodegradable Capable of being broken down by living organisms; usually used in reference to manmade organic compounds such as pesticides.
Biofilm A thin layer of living cells, usually microorganisms, coating a surface.
Biogeochemistry The study of the combined effects of biological and geological activities on chemical changes observed in nature.
Biomass The total amount of living material present in a sample or system.
Biosphere The total assemblage of all living organisms on earth.
Biosynthesis The synthetic reactions of cell metabolism; see *anabolism*.
Biotechnology Applications of modern microbiological techniques, usually genetic engineering techniques, to produce useful products.
Blastomycosis A systemic fungal disease originating in the lungs, cause by the mold *Blastomyces dermatitidis*.
Blue-green algae See *cyanobacteria*.
Botulism A neurologic disease caused by a potent exotoxin produced by the bacterium *Clostridium botulinum*; one form is a type of food poisoning, whereas infant botulism is an infection of the intestines leading to toxin production and disease.
Bronchopneumonia Inflammation of the lungs following infection of the bronchi.
Brownian motion Vibratory, haphazard motion observed in microscope preparations (as distinguished from true motility).
Budding The process of cell division in which the mother cell retains its identity, and the daughter cell forms by growth of a new cell upon one part of the mother cell.
Calorie A unit of heat or energy; that amount of heat required to raise the temperature of one gram of water by 1°C.
Cancer A type of disease in which normal cells lose *contact inhibition*, and grow profusely to form a tumor, which invades surrounding tissue and sometimes spreads to other locations; there are many different causes including carcinogenic chemicals, certain forms of radiation, and some viruses.
Capsule Gummy material secreted in a compact layer outside the cell wall.
Carbohydrate A polymer composed of sugar monomers; see *polysaccharide*.
Carcinogenic Able to cause cancer.
Carcinoma A malignant, often metastasizing tumor.
Cariogenic Involved in causing tooth decay.
Carrier An individual who harbors a virulent pathogen but does not exhibit symptoms of disease; may be a source of infection.
Catabolism The biochemical processes involved in the breakdown of organic compounds, usually leading to the production of energy.
Catalysis The speeding up of a chemical reaction by lowering the *activation energy*.
Catalyst A substance that promotes a chemical reaction without itself being changed in the end.
Cation A chemical species bearing positive ionic charge; cations interact with anions in ionic bonding reactions.
Cell An individual biological unit, capable of independent function and able to divide to form two new identical cells.
Cell-mediated immunity The portion of acquired immunity due to the activation of T cells.
Cell membrane A thin envelope surrounding the cell through which food materials pass in and waste materials and other metabolic products pass out.
Cellulose A polymer composed of repeating units of the sugar glucose linked in a specific way (β-1,4 linkage); often a main component of plant and algal cell walls
Cephalosporins Antibiotics produced by molds of the genus *Cephalosporium* which inhibit the synthesis of the bacterial cell wall.
Chemical feedstock A source of chemicals used in the industrial synthesis of organic chemical polymers, for example, petroleum or specific chemicals produced by microorganisms.
Chemoautotroph An organism that utilizes CO_2 as its sole carbon source and an inorganic compound as its energy source.
Chemoprophylaxis The use of a drug or antibiotic to prevent future infections in people who are unusually susceptible.
Chemotaxis Movement of a cell in response to a chemical; movement may be toward an attractant chemical or away from a repellent chemical.
Chemotherapeutic agent A chemical agent which is harmful to microorganisms.
Chemotherapy Treatment of an infectious disease by drugs that act against the pathogen but do not harm the host.
Chlamydia Small bacteria which have an obligate dependence on host cells and are thus obligate intracellular parasites; cause of nongonococcal urethritis (NGU) and trachoma.
Chloramphenicol A broad-spectrum antibiotic produced by streptomycete bacteria, which inhibits protein synthesis in bacteria.
Chlorophyll A green pigment used by plants, algae, and cyanobacteria to convert light energy into chemical energy during photosynthesis.
Chloroplast The green, chlorophyll-containing organelle in photosynthetic eucaryotes; site of photosynthesis.
Chlortetracycline A chemical derivative of tetracycline, a broad-spectrum antibiotic produced by streptomycete bacteria, which inhibits bacterial protein synthesis.
Chromoblastomycosis A subcutaneous fungal disease which causes cauliflower-like growths on the skin.
Chromosome The structure that contains the DNA in eucaryotes, usually complexed with proteins called histones.
Ciliate A type of protozoa which moves by means of cilia; (see *cilium*).
Cilium A short, filamentous structure that beats with many others to make a cell move; found only in certain protozoa.

Clone A collection of identical entities all derived by duplication of one original entity; a pure culture of an organism derived from the growth of a single cell; a collection of identical DNA molecules all derived from replication of one original copy.
Coagulase A substance produced by *Staphylococcus aureus* that causes plasma to clot.
Coccus A spherical bacterium.
Codon A sequence of three nucleic acid bases on messenger RNA which codes for a specific amino acid; (see *anticodon*).
Coliform A Gram-negative rod-shaped bacterium that ferments lactose with gas production within 48 hours at 35°C; common inhabitants of the gastrointestinal tract; used as indicator organisms for contamination of water by feces.
Coliform test Prescribed methods for analysis of water for the presence of *coliform* bacteria. May use either the *most-probable-number* or the *membrane filtration* procedure.
Collagenase An enzyme which breaks down the protein collagen, a type of cell-cementing substance found in the body.
Colony A population of cells arising from a single cell, growing on solid medium.
Combined nitrogen Nitrogen in a form in which it is bonded to other elements; for example, ammonia (NH_3), nitrate (NO_3^-), and nitrite (NO_2^-).
Commensalism A relationship between two organisms in which one benefits without harm or benefit to the other.
Complement A complex of proteins in blood serum that acts in concert with specific antibody in certain kinds of antigen-antibody reactions.
Complementarity The exact pairing of the nucleic acid base sequence of two single strands of DNA or RNA; based on the hydrogen bonding of adenine with thymine (or uracil) and guanine with cytosine.
Complement fixation The utilization of complement during an antigen-antibody reaction. Removal of complement can be measured and is a sensitive indicator of occurrence of an antigen-antibody reaction.
Compost A mixture of organic materials used as a soil conditioner and fertilizer; prepared by microbial degradation of plant materials.
Compound A specific association of atoms held together by strong chemical bonds; for example, oxygen gas (O_2) and water (H_2O).
Conidia Asexual reproductive structures of fungi, formed at the tip of aerial hyphae; spread by air currents.
Conjugation The transfer of genetic information from one cell to another by cell-to-cell contact.
Constitutive Always synthesized; a type of enzyme whose synthesis is not regulated.
Contact inhibition The ability of a cell to stop growing when it contacts adjacent cells.
Contagious (infectious) disease An *infection* transmitted through the air, by water, food, objects, direct contact, or by insects or animals.
Corepressor A chemical which, when bound to a repressor protein, shuts off expression of a gene; usually the chemical produced by the enzymes coded for by that gene.
Culture A particular strain or kind of organism growing in a laboratory medium.
Cyanobacteria Photosynthetic procaryotes which contain chlorophyll and phycocyanin pigment.
Cycloheximide An antibiotic synthesized by streptomycete bacteria which inhibits protein synthesis in eucaryotic cells.
Cysteine A sulfur-containing amino acid; commonly used as a reducing agent in anaerobic culture media.
Cytochrome An iron-containing electron carrier common in the electron-transport systems of many respiring organisms.
Cytoplasm Cellular contents, excluding the nucleus, inside the plasma membrane.
Cytotoxic T cell A T lymphocyte which is able to recognize foreign antigens on other cells and then destroy the other cell; also called *killer T cells*.
Dehydrogenase A class of enzymes which oxidize substrates in the process removing electrons and hydrogen atoms.
Denitrification The conversion of nitrate into nitrogen gases under anaerobic conditions, which results in loss of nitrogen from ecosystems.
Dental caries Tooth decay.
Dental plaque A coating on teeth that consists of organic materials and bacteria.
Deoxyribonucleic acid (DNA) A polymer composed of the sugar deoxyribose and of nucleic acid bases arranged in such a fashion as to encode the inheritable traits of an organism; the genetic material of an organism.
Dermatophytes Fungi that cause superficial mycoses.
Diarrhea Rapid and watery discharge of intestinal contents through the anus.
Differential medium A culture medium which can be used to distinguish between different microorganisms.
Differential stain A staining technique which stains different organisms or tissues in different ways so that they may be distinguished; for example, the Gram stain.
Differentiated Cells with identical, or nearly identical genotype, which have become distinct through developmental changes; for example, muscle cells and liver cells have differentiated from common precursor cells but are phenotypically distinct.
Dimorphic fungus A fungus which can exist either in unicellular yeast form or as a filamentous mold.
Diphtheria An infectious disease caused by the bacterium *Corynebacterium diphtheriae,* which establishes itself in the upper respiratory tract and elicits a potent exotoxin which spreads in the blood and may cause death.
Direct count Enumeration of microorganisms by counting them in a defined area or volume viewed under a microscope.
Disinfectant An agent that kills microorganisms; may be harmful to human tissue, and thus should be used on inanimate objects only.
Dispersal The spread of organisms through the environment by air, water, food, animals, or human contact.
Disulfide bond A bond formed between the sulfur atoms of two different organic compounds, such as cysteine; often

responsible for joining different proteins or bending the primary amino acid sequence into a more complex structure.
DNA See *deoxyribonucleic acid*
DNA virus A virus which contains DNA as its genetic material.
Dysentery An infectious disease resulting in watery, bloody diarrhea and abdominal pain; amoebic dysentery is caused by an amoeboid protozoan, whereas bacterial dysentery is caused by *Shigella dysenteriae.*
Ecologic Pertaining to the relationship between an organism or organisms and features of the environment.
Ecosystem The total community of organisms living in a particular habitat, together with their physical and chemical environment.
Electron acceptor A substance that accepts electrons during an oxidation-reduction reaction.
Electron donor A substance that gives up electrons during an oxidation-reduction reaction; an energy source.
Electron-transport (oxidative) phosphorylation The synthesis of high-energy phosphate bonds within the electron-transport system.
ELISA Enzyme Linked ImmunoabSorbant Assay; a diagnostic test which employs antibodies linked to an enzyme whose activity can be determined for the quantitative determination of the antigen which reacts with that antibody.
Endemic disease A disease that is constantly present in a population. Compare *epidemic disease.*
Endospore A structure produced inside some bacteria, highly resistant to heat and chemicals; can germinate to form a new cell.
Endothermic reaction A chemical reaction that requires input of energy to proceed.
Endotoxin A cellular component of certain pathogens released upon cell death, which then attacks and damages tissue cells. Compare *exotoxin.*
Enteric Intestinal.
Enterotoxin A toxin affecting the intestine.
Enterovirus A virus which causes infections of the gastrointestinal tract; sometimes this term is used to indicate a group of picornaviruses which cause such infections as polio and infectious hepatitis.
Envelope The outermost layers of a cell or the membrane of a virus.
Enzyme A protein functioning as the catalyst of living organisms, which promotes specific reactions or groups of reactions.
Epidemic A disease occurring in a greater-than-usual number of individuals in a community at the same time. Compare *endemic disease.*
Epidemiology The study of how disease spreads in a population, including sources of infection, modes of transmission, the incidence of disease, and prevention and control of disease.
Epithelium The layers of cells that make up the surfaces of the body, such as skin and linings of the interior openings of the body (mouth, upper respiratory tract, gastrointestinal tract, genitourinary tract).
Ergotism A type of poisoning caused by a mycotoxin produced by the fungus *Claviceps purpurea*; the toxin is found in contaminated grain and may cause dramatic neurologic effects as well as decreased blood flow to the extremities.
Erythrocyte Red blood cell.
Eubacteria One of the three major evolutionary lines of cellular life forms, primarily procaryotic microorganisms which possess a peptidoglycan cell wall (mycoplasma are the only exception).
Eucaryote One of the three major evolutionary lines of cellular life forms, including plants, animals, fungi, protozoa, and algae, all of which possess a nucleus and organelles.
Eutrophication Nutrient enrichment of natural waters, usually from manmade sources, which frequently leads to excessive algal growth.
Exergonic See *exothermic reaction.*
Exon A segment of eucaroytic DNA which codes for protein; See *intron.*
Exothermic reaction A chemical reaction that proceeds with the liberation of heat.
Exotoxin An extracellular product of some pathogens; spreads to distant tissues where it causes cellular damage. Compare *endotoxin.*
Exponential growth The phase of most rapid growth, in which the progressive doubling of cell number results in a continually increasing rate of growth.
Expression The synthesis of a protein or proteins encoded by a particular gene or operon.
Facultative A qualifying adjective indicating that an organism is able to grow either in the presence or absence of an environmental factor; e.g., "facultative anaerobe."
Fecal Pertaining to intestinal material excreted via the anus.
Feedback inhibition The inhibition by an end product of the biosynthetic pathway involved in its synthesis.
Fermentation The oxidation of organic compounds occurring in the absence of any external electron acceptor.
Fermentation (industrial) A large-scale microbial process.
Fermenter A large vessel (5,000 to 50,000 gallons) in which industrial fermentations are carried out.
Fibrin The protein which composes blood clots; formed from the blood protein fibrinogen.
Fibroma A benign tumor of connective tissue.
Filamentous Threadlike; many times longer than wide.
Fimbria Tube-like protein structures on the exterior of some bacteria, which may be involved in attachment to surfaces; *pilus.*
Flagellum A long whiplike organ of motility. The flagella of procaryotes and eucaryotes differ in structure.
Fluorescent antibody An antibody labelled with fluorescent dye; useful in staining procedures to identify bacteria or viruses that have particular antigens.
Food chain The path of food consumption in a system as different consumer groups sequentially prey upon other organisms.
Food infection Illness arising from ingestion of food containing pathogenic organisms such as some salmonellas which grow in the gastrointestinal tract.

Food poisoning Illness arising from ingestion of food containing toxins from the growth of organisms such as some staphylococci or clostridia.
Free-living Growing outside of and not directly dependent on other organisms; *saprophytic*.
Freeze-drying See *lyophilization*.
Fungi Eucaryotes that often show mycelial, spreading growth; included are the molds, yeasts, and mushrooms.
Gangrene Dead tissue resulting from bacterial invasion and interference with nutrition of tissue.
Gastroenteritis A food-borne or water-borne infection causing inflammation of the lining of the stomach and intestines and resulting in vomiting and diarrhea.
Gene A unit of heredity; a segment of DNA specifying a particular protein or polypeptide chain.
Generation time The time required for formation of two cells from one, or for the microbial population to double.
Genetic engineering The combining of DNA from two different organisms, primarily so that a product usually made by one organism can be more efficiently made by the other; *recombinant DNA technology*; see *biotechnology*.
Genetic marker Any mutant gene useful in genetic analysis.
Genitourinary Pertaining to the reproductive and urinary secretion systems of the body.
Genome The complete set of inheritable characteristics of an organism.
Genotype The genetic complement of an organism.
Genus A group of related species.
Germicide An agent that kills microorganisms. See also *antiseptic* and *disinfectant*.
Glycolysis The stepwise breakdown of glucose by fermentative reactions.
Gonococcus Gram-negative coccus-shaped bacteria classified as *Neisseria gonorrhoeae*; the cause of gonorrhea.
Gonorrhea A sexually transmitted disease of very high incidence, caused by the bacterium *Neisseria gonorrhoeae*.
Gram-negative bacteria Bacteria which stain pink in the differential stain, called the *Gram stain*; contain an outer membrane outside of the cell wall.
Gram-positive bacteria Bacteria which stain blue in the differential stain, called the *Gram stain*; possess a thick cell well with no outer membrane.
Growth An increase in the number of cells or in microbial mass.
Growth factor An organic compound that is required in very small amounts as a nutrient.
Growth rate The amount of increase in cell number or mass per unit time.
Halophile Any organism that grows best at high concentrations of salts.
Hapten A chemical which can combine with antibodies or T-cell receptors, but which does not by itself stimulate an immune response.
Hemagglutinin A substance which causes red blood cells to agglutinate; a protein of the influenza virus envelope which may help the virus adhere to cells of the upper respiratory tract.
Hemolysin See *hemolysis*
Hemolysis Dissolution of red blood cells, caused by specific bacterial toxins (*hemolysins*).
Hepato Pertaining to the liver.
Heterotroph An organism that uses an organic compound as carbon source.
Histocompatibility antigens The substances on the exterior of cells that determine tissue type.
Homology Similarity in the sequence of monomers which make up different polymers; for example, similarity in the base sequences of two different DNA or RNA molecules.
Host An organism that supports the growth of a parasite; an organism used to replicate or express recombinant DNA molecules.
Hybridoma A hybrid cell formed by the fusion of a B cell and a myeloma (cancer) cell, used to produce monoclonal antibody.
Hydrophilic Able to dissolve in or associate with water.
Hydrophobic Not soluble in water.
Hyperimmune Having a very high antibody titer.
Hypersensitivity Unusual and extreme allergic reaction to some antigen to which the host was previous exposed.
Hypha A filament that is the usual form of cellular structure in molds.
Icosahedral Many-sided; a type of virus which does not possess an envelope and has a crystalline appearance; naked virus.
Immune system The system in mammals which responds specifically to the presence of foreign antigens to aid in eliminating them from the body; provides *acquired immunity*.
Immunity Defense system against disease, including innate host defense mechanisms and acquired defense, such as antibodies.
Immunization The induction of specific immunity by injecting antigen or antibodies.
Immunodeficiency Incomplete immune system; either antibody-mediated, cell-mediated immunity, or both may be abnormal.
Immunogenic Able to stimulate an immune response; see *antigen*.
Immunoglobulin Proteins in blood serum that are antibodies.
Immunosuppression Drug treatment which temporarily shuts off the immune system; often used during organ transplants to prevent tissue rejection.
Immunosurveillance Recognition and elimination of cancer cells by killer T cells of the immune system.
Impetigo An infectious disease of the skin caused by staphylococci or streptococci.
Incineration Killing of organisms by burning.
Incubation period The time between encounter with a pathogen and the appearance of symptoms resulting from infection.
Indicator organism An organism associated with the intestinal tract whose presence in water indicates fecal contamination.

Inducible enzyme An enzyme whose synthesis does not occur unless a specific chemical, called an inducer, is present; the inducer is often the substrate of that enzyme.
Induction The process by which an enzyme is synthesized in response to the presence of an external substance, the inducer.
Infection Growth of a microorganism in a host, often but not always causing harm (disease) in the host.
Inflammation A response of host tissues to injury, microbial infection, or presence of foreign matter, characterized by swelling, redness, pain, and accumulation of phagocytes at the site.
Innate resistance The nonspecific defenses already present in the body which are stimulated by infection or trauma; for example, phagocytosis and inflammation.
Inoculum Material used to initiate a microbial culture.
Inorganic compound A compound which does not contain carbon atoms (the only exception being carbon dioxide, CO_2).
Interferon A naturally occurring antiviral agent, produced as a result of virus infection or the presence of some nucleic acids, that interferes with virus multiplication.
Intron A eucaryotic DNA segment that does not code for protein but may interrupt segments which do encode protein (see *exon*).
Invasiveness The degree to which an organism is able to spread through the body from a focus of infection.
In vitro In glass; in culture.
In vivo In the body; in a living organism.
Ionizable A chemical which is able to become charged with either positive or negative ionic charge.
Isolation The procedures by which a pure culture is obtained; confinement of a hospital patient to reduce the spread of infectious disease either from, or to other patients.
Keratin A protein found in large amounts in the outermost cells of the skin, and also the nails and hair, which imparts rigidity to these structures.
Koch's postulates Experimental criteria, devised by Robert Koch, to demonstrate that a specific disease is caused by a specific organism.
Lag phase The earliest phase of growth, during which the cell number stays constant while cells adjust to the new medium.
Latent virus A virus present in a cell but not causing any detectable effect.
Legume A plant which contains root nodules and is capable of symbiotic nitrogen fixation with bacteria of the genus *Rhizobium*.
Leukemia A condition in which large numbers of abnormal white blood cells (leukocytes) develop within the blood and blood filtering organs.
Leukocidin A microbial substance able to destroy phagocytes.
Leukocyte A white blood cell, usually a phagocyte.
Lichen A regular association of an alga with a fungus, usually leading to the formation of a plant-like structure.
Lipid Fat; found mainly in membranes of microbial cells.
Lithotroph An organism that can obtain its energy from oxidation of inorganic compounds.
Lymph A fluid similar to blood but lacking the red-cell component; flows within the lymphatic system.
Lymphocyte A cell of the immune system; B-lymphocytes make antibody; T-lymphocytes are involved in cell-mediated immunity.
Lymphoma Malignant cancer of the lymphatic system; often lymph nodes are involved.
Lyophilization The preservation of a food or culture of organisms by drying under vacuum while the product is frozen.
Lysis Rupture of a cell, resulting in loss of cell contents.
Lysogenic Describing a bacterium that contains a temperate bacteriophage integrated into the cell DNA.
Lysozyme An enzyme in body fluids, active in host defense; able to kill invading organisms by digesting the bacterial cell wall.
Macromolecules Very large polymers in cells, including proteins, polysaccharides, and nucleic acids.
Macrophage A type of phagocyte; includes *wandering cells* in the blood and lymph, and *histiocytes* in the reticuloendothelial system.
Malignant A type of tumor which invades surrounding tissue; contrast with *benign*.
Mastitis Inflammation of the udders of a cow resulting from bacterial infection.
Membrane filter (MF) method A method for counting organisms, especially in dilute samples; the liquid sample is passed through a membrane filter, which is placed on agar medium to permit the organisms to grown and be counted.
Meningitis Infection accompanied by inflammation of the meninges, the membranes surrounding the brain and spinal cord.
Meningococcus Gram-negative coccus-shaped bacterium classified as *Neisseria meningitidis*, one important cause of meningitis.
Mesophile An organism that grows best at temperatures between room temperature (18–25°C) and body temperature (37°C).
Messenger RNA (mRNA) An RNA molecule containing a base sequence complementary to DNA; directs the synthesis of protein.
Metabolic intermediate A compound produced during a series of enzymatic reactions, which is altered by the subsequent enzymatic reactions.
Metabolism All biochemical reactions in a cell, both anabolic and catabolic, by which an organism converts nutrients into cell material and energy.
Metabolite A product of cellular metabolism.
Metastasis The spreading of a malignant tumor from its initial location to other places within the body.
Methanogen A type of archaebacterium which is capable of producing methane gas.
Microaerophile An aerobic organism which is poisoned by oxygen when it is present in normal amounts (such as in air), and which requires reduced oxygen levels for growth.
Microbial leaching The extraction of minerals from low-grade ores by action of bacteria, especially *Thiobacillus ferrooxidans*.

Microenvironment The environmental conditions in the immediate microscopic surroundings of a microbial cell; often these conditions are different from the larger environment in which the organism grows.

Micrometer One-millionth of a meter, or 10^{-6} m (abbreviated μm), the unit used for measuring microbes. Formerly called micron.

Mineralization The conversion of organically bound elements into inorganic forms, which are then available for metabolism.

Minimum inhibitory concentration (MIC) The lowest concentration of an antibiotic that inhibits growth of a test organism.

Mitochondrion Intracellular organelle in eucaryotic cells; the site of respiratory activity and energy production.

Mold A fungus which grows as hyphae interwoven into an extensive mycelium, with aerial hyphae and conidiospores.

Molecular biology The study of the major macromolecules of living cells, including DNA, RNA, protein, and others.

Molecule See *compound*.

Monoclonal Produced from a single genetically pure cell type; often referring to antibody produced from a hybridoma cell made from a single type of B cell.

Monocyte See *macrophage*.

Monomer A small unit that is repeated many times in the formation of a polymer.

Mononucleosis An infectious disease caused by viruses, which is usually transmitted by direct contact, such as kissing, and is associated with an abnormal increase in phagocytic cells (monocytes).

Morbidity The incidence of diseases in a population, including fatal and nonfatal cases.

Mortality The incidence of diseases causing death in a population.

Most-probable-number (MPN) method A method for counting organisms, using liquid media; statistical evaluation of growth allows estimates of cell number.

Mucous membrane Interior epithelial layers across which mucus flows; not actually a membrane, but a layer of epithelial cells.

Mutagen Any agent that induces mutation.

Mutant A strain genetically different from its parent because of mutation.

Mutation A sudden inheritable change in the genotype of an organism.

Mutualism Relationship between two organisms in which both organisms benefit.

Mycelium The spreading, often highly branched, mass of hyphal filaments typical of vegetative fungal growth.

Mycorrhiza A symbiotic association between plant roots and fungi.

Mycosis Fungal disease of humans; may be widely disseminated in the host (*systemic*) or restricted to the skin, hair, or nails (*superficial*).

Mycotoxin Toxin produced by fungus; formed when grains or some other products become moldy.

Nanometer One trillionth of a meter, or 10^{-9} m (abbreviated nm); a unit used to measure viruses or individual features of microbial cells.

Nasopharyngeal Pertaining to the part of the upper respiratory tract above the mouth.

Nephritis Inflammation of the kidneys; glomerulonephritis.

Neuraminidase An enzyme which degrades neuraminic acid, a type of cell cement; a component of influenza virus which aids in the release of virus from host cells.

Neurotoxin A toxin which acts against the central nervous system; for example, botulinum, the botulism toxin.

Neutralization Reaction between antibody and soluble antigen, such as toxin. Viruses are also neutralized by antibodies.

Nitrification Conversion of ammonia to nitrite and the further conversion of nitrite to nitrate.

Nitrogen fixation The conversion of nitrogen gas into combined form; it may be done by either *free-living* or *symbiotic* microorganisms.

Nodule A tumor-like growth on the roots of leguminous plants caused by infection with nitrogen-fixing bacteria of the genus *Rhizobium*.

Nongonococcal urethritis (NGU) Inflammation of the urethra (the duct through which urine flows) which is caused by microorganisms other than gonococci, such as herpesvirus, chlamydia, and a trichomonad protozoan.

Normal flora The microorganisms which ordinarily grow on the various surfaces of a plant or animal.

Nosocomial infection An infection which is acquired in a hospital.

Nucleic acid A polymer of nucleotides, either deoxyribonucleic acid (DNA) or ribonucleic acid (RNA).

Nucleoside An individual nucleic acid base (adenine, cytosine, guanine, uracil, or thymine) bonded to a single ribose or deoxyribose sugar molecule.

Nucleotide An individual unit of a nucleic acid, composed of a nucleic acid base bonded to a ribose or deoxyribose sugar molecule which is itself bonded to one, two, or three phosphate groups.

Nucleotide sequence The order of nucleic acid bases in a DNA or RNA strand.

Nucleus Membrane-enclosed structure within eucaryotic cells containing the genetic material (DNA) organized in chromosomes.

Nutrient A chemical which an organism obtains from its surrounding environment and uses either as an energy source or as a source of the elements needed to biosynthesize cell constituents.

Oligonucleotide A nucleic acid composed of several or many nucleotides bonded together.

Oncogene A gene which codes for a product which causes transformation of a normal cell into a cancer cell.

Operon A cluster of related genes on a DNA molecule all under the control of one regulatory region.

Opportunistic pathogen An organism that only causes disease when host resistance is abnormally low.

Opsonization Combination of antibody with a cell in the presence of complement, increasing susceptibility of the cell to phagocytosis.

Optical density (O.D.) A means of expressing numerically the turbidity of a suspension, such as a bacterial culture. (As cell number increases, so does O.D.)

Organelle Internal membrane-bound structures of eucaryotic cells; examples include mitochondria and chloroplasts.

Organic compound A compound, other than carbon dioxide (CO_2), which contains one or more carbon atoms; when there is more than one carbon atom, carbon atoms are usually bonded to each other.

Osteomyelitis Infection of the bones.

Oxidation The process by which a compound gives up electrons, acting as an electron donor, and becomes oxidized.

Oxidation-reduction (redox) reaction A coupled pair of reactions, in which one compound becomes oxidized while another becomes reduced and takes up electrons released in the oxidation reaction.

Oxygenic Oxygen-producing; a type of photosynthesis carried out by plants, algae, and cyanobacteria in which the electron donor, water, is split and oxygen is produced.

Pandemic A worldwide epidemic.

Papilloma A benign tumor or overgrowth of epithelial tissue such as skin; for example, warts, corns.

Parasite An organism that lives in and damages another organism (its host); all of its nutrients come from the host.

Pasteurization A mild heat treatment of liquids to kill all pathogens and reduce the total bacterial count.

Pathogen A parasite that causes disease.

Pathogenicity Ability of an organism to cause disease.

Pathological Pertaining to disease; *pathological tissue* or *pathological samples* are those derived from a diseased individual.

Penicillin A broad-spectrum antibiotic produced by the mold *Penicillium chrysogenum*, which inhibits the formation of the bacterial cell wall.

Peptide Two or more amino acids bonded together by peptide bonds between amino ($-NH_2$) and carboxyl ($-COOH$) groups.

Peptidoglycan The chemical component of bacterial cell walls, composed of alternating sugar units, N-acetyl glucosamine and muramic acid, cross-linked by short peptides.

Peritonitis Inflammation of the peritoneum (the membrane lining the walls of the abdomen and pelvis).

Peritrichous flagellation Having flagella attached to many places on the cell surface.

Pesticide A chemical used to control weeds, insects, fungi, or other pests.

pH A measure of the degree of acidity or alkalinity of a solution: pH 7 is neutral; acids have pH less than 7; bases have pH between 7 and 14.

Phage See *bacteriophage*.

Phagocytosis The ingestion of particles or parasites by cells (phagocytes) of the animal body.

Pharmaceutical A chemical produced industrially, which is useful in preventative or therapeutic treatment of disease.

Phenotype The collection of characteristics which an organism expresses at a given time; not the same as *genotype* since not all genetic characteristics may be expressed at the same time.

Phospholipid A type of membrane lipid containing phosphorus.

Phosphorylation The addition of phosphate to a molecule, often to activate it prior to oxidation.

Photoautotroph An organism that uses CO_2 as sole carbon source and light as its energy source.

Photophosphorylation The synthesis of high-energy phosphate bonds as ATP, using light energy.

Photosynthesis The conversion of light energy into chemical energy that can then be used in the formation of cellular constitutents from CO_2.

Phototroph An organism that obtains its energy from light.

Pilus A tube-like protein structure on the exterior of many bacterial cells, which plays a role in the attachment of a cell to other cells.

Placenta The organ through which nutrients and some antibodies pass from mother to fetus.

Plasma The blood fluid that remains after cellular components are removed; contains antibodies.

Plasma cell A B-lymphocyte that produces antibodies.

Plasmid A extrachromosomal genetic element not essential for growth.

Plaque A localized circular area of viral lysis of host cells on a lawn of such cells. See also *dental plaque*.

Pleomorphic Able to undergo change in shape; existing in more than one form.

Pneumonia An infection that results in inflammation of the lungs.

Polar flagellation The condition of having flagella attached at one or both ends of the cell.

Polymer A large molecule made up of repeating small units (monomers).

Polymorphonuclear leukocyte (PMN) One type of phagocyte, numerous during the acute phase of infection.

Polypeptide A *peptide* composed of many amino acids; a protein.

Polysaccharide See *carbohydrate*.

Precipitation An antibody-antigen reaction that results in formation of an aggregate or precipitate.

Primary amino acid sequence The order of amino acids in a peptide or protein; determines the active three-dimensional structure of a protein.

Primary production The initial synthesis of organic matter from inorganic nutrients in an ecosystem; carried out by either phototrophic or lithotrophic autotrophs.

Prion Protein infectious agent; a subcellular entity composed only of protein and containing no DNA or RNA, which causes infectious diseases such as scrapie in sheep; prions are reproduced during the infection.

Procaryote A simple type of cellular organism which lacks a nucleus, organelles, chromosomes, and is unable to carry out sexual reproduction; bacterium (both archaebacteria and eubacteria).

Prodromal The period during infectious disease when symptoms appear.
Prophage A lysogenic bacterial virus which is integrated into the host cell DNA.
Prophylaxis Treatment, usually immunologic, designed to protect an individual from a future attack by a pathogen.
Protease An enzyme which degrades proteins into their component amino acids
Protein synthesis The production within a cell of proteins encoded by information in the DNA; involves the processes of *transcription* and *translation*.
Proton gradient A separation of protons (H^+) such that there are more on one side than on the other side of a membrane.
Proton motive force (PMF) The force established by a proton gradient; the mechanism by which energy released in respiration reactions is stored before being converted to chemical energy in the form of ATP.
Protoplasm The complete cellular contents: plasma membrane, cytoplasm, and nucleus: usually considered to be the living portion of the cell, thus excluding those layers peripheral to the plasma membrane.
Protozoa Eucaryotic single-celled animal-like microorganisms including amoebas, flagellates, ciliates, and sporozoans.
Psychrophile Any organism that grows best at cold temperatures.
Psychrotroph An organism which is able to grow at low temperatures, though its optimum is at a moderate temperature.
Pure culture An organism growing in the absence of all other organisms.
Pus The fluid containing cellular debris, pathogens, and phagocytic cells which accumulates in some areas of inflammation and is characteristic of pyogenic infections.
Pyogenic Pus-forming; causing abscesses.
Pyrogenic Fever-inducing.
Quarantine Restriction of movement of individuals with very infectious diseases to prevent spread of disease to others in a population.
Radioimmunoassay (RIA) A diagnostic test employing antibodies which interact with radioactive antigens, used to detect the concentration of that antigen in test samples.
Recombinant An organism which has acquired DNA from another donor organism either naturally or by artificial gene transfer using genetic engineering.
Recombinant DNA technology See *genetic engineering* and *biotechnology*.
Recombination The process in which genetic elements from two parents are brought together in one unit.
Redox See *oxidation-reduction reaction*.
Reducing agent A chemical which readily gives up electrons and reduces more oxidized compounds; used to remove oxygen from anaerobic culture media.
Reducing power Reduced chemicals which serve as electron donors in reduction reactions; often used to mean the source of electrons for reduction of carbon dioxide (CO_2) during photosynthesis.
Reduction The process by which a compound accepts electrons to become reduced.
Replication The exact duplication of a DNA (or occasionally an RNA) nucleotide sequence.
Reportable disease A disease which must or should be reported to public health authorities; notifiable disease.
Repression The process by which the synthesis of an enzyme is inhibited by the presence of an external substance, the repressor protein.
Reproduction The multiplication of one cell into more than one offspring cell.
Respiration Oxidation of organic compounds in which molecular oxygen serves as the electron acceptor.
Restriction enzyme An enzyme which cuts DNA at a specific site, often leaving single stranded "sticky ends."
Reticuloendothelial system The system of fixed phagocytes (histiocytes) able to remove particles or microorganisms from the blood stream.
Retrovirus A type of RNA virus which can make DNA from its RNA using the enzyme reverse transcriptase; the DNA can then integrate with the host cell DNA; cause of some cancers and AIDS.
Revertant An organism which has regained a characteristic either through reversal of a mutation or because a new mutation cancels the effect of the original one.
Rhizosphere The region around the plant root immediately adjacent to the root surface where microbial activity is unusually high.
RIA See *radioimmunoassay*.
Ribonuclease An enzyme which degrades ribonucleic acid (RNA) into its component nucleotides.
Ribonucleic Acid (RNA) A polymer composed of the sugar ribose and nucleic acid bases arranged in a specific sequence; see *messenger RNA*, *transfer RNA*, and *ribosome*.
Ribose A five-carbon sugar (pentose) found primarily in ribonucleic acid (RNA).
Ribosome A cytoplasmic particle composed of RNA and protein; part of a cell's protein-synthesizing machinery.
RNA See *ribonucleic acid*.
RNA virus A virus which contains RNA as its genetic material.
Rumen A special organ in ruminant animals in which initial digestion occurs by anaerobic fermentation by bacteria and protozoa.
Saprophytic Free-living; obtaining all nutrients from nonliving sources.
Sarcoma A malignant tumor of connective tissue.
Serology The study of antigen-antibody reaction in the laboratory.
Serotype A subgroup of a particular species unique because of the antigenic properties of some cellular component; a given species may contain many different subgroups of different serotype.
Serum The fluid portion of the blood that remains after the blood cells and materials responsible for clotting are removed; contains antibodies.
Sexually transmitted disease A disease spread primarily by sexual intercourse.

Single-cell protein Protein produced by microbes for use as food by humans or animals.
Sludge digestion An anaerobic system for treatment of solid wastes from sewage; the material is degraded by fermentation reactions.
Species A collection of closely related strains; one member of a genus.
Spirochete A coiled bacterium.
Spore A resistant structure formed by a cell, including endospores, conidia, and cysts.
Sporozoan A type of protozoan which is usually nonmotile; many cause important intracellular infections, for example, malaria.
Starter culture A pure culture, available commercially, for use as a direct inoculum in production of fermented milk and meat products.
Stationary phase Growth-cycle phase when rapid cell division has ceased; cells divide slowly or not at all.
Sterile Free of living organisms.
Sterilization Treatment resulting in death or removal of all living organisms in a material.
Stock culture A culture maintained in a culture collection for future study or reference.
Strain A member of a species which has unique properties and can be distinguished from other members of the same species.
Streptomycete A filamentous eubacterium which forms extensive mycelium, aerial hyphae, and conidiospores in a manner similar to molds; many produce antibiotics.
Substrate The compound undergoing reaction with an enzyme.
Substrate-level phosphorylation The synthesis of high-energy phosphate bonds with an activated (usually) organic substrate.
Symbiont An organism living in association with another organism; the relationship may be mutualistic.
Symbiosis Association of two organisms in which the relationship is often mutually beneficial.
Systemic Not localized; an infection disseminated widely throughout the body is said to be systemic.
T cell A type of lymphocyte which is involved in cell-mediated immunity; some T cells aid in the production of antibody.
Temperate virus A virus that, upon infection of a host, does not necessarily cause lysis but may become integrated into the host genetic material. See also *lysogenic*.
Thermal death time The heating time required (at a given temperature) to kill *all* the organisms present.
Thermoduric Able to survive specific high temperatures but not grow.
Thermophile Any organism that grows best at hot temperatures.
Tissue A collection of cells which compose a defined region of the anatomy of a plant or animal.
Tonsillitis Inflammation of the tonsils; sometimes due to bacterial infection.
Toxigenicity The ability of an organism to elicit toxins causing disease symptoms.
Toxin A microbial substance able to induce host damage.
Toxoid A toxin modified so that it is no longer toxic but is still able to induce antibody formation.
Transcription The synthesis of a messenger RNA molecule complementary to one of the two double-stranded DNA molecules.
Transduction The transfer of genetic information via a virus particle.
Transfer RNA (tRNA) A type of RNA involved in the translation process of protein synthesis; each amino acid is combined with a specific transfer RNA molecule.
Transformation The transfer of genetic information via free DNA; the conversion of a normal cell into a cancer cell by a carcinogen or a virus.
Translation The process during protein synthesis in which the genetic code in messenger RNA is translated into the polypeptide sequence in protein.
Transposon A genetic element which is able to move from one location to another on chromosomal or plasmid DNA.
Tricarboxylic acid (TCA) cycle (citric acid cycle) The series of steps by which pyruvate is oxidized completely to CO_2, also forming NADH which fuels ATP production.
Trichomonad A type of flagellated protozoan which inhabits the gastrointestinal or urinary tract; one type causes an extremely common sexually transmitted disease called trichomoniasis.
Trickling filter An oxidative process for the treatment of sewage; liquid sprayed onto a rock bed is oxidized by microbes growing on the rock surface.
Trypanosome One type of flagellated protozoan; includes the causal agent of African sleeping sickness, an infection of the bloodstream.
Tumor A collection of cells which have lost their ability to stop dividing when in contact with adjacent cells, and pile up; may be either benign or malignant.
Tumorigenic Able to induce the formation of tumors.
Vaccination Treatment to render an individual resistant or immune to a particular infectious disease.
Vaccine Material used to induce active immunity; a harmless preparation containing antigens from the pathogen, often as a suspension of killed or attenuated microorganisms.
Vaginitis Inflammation of the vagina, often the result of microbial infection.
Vector An agent, usually an insect or other animal, able to carry pathogens from one host to another; a unit of DNA, such as a plasmid or bacteriophage, which is recombined with foreign DNA in genetic engineering.
Venereal disease See *sexually transmitted disease*.
Viable count The enumeration of living organisms in a given sample, usually by plating in agar.
Viroid An infectious agent composed only of a small amount of RNA and containing no protein; the RNA is reproduced during infection.
Virulence The degree to which a given pathogen can cause disease; it is a function of both *invasiveness* and *toxigenicity*.
Virus A noncellular, parasitic, infectious agent able to reproduce only in living host cells; too small to be seen in the light microscope.

Vitamins Compounds required in small amounts for growth; some microorganisms cannot synthesize all those required and must have them added to the medium.

Water activity (a_w) A measure of water content equivalent to percent relative humidity divided by 100.

Xerophile An organism which can grow at very low moisture levels.

Zoonoses Diseases transmitted from animals to humans.

Credits and Acknowledgements

We would like to express our heartfelt thanks to all those who have generously provided illustrative material for use in this book. Without their kind cooperation, the completion of this book would not have been possible.

Chapter 1 **Figure 1.1** National Center for Health Statistics.

Chapter 2 **Figure 2.1d** Fleischmann Laboratories. **Figure 2.3** Councilman Morgan. **Figure 2.6** Carl Zeiss Inc. **Figure 2.9a** Stanley Holt. **Figure 2.10** Arthur Kelman. **Figure 2.12b** S.L. Tamm, *Journal of Cell Biology*, Volume 55, 250, 1972. **Figure 2.12c** R.G. Kessel and C.Y. Shih, *Scanning electron microscopy in biology*, Springer-Verlag, New York, 1974. **Figure 2.12d** Melvin Fuller. **Figure 2.15c** Norbert Pfennig. **Figure 2.15d** Ercole Canale-Parola et al., *Archiv f. Mikrobiologie*, Volume 59, 41, 1967. **Figure 2.15e** J.T. Staley, *Journal of Bacteriology*, Volume 95, 1921, 1968. **Figure 2.16** Wellcome Research Laboratories, Beckenham, Kent, England. **Figure 2.17** J.F.M. Hoeniger and E.L. Headley, *Journal of Bacteriology*, Volume 96, 1835, 1968. **Figure 2.18a** Hubert and Mary P. LeChevalier. **Figure 2.18b** Peter Hirsch. **Figure 2.19a** Hans Reichenbach. **Figure 2.19b** K. Stephens et al., *Journal of Bacteriology*, Volume 149, 739, 1982. **Figure 2.20** Elliot Juni. **Figure 2.21** Einar Leifson. **Figure 2.24** Fleischmann Laboratories. **Figure 2.33** J.W. Schopf, *Journal of Paleontology*, Volume 42, 651, 1968.

Chapter 3 **Box, page 76** Holger Jannasch, *Symposium Society for General Microbiology*, Volume 36, 97, 1984. **Figure 3.10a** C.H.W. Hirs et al., *Journal of Biological Chemistry*, Volume 235, 633, 1960. **Figure 3.10b** National Biomedical Research Foundation, Washington, D.C. **Figure 3.26** Norbert Pfennig.

Chapter 4 **Figure 4.8b** American Sterilizer Co., Erie, PA. **Figure 4.14** Contamination Control, Inc. Kulpsville, PA.

Chapter 5 **Figure 5.7** National Bureau of Standards, Washington, D.C. **Figure 5.9** Eli Lilly and Co., Indianapolis, IN. **Figure 5.16a** S.B. Levy, *The Lancet*, Volume II, 83, 1982.

Chapter 7 **Figure 7.3** R. Sivendra et al., *Journal of General Microbiology*, Volume 90, 21, 1975. **Figure 7.4** Plates prepared by Diane Derouen. **Figure 7.11**

Charles Brinton et al., *Proceedings of National Academy of Sciences*, Volume 52, 776, 1964. **Figure 7.12** Huntington Potter and David Dressler, Harvard Medical School. **Figure 7.16** Barbara Bachmann et al., *Bacteriological Reviews*, Volume 40, 116, 1976.

Chapter 8 **Figure 8.3** C. Lai et al., *Infection and Immunity*, Volume 11, 200, 1975. **Figure 8.5** Dwayne C. Savage. **Figure 8.6** M.R. Neutra, *Journal of Ultrastructure Research*, Volume 70, 186, 1980. **Figure 8.8** Charles River Breeding Laboratories, Inc. **Figure 8.10** C.L. Wisseman et al., *Infection and Immunity*, Volume 14, 1052, 1976. **Figure 8.15** J.G. Hirsch, *Journal of Experimental Medicine*, Volume 116, 827, 1962.

Chapter 9 **Figure 9.18** J.J. Witte, *American Journal of Public Health*, Volume 64, 939, 1974. **Figure 9.21a** B.B. Bohlool and E.L. Schmidt. **Figure 9.21b** B.B. Bohlool and E.L. Schmidt, *Soil Science*, Volume 110, 229, 1970.

Chapter 10 **Figure 10.11** Bar graph from C.H. Andrewes, *The natural history of viruses*, W.W. Norton and Co., New York, 1966. **Figure 10.13** E.C. Lippy and S.C. Waltrip, *Journal of American Water Works Association*, Volume 76, 60, 1984. **Figure 10.15** George A. Schuler, University of Georgia, Athens.

Chapter 11 **Figure 11.1a** A. Umeda et al., *Journal of Bacteriology*, Volume 141, 838, 1980. **Figure 11.1b** Bryan Larsen, Marshall University School of Medicine. **Figure 11.4** Theodor Rosebury. **Figure 11.8** Royal College of Surgeons, Edinburgh. **Figure 11.9** S.S. Arnon et al., *Epidemiologic Reviews*, Volume 3, 45, 1981. **Figure 11.13** C.H. Binford in F.H. Top and P.F. Wehrle, (editors), *Communicable and infectious diseases*, 7th edition, C.V. Mosby Co., St. Louis, 1972. **Figure 11.16a** Theodor Rosebury, *Microbes and morals*, The Viking Press, New York, 1971.

Chapter 12 **Figure 12.2a and b** M.G. Rinaldi. **Figure 12.2c and d** P.M.D. Martin. **Figure 12.3** M.G. Rinaldi. **Figure 12.4** J.E. Cutler. **Figure 12.5** J.W. Rippon, *Medical Mycology*, W.B. Saunders Co., Philadelphia. **Figure 12.6** M.G. Rinaldi. **Figure 12.13** M.J. Friedman and W. Trager, *Scientific American*, Volume 244, number 3, 154, 1981.

Chapter 13 **Figure 13.1a** Virus Laboratory, University of California, Berkeley. **Figure 13.1b** W.F. Noyes, *Virology*, Volume 23, 65, 1964. **Figure 13.1c and d** Erskine Palmer. **Figure 13.2** Paul L. Kaplan. **Figure 13.3a** Y. Chardonnet and S. Dales, *Virology*, Volume 40, 462, 1970. **Figure 13.3b, c, and d** Robert Gallo et al., *Science*, Volume 224, 500, 1984. **Figure 13.7a** P.W. Choppin and W. Stoeckenius, *Virology*, Volume 22, 482, 1964. **Figure 13.8** W.P. Glezen, *Epidemiologic Review*, Volume 4, 25, 1982. **Figure 13.9** A.D. Langmuir, *American Review of Respiratory Diseases*, Volume 83, 2, part 2, 1961. **Figure 13.10** D.A.J. Tyrell and M.L. Bynoe, *British Medical Journal*, Volume 1, 393, 1961. **Figure 13.17** Paul L. Kaplan.

Chapter 14 **Figure 14.6b** T.R.G. Gray, University of Essex.

Chapter 15 **Figure 15.3** Billy Davis and the Courier-Journal and Louisville Times. **Figure 15.4** E.D. Kilbourne and W.G. Smillie, *Human ecology and public health*, 4th edition, The Macmillan Co., New York, 1969. **Figure 15.5b** M.J. Allen et al., *Journal of American Water Works Association*, Volume 72, 614, 1980. **Figure 15.6** Wisconsin Department of Natural Resources, Madison. **Figure 15.10** Richard Unz.

Chapter 16 **Figure 16.4** Dried Fruit Association of California. **Figure 16.10a and 16.11a** APV CREPACO, Inc. **Figure 16.14 and 16.15a and b** N.F. Olson, *Ripened semisoft cheeses*, Charles Pfizer and Co., Inc., New York, 1969. **Figure 16.16** Wisconsin Department of Agriculture.

Chapter 17 **Figure 17.6** F.J. Bergersen. **Figure 17.8b** U.S. Department of Agriculture. **Figure 17.9** Chester J. Mirocha. **Figure 17.11** John Andersen.

Chapter 18 **Figure 18.1, 18.3, and 18.8** Charles Pfizer and Co. **Figure 18.9** Courtesy of Food Engineering. **Figure 18.10** Lee C. Schisler.

Index

ABO system, 241–242
Absorbance, 108
Acetic acid bacteria, 451
Acetobacter, 18, 517, 526–528
Acetone, commercial production, 528
Acetyl-CoA, 69–70
N-Acetyl glucosamine, 53
Acidity, food, 439–440
Acid mine drainage, 400–402
Acidophile, 86, 389
Acinetobacter, 33
Acne, 183, 193, 300
Acquired immunity, 192, 197, 205, 217
Acquired immunodeficiency syndrome (AIDS), 2–3, 236–237, 335, 351, 354, 370–372
  control, 281–282
  incidence, 370
  screening, 372
  transmission, 262
Actinomycetes, 32–33
Activated-sludge system, 422, 425–427
Activation energy, 59–60
Active immunization, 244–247
Active site, 55–56, 60–61
Acute illness, leading causes, 266
Acute period of disease, 207
Acylovir, 369
Adenine, 55–57, 132
Adenosine diphosphate (ADP), 64
Adenosine triphosphate (ATP), 64
  from fermentation, 66–68
  from photosynthesis, 76–79
  from respiration, 68–72
Adenovirus, 20, 42, 351, 354
ADP, *see* Adenosine diphosphate
Aeration, industrial fermentation, 508
Aerobe, 86–89
  facultative, 86
  microaerophilic, 86
  obligate, 86
Aerophile, 86
Aerotolerant anaerobe, 86
Aflatoxin, 337–338, 490–492
African sleeping sickness, 41, 263, 339–341
Agalactia, 320
Agammaglobulinemia, 236
Agar, 13, 85
Agar diffusion method, 123–124
Agaricus bisporus, 530
Agglutination, antibody, 227
Agriculture, 4–5
  antibiotics, 121, 127
  microbiology, 476–501
AIDS, *see* Acquired immunodeficiency syndrome
Airborne transmission, 260–262
Air pollution, lichens and, 40
Alcohol
  as antiseptic, 115
  commercial production, 511, 518–519
  from fermentation, 66–68
Alcoholic beverages, 6, 511, 513–519
  distilled, 517–518
Aldrin, 480
Algae, 18–19, 27
  characteristics, 34–35
  energy metabolism, 73
  filamentous, 34
  lichen, 38–40
  photosynthesis, 77
  single-cell protein, 529
  soil, 477
  water pollution, 431–432
Algal bloom, 431–432
Algicide, 114–115
Algistatic agent, 114
Alkalophile, 86
Allergen, 237, 240
Allergy, 36, 228–229, 237–240
  antibiotic, 116
  fungi, 329, 336
  occupational, 336
  skin test, 240
Alternaria, 336
Alternative energy sources, 518–519
Amanita caesarea, 38
Amanita muscaria, 38, 337–338
Amanita phalloides, 337
Amanita verna, 38
Amanitin, 337
Amantadine, 363
Ames test, 155
Amikacin, 120
Amino acids, 54–55, 132
  codons, 137
  commercial production, 521–525
  industrial uses, 523
  tRNA, 137–139
p-Aminobenzoic acid, 121–122
Aminoglycosides, 120
6-Aminopenicillanic acid, 119
Ammonia
  as energy source, 73, 75
  fertilizer, 485–486
  nitrogen cycle, 395–398
  as nitrogen source, 63
Amoeba, 40–41
  diseases caused by, 342–343
Amoeba proteus, 40
Amoebic dysentery, 342–343, 410
Amphotericin, 118, 120
Ampicillin, 118–119, 125
Amylase, 525–526
Anabaena, 18–19, 78
Anabolism, 48–49, 58, 62–64
Anaerobe, 86–89
  aerotolerant, 86
  enrichment culture, 96
  facultative, 74
  intestinal, 185
  obligate, 86–87

Anaerobic jar, 87–88
Anaerobic respiration, 73–75, 398–399, 485
Anaerobic sludge digestion, 422–425
Anaphylactic shock, 229, 238–239
Ancalomicrobium adetum, 30
Animal-cell culture, 352–353
Animal feed
  antibiotics, 127, 471, 495, 499–500
  silage, 500–501
  yeast, 511, 513
Animals, *see also* Zoonoses
  diseases, 4–5, 491–497
  germ-free, 188–190
  slaughter, 469–470
Anopheles, 344–345
Anoxygenic photosynthesis, 78
Anthrax, 12, 32, 263, 305–306
  immunization, 306
  livestock, 493–494
  transmission, 305
Antibiotic resistance, 124–127, 500
  gonorrhea, 297
  hospital infections, 283
  multiple, 162, 166–167, 283
  plasmid-coded, 160–165
  Salmonella, 110, 471
  selection for, 154–155
  transposon-coded, 166–167
  uncontrolled antibiotic use, 126–127
Antibiotics, 114–118, *see also* specific antibiotics
  administration, 116–117
  agriculture, 121, 127
  allergy, 116
  animal feed, 127, 471, 495, 499–500
  broad-spectrum, 116–117
  commercial production, 4, 505–506, 519–522
  measuring antimicrobial activity, 122–124
  mode of action, 118–122
  narrow-spectrum, 116–117
  for plant disease, 488
  range of action, 116–117
  screening for, 127–128
  selective toxicity, 114–115
  sterilization of intestine, 186–187, 200
  therapeutic index, 117
  toxicity to host, 116
  uncontrolled use, 126–127
  zone of inhibition, 123–125
Antibiotic sensitivity, 211
  Kirby-Bauer method, 124
  testing, 123–125
Antibiotic therapy, 211–214
Antibody, 205, 217, 220–222
  antigen-antibody reaction, 225–227
  antigen-combining site, 223–224
  constant region, 223–224
  diagnostic tests using, 250–252
  diversity, 229–230
  kinds, 224–225
  monoclonal, 248–249
  production, 229–232
  structure, 223–224, 229–230
  variable region, 223–224
Antibody-mediated immunity, 218, 220–227
Antibody titer, 221
Anticoagulant, 198
Anticodon, 137–138
Antigen, 217–218
  antigen-antibody reaction, 225–227
Antigenic determinant, 218
Antigenic variation, 360–362
Antihistamine, 203–204, 238
Antimicrobial therapy, 211–214
Antiseptics, 114–115
Antiserum, 245
Antitoxin, 225–226, 245
Antiviral drugs, 357–358
Appendaged bacteria, 30
Appert, Nicholas, 10–11
Archaebacteria, 20, 42–43
Arenavirus, 351
Arthobacter, 32
Ascorbic acid, commercial production, 528
Aseptic technique, 11, 93–94
Asexual reproduction, 26
Ashbya gossypii, 522
Asian flu, 362
Aspartame, 6, 523
Aspergillosis, 330
Aspergillus, 335–336, 439
Aspergillus clavatus, 336–337
Aspergillus flavus, 337–338, 490–492
Aspergillus niger, 528
Aspergillus ochraceus, 337
Aspergillus versicolor, 337
Asthma, 336
Athlete's foot, 180, 330–332
Atmosphere, 383–384
Atom, 50
ATP, *see* Adenosine triphosphate
ATPase, 71–72
Atrazine, 480
Attack rate table, food-specific, 447
Attenuation, 246
Autoclave, 91–92
Autoimmune disease, 240–241
Autoimmune response, 293
Autotroph, 63–64, 75
Avian influenza, 494
Axial filament, 318
Azotobacter, 396–397, 482

Bacillus, 30, 32, 119, 458, 519
  diseases caused by, 305–306
  in food, 439
Bacillus anthracis, 12, 263, 305–306, 493–494
Bacillus cereus, 18, 306, 474
Bacillus megaterium, 29, 201
Bacillus polymyxa, 520
Bacillus popillae, 489
Bacillus sphaericus, 489
Bacillus subtilis, 25, 520
Bacillus thuringiensis, 489
Bacitracin, 499, 520
Bacon, 470, 472
Bacteremia, 193
Bacteria, 18–19
  characteristics, 29–33
  diseases caused by, 291–324
  energy metabolism, 73
  Gram stain, 23–24
  as insecticide, 489–490
  rumen, 497–499
  shape, 30
  single-cell protein, 529
  size, 20, 29
  water-borne, 410–411
Bactericide, 114
Bacteriochlorophyll, 78
Bacteriophage, *see* DNA virus
Bacteriophage lambda, 171–172
Bacteriostatic agent, 114
Bacteroid, 483–484
Bacteroides, 121, 185
Baker's yeast, 511–513
Bambermycin, 499
Basidiospore, 37
Batch fermenter, 509
Bats, rabies, 374
B cells, 229–232, 248–249
Beer, 511, 515–517
Bee sting reaction, 238
Berkeley, Miles Joseph, 11
Bifidobacterium, 186
Binary fission, 25
Binomial nomenclature, 29
Biochemical oxygen demand (B.O.D.), 420–422
Bioconversion, 506, 524
Biodegradable substance, 427
Biofuels, 6
Biofilm, 425
Bioform, 419–420
Biogeochemical cycles, 387
Biological containment, 172
Biomass, 6
Biopsy, 208
Biosphere, 383–384
Biosynthesis, *see* Anabolism
Biotechnology, 2, 7–8, 167–175
  United States corporations, 7–8
Black Death, 1
Blackleg, 493
Blanching, 453
Blastomyces dermatitidis, 331, 335
Blastomycosis, 267, 330–331, 335
Blepharisma, 40
Blindness, 297, 322–323
Blood, 198–200
  clotting, 198, 203, 205
  components, 198
  specimen, 209
Blood groups, 241–244, 346
Blood poisoning, 292
Blood typing, 241–244
Blue-green algae, *see* Cyanobacteria
B.O.D., *see* Biochemical oxygen demand
Boils, 32, 193, 195, 203, 262, 294–295
Booster shot, 221, 247–248
Bordeaux mixture, 487
Bordetella, diseases caused by, 299–300
Bordetella pertussis, 299–300
Botulism, 180, 191–192, 195–196, 258, 304–305, 442–443
  outbreak, 442
  treatment, 245
Bovine virus diarrhea, 493
Bread making, 68, 511–513
Breast feeding, 225, 245, 374
Breed test, 105
  milk, 461–462
Brevibacterium linens, 467
Brewing, 68, 515–517
Brucella, 493
  diseases caused by, 314–316
Brucella abortus, 191, 315, 458–459, 493
Brucella melitensis, 203, 276, 315
Brucella suis, 315
Brucellosis, 191, 263, 267, 315, 492–493
  control, 276
  transmission, 260
Buboes, 313
Bubonic plague, 1, 197, 263, 267, 278, 313–314
  transmission, 260
Budding, yeast, 36
Buffer, 86
Bulking, 427
Bunyavirus, 351
Burn patients, 193, 283, 316
Butanol, commercial production, 528
Buttermilk, 5, 463

Calcium propionate, 455
Calf scours, 493, 495–496, 500
Campylobacter, 410
  diseases caused by, 317–318
Campylobacter fetus, 317, 493
Campylobacter jejuni, 318, 442, 474
Cancer, 233, 237, 249, 351, 355, 376–379
  cervical, 369, 378
  liver, 378, 490
  nasopharyngeal, 378–379
  skin, 378
  viruses, 376–379
Candicidin, 120
Candida, 188, 335, 371
Candida albicans, 183, 186, 331, 333, 335
Candidiasis, 330–331
Canning, 109, 441–443, 452–455
  commercial, 452–455
  meat, 471
  quality control, 454–455
  spoilage, 454–455
Capsule, bacterial, 26, 33, 202, 219
Carbadox, 499

Carbohydrate, 53, 58
Carbon compounds, requirements of cells, 62–64
Carbon cycle, 390–395
Carbon dioxide
  carbon cycle, 391–392
  as carbon source, 62–63
  from fermentation, 66–68
  fixation, 77
  greenhouse effect, 392
  in photosynthesis, 76–79
  from respiration, 68–72
Carbuncle, 193
Carcinogen, 155, 376
Carnivore, 386
Carrier, disease, 257, 259, 276–277, 312
Casein, 457
Catabolism, 48–49, 59, 64–72
Catalysis, 59–60
Cattle, *see also* Livestock
  infectious abortion, 315–316
  tuberculosis, 309
cDNA, *see* Complementary DNA
Cell, 17
  composition, 58
Cell count
  total, 104–105
  viable, 105–108, 448, 462
Cell culture, 352–353
Cell mass, 108–109
Cell-mediated immunity, 218, 220–222, 232–236
Cell membrane, 18
  electron transport system, 71–72
  eucaryotic, 27
  inhibitors, 118–120
  procaryotic, 24
  structure, 55–58
Cell structure
  eucaryotic, 26–29
  procaryotic, 24–26, 28
Cellulase, 61
Cellulose, 53
  digestion, 497–499
Cell wall, 18
  algal, 35
  bacterial, 53
  eucaryotic, 27
  inhibitors, 118–119
  procaryotic, 24
  structure, 52–54
Cephalosporin, 116, 119
  commercial production, 520
Cephalosporium, 116, 119, 519–520
Cephalothin, 118, 125
Cephoxitin, 125
Cephamycin, 118–119
C gene, 229–230
Chagas' disease, 341
Chain, Ernst, 212
Chancre, 319
Chantrelle, 38
Cheese, 5, 465–468, 525–526
Cheese worker's lung, 336
Chemical bond, 50
Chemical feedstock, 6
Chemical preservative, 455–456
  meat, 471–472
Chemical reactions, 58–62
Chemoprophylaxis, 213
Chemotherapeutic agents, 114–118
  mode of action, 118–122
Chemotherapy, 211–214
Chicken pox, 208, 267, 275, 351, 364, 368
Childhood, virus diseases, 364–368
Chlamydia, 194
  chlamydial nongonoccal urethritis, 262, 274, 281
  diseases caused by, 322–323
  transmission, 261
Chlamydia trachomatis, 322–323
Chloramphenicol, 116–120, 125
  commercial production, 520
Chlordane, 480
Chlorinated insecticide, 480
Chlorination, 415–416, 428
Chlorine compounds, 115
Chlorobium limicola, 78
Chlorophyll, 76
Chloroplast, 27, 34
Chlortetracycline, 118, 120, 472, 499
Cholera, 266–267, 278, 316–317, 410
  epidemic, 417
  immunization, 247, 271–272, 317
  outbreak, 273–274
  transmission, 260, 262
  treatment, 317
Chromatium okenii, 78
Chromobacterium violaceum, 154
Chromoblastomycosis, 330–333
Chromosome, 26
Chrysosporum, 439
Cilia
  eucaryotic, 28
  protozoan, 41
Ciliates, 40–41, 140
  rumen, 497–499
Circulatory system, 198–203
Citric acid, commercial production, 6, 528
Citric acid cycle, *see* Tricarboxylic acid cycle
Citrinin, 337
Cladosporium, 336
Claviceps purpurea, 337–339
Clindamycin, 118, 121, 125
Clinical course of infectious disease, 206–208
Clinical microbiology, 208–211, 250–252
Clitocybe, 337
Clonal selection, 231, 236
Clone, 82, 170
Clostridium, 32, 185, 396–397, 458
  diseases caused by, 303–305
  nitrogen fixation, 482
  reservoir, 257–258
Clostridium acetobutylicum, 528
Clostridium bifermentans, 31
Clostridium botulinum, 195–196, 258, 304–305, 442–443, 447, 455, 472, 474
Clostridium chauvoei, 493
Clostridium perfringens, 190, 195, 263, 304, 443–444, 469, 474, 499
Clostridium tetani, 31, 191, 195–196, 258, 303–304
Coagulase, 203, 294
Coal, 391
  mining, 400–402
Coccidioides immitis, 261, 334
Coccidioidomycosis, 267, 330–331, 334–335
  geographic incidence, 334
Coccus-shaped bacteria, 30, 32
Codon, 136–140, 152–153
Coliforms
  as indicator organisms, 412–413
  milk, 462–463
Collagenase, 193, 205
Colony count, 105
Colony-forming unit, 107
Colorado tick fever, 351
Colorimeter, 108
Commensal relationship, 179–180
Common cold, 191, 267, 351, 363–364
  epidemic, 271–273
  incidence, 266
  transmission, 260–262, 364
Complement, 227–229
  activation, 227–229
  fixation, 250–252
Complementary bond, 133–134
Complementary DNA (cDNA), 173–174
Compost, 478–479
Compound, 50–51
Condenser, 21–22
Conidia, 35–36
Conjugation, 158, 160–165
Conocybe, 337
Constitutive enzyme, 141
Contact inhibition, 376
Contagion, 11
Contagious disease, *see* Infectious disease
Contaminated items, disposal, 93, 110–111, 283–285
Continuous fermenter, 509
Continuous sterilization, 507–508
Convalescent period, 208
Copper mining, leaching, 403–404
Copper sulfate, 115
  algae control, 431
Coprine, 337
Coprinus atrementarius, 337
Corepressor, 143
Coronavirus, 351–352
Cortinarius orellanus, 337
Corynebacterium, 183, 187, 493
  diseases caused by, 300–303
Corynebacterium acnes, 300
Corynebacterium diphtheriae, 191, 194–196, 276, 300–303, 458–459
Corynebacterium simplex, 524
Cosmic rays, 154
Coughing, 258, 261, 360
Counting chamber, 104
Covalent bond, 50–51
Cover slip, 22
Cowpox, 351
Coxiella burnetii, 321, 458–459
  pasteurization, 458–459
  transmission, 261
Cremation, 287
Creutzfeldt-Jakob disease, 147
Cross-matching, blood, 242–243
Cryptococcosis, 335
Cryptococcus neoformans, 331, 335, 371
Cryptosporidium, 371
Cryptostroma corticale, 336
Crystal violet, 22
Culture
  diagnostic, 211
  enrichment, 94–96
  inoculation, 94–95
  long-term storage, 96
  pure, 12–13, 82, 94–96
  starter, *see* Starter culture
  stock, 96
Culture medium, 83–85
  aeration, 86–89
  complex, 84–85
  differential, 96, 211
  industrial fermenter, 506
  liquid, 85
  pH, 86
  selective, 96, 154–155, 159, 164–165, 211
  semisolid, 13
  solid, 85
  sterilization, 90–91
  synthetic, 84–85
  temperature, 89
Curd, 465–468
Curing of plasmid, 162–163
Curvularia lunata, 524
Cyanobacteria, 18–19, 43, 78, 396–397
  nitrogen fixation, 482
  photosynthesis, 77
Cycles of disease, 274–275
Cycloheximide, 117–118, 121, 488
  commercial production, 520
  mode of action, 140–141
Cycloserine, 520
Cytomegalovirus, 369–371
Cytosine, 55–57, 132

2,4-D, 480
Dairy microbiology, 5, 457–468, 493
Dalapon, 480
Dark-field microscope, 23
DDT, 480–481
Dead, disposal, 287
Death cap, 337
Death phase, 104

Death rate, human, 3
Debaryomyces, 439
Decline period of disease, 207
Decomposition, 386–387, 390–392, 396, 420
  aerobic, 391–392
  anaerobic, 391
  substances resistant to, 427–428
Defined medium, *see* Synthetic medium
Deforestation, 392
Dehydrogenase, 61
Delayed-type hypersensitivity, 233–235, 237–240
Deletion mutation, 153
Dengue, 351
Denitrification, 74, 396, 398, 485
Denitrifying bacteria, 73–74
Dental caries, 184, 293
Dental plaque, 184
Deoxyribonuclease, 62
Deoxyribose, 55
Dermatitis, 240
Dermatophytes, 329–331
Destroying angel, 38
Desulfovibrio, 399
Detergent, 115
  enzymes in, 525
  nonbiodegradable, 428
Dextran, 528
D gene, 229–230
Diabetes, 241
Diacetyl, 464
Diarrheal disease, 310, 351
  infant, 310–311, 351
  transmission, 410
  virus, 373–374
Diazinon, 480
Differential medium, 96, 211
Di George's syndrome, 236–237
Dihydroxyacetone, 528
Dilution
  serial, 106–107
  tube, 122–123
Diphosphoglycerate, 67–68
Diphtheria, 191, 194–197, 257, 267
  control, 276
  food-borne, 458–459
  immunization, 247–248, 300–301
  incidence, 301
  transmission, 261–262
Direct microscopic count, 104–105
  milk, 461–462
Direct microscopic examination, 210–211
Disease, *see* Infectious disease
Disease symptoms, 192
Disinfectant, 114–115
Disinfection
  hands, 2
  hospital equipment, 284
Disulfide bond, 55
DNA
  eucaryotic, 26, 144–145
  hybridization, 134–135
  mitochondrial, 140
  mutation, 151–155
  per cell, 58
  procaryotic, 25
  radiation effects, 112–113
  recombinant, 151, 155–175
  replication, 134
  sticky ends, 170–171
  structure, 55–57, 132–135
DNA clone, 170
DNA ligase, 168, 170–171
DNA polymerase, 134
DNA virus, 41–42, 351
  reproduction, 145–146
Dog
  leptospirosis, 320
  rabies, 374–376
Double helix, 134
Doubling time, *see* Generation time
Drechslera, 336
Drilling fluid, 394
Drinking water, 342, 398, 408–409, 412
Drinking water standards, 412–413, 416–418
Drugs
  allergy, 238–240
  antiviral, 357–358
  in disease eradication, 281
Dry heat sterilization, 92–93
Drying
  control of growth, 111–112
  food, 438–439, 449–451
  meat, 471
Dry-slide method, cell count, 105
Duffy antigen, 346
Dunaliella, 40
Dyes, 22–23
Dysentery
  amoebic, 342–343, 410
  bacterial, 458–459, *see also* Shigellosis
  control, 276
  outbreak, 273–274
  transmission, 410

Ecosystem, 385–387
  measurement of microbial activity, 388
Electromagnetic radiation, 112–113
Electron, 50
Electron acceptor, 65
Electron carrier, 65
Electron donor, 65
Electronic cell counter, 105
Electron microscope, 21, 23–24
  scanning, 24
Electron-transport phosphorylation, 69, 71, 74
Electron transport system, 68–72
ELISA technique, 250–251
Embalming, 287
Encephalitis, 351
Endemic disease, 256, 270
  travel to areas with, 270–272
Endocarditis, 191
Endoplasmic reticulum, 27
Endospore, 30
  germination, 31
  heat resistance, 90–91, 109
Endothermic reaction, 59
Endotoxin, 194–196
Energy
  flow through ecosystem, 386–387
  generation, 66–72
  metabolism, diversity among microorganisms, 73–79
  source, 64–66, 83
Enrichment culture, 94–96
Entamoeba histolytica, 342–343, 410
Enteric bacteria, 469
  diseases caused by, 310–313
  as indicator organisms, 412–413
Enterobacter aerogenes, 412, 458
Enterotoxin, 195, 310, 317
  staphylococcal, 448
Enterovirus, 372–373
Envelope, viral, 41
Environment, 48
Environmental microbiology, 383–404
Enzymes, 55–56, 60–62
  commercial applications, 525
  commercial production, 505, 511, 524–526
  constitutive, 141
  genetic engineering, 168–170
  inducible, 141–143
  regulation, 143–144
  repressible, 141–143
  restriction, 168–171
Enzyme-substrate complex, 60–61
Epidemic, 256, 271–274
Epidemiology, 256–257, 277–278
Epidermophyton, 330–332
Epiglottitis, 299
Epithelial layer, 182
Epstein-Barr virus, 369–370, 378–379
Ergot poison, 337–339
Erysipelas, 292
Erythritol, 315
Erythroblastosis fetalis, *see* Rh disease of the newborn
Erythrogenic toxin, 195, 292
Erythromycin, 116, 118, 121, 125, 499
  commercial production, 520
Escherichia, 32
Escherichia coli, 29, 185, 191, 195, 282–283, 410, 412, 458, 493, 495
  diarrhea, 208, 310
  enterotoxigenic, 310–311
  in genetic engineering, 172
  genetic map, 165
Ethylene oxide sterilization, 93
Eubacteria, 19, 42–43
Eucaryote, 18–19
  structure, 26–29
Euglena, 28
Eurotium, 439
Eutrophication, 77, 431–432
Evolution, 41–44, 151
Exon, 144
Exothermic reaction, 59
Exotoxin, 300
Exponential growth, 102
Exponential phase, 102–103
Eyepiece, 21

Facultative aerobe, 86
Facultative anaerobe, 74
FAD, 70–72
Farmer's lung, 336
Favus, 330–331
Feces, 185–186, 258, 262, 281
  specimen, 209
Feedback inhibition, 143–144
Fermentation, 66, 72, 74
  industrial, 504–531
  silage, 500–501
Fermented food, 451
Fermenter, industrial, 506–510
Fertilizer
  compost, 478–479
  nitrogen, 398, 485–486
  sulfur, 399
Fetus, immunity, 224
Fever blister, 351, 356, 369
F factor, 162–165
Fibroma, Shope, 377
Filipin, 120
Filter sterilization, 91–92
Filtration of water, 415–416
Fish kill, 420–421
Flagella
  bacterial, 33
  eucaryotic, 27–28
  procaryotic, 25–26
Flagellar antigen, 227
Flagellates, 40–41
  diseases caused by, 339–342
Flash pasteurization, 110, 459
Flatus, 186–187
Flea, 313
Fleming, Alexander, 212
Flesh of the gods, 337–338
Floc, 425–427
Florey, Howard, 212–213
Fluorescence microscope, 23
Fluorescent antibody technique, 250–252
Fly, 431
Focus of infection, 193
Folic acid, 121–122
Food
  acidity, 439–440
  allergen, 240
  assessing microbial content, 447–448
  from microorganisms, 529–531
  moisture content, 438–439
  moldy, 338
  oxygen availability, 440–441

processing, 437, 442–443
sanitation, 456
spoilage, 436–441
storage, 438–439
water activity, 438–439
Food-borne disease, 260, 262, 314, 441–447
control, 281
outbreak, 441–442, 446–447
virus, 372–374
Food chain, 386
Food handler, 259, 262, 276, 312, 373, 445, 456, 471
Food infection, 441
Salmonella, 445–446
Food microbiology, 436–474
Food poisoning, 191–192, 195–196, 267, 304, 306, 312, 441–446, 474
Clostridium, 442–443
staphylococcal, 443–445
Food preservation, 2–3, 5, 11, 154, 448–456
canning, 441, 452–455
chemical, 455–456
drying, 111–112, 438, 449–451
low-temperature, 111, 439–440, 450–451
pickling, 451
radiation, 113
salting, 451–452
Foot-and-mouth disease, 175, 493, 496
immunization, 247, 357–358
Formaldehyde, 455
F protein, 367
Francisella tularensis, 263, 314
Frankia, 397
Free radicals, 113
Freeze-drying, food, 449
Freezing
food, 111, 439–440, 450–451
meat, 470–471
Fructose, 6
commercial production, 526
Fruiting body
bacterial, 33
mushroom, 36–37
Fuel, 6, 509
alcohol, 518–519
decomposition, 393
Fungi, 18–19
allergy, 329, 336
characteristics, 35–38
dimorphic, 333
diseases caused by, 328–338, *see also* Mycoses
energy metabolism, 73
evolution, 43
as insecticide, 489
lichen, 38–40
opportunistic, 335
toxins, 336–338
Fungicide, 114
Furazolidone, 499
Fusarium, 337–338, 439, 490–491

Galactosidase, 457, 511
Galerina sulciceps, 337
Gametocyte, 345
Gas sterilization, 93
Gas gangrene, 32, 195, 263, 304
Gastroenteritis, 312, 314, 317–318, 351
food-borne, 445, 458
incidence, 269
transmission, 410
Gastrointestinal tract, 184–187
disease transmission, 258
effect of antibiotics, 186–187, 200
exit of pathogens, 262
gas, 186–187
invasion by pathogen, 193
normal flora, 185–186, 189
Gene, 131, 133
amplification, 520
chemical synthesis, 173
eucaryotic, 144–145
expression, 141–143, 169
isolation, 168, 172–173
Generalized transduction, 160–161
Generation time, 102
Genetic code, *see* Triplet code
Genetic engineering, 2, 7, 98, 151–175, 506, 520
hazards, 168
uses, 174–175
virus vaccine, 357–358
Genetic exchange, 155–157
Genetic map
construction, 165
Escherichia coli, 165
Genetics, 151–175
Genital herpes, 262, 351, 369, 379
Genitourinary tract, 187–188
normal flora, 187–188
Genotype, 141–142
Gentamicin, 125
Genus, 29
German measles, *see* Rubella
Germ-free animals, 188–190
Germicidal lamp, 112–113
Germicide, 114
Germination, endospore, 31
Germ theory of disease, 11–14, 190
Giardia lamblia, 341–342, 410
Giardiasis, 341–342
prevention, 416
transmission, 410
Gibberellin, commercial production, 528
Gliding bacteria, 33
Global distribution of diseases, 270
Global microbiology, 383–404
Gloeocapsa, 78
Glomerular nephritis, 292–293
Glove box, 88
Glucoamylase, 525–526
Gluconic acid, commercial production, 528
Glucose, 53
fermentation, 66–68
oxidation, 65
Glucose isomerase, 526
Glucose oxidase, 525
Glycerol, commercial production, 511
Glycogen, 53
Glycolysis, 66–68
Gonococcus, see *Neisseria gonorrhoeae*
Gonorrhea, 32, 208, 267, 296–298, 323
control, 276, 281
incidence, 274, 297–298
penicillin-resistance, 126–127
transmission, 262, 297
treatment, 297
Grain, mycotoxin, 338–339, 490–492
Grain dust allergy, 336
Gram-positive bacteria, 43
Gram stain, 23–24
Granulosis virus, 489
Green bacteria, 77–78; *see also* Photosynthesis
Greenhouse effect, 392
Griseofulvin, 115, 118, 122, 488
commercial production, 520
Groundwater, 408–409
Growth, microbial, 101–128
control of, 109–122
measurement, 104–108
Growth curve, 102–104
Growth cycle, 102–104
Growth factor, 83–84
analogs, 118, 120–121
Growth hormone, production, 174–175
Growth rate, 101
Guanine, 55–57, 132
Gum-producing bacteria, 394–395
Gyromitra, 337
Gyromitrin, 337

Habitat, microorganism, 388–390
Haemophilus, diseases caused by, 299
Haemophilus influenzae, 29, 191, 261, 298–299
Halobacterium, 439
Halophile, 42–43, 389
Hallucinogens, fungal, 337–338
Hand washing, 274, 284–285
Hapten, 218
Hay fever, 240–241
Health statistics, 264–268, 277
Heat sterilization, 109
Helminthosporium, 336
Hemagglutinin, 360–363
Heme, in culture medium, 84
Hemolysin, 205, 294
Hepatitis, 267, 373
Hepatitis A virus, 373, 410, 442
Hepatitis B virus, 373, 378–379
Heptachlor, 480
Herbicide, 480
Herpes, 208
Herpes bovis virus, 493
Herpes simplex virus, 356, 369, 371, 378
Herpesvirus, 42, 351, 368–370
Heterotroph, 62–64, 73–74
Hexachlorophene, 115, 284
Hexose, 53–54
Hfr strain, 164–165
High-energy phosphate bond, 64
High-fructose corn syrup, 526
High-temperature short-time pasteurization, 460–461
Hirsutella thompsonii, 489
Histamine, 203, 229, 238–239
Histiocyte, 199
Histocompatibility antigen, 233–234, 244
Histoplasma capsulatum, 331, 334
Histoplasmosis, 262, 330–331, 334
geographic incidence, 334
Hives, 238
Homograft, 244
Homologous recombination, 156
Hong Kong flu, 361–362
Hooke, Robert, 9
Hops, 516–517
Hospital infection, 32, 282–286, 295, 310, 316
control, 283–285
fungal, 335
incidence, 282–283
pathogens, 282–283
pneumonia, 324
staphylococci, 283
Host
for pathogen, 179–180
for recombinant DNA, 168–169, 172
viral, 41
Host defense, 180–181, 197–205
immune system, 217–253
Host-parasite relationship, 204–208
Hot spring, 389
H protein, 360–363
Human contact, spread of disease, 262
Human T-cell leukemia virus, 378–379
Human T-lymphotrophic virus, 370–371
Humus, 391–392, 477–478
Hungate technique, 88
Hyaluronidase, 193, 205
Hybrid, 156
Hybridization, 134–135, 172–173
Hybridoma technique, 249
Hydrocarbons, 392–395
Hydrogen bond, 50–51, 133–134
Hydrogen peroxide, 115
Hydrogen sulfide
in anoxygenic photosynthesis, 78
as energy source, 75–76
formation, 73–74
sulfur cycle, 398–400
in water pipelines, 418–419
Hydrologic cycle, 408
Hydrolysis, 60–61
Hydrophilic bonding, 51, 57–58
Hydrophobic bonding, 51, 57–58
Hydrosphere, 383–384
Hydrothermal vent, ocean floor, 75–76

Hypersensitivity
delayed-type, 233–235, 237–240
immediate-type, 237–240, 336
Hypha, 35

Ibotenic acid, 337
Immediate-type hypersensitivity reaction, 237–240, 336
Immobilization of motile cells, 227
Immune serum, 244–245
Immune system, 192, 197, 206, 217–253
cells, 222–223
memory, 218, 221
nonmicrobial phenomenon, 241–244
specificity, 217–218
tolerance, 218
Immunity
acquired, 192, 197, 205, 217
antibody-mediated, 218, 220–227
cell-mediated, 218, 220–222, 232–236
passive, 245
Immunization, 219, 221–222, 244–248, 276–280, *see also* Vaccine
active, 244–247
adverse reactions, 302
infants and children, 247
passive, 244–245
travelers, 270–272
Immunocompromised patient, 335, 368
Immunodeficiency disease, 236–237, 286
Immunoglobulin-degrading enzymes, 205
Immunoglobulins, 223–225, 229
Immunosuppression, 244
Immunosurveillance, 233, 237
Impetigo, 292, 294
Incineration, 93, 110–111
Incubation period, 207–208, 258
Incubator, 89
Indicator organism, 412
Induced mutation, 152–154
Inducer, 142
Induction control, 142–143
Industrial microbiology, 504–531
Infant
congenital herpes, 369
congenital syphilis, 319
gonococcal eye infection, 296–297
premature, 286
Infant botulism, 304–305, 443
Infant diarrhea, 310–311, 351
Infantile paralysis, *see* Polio
Infection, 180
focus, 193
opportunistic, 236–237
persistent, 319, 356
pyogenic, 193, 195, 202
sequence of events, 192
Infectious abortion, cattle, 315–317
Infectious bovine rhinotracheitis, 493
Infectious disease, 179–214, *see also* specific diseases
cause of human death, 2–3
clinical aspects, 206–208
control, 3–4, 275–278
in developed and developing countries, 268–270
eradication, 279–280
germ theory, 11–14
historical impact, 1–2
outbreak, 271–274
prevalence, 264
prevention, 278–282
transmission, *see* Transmission of infectious disease
Infectious hepatitis, 267, 373
control, 276
epidemic, 373
immunization, 272
transmission, 373, 410–411
Infectious mononucleosis, 351, 369
Inflammation, 203–204, 206, 228–229
Influenza, 180, 208, 233, 351, 359–363
avian, 494
control, 276
epidemic, 2, 361–363
immunization, 247, 363
incidence, 266, 361
pandemic, 272, 362
seasonality, 361
transmission, 260–262, 360
Influenza virus, 359–360
antigenicity, 219–220, 360–363
size, 20
Inky cap, 337
Innate resistance, 192, 197–205
Inoculating loop, 94
Inoculation, industrial fermenter, 508
Inocybe, 337
Inorganic compounds, as energy source, 73, 75–76
Insecticide, 341, 480–481
microbial, 488–490
Insect pathogen, 488–490
Insertion mutation, 153, 156–157
Insertion sequence, 166
Insulin, 55, 241
production, 173–174
Intercellular infection, 194
Interferon, 204–205, 234
genic engineering, 173–175
Intervening sequence, 144
Intestinal flora, 32
Intestinal tract, *see* Gastrointestinal tract
Intracellular infection, 194
Intron, 144, 174
Invasion, 192–193
Invasiveness, 196–197
Invertase, 60–61, 511, 525
Iodine, 115, 284
Ionizing radiation, 112–113
Iridovirus, 351
Iron, 194
Iron-oxidizing bacteria, 419
Isolation of microorganisms, 94–96
Isolation of patient, 285–286
Isolator, 188–189, 286
Isomerization, 526
Isoniazid, 118, 121–122
Itaconic acid, 528

Jellies, 451–452
Jenner, Edward, 359
Jet fuel, 393
J gene, 229–230
Jock itch, 330–331

Kanamycin, 120, 125
commercial production, 520
Kaposi's sarcoma, 371
Kidney, nephritis, 191
Kirby-Bauer method, 124
Kitchen, sanitation, 456
Klebsiella pneumoniae, 412
Koch, Robert, 11–14, 306, 417
Koch's postulates, 14, 190
Koplik's spots, 365

β-Lactamase, *see* Penicillinase
Lactase, 61
Lactic acid, commercial production, 528
Lactic acid bacteria, 74, 185, 451, 458, 463–464, 473, 501
Lactobacillus, 32, 188, 439, 463–464, 501, 517
Lactobacillus acidophilus, 458
Lactobacillus bulgaricus, 464
Lactobacillus casei, 458
Lactobacillus plantarum, 472
Lactoperoxidase system, 184, 205
Lactose, 186–187, 457
Lager beer, 517
Lagoon, sewage, 422, 429–431
Lag phase, 102–103
Lake
algae, 431–432
as ecosystem, 385–386
water, 409
Lassa fever, 351
Latent infection, viral, 146–147, 356, 368
Leach dump, 403–404
Leavening, 511
Leeuwenhoek, Antoni van, 9–10
Leghemoglobin, 483–484
Legionella pneumophila, 84, 324
Legionellosis, 84, 324
seasonality, 324
treatment, 121
Legume, 4–5, 397, 482–485
Leprosy, 190, 235, 267, 310
Leptospira, 320, 493
Leptospirosis, 320
livestock, 493
Leptotrichia buccalis, 184
Leuconostoc, 463–464, 501
Leukemia, human T-cell, 351, 378–379
Leukocidin, 195, 202–203, 205, 294
Leukocyte count, 201
Lichens, 38–40, 477
and air pollution, 40
Light, energy from, 76–79
Light microscope, 21–22
Lincomycin, 118, 121, 499, 520
Lindane, 480
Linnaeus, 29
Lipids
per cell, 58
structure, 55–58
Lister, Joseph, 11
Lithosphere, 383–384
Lithotroph, 73, 75–76
Livestock, 315
diseases, 491–497
infectious abortion, 317
Mycoplasma disease, 320
Louse, 321
Low-temperature pasteurization, 459
Lupus erythematosus, 241
Lymph, 198–200
Lymph node, 198–199
Lymphocyte, 199
Lymphokine, 234–235
Lymphoma, Burkitt's, 378–379
Lyophilization, 96
food, 449
Lysis, 24, 118, 227–228
Lysogenic infection, 146–147, 355
Lysozyme, 183–184, 205
Lytic infection, 145–146, 356

Macrolides, 120–121
Macromolecule, 52
Macrophage, 199–201, 231–235
Macrophage activating factor, 234
Macrophage chemotactic factor, 234–235
Magnification, 21
Malaria, 41, 180, 194, 343–347
control, 276
eradication, 280, 345
geographic distribution, 270, 346
incidence, 271
natural resistance, 346–347
transmission, 260, 263
treatment, 345
Malathion, 480
Malt, 515
Maltase, 61
Maltster's lung, 336
Manganese-oxidizing bacteria, in water pipelines, 419
Maple bark stripper's lung, 336
Mash, 515–516
Mass vaccination, 279–280

Mast cell, 225, 238–239
Mastitis, livestock, 292, 458–459, 493, 495
Mating, yeast, 36
Measles, 208, 233, 351, 364–366
  control, 276
  epidemic, 366–367
  immunization, 247–248, 366–367
  incidence, 365
  outbreak, 274
Meat
  aging, 469–470
  antibiotic-resistant bacteria in, 470–471
  contaminated, 263
  microbiology, 468–473
  preservation, 471–473
  processing, 470–473
  spoilage, 438–441
  surface shine, 469
Meat-packing industry, 305–306, 315, 468–470
Medium, *see* Culture medium
Membrane, *see* Cell membrane
Membrane, internal
  eucaryotic, 27
  procaryotic, 24
Membrane filter
  growth on, 89–90
  microbiological assay of water, 413
  sterilization of liquids, 91–92
  viable count, 108
Memory cell, 231–232, 236
Meningitis, 191, 209, 294, 298–299, 310
  control, 276
  immunization, 247, 299
  incidence, 298–299
  outbreak, 298–299
  transmission, 261–262
  treatment, 298
Meningococcus, see *Neisseria meningitidis*
Mercurials, 115
Merozoite, 344–345
Mesophile, 89
Messenger RNA (mRNA), 135–139
Metabolic diversity, 73–79
Metabolism, 18, 48–80
  regulation, 141–144
Metabolite, secondary, 103–104, 509
Metal, corrosion, 399
Methane, 51
  rumen, 499
  sewage-plant, 423–424
Methanogens, 42–43, 73, 75
  culture, 88
  intestinal, 186
  rumen, 499
  sewage treatment, 423–424
Methicillin, 118–119, 125
Methylene blue, 22
Metritis, 493
MIC, *see* Minimum inhibitory concentration
Micrasterias, 34
Microaerophile, 86
Microbial antagonism, 115–116
Microbial leaching, 400, 402–404
Micrococcus, 32, 467
Microenvironment, 388
Microfossil, 44
Micrometer, 19–20
Microorganisms
  natural habitat, 388–390
  size, 19–20
Microscope, 20–24
  cell counting, 104–105
  dark-field, 23
  electron, 21, 23–24
  fluorescence, 23
  invention, 8–10
  light, 21–22
  phase-contrast, 23
Microsporum, 330–331
Migration inhibition factor, 234
Mildew, 111–112, 487
Milk, 457–468
  aflatoxin, 490–492
  bacterial count, 105
  bacteriological examination, 461–463
  composition, 457–458
  dried, 449
  fermented products, 463–468
  government regulation, 463
  microorganisms in, 263, 315, 457–458, 493
  pasteurization, 110, 276–277, 309, 311, 316, 458–461
  processing, 457
  shelf life, 461
Mineralization, 397, 478
Minimum inhibitory concentration (MIC), 122
Mining microbiology, 400–404
Mitochondria, 27, 71, 140
Mitomycin, 154
Mitosis, 26
Moisture content, food, 438–439
Molds, 35–36, 329
Molecular biology, 2, 131–148
Molecule, 50
Monoclonal antibody, 248–249
Monocyte, *see* Macrophage
Monomer, 52–58
Mononucleosis, infectious, 351, 369
Monosodium glutamate, 523
Moraxella, 187
Morbidity, 265–266
Morel, 38
Mortality, 265
  infant and child, 269
Mosquito, 344–345
Moss, 477
Most probable number (MPN), 108, 413
Mouse, nude, 237
Mouth, normal flora, 183
MPN, *see* Most probable number
M protein, 202, 205–206, 219–220, 225–226, 292
mRNA, *see* Messenger RNA
Mucociliary flushing, 187, 193, 205, 360
Mucor, 439
Multiple drug resistance, 162, 166–167, 283
Multiple sclerosis, 241
Mumps, 208, 233, 351, 367–368
  immunization, 247–248, 367–368
  incidence, 365
Muramic acid, 53
Muscarine, 337–338
Muscazone, 337
Muscimol, 337
Mushroom, 36–38
  commercial production, 38, 530–531
  edible, 38
  halucinogenic, 337
  poisonous, 38, 337–338
Mushroom picker's lung, 336
Must, wine, 514
Mutagen, 152–155
Mutants
  antibiotic-resistant, 125, 154–155
  kinds, 154
  selection, 154–155
Mutation, 151–155
Mutualism, 38–40, 179–180
Mycelium
  bacterial, 32–33
  fungal, 35, 37
Mycetismus, 337–338
Mycetoma, 333
Mycobacterium, diseases caused by, 306–310
Mycobacterium avium-intracellulare, 371
Mycobacterium bovis, 493
Mycobacterium leprae, 310
Mycobacterium tuberculosis, 13–14, 111, 122, 191, 203, 235, 306–309
  in animals, 493
  growth cycle, 103
  milk pasteurization, 458–459
Mycoplasma mycoides, 320
Mycoplasma pneumoniae, 191, 320
Mycorrhiza, 486
Mycoses, 329–335
Mycotoxin, grain, 337–338, 490–492
Myeloma cell, 248
Myxobacteria, 33
Myxococcus xanthus, 33
Myxospore, 33
Myxovirus, 493

NAD, *see* Nicotinamide adenine dinucleotide
NADP, *see* Nicotinamide adenine dinucleotide phosphate
Nanometer, 20
Nasopharyngeal swab, 209
Natural gas, 6
Natural killer cell, 233
Natural selection, 151
Negri body, 374
Neisseria, 32
  diseases caused by, 296–299
Neisseria gonorrhoeae, 127, 276, 296–298
Neisseria meningitidis, 276, 298
Neomycin, 120, 125, 499, 520
Nephritis, 191
Neuraminidase, 317, 360–363
Neurotoxin, 195
Neutralization of antigen, 225–226
Neutral spirits, 518
Neutron, 50
NGU, *see* Nongonnococcal urethritis
Nicotinamide adenine dinucleotide (NAD), 65–72
Nicotinamide adenine dinucleotide phosphate (NADP), 65–66, 77
Nitrate
  as electron acceptor, 73–74
  fertilizer, 485–486
  formation, 75
  nitrogen cycle, 395–398
  as nitrogen source, 63
Nitrification, 73, 75, 396–398, 486
Nitrifying bacteria, 398, 486
  enrichment culture, 95
Nitrite
  as energy source, 75
  nitrogen cycle, 395–398
Nitrofurantoin, 125
Nitrogen
  fertilizer, 485–486
  fixation, 4, 63, 395–397, 482–485
  requirement of cells, 63
Nitrogen cycle, 395–398, 482–486
Nitrogen dioxide, silage, 501
Nitrogen gas
  denitrification, 398
  nitrogen fixation, 395–397
  as nitrogen source, 63
Nitrosamines, 154
Nitrous acid, mutagenicity, 154
Nocardia, 32
Nomenclature
  enzyme, 61
  microorganism, 29
  virus, 351
Nomuraea rileyi, 489
Nongonococcal urethritis (NGU), 341
  chlamydial, 262, 281, 323
Normal flora, 182–190
Norwalk virus, 374, 410–411
Nosema locusteae, 489
Nosocomial infection, *see* Hospital infection
Novobiocin, 499
N protein, 360–363
Nuclear polyhedrosis virus, 489
Nuclear region, procaryotic, 24
Nuclease, 62
Nucleic acid
  hybridization, 172–173
  structure, 55–57
  synthesis inhibitors, 118, 121
Nucleic acid base, 55, 57, 132–134
Nucleus, 26–27

Nutrient, 48–49
  in culture medium, 83–85
Nystatin, 118–120, 520

Objective lens, 21–22
Obligate aerobe, 86
Obligate anaerobe, 86–87
Ocean, hydrothermal vent, 75–76
Ochratoxin, 337
Ocular lens, 21–22
Oil-immersion objective, 21–23
Oil spill, 393
Oil well, 394–395
Oleandomycin, 499
Oncogene, 378
Operator, 141–143
Operon, 133
Opportunistic pathogen, 182, 197, 236–237, 316, 335, 371
Opsonization, 227
Orbivirus, 351
Orellanine, 337
Organelle, 26–27
Organic compounds, 51–52
  commercial production, 505, 528
  energy from, 73
Organophosphate insecticide, 480
Orthomyxovirus, 42, 351, 359–360
Oscillatoria, 78
Otitis media, 294, 299
Oxidase, 62
Oxidation pond, 403–404, 429
Oxidation reaction, 64–66
Oxidation-reduction cycle, 387
Oxidation-reduction potential, food, 440–441
Oxygen
  from photosynthesis, 76–79
  requirement of microorganisms, 86–89
  in respiration, 68–72
Oxygenic photosynthesis, 77
Oxytetracycline, 120, 472, 499
Ozone, 428

Pandemic, 256, 272
Paneolus, 337
Papilloma, Shope, 377
Papilloma virus, 376, 378
Papovavirus, 351, 376
Parainfluenza virus, 500
Paramecium, 18–19, 28
Paramyxovirus, 351, 364, 367
Parasite, 179–180
Parathion, 480
Paratyphoid fever, 247, 276
Parvovirus, 42, 351
Passive immunity, 244–245
Pasteur, Louis, 10–11, 110, 374
Pasteurella haemolytica, 493
Pasteurella multocida, 500
Pasteur flask, 10–11
Pasteurization, 110
  milk, *see* Milk, pasteurization
  wine, 515
Pathogens, 179–180, 190
  harm to host, 194–196
  invasion and growth, 192–194
  isolation and identification, 210–211
  opportunistic, 182, 197, 236–237, 316, 335, 371
  precautions for working with, 96–98
  shipping, 97–98
  specificity, 191–192
  transmission, *see* Transmission of infectious disease
Patulin, 337
Pectinase, 525
Pectinolytic organisms, 441
Pediococcus acidilactici, 472
Pediococcus pentosaceus, 472
Penicillin, 36, 115–119, 125
  allergy, 241
  animal feed, 499
  commercial production, 505, 520
  history, 212–213
  inactivation, 125
  prophylactic, 213
  resistance, 119, 159–160
  therapeutic index, 117
Penicillinase, 125–126, 297
Penicillium, 115, 336
  antibiotic production, 519
  in food, 439
Penicillium chrysogenum, 119, 520
Penicillium graminearum, 337
Penicillium griseofulvum, 122, 520
Penicillium roqueforti, 467
Pentose, 53–54
Peptide, 54–55, 132
Peptidoglycan, 53, 118
Peritonitis, 191, 292
Peritrichous flagellation, 33
Pertussis, *see* Whooping cough
Pesticide, 479–482
  persistence, 480–481
Petroleum, 391
  decomposition, 393
  microbiology, 6, 392–395
  prospecting, 392–393
  recovery, 393–394
pH
  culture medium, 86
  scale, 86
  sterilization, 109
Phagocyte, 199–203
Phagocytosis, 199–203, 205, 227–228
Phallotoxin, 337
Phanerozoic period, 44
Pharmaceutical industry, 4, 127
Pharyngitis, 299
Phenols, 115
Phenotype, 141–142
Phosphoenol pyruvate, 67–68
Photoheterotroph, 79
Photosynthesis, 27, 34, 76–79, 385–388, 477
  anoxygenic, 78
  oxygenic, 77
Photosynthetic membrane, 25
Photosynthetic organisms, 43
  enrichment culture, 95
Phototroph, 48, 76–79
Phytophthora citropthora, 489
Pickles, 6, 451
Picornavirus, 42, 351, 363–364
Pilot plant, 510
Pilus, 162–163, 297
Pimples, 183, 193, 195, 203, 262, 294
Pityriasis versicolor, 329, 332
Plague
  bubonic, *see* Bubonic plague
  immunization, 247, 272
  pneumonic, 313
Plants
  disease, 4–5, 329, 486–488
  normal flora, 486
Palque, dental, 184
Plaque, virus, 353
Plasma, 198
Plasma cell, 231–232
Plasmid, 158, 160–167
  in genetic engineering, 170–172
Plasmodium, 194, 263, 343–347
Plasmodium falciparum, 344–347
Plasmodium malariae, 344
Plasmodium ovale, 344
Plasmodium vivax, 41, 276, 344–346
Plate count, 105
  food assay, 448
  milk, 462–463
Pleomorph, 32
Pleural fluid specimen, 210
Pleuropneumonia, 320
Pleurotus ostreatus, 336
PMN, *see* Polymorphonuclear leukocyte
Pneumocystis carinii, 371
Pneumonia, 191, 324
  bacterial, 196–197, 202, 292–294, 299, 310, 360
  immunization, 247
  livestock, 493
  mycoplasmal, 320
  pneumococcal, control, 277
  primary atypical, 320
  transmission, 262
  treatment, 121
  viral, 351
Poison ivy, 180
Polar flagellation, 33
Polio, 41, 351, 372–373
  control, 277
  immunization, 246–248, 372–373
  incidence, 246
  transmission, 260, 410
Poliovirus, 20, 41–42, 352–355, 410
Pollen, 240
Polyenes, 119–120
Polymer, 52–58
Polymixin, 118–119, 125, 520
Polymorphonuclear leukocyte (PMN), 199–201, 203
Polyoma virus, 377–378
Polysaccharide, 52–54
Pomace, 514–515
Potato famine, 1, 11, 486
Poultry, influenza, 494
Pour plate, 106
Poxvirus, 351, 359
Precambrian period, 44
Precipitation, antibody, 226–227
Pregnancy, rubella during, 366–367
Preservative
  food, 455–456
  meat, 471–472
Preserves, 451–452
Pressure filtration, 415
Primary production, 385–386
Prion, 147
Privy, 429–431
Procaryote, 18–19
  structure, 24–26, 28
Prodromal period, 207
Promoter site, 135, 141–143
Propazine, 480
Prophage, 146
Propionibacterium, 467–468, 522
Propionic acid bacteria, 451
Protease, 62, 525
Protein
  inducible, 141–143
  per cell, 58
  respressible, 141–143
  structure, 54–55, 132
  synthesis, 131, 136–141, 144–145
    inhibitors, 118, 120–121
    regulation, 141–143
Protein coat, viral, 29
Proteolytic organisms, 441
Proton, 50
Proton motive force, 71–72
Proto-oncogene, 378–379
Protoplasm, 18
Protozoa, 18–19
  characteristics, 40–41
  diseases caused by, 338–347
  energy metabolism, 73
  evolution, 43
  as insecticide, 489
  intracellular infection, 194
  water-borne, 410–411
Provirus, 146
Pseudomembrane, 300
Pseudomonads, 458
Pseudomonas, 32, 73, 119, 458
  diseases caused by, 316
Pseudomonas aeruginosa, 120, 191, 283, 529

Pseudomonas denitrificans, 522
Pseudopodium, 41
Psilocybe, 337–338
Psilocybin, 337–338
Psittacosis, 261–262
Psychrophile, 89
Psychrotroph, 450, 457–459, 469–470
Public health microbiology, 256–288
Puerperal sepsis, 292
Puff ball, 37
Pure culture, 12–13, 82, 94–96
  maintenance, 96
Purple bacteria, 77–78; *see also* Photosynthesis
Pus, 193, 202, 204, 206, 294–295
Putrefactive bacteria, 287
Pyogenic infection, 193, 195, 202
Pyruvate, 67–68

Q fever, 321, 458–459
Quarantine, 278

Rabbit fever, *see* Tularemia
Rabies, 208, 233, 258, 263, 351, 374–376
  eradication, 279
  immunization, 247, 374
  transmission, 260
Rabies virus, 20
Radiation, control of growth, 112–113
Radioactive material, 154
Radioimmunoassay, 250–251
Reading frame, 138–140
Recombinant DNA, 151, 155–175
Recombination, 151, 155–157
Red blood cell
  antigenicity, 241–244
  lysis, 206
Reducing agent, 87
Reducing power, in photosynthesis, 77
Reductase, 62
Reduction reaction, 64–66
Refrigeration, 111, 439–440, 450–451, 470–471
Rennin, 465, 525
Reovirus, 351
Replication, DNA, 134
Reportable disease, 266–268
Repression control, 143
Repressor protein, 142–143
Reproduction
  asexual, 26
  sexual, 26–27
Reservoir, water, 409
  algae in, 431–432
Reservoir of pathogen, 257–259
Resistance transfer factor, 500
Resolution, 21
Respiration, 68–72, 385–387, 391
  anaerobic, 73–75, 398–399, 485
Respiratory tract, 187, 257
  diseases, 191
    seasonality, 274–275
    transmission, 258
  exit of pathogens, 261
  invasion by pathogen, 193
  normal flora, 187
  specimens, 209
Restaurant, sanitation, 456
Restriction enzymes, 168–171
Reticuloendothelial system, 201
Retrovirus, 42, 148, 351, 355, 379
Reverse isolation, 286
Reverse transcriptase, 148, 173–174, 355, 378
Revertant, 155
Reye's syndrome, 363, 368
Rhabdovirus, 351–352
Rh disease of the newborn, 242–243, 245
Rheumatic fever, 180, 213, 240–241, 292–293
Rheumatoid arthritis, 241
Rhinosporidiosis, 332–333
Rhinovirus, 363–364
Rhizobium, 397, 482–485
Rhizopus, 335
Rhizopus nigrificans, 524
Rhizosphere, 486
Rhodospirillum, 30
Rh system, 242–243
Riboflavin, commercial production, 522
Ribonuclease, 56, 62
Ribose, 53, 55
Ribosome, 137–140
  eucaryotic, 140
  evolution, 42–43
  procaryotic, 24, 140
  structure, 55
Rice paddy, 397
Rickettsia, diseases caused by, 320–322
Rickettsia prowazekii, 321
Rickettsia rickettsii, 194, 263, 321–322
Rifampin, 118, 121, 140
Ringworm, 329–331
RNA, 135
  per cell, 58
  processing, 144–145
  structure, 55–57
  synthesis, 135–136
RNA polymerase, 135–136, 142–143, 354–355
RNA virus, 41–42, 148, 351
Rocky Mountain spotted fever, 194, 208, 263, 321–322
  control, 277
  distribution in United States, 322
  transmission, 260
Rodent
  Leptospira carrier, 320
  plague, 313–314
  tularemia, 263, 267, 314
Rod-shaped bacteria, 30, 32
Root-nodule bacteria, 4, 482–485
Rose-gardener syndrome, *see* Sporotrichosis
Rose spot, 311
Rotavirus, 374
Rous sarcoma virus, 377
Rubella, 233, 351, 366–367
  control, 276
  epidemic, 366–367
  immunization, 247–248, 366–367
  incidence, 365
Rumen, 4–5, 75, 497–499

Saccharification, 526
Saccharomyces bailii, 439
Saccharomyces carlsbergensis, 516
Saccharomyces cerevisiae, 36, 172, 511, 514, 517
Saccharomyces rouxii, 439
Safranin red, 22
Saint Anthony's fire, 339
Saliva, 183
Salmonella, 96, 311, 410, 442, 445–447, 458, 474, 493
  antibiotic-resistance, 471
Salmonella newport, 471
Salmonella paratyphi, 276, 410
Salmonella typhi, 103, 203, 259, 262, 311–312, 410, 458–459
Salmonella typhimurium, 110, 493
Salmonellosis, 208, 267, 311–312
  livestock, 493
  outbreak, 278, 471
  seasonality, 275
  transmission, 260, 410
Salted food products, 451–452
  meat, 472
Sand filter, 415–417
Sanitary sewer, 422
Sanitation
  cheese factory, 468
  food handling, 456
  livestock, 493
  slaughterhouse, 469–470
San Joaquin Valley fever, *see* Coccidioidomycosis
Saprophyte, 180
Sauerkraut, 5, 451
Sausage, 6, 154, 470, 472–473
Scale-up, 510
Scarlet fever, 195, 292, 301
Scenedesmus, 34
Schick test, 303
Schizont, 344–345
Scrapie, 147
Scrub typhus, 267, 321
Seafood, contaminated, 317
Seasonality of infectious disease, 274–275
Secondary metabolite, 103–104, 509
Sedimentation basin, 414–415
Selective medium, 96, 154–155, 159, 164–165, 211
Semiperishable food, 437
Semisolid medium, 13
Semmelweis, Ignaz, 11
Septic tank, 429–430
Serial dilution, 106–107
Serologic method, microbial products in food, 448
Serology, 250–252
Serotonin, 238–239
Serum, 198–199
Serum hepatitis, 267, 373
  immunization, 272, 357–358
Serum sickness, 238–239, 245
Settling basin, 415
Severe combined immunodeficiency, 236
Sewage, 274, 373, 420
  soluble components, anaerobic treatment, 425–427
  sterilization, 113
Sewage lagoon, 422, 429–431
Sewage-treatment system, 75, 312, 317, 410, 422–431
  aerobic, 425–427
  anaerobic, 422–425
  effluent, 428–429
  nitrification, 398
  substance resistant to, 427–428
Sewer pipelines, 409, 418, 422
Sexually transmitted disease, 2, 257–260, 262, 267–268, *see also* specific diseases
  control, 281
  incidence, 274
Sexual reproduction, 26–27
Shape, bacteria, 30
Sheep, *see* Livestock
Shigella, 96, 276, 312–313, 410, 442
Shigella dysenteriae, 458–459
Shigellosis, 267, 312–313
  transmission, 410
Shingles, 351, 368
Shipping container, for pathogens, 97–98
Shipping fever, 493, 500
Sickle cell anemia, 346–347
Silage, 500–501
Silver nitrate, 115
Simazine, 480
Single-cell protein, 529
Sinusitis, 299
Size
  bacteria, 29
  microorganism, 19–20
  virus, 351–352
Skin
  allergy, 237–240
  antisepsis, 183, 284–285
  barrier to infection, 182–183, 192–193, 205
  cleansing, 115
  lesions, 294
  normal flora, 183
Slime, bacterial
  extracellular, 202, 293, 425–427
  meat, 469
  in water pipelines, 419–420
Slime mold, 43
Sludge
  activated, 425–427
  anaerobic digestion, 422–425
  dewatering, 424–425

Smallpox, 41, 233, 351, 358–359
  epidemic, 272
  eradication, 279–280, 358
  immunization, 247–248, 359
  reservoir, 258
Smallpox virus, 20
Smear, cheese making, 467
Sneezing, 258, 261, 360
Soap, 114–115, 285
Sodium benzoate, 455
Sodium nitrite, 455
  meat preservation, 472
Sodium propionate, 455
Soft drinks, 6, 526
Soil, 476–479
  bacterial spores, 30–31
  fertility, 478
  formation, 477–478
  as microbial habitat, 33, 388–390
  nitrogen cycle, 74–75, 395–398
  pathogens, 258
Soil profile, 477–478
Sorbic acid, 455
Sorbose, 528
Sore throat, 32
  streptococcal, 180, 202, 205–206, 240, 292–293
    outbreak, 273
    transmission, 260–262
  viral, 351
Specialized transduction, 160–161
Species, 29
Spectrophotometer, 108
Spinal fluid specimen, 209
Spirochaeta plicatilis, 30
Spirochete, 30, 43
  diseases caused by, 318–320
Spoilage, food, 436–441
Spoil bank, 400–401
Spontaneous generation, 10–11
Spontaneous mutation, 152
Spore
  bacterial, 30, 32–33, *see also* Endospore
  fungal, 35–37, 336
  procaryotic, 26
Sporothrix schenckii, 331–332
Sporotrichosis, 330–333
Sporozoa, 343–347
Sporozoan, 41
Sporozoite, 343–345
Spread plate, 105–106
Sputum culture, 209–210
Stachybotrys, 337
Stachybotrytoxin, 337
Staining, 22–23
  acid-fast, 306
  differential, 23
  Gram, 23–24
  negative, 33
Staphylococcus, 32, 183, 187, 292, 447
  diseases caused by, 294–296
  in food, 439
Staphylococcus aureus, 111, 183, 186, 191, 195, 294, 442–445, 474, 495
  antibiotic sensitivity, 295
  habitat, 295
  hospital infection, 283
  transmission, 261–262
Staphylococcus epidermidis, 294
Starch, 53
Starter culture
  fermented food, 451
  fermented milk products, 464–465
  sausage, 472
Stationary phase, 102–104, 509
Statistics, health, 264–268, 277
Sterigmatocystin, 337
Sterile solutions, 284
Sterilization, 11, 90–93
  culture medium, 90–91
  dry heat, 92–93
  filter, 91–92
  gas, 93
  glassware, 92
  heat, 109
  hospital equipment, 283–284
  industrial fermenter, 507–508
  pH, 109
Steroid hormones, commercial production, 524
Stigmatella aurantiaca, 33
Stock culture, 96
Stomach
  acidity, 184–185, 205
  fluid specimen, 209
Streak plate, 95
Streptococcus, 32, 187, 463–464, 493, 501
  antibiotic sensitivity, 293
  antigenicity, 292
  diseases caused by, 291–294
  extracellular slime, 293
  pyogenic, 292
  transmission, 293
Streptococcus agalactiae, 458–459, 493, 495
Streptococcus faecalis, 185, 191
Streptococcus lactis, 458
Streptococcus mutans, 184, 293
Streptococcus pneumoniae, 29, 191, 196–197, 292–294, 298
  encapsulated, 202
  transmission, 261
Streptococcus pyogenes, 195, 202, 205–206, 301, 458–459
  antigenicity, 219–220
  transmission, 261
Streptococcus sanguis, 184, 293
Streptococcus thermophilus, 464
Streptokinase, 203, 205–206
Streptolysin, 195
Streptomyces, 32–33
  antibiotic production, 116, 119–121, 127–128, 519
Streptomyces erythreus, 520
Streptomyces fradiae, 520
Streptomyces griseus, 520
Streptomyces kanamyceticus, 520
Streptomyces lincolnensis, 520
Streptomyces noursei, 520
Streptomyces orchidaceus, 520
Streptomyces rimosus, 520
Streptomyces roseochromogenes, 524
Streptomycin, 116–120, 125
  animal feed, 499
  commercial production, 520
  mode of action, 140
  for plant disease, 488
Subcutaneous mycoses, 329–335
Substrate, 60–61
Substrate-level phosphorylation, 68, 70
Sudden infant death syndrome, 304
Sugars, 52–54
Sulfa drugs, 121–122
Sulfamethazine, 499
Sulfamethoxypyridazine, 499
Sulfanilamide, 118
Sulfate, sulfur cycle, 398–400
Sulfate-reducing bacteria, 73–74, 399
  water pipeline corrosion, 418–419
Sulfides
  acid mine drainage, 400–402
  as energy source, 75
  solubilization, 402–404
  sulfur cycle, 398–400
Sulfolobus acidocaldarius, 389
Sulfonamides, 117, 121–122
Sulfur
  as energy source, 75
  oxidation, 399–400
Sulfur cycle, 398–400
Sulfur dioxide, 455
Sulfuric acid
  acid mine drainage, 400–402
  formation, 75
  soil, 399–400
Sulfur-oxidizing bacteria, 73, 75
Superficial mycoses, 329–333
  treatment, 331
Surveillance-and-containment, 279–280
Sweat glands, 183
Sweetener, high-fructose, 525–526
Swine flu, 363
Symbiosis, 482–485
Syphilis, 191, 208, 267, 318–320
  congenital, 319
  control, 277, 281
  incidence, 274
  stages, 318–319
  transmission, 262
  treatment, 319
Systemic lupus erythematosus, 241
Systemic mycoses, 329–335

2,4,5-T, 480
Tanning industry, 305–306
Tannins, 514–515
TCA cycle, *see* Tricarboxylic acid cycle
T cells, 205, 217, 220–223, 232–236
  cytotoxic (killer), 232–234
  delayed-type hypersensitivity, 233–235, 239–240
  helper, 231–233
  proliferation, 235–236
  receptors, 232
  suppressor, 232–233
Temperate virus, 146–147
Temperature
  dependence of microorganisms, 89
  food storage, 439–440
  industrial fermenter, 508
Tetanus, 191, 195–197, 258, 303–304
  immunization, 247–248, 301, 304
  incidence, 301
  transmission, 263
Tetracycline, 116–118, 120, 125
  animal feed, 499
  commercial production, 520
  mode of action, 140
  for plant disease, 488
Thallus, lichen, 38–39
Therapeutic index, 117
Thermal death time, 109
Thermophile, 89, 95
Thiobacillus, 399
Thiobacillus ferrooxidans, 400–404
Thioglycollate, 87
Throat swab, 209
Thrush, 335
Thymine, 55–57, 132
Tick, 314, 321–322
Tinea capitis, *see* Ringworm
Tinea cruris, *see* Jock itch
Tinea favosa, *see* Favus
Tinea pedis, *see* Athlete's foot
Tissue culture, 352–353
Tissue transplantation, 233
Tobacco mosaic virus, 29, 41
Tobramycin, 118, 120, 125, 140
Togavirus, 351
Tonsillitis, 195
Torulopsis, 188
Total cell count, 104–105
Toxemia, 193
Toxic shock syndrome, 295–296
Toxigenicity, 196–197
Toxin, 194–196, 213
  antigenicity, 219
  erythrogenic, 195, 292
  in food, 442–445
  fungal, 329, 336–338
Toxoid, 246–247, 301
Toxoplasma, 371
Trachoma, 322–323
Transcription, 135–136, 144–145
Transduction, 158, 160–161
  generalized, 160–161
  specialized, 160–161
Transfer hood, 96–97

Transfer RNA (tRNA), 137–139
Transformation, 157–160, 169, 357, 377–378
Translation, 137–140
Transmission of infectious disease, 192, 256–264
  during infection, 258
  interruption, 281–282
Transplantation immunity, 244
Transposon, 166–167
Travelers' diarrhea, 310–311
Treponema pallidum, 23, 111, 191, 318–320
Tricarboxylic acid cycle (TCA cycle), 69–70
Trichomonads, 341
Trichomonas hominis, 341
Trichomonas tenax, 341
Trichomonas vaginalis, 341
Trichomoniasis, 341
  livestock, 493
Trichophyton, 330–331
Trichothecene, 337
Trickling filter process, 422, 425, 427
Trimethoprim-sulfamethoxazole, 125
Triplet code, 133, 136–137, 140, 152–153
tRNA, *see* Transfer RNA
Truffle, 38
Trypanasoma, 263
Trypanosoma brucei, 341
Trypanosoma cruzi, 341
Trypanosoma gambiense, 340–341
Trypanosomes, 41, 339–341
Tsetse fly, 340–341
Tsutsugamushi disease, *see* Scrub typhus
Tubercle, 307
Tuberculin test, 235, 307, 309, 493
Tuberculosis, 13–14, 191, 203, 235, 240, 267, 273, 306–309
  carrier, 259
  control, 277
  diagnosis, 209–210
  food-borne, 458–459
  immunization, 247, 308–309
  incidence, 269, 308
  livestock, 309, 492–493
  postprimary, 307–308
  primary, 307
  transmission, 262–263
  treatment, 120–122, 307–308
Tularemia, 263, 267, 314
Tumor, 376–379
Tumor virus, 148
Turbidity, and cell mass, 108
Tylosin, 499
Typhoid fever, 203, 257, 267, 311–312
  carrier, 259, 312
  control, 277
  immunization, 246–247, 272, 312
  incidence, 415–416
  stages, 207
  transmission, 260, 262, 410
  treatment, 311–312
Typhoid Mary, 259
Typhus fever, 321
  immunization, 247
  outbreak, 273–274

Ulothrix, 34
Ultra-high temperature pasteurization, 461
Ultraviolet radiation, 112–113, 154
  sewage treatment, 429
Undecylenic acid, 331
Undulant fever, 203, 315, 493–494
  food-borne, 458–459
  treatment, 315–316
Universal ancestor, 43
Uracil, 55–57, 135
Uranium mining, 404
Urea, fertilizer, 485–486
Ureoplasma urealyticum, 320
Urethritis, 320
Urinary tract infection, 282–283, 310
Urine specimen, 209

Vaccination, *see* Immunization
Vaccine, 219, 244–247
  adverse reactions, 302
  polyvalent, 363
  production, 174–175
  viral, 174–175, 357–358
Vaccinia virus, 358–359
Vacuole, 27
Vacuum-drying, food, 451
Vagina, normal flora, 188
Vaginitis, 335
Varicella-Zoster virus, 368
Vector
  control of, 280
  for genetic engineering, 168, 170–172
  insect, 258, 260, 263, 270, 313–314, 320–322, 340–341, 344–345
Verticillium lecanii, 489
V gene, 229–230
Viable count, 105–108
  food, 448
  milk, 462
Vibrio, diseases caused by, 316–317
Vibrio cholerae, 262, 316–317, 410, 417
Vibrio parahaemolyticus, 317, 442, 474
Vibriosis, livestock, 493
Villemin, Jean-Antoine, 13
Vinegar, 526–528
Viremia, 193
Viroid, 147
Virulence, 181, 196–197, 205
Virus, 19–20, 194
  and cancer, 376–379
  characteristics, 41–42
  control, 276, 357–358
  cultivation, 352–353
  diseases caused by, 350–380
  enveloped, 351–352
  in genetic engineering, 170–172
  icosahedral, 351–352
  as insecticide, 489–490
  interferon and, 204
  naked, 41, 351
  reproduction, 145–148, 353–357
  size, 20, 351–352
  structure, 29
  in treated sewage, 428–429
  water-borne, 410–411
Vitamin K, 189
Vitamins
  commercial production, 511, 521–525
  in culture medium, 83–84
Volvox, 34

Wallemia, 439
Wandering cell, 199
Wart, 351, 376
Waste product, metabolic, 49–50
Wastewater, 407
Water
  algal pollution, 421–432
  bacterial count, 108
  contaminated, 320
  distribution, 407
  emergency purification, 415–416
  fecal contamination, 412
  microbiological assay, 411–413
  microbiology, 407–432
  pollution, 420–422
    mining, 400–402
  purification, 274, 312, 317, 413–417
  self-purification, 420–421
  soil, 389–390
Water activity, food, 438–439
Water bath, 89
Water-borne disease, 260, 262, 341–342, 409–411, 417
  control, 281
  incidence, 274
  virus, 372–374
Water pipelines, 409
  corrosion, 419
  cross-contamination, 409–410
  microbiology, 418–420
Water supply, 276, 407–409
  urban, 409, 417–418
Water-treatment plant, 3–4, 414–415
Weathering, 477–478
Weight, cell mass, 108–109
Weil's disease, 320
Whey, 465–466
Whooping cough, 121, 299–300
  age distribution, 300
  control, 277
  epidemic, 302
  immunization, 246–248, 300–302
  incidence, 301–302
  transmission, 261
Wine, 511, 513–515
Witchcraft, 339
Wood pulp worker's lung, 336
Woolsorter's disease, 306
World health, 268–271
Wort, 516–517
Wound infection, 193, 260, 263, 303–304, 310

Xanthan gum, 394–395
Xanthomonas campestris, 394–395
Xeromyces bisporus, 439
Xerophile, 438

Yaws, 281
Yeast, 18–19, 21, 36
  active dry, 513
  alcohol production, 518–519
  alcoholic beverage production, 513–519
  animal feed, 511, 513
  brewing, 516–517
  cheese making, 467
  commercial production, 509, 512
  disease caused by, 333–335
  fermentation, 66–68
  food, 511, 513, 529
  industrial, 511–513
Yeast cake, 513
Yellow boy, 400
Yellow fever, 263, 267, 278, 351
  eradication, 280
  immunization, 247, 271–272
Yersinia enterocolitica, 314, 442, 474
Yersinia pestis, 196–197, 263, 313–314
Yogurt, 5, 32, 464

Zearalenone, 337
Zoogloea ramigera, 425–427
Zoonoses, 258, 260, 263–264